Aluminum Alloys—Contemporary Research and Applications

Treatise on Materials Science and Technology

VOLUME 31

ALUMINUM ALLOYS—CONTEMPORARY RESEARCH AND APPLICATIONS

EDITED BY

A.K. VASUDEVAN

Office of Naval Research
Arlington, Virginia

and

R.D. DOHERTY

Drexel University
Philadelphia, Pennsylvania

TREATISE ON MATERIALS SCIENCE AND TECHNOLOGY

VOLUME 31

ACADEMIC PRESS, INC.
Harcourt Brace Jovanovich, Publishers
Boston San Diego New York
Berkeley London Sydney
Tokyo Toronto

ACADEMIC PRESS, INC.
1250 Sixth Avenue, San Diego, CA 92101

United Kingdom Edition published by
ACADEMIC PRESS INC. (LONDON) LTD.
24–28 Oval Road, London NW1 7DX

LIBRARY OF CONGRESS CATALOGING-IN-PUBLICATION DATA

Aluminum alloys—contemporary research and applications/edited by
A.K. Vasudevan and R.D. Doherty.
p. cm.
Bibliography: p.
Includes index.
ISBN 0-12-341831-3 (alk. paper)
1. Aluminum alloys. I. Vasudevan, A.K. (Asuri K.) Date-.
II. Doherty, R.D. (Roger Davidge), Date-.
TA403.T74 vol. 31
[TA480.6]
620.1'1 s—dc20 89–15182
[620.1'86] CIP

ISBN 0-12-341831-3

89 90 91 92 9 8 7 6 5 4 3 2 1
PRINTED IN THE UNITED STATES OF AMERICA

Contents

PART 6: COMPETITIVE MATERIALS TO ALUMINUM

Contributors

Numbers in parentheses indicate the pages on which the authors' contributions begin.

C.C. BAMPTON (189), *Science Center, Rockwell International Corporation, 1049 Camino Dos Rios, PO Box 1085, Thousand Oaks, California 91360*

S.F. BAUMANN (65), *Alcoa Laboratories, Alcoa Center, Pennsylvania 15069*

ROBERT BUSK (663), *International Magnesium Consultants, Inc., PO Box 3163, Hilton Head, South Carolina 29928*

R.J. BUCCI (295), *Alcoa Laboratories, Alcoa Center, Pennsylvania 15069*

L. CHRISTODOULOU (463), *Martin Marietta Laboratories, Baltimore, Maryland*

C.E. CROSS (171), *IMAC Research & Consulting, PO Box 1989, Idaho Springs, Colorado 80452*

R.D. DOHERTY (445), *Department of Materials Engineering, Drexel University, Philadelphia, Pennsylvania 19104*

G.R. EDWARDS (171), *Center for Welding Research, Colorado School of Mines, Golden, Colorado 80401*

J.D. EMBURY (579), *Department of Metallurgy and Materials Science, McMaster University, Hamilton, Ontario, Canada*

D.J. FIELD (523), *Alcan International Limited, Southam Road, Banbury, Oxon OX16 7SP, England*

W.G. FRICKE (563), *Alloy Technology Division, Alcoa Laboratories, Alcoa Center, Pennsylvania 15069*

A.K. GHOSH (189), *Rockwell International Science, 1049 Camino de Rios, PO Box 1085, Thousand Oaks, California 91360*

D.A. GRANGER (109), *Molten Metal Processing Division, Alcoa Laboratories, Alcoa Center, Pennsylvania 15069*

J.W. Gunnink (295), *Delft University of Technology, Delft, The Netherlands*

S.J. Harris (255), *Department of Metallurgy and Materials Science, University of Nottingham, NG7 2RD Nottingham, England*

G.H. Hildeman (323), *Alcoa Laboratories, Alcoa Center, Pennsylvania 15069.*

N.J.H. Holroyd (463, 485), *Alcan International Limited, Banbury Laboratories, Banbury, Oxon OX1 67SB, England*

P.R. Howell (365), *Department of Materials Science and Engineering, Pennsylvania State University, University Park, Pennsylvania 16802*

J.A. Hunter (485), *Alcan International Limited, Banbury, Oxon, England*

S.C. Jha (409), *Materials Scientist, Marko Materials, Inc., 144 Rangeway Road, North Billerica, Massachusetts 01862*

F.R. Jones (605), *School of Materials, Division of Ceramics, Glasses, and Polymers, University of Sheffield, Elmfield, Northumberland Road, Sheffield S10 2TZ, England*

M.J. Koczak (323), *Department of Materials Engineering, Drexell University, Philadelphia, Pennsylvania 19104*

G.W. Kuhlman (651), *Alcoa Forging Division, Cleveland, Ohio*

D.J. Lloyd (579), *Alcan International Limited, Kingston Research & Development Centre, PO Box 8400, Kingston, Ontario K7L 4Z4, Canada*

M. Meshii (501), *Materials Research Center and Department of Materials Science and Engineering, Northwestern University, Evanston, Illinois 60208*

L.N. Mueller (295), *Alcoa Laboratories, Alcoa Center, Pennsylvania 15069*

G.H. Narayanan (219), *Boeing Commercial Airplane Company, PO Box 3707, M/S 73-43, Seattle, Washington 98124*

S.R. Nutt (389), *Division of Engineering, Brown University, Providence, Rhode Island 02912*

B.A. Parker (539), *Department of Materials Engineering, Monash University, Clayton, Victoria 3168, Australia*

M.A. Przystupa (563), *PO Box 39, Redondo Beach, California 90277*

W.E. Quist XXX, *Boeing Commercial Airplane Company, PO Box 3707, M/S 73-43, Seattle, Washington 98124*

T.R. Ramachandran (579), *Department of Metallurgical Engineering, Indian Institute of Technology, Kanpur, India*

R.E. Sanders, Jr. (65), *Alcoa Laboratories, Alcoa Center, Pennsylvania 15069*

T.H. Sanders, Jr. (409), *School of Materials Engineering, Georgia Institute of Technology, Atlanta, Georgia 30332*

J.M. Sater (409), *Institute for Defense Analysis, Alexandria, Virginia*

G.M. Scamans (485), *Alcan International Limited, Banbury Laboratories, Banbury, Oxon OX1 67SB, England*

J.T. Staley (3), *Alloy Technology Division, Building C, Alcoa Laboratories, Alcoa Center, Pennsylvania 15069*

Edgar A. Starke (35), *Department of Materials Science, University of Virginia, Charlottesville, Virginia 22901*

H.C. Stumpf (65), *Alcoa Laboratories, Alcoa Center, Pennsylvania 15069*

S. Suresh (445), *Division of Engineering, Brown University, Providence, Rhode Island, 02912*

A.K. Vasudevan (445, 463), *Office of Naval Research, Applied Research and Technology Directorate, 800 North Quincy Street, Arlington, Virginia 22217*

L.B. Vogelesang (295), *Delft University of Technology, Delft, The Netherlands*

J. Wadsworth (189), *Alloy Technology Division, Alcoa Laboratories, Alcoa Center, Pennsylvania 15069 or Lockheed Missiles and Space Company, Inc., 3251 Hanover Street, Palo Alto, California 94394-1187*

J.W. Watson (501), *Armco Inc., 703 Curtis Street, Middletown, Ohio 45043*

J.A. Wert (137), *School of Engineering and Applied Science, University of Virginia, Charlottesville, Virginia 22901*

D.B. Williams (365), *Department of Metallurgy and Materials Engineering, Lehigh University, Bethlehem, Pennsylvania 18015*

J.C. Williams (651), *General Electric Company, Cincinnati, Ohio*

M.A. Zaidi (137), *Alcoa Laboratories, Alcoa Center, Pennsylvania, 15069*

Preface

This volume describes (i) the processing, structure, and properties of current commercial aluminum alloys, (ii) the research and technology of the aluminum alloys that are being developed for future use, and (iii) the competitive materials that are challenging aluminum for its major application—aerospace engineering. Although aluminum has had, and continues to have, many varied non-aerospace applications since its first commercial development 100 years ago, the most indispensable application of aluminum has been its use as a light-weight structural metal for flight applications.

Since 1930, modern aircraft have been built almost universally from aluminum alloys—both the Al-Cu-Mg (2xxx) series and the Al-Zn-Mg-Cu (7xxx) alloys. Since their introduction, these alloys have undergone steady incremental improvements as described in the first chapter. These improvements were driven by engineering needs for high strength combined with resistance to fracture, fatigue, and both general corrosion and stress corrosion failure.

Recently, additional challenges have arisen from two factors. The first was the increase in oil prices during 1970–1980 that made the need for light-weight alloys more critical than ever as every kilogram of structural weight saved would reduce the amount of fuel necessary to provide the lift to offset the weight of a structure. In addition, because only about 10 percent of the total take-off weight of an aircraft is available for passengers, cargo, etc., a 10 percent weight savings in the structure offers the potential to double the aircraft's carrying capacity. The second factor to challenge the traditional alloys was the development of totally novel competitive materials. The most important has been fiber-reinforced polymeric materials using the high-strength, non-metallic fibers developed in the last 10 to 25 years, particularly graphite and the polymeric aramid fibers such as "Kevlar." Other competitive materials are the lighter, but weaker, magnesium alloys and the heavier, but stronger, titanium alloys. The former metal has its main application in shaped castings because the limited slip systems in magnesium create major difficulties in wrought applications; however, the chemical reactivity of magnesium also creates some additional problems. Titanium has its major applications in higher temperature use due to the much higher melting point of titanium, although even here its higher cost is a major difficulty. Chapters that deal with these competitive materials are included because they are interesting in their own right, but, more importantly, because they are necessary to show which properties of aluminum alloys must improve if aluminum is to retain its predominant role in aerospace applications.

The early chapters cover the historical development of aluminum physical metallurgy and discuss carefully the two main classes of aluminum alloys, conventionally described as "heat-treatable" and "non-heat treatable." This term refers to a *specific* heat treatment—precipitation hardening. This heat treatment starts with the dissolution of a range of soluble elements at a temperature close to the alloy solidus temperature, rapidly quenching to preserve the highest possible amount of solute in metastable solid solution, and finally "aging" the quenched material to precipitate the solute in fine second-phase "zones" or other finely dispersed, usually metastable, precipitates. This type of processing produces the highest strength aluminum alloys for aerospace and other applications. Aluminum alloys that are described as non-heat treatable are nevertheless processed in ways that use high temperatures for ease of shaping, for softening cold-worked materials, for changing the texture of wrought metal, and also for controlling the amount and distribution of soluble alloying additions and second phases-although the latter are much more coarsely dispersed than the precipitates formed in the final stage of the heat treatment of heat-treated alloys.

An initial thermal treatment applied to almost all aluminum alloys is the "homogenization" anneal. This is a lengthy anneal at the highest temperatures at which all trace of melting of the segregated casting can be avoided. Its object is to make the as-cast solute distribution more uniform by allowing diffusion of the faster diffusing solute elements—mainly magnesium, copper, zinc, and silicon. This thermal process also precipitates solutes that are slow diffusing and only slightly soluble in aluminum-high melting transition elements, mainly manganese, chromium and zirconium. These form the small (typically 0.1 μm but 10 nm for Zr) dispersoids at volume fractions of less than about 1 percent. These dispersoids have two functions. The first is to prevent high-temperature grain growth and in some cases also to prevent recrystallization of hot worked material during solution treatment. The second function of these fine particles is that they are believed to make dislocation slip less heterogeneous during cold working by providing non-shearable obstacles to dislocation motion. Iron, at

0.15–0.5 wt%, is an inevitable impurity in commercial aluminum alloys since its removal from primary and secondary aluminum is usually too costly to be practical in commercial alloys. It is mainly present as large "constituents," a few micrometers in size, that were produced during casting and which are only marginally changed (in their chemistry) by heat treatment although they are usually partially broken up during working. It should be recognized that homogenization processing and the thermal processing to softening wrought materials are applied to both classes of aluminum alloys, heat-treatable and non-heat treatable.

Subsequent chapters in the book deal with the research and technology of three new classes of aluminum alloys whose development has been the stimulus for this volume. The three classes are powder metallurgy (PM) aluminum alloys, mainly produced by rapid solidification; aluminum-based composite alloys, including the new alloys—"Arall" materials—that use aluminum as a "macro-composite" with aramid reinforced polymers; and finally, the new range of aluminum–lithium alloys.

The PM alloys take advantage of rapid solidification to reduce the size of the iron-rich constituents to make them capable of acting as dispersoids and, at high enough concentrations of iron, to provide dispersion hardening. Not only iron but other limited-solubility, slow-diffusing transition metal additions provide fine dispersions resistant to coarsening at moderate temperatures when processed by rapid solidification, thus providing an aluminum-based alternative to titanium for application at these temperatures.

The aluminum matrix composite alloys provide an alternative to other non-aluminum composites. These composites all provide the possibility of higher elastic stiffness and higher strength but at the cost of much reduced ductility (although not necessarily reduced toughness) compared with "monolithic" aluminum. Arall is a hybrid composite that consists of a core of Aramid fibers in a polymer matrix that is clad by (and sometimes contains) high-strength aluminum sheets to improve corrosion resistance and give more isotropic mechanical properties than would be possible by the fiber reinforced polymer. Both aluminum and non-aluminum composites require a whole new technology for design and production of the final structural parts. This problem provides an exciting technological challenge, some of the materials aspects of which are described in this volume.

Perhaps the most significant recent development in aluminum physical metallurgy has been the interest in Al-Li alloys. It has been known for many years that lithium was a valuable alloying addition in aluminum, giving a major reduction in density, additional precipitation hardening capacity, and, most remarkably, a significant increase in elastic stiffness.

The last phenomenon is not understood since usually the effect of a solute addition on the elastic stiffness of the matrix is related to the melting point of the added solute. For example, low melting magnesium reduces the stiffness of aluminum while high melting solutes such as manganese raise the stiffness of aluminum. Lithium, however, although it is a very low-melting point metal, appears to raise the bond strength in the aluminum matrix and thus its elastic modulus. Because the atomic weight of lithium is very low, small weight percentage additions of lithium (1–3 wt%) give large atomic fractions (3–10 %), so the effect of small amounts of Li can be very significant in reducing density and increasing elastic stiffness. The challenges presented by Al-Li are to develop alloys that can be produced despite the chemical reactive nature of lithium metal and to produce Al-Li alloys that have acceptable fracture toughness values. Difficulties, mainly with intergranular brittleness, delayed the introduction of this new class of alloys until recently when pressures for low-density, stiff aluminum alloys led to much of the successful research and development described here. Potentially of major importance is the improved fatigue resistance of many of the lithium containing alloys, which are apparently associated with highly branched fatique cracks seen under conditions where the alloys show much lower crack growth rates than alternative alloys. It is widely expected that successful translation of the research described here to technological application will largely determine the types of aerospace vehicles flying in the next century. Much still remains to be understood in the science of the Al-Li alloys and in the older alloys as well.

The editors of this volume are very grateful to the authors of the individual chapters for their contributions. We were delighted to have a mix of authors: 14 work in industry, 12 work at universities, and 7 of the authors work outside the United States. All are acknowledged experts in their various fields and have made major contributions to the developments they describe. As is inevitable in this type of volume, there is some duplication and, more seriously, some topics not covered to the extent that the editors would have liked, particularly in the area of shaped castings and high-temperature materials. However, only by this type of multi-authored work could we have produced a volume describing the current state of research in this exciting and fast-moving field. We intend the book to be of interest to several different types of readers; new and established workers in the field of aluminum metallurgy, users of aluminum alloys, materials engineers working with other materials who will be intested in the current developments in our field, and, most importantly, to students of all ages of materials science and engineering. We hope that all our readers, particularly the last group, will see in the present volume how the many different aspects of physical metallurgy must be brought together for successful technological application of a critical material.

We are very grateful to the authors for providing novel and stimulating discussions of their fields, to the reviewers of the chapters for their critical and helpful comments, to the staff at Academic Press for all their help, and also to Drexel University, Alcoa Technical Center, and the Office of Naval Research for providing facilities while we were editing this volume.

A. K. Vasudevan
R. D. Doherty

July 1989

Part I

Historical Background

1

History of Wrought-Aluminum-Alloy Development

J.T. STALEY

Alcoa Laboratories
Alcoa Technical Center
Alcoa Center, Pennsylvania

ISBN 0-12-341831-3

In almost every instance, aluminum and its alloys have replaced another material in an existing application. The driving force for the replacement depends on the application. The first aluminum alloys were developed when characteristics of unalloyed aluminum were insufficient to meet all of the needs of the prospective user. For structural use, the strongest alloy that meets the minimum requirements for properties such as corrosion resistance and ductility or toughness is usually selected if it is cost-effective. The continuing pursuit for cost-effective solutions to materials problems remains the driving force for alloy development today. Properties depend on the microstructure, which, in turn, is controlled by both composition and processing. Consequently, the development of tempers is as least as important as the establishment of the composition. This brief description of the history of aluminum alloys will confine itself to the development of wrought alloys and tempers used as structural materials. Development of the high-strength alloys will be described in greater detail because more is known about them.

I. Al-Mn Alloys

A. 3003

Unalloyed aluminum containing 0.95% Fe and Si as impurities is designated 1100 and develops a yield strength of 35 MPa in the annealed, O, temper. That can be increased to 150 MPa by cold-rolling 80% to the H18 temper. To increase strength, attempts were made to alloy aluminum with other elements. Most of the alloying additions decreased resistance to corrosion. Manganese was found to increase strength in both the O and H18 tempers, however, without sacrificing corrosion resistance, particularly when a small amount of copper was also present. An alloy containing 1.25% Mn and 0.12% Cu, now designated 3003, was introduced in the United States in 1906. Alloy 3003-O developed about 12-MPa yield strength and 30% elongation, while 3003-H18 developed about 180-MPa yield strength and 4% elongation. To assess its resistance to corrosion, specimens were exposed in the artificial tropical climate of Phipps Conservatory in Pittsburgh. In the 1970s, it was demonstrated that 3003 can develop yield strength up to 400 MPa by special thermal-mechanical treatments. This process has never been commercialized because ductility suffers when strength is increased by cold-working.

B. 3004

To meet the need for an economical sheet material that would develop higher strength than 3003 with excellent corrosion resistance, 3004 was commercialized in 1929. This alloy, containing 1.25% Mn and 1% Mg, developed yield and tensile strengths double those of 1100 in the annealed temper at a tensile elongation of 20%, and corrosion resistance was excellent. In the cold-worked H18 temper, 3004 developed tensile and yield strengths over 50% higher than that of 1100-H18 at the same tensile elongation value of 5%. In the 1980s, more 3004-H19 sheet is produced than any other aluminum alloy because it is the material of choice for the body of beer and beverage containers, the largest market for aluminum sheet. The greatest reasons for its popularity are ability of this severely cold worked sheet to be drawn into a cup, to be ironed to form a can, to maintain sufficient strength after pasteurization or paint baking at elevated temperatures, and to be compatible with the Al-Mg-Mn alloy used for the easy-open end so that cans are recyclable.

C. 3X05 and Others

Although 3003 and 3004 remain the premier Al-Mn alloys, a number of other alloys have been developed over the years for particular combinations of properties. Alloy 3005 is a hybrid between 3003 and 3004 both in composition and properties. Alloy 3105 has lower manganese and higher chromium. Two of the newer alloys, registered in 1978, contain zirconium. One of them (3009) contains 1.4% Si and 1.5% Mn; the other (3011) contains 1% Mn plus 0.12% Cu and 0.25% Cr.

II. Al-Cu Alloys

A. 2X17

In 1906, the same year that 3003 was introduced in the United States, Wilm discovered in Germany the phenomenon of precipitation-hardening at room temperature (natural aging) following quenching. The aluminum alloy contained about 4% Cu, 0.5% Mg, 0.5% Mn, and the usual Fe and Si impurities that were usually around 0.5% each. Experimental production of sheet was begun in 1908 at the Durener Mettwerke-A.G. plant, which licensed the patent and called the material duralumin. Vickers in England also became a licensee and produced the alloy in some quantity starting in 1911, the same year that Wilm published his work. Count von Zeppelin had been looking for an aluminum alloy with higher specific strength than the Al-Mn alloy available for his lighter-than-air vehicles, and provided impetus for the development of these materials. Germany produced almost 100 zeppelins for the first World War, and more than 1,650,000 pounds of duralumin were provided in one year. This alloy also began to replace the spruce and fabric construction in parts of some airplanes. The first all-metal airplane is believed to be a Junkers F-13, which first flew in 1919. As soon as word of Wilm's discovery reached this country, work began to develop an alloy of this type. Laboratory heats were made and tested in 1911, and Alcoa commercialized its version in 1916. The yield strength was 280 MPa, the ultimate strength was 430 MPa, and (in contrast to the low ductility associated with the cold-worked tempers of 1100 or 3003) tensile elongation was high. Aging at elevated temperatures also increased yield strength somewhat, but the accompanying loss in elongation values discouraged the use in such tempers.

To increase the corrosion resistance of 2017-T4 sheet, a clad product was developed for the aircraft industry and commercialized in 1928. This product, known as alclad 2017-T4 sheet, consists of a 2017 core covered on one or both sides with a surface layer of high-purity aluminum that is 2.5–10% of the total thickness of the product. The underlying alloy is cathodic relative to the pure metal cladding so it is galvanically protected not only where the cladding is continuous, but also where the core is exposed, as at sheared edges, rivet holes, and scratches through the surface layer. The Northrop Alpha, credited with being the first all-metal airplane in the United States, was constructed almost exclusively of this material.

Although 2017-T4 was a commercial success, it was not without its problems. Production personnel were not pleased with the fabricability of the new alloy. They found that 2017 was both hot and cold short, and that it was particularly difficult to forge. Users found that rivets made from 2017 sometimes cracked. Alloy 2117, a lower copper version of 2017, provided the answer to the rivet problem and is still used for rivets as of the late 1980s.

B. 2025

Alloy 2025-T6 was developed to provide improved forgeability. A large use was for propeller blades. This alloy contained no addition of magnesium, but had a controlled addition of 0.8% Si in addition to 4.25% Cu and 0.75% Mn. Unlike 2017, it did not age-harden at room temperature; it had to be age-hardened at elevated temperatures (artificial aging) to develop the strength of 2017-T4. Alloy 2025-T6 was susceptible to intergranular corrosion, but the degree of susceptibility did not prevent it from being used successfully in many applications.

C. 2014

This alloy, introduced in 1928, combined the magnesium addition in 2017 with the silicon addition in 2025. Yield strength in the T4 temper was only slightly higher than that of 2017-T4, but the alloy responded to artificial aging to such an extent that 2014-T6 developed a yield strength of 420 MPa, 50% higher than that of 2017-T4. Tensile-elongation values of 2014-T6 were 13%, significantly lower than the 22% developed by 2017-T4. Aircraft designers at the time were reluctant to use 2014-T6 sheet because of the lower tensile elongation, but did try it in forgings. The forged products performed successfully and led to the use of extrusions and sheet. 2014 forgings continue to be used for the manufacture of airplane wheels in the United States, and this alloy remains a general-purpose, high-strength forging alloy, but it is rarely manufactured in other product forms. In the U.K., however, 2014 in both T4 and T6 tempers is still used extensively in aircraft as flat-rolled products and extrusions as well as forgings.

D. 2X24

Naturally Aged Tempers. The continuing demand of the aircraft industry for lighter-weight structures led to the development of alloy 2024-T3 in 1931. This alloy contains higher magnesium than 2017 (1.5% versus 0.5%). It develops almost 20% higher yield strength than 2017-T4 along with a high ratio of ultimate to yield strength and high tensile-elongation values. The T3 temper designation recognizes the significant increase in strength attributable to the 1.5 to 3% stretch applied to rolled and extruded products after quenching to straighten and/or minimize effects of residual quenching stresses. Because of its higher strength, this alloy soon supplanted 2017-T4, and the alclad version remains the alloy of choice for the fuselage skin of every commercial passenger airplane. The venerable and extremely successful DC-3 was constructed almost exclusively from alclad 2024-T3. This was the first plane that was economical enough to fly that airlines could make a profit without carrying mail in addition to passengers, and the higher fuel economy made possible by the stronger alloy was a major factor in its success. It has also been noted, along with the jeep, as one of the critical vehicles in World War II. Alclad 2024-T3 is also used in most commercial passenger planes for the lower wing skins and is used almost exclusively by the builders of the smaller general aviation aircraft.

In the late 1970s, Boeing was designing the 757 and 767. They wanted to use a stronger alloy for the lower wing skin than the 2024-T3 plate and extrusions that they had been using. The goal was to increase strength without sacrificing other characteristics. The approach was to start with alloy 2024 and increase strength by increasing the dislocation density and to maintain toughness by reducing the number of coarse constituent particles. Details differed for plate and extrusions. For plate, the dislocation density was increased by cold-rolling; for extrusions, it was increased by ensuring that the dynamically recovered structure produced during extrusion was stable enough so that the product didn't recrystallize during solution heat treatment. Toughness was maintained at the higher strength level by decreasing the volume fraction of coarse intermetallic particles. Limiting the iron and silicon contents reduced the number of Al_7Cu_2Fe and Mg_2Si constituent particles, and adjusting the copper and magnesium minimized the volume fraction of Al_2Cu and Al_2CuMg particles that were not capable of being dissolved at the solution temperature. The new materials were designated 2324-T39 plate and 2224-T3 extrusions.

Artificially Aged Tempers. The yield strength of alloy 2024 can be increased by artificial aging, but aircraft designers were reluctant to try the material because of the small difference between yield and ultimate strengths. Initial results also indicated that strength could be increased even higher by cold-work prior to the artificial aging treatment (T8 temper). The increase in strength was accompanied by a decrease in corrosion resistance at short aging times, but aging at temperatures near 200°C could provide good resistance at peak strength. Some 2024-T8 was used in the

1940s, but the large increase came in the supersonic era, at which time designers selected it for areas that would experience aerodynamic heating.

When aircraft designers began to machine thick plate (100–150 mm) to make bulkheads to reduce the number of fasteners, the properties in the short-transverse direction (through the thickness) began to be important. The short-transverse ductility of 2024-T8 was unacceptable, so efforts to increase this property began in the 1960s. An early solution was to preforge the ingot prior to rolling to modify the intermetallic particles that produced stringers during rolling. This solution was uneconomical, however, so alloy 2124-T8 was developed. This alloy contained significantly less iron and silicon than 2024, and copper and magnesium were controlled to minimize the volume fraction of Al_2CuMg and Mg_2Si constituent particles that could not be dissolved during solution heat treating. Closely controlled thermal-mechanical treatments were employed to minimize the formation of stringers. Alloy 2124-T8 is the major material of construction in the space shuttle orbiter.

E. 2011

Although aluminum alloys are easily machinable, they form long chips during drilling or turning. Alloy 2011 was developed in 1934 to fill the need for a high-strength alloy that could be machined without forming the long chips. The addition of small amounts of lead and bismuth, which are virtually insoluble in aluminum, solved the problem.

F. 2X18

The National Physical Laboratories in the United Kingdom had been evaluating the addition of nickel to improve the elevated temperature properties of Al-Cu-Mg alloys since the early 1920s. Their Y alloy (containing 4% Cu, 2% Ni, and 1.5% Mg) developed strength at 200°C that was higher than that of 2017 at room temperature. The Al-Cu-Mg-Ni alloy, 2618-T61, was introduced in the United States in 1932. This was followed two years later by 2218-T72, a modified version combined with a new temper that provided more stable properties at elevated temperatures. New versions followed over the years, and the latest version, 2618, recognized that iron was a useful alloying element and not just an impurity. Iron and nickel in equal amounts combine with aluminum to form Al_9(Fe-Ni) and confer stable strength at elevated temperatures. This alloy (known as

RR58 in the U.K.) was selected as the major construction material for the Concorde because of its creep resistance. The main use of 2618-T6 in the United States is as a mold material for the manufacture of tires.

G. 2X19

This alloy was introduced in 1954. It developed mechanical properties higher than those of any aluminum alloy at temperatures above about 230°C. It contains nominally 6.1% Cu, 0.3% Mn, 0.15% Zr, and 0.1% V. The latter three elements all form dispersoids, high-temperature precipitates, which contribute to the strength at elevated temperatures. In addition, welding characteristics and weld strengths are excellent. Alloy 2219-T8 also possesses superior properties at cryogenic temperatures and is resistant to corrosion by a number of chemicals. It has been used for a number of applications where its weldability, mechanical properties, and corrosion resistance made it the best choice. A high-purity, controlled composition version, 2419, is used in the new BIB aircraft. A modification containing a small addition of magnesium offers higher strength, particularly in highly worked T8 tempers. This alloy, registered as 2519, may replace 2219 for welded tanks in the space shuttle and may be used for armor plate.

H. Al-Cu-Li

Alloy 2020 was the first commercial alloy to contain a significant amount of lithium. Developed in the late 1950s to provide a high ratio of strength to density, this alloy contained 4% Cu, 1.4% Li, .5% Mn, and 0.2% Cd. The cadmium was added to increase strength by its effect on nucleating the hardening phase. Alloy 2020-T6 developed strength equal to that of 7075-T6 with lower density and higher modulus. Although this alloy developed the lowest fracture toughness of any commercial aluminum alloy, it was used successfully as both upper and lower wing skins and part of the tail for the RA5C Vigilante aircraft. In-service corrosion resistance of this United States Navy plane was outstanding although it was used bare. Initial attempts to increase toughness without sacrificing the strength of 2020-T6 were unsuccessful. Furthermore, the production problems with the alloy were severe. Because of the combination of low toughness and production difficulties, the alloy was withdrawn in 1974. Interest in the use of lithium as an alloying element in aluminum alloys experienced a resurgence in the late 1970s as airframe builders looked for new materials to reduce weight. As of 1988, four Al–Li alloys have been registered. All contain zirconium to control grain structure of the product. Two of the alloys contain a higher weight percentage of copper than lithium and are designated as 2090 and 2091. The other two contain a higher weight percentage of lithium and are designated 8090 and 8091. Alloy 2090 does not contain magnesium as an alloying addition.

Alloy	Li	Cu	Mg	Zr
2090	1.9/2.6	2.4/3.0	0.25*	0.08/0.15
2091	1.7/2.3	1.8/2.5	1.1/1.9	0.04/0.16
8090	2.2/2.7	1.0/1.6	0.6/1.3	0.04–0.16
8091	2.4–2.8	1.6–2.2	0.5–1.2	0.08–0.16

* = maximum allowed.

Alloy 2090, developed in the U.S., is strengthened by Al_3Li, designated δ', and Al_2CuLi, designated T_1. In a T8 temper, it develops the strength of 7075-T6 with comparable toughness. It is being test flown and is under consideration for several applications. Because it is weldable, alloy 2090 is a candidate for a new space launch vehicle. Alloy, 2091, developed in France, is strengthened by metastable Al_2CuMg, designated S$'$, in addition to T_1 and δ'. It has been tested in a French fighter prototype. Because its properties are comparable to those of alloy 2024-T3 with lower density, it is a leading contender for the portion of the fuselage of a commercial airliner to be manufactured by Aerospatiale. Alloy 8090, developed in Great Britain, has the lowest density of the commercial alloys. It is also strengthened by S$'$, T_1, and δ'. This alloy is being used by Lockheed for a classified application and is a leading candidate for the portion of the fuselage of a commercial airliner to be manufactured in Great Britain. Alloy 8091, a higher copper version of 8090, is being evaluated for several applications.

I. 2036

This alloy was developed to fill a different type of need. In the 1970s, the United States auto industry was receptive to substitution of traditional steel alloys by lightweight materials in an effort to achieve government-mandated targets for fuel economy. They desired an economical, lightweight alloy that was as strong and formable as mild steel. A French alloy called Avional 21 had been used for packaging for some time. The alloy, which is similar to 2117-T4, developed medium strength and good formability. A modification of this alloy was registered in the United States in

1970 as 2036 to fill the need of the auto industry. It contains nominally 2.5% Cu, 0.5% Mg, and .25% Mn. Alloy 2036-T4 was used to solve particular weight problems in the 1970s, but its lower formability compared to 1010 steel and its higher cost have limited its use.

III. Al-Mg-Si Alloys

A. *6X51*

Wilm's discovery led to searches for other alloys that could age-harden. Although binary Al-Mg and Al-Si alloys could not be age-hardened, attempts to heat treat Al-Mg-Si alloys proved to be successful. Alloy 6051, a copper-free alloy containing 1% Si and 0.5% Mg, was introduced in the United States in 1921. Although not as strong as 2017, it was significantly stronger than 3003 and as fabricable as 1100. It was largely used for furniture in the T4 temper because of the unique combination of strength, corrosion resistance, and fabricability. It also found use in the T6 temper as forgings, extrusions, and tubing although it was also susceptible to intergranular corrosion. An improved version of 6051, 6151, was introduced in 1928. It developed higher yield strength along with higher elongation values. Other versions have been introduced over the years. Alloy 6351 was developed for its finishing characteristics.

B. *6061*

This heat-treatable alloy, introduced in 1935, has long set the standard for a lightweight, economical, medium-strength, corrosion-resistant material for general-purpose structural use. It has a balanced composition in that the 1% Mg and the 0.6% Si are in the ratio to form Mg_2Si. It also contains an addition of 0.25% Cu and 0.04 to 0.35% Cr. The lower amounts of chromium are usually used in the manufacture of extruded products that, for economic reasons, are not subjected to a separate solution heat treatment but are quenched as they leave the die.

C. *6X63*

The year 1944 marked the introduction of 6063, the most widely used extrusion alloy. It contains nominally 0.67% Mg and 0.4% Si, but more versions of this alloy are probably made for special customer use than there are versions of any other alloy. Its main virtue is that it can be extruded readily with low pressures at low temperatures and is quenchable at the press by forced air cooling. Moreover, it can be easily finished by almost any process. Most of the aluminum extrusions in the building and construction industry are versions of 6063.

D. 6066, 6070, and 6013

These alloys contain higher solute and develop higher strength than 6061. The corrosion resistance of 6066, registered in 1954, and the toughness and ductility of 6070, introduced in 1962, are inferior to these characteristics in 6061, so these alloys have not been used widely. Laboratory tests and initial field performance of alloy 6013, introduced in 1983, strongly indicate that this alloy can develop strength 25% higher than that of 6061-T6 with no sacrifice in other characteristics. Because this alloy is not susceptible to the development of Luders lines during stretching, it has been used for applications requiring stretch forming.

E. 6009 and 6010

These alloys were introduced in 1976 to fill the need for materials that would be highly formable in the soft T4 temper but could age-harden during the automotive paint bake cycle. At the time, alloy 5182-0 was being used for the inner panels, which require the highest formability. Alloy 2036-T4 was the choice for outer panels that require higher strength. The scraps of these alloys were incompatible, however, and the duo-alloy combination of 6009 and 6010 was developed as a replacement. Alloy 6009 is being used in the automotive industry in the late 1980s, and alloy 6010-T6 has found use as an alloy for canoes.

IV. Al-Mg Alloys

A. 5X54

Because of its low density, magnesium had been evaluated as an alloying element before the 1900s. The ultimate strength of Al-Mg alloys in the annealed temper increases progressively with rising magnesium content to about 15%, and the tensile yield strength continues to increase up to about 20% Mg. The yield strength of an annealed Al-10% Mg alloy is about 175 MPa, and the ultimate strength is about 370 MPa. Moreover, magnesium is an unusual, if not unique, alloying element in aluminum in that after an initial decrease with the first few percent addition, tensile elongation increases slightly with subsequent additions of up to 10%. This phenomenon is attributed today to the effect of magnesium on decreasing stacking-fault energy. This decrease in stacking-fault energy decreases the ability to cross-slip giving the observed high rate of work-hardening. The alloys are also resistant to corrosion, even in salt water, and are highly weldable. These benefits are not without drawbacks. Alloys containing more than about 3.5% Mg are susceptible to stress-corrosion cracking and exfoliation corrosion unless the microstructure is carefully controlled, and material in cold-worked tempers age-soften at room temperature. The challenge of alloy developers is to determine how to capture the maximum potential of such alloys by control of composition and microstructure. Development of tempers to solve these problems plays a significant role in the history of Al-Mg alloy development.

The high cost of magnesium, the tendency of aluminum alloys containing more than about 1% Mg to form a large amount of dross during melting, and the difficulty of fabrication limited the early development of Al-Mg alloys. Sheet of an alloy containing 4% Mg was produced in the United States in 1921, but the alloy was not commercialized. Because of the need for an alloy that would have high strength in the as-welded condition, development activity of Al-Mg alloys increased in the early 1930s. The alloys also contained small amounts of chromium, manganese, or both. Generally, chromium was preferred in the United States and manganese in Europe. Alloy 5154 was introduced in 1932. This alloy contained 3.5% Mg and .25% Cr. It developed the highest strength of all aluminum alloys in the annealed temper: 120 MPa yield and 240 MPa ultimate. The H38 temper was developed to solve the problem of age-softening. This stabilizing treatment includes exposure of a few hours at 120%C after cold-rolling to provide a stable structure with improved ductility. At about the same time in England, a 3.5% Mg alloy containing 0.5% Mn instead of chromium was introduced. Because of the opportunity for precipitation in strain-hardened Al-Mg alloys of anodic aluminum-magnesium compounds at temperatures between about 70 and 145°C, these alloys may become sensitized after exposure to temperatures in this range. To alleviate this potential problem for applications where the product may be exposed at these temperatures, a lower-solute version of 5154 was introduced in 1957 as 5454. The alloy contained 2.75% Mg, 0.8% Mn, and 0.1% Cr.

B. 5X52

The need for a less expensive, more fabricable alloy for sheet-metal parts led to the development of 5052 in the 1930s. This alloy contains 2.5%

Mg and 0.25% Cr and was available in the annealed and stabilized tempers. It has been used for packaging and general applications. A number of modifications of 5052 have been introduced over the years, often for finishing applications and in the chemical industry for storage tanks.

C. 5X05

Alloy 5005 was introduced in 1935 to fill the need of the budding mobile-home industry for a lightweight, inexpensive, workable, corrosion-resistant siding material. It is essentially a binary alloy with 0.85% Mg with maximum levels on copper, chromium, and manganese. Alloy 5205, registered in 1967, has a controlled amount of copper with lower limits on chromium and manganese.

D. 5X56, 5X83, and 5X82

In the early thirties, there was an opportunity for an aluminum alloy that could be drawn into fine wire for several applications. An alloy containing 6% Mg was tried initially, and the magnesium level was subsequently lowered to a nominal 5.25% with an addition of 0.25% Cr. A problem developed when the chromium separated as coarse, primary $CrAl_7$ crystals during solidification. These brittle intermetallic particles caused breakage during the wire-drawing process. Substituting 0.1% Mn plus 0.1% Cr for the 0.25% Cr solved the problem, and 5056 wire was commercialized in 1936. This product performed creditably for armoring cable in aircraft, for marine service, and for hookless fasteners. After a few years of service as insect screening, however, the fine, severely cold worked wire was found to be susceptible to intergranular corrosion. Alclad 5056 wire was developed to solve the problem. Continuing research and development led to incremental improvements in the base alloy and tempers for use as rolled products and extrusions. This work led to 5156 and 5256. In the fifties, corrosion problems with the high-solute alloys led to the introduction of H2 and H3 tempers. The H2 tempers were obtained by cold-rolling followed by a high-temperature stabilizing treatment. The H3 tempers utilized cold-rolling followed by a low-temperature stabilizing treatment. Alloy 5356-H321 was one of the first applications of an H3 temper. Another modification of alloy 5056 appeared in 1956 with the introduction of 5456. This alloy contains 0.8% Mn along with a nominal 5.1% Mg and 0.1% Cr. Alloy 5456 is the strongest non–heat-treatable alloy commercially available for welded structures. Alloy 5083 and 5183 are similar to 5456, but the nominal magnesium levels are less than 5%. The H323 and

H343 tempers for 5X56 and 5X83 alloys were introduced in 1960. Corrosion of military vehicles in the Vietnam War and with crab boats due to uncontrolled precipitation of Mg_2Al_3 led to the development of the H11X tempers for these alloys. These tempers can be made by alternate processes. In one, the plate is cooled rapidly from the hot-rolling temperature to prevent precipitation of magnesium. The plate is then cold-rolled to develop required strength. The other approach is to provide a partial annealing treatment after cold-rolling to precipitate the magnesium in a controlled manner and then cold-roll to final gage. Many of the armored vehicles are made from 5083-H131 because of its low density, weldability, strength, resistance to exfoliation-corrosion and stress-corrosion, and ability to resist penetration of projectiles. Alloy 5082, introduced in 1963, is essentially a binary 4.5% Mg alloy. The low number of constituent particles provides increased formability. A modification, 5182 contains 0.4% Mn for higher strength. In the hard-rolled H19 temper, this alloy possesses the unique characteristics required for the easy-open end for beer and beverage containers.

E. Higher-Mg Alloys

In the late twenties, a sheet alloy with 10% Mg was evaluated for saucepans. When it was tested for resistance to corrosion by tap water, the pans began to corrode by a process poetically described as "like the unfolding of rose petals." These observations led to extensive research into the mechanism of corrosion in these alloys. Briefly, precipitated particles of the aluminum-magnesium compound are strongly anodic to the depleted solid solution and, hence, are selectively corroded by electrolytic action. The corrosion problems plus the tendency to age-softening that increases with rising magnesium level have thwarted efforts to develop commercial alloys with higher magnesium contents than the maximum 5.5% in 5456. Alloy containing up to 9% Mg with small amounts of magnesium were introduced both in England and in Germany. When the Germans attempted to introduce them into the United States in the 1930s, they were informed of the United States experiences. They withdrew the alloys and began their own research. Despite the problems, the potential benefits of high strength, low density, and weldability attract periodic interest. Several experimental alloys containing nearly 7% magnesium were developed in the early 1980s for use in armored vehicles. They are being evaluated by vehicle manufacturers. United States Navy interest in lightweight alloys for ships led to the development of an alloy containing 8% Mg in the 1970s. The alloy was not commercialized, however, because of production difficulties in both hot- and cold-rolling. Research in the 1980s with alloys containing 10% Mg indicates that the corrosion problems can be solved successfully by thermal-mechanical treatments that modify the dislocation substructure and the Al_3Mg precipitates. Probability that these treatments can be applied in a commercial facility are low at this time.

V. Al-Zn-Mg-Cu Alloys

A. 7075-T6

German investigators had shown in the early 1920s that Al-Zn-Mg alloys developed exceptionally high tensile properties. These ternary alloys, however, were susceptible to stress-corrosion cracking. Much research was done in the United States, Europe, and Japan in the late 1920s and 1930s to understand the phenomenon and to develop commercial alloys. The first continuously produced Al-Zn-Mg alloy in the United States was 7076-T6. It was introduced in 1940 and contained 7.5% Zn, 1.6% Mg, 0.7% Cu, and 0.6% Mn. It was developed as a high-strength forging alloy for the production of propeller blades that were more resistant to abrasion and fatigue than alloy 2025.

By 1938, laboratory stress-corrosion tests indicated that the stress-corrosion problem in sheet had been solved, and parts of an airplane wing were fabricated in an alloy known at the time in the United States as X74S. The alloy contained 5.2% Zn, 2.1% Mg, 1.5% Cu, and 0.4% Mn. The laboratory tests used to evaluate X74S and other candidates, however, did not include tests of sheet plastically deformed in the heat-treated temper. Field experience soon showed that X74S sheet was highly susceptible in this condition, so the search for an alloy that would be resistant as formed sheet components continued. The base composition of X74S was retained, and effects of small additions of high-melting elements were evaluated on plastically deformed sheet. Alloys containing 0.2 to 0.35% Cr developed high resistance to stress-corrosion. To meet the demands for higher strength, the zinc, magnesium, and copper contents were increased slightly, and when tests confirmed that the more highly alloyed composition developed adequate resistance, the alloy known today as 7075-T6 was introduced in 1943. Interestingly, chemical analysis of sheet from a downed Japanese Zero fighter aircraft disclosed that the composition was almost the same as that of 7075. The new alloy was quickly introduced into designs on the drawing boards. Alloy 7075-T6 sheet was used as the skin and stringers for the upper wing skin of a later version of the B-29 Super

Fortress with a saving of over 180 kg per airplane. Extrusions soon followed sheet, and forgings were produced in mid-1945 and plate in 1946. The adverse effects of chromium on the ability to develop high strength in the center of thick products and in products quenched at a low rate were noted early.

Why additions of chromium affected resistance to stress-corrosion and quench sensitivity were unknown at the time. Today we know that the chromium remains in solid solution during ingot casting and precipitates during the ingot preheat treatment as dispersoid particles of $Al_{12}Mg_2Cr$ known as E phase. During hot-working, the product dynamically recovers. This recovered structure recrystallizes during solution heat treatment of material not containing dispersoids. The dynamically recovered structure is retained in 7075 when the hot-worked plate, forging, or extrusion is solution heat treated, however, because the E-phase dispersoids pin the subgrain boundaries. The higher dislocation density in cold-worked 7075 sheet provides enough stored energy to overcome the pinning effect of the dispersoids, so the sheet recrystallizes during solution heat treatment. The dispersoids are still effective in pinning high-angle grain boundaries, however, so the sheet recrystallizes to an "elongated-pancake" grain structure. Stress-corrosion in aluminum alloys is almost invariably intergranular. The elongated grain structure prevents the development of high stress along grain boundaries, thus improving resistance to stress-corrosion. The effect of chromium on quench sensitivity is attributed to the $Al_{12}Mg_2Cr$ dispersoid particles. During all but the most rapid quench from the solution-heat-treatment temperature, the solid solution is depleted by the precipitation of the solute elements as Mg(ZnCuAl), eta phase. In an alloy containing only aluminum, zinc, magnesium, and copper, the precipitates form only at grain boundaries. In 7075, however, this precipitate nucleates as well on the E-phase particles. This incoherent precipitation depletes the matrix for subsequent precipitation of coherent Guinier-Preston (GP) zones and partially coherent eta prime during artificial aging, so lower strength is developed. The precipitation of E-phase dispersoids may also contribute to the improved resistance to stress-corrosion cracking by developing local sites anodic to the adjacent matrix that would be depleted of the solute elements.

B. 7178 and 7001

The incessant demand of the aircraft industry for lighter weight structure led to the development of 7178-T6 in 1951. This alloy contained higher zinc, magnesium, and copper than 7075, and offered 35 MPa higher strength. It was used in the United States and the United Kingdom in sheet and extruded form for skin and stiffeners for the upper wing skins of several airplanes and was even used as fuselage sheet in a military plane because of the weight advantage over 7075-T6. Service failures of components of 7075-T6 and 7178-T6 in planes built in the United Kingdom in the 1950s led to a retraction of the use of 7XXX alloys in that country for many years. The alloy gradually fell into disfavor when considerations for fracture toughness and damage tolerance became prominent. Alloy 7001, introduced in 1955, contained even more solute than 7178 and developed higher strength. Because of increasing emphasis by aircraft designers on toughness, this alloy was not used to any extent. It is interesting historically, however, because it marked the peak in strength of aluminum alloys for aircraft. Future alloys would be developed to meet other criteria and to solve deficiencies in older alloys that developed as requirements changed.

C. 7079-T6

In the 1950s, aircraft designers wanted to use more massive forgings. Attempts to use 7075-T6 were disappointing because of the alloy's quench sensitivity. This was aggravated in forgings because they are usually quenched in hot water to minimize residual quenching stresses. Furthermore, field experience with 7075-T6 products that were thick enough to develop stresses through the thickness revealed that they were susceptible to stress-corrosion cracking in the short-transverse direction. Laboratory tests in 3.5% NaCl solution by alternate immersion showed that they could develop cracks at stresses as low as 70 MPa. In addition, ductility in the short-transverse direction was lower than in the other principal directions. To fill the need for an improved forging alloy, Hy43 (an alloy developed in Germany) was evaluated in the United States. Tests were encouraging. Ductility in the critical short-transverse direction was superior to that of the other alloys; strength remained high in slowly quenched sections; and accelerated stress-corrosion in the NaCl solution indicated that performance would be superior to that of the alloys currently in use. This alloy was commercialized in the United States in 1954 as 7079-T6. It contained 4.3% Zn, 3.3% Mg, 0.6% Cu, 0.2% Mn, and 0.2% Cr. The alloy was widely specified in the late 1950s and early 1960s for a variety of new and old designs. Despite the good resistance predicted by the accelerated testing, however, service failures by stress-corrosion cracking began in the late 1950s. The problem became so severe that some 7079-T6 forgings failed in the forge plant before they could be shipped. The experience so affected materials people in the aircraft industry that even in the late 1980s, when a new aluminum alloy is proposed, they ask: "How do we know this won't be another 7079-T6?" We know now that, although 7079-T6 is resistant to the initiation of stress-corrosion, the rate of growth of stress-corrosion cracks in this alloy is orders of magnitude faster than that in 7075-T6. Testing of precracked fracture-mechanics–type specimens was unknown at the time that 7079-T6 was first evaluated, so the deficiency in its characteristics was not discovered.

D. 7075-T73

The aircraft industry stimulated aluminum producers to solve the problem of stress-corrosion cracking in the short-transverse direction. Besides the problems with 7079-T6 in the field, the new wide-body aircraft were being designed, and designers wanted to use thick forgings and plate. The initial response to the challenge was the development of the stress-corrosion–resistant T73 temper for 7075. Experiments showed that aging for sufficient time at temperatures above about 150°C significantly increased the resistance to stress-corrosion cracking in the short-transverse direction. Preceding the overaging treatment by a first-step near 120°C minimized the strength loss during aging above 150°C. Although products processed using this overaging treatment developed strengths about 15% lower than 7075-T6 and 20% lower than those of 7079-T6 in thick sections, the aircraft industry welcomed the new temper. One of the first applications for 7075-T73 was in the DC10. These were the largest aluminum forgings produced. The reason for the pronounced increase in resistance to stress-corrosion cracking with aging above 150°C has been studied by many investigators over the years, but no explanation is universally accepted. Regardless of the mechanism, products of 7075-T73 have been in service for over 20 years, and no known instance of stress-corrosion cracking has been reported.

E. 7075-T76

Another problem with the corrosion performance of 7075-T6 surfaced in the early 1960s. Light-gauge plate and extrusions began to suffer exfoliation corrosion. In this form of corrosion, attack proceeds mainly along grain boundaries. The corrosion product wedges open the grains causing both esthetic and mechanical degradation. This problem, combined with the stress-corrosion problem in the short-transverse direction of 7075 and 7079 products, caused the United States Air Force to bar the use of T6 tempers in products of these alloys that were thicker than about 60mm. Alloy 7075-T73 was not prone to exfoliation corrosion, but designers did not want to accept the 15% strength loss. Alcoa developed the mildly overaged T76 tempers for 7075 and 7178 to solve this problem. One of the first uses of the new temper was as one-side alclad 7075-T76 for the wing skins of the L1011. A new cladding alloy had to be solved to make the application successful. These wings are slightly gull-shaped, and peening with glass shot was used to plastically deform them to conform to the intended shape. When 7075-T76 clad with the conventional low-strength 1% Zn cladding alloy was tried, several problems arose. First, the soft cladding deformed to such an extent that forming was difficult. Second, the peened cladding was esthetically displeasing. Third, near the wing tips, the cladding became a significant fraction of the thickness, so strength of the section was decreased substantially. An Al-Zn-Mg heat-treatable cladding alloy, 7008, was developed and solved the problem.

F. 7175-T736

Alloy 7175–T36 forgings were also developed in the mid-1960s. The work began by a production forging metallurgist who carefully documented the smallest details concerning the composition and processing of 7075 forgings and correlated them with mechanical properties. He found that certain combinations of compositions and practices resulted in properties of forgings that were consistently in the upper half of the range for 7075-T6. Applying an overaging treatment to forgings processed in this manner provided a product that had strength approaching that of 7075-T6 with substantially higher resistance to stress-corrosion cracking. Since its introduction in 1968, many 7175–T736 forgings have been used successfully in a number of programs.

G. 7X49-T73

Although 7175-T736 was used successfully, it was not useful for sections thicker than about 75 mm because of its quench sensitivity. Several programs were initiated in the 1960s that had the goal of developing an alloy that would provide the strength of 7079-T6 in sections of 75 mm and greater thickness with high resistance to stress-corrosion cracking. Alloy 7049-T73, registered in 1968, was the first of several commercialized. Compared with 7075, the alloy contains higher zinc for higher strength and lower chromium for lower quench sensitivity. It has been used as forgings in a number of airplanes both in new design and as a retrofit for 7079-T6. Alloy 7049-T73 has been used in a few applications as extrusions, but has not found a use as plate. A high-purity version was introduced later to provide higher toughness.

H. AZ74

At about the same time that alloy 7049-T73 and 7175-T736 were being developed, claims were made that additions of silver to Al-Zn-Mg-Cu alloys would provide high strength after aging in the temperature range found to confer high resistance to alloys such as 7075. Careful investigation in the United States of matched pairs of silver-free and silver-bearing Al-Zn-Mg-Cu alloys containing either Cr, Mn, or Zr revealed, however, that silver showed an advantage only when it was compared with product rapidly heated to the artificial aging temperature. Silver-free material that was preaged near 120°C prior to aging above 150°C (a standard practice for 7075-T73) developed the same strength as the silver-bearing counterpart. Stress-corrosion tests showed that the pairs developed the same resistance when compared at the same strength levels. As a result of this work, silver has not been used as an alloying element in Al-Zn-Mg alloys in the United States. An alloy known as AZ74, however, has been commercialized in Germany and has found use in a few applications in Europe in forgings.

I. X7080-T7

One candidate alloy to develop high strength in thick sections of slowly quenched forgings was registered in 1965 as X7080-T7. It contained nominally 6% Zn, 2.5% Mg, 1% Cu, and 0.35% Mn. Initial results were extremely promising. Forgings developed high strength after being quenched at low rates, and results of stress-corrosion tests of smooth, short-transverse specimens exposed in the standard 3.5% NaCl solution by alternate immersion indicated that the alloy would be resistant at high stress levels. Because of the experience with 7079-T6, however, the alloy developers had exposed replicate sets of test specimens in outdoor environments. Despite strong customer interest, none of the new material was sold except for test and evaluation pending performance of specimens that had been exposed outdoors for sufficient time. The conservatism was shown to be justified when specimens began to fail in the New Kensington atmosphere, and registration of the alloy was withdrawn.

J. 7X50 and 7010

Investigations to study the effects of aging practice on the stress-corrosion resistance of 7075 provided an indication that copper remained in solution during conventional T6 aging treatments near 130°C, but precipitated as eta phase during the second aging step of the T73 treatment near 160°C. This result, combined with the lack of success of aging treatments in this temperature range to significantly increase the stress-corrosion resistance of 7079 (0.6% Cu) and X7080 (1% Cu), indicated to alloy developers that an alloy containing copper above the 1.6% nominal amount in 7075 may provide more attractive combinations of strength and resistance to stress-corrosion cracking. Two alloys were selected for the initial evaluation. Both contained zinc and magnesium near the nominal levels in 7075, but copper was raised to about 2.5%. One alloy contained chromium and manganese, the other only manganese. Both were aged 24 hours at 120°C prior to a series of aging treatments near 160°C. Unlike behavior of all other Al-Zn-Mg-Cu alloys evaluated before, strength of the chromium-free alloy initially increased with aging time at 160°C before it decreased. This observation was initially treated with the same combination of excitement and skepticism that Wilm gave to the initial hardness tests of duralumin, but retests confirmed the phenomenon. Significantly, initial stress-corrosion tests indicated that this alloy could develop higher strength while maintaining the excellent resistance to stress-corrosion cracking of 7075-T73. Analysis of other tests had indicated that while the 3.5% NaCl test overrated performance of alloys containing less than about 1.25% Cu, it was predictive of outdoor performance of Al-Zn-Mg alloys containing at least 1.5% Cu. Nevertheless, specimens were exposed in several natural environments to prevent another situation like that of X7080-T7. Meanwhile, other work had demonstrated that additions of 0.1% Zr were about as effective as 0.25% Cr in preventing recrystallization of hot-worked products during solution heat treatment and that additions of 0.5% Mn were not as effective. Furthermore, alloys containing zirconium were not as quench-sensitive as those containing 0.5% Mn and were significantly less quench-sensitive than alloys containing 0.25% Cr. This was attributed to the difficulty of eta-phase nucleating on the coherent Al_3Zr dispersoids compared to the ease of nucleating on the incoherent $Al_{12}Mg_2Cr$ and $Al_{20}Mn_3Cu_2$ particles. In addition, the alloys containing zirconium developed higher toughness. Consequently, alloys containing 0.1% Zr with over 2% Cu were also evaluated. Development proceeded under sponsorship of the United States Naval Systems Command and, later, the United States Air Force Materials Lab. Selection of the optimum combination of zinc, magnesium, and copper required development of a new analytical tool. Because of the different programs involved, one-to-one comparisons were possible on only a few of the many compositions evaluated. In particular, the number of specimens exposed at each stress level and the stress levels employed differed widely. Results of hundreds of specimens from 28 candidate alloys had to be analyzed. The problem was solved using the statistical procedure known as probit analysis. The effects of zinc, magnesium, and copper on the strength that would provide equivalent resistance to stress-corrosion were established. The composition providing the highest strength was 6.2% Zn, 2.3% Cu, and 2.25% Mg. Impurity elements were established at low levels for high fracture toughness, and 0.1% Zr was added because it provided superior toughness and quench sensitivity. The alloy was registered in 1971 as 7050, but was tried before that by the United States Naval Air Systems Command to retrofit the wing carry-through fitting of the A6 plane, which had been made from 7079-T6 and was suffering from stress-corrosion cracks. Development of a casting practice for the large ingots needed for thick plate, forgings, and extrusions required extensive development. Alloy 7050-T736 is the main material of construction for the F18 and is used in many planes built since its introduction. It is also used widely as rivets because of its combination of strength and resistance to stress-corrosion cracking. The European versions contain somewhat less copper for improved ingot castability and are called 7010 in the United Kingdom and Zergal in Italy.

At the same time that Boeing wanted a higher-strength alloy to replace 2024-T3 as the lower wing skin materials, they wanted a higher-strength material to replace 7075-T6 for the upper wing skin. A minor modification to 7050 proved to be successful, and this material, known as 7150-T6, was made as extrusions and plate for the 757 and 767 aircraft.

K. 7475

Another 7XXX-alloy–development program was underway in the sixties concurrent with the programs to develop material with high resistance to stress-corrosion cracking. The goal of this program was to develop a material that was as resistant to fracture as 2024-T3 at significantly higher strength levels. Fracture toughness was just beginning to get attention from the scientific community, and no experience was available. Consequently, the program began by analyzing the relationships between strength and fracture toughness of commercial aluminum alloys and tempers. Although 2024-T3 developed higher toughness than 7075-T6, the 7XXX alloys developed toughness superior to that of the 2XXX alloys when compared at the same strength level. This analysis was followed by experiments with 7XXX alloys. Tests of laboratory-fabricated sheet aged by practices simulating plant-heating rates revealed that alloys of a particular solute level at peak strength developed higher toughness than higher-solute alloys either underaged or overaged to the same strength level. Alloys aged to peak strength also developed higher toughness than alloys that developed strength by combinations of aging followed by cold-working. These results were interpreted as an indication that a partially coherent precipitate provided the most effective way to strengthen these alloys with the least sacrifice in toughness. Experimental programs comparing effects of the major solute elements revealed that alloys containing the ratios of zinc, magnesium, and copper near those of 7075 developed the most attractive combinations of strength and toughness. Taken as a whole, the broad investigation indicated that modifications in alloy 7075 could provide a material that would meet the target, so alloy development proceeded in that direction.

The microstructure of 7075 is characterized by coarse constituent particles, fine chromium dispersoid particles, the semicoherent hardening precipitate in the matrix, the equilibrium precipitate that can be present at grain boundaries and other free surfaces, and the grain structure. These features are affected by ingot casting and preheat practices, slab reheat and rolling temperatures, in-process annealing practices, amount of cold-work, solution-heat-treatment time and temperature, quench practice, time between quenching and stretching and between quenching and aging, amount of stretching, rate of heating to aging temperature, and aging time and temperature. Effects of modifying these microstructural features on fracture topology and toughness were systematically evaluated, and all were found to have a significant effect. Although even large changes in the amounts of the impurity elements iron and silicon have no detectable effect on elongation of 7075 sheet, examination of fractured and cracked specimens of 7075-T6 sheet indicated that Al_7Cu_2Fe and Mg_2Si intermetallic constituent particles were playing a major role in the fracture of precracked or notched specimens. Consequently, effects of decreasing the iron and sillicon were examined, and the results showed that decreasing the volume fraction of these particles improved toughness without affecting strength. At the levels of iron and silicon deemed to be commercially feasible, toughness was still not at target levels, so other ways of improving fracture resistance were pursued. Examinations of fractured surfaces revealed two other features besides fractured constituent particles. Portions of the surfaces were intergranular, and others showed evidence of dimpled rupture. Research established a strong effect of grain morphology and quench rate on percent intergranular fracture and toughness. For thin sheet, which can be rapidly quenched, fracture mode is transgranular, regardless of grain morphology, and resistance to crack propagation increases with decreasing grain size. In either slowly quenched sheet or thick products that cannot be rapidly quenched, however, a significant portion of the fracture is intergranular when the structure is recrystallized, and toughness suffers. Grain-boundary precipitates and precipitate-free zones along boundaries have been observed in slowly quenched recrystallized material, and these have been correlated with intergranular fracture. In unrecrystallized structures, however, the precipitation occurs on subgrain boundaries as well as on grain boundaries that show fewer precipitates and narrower precipitate-free zones. This work showed that development of a transgranular fracture was a necessary feature for high toughness, but additional improvements were required to meet the toughness target. Consequently, details of the formation of the transgranular dimples were studied. The scanning electron microscope was not yet developed, so anodic replicas were prepared and examined in the transmission electron microscope. Painstaking examination of mating fracture surfaces revealed an $Al_{12}Mg_2Cr$ dispersoid particle at the base of each dimple. Tests of experimental alloys with differing amounts of chromium showed that toughness increased with decreasing amounts of chromium below the .25% nominal in 7075. In addition, special thermal-mechanical treatments were developed that increased the interparticle spacing of the dispersoids. As particle spacing increased, toughness increased. The final optimization was to modify the proportions of the major alloying elements that affect the matrix and grain-boundary precipitates. Work with a number of experimental compositions revealed that alloys containing a higher ratio of zinc to magnesium than is standard for 7075 provided the highest toughness at a particular strength level. As a result of these investigations, an alloy with higher zinc and lower magnesium, chromium, iron, and silicon was registered as 7475. The processing for the alloy includes thermal-mechanical treatments to modify the chromium dispersoids and control grain structure. Alloy 7475 is available in T6, T76, and T73 tempers. It is the first aluminum alloy developed to provide high and controlled fracture toughness. It was first used on the European military plane called the MRCA Tornado and then on the F16 United States Air Force fighter. Alloy 7475 has been used on other high-performance aircraft and is being used as the fuselage skin on the C5B cargo plane. In the late 1970s, it was discovered that 7475 is superplastic when grain size is about 10 micrometers. A commercial practice was developed to produce the fine grains, and superplastic 7475 sheet is being considered for several applications in military aircraft.

L. 7090 and 7091

These are wrought powder metallurgy alloys. Much higher levels of alloying elements can be retained in solid solution during the rapid solidification of powder atomization than during conventional solidification processes. Consequently, new compositions can be explored. Results of a program in the 1960s sponsored by the United States Army indicated that Al-Zn-Mg-Cu alloys containing additions of cobalt developed combinations of strength and resistance to stress-corrosion cracking superior to that of 7075. Cobalt cannot be added successfully to aluminum alloys cast conventionally because of its low solid solubility. Continuation of this work, under partial sponsorship of the United States Air Force, led to the registration in 1980 of 7090 and 7091. These alloys develop combinations of properties that are more attractive than those of conventional alloys for several applications, but they cost more because of the special processing. Consequently, although they have been extensively evaluated by airframe builders, attempts to find applications where they are cost-effective have been largely unsuccessful. An alloy-development program initiated in the early 1980s has combinations of properties as targets that several airframe builders believe will be cost effective if achieved.

VI. Al-Zn-Mg Alloys

A. *7005 and 7004*

Although copper additions were found to be essential in developing a high resistance to stress-corrosion in the high-strength alloys, the addition renders the products unweldable using conventional techniques. Alloy developers in the 1960s, however, discovered that Cu-free alloys containing about 4.5% Zn and 1.5% Mg with small amounts of Cr, Mn, and Zr were both weldable and resistant to stress-corrosion cracking. They develop strength intermediate to that of 6061-T6 and 2219-T8. These alloys have the ability to age-harden in the heat-affected welded zone up to about 80% of the parent metal by natural aging and can recover 100% of the strength if artificially aged after welding. Consequently, the alloys are used for welded structures.

B. *7039*

This alloy contains about 4% Mg and 2.8% Mg plus small amounts of Cr and Mn. It has found use as material for armored vehicles because it is superior to 5083-H116 in ability to protect against projectiles. Problems with stress-corrosion cracking, however, have been so severe that it will be replaced by another alloy in new designs.

C. *7X21, 7X16, 7X29, and 7X46*

These alloys were developed in the 1970s in response to a need for economical, lightweight materials for auto-bumper face plates and backup bars. The properties are sensitive to the minor alloying elements and to the processing that control the degree of recrystallization. The high strength, weldability, and ability of the high-purity versions to be bright anodically finished has led to their acceptance as face plates. Alloys 7016 and 7029 contain moderate copper (0.5–1.0%) for finishing characteristics and no recrystallization retardants.

VII. Overview

Aluminum-alloy development began when the properties of unalloyed aluminum were insufficient to meet the needs of potential customers. The earliest approach was strictly empirical. Alloying elements were added, and the products were evaluated by the user. If an alloying element provided improvement in a particular property, more was added until other properties of interest became unsuitable for the application. With time, customer needs became more sophisticated in that combinations of properties were needed; the science of metals grew; instruments and procedures to characterize microstructure and properties were improved; and producers developed many tests in an attempt to predict service performance. As the science progressed, alloy developers began to understand why certain alloying elements produced the property changes, but reasons for the effects that they observed usually were not understood by the scientific community for many years after the initial observation. The dictum that "if a little is good, add more until something bad happens" was still used. The realization that properties were controlled by the microstructure (which was in turn controlled by the composition, casting, fabricating, and heat-treating practices) came considerably before the ability to identify the particular microstructural features that controlled properties. Consequently, the alloy developer created a host of heuristic rules relating composition and process to properties. Although the relationship between structure and properties is better known today than it has ever been, it is still insufficient to serve as more than a guideline to alloy developers. Attempts to design an artificial intelligence system for alloy development have emphasized that models of the relationship between microstructure and even a relatively simple property such as yield strength are significantly less accurate than the magnitude of the improvements targeted for in a typical alloy-development program. Consequently, aluminum-alloy development proceeds by a process of "enlightened empiricism." In a typical program, the alloy developer determines which current alloy-process combination develops the properties closest to the target. Then, a judgment is made regarding which microstructural features are controlling the property of interest. Changes in composition and practice designed to provide the desired microstructure are tried using both experience and science as guidance. The microstructure and properties obtained are then compared with the targets, and the process is iterated until the target is reached. As greater understanding is gained of the relationships between structure and properties, alloy developers can rely more on science and less on lore and heuristics.

References

1. *75 Years of Alloys from Alcoa: A Record of Alloy and Metallurgical Development by Aluminum Company of America* (1962), Aluminum Company of America, Pittsburgh, PA.
2. *Age Hardening of Metals* (1940), Tower Press, Cleveland, Ohio.
3. *Alcoa Aluminum and Its Alloys* (1950). Aluminum Company of America.
4. Aluminum Association (1982). *Aluminum Standards and Data.*
5. Aluminum Association (1984). *Registration Record of Aluminum Association Designations and Chemical Composition Limits for Wrought Aluminum Alloys.*
6. "Aluminum in Aircraft" (1930, 1941, 1943) Aluminum Company of America.
7. Anderson, R.J. (1925). *The Metallurgy of Aluminum and Its Alloys.* Henry Carey Baird.
8. Anderson, W.A. (1957). "Precipitation Hardening Aluminum-Base Alloys," *Precipitation from Solid Solution,* American Society for Metals, Cleveland, Ohio.
9. Anderson, W.A. (March 1967). "New Alloys for Rigid Containers, "*Modern Packaging.*
10. Brown, H. (1948). *Aluminum and Its Applications*, Pitman Publishing Corporation, New York.
11. Budgen, N.F. (1947). *Aluminum and Its Alloys*, 2nd ed. Sir Issac Pitman and Sons.
12. Carr, C.C., (1952), *Alcoa, An American Enterprise.* Rinehart, New York.
13. Collins, F.R., Nelson, F.G., and Robinson, I.B. (November 1961). "Aluminum 2219 . . . New Alloy for Welded Structures," *Metal Progress.*
14. Davis, W.H. (1967). *A Hundred Years of Metallurgy.* Aldins.
15. Develay, R. (February 1974). "Current Trends in the Development of the Composition of Wrought Aluminum Alloys," *Revue de l'Aluminum.*
16. Deville, H. (1933). *Aluminum*, R.A. Anderson trans. Sherwood.
17. Dix, E.H., Jr. (1949). "Aluminum-Clad Products" in *Engineering Laminates.* John Wiley, New York.
18. Dix, E.H., Jr. (October 1950). "Aluminum Alloys—1940 to 1950," *Metal Progress.*
19. Dix, E.H., Jr. (1950). 24th Campbell Memorial Lecture, "Aluminum-Zinc-Magnesium Alloys—Their Development and Commercial Production," *Trans. ASM*, **42**.

20. Dix, E.H., and Bowman, J.J. (February 1936). "Fifty Years of Aluminum Alloy Development," *Metals and Alloys.*
21. Dix, E.H., Jr., Anderson, W.A., and Shumaker, M.B. (1958). *Development of Wrought Aluminum-Magnesium Alloys*, Alcoa Tech. Paper no. 14, Aluminum Company of America, Pittsburgh, PA.
22. Edwards, J., Frary, F.C., and Jeffries, Z. (1930). *Aluminum Products and Their Fabrication.* McGraw-Hill.
23. Fischer, G., Linker, and Markworth, M. (1974). "A Non-Quench Sensitive Al-Zn-Mg-Cu Alloy with Additions of Chromium, Zirconium, and Silver," *Aluminum* **50**.
24. Fuss, V. (1936). *Metallography of Aluminum and Its Alloys*, Sherwood Press, Cleveland, Ohio.
25. Grard, C. (1921). *Aluminum and Its Alloys* C.M. Phillips and H.W.L. Phillips, Constable. trans.
26. Gregory, E., and Simons, E.N. (1948). *Non-Ferrous Metals and Alloys*, P. Elek, London, England.
27. Hatch, J., ed. (1984). *Aluminum: Properties and Physical Metallurgy.* ASM.
28. Hobbs, D.B. (1938). *Aluminum History, Metallurgy, and Uses.* Bruce.
29. Hoff, N.J. (1984). *Innovation in Aircraft Structures—Fifty Years Ago and Today.* American Inst. Aeronautics and Astronautics.
30. Hunsicker, H.Y. (1976). "Development of Al-Zn-Mg-Cu Alloys for Aircraft" in *Proceedings of the 1975 Rosenhain Centenary Conference*, Metals Society, London, England.
31. Hunsicker, H.Y., and Stumpf, H.C. (1963). "History of Precipitation Hardening" in *The Sorby Centennial Symposium on the History of Metallurgy.* Gordon and Breach.
32. Law, E.F. (1919). *Alloys and Their Industrial Applications.* Charles Griffin and Company, London, England.
33. Luhan, J.V., and Summerson, T.J. (1970). "Development of 7049-T73 High-Strength, Stress-Corrosion Resistant Aluminum Alloy Forgings," *ASM Met. Eng. Quart.*, **10** no. 4.
34. Mondolfo, L.F. (1976). *Aluminum Alloys Structure and Properties.* Butterworth Woburn, Mass.
35. Mori, L. (1974). "Aluminum Alloys: Characteristics, Selection and Uses," *Aluminio*, **43**.
36. Nock, J.A., Jr., (October 1944). "75S—Alcoa's New High-Strength Aluminum Alloy." *Metals and Alloys.*
37. Nock, J.A., Jr., (1947). "Commercial Wrought Aluminum Alloys," *Physical Metallurgy of Aluminum Alloys*, American Society for Metals, Cleveland, Ohio.
38. Nock, J.A., Jr. (1953). "Today's Aluminum Aircraft Alloys," *SAE Transactions*, **61**.
39. Nock, J.A., Jr., Holt, M., and Sprowls, D.O. (September 1961). "A New High-Strength Aluminum Alloy," *Metal Progress.*
40. Perry, J. (1947). *The Light Metals Industry.* Longmans, Green,
41. Pruitt, B., and Smith, G.D. (1986). "The Corporate Management of Innovation: Alcoa Research, Aircraft Alloys, and the Problem of Stress-Corrosion Cracking," in *Research On Technological Innovation, Management, and Policy*, R. Rosenbloom, ed. JAI Press, Greenwhich, CT.
42. Reynolds, M.A., and Harris, J.G. (1974). "Development of a Tough, High-Strength Aluminum Alloy with Improved Stress-Corrosion Resistance," *Aluminium*, **50**.
43. Staley, J.T. (1976). "Microstructure and Toughness of High-Strength Aluminum Alloys," *ASTM STP 304*, American Society for Testing and Materials. Philadelphia, PA.

44. Staley, J.T., Hunsicker, H.Y., and Schmidt, R. (1971). "New Aluminum Alloy X7050." *AIME-TMS Paper F71-7*, Metallurgical Society of AIME, Warrendale, PA.
45. Starke, E.A., Jr. (1977). "Aluminum Alloys of the 1970's: Scientific Solutions to Engineering Problems," *Materials Science and Engineering*, **29**.
46. Van Horn, K.R., ed. (1967). *Aluminum: Properties, Physical Metallurgy, and Phase Diagrams*. ASM,
47. Vickers, C. (1923). "*Metals and Their Alloys*. Henry Carey Baird,
48. Wilhelm, D. (1932). *The Book of Metals*. Harper,
49. Yoshida, Y., and Sasajima, M. (December 1983). "Current Status of Aluminum Alloys as Aircraft Structural Material and Fabrication Methods," *J. Jap. Inst. Light Met.*, **33**, no. 12.

Part II

Conventional Aluminum Alloys

2

Heat-Treatable Aluminum Alloys

EDGAR A. STARKE, JR.

Department of Materials Science
University of Virginia
Charlottesville, Virginia

I. Introduction

Aluminum alloys are classified as heat-treatable or non–heat-treatable, depending on whether or not they respond to precipitation hardening. The heat-treatable alloys contain elements that decrease in solubility with decreasing temperature and in concentrations that exceed their equilibrium solid solubility at room and moderately higher temperatures. The most important alloying elements in this group include copper, lithium, magnesium, and zinc. A normal heat-treatment cycle includes a solutionizing soak at a high temperature to maximize solubility, followed by rapid cooling or quenching to a low temperature to obtain a solid solution supersaturated with both solute elements and vacancies. During aging at either room or some intermediate temperature, the solute atoms cluster together, forming preprecipitates which are coherent with the matrix and often referred to as Guinier-Preston zones after the two researchers who confirmed their existence in 1938. The research by Guinier and Preston followed the discovery of age hardening in 1906 by Alfred Wilm and the explanation of the phenomenon by Merica, Waltenberg, and Scott in 1919.

ISBN 0-12-341831-3

II. Precipitation in Age-Hardening Systems

Classical nucleation theory, originally developed for vapor-to-liquid changes, is inadequate in describing "homogeneous" precipitation that occurs in an age-hardening alloy since this process can be varied over a wide range by seemingly insignificant changes in heat-treatment procedure. Theories that adequately describe the nucleation of precipitates from supersaturated solid solutions have been developed by Lorimer and Nicholson as well as by Pashley, Jacobs, and Vietz. These theories explain a number of important observations on the precipitation behavior of alloys, viz.:

(a) the importance of a certain critical temperature that lies below the normal solvus temperature,
(b) the effect of excess vacancies on precipitate nucleation,
(c) the observation of precipitate-free zones (PFZs) at grain boundaries,
(d) the response of alloys to double-aging treatments, and
(e) the sensitivity of alloys to small changes in heat treatment.

The type and rate of the decomposition process depend on whether aging is carried out below or above a critical temperature defined by a metastable solvus line on the equilibrium diagram, as in Fig. 1. The solid

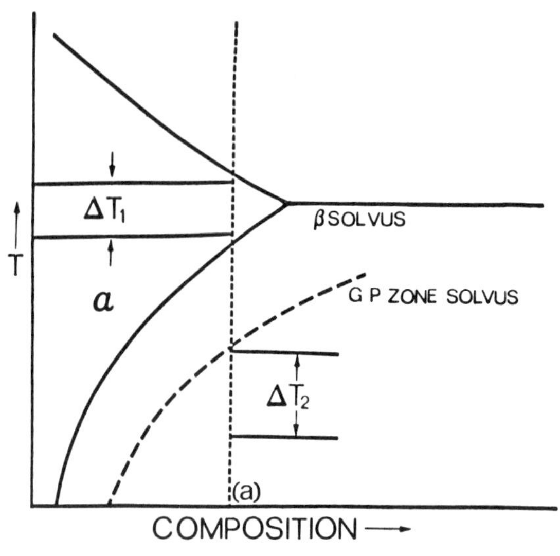

FIG. 1 Phase diagram of a hypothetical system A-B showing the β solvus and the GP zone solvus. For composition (a), ΔT_1 temperature range for solution heat treating, ΔT_2 temperature range for precipitating heat treatment.

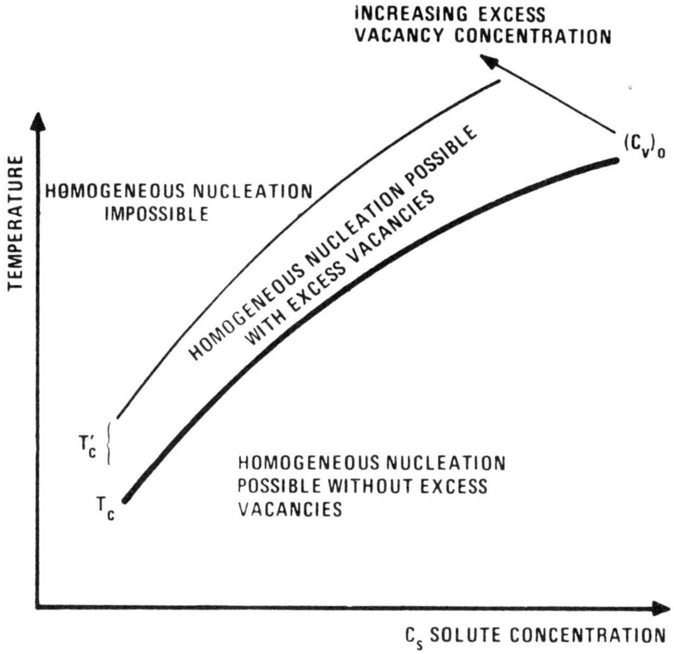

FIG. 2 Critical temperature versus solute concentration for equilibrium and excess vacancy concentrations. (After Jacobs and Pashley)

solution is metastable with respect to Guinier-Preston (GP) zones below this temperature and the zones do not form above the critical temperature. The critical temperature may be displaced to higher temperatures when the vacancy concentration is increased, as in Fig. 2. The vacancies are believed to accelerate the formation of clusters during the quench and/or act as aggregates that nucleate the new phase. The decomposition of supersaturated solid solutions below either critical temperature normally occurs by the following sequence: supersaturated solid solution → clusters → transition structure → final structure.

Often a PFZ is observed adjacent to a high-angle grain boundary, as shown in the transmission electron micrograph (TEM) of Fig. 3. The PFZ may be caused by two different phenomena:

1. The aging temperature is above the critical temperature determined by the equilibrium concentration of vacancies but below that determined when excess vacancies are present, and vacancies have been lost to the grain boundary prior to aging. In this case, the PFZ is called a vacancy-depleted PFZ.

2. Normal nucleation and growth of the equilibrium phase occur at the grain boundary, depleting the adjacent region of solute, and thus producing a solute-depleted PFZ. This is the type shown in Fig. 3.

Both types of PFZs may be minimized by lowering the aging temperature, which increases the solute supersaturation (and thus the driving force for homogeneous decomposition) and decreases diffusion rates, which decreases the nucleation and growth of the equilibrium precipitates. Consequently, age-hardening alloys are frequently given double-aging treatments: a low temperature age followed by a higher temperature age. The low temperature age increases the number density of preprecipitates and minimizes the formation of PFZs, and the higher temperature age accelerates the growth of the precipitates and often their transition to an intermediate precipitate. This allows the desired strength level to be obtained in the minimum amount of aging time.

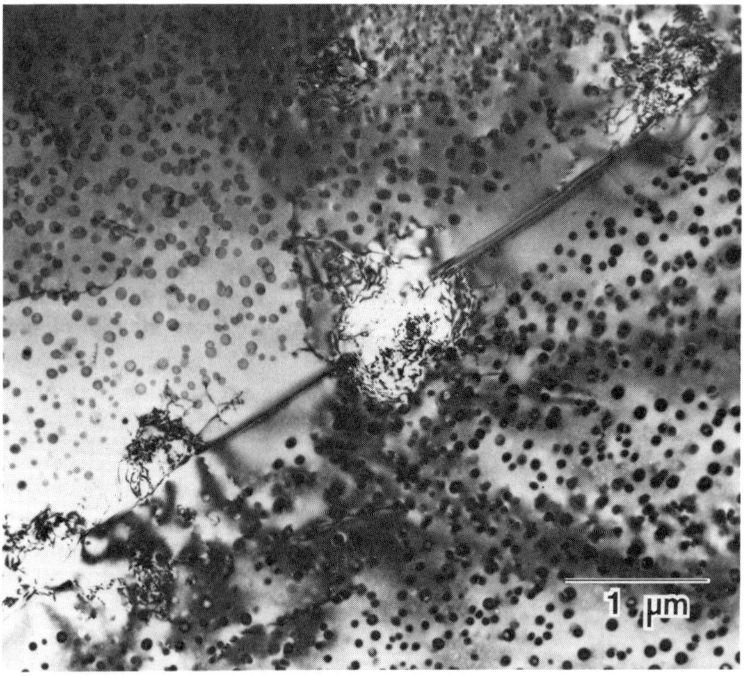

FIG. 3 TEM showing precipitate-free zone (PFZ) adjacent to a high-angle grain boundary in an Al-2 wt % Li alloy. The matrix precipitates are the metastable phase Al_3Li and the large strain fields at the grain boundary are produced by the equilibrium phase AlLi. (Courtesy of W.A. Cassada)

PFZs almost always have a deleterious effect on the properties of aluminum alloys. Since they are free of precipitates, they are softer than the surrounding matrix and, under an applied stress, can be regions of strain localization. This may lead to premature fracture under either monotonic or cyclic loading. In addition, PFZs are believed to increase the susceptibility of age-hardenable aluminum alloys to stress corrosion cracking. The grain boundary precipitates that produce the PFZs can also play an important role in the failure process of both mechanically and environmentally induced fracture. Consequently, processing procedure (e.g., low temperature aging and/or deformation prior to aging) are frequently employed to minimize grain boundary precipitation and PFZ formation. The dislocations induced during the deformation often serve as nucleating sites for precipitates in competition with the grain boundaries.

Trace elements may often be present in aluminum alloys, either as impurities or by design, and can frequently have an effect on the precipitation process. Although the influence of trace elements depends on the particular element and alloy system, they usually modify the precipitation process by one or more of the following mechanisms:

(a) They interact with vacancies and alter the kinetics of the precipitation process.

(b) They change the interfacial energy of the clusters and/or precipitates and may raise or lower the critical temperature and alter the nucleation frequency.

(c) They may alter the type of precipitate formed.

Although trace elements have been added to some alloy systems to improve machinability, they are normally added to aid in the nucleation of the strengthening precipitates. The effect of cadmium on the nucleation of the θ' precipitate in an Al-Cu alloy is shown in the TEMs of Fig. 4.

Most commercial alloys are designed to have precipitates that control the grain structure. Aluminum combines readily with the transition metals, chromium, manganese, zirconium, etc., to form intermetallic phases with little or no solubility in the aluminum matrix. Because of their slow diffusivity in aluminum, these alloying additions form very small precipitates, less than one micron in size, either during solidification or during ingot preheat. This fine distribution of precipitates, called dispersoids, delays or prevents static recystallization during processing, and, because they are strung out in the working direction (a process called mechanical fibering), they aid in retaining the elongated or "pancake"-shaped grains that develop during working. The mean free path of the dispersoids is greater in the working direction than in directions perpendicular to it. Thus, recrystallized grains can grow faster in the working direction. Such

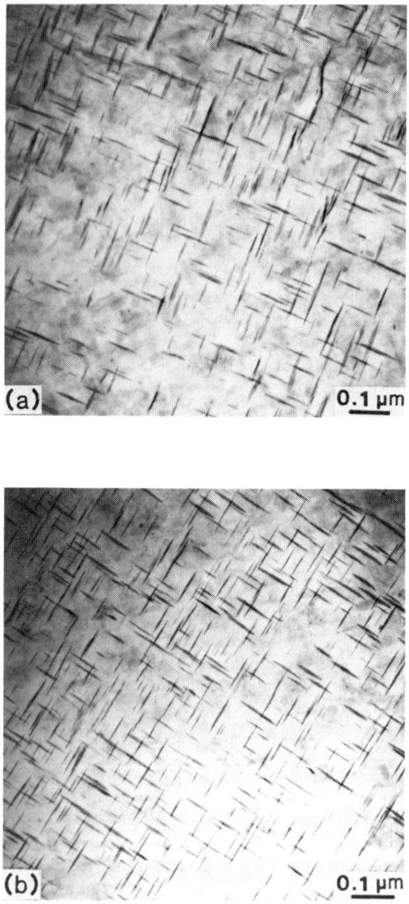

FIG. 4 TEMs showing the effect of Cd on the precipitation of θ' (Al_2Cu) in an Al-Cu alloy aged to peak strength. (a) No Cd. (b) 0.2 wt % Cd.

elongated grains impart anisotropic behavior in the material, as discussed in Section I. The TEM of Fig. 5 shows the Cr-rich $Al_{12}Mg_2Cr$ dispersoids in an Al-Zn-Mg alloy.

The effectiveness of a particular dispersoid in controlling the grain structure depends on its size, spacing, and coherency. Small particles (normally less than 0.4μm in diameter) retard continuous recrystallization by pinning subgrain boundaries and preventing subgrain coalescence. They retard discontinuous recrystallization by pinning migrating high-angle boundaries—a process described by Zener. Zener suggested that a grain

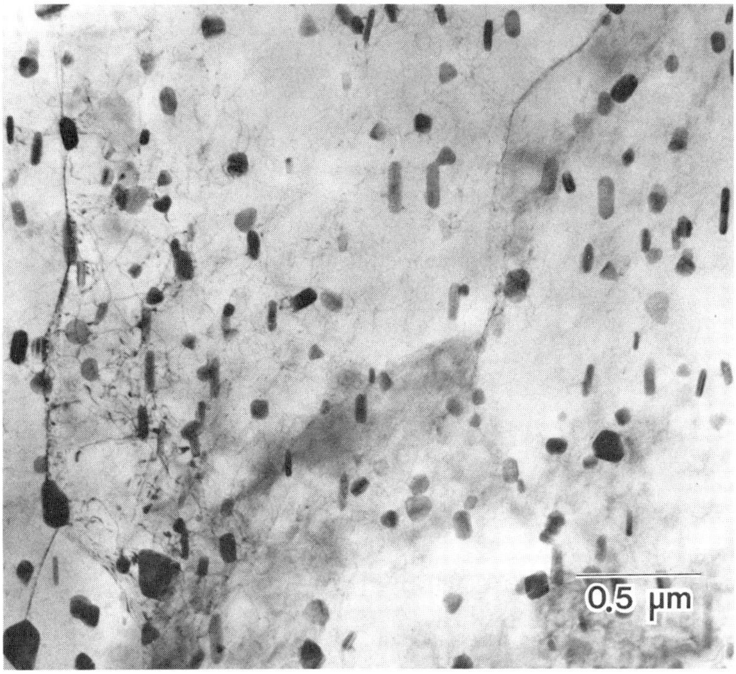

FIG. 5 TEM showing the $Al_{12}Mg_2$ Cr dispersoids used for grain-structure control in many Al-Zn-Mg alloys. (Courtesy of John Wert)

would cease to grow when its radius, R, was approximately equal to the ratio of the particle radius, r, divided by the particle volume fractional f, i.e.,

$$R(\text{crit}) = (3\beta/4)\ (r/f), \tag{1}$$

where β is a constant approximately equal to one. Eq. (1) suggests that small, closely spaced particles have the greatest effect in retarding recrystallization. Coherent dispersoids (e.g., Al_3Zr) are more effective than the incoherent dispersoids (e.g., $Al_{12}Mg_2Cr$ and $Al_{20}Mn_3Cu_2$) in inhibiting boundary migration, because the precipitate–matrix interface must either change from coherent to semicoherent or incoherent as the boundary passes the dispersoid or, alternatively, dissolve and reprecipitate after the boundary passes through its local. Both processes require considerable energy.

The type of dispersoid—i.e., coherent or incoherent—can also effect the "quench sensitivity" of an age-hardenable alloy. As mentioned previously, age-hardenable alloys must be quenched from a relatively high temperature to maximize the supersaturation of the alloying elements that will participate in formation of the strengthening precipitates during subsequent aging. The quenching medium may be forced air, cold or boiling water, or some other liquid. The quench sensitivity describes the tolerance of a particular alloy to the slower quenching mediums. During slow cooling, some of the solute may precipitate out as coarse particles, which reduces the degree of supersaturation and adversely affects the alloy's fracture toughness and ductility. Quench sensitivity is higher in the more concentrated alloys and enhanced in alloys containing incoherent particles whose interfaces serve as nucleation sites for heterogeneous precipitation. Consequently, alloys that contain chromium and manganese as dispersoid-forming elements are more quench-sensitive than those that contain zirconium, which forms a coherent dispersoid.

III. Strengthening in Age-Hardening Systems

The increase in strength of age-hardenable aluminum alloys over that of the pure metal or solid-solution alloys is due to the interaction of dislocations with preprecipitates and transition precipitates formed during aging below the critical temperature. Being coherent and/or partially coherent with the matrix, they are penetrated by dislocations during plastic flow. For this cutting mechanism, the critical resolved shear stress, τ, due to the interaction of a gliding dislocation with the dispersion, can be represented by an equation of the form

$$\tau = \mathrm{c} f^{\mathrm{m}} r^{\mathrm{p}} \tag{2}$$

where c is an alloy constant that depends on the particular strengthening mechanism (i.e., coherency, surface, chemical, stacking fault, and/or modulus hardening), f is the volume fraction, and r is the particle radius. Coherency hardening is associated with the elastic coherency stresses surrounding a particle that does not exactly fit the matrix. Surface hardening is associated with the energy necessary to create additional particle–matrix interfaces when the particle is sheared by the dislocation. Chemical or order hardening is associated with the additional work required to create an internal interface or antiphase boundary in the case of internally ordered particles. Stacking fault hardening is associated with the difference in stacking fault energy between the matrix and the particle, and modulus hardening is associated with the difference in modulus between the matrix and the particle. The exponents m and p of Eq. (2) are always positive and the strength increases with both volume fraction and particle size.

The particles may form in a variety of shapes (e.g., spheres, cubes, discs, and needles), depending on the misfit between particle and the matrix and the interfacial energy. The shape that occurs generally minimizes the total energy associated with the formation of the particle. For example, when the atomic misfit is very small, spherical particles are formed since this shape has the lowest surface-area–to–volume ratio. However, when the atomic misfit is large, the particles generally assume the shape of discs or needles in order to minimize the strain energy. Cubes are an intermediate state and have not, as of early 1988, been observed in aluminum alloys.

As the particles grow, coherency strains build up and are finally released when the particles become incoherent, e.g., during the transformation from the transition phase to the equilibrium phase. This is accompanied by a change in particle-dislocation interaction from one of shearing to that of dislocation looping or bypassing the particles. This mechanism is called Orowan hardening, and the critical resolved shear stress can be represented by

$$\tau = \frac{0.8G\mathbf{b}(f)^{1/2}}{2\pi[2(1 - \nu)r]^{1/2}} \ln \frac{1.6}{r_0} r \tag{3}$$

where **G** is the shear modulus; **b**, the Burgers vector; ν, Poisson's ratio, and r_0, the inner cut-off radius of the dislocation. Here the strength decreases with increasing particle size. A schematic representation of the increase in critical resolved shear stress associated with particles as a function of particle size is given in Fig. 6.

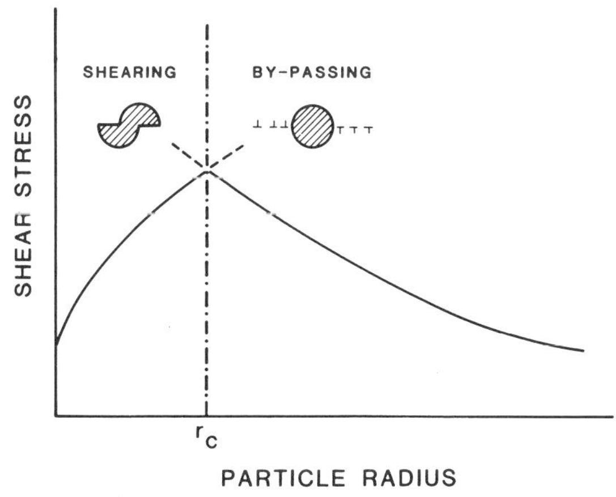

FIG. 6 Schematic of the resolved shear stress as a function of particles radius when the particles are sheared or bypassed by dislocations.

As mentioned previously, the strength of the age-hardened alloy is due to the interaction of dislocations with preprecipitates and precipitates that may be coherent, partly coherent, or incoherent with the matrix. Depending on the size, spacing, and degree of coherency, the precipitates are either sheared or looped and bypassed by the dislocations during plastic deformation. The particular deformation mechanism controls the distribution of strain and can affect the fracture properties, both monotonic and cyclic, of aluminum alloys. When particles are sheared by moving dislocations, the strengthening mechanisms associated with the particles are reduced. (The size of the particle on the glide plane is reduced.) This results, successively, in a local decrease in resistance to further dislocation motion, concentration of slip, and destruction of the strengthening agents. An example of this type of strain localization is shown in the TEM of Fig. 7. Such concentrated slip may lead to early crack nucleation in slip bands under both monotonic and cyclic loading. When these bands impinge on grain boundaries, they can aid in low-energy intergranular failure, including environmentally assisted stress corrosion cracking. However, the planar slip and strain localization associated with shearable precipitates are beneficial for fatigue crack propagation resistance since slip band cracking enhances zigzag crack growth and crack branching, both of which increase the total crack path and lower the effective stress intensity at the main crack tip. In addition, slip band cracks must usually be renucleated at grain boundaries. These factors, along with a higher degree of slip reversibility,

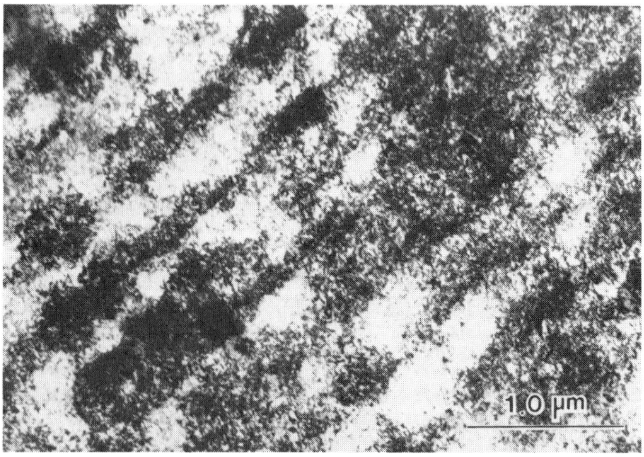

FIG. 7 TEM showing intense slip bands in an Al-Mg-Si alloy aged to produce shearable precipitates. (Courtesy of Anil K. Sachdev)

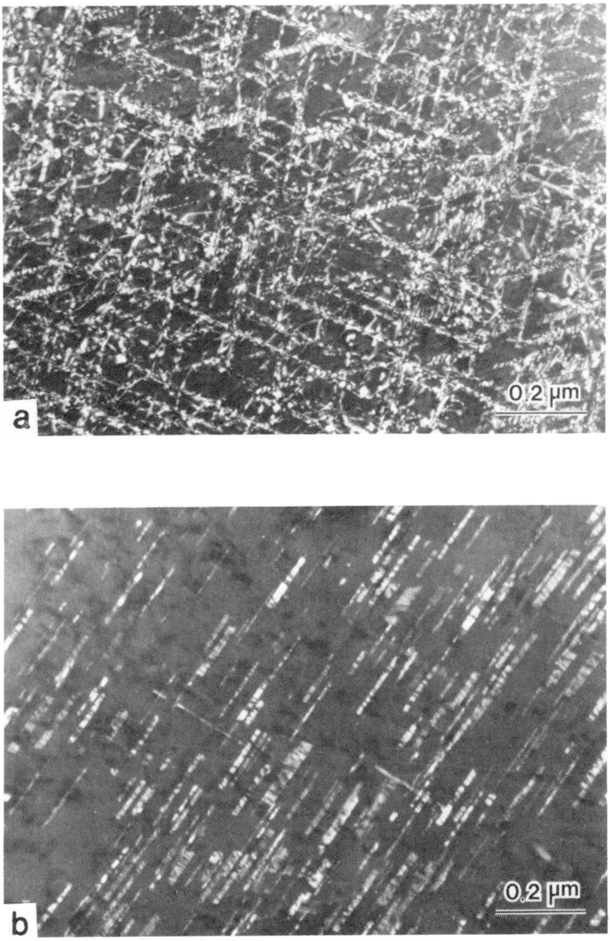

FIG. 8 TEM showing homogeneous deformation in an Al-Mg-Si alloy aged to produce nonshearable precipitates. (a) Weak beam. (b) Dark field TEM of nonshearable precipitates following 4% tensile strain. The fringes on the rod-shaped precipitates are loops left behind during dislocation bypass by the Orowan mechanism. (Courtesy of Anil K. Sachdev)

result in lower fatigue crack growth rates for alloys containing shearable precipitates.

Incoherent strengthening precipitates that are looped and bypassed by dislocations tend to homogenize deformation, as illustrated in the TEM of Fig. 8. Increasing the homogeneity of deformation normally improves ductility and stress corrosion resistance, but it usually decreases fracture toughness and fatigue crack propagation resistance.

By considering the relationships expressed in Eqs. (2) and (3), one can optimize the strength of age-hardenable aluminum alloys by maximizing those elements that participate in the aging sequence and can be put in solid solution at the solutionizing temperature. Solubilities of these elements should not be greatly exceeded since all excesses form coarse precipitates that do not add significantly to the strength and that adversely affect the fracture toughness. However, commercial heat-treatable aluminum alloys are designed to precipitate out a sizable volume fraction of a second phase. Most of these alloys contain two or more solute additions since the addition of a second alloying element to a binary system reduces the solubility of the first alloying element if an intermetallic compound between the two alloying elements forms. In addition, the second alloying element often increases the driving force for nucleation and/or reduces the activation energy for nucleation. For example, if the first alloying element has a smaller atomic radius than aluminum, the second alloying element should have a larger atomic radius, and vice versa, since this would tend to decrease the strain energy in the lattice, thus reducing the activation energy required for nucleation. Most commercial age-hardenable aluminum alloys (e.g., Al-Mg-Zn, Al-Mg-Cu, Al-Mg-Si, and Al-Li-Cu) satisfy this criterion.

IV. Alloy and Temper Designations

The Aluminum Association of the United States uses a four-digit numerical system to identify aluminum alloys. The system used for wrought alloys is slightly different from that used for cast alloys; however, the digit that designates the alloy group is essentially the same. The nomenclature for wrought alloys has been accepted by most countries and is now called the International Alloy Designation System (IADS). The first digit indicates the alloy group and the last two digits identify the aluminum alloy or indicate the aluminum purity. The second digit indicates modifications of the original alloy or impurity limits. Experimental alloys are also identified in accordance with this system, but they are indicated as experimental by the prefix X. The designation system for wrought aluminum alloys is given in Table I.

The heat-treatment or temper-nomenclature system developed by the Aluminum Association has also been adopted as part of the IADS by most countries. It is used for all forms of wrought and cast aluminum alloys with the exception of ingot. The system is based on the treatments used to develop the various tempers and takes the form of letters added as suffixes

TABLE I

Aluminum Alloy Designation for Wrought Alloys

Four-digit series	Al content or main alloying elements
1XXX	99.00% minimum
2XXX	Copper
3XXX	Manganese
4XXX	Silicon
5XXX	Magnesium
6XXX	Magnesium and Silicon
7XXX	Zinc
8XXX	Others
9XXX	Unused

TABLE II

Temper Nomenclature: Symbols Added as Suffix Letters or Digits to the Alloy Number

Suffix letter F, O, H, T, or W indicates basic treatment or condition.	First suffix digit indicates secondary treatment used to influence properties.	Second suffix digit for condition H only indicates residual hardening.
F—As-fabricated		
O—Annealed, wrought products only		
H—Cold-worked, strain-hardened	1—Cold-worked only	2—1/4 hard
	2—Cold-worked and partially annealed	4—1/2 hard
	3—Cold-worked and stabilized	6—3/4 hard
		8—hard
		9—extra hard
W—Solution heat-treated		
T—Heat-treated stable	1—Partial solution plus natural aging	
	2—Annealed cast products only	
	3—Solution plus cold-work	
	4—Solution plus natural aging	
	5—Artificially aged only	
	6—Solution plus artificial aging	
	7—Solution plus stabilizing	
	8—Solution plus cold-work plus artificial aging	
	9—Solution plus artificial aging plus cold-work	

to the alloy number. Subdivisions of the tempers, when they significantly influence the characteristics of the alloy, are indicated by one or more digits following the letter. Alloys supplied in the as-fabricated or annealed condition are designated with the suffixes F and O, respectively. Those supplied in the solution heat-treated condition are designated by the letter W. Alloys supplied in the strain-hardened condition are designated with the letter H, and those in the heat-treated condition with the letter T. Digits following H represent the degree of strain hardening and those following T the type of aging treatment. Stress relieving is indicated by the number 5 following the first digit for age-hardening tempers. The temper-designation system is given in Table II.

V. 2XXX Alloys

As mentioned previously, the first heat-treatable aluminum alloy was based on the Al-Cu system, and wrought aluminum alloys having copper as the main alloying element are designated as 2XXX alloys using the four-digit numerical system developed by the Aluminum Association. During aging of Al-Cu alloys, copper atoms cluster into two dimensional disc-shaped Guinier-Preston zones on {100} matrix planes. These zones are usually 30-50A in diameter and are shown in the TEM of Fig. 9. At temperatures of 373K and higher, three-dimensional zones are formed.

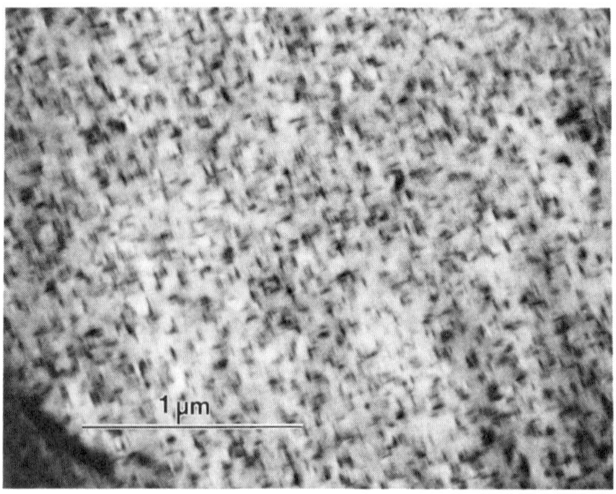

FIG. 9 TEM showing GP zones in an aged Al-1.9 wt % Cu alloy. (Courtesy of J.A. Hawk)

Depending on the aging time and temperature, they may transform into the transition phase θ', which has the same chemical composition as the equilibrium phase Al_2Cu, but which has a tetragonal crystal structure and is only partially incoherent with the matrix. θ' is fully coherent with the matrix along the broad faces, and the {001} planes and ⟨001⟩ directions of the θ' are parallel to the {001} planes and ⟨001⟩ direction of the matrix. θ' often nucleates directly on dislocations since the dislocations can act to relieve the misfit associated with the precipitate. The distribution of θ' may be very heterogeneous and reflects the distribution of dislocations. Deformation prior to aging can increase the dislocation density and thereby the number of nucleating sites for heterogeneous precipitation of θ'. The aging sequence in Al-Cu alloys is usually described as

$$\text{supersaturated } \alpha \rightarrow \text{GP zones} \rightarrow \theta' \rightarrow \theta\ (Al_2Cu).$$

However, deformation prior to aging can accelerate the formation of θ' at the expense of the GP zones.

The addition of magnesium to Al-Cu alloys enhances both the rate and magnitude of natural aging. The enhancement probably results from the complex interactions between the solute elements and the vacancies. For high magnesium concentrations, the decomposition process may also be modified and can be described as

$$\text{supersaturated } \alpha \rightarrow \text{GP zones} \rightarrow S' \rightarrow S\ (Al_2CuMg).$$

The transition phase S′ has the same chemical composition and an orthorhombic crystal structure very similar to the equilibrium S phase. The [100], [010], and [001] directions of S′ are respectively parallel to the [100], [021], and [012] directions of the matrix. The S′ precipitates normally nucleate on dislocations and grow as laths on {210} planes along ⟨100⟩ directions. As growth continues, the laths of S′ form corrugated sheets on {210} planes with {110} composite planes as shown in the TEMs of Fig. 10.

Significant age hardening occurs in the 2XXX during aging at room temperature, i.e., natural aging. Although the rate and extent of natural aging varies from one alloy to another, most of the strengthening occurs during the first 24 hours, and stability is reached after four days. Consequently, many 2XXX alloys are used in the naturally aged tempers (e.g., T3 and T4). Although the maximum strength cannot normally be realized during natural aging, the T3 and T4 tempers possess good combinations of strength, ductility, fracture toughness, and fatigue crack growth resistance. The good fracture properties can be related to the lack of precipitation of coarse grain boundary phases and associated development of PFZs during room-temperature aging. Maximum strength in 2XXX alloys is obtained during artificial aging at temperatures ranging from approximately 368

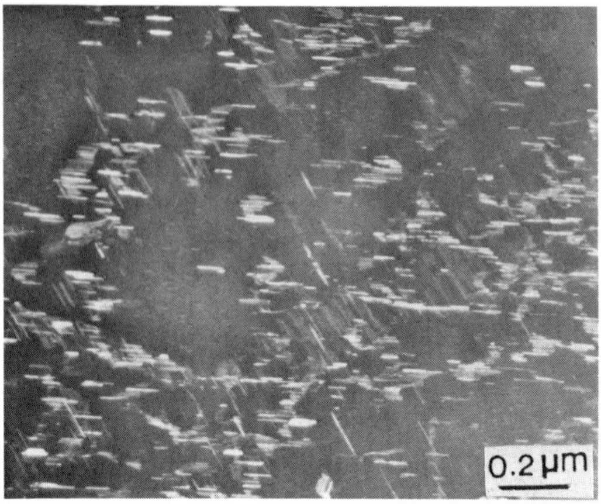

FIG. 10 Dark field TEM of an Al-Cu-Mg alloy showing corrugated sheets of S′ on {110} planes. [112] foil normal.

to 478K. The increase in strength is due to the precipitation of the transitional phases, θ', and S′ (depending on alloy composition). During artificial aging, the yield strength normally increases more than the tensile strength. Ductility and fracture toughness decrease disproportionately with increasing strength so that strength-ductility combinations are not as attractive as those obtained during natural aging. Although some ductility may be recovered during overaging, accompanying strength reductions are proportionately larger. 2XXX alloys respond very well to deformation prior to aging, and improvements in strength-toughness combinations are observed for the T8 temper. This improvement is related to the refinement in size and distribution of the strengthening precipitates and the decrease in size of grain boundary precipitates and PFZs. An illustration of this effect is given in the TEMs of Fig. 11. Precipitation-hardening curves have been established for most commercial alloys, from which one can extract time–temperature–strength relationships.

The high strength-density ratios of the 2XXX and 7XXX (to be discussed later) alloys have made them extremely attractive for use in aircraft, and subsonic aircraft structures of the late 1980s contain on average 80 wt % aluminum. The primary structural components, wing spars, fuselage, and landing gear, are made from extrusions or forgings of either 2XXX or 7XXX alloys. The wings and fuselage are skinned with plate of similar alloys. The particular design criteria usually dictate the alloy selection. The 7XXX alloys have higher strength than the 2XXX alloys, but the latter have a higher temperature stability, are generally more resistant

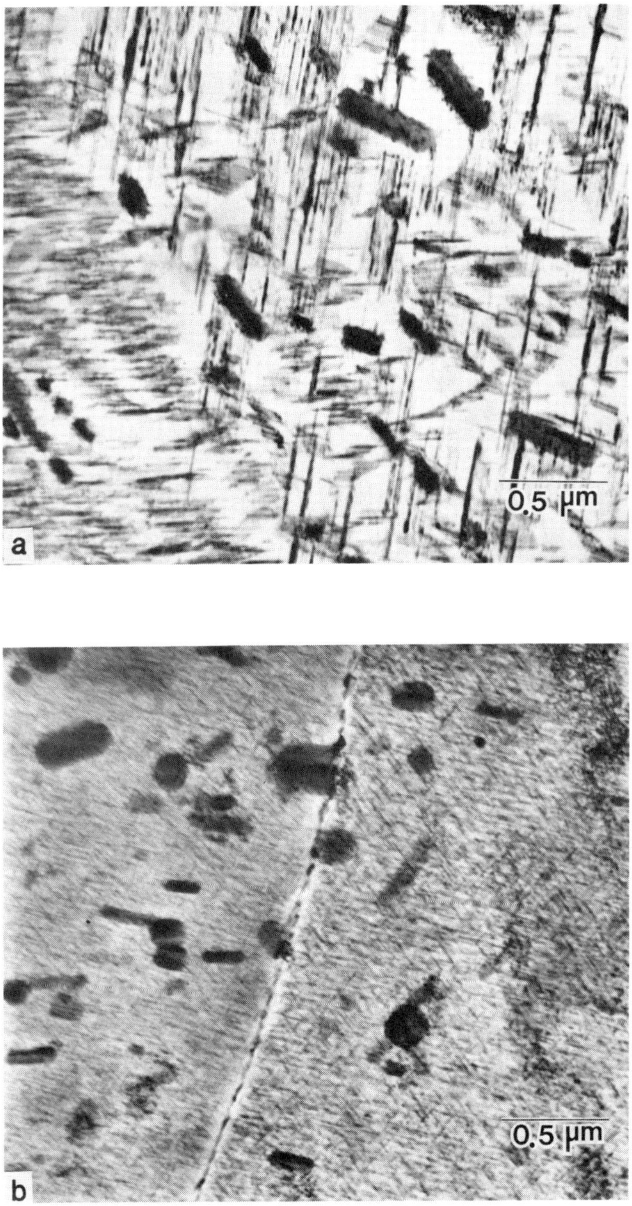

FIG. 11 TEMs showing the effect of plastic deformation on the S′ precipitate distribution in a commercial Al-Cu-Mg-Mn alloy. (a) Undeformed, aged 12h at 190°C. (b) Stretched 6% and aged 12h at 190°C. The large precipitates are the $Al_{20}Cu_2Mn_3$ dispersoids. (Courtesy of ALCOA)

TABLE III

CHEMICAL COMPOSITION OF SOME 2XXX ALUMINUM ALLOYS

Alloy	Si	Fe	Cu	Mn	Mg	Cr	Zn	Ti	Zr	Others (total)	Aluminum
2117	0.8	0.7	2.2–3.0	0.2	0.2–0.5	0.1	0.25	—	—	0.15	Remainder
2219	0.20	0.30	5.8–6.8	0.20–0.40	0.02	—	0.10	0.02–0.10	0.10–0.25	*	Remainder
2419	0.15	0.18	5.8–6.8	0.20–0.40	0.02	—	0.10	0.02–0.10	0.10–0.25	*	Remainder
2024	0.50	0.50	3.8–4.9	0.30–0.9	1.2–1.8	0.10	0.25	—	—	0.15	Remainder
2124	0.20	0.30	3.8–4.9	0.30–0.9	1.2–1.8	0.10	0.25	0.15	—	0.15	Remainder
2036	0.5	0.5	2.2–3.0	0.1 –0.4	0.3–0.6	0.1	0.25	0.15	—	0.15	Remainder
2048	0.15	0.20	2.8–3.8	0.20–0.6	1.2–1.8	—	0.25	0.10	—	0.15	Remainder

* 0.05–0.15V

to fracture, and are frequently used in fatigue and fracture-toughness critical components. 2XXX and 7XXX alloys have both seen recent use in the auto industry as body sheet and bumpers, respectively.

The chemical-composition limits of some of the 2XXX alloys are listed in Table III. The composition is in maximum percentage unless shown as a range. The roles of most of the alloy additions in Table III have been described previously; however, some special note should be made of iron, silicon, and titanium. Silicon normally increases fluidity, feedability, and resistance to hot cracking during casting. Iron reduces hot cracking in Al-Cu alloys, but both iron and silicon can form coarse constituent phases (e.g., Al_7Cu_2Fe and Mg_2Si), which reduce the fracture toughness. Consequently, when this property is of prime importance, iron and silicon contents must be kept to a minimum. For example, 2124 is a higher-purity version of 2024. (See Table III.) Titanium and titanium plus boron refine the cast grain structure of aluminum alloys (as does zirconium) and improve the resistance to hot cracking. However, these additions also reduce fluidity.

VI. 6XXX Alloys

The second class of age-hardenable aluminum alloys developed for commercial utilization contained magnesium and silicon as the major alloying additions and are designated 6XXX alloys. These are medium-strength alloys having good weldability and corrosion resistance, and they can strengthen appreciably during room-temperature aging. They can be cheaply formed into a variety of shapes and are extensively used in the building, construction, and transportation industries. 6XXX alloys are used for many architectural products (including window and door frames, lightning poles, and fixtures), for ladders, and for welded structures. Their low cost and attractive combination of properties have, in some instances, made them an attractive replacement for steel in auto bodysheet and bumpers. Some 6XXX aluminum alloys, which have excellent combinations of strength and electrical conductivity, have virtually replaced copper for transmissions lines.

Silicon and magnesium are often added in balance to form the quasi-binary Al-Mg_2Si, although sometimes silicon is added in excess of that needed to form Mg_2Si. The quasi-binary has a eutectic at 868K and a maximum solubility of Mg_2Si of 1.85 wt %. The precipitation sequence is supersaturated solid solution → GP zones → β' → β (Mg_2Si). The GP zones are needle-shaped and form along the ⟨100⟩ direction of the aluminum matrix. The β' transition phase is rod-shaped and also lies along the

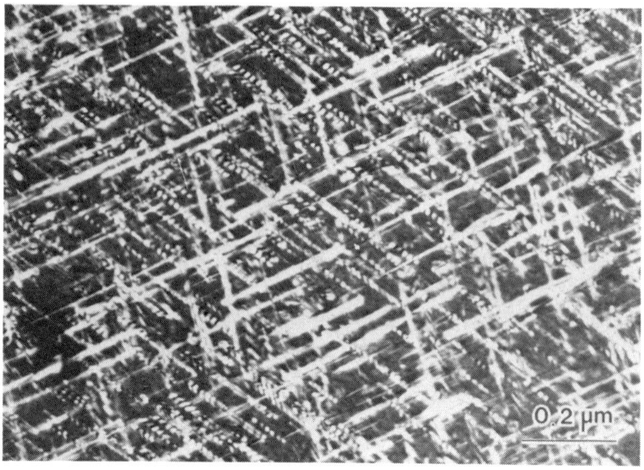

FIG. 12 Weak-beam TEM showing the rod-shaped precipitate in an Al-Mg-Si alloy. The fringes are due to oscillary contrast from the precipitates being observed at an angle to the plane of observation. (Courtesy of Anil K. Sachdev)

⟨100⟩ direction of the matrix. The β' rods have a hexagonal crystal structure with the {001} planes and ⟨100⟩ directions parallel respectively to the {100} planes and ⟨011⟩ directions of the aluminum matrix. These precipitates are shown in the TEM of Fig. 12. The equilibrium β phase is face centered cubic with the CaF_2 structure and forms as platelets on the matrix {100} planes. The {100} planes and ⟨110⟩ directions of the β phase are parallel respectively to the {100} planes and ⟨100⟩ directions of the aluminum matrix.

Most 6XXX alloys are used in the artificially aged, maximum-strength temper, T6, obtained by aging for various times at temperatures ranging from 433 to 463K. However, some 6XXX alloys, such as those used for automotive applications, may be used in the T4 temper. The chemical compositions of a selected number of 6XXX alloys are listed in Table IV.

TABLE IV

CHEMICAL COMPOSITION OF SOME 6XXX ALUMINUM ALLOYS

Alloy	Si	Fe	Cu	Mn	Mg	Cr	Zn	Ti	Zr	Others (total)	Aluminum
6009	0.6–1.6	0.50	0.15–0.6	0.2–0.8	0.40–0.8	0.1	0.25	0.10	—	0.15	Remainder
6010	0.8–1.2	0.50	0.15–0.6	0.2–0.8	0.6 –1.0	0.1	0.25	0.10	—	0.15	Remainder
6151	0.6–1.2	1.0	0.35	0.2	0.45–0.8	0.15–0.35	0.25	0.15	—	0.15	Remainder
6061	0.4–0.8	0.7	0.15–0.4	0.15	0.8 –1.2	0.04–0.35	0.25	0.15	—	0.15	Remainder

VII. 7XXX Alloys

The aluminum alloys having the greatest response to age hardening are based on the Al-Zn-Mg system and are designated 7XXX alloys by the Aluminum Association. As mentioned previously, their excellent combination of low density and high strength make them very attractive materials for the transportation industry. The alloys of commercial interest form spherical GP zones after quenching and during aging from room temperature up to approximately 435K. An example of these zones is shown in the TEM of Fig. 13. Extended warm aging of alloys with high Zn : Mg ratios produces the transition precipitate η', the precursor of the equilibrium $MgZn_2$, η phase. The basal planes of the hexagonal η' transitional phase are coherent with the (111) matrix planes, but the interface between the matrix and the c direction of the precipitate is incoherent. For low Zn : Mg ratios, the $Al_2Zn_3Mg_3$, T phase may form. The decomposition process may be described as supersaturated solid solution → GP zones → η' → η or as supersaturated solid solution → GP zones → T′ → T.

Copper additions considerably increase the strength of Al-Zn-Mg alloys. Concentrations up to one wt % do not appear to alter the basic precipitation mechanism and in this range probably add a component of solid-solution strengthening. Low–copper-content alloys are readily weldable but are not as resistant to stress corrosion cracking as the higher-content alloys. When in excess of one percent, copper participates in the precipitation process during aging and decreases the coherency of the precipitate

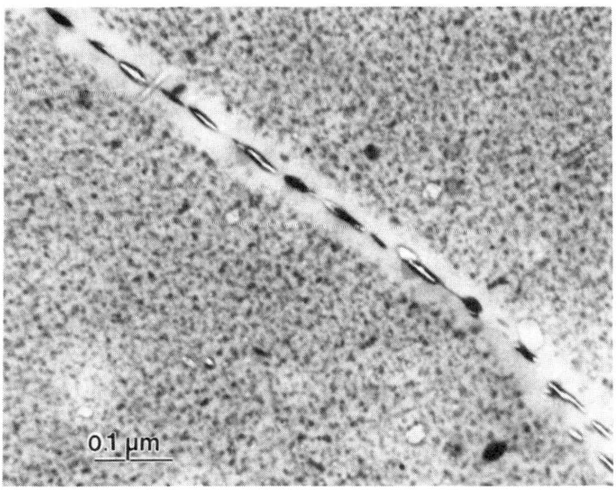

FIG. 13 TEM showing spherical GP zones in an aged Al-Zn-Mg alloy. The equilibrium η phase has precipitated at the high-angle grain boundary producing a solute-denuded PFZ. (Courtesy of J.A. Hawk)

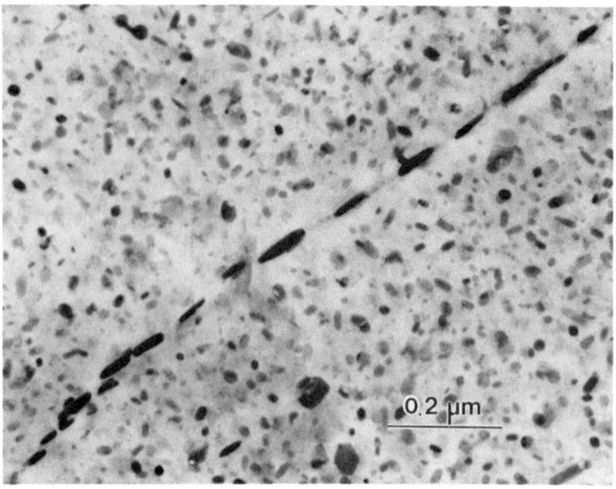

FIG. 14 TEM showing the metastable η' phase in the matrix and the equilibrium η phase along a high-angle grain boundary in an Al-Zn-Mg-Cu alloy. (Courtesy of J.A. Hawk)

when aged to peak strength. The η' phase in an Al-Zn-Mg-Cu alloy is shown in Fig. 14. In the quaternary Al-Zn-Mg-Cu system the phases $MgZn_2$ and AlMgCu form an isomorphous series, with aluminum and copper substituting for zinc in $MgZn_2$. The improvement in resistance to stress corrosion cracking of Al-Zn-Mg-Cu alloys when compared to Al-Zn-Mg alloys may be related to a reduction in the electrochemical activity of the precipitates as their copper content is increased. The improvement may also be somewhat related to the effect that copper has on the coherency of the transition precipitate, η', and the associated effect of homogenizing deformation.

Chromium was originally added to Al-Zn-Mg alloys used for sheet product applications to improve their resistance to stress corrosion cracking. The $Al_{12}Mg_2Cr$ dispersoids aid in retaining the directional grain structure developed during processing of wrought products and also prevent excessive growth of recrystallized grains which form during subsequent heat treatments. However, as mentioned earlier, these incoherent dispersoids increase the quench sensitivity of high-copper Al-Zn-Mg alloys since they act as nucleating agents for solute-rich precipitates during quenching. This process eliminates some of the copper and magnesium from participating in the low-temperature aging sequence, and it leads to a lower strength than would be predicted from consideration of the alloy chemistry alone. Zirconium may be substituted for chromium and offers the advantage of forming a coherent dispersoid which minimizes quench sensitivity. Zirconium also seems to be the most effective addition for inhibiting

TABLE V

CHEMICAL COMPOSITION OF SOME 7XXX ALUMINUM ALLOYS

Alloy	Si	Fe	Cu	Mn	Mg	Cr	Zn	Ti	Zr	Others (total)	Aluminum
7016	0.1	0.1	0.6 –1.4	0.03	0.8–1.4	0.3	4.0–5.0	—	—	0.15	Remainder
7029	0.1	0.12	0.5 –0.9	0.03	1.3–2.0	0.03	4.2–5.2	—	—	0.15	Remainder
7039	0.30	0.40	0.10 max	0.10–0.4	2.3–3.3	0.10 max	3.5–4.5	0.10	—	0.15	Remainder
7046	0.2	0.4	0.25	0.3	1.0–1.6	0.2	6.6–7.6	0.06	—	0.15	Remainder
7146	0.2	0.4	0.25	0.03	1.0–1.6	0.03	6.6–7.6	0.06	—	0.15	Remainder
7049	0.25	0.35	1.2 –1.9	0.20 max	2.0–2.9	0.10–0.22	—	0.10	—	0.15	Remainder
7050	0.12	0.15	2.0 –2.6	0.10	1.9–2.6	0.04	5.7–6.7	0.06	0.08–0.15	0.15	Remainder
7075	0.40	0.50	1.2 –2.0	0.30	2.1–2.9	0.18–0.35	5.1–6.1	0.20	—	0.15	Remainder
7175	0.15	0.20	1.2 –2.0	0.10	2.1–2.9	0.18–0.30	5.1–6.1	0.10	—	0.15	Remainder
7475	0.10	0.12	1.2 –1.9	0.06	1.9–2.6	0.18–0.25	5.2–6.2	0.06	—	0.15	Remainder
7178	0.40	0.50	1.6 –2.4	0.30	2.4–3.1	0.18–0.35	6.3–7.3	0.20	—	0.15	Remainder
7079	0.30	0.40	0.40–0.8	0.10–0.30	2.9–3.7	0.10–0.25	3.8–4.8	0.10	—	0.15	Remainder

recrystallization. The chemical compositions of a selected number of 7XXX alloys are listed in Table V.

Although 7XXX alloys respond to natural aging, they are not normally used in that condition because they age indefinitely and are therefore unstable. Since the GP zones that form during aging of Al-Zn-Mg and Al-Zn-Mg-Cu alloys have very low coherency strains, deformation prior to aging has little or no effect on the precipitation behavior. Consequently, there is no T8 temper, and most 7XXX alloys are used in either the T6 or T7 tempers. Peak strength is normally obtained by aging for various times, usually up to 24 hours, at temperatures ranging from 388 to 403K. The T73 and T76 overaged tempers are primarily used for applications requiring a high resistance to stress corrosion cracking and exfoliation, respectively. However, these overaging tempers lead to a sacrifice in strength, and new methods to improve the resistance to stress corrosion cracking with no sacrifice in strength are constantly being pursued. One such method, called retrogression and reaging (RRA), produces a several-fold increase in the threshold stress for SCC in 7075 over that for the T6 temper with no significant sacrifice in strength. The RRA treatment involves a short time retrogression anneal at a temperature in the two-phase ($\alpha + \eta$) field of the phase diagram, followed by water quenching and a final reaging treatment equivalent to the original T6 temper. Many variables affect the aging response of 7XXX alloys, including composition, quench rate, time between quench and artificial aging, and heat-up rate to aging temperature. As with the 2XXX alloys, precipitation-hardening curves have been established for most of the commercial 7XXX alloys for several temperatures.

VIII. Properties of Heat-Treatable Aluminum Alloys

Chemical composition and processing control the microstructure and thus the physical, mechanical, and corrosion properties of heat-treatable aluminum alloys. Since these can be varied greatly, it is not feasible to cover all possibilities here, and the reader is referred to *Aluminum Standards and Data* published by the Aluminum Association. Typical mechanical properties of a selected group of heat-treatable alloys are listed in Table VI. A schematic showing the change in strength with particle radius, and thus with increasing aging time, has been given in Fig. 6. The toughness levels of age-hardenable aluminum alloys generally decrease as the strength level is raised by alloying and heat treatment. This trend is illustrated in Fig. 15a for the 2XXX and 7XXX alloys. Once peak strength

TABLE VI

Typical Mechanical Properties of a Selected Group of Heat-Treatable Alloys

Alloy and temper	Tension				Hardness	Shear	Fatigue	Modulus
	Strength ksi		Elongation percentage in 2 in.		Brinell Number	Ultimate Shearing Strength	Endurance Limit	Modulus of Elasticity
	Ultimate	Yield	1/16-in.–thick specimen	1/2-in.–diameter specimen	500 kg load 10 mm ball	ksi	ksi	ksi $\times 10^3$
2024-0	27	11	20	22	47	18	13	10.6
2024-T3	70	50	18	—	120	41	20	10.6
2024-T4, T351	68	47	20	19	120	41	20	10.6
2036-T4	49	28	24	—	—	—	18	10.3
2117-T4	43	24	—	27	70	28	14	10.3
2219-0	25	11	18	—	—	—	—	10.6
2219-T81, T851	66	51	10	—	—	—	15	10.6
6061-0	18	8	25	30	30	12	9	10.0
6061-T6, T651	45	40	12	17	95	30	14	10.0
7049-T73	75	65	—	12	135	44	—	10.4
7050-T7651	80	71	—	11	—	47	—	10.4
7075-0	33	15	17	16	60	22	—	10.4
7075-T6, T651	83	73	11	11	150	48	23	10.4
7178-0	33	15	15	16	—	—	—	10.4
7178-T6, T651	88	78	10	11	—	—	—	10.4
7178-T76, T7651	83	73	—	11	—	—	—	10.3

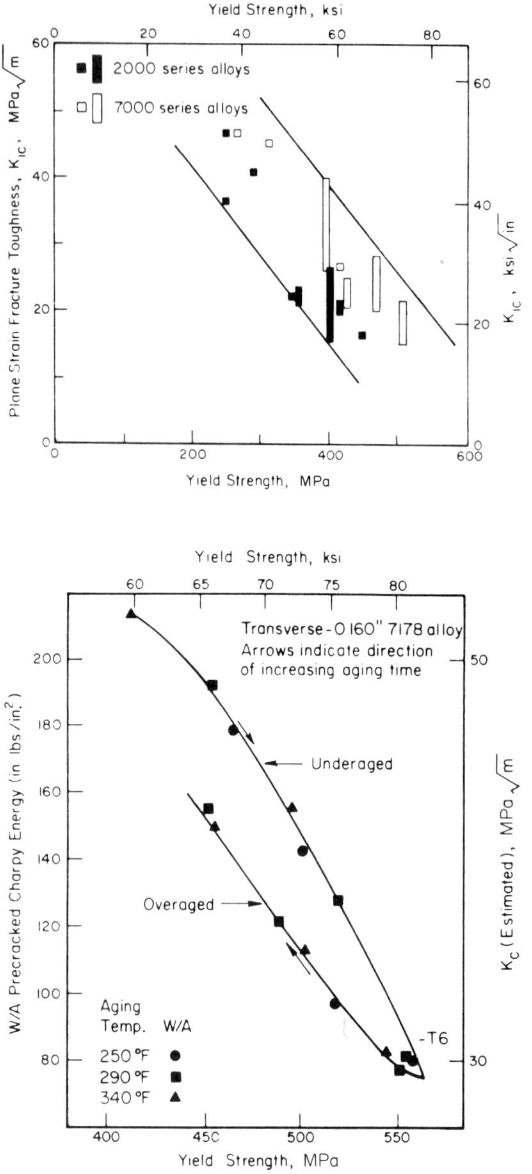

FIG. 15 The relationship between strength and toughness. (a) For various 2XXX and 7XXX alloys. (b) For 7178 as a function of aging. After Hahn and Rosenfield.

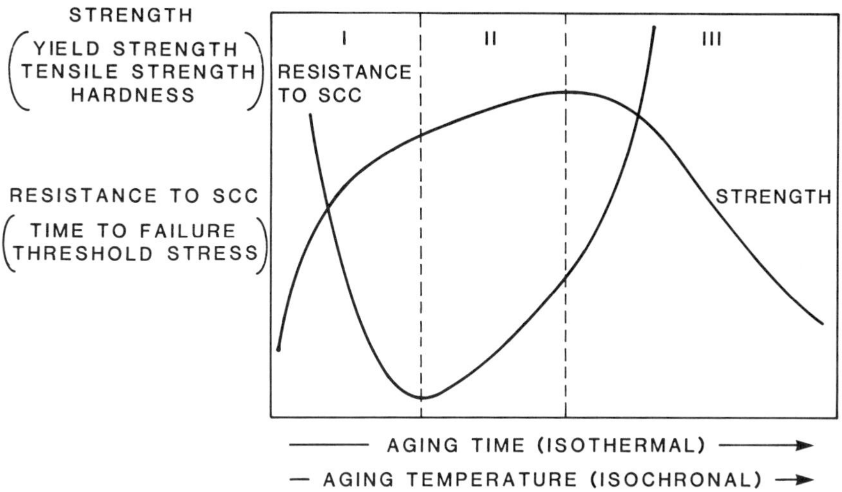

FIG. 16 Schematic representation of the change in resistance to SCC and strength as a function of aging time or aging temperature. After Speidel.

is reached for a particular alloy, some improvement in toughness may be obtained by overaging. However, the overaged strength–toughness curves always lie below the underaged strength–toughness curves as illustrated in Fig. 15b. This is most likely due to the presence of coarse grain boundary precipitates and precipitate-free-zones in the overaged condition which lead to an increased incidence of low-energy intergranular fracture.

Aluminum and its alloys may fail by intergranular stress corrosion cracking while under stress and exposed to aggressive environments, which may include water vapor, aqueous solutions, organic liquids, and liquid metals. Stresses sufficient for the initiation and growth of stress corrosion cracks may be well below those necessary for gross yielding. In general, aluminum alloys are most susceptible near or slightly below peak strength, as shown schematically in Fig. 16, when planar slip and strain localization are most pronounced. Overaging is often used for improving the stress corrosion resistance of some 7XXX alloys, but is impractical for 2XXX alloys. For these materials, stress corrosion must be controlled by alloy chemistry or by using a barrier that isolates the environment from the susceptible material. The resistance to stress corrosion cracking increases in the Al-Zn-Mg 7XXX alloys when copper is added, as shown in Fig. 17a. Although all of these alloys show an improvement in stress corrosion resistance with overaging, only those having an excess of one wt % Cu become completely immune, as seen in Fig. 17b.

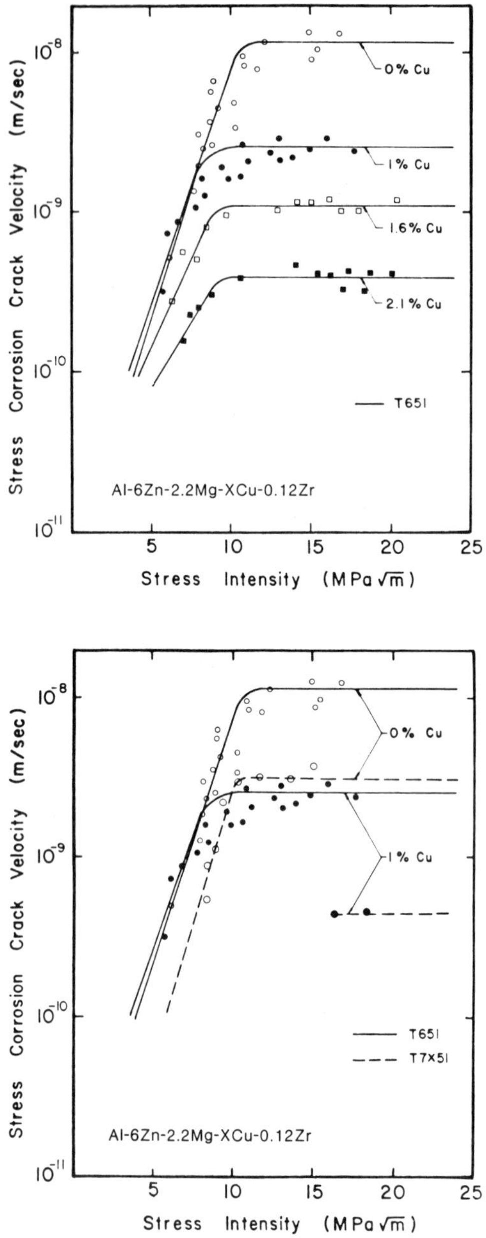

FIG. 17 Effect of stress intensity on stress corrosion crack velocity of an Al-Zn-Mg alloy in 3.5 pct NaCl. (a) With varying copper contents. (b) With overaging.

References

1. *Aluminum Standards and Data 1978 Metric SI*. Aluminum Association, Washington, DC.
2. Hahn, G.T., and Rosenfield, A.R. (1975). "Metallurgical Factors Affecting Fracture Toughness of Aluminum Alloys," *Met. Trans. A* **6A**, 653–670.
3. Hatch, J. E., ed. (1984). *Aluminum; Properties and Physical Metallurgy*. American Society for Metals, Metals Park, Ohio.
4. Kelly, A., and Nicholson, R.B. (1963). "Precipitation Hardening" in *Progress in Materials Science*, **vol. 10**, B. Chalmers, ed. Pergamon Press, New York, pp. 151–391.
5. Lorimer, G.W. (1978). "Precipitation in Aluminum Alloys" in *Precipitation Processes in Solids*, K.C. Russell and H.I. Aaronson, eds. The Metallurgical Society of AIME, New York, pp. 87–119.
6. Mondolfo, L.F. (1976). *Aluminum Alloys: Structures and Properties*. Butterworth, London, England.
7. Polmear, I.J. (1981). *Light Metals: Metallurgy of the Light Metals*. Edward Arnold, London, England.
8. Ryum, N. (1986). "Physical Metallurgy of Heat Treatable Alloys" in *Aluminum Alloys: Their Physical and Mechanical Properties*, **vol. III**, E.A. Starke, Jr., and T.H. Sanders, Jr., eds. Engineering Materials Advisory Services Ltd., West Midlands, United Kingdom, pp. 1511–1545.
9. Sarkar, B., Marek, M., and Starke, E.A., Jr., (1981). "The Effect of Copper Content and Heat Treatment on the Stress Corrosion Characteristics of Al-6Zn-2Mg-X Cu Alloys," *Met. Trans. A* **12A**, 1939–1943.
10. Speidel, M.O. (1975). "Stress Corrosion Cracking of Aluminum Alloys," *Met. Trans. A* **6A**, 631–642.
11. Starke, E.A., Jr. (1970). "The Causes and Effects of Denuded or Precipitate Free Zones at Grain Boundaries in Aluminum Base Alloys," *J. of Met.* **22**, no. 1, 54–63.
12. Starke, E.A., Jr. (1977). "Aluminum Alloys of the 70's; Scientific Solutions to Engineering Problems," *Mater. Sci. Eng.* **29**, 99–115.

3

Wrought Non–Heat-Treatable Aluminum Alloys

R.E. SANDERS, Jr., S.F. BAUMANN,
and H.C. STUMPF (Retired)

Alcoa Laboratories
Alcoa Center, Pennsylvania

ISBN 0-12-341831-3

I. Introduction

A. Non–Heat-Treatable Alloys

The manufacture of strain-hardened or non–heat-treatable (NHT) aluminum-alloy sheet represents a major portion of the aluminum industry of the late 1980s. In 1984, the U.S. aluminum industry shipped approximately 3.2 billion kilograms of NHT alloy sheet, plate, and foil. This figure represents approximately 95% of all (heat-treatable and NHT) flat-rolled products shipments [*1*]. The rigid-container-sheet (RCS) market for production of aluminum beverage containers comprised over 40% of this total. Other large-volume applications of NHT alloy sheet include household foil, architectural sheet, lithographic sheet, residential siding, fin (heat-exchanger) stock, and formed container stock.

The Aluminum Association has assigned the 1XXX-, 3XXX-, and 5XXX-series alloy designations to the major non–heat-treatable aluminum alloys. Unlike the heat-treatable alloys, which derive strength from precipitation-hardening, the NHT alloys are strengthened by elements in solid solution and dislocation structures introduced by cold-rolling. The 1XXX alloys are defined as those that are 99.0 wt % or greater aluminum with iron and silicon as the major alloying elements. The 3XXX alloys have manganese as the primary alloying element, while the 5XXX alloys are alloyed principally with magnesium. A number of 8XXX-series NHT alloys, mostly Al-Fe-X compositions with Fe + X > 1.0%, have been developed over the years, but are not used in large volumes.

The temper designations used for the NHT aluminum alloys are based upon the amount of strengthening due to work-hardening that remains in the sheet after processing. Cold-working operations, recovery, and recrystallization treatments are manipulated during fabrication to provide the desired combination of final properties in the sheet product. A summary of the Aluminum Association NHT alloy temper designations is contained in Table I.

This summary will describe the important aspects of microstructure development and strengthening mechanisms in the three major NHT alloy systems. Phase transformations during solidification and solid-state processing will be discussed in view of their important effects on strain-hardening, recovery, and recrystallization. Microstructure-property relationships pertinent to selected structural applications of NHT alloy sheet will be emphasized.

Much of the data presented in this chapter are from previously unpublished research performed at Alcoa Laboratories from 1930 through 1985.

TABLE I

TEMPER DESIGNATIONS FOR NON–HEAT-TREATABLE ALUMINUM ALLOYS

F - *As-fabricated*: no special control over thermal or working operations; no mechanical property limits.
O - *Annealed*: to obtain the lowest strength temper.
H - *Strain-hardened*: applies to products that have had their strength increased by strain-hardening, with or without subsequent thermal operations; always followed by two digits.
H1X - Strain-hardened only, no subsequent thermal operation.
H2X - Strain-hardened and partially annealed.
H3X - Strain-hardened and stabilized.
X indicates the degree of strain-hardening remaining after rolling and/or partially annealing; except for H1X tempers, it does not directly correspond to the cold-reduction experienced by the metal.

EXAMPLES:
H18 - Corresponds to UTS achieved by equivalent cold-reduction of 75% after full annealing.
H14 - Applies to material strengthened to one-half UTS level of H18.
H34-Applies to material with strength of one-half UTS level of H18 but achieved by stabilizing after cold-rolling.
H19 - Applies to material with UTS of 2.0 ksi (14 MPa) greater than that of H18.
H26 - Applies to material strengthened to three-quarters UTS level of H18 by cold-rolling and partial annealing.

Source: *Aluminum Standards and Data* (1982). The Aluminum Association, Washington, D.C.

This chapter summarizes much of the basic and applied research by workers such as Dr. W.A. Anderson, Dr. H.C. Stumpf, Mr. J.A. Nock, Jr., and Mr. J.K. McBride. Their efforts have, to a large extent, resulted in the development and widespread application of many of the NHT alloys in use today.

II. 1XXX Alloys

A. *Background*

Following the development of the Hall cell, researchers found that the metal from the pot was approximately 99% (by weight) pure aluminum, with Fe and Si as the major contaminants. This "commercial-purity" aluminum, designated as Alcoa Alloy 2S in 1888 [2] and as the present Aluminum Association Alloy 1100, is the oldest aluminum alloy in continuous use.

Improvements in refining and smelting technology have led to a wide variety of 1XXX alloys ranging in composition from 99.0 to 99.99% aluminum. Minor elements as well as Fe and Si are controlled to obtain microstructures or properties for specific end uses. Small amounts of copper in solid solution are particularly effective in increasing the H1X and O-temper yield strengths of 1XXX alloys. Alloys of >99.99% aluminum are produced for some very specialized applications. Refinement to this purity level is very expensive and alloys of this type are not covered in this review.

B. Phase Reactions in 1XXX Alloys

The relevant phase reactions in the 1XXX alloys are those of the Al-Fe-Si system. The maximum solid solubility of Fe in Al is 0.05%. Any Fe in excess of this amount will form constituents at grain and cell boundaries by a eutectic reaction during solidification. The equilibrium diagram, Fig. 1, predicts that the Fe will be present in the form of the Al_3Fe constituent, but in most commercial solidification processes such as direct chill (DC) or roll casting of 1XXX alloys, metastable phases Al_6Fe or Al_m Fe, will be observed [*3*]. When Si is present, $Al_{12}Fe_3Si$(α-phase) or $Al_9Fe_2Si_2$ (β-phase) may be found in the 1XXX alloy sheet.

High-temperature thermal treatments transform the Al_6Fe or Al_mFe to the stable Al_3Fe phase occupying approximately $\frac{1}{2}$ the volume of the Al_6 phase.

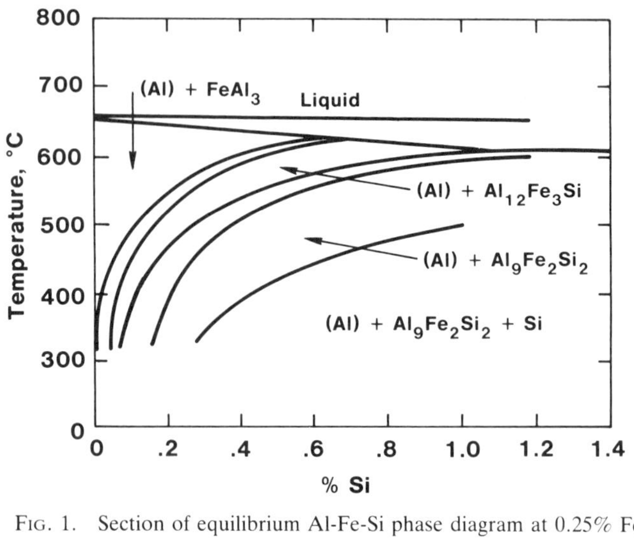

FIG. 1. Section of equilibrium Al-Fe-Si phase diagram at 0.25% Fe.

C. Strength Development

Since the strength levels of the 1XXX alloys are very low, 14–28 MPa in the "O" or fully annealed temper, these alloys are generally strengthened through strain-hardening. This is accomplished through cold-rolling for sheet materials, but deformation during forming of softer 1XXX alloys can impart some benefit.

The influence of microstructure on strain-hardening of the 1XXX series alloys is demonstrated in an experiment in which two series of DC-cast Al-Fe-Si alloys were preheated, hot-rolled, and then cold-rolled 67% (H1X-temper) prior to annealing [*4*,*5*]. In the first series, Fe was varied from 0.15 to 1.2% with a 0.05 Si base. Preheating conditions were varied in order to produce either the Al_3Fe phase (16 hours/607°C) or the Al_6Fe phase (16 hours/482°C). Yield strength of the H1X sheet varied with the calculated constituent volume fraction as shown in Fig. 2a. Increased constituent content increased strength about 7 MPa per volume percent for both the H1X and annealed ("O") tempers. Thus, increasing constituent yielded no additional benefits in subsequent strain-hardening during rolling. The 607°C H1X material retained higher strength after rolling due to the larger amount of Fe retained in solution. The constituent–yield strength dependence was similar in magnitude regardless of the type of constituent present: Al_6Fe or Al_3Fe.

When the H1X alloys were cold-rolled an additional 90% without an anneal (H19), the lower-Fe alloys strengthened more than the high-Fe alloys. The strength increase ($\Delta = YS_{H19} - YS_{H1X}$) after 90% cold-reduction in the Fe alloy series varied with the mean interparticle spacing of the constituent, as shown in Fig. 2b. The lattice strain, as calculated from X-ray line breadth measurements, actually decreased with increasing Fe content, indicating that the additional volume fraction of Fe bearing constituents promotes particle-induced dynamic recovery for rolling to very high strains. This interpretation is reasonable since the cold-rolling of the Al-Fe alloys promotes the accumulation of dislocations in cellular structures (Fig. 3). In the second series of alloys, Si was varied from 0.05 to 0.60% with a 0.27% Fe base. Strengthening in these alloys could be predicted from a knowledge of the effect of Si in solid solution on increasing strength in strain-hardened Al alloys (Fig. 4). The higher strength with increased Si in solid solution is a typical solute atom strengthening effect in that the Si atoms impede dynamic recovery during rolling and promote strain-hardening through dislocation multiplication. This effect is illustrated in Fig. 5, which compares the dislocation structures of pure Al and Al-1.0Si after 90% cold-reduction by rolling.

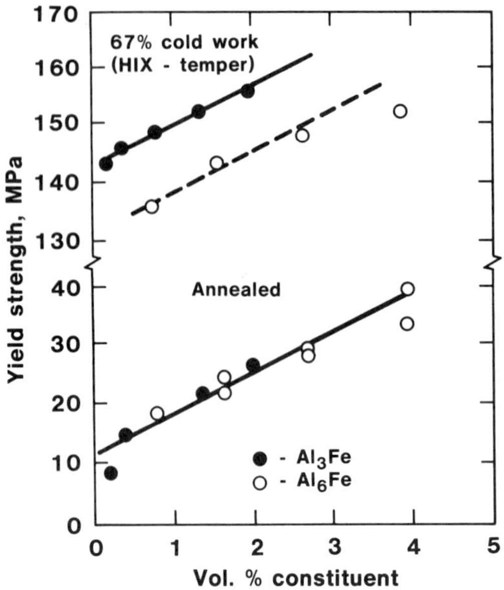

(a)

FIG. 2(a). Yield strength versus volume fraction constituent (Al_3Fe or Al_6Fe) for Al-0.05 Si-x Fe series of lXXX alloys.

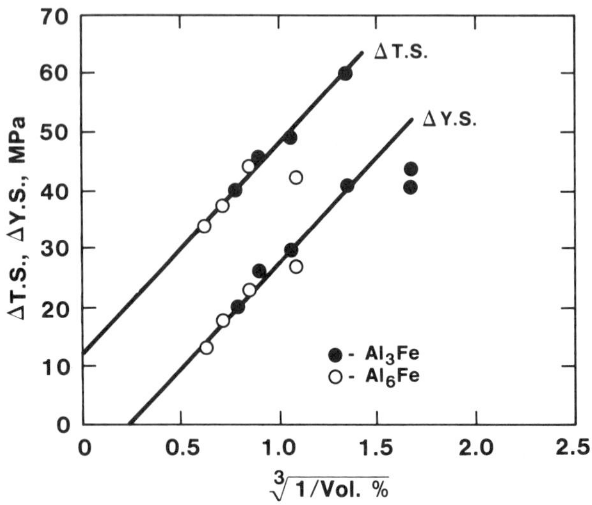

(b)

FIG. 2(b). Increases in yield and tensile strengths due to additional 90% cold-rolling of HlX temper (67% cold-rolling) Al-0.05 Si-x Fe sheet.

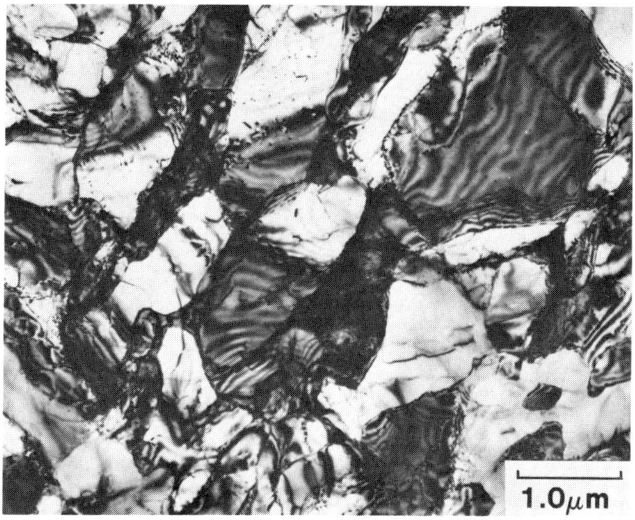

FIG. 3. Transmission electron micrograph illustrating the dislocation structure in alloy 1145-H19 foil.

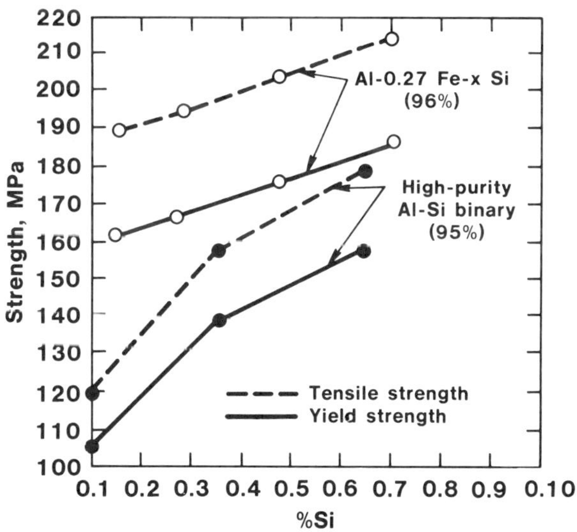

FIG. 4. Strength versus Si content for Al-0.27 Fe-x Si and high-purity Al-Si binary alloys after cold-rolling.

(a)

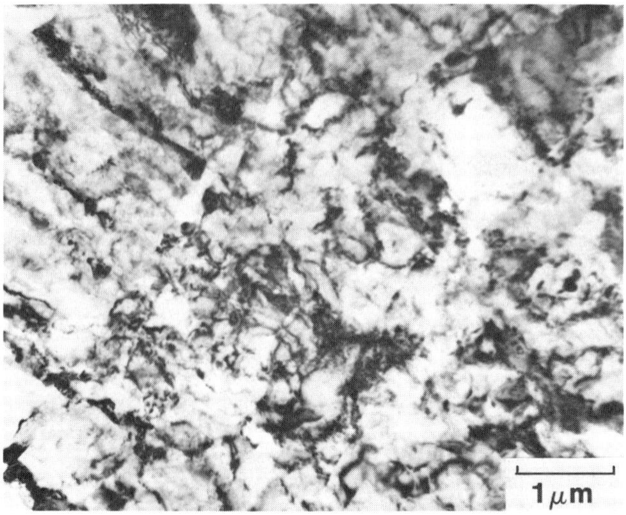

(b)

FIG. 5. Transmission electron micrographs comparing the dislocation structures of (a) pure Al and Al-1.0% Si after 90% reduction by rolling.

D. Recovery and Recrystallization

Recovery in 1XXX-H1X sheet is thermally activated, making it both time- and temperature-dependent. Recovery will typically begin immediately upon heating to temperatures above that at which the rolling deformation was performed. However, recovery can take place at room temperature for long-term exposures (>10 years) of 1100-H18 sheet [*6*]. Figure 6 illustrates the changes in mechanical properties that occur upon heating of alloy 1100-H19 sheet to progressively higher temperatures. Strength is lost through the initial annihilation of free dislocations and later through the growth of the subgrain structure. During recovery, a minimum in the tensile elongation is typically observed in the 180 to 240°C temperature range.

The onset of recrystallization is accompanied by a rapid drop in strength and increase in ductility as strain-free high-angle grain boundaries are formed. The concepts of particle-stimulated nucleation (PSN) and Zener drag describe the recrystallization behavior of cold-rolled 1XXX sheet quite adequately [*7–9*]. The reduction in grain size with increasing Fe content (Fig. 7) can be explained by an increase in the volume fraction of critical-sized Fe-bearing particles that can act to nucleate recrystallized grains [*10,11*]. The deleterious effect of Si on final grain size is caused by

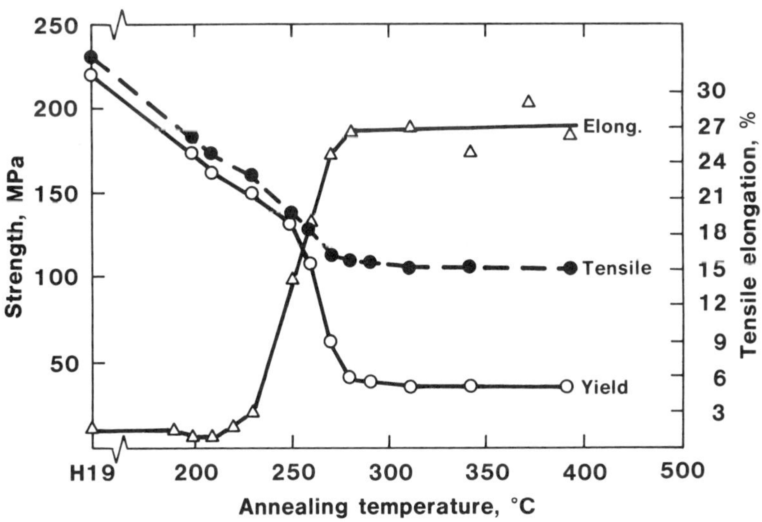

FIG. 6. Isochronal (one-hour soak) annealing curve for alloy 1100-H19 sheet, 0.2 mm thick, produced from roll-cast strip.

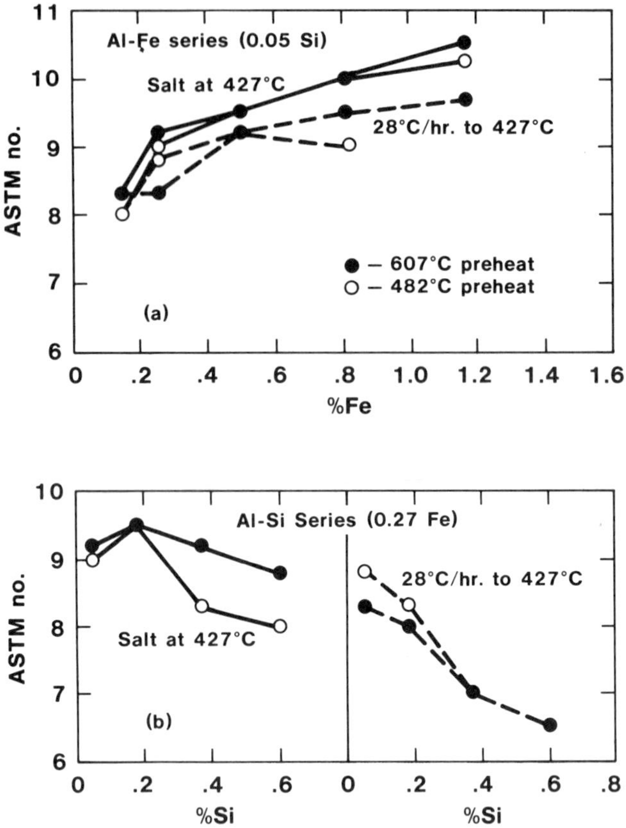

FIG. 7. Effects of variations in Fe and Si content on grain size of sheet rolled 96% prior to annealing.

precipitation of Si from solid solution prior to or during the annealing operation. The fine Si particles impede early stages of recrystallization within deformation zones and lead to a reduction in the number of deformation zones that can be effective nuclei for new grains [*12*]. The effect of cold-reduction on grain size is also consistent with PSN as the increased intensity of the deformation zones adjacent to particles at higher cold-reductions serves to activate additional particles of smaller sizes as nucleation sites for recrystallization. A more complete description of recrystallization and grain-size control is undertaken in the discussion of 3XXX alloys.

The presence of Fe in solid solution can affect the onset of recrystallization. The effect is most noticeable in high-purity alloys [*13*] (see Fig. 8), but

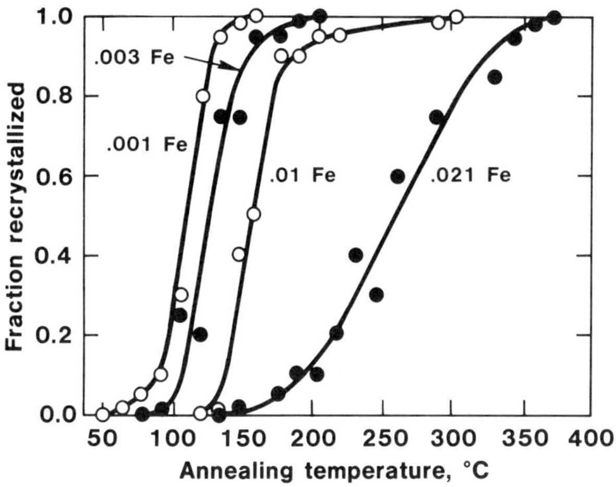

FIG. 8. Effect of Fe in solid solution on recrystallization behavior of high-purity lXXX sheet.

can also be observed in commercial-purity alloys. The retardation of recrystallization by Fe in solution is caused by precipitation of fine Al–Fe particles during the anneal.

III. 3XXX Alloys

A. *Background*

The first commercial 3XXX alloy, 3S, resulted from research in 1906 by Dr. Earl Blough [2], who was looking for a corrosion-resistant aluminum sheet alloy for camera cases. He found that Mn additions to aluminum resulted in a moderate-strength, formable sheet with excellent corrosion resistance. His work resulted in what is now substantially the composition of Alloy 3003: 1.25% Mn, 0.13% Cu. Alloy 3003 is still one of the most widely used of all aluminum alloys, with many applications as light gauge sheet and foil.

In the 1920s, the need arose for an alloy stronger than 3S but not as difficult to cold-roll on existing equipment as the Al-Mg alloys. In 1929, Alcoa Alloy 4S, essentially 3S + 1%Mg, first came into commercial use. The alloy was supplanted temporarily by 5XXX alloys in the 1940s, but was later returned to production. As of the late 1980s, Alloy 3004 is the primary alloy used for can body stock, making it the largest-volume aluminum alloy in production.

B. Solidification–Phases

The Al-Fe-Mn-Si system serves as the basis for the 3XXX alloys. While Mn is the major alloying element, all the commercial alloys from this group contain controlled amounts of Fe and Si that play a vital role in the phase reactions and microstructural development.

In as-cast 3XXX ingot, the Al matrix phase is dendritic. Between dendrite arms is the eutectic mixture, which contains the coarse Fe-rich intermetallic constituent particles. The particles are of two major types: the orthorhombic Al_6 (Fe,Mn) phase and the cubic Al_{12} $(Fe,Mn)_3$ Si. The relative amounts of these phases are determined by the specific alloy composition and the solidification rate. Fe, being highly insoluble in Al, has the largest effect on the volume fraction of constituent. Most commercial ingot cooling rates favor the Al_6 phase over Al_{12} for the as-cast constituent. In roll-casting of 3003, however, where the solidification rates are much higher, Al_{12} is the predominant constituent phase.

C. Preheating

During solidification, substantial amounts of Mn are retained in solution in the Al matrix. Typical as-cast 3003 ingot will contain 0.7–0.9 wt % Mn in solution [*14,15*]. During preheating, this supersaturation is relieved by the precipitation of a fine dispersoid of Al_{12} $(Mn,Fe)_3Si$, or Al_6 (Mn,Fe) if the Si content is very low (i.e., less than approximately 0.07 wt %). The dispersoid is submicron in size and forms within the Mn-rich dendrites. Coring of the Mn in the dendrites during solidification results in heavier dispersoid populations within the dendrites as seen in the micrograph of a preheated 3004 ingot in Fig. 9. The constituent particles are visible between dendrites and dendrite arms.

The preheat practices for 3XXX alloys are designed to produce a coarse, widely spaced dispersoid distribution with minimum Mn remaining in solid solution. The importance of this microstructure to the control of recrystallized grain size during subsequent annealing treatments will be discussed in a later section. The precipitation of Mn during the preheat is found to follow C-curve–type kinetics, as shown in Fig. 10 [*14*], which plots percentage of Mn + Fe in solution in a 3003 ingot as a function of time at temperature. An initial heating rate of approximately 28°C/h was used to simulate heat-up of a large commercial ingot. Early during heat-up, a fine dispersoid forms, but at temperatures above about 480°C, many of the smaller particles revert, causing the matrix Mn level to rise while the number of stable dispersoid particles decreases. With time at the soak

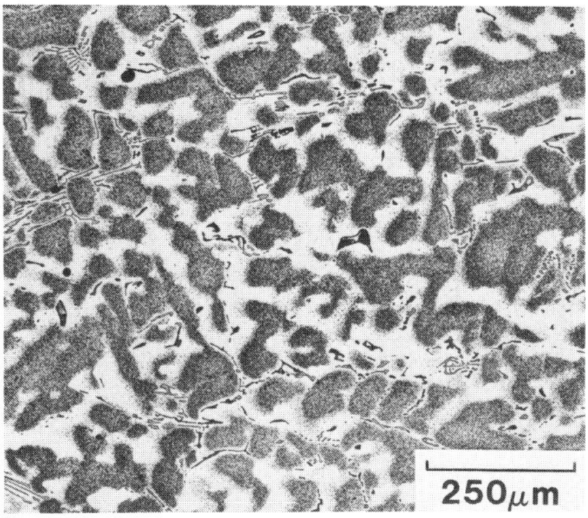

FIG. 9. Optical micrograph of a preheated 3004 ingot etched to show dipersoids within the cored dendrites. Constituent particles are between dendrites and dendrite arms.

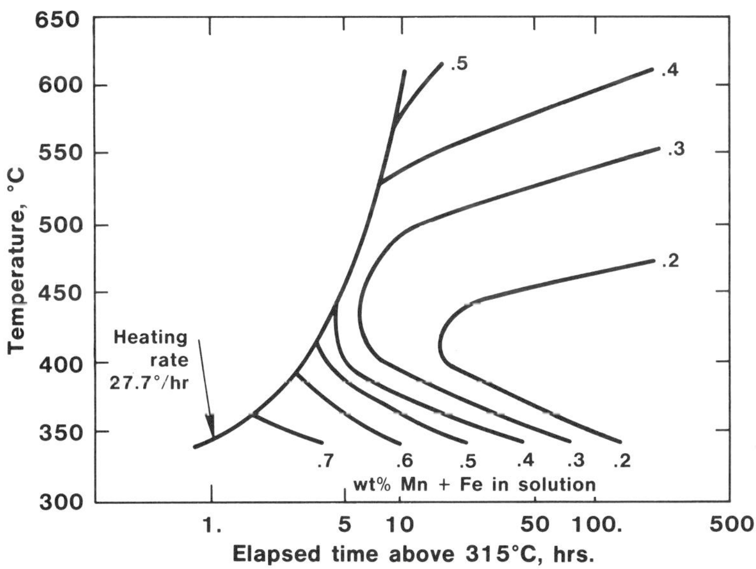

FIG. 10. Change in matrix solute level (Mn + Fe) with time during isothermal soak on preheating 3003 ingot. A slow heat-up to the soak temperature was used as indicated.

temperature, there is continued growth and coarsening of the surviving dispersoids while the Mn in solution continues to drop. The cool from the soak effectively removes additional Mn from solution through growth of the existing dispersoids rather than nucleation of new ones.

The use of a high soak temperature and slow cool is the most effective way of obtaining a coarse dispersoid distribution since the low diffusivity of Mn does not allow for significant particle coarsening at low temperatures. In well-homogenized alloys, the Mn dispersoids will be approximately 0.25 μm in diameter and spaced approximately 0.5 μm apart. Total relief of the supersaturation of Mn is never achieved during the preheat because of the slow precipitation kinetics (governed by Mn diffusion). The level of residual Mn in solution should be low enough (approximately 0.25%) to minimize any additional precipitation during subsequent hot-rolling and annealing.

The high-temperature, slow-cool preheating route just described is the most effective way to produce the desired coarse dispersoid distribution in modern DC ingot. However, during the 1930s, fine-grained 3003-0 was routinely produced with much lower temperature homogenization treatments. This was possible due to the old "tilt" mold technology in which molten aluminum was poured directly into a steel mold. The extremely slow rates of cooling experienced by these ingots after casting effectively produced a coarse dispersoid structure and low residual Mn in solution. When DC casting was introduced, the higher solidification and cooling rates produced dramatic differences in the microstructures of the as-cast 3003 ingot. As early as 1937, the higher Mn in solution and finer dispersoids present after preheating the "new" DC ingots were documented at Alcoa Laboratories [*16*]. Extensive development work was required to redesign preheating practices for obtaining fine-grained 3003-0 sheet from DC ingot [*17,18*].

The kinetics of the Mn precipitation reaction can be markedly affected by alloy composition. For instance, it has been shown that Si accelerates the precipitation of Mn and results in a denser distribution of dispersoids and lower matrix Mn levels after homogenization [*5,19*]. Furthermore, Fe decreases the solubility of Mn in Al and can lead to more rapid precipitation kinetics [*5,20*].

Preheating also serves to homogenize the distribution of the more mobile atomic species such as Mg and Cu, if present. If sufficient Si is present, some of the Al_6 (Fe,Mn) constituent transforms to the Al_{12} $(Fe,Mn)_3Si$ phase [*15*]. Some spheroidizing of the constituent phases also occurs during high-temperature preheating.

After preheating, hot-rolling progressively makes the microstructure more homogeneous by breaking up the cell structure, fracturing the con-

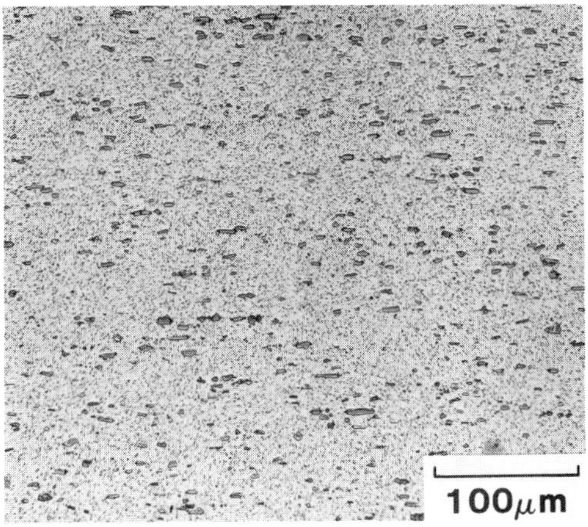

FIG. 11. Longitudinal cross section of 3004 after cold-rolling to .31-mm sheet. As compared to Fig. 9, the constituents and dipersoids have been uniformly redistributed.

stituent particles, and distributing the constituent and dispersoid more uniformly throughout the microstructure. Recrystallization even after extensive hot-rolling results in flat, elongated grains indicating that the microstructure is still layered with the remnants of the ingot structure. Final cold-rolling to thin gauge sheet results in a more homogeneous microstructure with uniformly distributed constituents and dispersoid. Figure 11 shows a typical microstructure for Alloy 3004 after rolling to H19 sheet. The duplex particle distribution of coarse constituents and fine dispersoids is seen.

D. Strength Development

Mn produces a modest solid-solution strengthening, but is not very effective by itself in promoting work-hardening during cold-rolling. Figure 12 [*21*] shows O temper and H19 yield and tensile strengths for Al-Mn binary alloys where the Mn has been retained in solution by a high-temperature anneal and a rapid quench. Annealed yield strength increases approximately 21 MPa and tensile strength about 42 MPa per wt % Mn in solution. The H19 yield strength only increases about 35 MPa with 1 wt % Mn in solution. This indicates that on a weight basis, Mn is not as effective in promoting work-hardening as is Mg, which raises the H19 yield strength of Al by about 70 to 85 MPa at the 1 wt % level. Thus, many commercial 3XXX alloys such as 3105 and 3004 contain Mg in order to achieve higher strength.

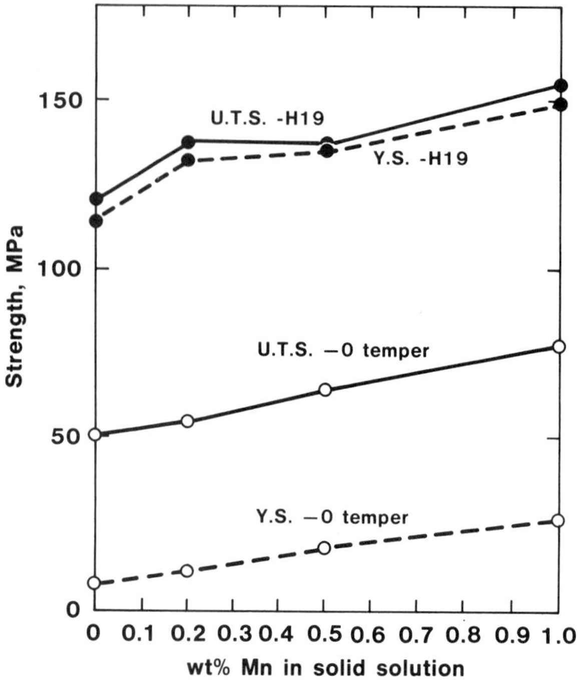

FIG. 12. Effect of manganese in solution on strength of annealed and fully hardened sheet.

Structural studies of deformed binary Al-Mn alloys have shown that cell formation follows the initial increase in dislocation density. This is observed both when Mn is retained in solution and when it is present as a dispersoid [22,23]. The development of the cell or subgrain structure during rolling indicates that dynamic recovery is occurring. With increased rolling, the subgrain structure becomes more well defined and the subgrain size decreases, resulting in an increase in strength. Figure 13a shows the microstructure of a 1 wt % Mn alloy (3003) after cold-rolling 90%. The matrix is segmented into subgrains of approximately 1.5 μm by fairly well defined subgrain boundaries. The subgrain interiors are relatively dislocation-free. When Mg is added for strengthening, the tendency toward cell formation is decreased and the overall dislocation content of the alloy increases with increasing Mg. Figure 13b shows the substructure present in 3004-H19 (1% Mn, 1% Mg). Compared to the microstructure of the Mg-free alloy (Fig. 13a), the Mg-bearing alloy has a finer substructure with an overall higher dislocation content and, hence, a higher strength.

(a)

(b)

FIG. 13. TEM micrographs showing dislocation and subgrain structure of (a) alloy 3003 (1% Mn) and (b) alloy 3004 (1% Mn, 1% Mg) after 90% cold-rolling. The addition of Mg reduces the extent of dynamic recovery that occurs during rolling.

E. *Recovery and Recrystallization*

The early stages of annealing cold-worked 3XXX alloys involves recovery processes where dislocation cells and subgrain boundaries become better defined. Subgrain growth occurs and eventually leads to the onset of recrystallization. As with the 1XXX alloys, lower strength and increased ductility usually accompany these microstructural changes. (Mg-free 3XXX alloys, however, partially annealed from an H19 temper at low temperatures, are known to display an anomalous decrease in tensile ductility similar to that for 1XXX alloys shown in Fig. 6). In applications where formability is important, it is generally found that formability can be optimized at a given strength level by partial annealing back from an initially higher strength level as opposed to directly cold-rolling to that strength level [*24*].

Recrystallization behavior and grain-size control in these alloys is best understood in terms of the recrystallization model developed by Nes [*9*] for alloys with a bimodal particle distribution. During deformation, the matrix adjacent to large particles undergoes large lattice rotations [*25*]. Subgrain growth in these deformation zones can result in subgrains with large misorientation to the rest of the deformed matrix, making them potential recrystallization nuclei. This is commonly referred to as particle-stimulated nucleation. The large constituent particles assist recrystallization nucleation and help refine grain size as suggested by Fig. 7a for the 1XXX alloys. The submicron dispersoids impede grain-boundary migration through a Zener drag mechanism as shown in Fig. 14 for 3004 alloy. Zener drag enables dispersoids to impede grain growth in recrystallized materials. However, by impeding boundary migration, dispersoids also interfere with the recrystallization nucleation process. When subgrain boundaries are pinned, subgrain growth is retarded and fewer recrystallization nuclei are formed. Furthermore, many recrystallization nuclei can have their growth stunted by dispersoids, decreasing the number of successful nucleation events and increasing the final grain size. To minimize this Zener drag effect, the dispersion parameter, f/r (f = volume fraction dispersoid, r = average particle radius) should be minimized [*9*]. It is clear that the ingot preheat practice must establish the appropriate dispersoid distribution to allow grain-size control in the sheet. Residual Mn in solution can also interfere with recrystallization by segregating to subgrain boundaries and precipitating as fine particles, which increase the Zener drag. (See reference [*26*], for more details on concurrent recrystallization and precipitation.)

Grain-size control in roll-cast 3XXX O-temper alloys continues to be a problem since the roll-casting process results in a much finer as-cast consti-

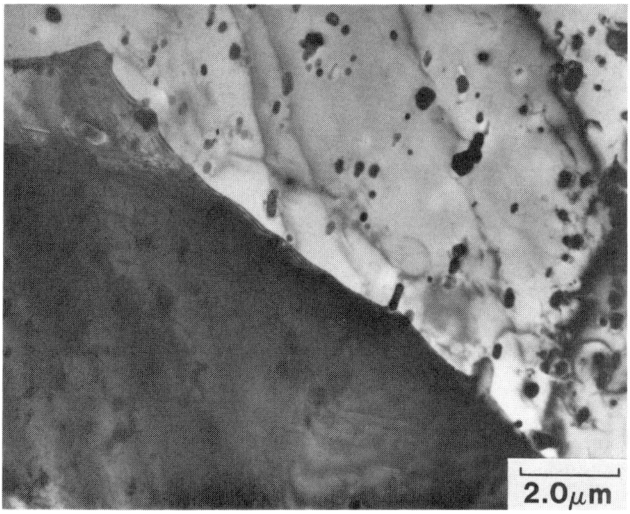

FIG. 14. Pinning of a grain boundary by dispersoids in a recrystallization 3004 alloy.

tuent size and higher amounts of Mn retained in solution. As a consequence, even the largest particles become very ineffective in stimulating nucleation of recrystallization in the presence of the dense dispersoid distribution. Roll-cast 3XXX alloys are typically used only for applications that require H1X or lightly annealed H2X tempers.

IV. 8XXX Alloys

Another group of alloys (usually grouped into the 8XXX classification) are centered around the microstructures achievable in eutectic or near-eutectic compositions where a nearly uniform dispersion of fine particles is used to stabilize a very fine grain structure. The volume fraction of particles is high, in the range of 4 to 10%. The fine grain size promotes a unique combination of high strength (through grain-boundary strengthening) with good ductility and formability.

Numerous eutectic systems have been used to produce particle-stabilized grain structures, including Al-Ni, Al-Fe and Al-Si binaries and Al-Fe-Ni, Al-Fe-Si, and Al-Fe-Mn ternary systems [27]. With these alloys, it is crucial that the as-cast structure be controlled to give the necessary particle size and distribution. Eutectics that solidify in a regular coupled growth

FIG. 15. TEM micrograph showing grain structure of an experimental alloy (1.36% Fe, 0.6% Mn) that was roll-cast, preheated, cold-rolled to .13 mm, and partially annealed. A well-defined subgrain structure exists with particles pinning many of the boundaries.

manner with a rod-shaped compound phase (e.g., Al-Ni) can be DC cast. Those alloy systems whose eutectic compounds tend toward a coarse, blocky morphology (e.g., Al-Fe-Mn, Al-Fe-Si) have to be cast with a high thermal gradient to avoid formation of coarse primary constituents. This typically requires a thinner DC ingot or roll casting.

The cast structure is broken up by appropriate combinations of preheating, and hot and cold working. The fully hardened H19 sheet or foil possesses high strength and limited ductility. The desired level of properties is obtained by partially annealing the fully hardened material. On annealing, a fine grain structure is established with the particles pinning the grain boundaries. (See Fig. 15.) The grain size is, therefore, influenced by the interparticle spacing, which is on the order of about 0.5μm.

An important advantage of these particle-stabilized alloys is the ease of strength control they afford. Properties of sheet or foil are controlled by the partial annealing temperature with strength dropping off gradually with increasing temperature. Figure 16 illustrates this with a partial annealing curve for a 0.8Fe-0.9Mn alloy [*28*]. The rapid drop in strength that accompanies recrystallization in 1XXX alloys (Fig. 6) is not observed. The desired annealing behavior results in no clear recrystallization or discontinuous grain growth, but rather a continuous grain growth controlled by the particle distribution. After partial annealing, grain boundaries are well

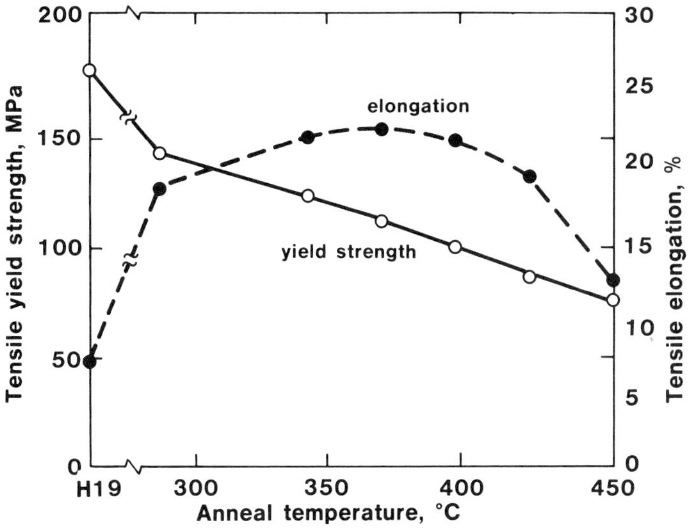

FIG. 16. Change of tensile properties of a 0.8Fe = 0.9Mn alloy with annealing temperature. Note the gradual decrease in yield strength with temperature compared to the response of alloy 1100 (Fig. 6).

defined and grain interiors are essentially dislocation-free. Thus, the strength is supplied chiefly by grain-boundary hardening.

The fine grain structure of these alloys gives them some unique tensile properties. It is commonly observed that yield strength varies with the reciprocal of grain size (d^{-1}) in these materials. This is a stronger grain-size dependence than observed in coarser-grained materials that follow a Petch-Hall ($d^{-1/2}$) type of relationship. The strong grain-boundary hardening effect is, unfortunately, accompanied by a loss of further hardening ability during tensile deformation [*29*]. This is understood in terms of the small mean free path for dislocations (limited by the grain size) resulting in less dislocation–dislocation interaction. At very small grain sizes (approximately 1 μm), this lack of work-hardenability can severely limit tensile ductility at room temperature.

V. 5XXX Alloys

A. *Background*

The 5XXX alloys were initially developed in the 1930s in response to demands for sheet materials with higher strengths, good formability, and high levels of corrosion resistance and weldability. The first recorded uses of 52S (later 5052), 56S (5056), and X54S (5154) were in the period 1932–1934. However, problems with fabrication of these alloys and the instability of their strain-hardened tempers (discussed in a later section) prevented their widespread use for a number of years. Shortages of Mg during World War II also curtailed the development of Al-Mg alloys. The introduction of improved metal-filtration technology, DC casting, and the development of more powerful rolling mills largely solved the fabrication problems for alloys up to approximately 5.5% Mg. Thus, the 1950s and 1960s saw the development and registration of numerous new 5XXX compositions. Alloys 5082 and 5182, produced in large quantities for the late 1980s beverage can market, were introduced in 1963 and 1967, respectively.

The attractive properties of the 5XXX alloys make them useful in the form of castings, extrusions, plate, sheet, or wire. In fact, the 5XXX alloys are the only NHT aluminium alloys with extensive uses as castings or plate. Over 30 5XXX alloys are currently registered with the Aluminum Association and are utilized in applications as diverse as rigid container sheet, screen wire, and ballistic armor plate.

B. Phases and Preheating

Magnesium has a very high solid solubility in Al (up to 14.9 wt % at 451°C, the binary eutectic temperature). Its solubility decreases to approximately 1.7 wt % at room temperature (extrapolated from Murray [*30*], but the rate of decomposition of the supersaturated solid solution is very slow. In commercially available alloys, essentially all of the Mg can be retained in solution without the need for extraordinary quenching procedures. Mg decreases the density of Al, but the most important influence of Mg is its ability to promote the strengthening of Al through work-hardening.

Commercial 5XXX alloys fall within a range of 0.5 to 6.0 wt % Mg. The constituent phases in 5XXX alloys can be of various types depending upon the specific composition. Due to the very low solubility of Mg_2Si in Al at high Mg contents, it is often present in the microstructure as a major constituent phase. Constituent phases containing Fe, Mn, and Si are also present. These can be $Al_{12}(Fe, Mn)_3$ Si, $Al_6(Fe, Mn)$, or Al_3Fe. Commercial alloys can contain up to 1% Mn. When Mn is precipitated as a dispersoid during preheating, the $Al_6(Mn, Fe)$ phase is favored since the Si has been tied up with Mg as Mg_2Si. Preheating of 5XXX ingot also eliminates the Mg coring of the dendrites that occurs during solidification. Cr, which is often added for corrosion protection, has a very low solubility in Al. During solidification and preheating, ternary compounds of Al, Mn, and Cr can form in alloys containing both Cr and Mn. In wrought sheet products, Cr is often present as the E phase ($Al_{12}\ Mg_2\ Cr$) as a result of precipitation during the preheat.

C. *Strength Development in Wrought 5XXX Alloys*

Magnesium in solution imparts a limited solution-hardening effect on the aluminum matrix. On a weight basis, it is even more effective as a solution-hardening element than is Cu (although, on an atomic basis, Cu is more potent). Figure 17 [*31*] illustrates the effect of Mg in solution on the tensile properties of annealed high-purity Al sheet. Both yield strength (YS) and ultimate tensil strength (UTS) increase with increasing Mg while elongation drops sharply with even small Mg additions. Careful analysis of the data shows that on an atomic basis, Mg is more efficient in hardening in the very dilute range (<1 at %) than in the more concentrated alloys. At Mg levels of <1 at %, the incremental increase in yield strength (Δ YS) was found to be proportional to (at % Mg)$^{2/3}$, whereas, in the more concentrated alloys, Δ YS increased linearly with (at % Mg)$^{1/3}$. This result suggested that the solution-hardening effect of Mg differs from dilute to concentrated alloys, but the mechanism responsible could not be defined.

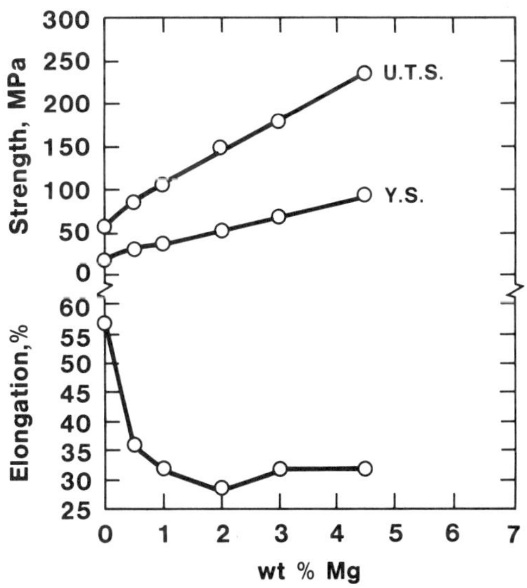

FIG. 17. Effect of Mg in solution on the tensile properties of annealed Al-Mg binary alloys.

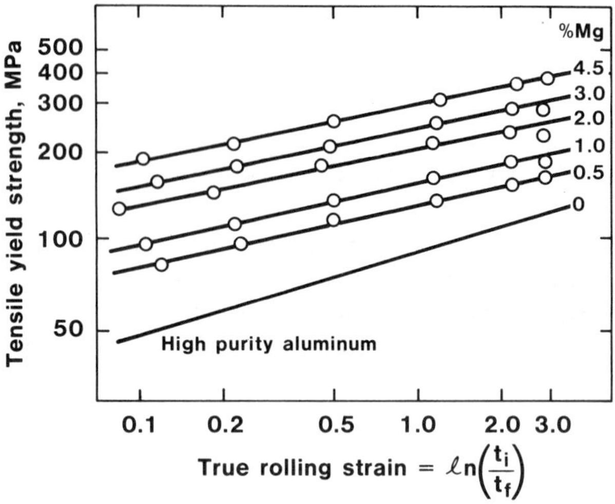

FIG. 18. Strain-hardening response of high-purity Al and Al-Mg binary alloys during cold-rolling.

While Mg does provide some solution-strengthening, it is most important as a strengthening element through its effect on work-hardening. The high-purity Al-Mg binary alloys of Fig. 17 show a more rapid rise in tensile strength than in yield strength, indicating that Mg is promoting work-hardening. The effect of Mg on work-hardening is seen more clearly in Fig. 18, which shows the increase in yield strength (Y.S.) as a function of true thickness strain during cold-rolling. The data is plotted as log Y.S. versus log strain to show the good agreement with the Hollomon equation $\sigma = k\epsilon^n$. The addition of Mg to pure Al results in a decrease in the strain-hardening exponent n, but an increase in k. With increasing Mg, n remains constant, but k continues to increase, indicating that Mg is increasing the rate of work-hardening.

This influence of Mg on enhancing strengthening during cold-deformation is microstructurally related to its effect on development and retention of lattice strain. Because Al has a high stacking fault energy, dislocation cross-slip is easy, and recovery processes occur readily in Al during both deformation and low-temperature annealing. The microstructure of deformed Al shows evidence of mutual dislocation annihilation and extensive dislocation rearrangement into cells and subgrains. X-ray line broadening studies [*31*] have shown that, as Mg is added to Al, the amount of retained lattice strain after deformation increases. Higher Mg contents result in a more rapid increase in lattice strain and higher amounts of

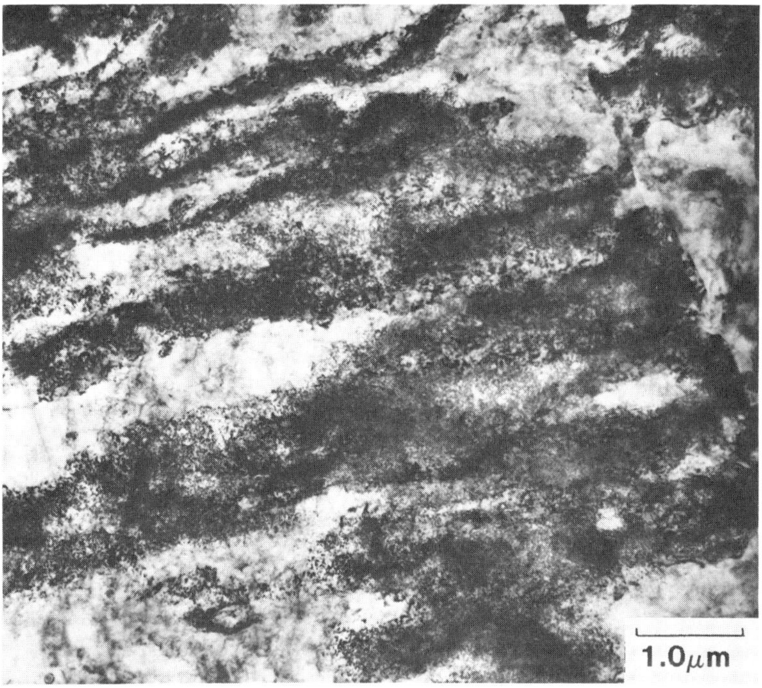

FIG. 19. TEM micrograph showing the dense, entangled dislocation structure present in fully hardened (H19) 5182 alloy (4.5% Mg, 0.35% Mn).

retained strain for a given amount of deformation. Thus, by decreasing the stacking fault energy, or by some other means restricting dislocation mobility, Mg decreases the tendency for dynamic recovery and dislocation rearrangement into lower energy configurations. The net result is an increase in the dislocation content of the alloy and a more uniform dislocation distribution. This can be seen by comparing Fig. 19 with Fig. 3. These figures show the H19 dislocation structure found in alloys 5182 (4.5% Mg) and 1145 (0% Mg), respectively. While the 1145 alloy shows clear evidence of subgrain formation, alloy 5182 shows little evidence of dynamic recovery with the retention of a dense uniform dislocation network. Thus, by suppressing recovery, Mg increases both the dislocation multiplication and storage rates leading to increased strength.

The combination of Mg in solution and cold-working is extremely effective in bringing about high strength. Unfortunately, there are practical limitations that restrict how far this combination can be taken in producing commercial wrought products. Included in these problems are fabrication

TABLE II

EFFECT OF MN ON TENSILE PROPERTIES AND GRAIN SIZE OF A 3.54% MG ALLOY

		H19 + 2 hrs @ 121°C			O-temper (2 hrs @ 343°C)			"O" Grain Size
Mg (wt %)	Mn (wt %)	Yield (MPa)	Tensile (MPa)	El (%)	Yield (MPa)	Tensile (MPa)	El (%)	(grains/mm^3)
3.54	0	314	349	5.0	103	230	—	1.05×10^5
	0.05	318	351	6.0	109	237	—	1.93×10^5
	0.10	324	356	6.0	114	240	—	2.79×10^5
	0.24	340	374	5.0	125	248	20.5	4.61×10^5
4.37	0	344	386	6.2	118	260	24.0	1.06×10^5

Note: All alloys contained 0.04Fe, 0.05Cu.

difficulties such as alligatoring and edge cracking during rolling and susceptibility of high-Mg alloys to intergranular corrosion or stress corrosion cracking (SCC). Commercially, higher strengths are achieved by the addition of other solid-solution elements. A requirement of such an addition will be that the element provides additional work-hardening but will not chemically interact with Mg and remove it from solid solution. Mn is often added to 5XXX alloys and provides additional strengthening at lower Mg levels. Table II [*32*] illustrates that relatively small additions of Mn to a 3.5% Mg binary can be more efficient in promoting strengthening than further Mg additions. Its effect on decreasing recrystallized grain size is also evident. Conductivity measurements showed that the Mn was retained in solution in these alloys so that results are attributable to solution effects. At higher levels, the Mn precipitates out as a dispersoid during preheating.

D. *Recovery and Recrystallization*

The strain-hardened tempers of the 5XXX alloys are of obvious commercial interest, as they offer the highest strengths possible of all NHT Al alloys. The strain-hardened condition, however, is a metastable state in which there is a driving force for internal strain to relax through the process of static recovery. In applications where long service times are expected, the time dependency of properties is important. Figure 20 [*33*] illustrates the gradual drop in strength and increase in tensile elongation with time at ambient temperature measured for an Al-6% Mg alloy. The data show a good linear fit of change in properties with log (time) characteristic of the recovery process. Within the first year, the yield strength of the H18 temper sheet dropped below the initial yield strength of the H16 temper.

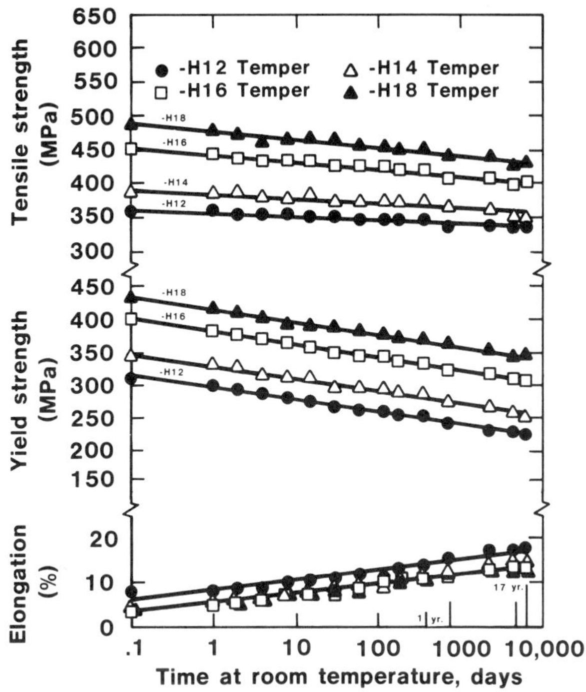

FIG. 20. Age-softening of an Al-6% Mg alloy cold-rolled to various initial HIX tempers. A linear change in tensile properties with log time is observed out to 17 years at room temperature. Legend indicates materials original temper.

The rate of property change is greatest in the more highly strain hardened tempers where the driving force for recovery is greatest.

This behavior is sometimes referred to as "age softening." Being basically a metallurgical recovery process, it is both temperature- and time-dependent. Exposure to higher service temperature allows thermally activated recovery mechanisms to proceed at a faster rate and increases the rate of strength loss.

Greater stability of properties with time is achieved by stabilizing strain-hardened 5XXX alloys after cold-working. "Stabilization" involves a low-temperature thermal operation, typically 120–180°C, which allows a controlled amount of recovery to occur. Strength drops and ductility is increased by the stabilization treatment. Since the internal structure has relaxed to a state of lower energy, the driving force for further recovery is reduced and properties are less susceptible to change with time at the lower service temperature. Figure 21 shows the influence of a stabilization treatment on the time dependency of properties for 5052 sheet [33]. The

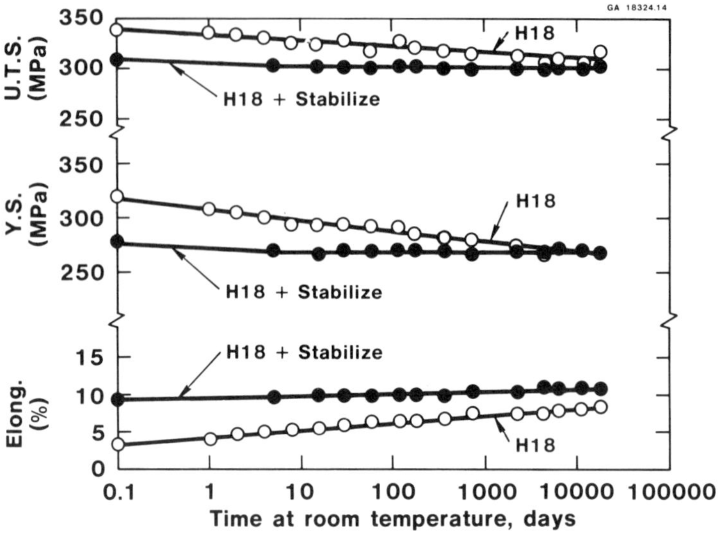

FIG. 21. Age-softening of 5052-H18 alloy with and without stabilization treatment (eight hrs at 143°C). Properties of stabilized material remain nearly constant while those of the H18 sheet show a gradual but continued change. Final data points represent properties after 52 years at room temperature.

strain-hardened sheet shows the expected gradual change of properties with time while the stabilized sheet shows properties that are nearly independent of time. Stabilized tempers are identified commercially, with the H3X designation. [Table I]

Partial annealing of 5XXX alloys to useful commercial tempers may involve recrystallization as well as recovery of the work-hardened structure. Factors that influence microstructure development during partial annealing include temperature, initial degree of work-hardening, and the presence of solute elements and second-phase particles. Figure 22a [*32*] contains partial annealing curves for a range of high-purity Al-Mg alloys that had initially been rolled 70%. During isochronal annealing, higher-Mg alloys lose proportionately more of their strength through recovery than do lower Mg alloys, which lose more of their strength through recrystallization. Elongation is recovered during annealing as shown for these same alloys in Fig. 22b. This figure shows that, for alloys partially annealed back to the same strength level from the fully hard condition, the higher Mg alloy will have the higher elongation because of the greater extent of recovery they have experienced. (The minimum in O-temper elongation in the 1 to 2 wt % Mg composition range shown in Fig. 17 is also seen in these data.)

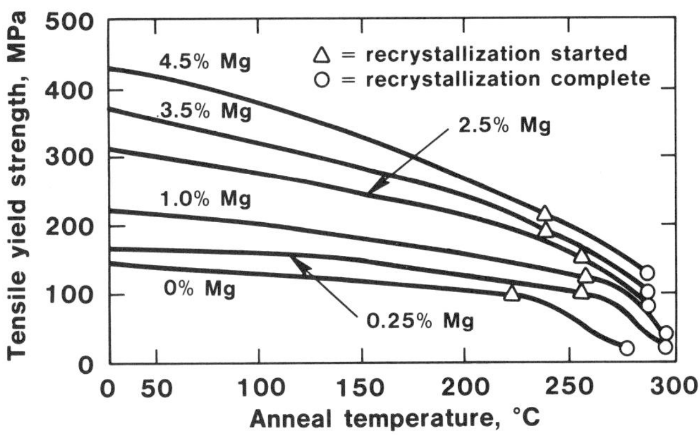

FIG. 22a. Partial anneal curves for a series of binary Al-Mg alloys cold-rolled 70% and annealed for two hours at the annealing temperatures.

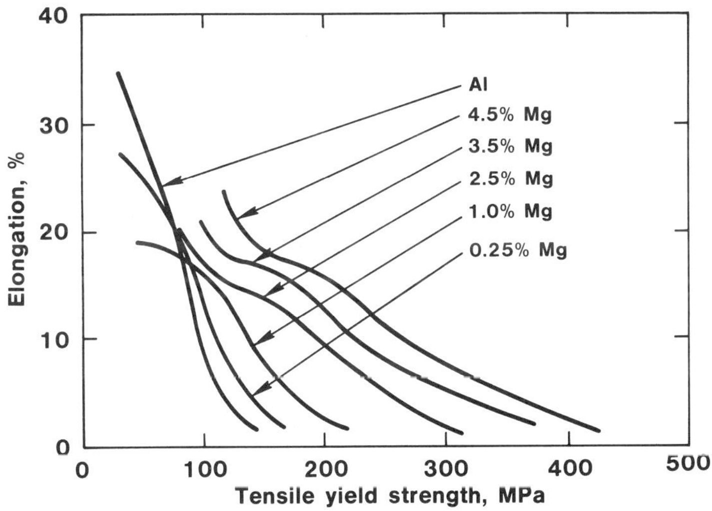

FIG. 22b. Recovery of elongation with decreasing strength for the partial annealed alloys of Fig. 22a.

Structural rearrangements during recovery were studied through the analysis of X-ray line-breadth measurements [*34*]. For the alloys of Fig. 22, it was found that increasing Mg and decreasing annealing temperature decreased the size of the recovered subgrains, resulting in higher retained strength. Other elements present in solution will also exert an influence

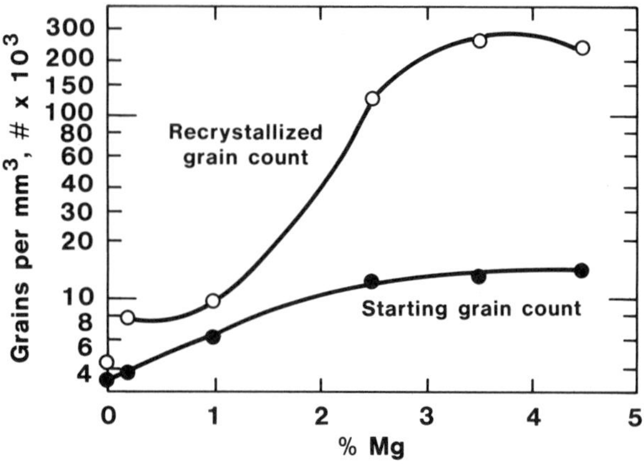

FIG. 22c. Initial and recrystallized grain count for the alloys of Fig. 22a.

on the recovery behavior, but Mg, because of its strong influence on dislocation mobility, is dominant.

Additions of Mg decrease the recrystallized grain size by increasing the driving force for recrystallization in the cold-worked material. Figure 22c shows annealed grain size as a function of Mg level for the same series of binary alloys. The starting grain size is also indicated. All samples were rolled to a true strain of 2.15 prior to annealing. The recrystallized grain size corresponds to that observed at the lowest temperature for complete recrystallization. The trend toward decreased grain size with increase in Mg content is evident up to a Mg level of 3.5%.

E. Precipitation of Al_3Mg_2

Many commercial 5XXX series alloys are produced and used in the condition where the matrix is supersaturated with Mg. Thus, there is a driving force for the precipitation of Mg from solution. Our level of understanding of Mg precipitation in commercial alloys is incomplete and qualitative at best.

Long-term studies have shown that the Al-Mg supersaturated solid solution is very stable with respect to decomposition at room temperature. Precipitation is accelerated by deformation and elevated temperatures and occurs most readily on gross structural defects such as grain boundaries and shear bands. (See Fig. 23.) Unless precipitation occurs at high temperatures (i.e., above approximately 260°C), the precipitate phase that forms

FIG. 23. Optical micrograph showing precipitation of β' on shear bands and grain boundaries in an Al-6.9% Mg alloy rolled to H18 and aged 52 years at room temperature.

first is not the equilibrium β phase but a metastable phase generally referred to as β'. Some investigations have suggested that more than one metastable precipitate phase can form depending upon the aging temperature [35,36]. The metastable precipitate is very stable at low temperature and transformation to the equilibrium phase is not typically observed even after very long aging times.

Precipitation is not an effective strengthening mechanism for wrought commercial 5XXX-H1X alloys. Precipitation of Mg in the cold-worked structure might accelerate recovery and result in a net softening of the alloy. Precipitation on grain boundaries or shear bands has detrimental effects on corrosion resistance stemming from the anodic nature of this phase with respect to the matrix.

High Mg alloys (>7 wt %) exhibit some degree of age-hardening behavior. (Note: Casting alloys 520 [9.5–10.6 Mg] and 535 [6.2–7.5 Mg] make use of precipitation-hardening to obtain useful properties in aged [T4 or T5] tempers. However, no wrought 5XXX alloys currently exploit age-hardening to obtain higher levels of strength.) Near ambient temperatures, this age-hardening behavior has been associated with a fine-scale precipitation (disordered and ordered G.P. zones as well as possible spinodal decomposition) [37,38]. Precipitation of β' during artificial aging can also harden high-Mg alloys [35,39]. Increased corrosion susceptibility and decreased ductility, however, are problems that often accompany the precipitation strengthening.

F. *Ludering Behavior during Deformation*

Many of the 5XXX alloys offer a high level of formability, the lower Mg alloys and 0-temper condition being the most formable. During forming or stretching operations, however, Al-Mg alloys can exhibit ludering behavior. Two types of luder bands have been observed. The first type (Type A or "flamboyant" bands) occur only in annealed alloys and are associated with a yield point and yield-point extension. The severity of this type of ludering increases with increasing Mg, up to approximately 3 wt % [*40*], and decreasing grain size. Phillips et al. [*41*] relate this behavior to the low availability of matrix dislocations in annealed samples and the difficulty in transferring strain across grain boundaries when Mg is present. These luder bands are objectionable mostly for cosmetic reasons and can be eliminated by introducing dislocations into the microstructure by stretching the piece past the yield point or by a rolling or roller leveling operation.

The other type of ludering (type B or the Portevin–Le Chatelier effect) involves repeated, discontinuous localized yielding events. This behavior is found to occur in strain-hardened as well as annealed tempers and is associated with a serrated stress–strain curve and a condition of negative strain-rate sensitivity [*42*]. The surface markings associated with this yielding are true diffuse necks, but are less visible than the type A bands. Only small amounts of Mg are required to produce serrated flow behavior in Al and all commercial 5XXX alloys have the potential to show this behavior. While the serrated flow behavior is itself not desirable, the negative strain-rate sensitivity that supports this phenomenon is detrimental from strain-localization and formability standpoints. Apart from forming at temperatures or strain rates where this behavior is not encountered, there is currently no clear understanding of how to avoid this serrated yielding behavior.

VI. Special Applications

A. *Finishing Characteristics*

Because the soft surfaces of 1XXX alloy sheet are susceptible to abrasion and scuffing, many uses require the application of hard anodic coatings. Other decorative uses such as trim for appliances or automobiles call for surfaces with highly reflective, specular finishes. Chemical composition and processing play important roles in determining an alloy's response to surface treatments.

Some 5XXX alloys such as 5252 and 5657 have been designed for bright finishing applications. Specular surfaces are produced by a combination of buffing (if necessary) and chemical or electrobrightening to improve surface smoothness. These surfaces are then protected, usually by growth of an anodic coating. Finishing of these alloys is typically done in H2-type tempers, where no recrystallization has occurred. The recovered structure assures a good combination of strength and ductility and promotes a good surface finish by minimizing surface relief associated with recrystallized grains. High alloy purity (low Fe and Si) minimizes both pitting during brightening and constituent inclusion in the anodic coating that detract from surface reflectivity. Cu and Si are also minimized to avoid precipitation of the S phase (Al_2CuMg) or Mg_2Si during partial annealing since these phases detract from image clarity. Magnesium, in solution, however, does not appreciably decrease reflectivity.

The type of constituent phase, Al_3Fe versus Al_6Fe or Al_mFe, will also influence the appearance or color of the oxide. The Al_mFe phase results in a dark anodized coating while the Al_3Fe phase leads to a bright aluminum-colored coating. Casting and homogenization practices must be designed to form specific phases for eliminating streaking in sheet products with anodic coatings [*43*,*44*]. Composition and processing can also be controlled to obtain unique coloration of anodic coatings in architectural sheet [*45*].

B. Al-Mg-Mn Alloys for Rigid Container Sheet

The most extensive use of NHT alloys by far is in the area of rigid container sheet for beer and beverage cans. Some limited food-can applications exist, but the volume is small compared to the overall RCS market. Over 1.36 billion kilograms of aluminum RCS were produced in the United States in 1984. The unique technological requirements for this application make the production of high-quality RCS one of the most challenging in the aluminum industry. The challenge of forming the necessary end, body, and tab configurations at high production rates from highly strain hardened (H19) sheet has been met by metallurgists and production personnel by controlling the microstructure and properties of the sheet.

Typical alloys in use for RCS include 3004 (body), 5082 or 5182 (end), and 5082 or 5042 (tab). The selection of alloys depends somewhat upon the can maker's need for strength and formability. But scrap compatibility determines to a large extent what alloys are available for this market since 50–55% of the beverage cans (500,000,000 kg) produced in the United States are recycled. All of the alloys are basically Al-Mg-Mn compositions with the exception of 5082, which has no Mn. Ancillary elements such as Zr or Cr are kept to a minimum since recycling assures that their addition to one of the alloys will make the alloy incompatible with the RCS scrap loop.

C. *Alloy 3004 Body Stock*

The primary alloy used for the body of aluminum beverage cans is 3004-H19. The composition of the alloy—nominally 1.1 Mn, 1.0 Mg, 0.4 Fe, 0.18 Si—yields a compromise among formability, high strength to obtain adequate container performance, and recycling considerations. The most notable feature of the 3004-H19 microstructure (see Fig. 11) is the approximately 2.5 vol % of constituent particles. These particles are the remnants of the eutectic structure formed at cell boundaries during casting. Particle fracture during hot- and cold-rolling results in a distribution of particles typically 5–10 μm in length by 2 μm in thickness.

The fine dispersoids result from precipitation and coarsening during preheating, hot-rolling, and annealing operations. After casting and processing to sheet gauge, about 0.2–0.3 wt % Mn remains in solid solution in typical 3004-H19. Preheating dissolves any Mg_2Si present in the cast ingot and the dispersoid formation ties up most of the Si, making it unavailable for reprecipitation. Very small amounts of Mg_2Si may precipitate during intermediate anneals and can be present in 3004-H19. Any Mg_2Si present in conventionally processed 3004-H19 is not visible optically and has little or no influence on the mechanical properties or performance of the sheet.

Stumpf [*46*] isolated the strengthening influence of each of the microstructural components (matrix, dispersoid, and constituent) in 3004. (See Table III.) By careful selection of alloy composition and processing, alloys consisting only of the 3004 matrix, matrix and constituent only, and matrix and dispersoid only were produced. Although the amount of Mn in solution varied slightly among the samples, the H19 tensile data still indicated substantial strengthening effects of both the constituent and the dispersoid phases in alloy 3004-H19 sheet.

Alloy 3004-H19 can stock is formed by the can maker by a sequence of drawing and ironing (D&I) operations. Cups are blanked and drawn initially 35–40%. The cups are then redrawn and ironed in a series of ironing rings from sheet gauge, approximately 0.33 mm to 0.12 mm, representing an additional thickness reduction of 65 to 70%. Necking, doming, and flanging complete the can-making process. The cans are formed at a rate of roughly 200 cans per minute; a tear-off (or fractured can rate) of 1 per 10,000 is considered marginally acceptable.

TABLE III

CONTRIBUTION OF VARIOUS MICROSTRUCTURAL COMPONENTS TO STRENGTH OF ALLOY 3004

	Mn (ss) wt%	H19 YS (MPa)	H19 UTS (MPa)
Matrix only	0.27	254	267
Matrix and dispersoid	0.36	283	295
Matrix and constituent	0.25	290	308
Matrix, dispersoid, and constituent	0.43	305	328

Normally, the entire can-making operation takes place prior to any opportunity for recovery to decrease strength and increase elongation of the sheet. Thus, the D&I process must take place using thin sheet with typical tensile elongations of 2 to 4%. The ironing process itself minimizes tensile stresses in the can sidewall, but the unique microstructure and mechanical behavior of 3004-H19 is also necessary for successful D&I can making. Figure 24 compares the stress–strain behavior of 3004-H19 and an Al-2.0 Mg alloy (5352-H19) of similar strength [*47*]. The strain-hardening exponent of the 3004 is lower than that of the 5352, but the ability to maintain strain-hardening and prevent strain-localization extends to much

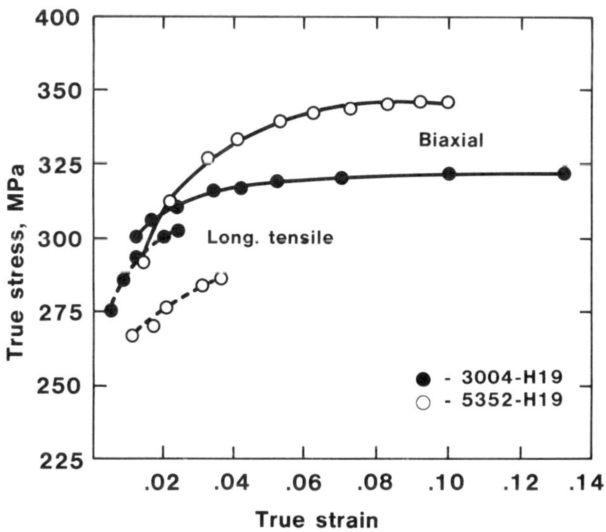

FIG. 24. Biaxial bulge and tensile test results comparing strain-hardening behavior of alloy 3004 and 5352-H19 sheet.

higher strain. Microstructural studies showed that high strain deformation (drawing or biaxial bulge testing) produced polygonization in alloy 3004-H19 but not for the 5352 alloy. The balance between the strain-hardening due to Mg atoms and the deformation-induced recovery in 3004-H19 must be an important factor in achieving the necessary high ironing reductions without fracture during can making.

Controlling recovery in 3004 sheet becomes important when the formed cans are subjected to a baking operation in the 180 to 200°C range to cure the interior lacquer coatings and decorative exterior paint applied to the finished cans. The strength and thickness of the sheet in the bottom (dome) portion of the can determine the pressure at which dome reversal (i.e., failure) will occur once the can is filled and sealed. Thus, minimizing the loss in strength during the can bake is important to enable the use of thin sheet gauges for D&I stock. Additional solid-solution elements (Cu or Mg) will impede recovery and act to strengthen the sheet as discussed previously. However, extra solute can be detrimental to formability as the H19 yield strength is increased considerably.

A clever way to resist the loss in strength during the can bake is to adopt some of the principles of heat-treatable alloys: High-temperature annealing and rapid cooling can be used to dissolve and "quench" in solute, which will reduce the strength loss due to recovery. This effect was first noted at Alcoa Research Laboratories in 1943 [*18*] for alloys 4S (3004) and 52S (5052) in the H14 temper (Fig. 25). The high-temperature anneal (482°C) was originally thought to dissolve Mg_2Si and make precipitation-hardening possible during the subsequent thermal treatment. Subsequent work, however, showed that the hardening effect was due to the inhibition of recovery by the additional solute [*48*]. Kobe Steel has patented an alloy and process for a "bake-hardening" D&I alloy [*49*]. The product produced by this method does not actually get stronger after baking, but does resist softening to approximately 200°C.

Controlling the fabrication sequence to produce 3004-H19 sheet with a low degree of deformation texture to minimize earing during deep drawing is a matter of concern to aluminum sheet manufacturers. The desired final texture is a balance between cube texture developed during the anneal and the deformation textures introduced during cold-rolling.

D. End Stock

The formability requirements for producing ends successfully from a strain-hardened 5XXX alloy such as 5182 represent a severe challenge for the material. As with body stock, the forming must be done at very high

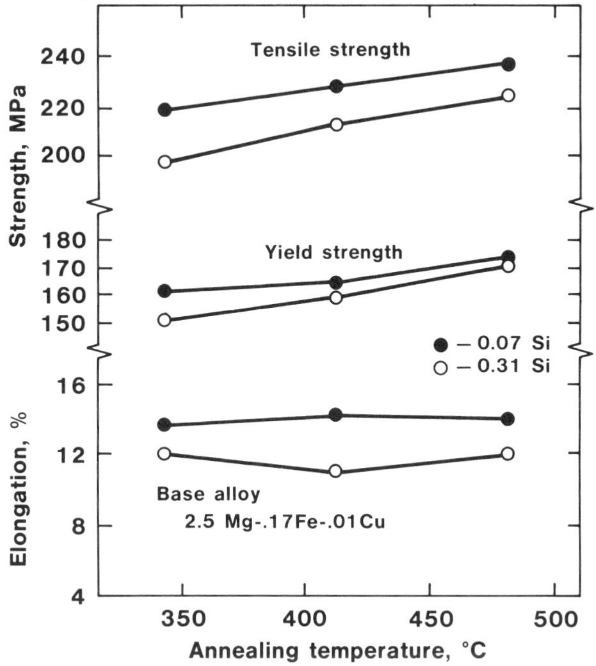

FIG. 25. Effect of annealing prior to cold-rolling and Si content on the mechanical properties of alloy 52S (5052)-H14 sheet after heating for one hour at 177°C.

production rates with a low incidence (fewer than 1 per 100,000) of fractures during the operation. However, integral rivet formation and scoring operations that are necessary for the "easy-opening" feature of the end represent unique forming problems quite unlike those encountered in drawing and ironing. Figure 26 illustrates the severe metal-working encountered during rivet formation and scoring of 5182 end stock. (The integral rivet is the feature of the aluminum end that holds the opening tab in place. The score is the groove along which the end fractures during opening to form the pour spout. The score depth is typically 50–70% of the metal thickness.)

The microstructures of the 5XXX alloys, 5182 and 5082, are uniquely suited to the metallurgical challenge presented by end-forming. The higher Mg content (4.5%) of these alloys relative to alloy 3004 provides a higher level of strain-hardening during forming. While the alloys are typically rolled to initially "hard" H18 or H19 tempers, they are actually formed after the application and curing of an organic coating. The coating cure, which may be several minutes in the 180 to 200°C range or as little as

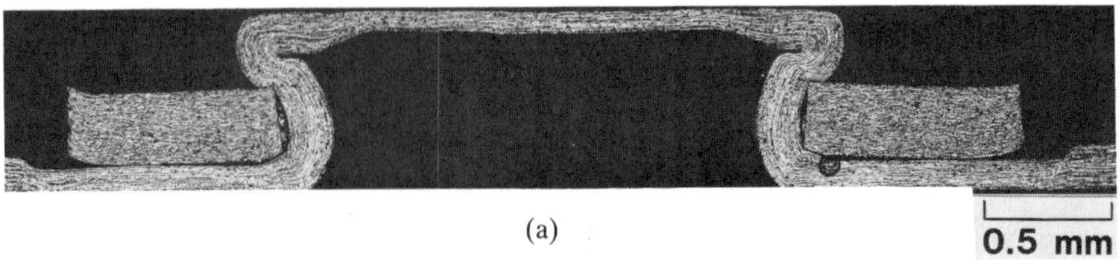

(a)

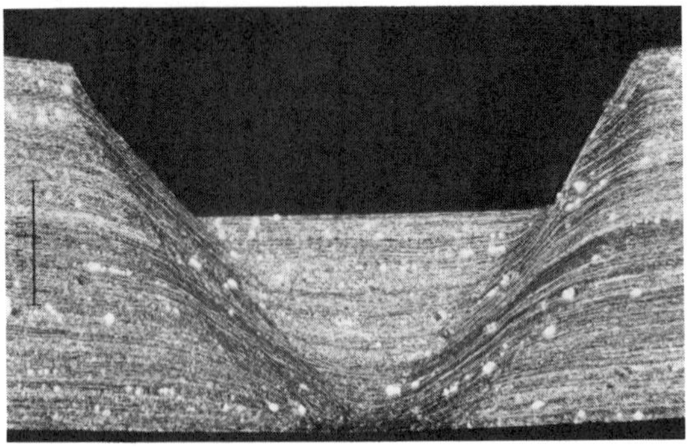

(b)

FIG. 26(a). Cross section of integral rivet and stay-on tab from alloy 5182 end produced for aluminum beverage container. (b) Cross section of score region of alloy 5182 end.

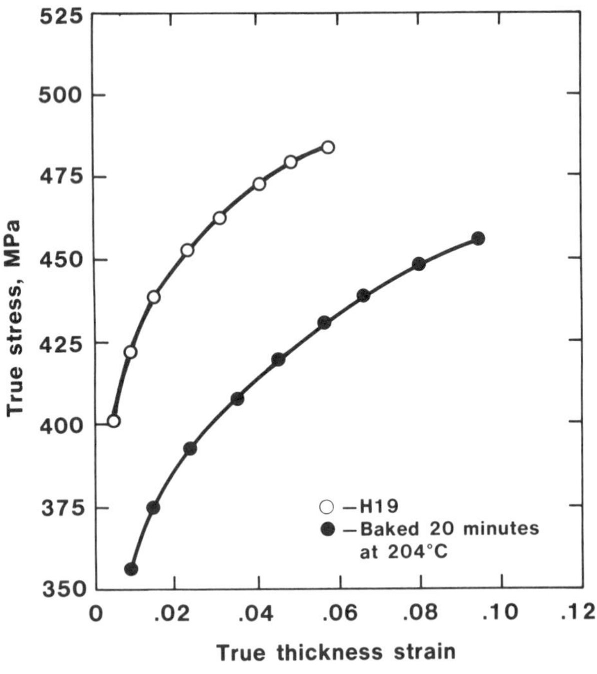

FIG. 27. Biaxial bulge test results for alloy 5182 sheet, 0.29 mm, illustrating the effects of recovery on subsequent strain-hardening ability during forming.

several seconds at 220 to 250°C (for high-speed coil-coating), reduces the yield strength of the 5182-H19 sheet from approximately, 360 MPa to approximately 320 MPa. While the coating itself and external lubricant assist the forming process, recovery and polygonization of the cold-worked structure during the coating cure are also crucial to the ductility and work-hardening ability of the sheet. (See Fig. 27.) The Fe and Si levels of the 5182 sheet are maintained at lower levels than those 3004 in order to reduce the volume fraction of insoluble Fe-bearing and Mg_2Si constituents. These particles can lead to poor formability and low fracture resistance during the strenuous end-forming processes.

VII. Summary

This chapter has attempted to provide a fundamental understanding of the metallurgy of the three major non–heat-treatable aluminum alloy systems. Microstructure development during work-hardening, recovery, and recrystallization were shown to be influenced strongly by elements in solid solution and phases formed during casting (constituents) and preheating (dispersoids). In the 1XXX (Al-Fe-Si) alloys, the roles of Fe-bearing constituent particles and Si in solid solution on strain-hardening and nucleation of recrystallized grains during annealing were discussed. The effects of cast microstructure and preheating in determining the size and distribution of $Al_{12}(Mn,Fe)_3Si$ dispersoids were shown to be critical for control of mechanical properties and grain size of 3XXX (Al-Mn-Fe-Si) alloys. The interaction of Mg atoms and dislocations during cold-rolling of the 5XXX (Al-Mg) alloys inhibit dynamic recovery and provide the basis for the highest aluminum alloy strengths attainable without heat treatment. Precipitation of Al_3Mg_2 or a related metastable phase, β', does not provide useful strengthening in the commercial wrought 5XXX alloys and may lead to corrosion problems due to its tendency to precipitate on grain boundaries.

The Al-Mg-Mn (3XXX and 5XXX) alloys are the basis for the largest-volume application for aluminum alloys, rigid container sheet, which consumes over 1.25 billion kilograms of metal annually. The microstructures and mechanical behavior of the important rigid container sheet alloys were described. The ability of alloy 3004-H19 to dynamically recover during high-strain deformation allows it to be drawn and ironed reliably into can bodies at extremely high production rates. After recovery induced by the coating bake, alloy 5182 is formed by a complex process into the can end.

Acknowledgments

The authors are indebted to Messrs. J. Paul Lyle, Jr., J.T. Staley, T.L. Hartman, and R.D. Doherty for critical reviews of this manuscript.

References

1. Hazzard, L.W. (1984). Aluminium Company of America, Pittsburgh, U.S. Dept. of Commerce Data.
2. Dix, E.H., Jr., and Bowman, J.J. (February 1936). *Metals and Alloys.*, **7**, pp. 29–34.
3. Westengen, H. (1982). *Aluminium*, **58**, 398–401.
4. Stumpf, H.C. (1969) Progress Report, Alcoa Laboratories.
5. Stumpf, H.C. (1970) Progress Report, Alcoa Laboratories.
6. Nock, J.A., Jr. (1947) Division Technical Report, Alcoa Laboratories.
7. Mould, P.R., and Cotterill, P. (1967). *J. Materials Science*, **2**, 241–255.
8. Humphreys, F.J. (1980). in *Recrystallization and Grain Growth of Multi-Phase and Particle-Containing Materials*, 1st Risø International Symposium on Metallurgy and Materials Science, Risø National Laboratory Roskilde, Denmark, pp. 31–44.
9. Nes, E. (1976). *Acta. Met.*, **24**, 391–398.
10. Bay, B., and Hansen, N. (1979). *Met. Trans. A*, **10A**, 279–288.
11. Bay, B., and Hansen, N. (1983). In *The Metallurgy of Light Alloys.* Institution of Metallurgists, London pp. 231–235.
12. Chan, H.M., and Humphreys,, F.J. (1984). *Acta. Met.*, **32**, 235–243.
13. Anderson, W.A. (1947). Division Technical Report, Alcoa Laboratories.
14. Stumpf, H.C., (1974) Division Technical Report, Alcoa Laboratories.
15. Warlimont, H. (1977). *Aluminium*, **53**, 171–176.
16. Keller, F.G. (1937) Letter Report, Alcoa Research Laboratories.
17. Nock, J.A., Jr. (1937) Progress Reports, Alcoa Laboratories.
18. Nock, J.A. Jr. (1943) Division Technical Report, Alcoa Laboratories.
19. Watanabe, H. Ohori, K., and Takeuchi, Y. (1984). *Aluminium*, **60**, E310–313.
20. Theler, J.J., and Furrer, P. (1974). *Aluminium*, **50**, 467.
21. Stumpf, H.C. (1971) Progress Report, Alcoa Laboratories.
22. Conserva, M., and Gatto, F. (1965). *Aluminio*, **34**, 401–407.
23. Gatto, F., Camona, G., Conserva, M., and Fiorini, P. (1968–1969). *Mat. Sci. Eng.* **3**, 56–61.
24. Altenpohl, D. (1982).*Aluminium Viewed from Within*, Aluminium-Verlag, Dusseldorf, West Germany, pp. 139–140.
25. Humphreys, F.J. (1979). *Metal Sci.*, 136–145.
26. Nes, E., and Embury, J.D. (1975). *Z. Metallk.* **66**, 589–593.
27. Morris, L.R. (July 1977). In *Proc. Conf. Solidification and Casting of Metals*, Univ. of Sheffield Metals Soc., London, pp. 18–21.
28. Baumann, S.F. (1985) Progress Report, Alcoa Laboratories.
29. Westengen, H. (1982b). In *Proc. 6th ICSMA*, Pergamon Press, New York vol. 1, pp. 461–466.
30. Murray, J.L. (1982). *Bulletin of Alloy Phase Diagrams*, **3**, no. 1, 60–74.
31. Stumpf, H.C. (1966) Progress Report, Alcoa Laboratories.

32. Staley, J.T., and McBride, J.K. (1965) Division Technical Report, Alcoa Laboratories.
33. Nock, J.A. Jr. (1951) Division Technical Report, Alcoa Laboratories.
34. Stumpf, H.C. (1964) Progress Report, Alcoa Laboratories.
35. Perryman, E.C.W., and Brook, G.B. (1949). *Res. Rep. #A. 833*, Brit. Non-Ferrous Metals Res. Assn., London.
36. Stumpf, H.C. (1959) Progress Report, Alcoa Laboratories.
37. Osamura, K., and Ogura, T. (May 1984). *Met. Trans. A*, **15A**, 835–842.
38. Raynal, J.M., and Roth, M. (1975). *J. Appl.Cryst.*, **8**, 535–537.
39. Fink, W.L., and Smith, D.L. (1936). *AIME Tech. Publication no. 760.*, AIME, New York.
40. Rossig, H.H., Ibe, G., and Gruhl, W. (1974). *Metall.*, **28**, 232.
41. Phillips, V.A., Swain, A.J., and Eborall, R. (1952–1953). *J. Inst. Metals*, **81**, 626–647.
42. Mulford, R.A., and Kocks, U.F. (1979). *Acta. Met.*, **27**, 1125–1134.
43. Westengen, H. (1982). *Z. Mettalkunde*, **73**, 360–368.
44. Granger, D.A., and Jensen, C.L. (1984). In *Light Metals 84*, 113th Annual Meeting of TMS-AIME, Los Angeles, Calif. AIME, Warrendale, PA pp. 1249–1263.
45. Cochran W.C. (1967). In *Aluminum, Vol. III, Fabricating and Finishing*, K.R. Van Horn, ed. American Society for Metals, Metals Park, Ohio. pp. 641–684.
46. Stumpf, H.C. (1973) Progress Report, Alcoa Laboratories.
47. Anderson, W.A., and Stumpf, H.C. (May 1981). *Metals Progress*, 60–66.
48. Doherty, R.D. (1985) Divisional Technical Report, Alcoa Laboratories.
49. Taii, E., Inaba, T.,Kitao, Y., and Anbai, M. (July 28, 1983). Japan Patent Application no. 57(1982)-9340. Kobe Steel Co., Ltd.

Part III

Processing of Aluminum Alloys

4

Ingot Casting in the Aluminum Industry

D.A. GRANGER

Molten Metal Processing Division
Alcoa Laboratories
Alcoa Center, Pennsylvania

I. Introduction and Background

Ingots are cast shapes produced with a cross-section that is suitable for fabrication processing using such methods as rolling, extrusion, and forging.

At the outset, some 100 years ago, the industry used ingot-casting techniques commonly employed in the well-established steel and copper

ISBN 0-12-341831-3

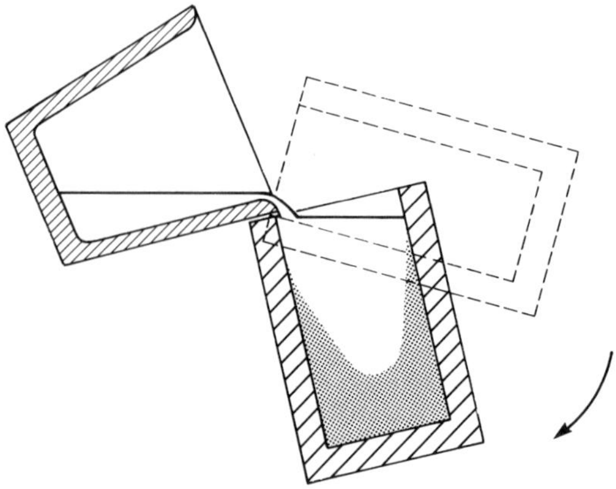

FIG. 1. Early permanent mold method of casting aluminum-alloy ingots: the tilting-mold technique.

industries. The reactivity of molten aluminum with moisture in the atmosphere soon forced a change from the usual pouring methods employed for making permanent (or book-mold) ingot. A more quiescent mode of transferring the molten metal to the mold, as exemplified by the tilting mold method, illustrated in Fig. 1, was adopted [*1*].

In the 1920s, the demand for large wrought aluminum components for aircraft structures was to have a dramatic impact on the ingot casting of aluminum alloys. The increased ingot dimensions, particularly in thickness, resulted in an unacceptable coarse metallurgical structure. In alloys such as 24S, large intermetallic particles were a source of fracture during rolling and forging. (24S composition is equivalent to the AA2024 alloy of the late 1980s.) These coarse structures were a direct consequence of the poor heat removal obtained from a permanent mold once the initial ingot shell setup and a gap formed between the casting and the mold. To overcome this problem, a more efficient method of heat removal had to be found. This need resulted in the development of the revolutionary direct-chill (DC) ingot-casting process [*2*]. The simplicity of the molten-metal transfer and mold arrangement illustrated in Fig. 2, along with the fact that great efficiency can be realized by making multiple ingot at one time, ensured the rapid replacement of book-mold casting for all ingot-making requirements in the industry.

Despite the advantages, both economic and metallurgical, that may be enumerated for continuous casting thinner cross-section ingot [*3*,*4*], the DC casting process and its derivatives remain the dominant ingot-casting method employed in the late 1980s, accounting for at least 90% of the ingot made worldwide.

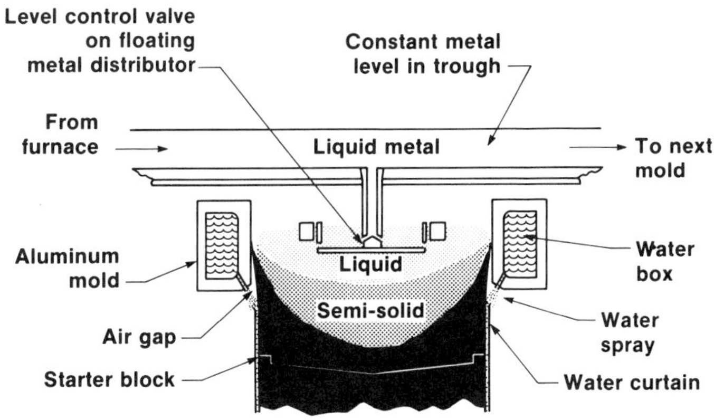

FIG. 2. Schematic of the vertical direct-chill casting process.

II. DC Casting Process

The DC process is illustrated schematically in Fig. 2. At start-up, the open end of the mold is closed off with the bottom block. As the mold begins to fill with metal, the bottom block is lowered at a rate that allows the molten metal level to be controlled at some predetermined distance above the lower end of the mold. (This distance is referred to as the metal head.) Thus, during casting, the metal entering the mold exactly equals the metal exiting. As shown, solidification commences with the formation of a solid shell in the water-cooled mold. This shell shrinks away from the mold and a gap forms which retards heat transfer to the mold. Most of the heat of solidification is removed from the ingot as it emerges from the mold into a water curtain which impinges directly onto the surface. Hence, the term *direct-chill casting*.

The first DC molds were simple and somewhat flimsy in construction, which made them inexpensive to fabricate, but subject to distortion, which, in turn, resulted in nonuniform water distribution. Molds in use today are still fabricated from aluminum alloy, typically 5052 alloy, for good thermal conductivity, but are much more rigid in construction (see Fig. 3) and provide a consistently uniform distribution of water. Most molds now incorporate a method of distributing lubricating oil, usually castor oil, to reduce friction between the ingot surface and the mold. This is in place of a heavy grease which was used to coat the surface of the mold prior to casting an ingot.

It was soon appreciated that surface and subsurface defects found on DC cast ingot were associated with the gap that forms after the initial shell is established.

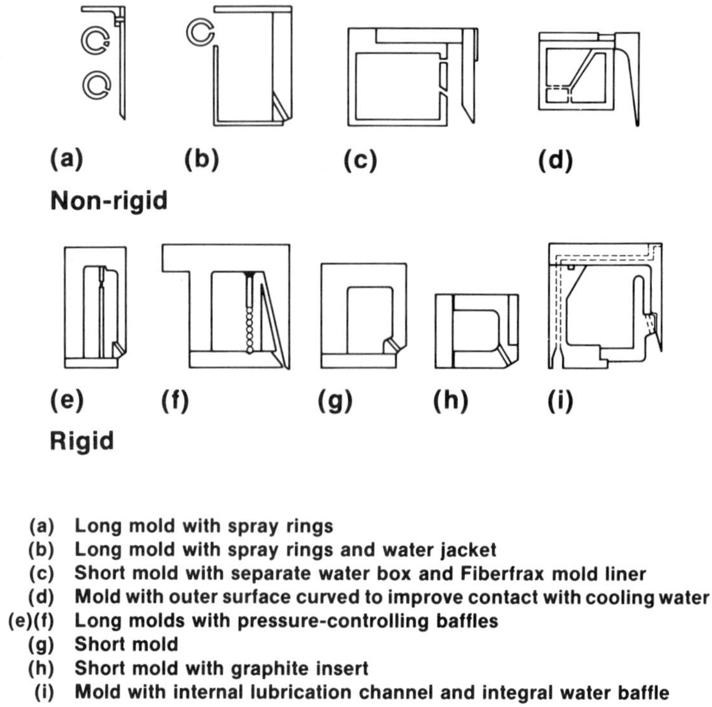

FIG. 3. A variety of DC mold designs.

This gap retards heat removal to the mold and the solidified metal reheats causing defects (as illustrated in Fig. 4), which have been reviewed in detail by Emley [4]. These defects necessitate the surface machining (or scalping) of ingot prior to fabrication and provide one of several incentives to modify the DC process to give an ingot of superior quality that, depending on the alloy being cast and the application to which the material is to be applied, requires little or no scalping.

A. *Mathematical Modelling of the DC Casting Process*

Attempts have been made to model mathematically the freezing of a DC cast ingot [5–8], but any analytical solution is seriously impeded by an inability to assign an accurate heat-transfer coefficient to the various stages of solidification as a direct consequence of the complex and variable boundary conditions. It is generally agreed that gap formation in the mold not only is the cause of surface and subsurface disturbances to the structure

FIG. 4. Surface appearance of "bleed bands" on DC cast 3003 alloy.

but is the reason why most of the heat in DC casting is removed by the water curtain that impinges directly on the ingot as it emerges from the mold.

One of the more successful models (and one for which most information has been published) was that developed by Peel and Pengelly [6]. In the later modified version [7], which incorporated experimental measurements of heat flux through the mold and in the region of the water cooling, the model was able to predict with reasonable accuracy (Fig. 5) sump shape, cell-size distribution, and scalping depth for a given set of casting parameters.

B. Structural Characteristics of DC Cast Ingot

Before proceeding further with the discussion of the evolution of the DC process and the introduction of close-to-final size casting methods, it is necessary to give an overview of the important structural characteristics of ingot. These metallurgical features are:

1. Grain size and shape
2. Cell size
3. Segregation (both micro-and macrosegregation)
4. Constituent type and distribution

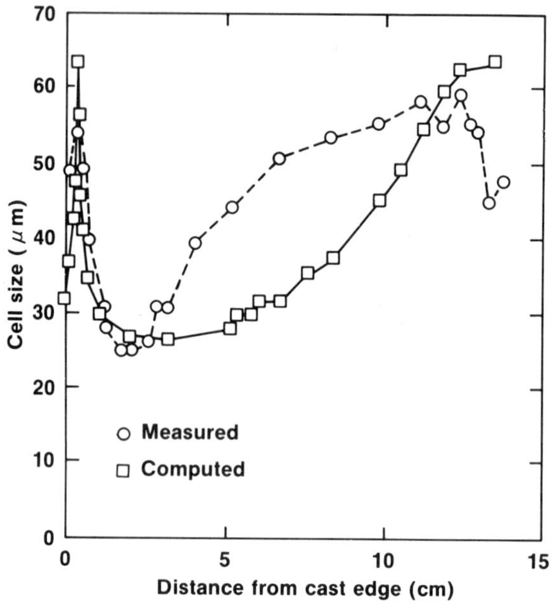

FIG.5. Cell-size distribution in a DC cast 300 × 970-mm pure aluminum ingot cast at 76 mm per minute with a 76-mm metal head. (After Beattie [7])

1. Grain Structure

The grain size ranges from about 50 to 500 microns in a grain-refined DC cast ingot. Master alloys containing heterogeneous nuclei are used to obtain an equiaxed grain structure. Grain size is a function of the type and amount of inoculant used, the alloy, ingot size, and casting variables such as melt superheat. In contrast with a grain-refined ingot (Fig. 6a), an unrefined ingot (Fig. 6b) will contain a columnar crystal structure usually highly detrimental for mechanical properties and surface finishing characteristics in wrought material.

2. Cell Size (or Dendrite Arm Spacing)

This structural feature illustrated in Fig. 7 can range from about 10 to 100 microns in DC ingot and is inversely proportional to the solidification rate. An empirical equation has been developed [9] that relates cells size to solidification rate:

$$d = k\theta^{-n}$$

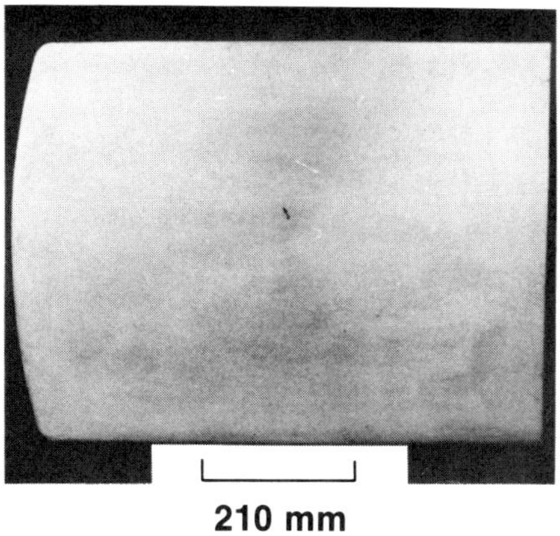

(a)

FIG. 6a. Fine equiaxed grain structure in 3003 alloy ingot grain refined using Al-5% Ti-0.2% B rod. Etch: Tucker's Reagent.

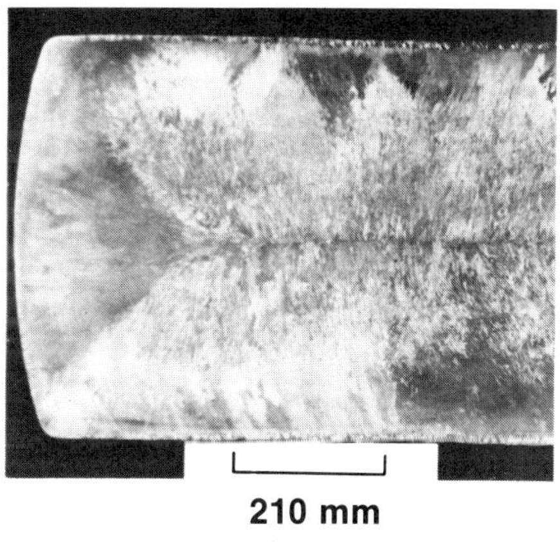

(b)

FIG. 6b. Coarse grain structure in DC semicontinuous cast 3003 alloy ingot that has not been grain-refined. Etch: Tucker's Reagent.

where d = cell size (in microns), k is a constant, $n = 0.33$, and θ is the solidification rate in °C/sec.

3. Segregation

Microsegregation is the source of the "cored" appearance of an etched cast structure as shown in Fig. 7. In all commercial aluminum alloys, the

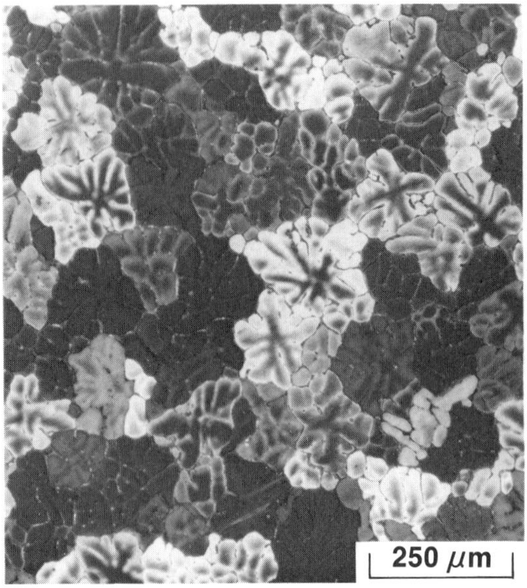

FIG. 7. The branched dendrite cell structure in equiaxed grained 2124 alloy electropolished and photographed under polarized light.

first metal to solidify is the most pure. As freezing continues, there is a buildup of solute in liquid and the solid gradually becomes richer in the alloying elements. This is shown by an electron probe microanalysis trace across a dendrite cell as shown in Fig. 8. The Scheil equation [10] gives a good semiquantitative description of solute redistribution:

$$C_s = C_o\,[1 - f_s]^{k-1}$$

where C_s = solute concentration in solid, C_o = average alloy composition, f_s = fraction solid, and k = partition coefficient.

Microsegregation is on the scale of the cell spacing but macrosegregation describes differences in composition over distances of centimeters. A typical composition profile across a thick DC cast ingot of 2024 alloy is shown in Fig. 9 taken from the work of Yu and Granger [11].

4. Primary Constituents

There are two types of intermetallic phases that form during solidification:

- the insoluble phases (which are usually iron-bearing), such as Al_3Fe
- the soluble phases, such as $CuAl_2$

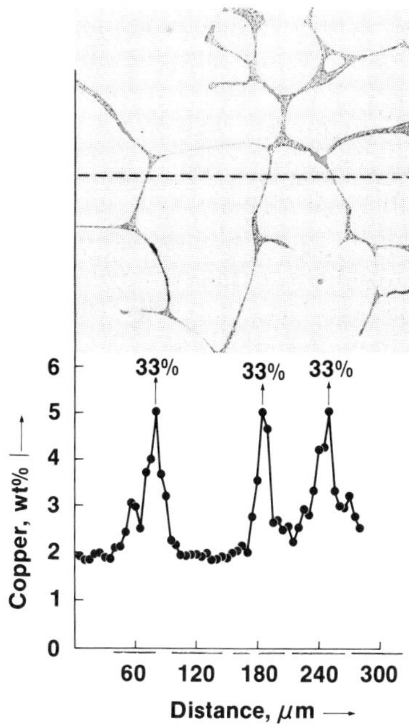

FIG. 8. Copper distribution between dendrite arms in DC cast 2124 alloy.

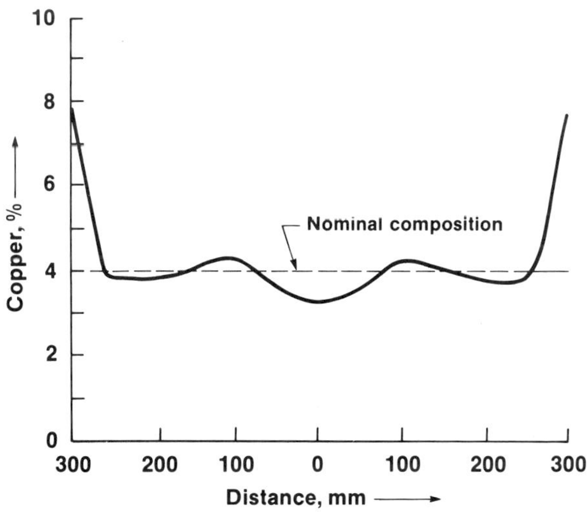

FIG. 9. Variation in copper concentration across a 600-mm–thick DC cast ingot of Al-Cu-Mg alloy.

The former can change in morphology, crystal type, and composition upon heating, but the latter may go into solution and disappear. Like dendrite cells, the size of a constituent particle is proportional to the solidification rate, with the coarser constituents being located at the center of an ingot where the cooling rate is slowest.

Both stable and metastable constituents may be found in an ingot depending upon composition and solidification rate. For instance, several forms of the insoluble iron bearing phase may be found in a large ingot: the metastable phases Al_6Fe and Al_mFe close to the surface, where cooling rates are high, and Al_3Fe toward the more slowly cooled center of the ingot [12–14].

III. Control of DC Ingot Structure

As will be shown, ingot structure has a direct impact on the quality of the wrought product. Therefore, it is of paramount importance to understand what can be done to control DC ingot structure.

A. *Charge Makeup—Impurities*

Ingot-structure control starts with the selection of materials from which the alloy is to be prepared. For the economic production of commercial aluminum alloy ingot, the charge is usually made up from recycled scrap metal, master alloy additions as well as smelter metal. Even high-grade smelter metal contains a variety of impurities besides iron and silicon, as shown in Table I. Some of these impurities (such as Na, Ca, Li, H, and C) will be removed or reduced in the course of molten-metal fluxing and filtration. Others (such as the peritectic forming elements: V, Ti, and Zr) may be precipitated with boron for special applications such as electrical conductor wire, but will normally be found in the cast ingot where their influence on constituent formation may be detrimental.

Major alloy additions are made using either aluminum master alloys, pure metal, or briquetted powder depending on economic factors and production issues such as solution rate and recovery. In general, the lower melting point additions such as Mg, Zn, Si, and Cu are added as pure elements while high melting solute such as Cr, Zr, Mn, and V are added as master alloys. (See Table II.) Also listed in Table II are the master alloys commonly used for purification, modification (i.e., refinement of the eutectic structure in hypoeutectic Al-Si alloys), and grain refining. These

TABLE I

IMPURITIES IN ALUMINUM FROM THE HALL-HEROULT CELL

Impurity	Usual concentration range, %
Fe	0.06–0.35
Si	0.05–0.15
H	ca. 3×10^{-5}[a]
Na	0.002–0.004
Ca	0.002–0.004
Li	0.002–0.004
Ti, Zr	0.005–0.020
V, Cr	0.005–0.020
Al_2O_3	Variable
Al_2MgO_2 (Spinel)	Variable
Al_4C_3	Variable
TiB_2, etc.	Variable

[a] 0.20–0.60 cc/100 g Al.

TABLE II

ALUMINUM MASTER ALLOYS

Purpose	Composition	Form[a]	Principal constituents
Alloying	Al-25% Mn	Waffle	$MnAl_3 + MnAl_4 + MnAl_6$
	Al-75% Mn	Briquette	Mn + Al powder
	Al-5% Fe	Waffle	$FeAl_3$
	Al-75% Fe	Briquette	Fe + Al powder
	Al-10% Cr	Waffle	$CrAl_7$
	Al-6% Zr	Waffle	$ZrAl_3$
	Al-5% V	Waffle	$VAl_3 + VAl_{10}$
Purification	Al-3% B	Waffle and rod	AlB_2 (major) + AlB_{12}
Modification	Al-10% Sr	Waffle and rod	Al_4Sr
	Al-16% Si-10% Sr	Waffle	Al_2SrSi_2
Grain refining	Al-10% Ti	Waffle	$TiAl_3$
	Al-6% Ti	Rod	$TiAl_3$
	Al-5% Ti-1% B	Rod	$TiAl_3 + TiB_2 + (TiAl)B_2$
	Al-5% Ti-0.2% B	Rod	$TiAl_3 + TiB_2 + (TiAl)B_2$

Note: Table lists the recommended forms in which to use the master alloy.
[a] Waffle: 16-lb. cast slab.
Rod: 3/8-in. diameter rod.
Briquette: 1/4-lb. powder compact.

master alloys may be a source of undesirable contaminants so, as with scrap and smelter metal, it is important to pay close attention to charge makeup.

B. Molten Metal Treatment

The mechanical properties and finishing characteristics of aluminum alloys are deleteriously affected by nonmetallic inclusions such as oxides, borides and carbides. Alkali elements, particularly sodium and calcium, reduce the hot-workability of aluminum alloys, especially those containing appreciable amounts of magnesium. Hydrogen, the only gas soluble in aluminum alloys, is the source of porosity in ingot [15] and is the primary cause of defects such as blisters on sheet and flakes in forgings [16].

In order to avoid these problems, metal fluxing and filtering methods have been developed to reduce hydrogen [17], alkali elements [18], and nonmetallic inclusions [19]. Fluxing is usually achieved by treating the metal with a reactive gas such as chlorine, as it flows from the furnace to the casting unit, in a reaction chamber where the gas phase is well dispersed, as illustrated in Fig. 10.

A variety of filter systems have been developed to remove solid particulate ranging from sophisticated deep-bed filters with a counter flow of inert gas [20] to throwaway foam ceramic filters that depend to a large extent on cake mode filtration [21]. Choice of metal-treatment system is dictated by product-quality requirements and economics.

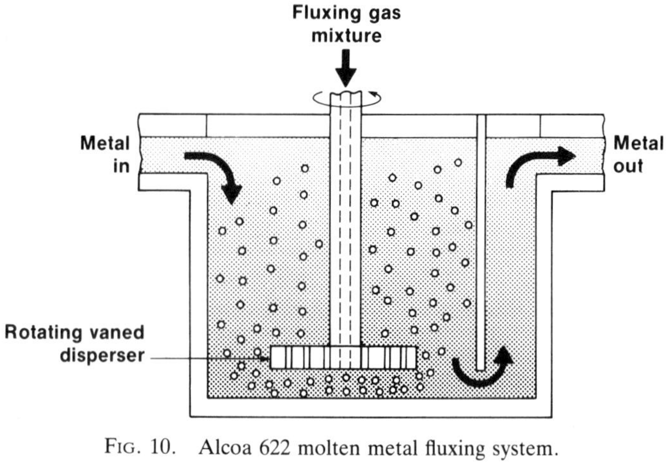

FIG. 10. Alcoa 622 molten metal fluxing system.

C. Grain Refinement

Control of grain structure in aluminum alloys ingot is achieved by providing the right combination of nucleation and growth conditions [22]. Nuclei are found in the form of aluminum master alloys; the most commonly used because of their effectiveness are those based on the Al-Ti-B alloy system where the titanium content may range from 3 to 10% and the titanium-to-boron ratios is anywhere from 3 to 50:1. A popular master alloy widely used in the form of 3/8-inch–diameter rod which may be added to the reaction chamber of an Alcoa 622 unit (see Fig. 10) is Al-5% Ti-0.2% B. The structure of this rod is illustrated in Fig. 11. The large $TiAl_3$ aluminides are soluble in molten aluminum, but the small boride particles are insoluble and provide the long-lasting grain-refining nuclei which are dispersed into the molten metal alloy stream when the aluminum matrix of the master alloy dissolves.

Addition levels range from 0.003 to 0.015% titanium depending on the alloy, ingot size, and casting conditions. In order to provide the best environment for growth of insoluble nuclei, introduced by a grain-refining master alloy, it is desirable to maximize the zone of residual constitutional supercooling that exists ahead of the advancing solidification front. Full constitutional supercooling only occurs with a planar solid–liquid interface, but in ingot casting it is so severe that breakdown to form cells and dendrites always takes place. However, the small amount of residual supercooling that remains ahead of the dendrites is maximized, as illustrated

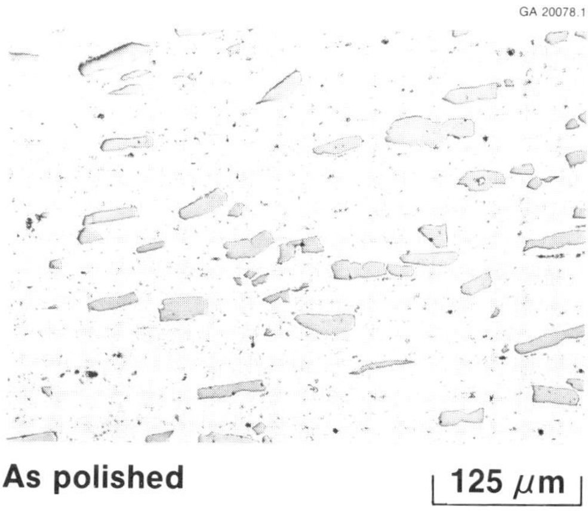

FIG. 11. Longitudinal cross-section through Al-5% Ti-0.2% B rod showing large $TiAl_3$ and small, insoluble TiB_2 particles.

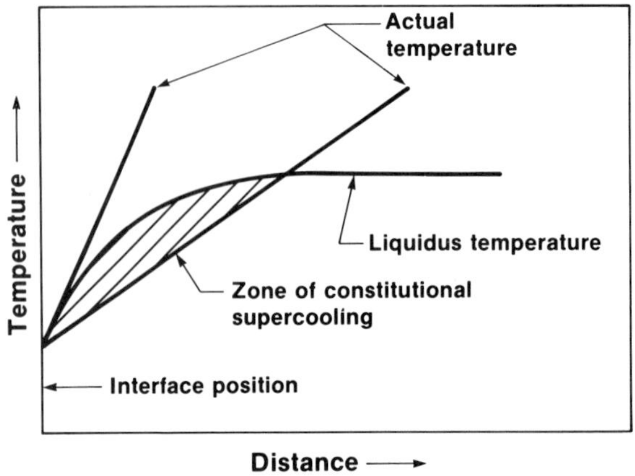

FIG. 12. Schematic illustration of constitutional supercooling in alloy solidification.

in Fig. 12, by minimizing the temperature gradient in the head of the ingot. In practice, this is accomplished by casting with the lowest melt superheat compatible with avoiding the formation of coarse intermetallic particles [23].

The principal reason for grain refining is to achieve an equiaxed structure, which increases the resistance of the alloy to hot-cracking during casting by lowering the coherency temperature [24]. That is, it lowers the temperature in the freezing range at which the alloy can sustain a load, and thus it reduces the stress level in the ingot on completion of solidification [25]. An equally important reason for controlling grain structure is to avoid nonuniform segregation and constituent distribution associated with coarse columnar grains, since these heterogeneities can lead to inferior properties in the wrought material [26].

D. Casting-Process Variables

To a great extent, the structure of a DC cast ingot is determined before the actual solidification event. The most important factor in the casting process is to establish the variables (such as water flow rate and distribution, metal level in the mold, casting speed, and lubricant flow rate) that will allow a crack-free ingot to be produced. As just mentioned, metal superheat has a direct impact on grain-structure control, but if too low, it may give rise in some alloys of the 7XXX, 3XXX, and 2XXX alloy series to coarse intermetallic particles. However, once the process and ingot size are selected, little further can be done to influence the cast structure without changing the process itself, as will be discussed later.

IV. Deficiencies of the DC Casting Process

While the DC casting process solved the immediate problem of producing hard alloy rolled plate for aircraft manufacturers, it did not provide a satisfactory ingot for forged parts. There were several reasons for this, but the principal difficulty was associated with the bilevel transfer system for delivering molten metal from the trough into the mold. (See Fig. 2.) Turbulence in this transfer scheme causes disruption of the oxide film on the molten metal surface, allowing freshly exposed metal to react with moisture in the atmosphere, resulting in the generation of oxide films and the absorption of hydrogen according to the reaction:

$$2Al + 3H_2O_{(v)} = Al_2O_{3(s)} + 6H_{(soln)}$$

This degradation in metal quality was found to be the source of defects, on broken tensile bars, known as "bright" and "dark" flakes, which are a source of low ductility and toughness in forged parts [16].

Compared with book-mold castings, ingots produced by the DC process are subject to much steeper thermal gradients. As a consequence, the thermal stress level is greater and special casting practices had to be developed in order to produce crack-free ingot. At first, the most crack-sensitive alloys could only be made in smaller diameters and thicknesses. Improvements in casting technology such as greater uniformity of molten metal distribution, control of water application to the ingot surface, speed control, and grain refinement allowed increasingly large ingots to be made until nowadays ingots 40 inches in diameter and up to 24 inches thick are routinely cast. Nevertheless, there remain some highly crack-sensitive alloys in the 2XXX and 7XXX series that can only be cast crack-free by "reducing" the quench of the direct water or by affecting a reduction in the thermal stress level in the ingot by removing the water from the ingot surface once solidification is complete [27].

V. DC Ingot Casting Process Development

Since the inception of DC casting, there have been continuing efforts to improve both the productivity and quality of DC cast ingot. Of principal concern here are the developments affecting ingot quality. In the broadest sense, these may be separated into internal and surface quality issues.

A. *Level Transfer Method*

To overcome the problem of oxide generation and hydrogen pickup in the casting process, a method had to be devised to avoid the bilevel transfer pouring and delivery of the molten metal to the mold without disturbing the oxide skin on the molten metal surface. If this could be accomplished, then only the upstream molten metal treatment (fluxing and filtering operations) would limit the internal quality because the oxide skin forms an effective barrier between the molten metal and the moisture in the atmosphere [15]. The solution to the problem was level transfer casting, illustrated in Fig. 13. In this process, the integrity of the oxide skin is retained as molten metal is introduced through the side of a refractory-lined basin, located above the water-cooled mold, with no change in level.

This mold assembly is clearly more complex and expensive than the simpler DC mold package; furthermore, it proved to be more difficult to get satisfactory surface and subsurface qualities. Many variables are involved in optimizing the surface of an ingot produced by the level transfer method. This topic has been addressed at length by Bergman [28].

B. *Low-Heat Casting*

Where internal quality is not of paramount importance, there is an incentive to obtain an improved surface structure in order to minimize, if not eliminate, scalping. It was clear that the primary cause of surface defects was reheating of the ingot shell as it contracted from the mold wall. A way to improve surface quality then would be to reduce the time that

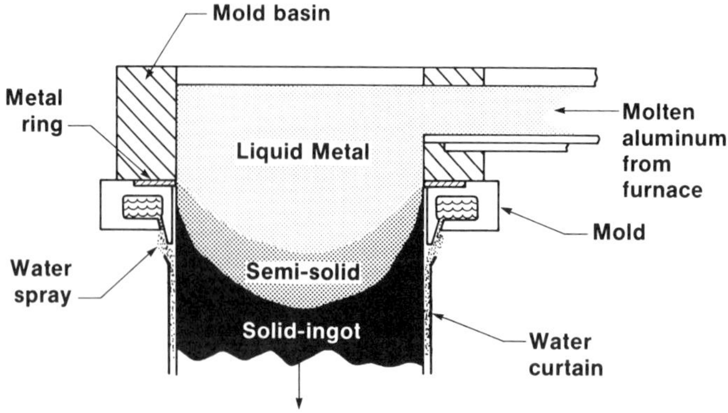

FIG. 13. Level transfer mold assembly.

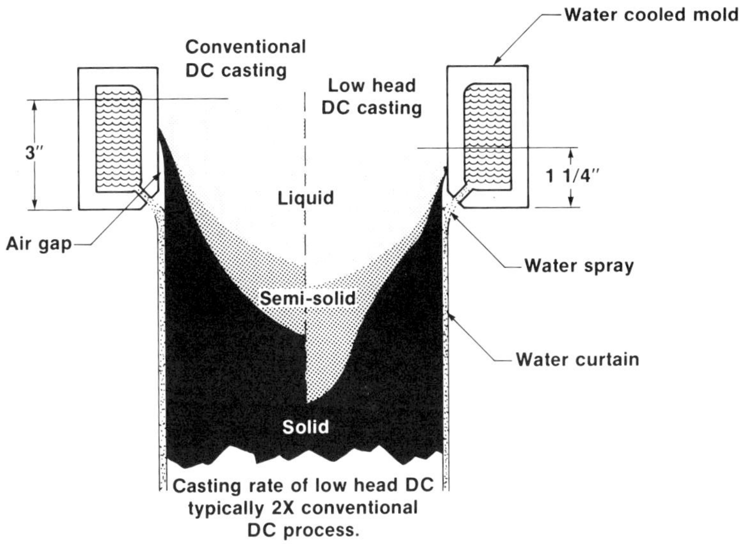

FIG. 14. Schematic of the low-head DC casting process.

reheating could occur by reducing the effective mold length (see Fig. 14) and increasing the casting speed. While simple in concept, this reduction in mold length is not easy to accomplish in practice. This is due to the need to accommodate the molten metal delivery hardware close to, but not actually touching, the starting block. Most importantly, there is the need to avoid molten metal bleed-outs at start-up, when thermal stresses in the first formed ingot cause it to curl [29]. In effect, the metal level in the mold was set by start-up requirements. One of several solutions was the Isocast process [30] in which the mold assembly and delivery system are raised after start-up, effectively reducing the liquid level in the mold. Another technique, the Alcoa 729 process, reduces the quench at the start of the drop by the introduction of CO_2 gas into the cooling water, thus minimizing curl and allowing the ingot to be started with a low head [29, 31].

The advantages of the low-head DC casting process are, in addition to a superior surface quality and higher casting rate, a more uniform and finer subsurface structure than can be used to advantage in certain applications requiring a uniform surface finishing response [32].

1. Electromagnetic (EM) Castings

This seemingly revolutionary method was a Russian invention [33] in which an electromagnetic field is used to contain the molten metal rather

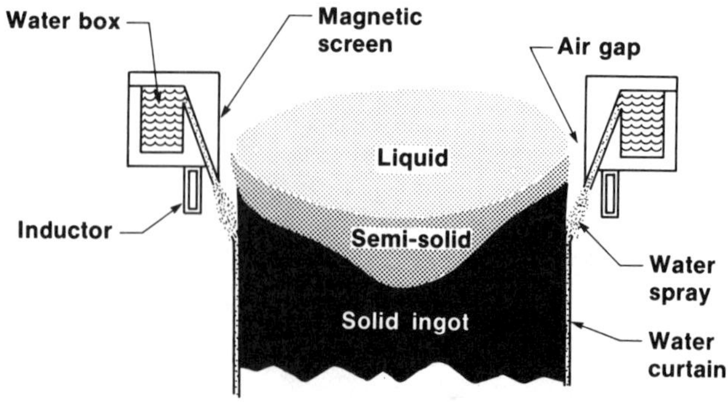

FIG. 15. Schematic of the electromagnetic (EM) casting process.

than a water-cooled mold as shown in Fig. 15. In essence, EM (or moldless casting as it is sometimes described) provides the closest approach to continuous heat removal from the solidifying ingot and (as expected) the surface defects characteristic of shell reheating are largely eliminated. Since the early 1980s, this method has been gaining increased acceptance because it allows scalping to be either eliminated or greatly reduced, and it markedly reduces the need for edge trimming during rolling to plate and sheet. The method is still largely confined to the non–heat-treatable alloys because cracking problems associated with the more crack-sensitive heat-treatable alloys have not been overcome yet.

2. Showa Denko Process

A limitation of the level transfer (or hot top process) for circular cross-section ingot is the inability to achieve the equivalent of a low-head condition because the start-of-freeze location is fixed immediately below the refractory-lined basin as shown in Fig. 13. The Showa Denko process overcomes this problem by the use of inert gas introduced at the lubricating ring where it forms a bubble. By use of gas-pressure control, it is possible to push the molten metal away from the oil ring, located below the refractory basin, and cause freezing to start closer to the mold exit. In this way, superior surface quality ingots, requiring little or no scalping, have been produced in a wide variety of alloys and ingot sizes [34].

3. Horizontal DC Casting

The process shown schematically in Fig. 16 has several attractive features, but they are offset by some significant metallurgical and production

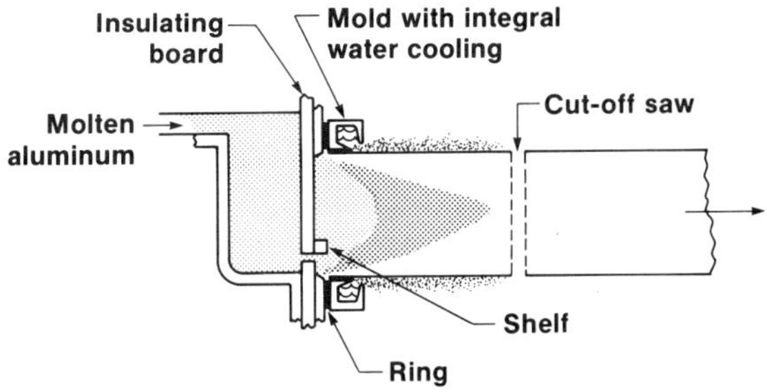

FIG. 16. Schematic of the horizontal DC casting process.

constraints. Of the DC casting derivatives, this is the only fully continuous process; it provides ingot of high internal quality as a result of the level transfer [35]. However, there are some structural features of a less desirable nature that arise from the molten metal distribution in the ingot crater. This flow pattern results in an asymmetry in the structure (in rectangular sheet ingot, it is observed that the thermal centerline is above the geometric centerline of the ingot), but the most detrimental consequence is the heavily liquated lower surface which usually requires doubling the scalping depth. Use of the method is normally confined to common alloy sheet ingot (e.g., 5182 and 3004 alloy, 6XXX extrusion ingot, and Barlock foundry ingot of the A356 type). Because of the complexity of the mold package and time required to change out a casting complex, the method is only economically viable where both ingot size and alloy types are minimized and large production runs of a single alloy of one ingot size are common.

VI. Evolution of Ingot Structure

The impact of ingot structure on the fabricability and final product performance is not well documented. Part of the problem may have arisen because of the belief that the combination of heat treatment and deformation would largely obliterate all traces of the cast structure. Nothing could be further from the truth. The cast structure places an indelible imprint on the alloy and must be controlled in a manner conducive with the desired fabricating practices and end uses.

Many examples could be given of how ingot structure impacts on wrought material properties, but only two cases will be given here for illustrative purposes.

A. *Twinned Columnar Growth (TCG) Structure*

The occurrence of this structure is the direct consequence of a poor grain-refining practice (i.e., use of an inadequate grain-refiner master alloy or lack of control of the growth conditions, possibly a combination of both). It is well known that TCG severely reduces the fabricability of ingot. Figure 17 illustrates splitting that occurs in 2024 forged stock during the first upset when the coarse grain structure is present. Less well known is the consequence of TCG on the mechanical properties of plate.

Granger and Liu [26] demonstrated that segregation associated with the TCG structure is not only coarser than that found in normal 7050 (and 2219) ingot, but also takes on a more oriented, less random distribution (Fig. 18). As a consequence, the ductility of a tensile test bar oriented normal to the grain-growth direction is markedly reduced, as demonstrated in Fig. 19.

B. *Streaking in Architectural Products*

The nonuniform heat removal that occurs in the mold during DC casting (see Section II) results in a variation in both cell size and constituent type. For instance, it has been shown that the iron-bearing phases formed in both 5005 and 1100 alloy are a function of the solidification rate [13 and 14]. As the freezing rate increases, phases of greater metastability are

FIG. 17. Cross-section through a forging showing a crack associated with a region of twinned columnar growth (45% of original size).

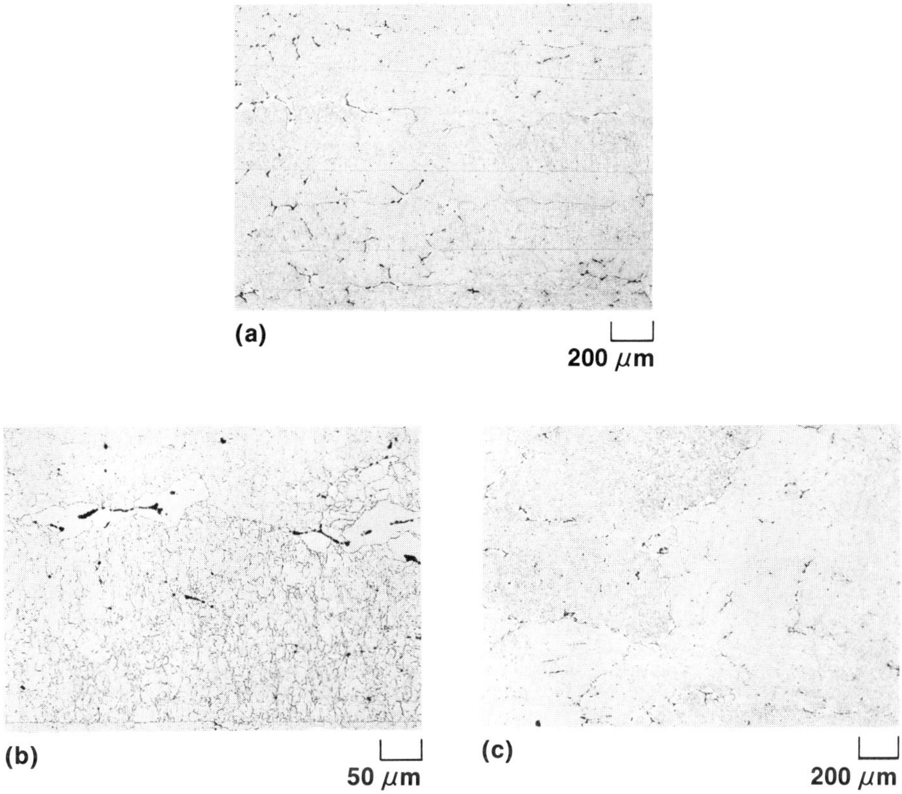

FIG. 18. Photomicrographs of TCG-bearing 7050 alloy plate. (a) Traces of [111] twin planes and intergranular segregation. (b) Higher magnification of (a). (c) More randomly oriented segregation in a TCG-free plate. Etch: Keller's Reagent.

formed. Since the different phases respond differently to etching and anodizing treatments, it is important that there is a single phase present in the ingot surface revealed by scalping if a uniform finishing reponse in the sheet surface is to be obtained.

In order to better ensure freedom from structural streaking, Granger and Jensen [32] recommended the adoption of a low-head DC casting practice so that ingot reheat during casting is kept to a minimum. Such a practice allows the production of only the metastable phase Al_mFe to be formed close to the surface of 5005 alloy ingot. A low-temperature preheat results in the conversion of some of the m-phase to $Al_{12}Fe_3Si$, but this phase responds identically with the Al_mFe during etching and anodizing and thus no streaking appears on the sheet surface.

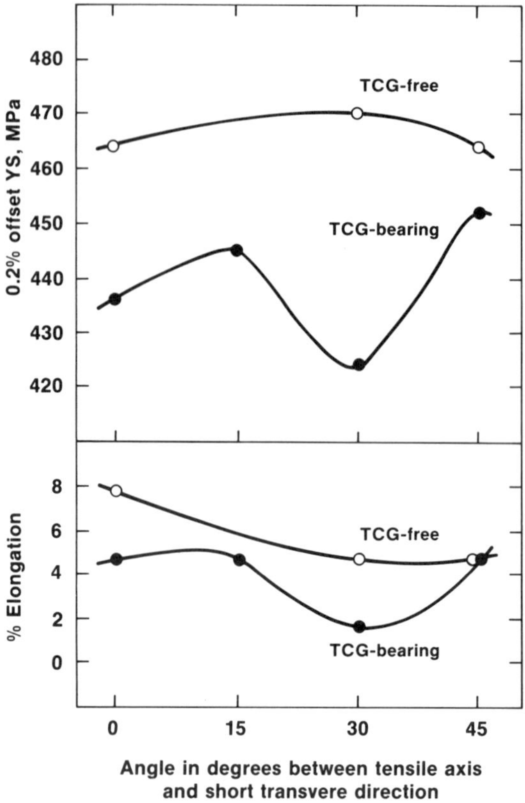

FIG. 19. Tensile properties as a function of orientation in 144-mm–thick 7050-T73651 plate.

VII. Continuous (Near-Net Shape) Casting Processes

The concept of continuous casting can be traced back to Bessemer, who, in the nineteenth century, cast steel strip between water-cooled drums [36]. However, with the exception of high-purity aluminum for electrical conductor cables, truly continuous casting methods have not been widely adopted. This must, in part, be attributed to the huge success of the DC casting of large ingot which could then be fabricated economically in large rolling and extrusion facilities. It is only when the metallurgical advantages that derive from the more rapid solidification of a smaller cross-section ingot are realized that the possibility exists to readily displace the incumbent cost-effective methods. This requires the development of new alloys with unique property combinations derived from the rapid freezing of a process such as the Hunter-Engineering method, which produces 1/4-inch–(6 mm–) thick reroll stock with a cell size ranging from 5 to 12 microns. In the mid-1970s, alloys (AA8006/8007) were developed for this purpose, [37] (see page 83), but there has been little published activity since that time and the continuous casting techniques still remain to be fully exploited.

A. *Processes for the Production of Wire Products*

The most successful methods for the production of conductor wire have been the Properzi process [38] and the analogous Southwire method [39]. These methods employ a water-cooled copper-alloy wheel, with a peripheral groove which is closed by a steel belt to form the mold as illustrated in Fig. 20. Following casting, the solid bar is fed into a multistand mill, where it is hot-rolled to rod typically 3/8 inch (9 mm) in diameter. Both Properzi and Southwire processes have been largely confined to the production of conductor wire in pure metal such as 1450 alloy or 6201 alloy where greater creep resistance was required.

Some expansion in the capability of the continuous casting of alloy rod has been achieved with the introduction of the Secim method [40]. Using this method, alloys of the 2XXX, 4XXX, and 5XXX series have been successfully cast and rolled to rod for use in such applications as welding wire, zippers, rivets, and machining rod. The principal differences between the Secim process and the earlier wheel and belt casting methods that allow this to be achieved are (1) a Y-shape copper mold for symmetrical cooling, (2) a tensionless copper belt for improved surface quality, (3) variable mold length, (4) scalping between the first and second rolling stands, and (5) an eight-stand grooved mill oval-round.

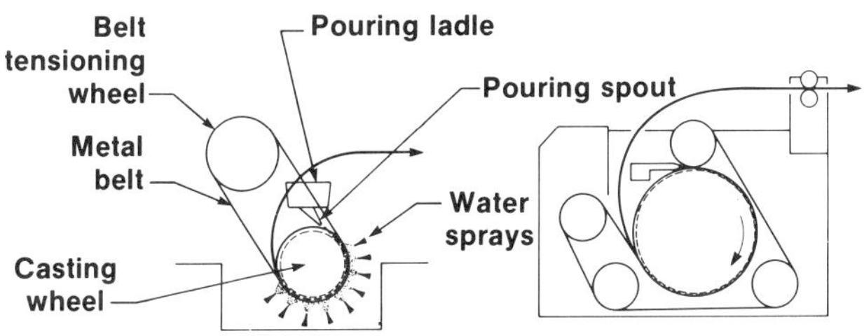

(a) Properzi equipment (b) Southwire Casting System

FIG. 20. Wheel-and-belt casting methods (schematic).

It appears that success of the Secim method will depend on the economics of the process rather than the possibility of generating any unique property combinations from the generally faster freezing rates achieved from the smaller cross-section bar, compared with 6-in–diameter ingot.

B. Processes for the Production of Reroll Stock

There have been numerous processes developed for making reroll stock using the moving-mold principle. The most successful are those using articulated blocks such as the Hunter-Douglas [41] and the Alusuisse Caster II [42], moving belt, (as exemplified by the Hazlett process [43]) and Hunter-Engineering [44] and Pechiney III C [45], which employs water-cooled rolls. Of all these processes, it seems that the latter of two methods have most successfully combined the economics of near-net casting in addition to the metallurgical advantages to be obtained from the more rapid solidification rate achieved by casting thin section strip.

Hunter-Engineering equipment is illustrated schematically in Fig. 21. Originally designed to cast vertically upward, it is now most commonly employed with a horizontal (or close to horizontal) casting direction. The method has been used most successfully to cast common alloys such as 1100 and 3003, though its productivity has always fallen short of expectations due to sticking problems. Other difficulties that have limited its full exploitation have been heat lines and centerline segregation [46]. The usual heterogeneities of an as-cast surface have prevented the use of "roll-cast strip" for surface critical applications such as lithographic and architectural sheet.

Morris [37] reported the invention of alloys 8006 and 8007 which were specifically designed to take advantage of the high solidification rates

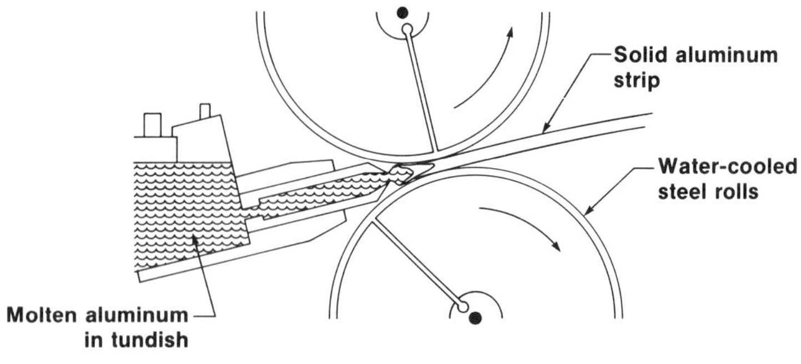

FIG. 21. Schematic of the twin-roll Hunter-Engineering roll-casting process.

associated with roll casting. These alloys were reported to provide a unique combination of strength and formability when produced from roll cast stock and have been successfully used for making such items as formed containers and heat exchangers. The difficulties associated with the casting of longer-freezing-range alloys with this type of equipment probably accounts for the limited progress made in the development of other alloys that would benefit from the high solidification rates of the roll-casting processes.

VIII. Summary and Future Trends

The various structural elements of an ingot are described and the means available to control them have been discussed. In doing so, an attempt has been made to emphasize the importance of factors such as alloying, molten-metal treatment, and grain refinement, in addition to the casting process, on final product quality. Nevertheless, the most important contribution to the final structure comes from the ingot-casting process. There are numerous processes ranging from the DC method, utilized nowadays for large cross-section ingot, to near-net shape casting for producing reroll stock and bar.

The DC method and its derivatives have been the dominant casting technique in the industry since the 1940s, but powerful economic and market forces are creating a situation in which other methods are becoming more widely exploited. There is an increasing demand for processes that will provide:

- more consistent ingot structure
- increased ingot quality
- reduced overall fabrication costs

For sheet ingot, these needs are met in large part by EM casting, which gives a relatively uniform ingot structure and substantially reduced fabrication costs. The Showa Denko and similar methods, which give superior surface quality, will be more widely exploited for the production of extrusion and forging ingot. Further improvements in structural consistency and quality are being obtained by the use of computer-controlled casting units and the development of sophisticated process-control models.

Strip- and bar-casting processes are being used to cast an increasingly large number of alloys principally as a consequence of the economic benefits of near-net shape casting.

Progress in ingot-casting methods has been accompanied by significant developments in both the techniques and the control of molten-metal degassing, filtration, in-line alloying, and grain refinement. It is anticipated that continued improvements will occur in this area since they are essential to the goal of higher ingot quality and consistency.

No mention has been made of the problems associated with the production of Al-Li alloy ingot. This occurred for two reasons: (1) space limitation would permit only a superficial treatment of this complex subject and (2) many of the unique solutions that have been devised to produce quality ingot remain proprietary. Nevertheless, it would be remiss to omit all aspects of what has been a remarkably successful and challenging event for the aluminum industry.

The introduction of lithium into aluminum creates a highly reactive molten alloy which requires novel approaches in the ingot plant including special refractories for containment, inert gas protection during melting and transfer to avoid excessive hydrogen pickup and oxide generation, superior degassing procedures, special mold lubricants, and extraordinary care in ingot casting because of the reactivity of molten Al-Li alloys with water. The latter problem has caused aluminum companies to explore alternative coolants.

The commercial success of the Al-Li alloys developed during the 1980s will be predicated, in large part, by the ability of the ingot plant to recycle lithium containing scrap, to manufacture a quality product, and to do so at an acceptable price.

References

1. Bower, T.F., Granger, D.A., and Keverian, J. (1971). *Solidification*. ASM, Metals Park, Ohio, pp. 394–396.
2. Ennor, W.T. (1942). U.S. Patent 2, 301, 027.
3. Clyne, T.W. (1984). *Metals Technology*, **11**, 350–357.
4. Emley, E.F. (1976). *Int. Met. Rev.*, **21**, 75–115.
5. Adenis, D., Coats, K.H., and Ragone, D.V. (1962). *J. Inst. Metals*, **91**, 395–403.
6. Peel, D.A., and Pengelly, A.E. (1970). *Mathematical Models in Metallurgical Process Development*. Iron and Steel Institute, London, p. 186.
7. Beattie, D.D. (1977). *Metals Technology*, **4**, 147–152.
8. Weckman, D.C., and Niessen, P. (1982). *Met. Trans.*, **13B**, 593–602.
9. Spear, R.E., and Gardner, G.R. (1963). *Trans. Amer. Foundrysmen's Soc.*, **71**, 209–215.
10. Flemings, M.C. (1974). *Solidification Processing*. McGraw-Hill, New York.
11. Yu, H., and Granger, D.A. (1984). *Proc. of Conf. on Fundamentals of Alloy Solidification Applied to Industrial Processes*. NASA Conf. Publication 2337, pp. 157–168.

12. Hollingsworth, E.G., Frank, G.R., and Willett, R.E. (1962). *Trans. Met. Soc. AIME*, **224**, pp. 188–189.
13. Kosuge, H., and Mizukami, I. (1975). *J. Jap. Int. Light Met.*, **28**, 321–327.
14. Westengen, H. (1982). *Z. Metallkunde*, **73**, 360–368.
15. Talbot, D.E.J. (1975). *Int. Met. Rev.*, **20**, 166–184.
16. Hess, P.D., and Turnbull, G.K. (1974). Int. Conf. on Hydrogen in Metals. ASM, Metals Park, Ohio, pp. 277–287.
17. Szekely, A.G. (1976). *Met. Trans.*, **7B**, 259–270.
18. Miller, R.E., Blayden, L.C., Bruno, M.J., and Brooks, C.E. (1978). *Light Metals*. AIME, Warrendale, PA., pp. 491–505.
19. Emley, E.F. (1976). *Metal Technology*, **3**, 118–127.
20. Brondyke, K.J., and Hess, P.D. (1964). *Trans. AIME*, **230**, 1553–1557.
21. Dore, J.E., and Yarwood, J.C. (1977). *Light Metals*, AIME, New York, pp. 171–189.
22. Granger, D. A. (1985), *Int. Seminar on Refining and Alloying of Liquid Aluminium and Ferro-Alloys*. Trondheim, Norway, pp. 229–241.
23. Yoshikawa, K., Sakamoto, T., and Takada, H. (1982). *Light Metals*. AIME, Warrendale, PA., pp. 817–834.
24. Claxton, R.J. (1973). *Proc. of the Continuous Casting Symposium*. AIME, New York, pp. 341–352.
25. Flemings, M.C., and Metz, S. (1970). *AFS Transactions*, **78**, 453–460.
26. Granger, D.A., and Liu, J. (1983). *Journal of Metals*, **6**, 54–59.
27. Dassell, J.L., and Zinniger, T.C. (1982). *Light Metals*, AIME, Warrendale, PA., pp. 793–801.
28. Bergmann, W.J. (1970). *Met. Trans.*, **1**, 3361–3364.
29. Spear, R.E., and Yu, H. (1984). *Aluminium*, **60**, 440–442.
30. Altenpohl, D. (1968). *Proc. 5th Light Metals Congress*. Leoben, p. 372.
31. Yu, H. (1979). U.S. Patent 4, 166, 495.
32. Granger, D.A., and Jensen, C.L. (1984). *Light Metals*. AIME, Warrendale, PA., pp. 1249–1263.
33. Getselev, Z.N. (1971). *Journal of Metals*, **23**, 38.
34. Mitamura, R., Ito, T., Takahaski, Y., and Hiraoka, T. (1978). *Light Metals*. AIME, Warrendale, PA., pp. 281–291.
35. Powers, R.B. (1975). *Continuous Casting Seminar*. Aluminum Association, Kansas City, Missouri, **111–5**, pp. 1–7.
36. Bessemer, H. (1846). British Patent 11, 317.
37. Morris, L.R. (1976). U.S. Patent 3, 989, 548.
38. Russell, J.B., and Nichols, F.R. (1958). *J. Inst. Metals*, **87**, 209–219.
39. Bonnamour, Y.B. (1975). *Continuous Casting Seminar*. Aluminum Association, Kansas City, Missouri, **11–3**, pp. 1–29.
40. Figueres, R. (1975). *Continuous Casting Seminar*. Aluminum Association, Kansas City, Missouri, **11–2**, pp. 1–21.
41. Hunter, J.L., and Quadt, R.A. (1952). *Mod. Met.*, **8**, 45.
42. Odok, A.M., and Gyongyos, I. (1975). *Continuous Casting Seminar*. Aluminum Association, Kansas City, Missouri, **1–4**, pp. 1–29.
43. Hazelett, R.W., and Swartz, C.E. (1964). AIME Presentation, New York.
44. Stephens, W.E., and Vassily, G. (1971). *Light Metals*. AIME, New York, pp. 535–547.
45. Crouzet, P. (1975). *Continuous Casting Seminar*. Aluminum Association, Kansas City, Missouri, **1–2**, pp. 1–7.
46. Bagshaw, M.J., Hunt, J.D., and Jordan, R.M. (1986). *Proc. of the Third Conf. on Modeling of Casting and Welding Processes*. AIME, Warrendale, PA., pp. 135–147.

5

Thermomechanical Processing of Aluminum Alloys

M.A. ZAIDI

Alcoa Laboratories
Alcoa Center, Pennsylvania

and

J.A. WERT

School of Engineering and Applied Science
University of Virginia
Charlottesville, Virginia

I. Introduction

The basic engineering properties of importance in wrought aluminum alloys are elastic modulus, strength, ductility, fracture-toughness, creep, fatigue, corrosion, and stress corrosion resistance. Historically, most if not

ISBN 0-12-341831-3

all of these properties have been controlled by changes in alloy composition. Though it was realized at the begining of the twentieth century [*1*,*2*] that for any alloy composition, significant property variations could be obtained by altering process parameters such as strain, strain rate, and temperature, the reasons for these variations were not clear. Hence, deformation processing primarily involved utilization of appropriate equipment to obtain the desired shape changes. With the advent of X-ray analysis [*3*] and later electron microscopy [*4*] and microanalysis [*5*] techniques, it was realized that the microstructure and constitutive nature of aluminum alloys changed with the degree and rate of deformation in conjunction with the deformation temperature. Microstructural development during postdeformation thermal treatments could also be related to the deformation-processing conditions. These changes in microstructure were found to affect strongly the properties of an alloy and thus the stage was set to manipulate the microstructure by thermomechanical processing (TMP) in order to alter the properties.

Perhaps the earliest utilization of thermomechanical processing was made by Anderson [*6*] in 1918 to improve the formability of aluminum sheets by controlled cold-rolling and annealing practices. Anderson, however, did not clarify the reasons for improved formability. It is now well known [*7*] that a decrease in grain size improves formability and methods are now being developed to produce superplastic aluminum alloys through development of ultrafine grain sizes by TMP [*8*]. Between 1930 and 1960, research was concentrated strongly on utilizing the growth of fundamental knowledge in developing new alloys and studying the effects of alloying additions on microstructure and properties. It was not until the 1960s that the metallurgists began to observe significant improvements in properties by careful control of combined thermal and mechanical treatments [*9*,*10*]. It was realized that a broader range of microstructures could be obtained by utilizing thermomechanical processing. Since the mid-1960s, TMP has been shown to improve strength, fracture-toughness, fatigue, stress corrosion cracking resistance, and superplastic formability in modern aluminum alloys [*11–19*]. Extensive use of TMP, however, has been hampered due to a lack of clear understanding of the influence of thermomechanical history and the process parameters on the microstructure and properties of aluminum alloys. Considerable efforts are now under way to understand these relationships to a large extent through careful simulation of deformation processes in the laboratories [*20–22*]. For example, it has been clearly shown that different thermomechanical processing histories can be utilized to produce similar microstructures [*23*]. Models are also being developed (and experimentally verified) that predict the evolution of microstructure with variations in process parameters

[*22,24,25*]. Such models also aim at predicting the most efficient processing route to obtain a desired microstructure and hence provide the vital link between research and extensive use of TMP by industry. The next two sections of this chapter describe the microstructural evolution during hot-deformation and postdeformation heat-treatments and its relationship to the process parameters. This is followed by a section on some applications of TMP for improvements in properties of aluminum alloys.

II. Microstructural Development during Thermomechanical Processing

Most aluminum alloys are fabricated by utilizing hot-rolling, extrusion, or forging cast and homogenized ingots into desired shapes and sizes. Some of these products are subsequently annealed and further deformed while other products are given a precipitation-hardening treatment which involves solutionizing at a high temperature, quenching in water, and stretching followed by an age-hardening treatment. Modern aluminum alloys such as advanced powder metallurgy alloys are fabricated by consolidation of powders into billets followed by extrusion, forging, or rolling.

In order to obtain a broad range of microstructures, it is desirable first to understand the mechanism of microstructural development during both deformation and postdeformation treatments. This section deals primarily with microstructural development during thermomechanical processing and the effect of process parameters on the microstructural characteristics and properties.

A. *Microstructural Development during Hot-Deformation*

During hot-deformation, grains change shape, subgrains form, constituent particles fracture, deformation bands form, precipitation may occur at subboundaries, recrystallization may occur, and texture changes take place. All of these factors are affected by changes in process parameters and have an effect on the properties of the material. These are briefly described next.

1. Subgrain Formation

In the early stages of hot-deformation, dislocations are generated that move in response to the difference between the applied and internal stress. Regions of internal stress exist in materials whenever there is an atomic mismatch in the structure (e.g., between the parent lattice and some impurity atmosphere, at clusters of dislocations, adjacent to large particles, and at dispersoids and fine precipitates). Since the velocity of a dislocation is proportional to the effective stress (applied stress minus internal stress) acting on it, dislocations slow down considerably at regions of high internal stress or they may get pinned at precipitates, dispersoids and other boundaries. By interaction of more dislocations with the "slowed down" or pinned dislocations, clusters are formed, thereby distorting the slip plane and creating small misorientations. As the temperature during hot-working is generally high enough for climb, some of the clustered dislocations climb perpendicular to the glide plane while others may cross-slip, eventually forming a three-dimensional network of relatively stable subboundaries. With further deformation, more dislocations are generated, the generation rate depending on the strain rate and the associated effective stress. These dislocations get attracted to the existing subboundaries and some get annihilated by interacting with dislocations of the opposite signs. The rate of annihilation depends on the dislocation density and on the ease of operation of the recovery mechanisms, such as climb, cross-slip, and node unpinning [*9*]. In aluminum alloys, a state is generally reached whereby the dislocation-generation rate is balanced by the dislocation-annihilation rate. Under these steady-state conditions, the alloy can be deformed to large strains without any significant change in the flow stress and average subgrain size. Figure 1 provides a sequence of micrographs showing the development of subgrains in the roll gap during hot-rolling of commercial-purity aluminum. Figure 1a shows the clustering of dislocations at grain boundaries and adjacent to particles in the early stages of deformation which is occuring before the well-annealed metal reaches the roll gap. Figure 1b shows formation of well-defined subgrains adjacent to grain boundaries with increasing deformation. Figure 1c shows the presence of cells and subgrains during the positive strain-hardening phase. Figure 1d shows the steady-state subgrain structure. Note that in the particular example shown in Fig. 1, the average subgrain size attained a (steady-state) constant value at approximately one-third of the projected roll bite length. Also note that the subgrains are essentially equiaxed. This is because the formation of subgrains is not related to the shape change of the work piece. It is related to the motion of dislocations through glide. cross-slip, and climb on slip planes. Even after very large strains (as high as $\epsilon = 40$), the subgrians are known [*26,27*] to remain equiaxed. During hot-deformation, all aluminum alloys develop subgrains whose size depends partly upon the deformation conditions and partly upon certain characteristics of the material (e.g., composition, stacking fault energy, interparticle spacing). For any specific alloy, the subgrain size is determined

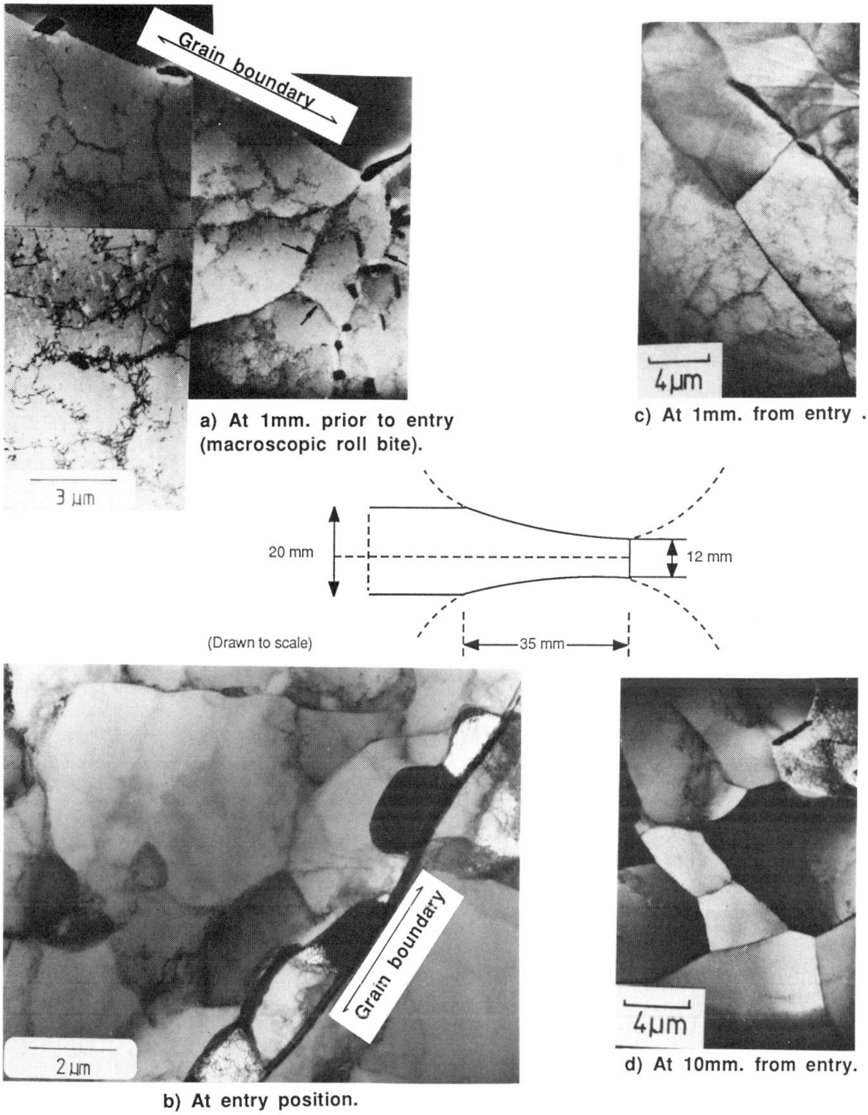

FIG. 1. TEM micrographs showing the development of substructure in the roll gap during hot-rolling of commercial purity aluminum at 500°C (a) At 1 mm prior to entry (macroscopic roll bite). (b) At entry position. (c) At 1 mm from entry. (d) At 10 mm from entry.

by the deformation variables. The subgrain size has a strong influence on the strength of the as-deformed material and on the postdeformation annealing and precipitation-hardening treatments. Some of these aspects will be discussed later.

2. Deformation and Shear Banding

The grains change shape in accordance to the shape change of the workpiece. Most aluminum alloys, during hot-working, develop bands of high misorientation within each grain which are known as deformation bands. Dillamore [28] and others [29,30] have suggested that deformation bands form whenever local differences in stress state activate different combinations of slip systems to produce the same imposed strain in different parts of the crystals. Different slip rotations will then occur and a relative misorientation is produced. The deformation-banding propensity rises with increase in strain and initial grain size. With increasing strain, the number of bands within a grain tends to saturate. Beyond this strain, the misorientations across the bands can substantially increase, thereby generating medium- or high-angle boundaries. Such highly misoriented regions in the deformed structure have a significant effect on the nucleation of recrystallization. This aspect has been studied in detail by Doherty and his coworkers [31,32]. It has also been reported [33] that the orientation of nuclei formed adjacent to deformation-band boundaries is similar to the orientation of the prior deformed region. This particular mechanism has a profound effect on the development of recrystallization textures [34,35] and subsequent formability of sheet metals. At large strians, the deformation bands can extend beyond individual grains and develop into macroscopic shear bands [36–38] running across many grains and even across the entire workpiece. These shear bands are known to deteriorate a material's workability [39] and are an unacceptable feature of the product microstructure.

3. Effect of Large Particles

Constituent particles based on Al,Fe, and Si are present in all aluminum alloys in the as-cast state. These particles are generally large (>10 microns) and brittle. At plastic strains of less than 0.1, most of the larger particles fracture and with further deformation fragment into relatively smaller particles (<5 microns) [40,41]. The fracture of particles generally leads to the presence of voids in the microstructure. Controlled thermomechanical processing that promotes extensive matrix flow can promote void closure as well as prevent further fracture by enhancing relaxation processes at the particle–matrix interfaces. Large, uncracked particles are indeed observed in hot-worked aluminum alloys, indicating that extensive matrix flow must occur to maintain compatibility between the hard, nondeforming particles and the softer matrix. The maintenance of compatibility can lead to high dislocation densities and large lattice misorientations (>10 deg.) adjacent to the particles. The dislocation densities and the misorientations are, however, dependent upon the processing parameters. Generally, in aluminum alloys, deformation at very high temperatures (>0.8 T_m) and low strain rates produces small misorientations and low dislocation densities adjacent to particles due to the rapid dynamic relaxation processes operative at the high temperatures [*42,43*]. The details of the structure adjacent to large particles, generally referred to as deformation zone structure, is discussed in detail by Humphreys [*44*] and others [*45,46*]. The structure in the deformation zones adjacent to large particles has a significant effect on postdeformation changes, especially on nucleation of recrystallization during annealing [*47,48*]. This aspect is described in detail in Section III.

4. Dynamic Precipitation

Precipitation during hot-deformation of aluminum alloys has not been studied in any detail, primarily due to difficulties in experimental design to separate static and dynamic effects. Observations [*49,50*] of precipitate morphologies have, however, indicated that dynamic precipitation does occur in some aluminum alloys. In general, dynamic precipitation has been related to the large point defect densities and the high recovery rates operative during hot-deformation of aluminum alloys. Guyot and Wintenberger [*49*] have proposed that during climb, the dislocations drag solute atoms from the solid solution, which by rapid diffusion along the core nucleate precipitates on the dislocations. More detailed analysis of dynamic precipitation has been published by Dahmen and Westmacott [*50*]. The effect of deformation substructure on precipitation after deformation is well documented and is known to exert considerable influence on the microstructure and properties. These aspects will be discussed later.

5. Dynamic Recrystallization

The occurence of recrystallization during hot-working in copper [*51,52*], steels [*53*], and nickel alloys [*54*] has been well established since the early 1960s, but only since the early 1980s has it been observed [*55–57*] in aluminum alloys. The major difficulty in the study of this phenomenon is that the presence of equiaxed grains in hot-worked material examined after deformation does not prove that recrystallization took place during deformation. Static recrystallizaton can occur fairly rapidly after heavy deformations and at high holding temperatures. Usually, dynamic recrystallization can only be positively identified by the presence of well-defined substructure within fine equiaxed grains is hot-deformed material. Dynamic recrystallization has been reported only in those aluminum alloys in which the recovery rates are low at high temperatures due to certain solute additions and the alloy contains a large amount of heterogeneities such as large particles, dispersoids, and grain boundaries. Under these conditions, the dislocation density builds up at existing subboundaries, giving rise to a high-angle boundary which is mobile at the elevated temperatures. An aluminum alloy that recrystallizes completely during hot-deformation has not yet been reported. It is however conceivable to develop an alloy containing recovery-inhibiting additions, control cast to produce a fine grain size and using appropriate combinations of strain rate and temperature to completely recrystallize and produce an ultrafine grain-size material. Such an alloy can be of substantial practical importance.

6. Texture Development

During deformation, there is a tendency for certain crystallographic planes and directions to orient themselves parallel to the rolling plane and rolling or extrusion direction. Such a tendency is primarily due to constraints on available slip systems for deformation. Thus, in a rolled sheet or extrusion, certain preferred orientations or textures develop. Though the development of textures during cold-rolling has been extensively studied [*58,59*], work on texture development during hot-rolling and extrusion has been very limited [*60–62*]. Hot-rolled aluminum alloys generally exhibit a range of orientations from {110} ⟨112⟩ to {112} ⟨111⟩ along with a large spread in orientations depending upon the deformation variables. Extrusions generally exhibit a strong ⟨111⟩ and a weaker ⟨100⟩ fiber texture. The intensity of the different texture components is strongly dependent on the thermomechanical process parameters. In commercial aluminum-alloy sheets, the hot-deformation texture plays an important role in the formability of final processed sheet [*63,64*].

B. *Effect of Process Parameters on the Hot-Work–Induced Microstructure*

The ability to obtain a broad range of microstructures is fundamental to thermomechanical processing. The most important (and perhaps the most neglected) way to obtain a broad range of microstructures is through the variation of process parameters. The relationship between microstructure and process parameters is discussed next. As was previously indicated, the subgrain size is related to the alloying additions (affecting the dislocation obstacle density) and the deformation variables (affecting the obstacle density and recovery rates). The deformation variables involved in hot-deformation are the imposed strain ϵ, strain rate $\dot{\epsilon}$, and temperature T. Since the measured subgrain size d corresponds to the size established during steady-state deformation, the effect of imposed strain can be ignored. The relationship generally observed [*65*] is of the form

$$d^{-\mathrm{m}} = \mathrm{a} + \mathrm{b} \ln[\dot{\epsilon} \exp(\mathrm{Q}/\mathrm{R}T)] \tag{1}$$

where a, b, m, and R are constants, Q is the activation energy (assumed constant) whose value is generally close to the self-diffusion activation energy, and T is the absolute temperature. The term in the outer parenthesis is known as the Zener-Holloman parameter, Z. Thus, Eq. 1 can be written as:

$$d^{-\mathrm{m}} = \mathrm{a} + \mathrm{b} \ln Z \tag{2}$$

From the preceding equations, it should be obvious that similar subgrain sizes can be obtained by choosing various combinations of deformation-process parameters as long as Z is constant. Alternatively, by keeping a process parameter such as temperature constant, significantly different subgrain sizes can be obtained by varying the strain rate. An example of the preceding two cases is shown for pure aluminum in Figs. 2a–c. Figure 3 shows an example of the relationship between d and Z for an Al-Mg alloy and commercial-purity aluminum. It is generally found [*66*] that good correlation can be obtained for values of m between 0.5 and 1.0. Figure 3 also shows that the subgrains for the same processing parameters (constant Z) are much finer in the Al-Mg alloy. This is due to the higher density of obstacles in the alloy relative to the commercially pure aluminum. Equation 2 is generally valid for most aluminum alloys irrespective of the mode of deformation [*66–69*].

The control of subgrain size by thermomechanical processing is extremely effective in imparting strength, especially for hot-deformed, non–heat-treatable aluminum alloys. The mechanism of substructural strengthening has been discussed by many workers [*70*]. It has generally been found [*69–71*] that a modified Hall–Petch type of relationship is obeyed. For a subgrain size d, the strength at room temperature σ is given by:

$$\sigma = \sigma_0 + kd^{-\mathrm{p}} \tag{3}$$

where σ_0 is the strength of the annealed material, k is the subboundary strength, and p is a constant. Figure 4 shows the variation in room-temperature strength with the parameter Z for two non–heat-treatable

a) T=250 C ; ε= 0.01 s^{-1} ; ε=1.2;
Z=5.0 x10^{15}; d=2.3 um.

b) T=350 C ; ε= 10.0 s^{-1} ; ε=1.2;
Z=7.2 x10^{15}; d=2.2 um.

c) T=350 C ; ε= 0.01 s^{-1} ; ε=1.2;
Z=7.2 x10^{13}; d=6.2 um.

FIG. 2. TEM micrographs showing the effect of process parameters on substructure development. Note that similar or different subgrain sizes (d) can be produced by widely different combinations of deformation-process parameters.
(a) T = 250°C; $\dot{\epsilon} = 0.01\ s^{-1}$; $\epsilon = 1.2$; $Z = 5.0 \times 10^{15}$; $d = 2.3\ \mu m$.
(b) T = 350°C; $\dot{\epsilon} = 10.0\ s^{-1}$; $\epsilon = 1.2$; $Z = 7.2 \times 10^{15}$; $d = 2.2\ \mu m$.
(c) T = 350°C; $\dot{\epsilon} = 0.01\ s^{-1}$; $\epsilon = 1.2$; $Z = 7.2 \times 10^{13}$; $d = 6.2\ \mu m$.

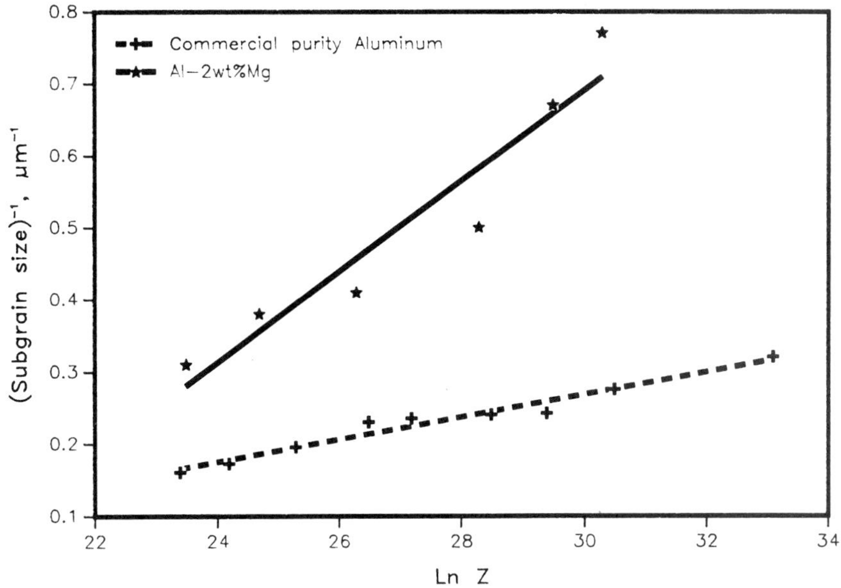

FIG. 3. Effect of hot-deformation-process parameters (defined in Z) on subgrain size.

aluminum alloys. Since the subgrain size is dependent upon the process parameters defined by Z (see Figure 3), it should is clear (from Fig. 4) that strength increments of over 100% are possible by producing a fine, stable subgrain-size material by thermomechanical processing.

Since the stress needed to deform the material depends basically on the instantaneous structure, strain rate, and temperature of deformation, the elevated-temperature flow stress is generally [72] related to the parameter Z as

$$Z = \dot{\epsilon} \exp(Q/\mathrm{R}T) \; \mathrm{A} \sinh(\alpha\sigma)^{\mathrm{n}} \tag{4}$$

where A, α, n, and R are constants. Since the subgrain size also depends upon Z, a relationship between elevated-temperature flow stress and subgrain size is generally observed in a large number of aluminum alloys. Such a relationship is shown in Fig. 5. Prediction of flow stress and its relationship to structure is extremely important for thermomechanical-process optimization of both the process and properties. Note that under practical conditions, the optimum is to fabricate the product having the desired properties at high productivity rates by expending the least amount of energy (which is related to the flow stress).

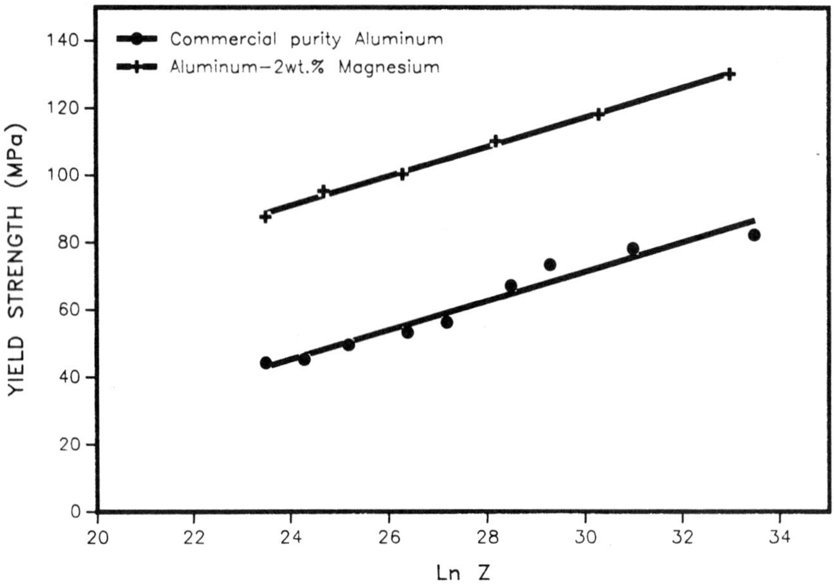

FIG. 4. Effect of process parameters (defined in Z) on the room-temperature strength of two hot-deformed aluminium alloys.

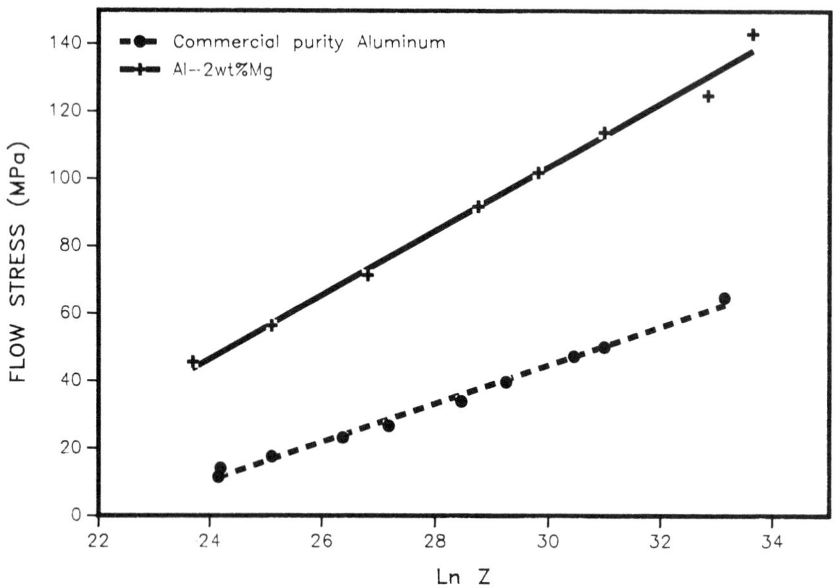

FIG. 5. Effect of hot-deformation–process parameters (defined in Z) on the elevated-temperature flow stress of two aluminum alloys.

The effect of processing parameters on deformation and shear banding has not been studied in any detail. Semiatin et al. [*73*] have shown that shear banding in torsion testing of 6061 Al alloy is promoted at high strain rates and temperatures between 316 and 427°C. At low strain rates and high temperatures (>427°C), no shear banding was observed. The authors attributed the occurrence of shear bands to thermal gradients set up at the high strain rates. In general, it is known that inhomogeneous deformation promotes banding. The degree of banding also rises with increase in strain and initial grain size. Thus, by utilizing thermomechanical processing that promotes recrystallization to a fine grain size between successive deformations, it may be possible to suppress shear banding.

The effect of processing parameters on dynamic precipitation is not at all clear. It is most likely that dynamic precipitation occurs if the processing path is such that the temperature falls during deformation from a single-phase region to a two-phase region (e.g., during hot continuous rolling of some Al alloys).

Perhaps, a future application of thermomechanical processing that can have a profound commercial impact relates to controlling damage and recrystallization adjacent to coarse second-phase particles. This would involve a systematic study of interactions between process parameters and their effect on stress-relaxation processes at particle–matrix interfaces. In general, hot-working–process parameters that promote high dislocation densities and misorientations adjacent to particles enhance the probability of particle-stimulated nucleation (PSN) during annealing. Alternatively, process parameters that promote extensive relaxation at particle–matrix interfaces can suppress PSN (and possibly void nucleation at particles). Examples of these two cases are shown in Figs. 6a and 6b. Note that deformation in the two cases was carried out at the same temperature but at different strain rates. However, only the specimen deformed at high Z exhibits PSN at the same annealing conditions. Thus, thermomechanical processing can be utilized to either promote or suppress particle-stimulated nucleation of recrystallization. It should be emphasized that since most commercial aluminum alloys contain coarse particles, PSN is generally difficult to avoid during annealing (or just after hot-deformation) unless appropriate paramaters are utilized during deformation processing. Control of PSN is important if unrecrystallized structures are desired. Furthermore, the orientations of nuclei formed adjacent to coarse particles are undesirable for subsequent cold-forming of some Al alloys.

It is known that in those aluminum alloys where substantial dynamic recrystallization is observed, the grain size has been reported to depend upon the extrusion temperature with the grain size rising with increasing deformation temperature [*74*]. Such a relationship has been ascribed to a

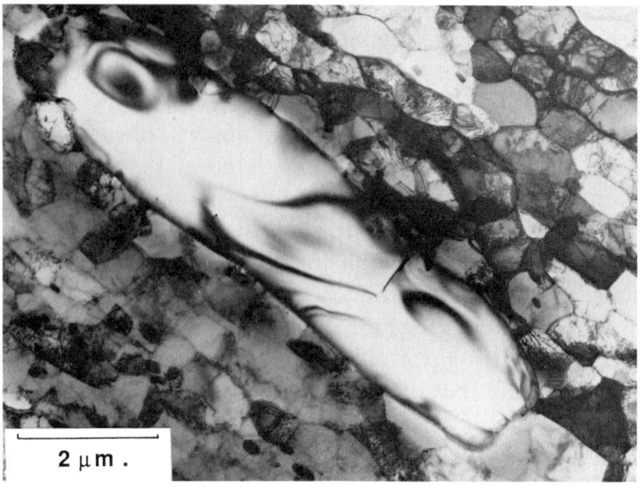

FIG. 6a. TEM micrograph of a partially annealed specimen showing the structure adjacent to a large particle. (Alloy Al-1Mg-Mn hot-deformed at Z = 3.5×10^{18}; annealed at ~315°C for 30 mins.)

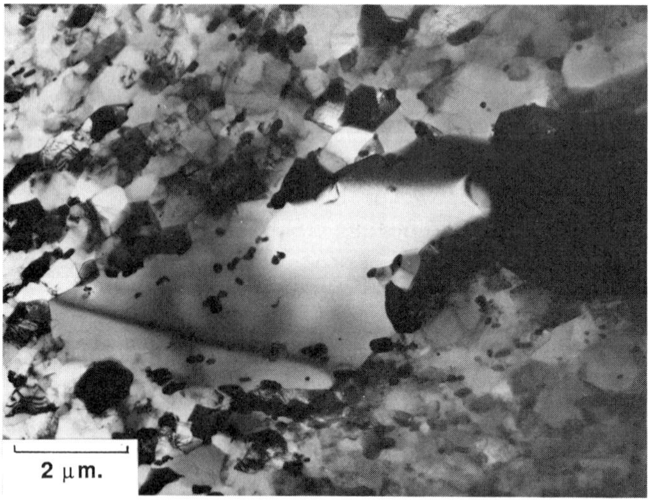

FIG. 6b. TEM micrograph of a partially annealed specimen exhibiting nucleation of recrystallization adjacent to a large particle. (Alloy Al-1Mg-1Mn hot-deformed at Z = 8.7×10^{19}; annealed at ~315°C for 30 mins.)

lower nucleation frequency and a higher growth rate at the high temperatures. The combined effects of strain rate and temperature on the dynamically recrystallized grain size of aluminum alloys has not been studied in any detail [*75*]. However, in other alloys, a number of workers [*53,76,77*] have shown that the occurrence of dynamic recrystallization, the flow behavior, and the recrystallized grain size are strongly dependent upon the initial grain size and the parameter Z.

The effect of process parameters on texture development during hot-deformation is another area that has not been investigated in much detail. Work on a commercial Al-Mg-Mn alloy and an Al-Mg alloy has shown [*61,64*] that though the major texture components are similar, the intensities vary with process conditions. A decrease in the rolling temperature, for instance, has been shown to increase the sharpness of certain orientations [*64*]. Sheppard et al. have, however, shown in an Al-Mg-Mn alloy that the rolling-texture intensities actually rise with an increase in rolling temperature [*61*]. Panchanadeswaran [*62*] has shown that in extruded pure aluminum, the intensity of the $\langle 010 \rangle$ fiber increases with decrease in extrusion temperature. At all temperatures, however, the $\langle 111 \rangle$ and $\langle 010 \rangle$ components dominate the texture.

C. Postdeformation Microstructural Changes

The microstructural changes prior to and during thermal treatments after deformation can be significantly influenced by the thermomechanical history of the material. Subgrain growth, recrystallization, and precipitation are known to occur after or between deformations prior to any static thermal treatments. All these effects are extremely important since they affect the final properties of the material. Zaidi [*79*] has reported significant increases in subgrain size with interrupted quenching time after hot rolling, (depending upon the process parameters), which leads to reduced room-temperature strength in non–heat-treatable aluminum alloys. Chia and Starke [*80*] have shown that delayed cooling of an Al-Fe rod upon exit from the hot mill causes localized recrystallization and subgrain growth. The resulting inhomogeneous microstructure has a deleterious effect on the physical and mechanical properties of the rod and the finished wire. Tutcher [*81*] has reported complete surface recrystallization in extruded Al-Mg alloys. He has shown that the depth of the recrystallized layer increases linearly with decrease in the logarithm of the parameter Z where $Z = \dot{\epsilon} \exp(Q/RT)$. Thus, to control such undesirable features or to improve product homogeneity, it is essential to understand the effect of process parameters on the postdeformation changes. Such an understanding can also be very useful in choosing the appropriate process parameters to promote complete recrystallization in certain products without using a static anneal.

The main thermal treatments given after hot-deformation are annealing and precipitation-hardening treatments. Annealing is usually carried out to recrystallize the alloy in order to control the grain size and texture, and/or to soften the material for further processing. Precipitation-hardening treatments are given to heat-treatable alloys primarily to increase the strength and obtain good combinations of other mechanical properties. The mechanisms of recrystallization and precipitation-hardening have been extensively studied [*82–84*] and will not be discussed here. The rest of this section will deal briefly with the effect of deformation-process parameters (and hence deformation structure) on the recrystallized and precipitation-hardened structure.

Work on steels by Sellars et al. [*85*] has shown that the kinetics of recrystallization during annealing after hot-deformation are strongly dependent upon the process parameters (defined in parameter Z), strain ϵ, and the initial grain size D_0. Based on their data, they proposed the following relationship:

$$t_{0.5} = \mathrm{A}\, D_0{}^{\mathrm{b}} \epsilon^{\mathrm{c}} \mathrm{Z}^{\mathrm{d}} \exp(Q_{\mathrm{rec}}/\mathrm{R}T_{\mathrm{rec}}) \qquad (5)$$

where A, b, c, d, and R are constants Q_{rec} is the activation energy for recrystallization, T_{rec} is the annealing temperature, and $t_{0.5}$ is the time for 50% recrystallization. Work on aluminum alloys, though extensive, has concentrated primarily on the effect of cold-work on recrystallization. The majority of industrial applications, however, utilize annealing (to soften) or solution heat-treating (to dissolve precipitates) after hot-deformation in order to further process the material. For non–heat-treatable Al alloys, it would be very useful to know the minimum annealing temperature, as a function of thermomechanical history, which would completely recrystallize the alloy. For heat-treatable alloys, it would be useful to know the thermomechanical process parameters that, if utilized during deformation processing, would prevent recrystallization during the solution heat-treating practice. In general, very little work on aluminum alloys has been done to predict relationships between process parameters and recrystallization aspects. However, Raghunathan et al. [*86*] have shown their data to fit Eq. (5) and have demonstrated very pronounced effects of process parameters on both recrystallization kinetics and recrystallized grain size in two Al-Mg alloys. They have shown that at constant strain and initial grain size, the recrystallization kinetics are strongly dependent upon the temperature-compensated strain rate (i.e., parameter Z). Furthermore, a nonlinear relationship exists between the rate of recrystallization and the process parameters described by the parameter Z. At low values of Z, the recrystallization kinetics are extremely sluggish. The relationships presented indicate quantitative means of suppressing or accelerating recrystallization during annealing by varying the thermomechanical process parameters. They also provide means of optimizing batch-annealing time by taking into account the variations in the hot-deformation–process parameters. It is known [*88*] that the degree of recrystallization induced during solution-treatment of heat-treatable (age-hardenable) alloys can have a profound effect on the strength and toughness of the aged alloys. Since the degree of recrystallization depends upon the thermomechanically produced microstructure (subgrain size, deformation bands, structure adjacent to large particles, etc.), variations in process parameters should have a significant effect on the final properties of the heat-treated material. Raghunathan et al. have also shown that the recrystallized grain size is a strong function of the alloy and the parameter Z. In separate work [*79*], it has also been shown that variations in subgrain size (produced by varying process parameters) have a strong effect on the recrystallized grain size of the annealed material. This is shown in Fig. 7 for an Al-2 Mg alloy. Since grain-size

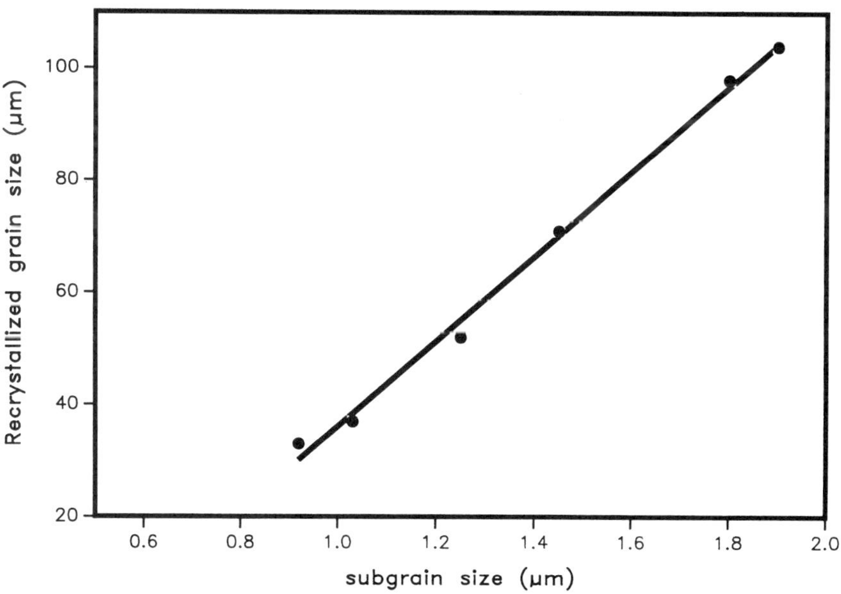

FIG. 7. Effect of hot-work induced subgrain size on the recrystallized grain size of an annealed Al-2Mg alloy.

control is essential for many applications, optimization of process parameters can have a significant impact on the recrystallization behavior and the final grain size.

It should also be pointed out that the preheating homogenization treatment prior to hot-deformation can significantly affect the material's recrystallization behavior during annealing after hot-deformation. This is because the homogenization treatment can be used to alter the size and distribution of dispersoids, which influences both the nucleation and growth of recrystallization. Utilization of high preheating temperatures followed by very slow furnace cooling can promote recrystallization at lower annealing temperatures in similarly processed material [*78*]. This is shown in Fig. 8 for two commercial Al-Mg-Mn alloys [*78*]. Such results indicate that the homogenization treatment is an important process variable which may be used in the thermomechanical process either to inhibit or enhance recrystallization.

The effect of hot-deformation–process parameters on recrystallization textures has not been reported in any detail. It is, however, known that at constant strain, variations in the deformation temperature and strain rate can significantly affect the recrystallization texture upon annealing [*60*]. It is generally found that as the hot-deformation temperature increases, the

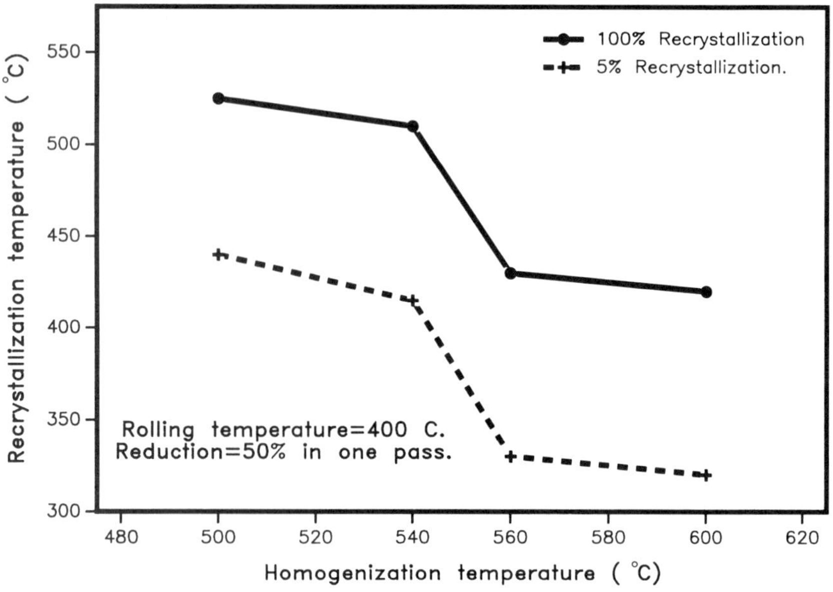

FIG. 8. Effect of ingot homogenization temperature on recrystallization temperature of a hot-rolled Al-2.7 Mg-0.8 Mn alloy (AA5454).

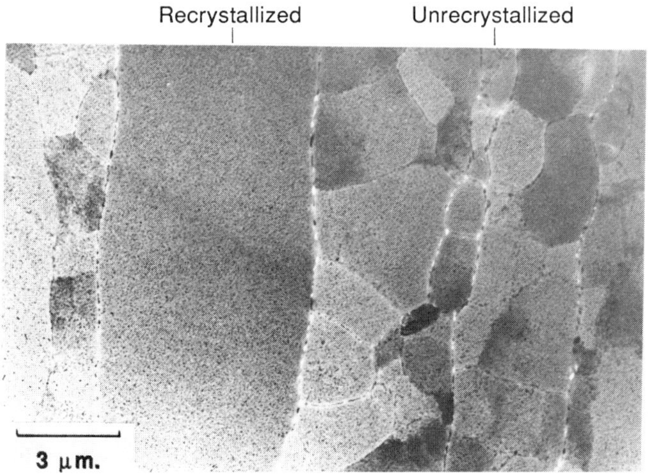

FIG. 9. TEM micrograph showing the distribution of precipitates in a partially recrystallized overaged Al-6.5Zn-2.3Mg-2Cu alloy (AA7150).

recrystallization-texture components weaken. Variations in thermomechanical processing are extensively utilized by the aluminum industry to balance deformation and recrystallization textures in order to reduce strength anisotropy. The reduction in anisotropy is extremely important in subsequent forming operations.

Though detailed studies to evaluate the effect of hot-working–process parameters on precipitation during aging have not been carried out, it is well known [*87*] that subboundaries are active nucleation sites. Figure 9 shows the microstructure of an averaged (T7 temper) 7150 Al alloy that was partially recrystallized during solution-treatment. Note that the precipitate size and distribution are more uniform in the recrystallized region. In the unrecrystallized region, the subboundaries exhibit relatively coarse precipitates (coarse compared to the matrix precipitates) and adjacent precipitate-free zones. Thus, any thermomechanical process that leads to reduction in subgrain size is likely to promote more inhomogeneous distribution of precipitates. It is, however, known [*88*] that the toughness of unrecrystallized 7xxx alloys is higher than that of the recrystallized material with no loss in strength. It is likely that the subboundary precipitation and the associated precipitate-free zones promote crack branching and hence lead to increased toughness. Thus, thermomechanical processing can be utilized to enhance properties of unrecrystallized heat-treatable aluminum alloys by developing fine subgrains during hot-deformation. It can also be utilized to suppress recrystallization during solution-treatment.

The previous sections have clearly indicated that wide variations in microstructures can be achieved by thermomechanical processing of a material for any desired shape change. Such variations in microstructures can have a profound influence on the final properties of the material. The thermomechanical processing aspects just described, however, relate to process modifications in the conventional bulk deformation processes (e.g., rolling, extrusion, and forging). It has also been found that certain thermomechanical treatments given to semifinished products can produce rather special microstructures. Such treatments are generally specific to certain systems, but nevertheless have a profound effect on the material's properties and performance. The next three sections describe the basis for such thermomechanical treatments and provide examples of applications to commercial aluminum alloys.

III. TMP for Grain-Size Control in Aluminum Alloys

All TMPs for grain-size control in aluminum alloys rely on particles to influence recrystallization in such a way that the desired grain size is produced. The effects of particles on recrystallization are so significant that many basic studies have been performed to clarify how particle size and spacing influence recrystallization. These studies have shown that particles can have two different (in many respects, opposite) effects on recrystallization.

A. *The Influence of Large Particles on Recrystallization*

One effect is due to nucleation of recrystallized grains in the neighborhood of large particles that were present before deformation of the alloy, often termed "particle-stimulated nucleation" [*47,89,90*]. This occurs because strain is concentrated near particles during deformation, creating an envelope of strongly deformed material around each particle. The envelope of highly strained material, called the deformation zone of the particle, is a favorable site for nucleation of recrystallized grains. Figure 10a shows a deformation zone formed around an M-phase particle in 7075 Al alloy. Detailed investigations of the dislocation cell structure in such deformation zones have shown that the dislocation cell boundaries have much higher misorientations in the deformation zone than are commonly found away from large particles. During the early stages of recrystallization, these cell walls transform into high-angle recrystallization fronts

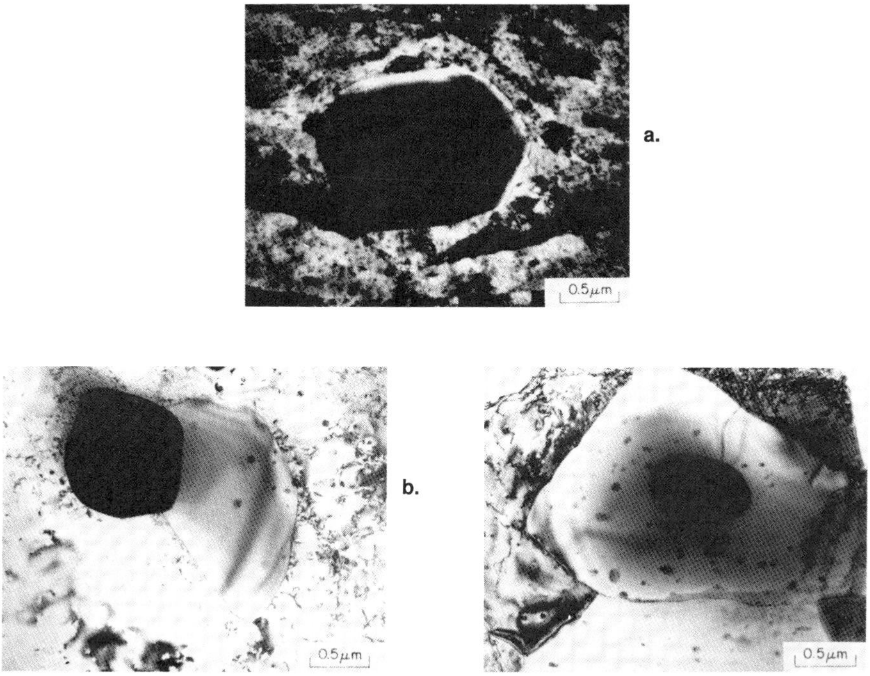

FIG. 10. TEM micrographs showing a 7075 Al alloy microstructure at several TMP stages.
(a) Deformation zone around an M-phase particle.
(b) 80-sec. recrystallization anneal.
(c) 100-sec. recrystallization anneal.

which sweep through the matrix, driven by the stored energy of deformation. Figures 10b and 10c show newly formed recrystallized grains consuming deformation zones and growing away from the particles where they nucleated.

Studies have shown that the degree of misorientation between adjacent dislocation cells in a deformation zone increases with particle diameter [*47*]. This result suggests that only particles larger than a certain minimum diameter would form deformation zones with sufficiently large dislocation cell misorientations to favor nucleation of a recrystallized grain. Experimental observations confirm existence of a minimum particle diameter for particle-stimulated nucleation [*47,90,91*]. The critical particle diameter, η_c, depends on the type and severity of deformation, but it is often near 1 μm for the heavy rolling reductions normally used for TMPs designed to control grain size.

In the presence of a dispersion of particles with diameters larger than η_c, recrystallization is accelerated because additional nucleation sites for recrystallization are available [*90*]. Moreover, an alloy containing particles larger than η_c recrystallizes to a finer grain size than an alloy without any particles larger than η_c. However, these effects are only obtained if particles are present before deformation. Particles larger than η_c that are precipitated after deformation are not surrounded by deformation zones and do not create preferential sites for nucleation of recrystallized grains.

B. *The Influence of Small Particles on Recrystallization*

The second effect that particles can have on recrystallization is due to retardation of grain-boundary migration by particle dispersions. Recall that a recrystallization front sweeping through a defect structure left by deformation is simply a grain boundary between the deformed matrix and a recrystallized grain. The drag pressure (P_Z) that a particle dispersion exerts on a boundary is given by [*92,93*]

$$P_Z = \frac{3}{2}\beta\gamma(f/r) \qquad (6)$$

where γ is the specific grain-boundary energy ($\sim$0.3 J/m^2 for Al), f is the volume fraction of particles, r is the average particle radius, and β is a constant having a value near 1/2.

Boundary drag pressure due to a dispersion of small particles makes nucleation of recrystallizing grains more difficult, thereby retarding recrystallization and producing larger recrystallized grain size than is obtained in an alloy without small particles [*82*]. Particles need not be present prior to deformation to create a drag pressure during recrystallization. Many examples are known where recrystallization is retarded by particles that precipitate concurrently with recovery and recrystallization [*94,95*].

In addition to impeding nucleation of recrystallizing grains, the boundary drag pressure from small particles also prevents grain growth after recrystallization [*92,93,96*]. The effect of particle dispersions on grain growth can be understood by considering that grain growth is driven by the energy stored in grain boundaries. Migration of grain boundaries in a direction that causes grain-boundary area to be reduced lowers the total energy of the metal. However, as grain growth proceeds, the driving pressure for continued grain growth is reduced because fewer grain boundaries are present. If particle dispersion is present during grain growth, the driving pressure for continued grain growth may fall to the same level as the drag pressure from the particle dispersion. When this situation occcurs, grain growth ceases because the net driving pressure has fallen to zero.

Zener [*92*] calculated the grain diameter at which the driving pressure and drag pressure are equal (D_Z) and found

$$D_Z = \frac{4}{3}\left(\frac{r}{f}\right). \tag{7}$$

As an example of the limiting grain diameter, consider the case of 7475 Al, an alumimum alloy sometimes used in superplastic-forming applications. The $Al_{18}Mg_3Cr_2$ particles in 7475 Al provide the only significant contribution to drag pressure in this alloy. We can estimate that $f = 0.005$ and $r = 0.05\ \mu m$ for these particles [*97*]. The limiting grain diameter calculated from Eq. 7 is 13 μm, which is comparable to experimental observations of the limiting grain size for this alloy.

C. *Recrystallization in Alloys with Large and Small Particles*

The results just described were obtained using model alloys containing, as far as possible, particles of one type and one size. Of course, this is not the case for commercial alloys, which generally contain several types of particles—some with diameters larger than η_c, others with diameters smaller than η_c. While the recrystallization behavior of commercial alloys can be complex, the following ideas have emerged from research on alloys containing particles of several types [*98–102*]. The concept of a deformation zone around a particle remains valid, and only particles larger than a critical diameter provide sufficiently intense deformation zones for nucleation of recrystallizing grains. However, the critical particle diameter required for nucleation of a recrystallizing grain is larger when both large and small particles are present. This occurs because the drag pressure from small particles lowers the net driving pressure for nucleation of recrystallizing grains. Large particles that would provide marginal nucleation sites in the absence of small particles become ineffective when small particles are present. Thus, one effect of small particles is to block some of the potential nucleation sites for recrystallizing grains, promoting a coarser recrystallized grain size. This effect has been experimentally confirmed by Nes [*99*] and by Chan and Humphreys [*102*].

However, recall that small particles also help retain fine grain size by preventing grain growth. Thus, we see that dispersions of small particles have two counterposed effects: they prevent nucleation of some grains, but they retain a fine grain size after recrystallization is complete. Theoretical considerations suggest that an optimum level of drag pressure from small particles may be found where only a few nucleation sites are blocked but grain growth is completely suppressed.

D. Application to Design of Thermomechanical Processes

Now let us consider how these basic concepts of particle effects on recrystallization can be used to design TMPs for grain-size control. The principles previously described can be used to design TMPs for either heat-treatable alloys or non–heat-treatable alloys that contain insoluble particles. We know that particles of two sizes should be present in the alloy: particles larger than η_c to stimulate nucleation of recrystallizing grains and some particles small enough to provide a significant boundary drag pressure. The spacing of particles larger than η_c should be smaller than the desired grain size since one particle-stimulated nucleation event is needed for each grain in the final microstructure. The dispersion of small particles must provide enough drag pressure to suppress grain growth at the anticipated grain size, but higher drag pressures will be detrimental to recrystallization.

The processing stages in a generalized TMP for grain-size control are simply:

1. Heat-treatment to produce the required particle dispersions
2. Deformation below the recrystallization temperature
3. Recrystallization anneal

Although TMPs designed for grain-size control in commercial aluminum alloys may consist of many individual steps involving complex sequences of isothermal annealing and controlled cooling or heating treatments, consideration of the function of each step will normally allow a TMP to be separated into the three principle stages just listed. An example of how TMP for grain-size control may be designed is described in Section V.

IV. TMP for Precipitate-Size and Distribution Control

TMPs designed to control grain size and texture have been described in previous sections of this chapter. These microstructural characteristics primarily influence fabrication of components from aluminum alloys. In this section, we shall briefly consider TMPs designed to control precipitate-dispersion characteristics in heat-treatable alloys, a microstructural factor that dominates service properties such as strength and fracture resistance.

Aging of heat-treatable aluminum alloys can produce different types of precipitate particles [*83*,*84*]. The first metastable precipitates to form during aging at low temperatures are coherent with the face centered cubic (fcc) matrix. Prolonged aging at a low temperature or aging at a higher temperature causes the metastable coherent precipitates to dissolve and the same solute to be precipitated as metastable semicoherent precipitates. The semicoherent precipitates have dislocations at some interfaces where the precipitate and matrix lattices do not match.

Since the coherent precipitates are smaller and closer together than the semicoherent precipitates, they generally produce higher strength levels than the semicoherent precipitates. However, they often promote localized slip which can lead to poor ductility [*103*] and possibly to stress-corrosion-cracking (SCC) susceptibility [*104*]. In many applications, some strength must be sacrificed to gain ductility and a degree of SCC resistance. Thus, precipitation-hardening aluminum alloys are often deliberately overaged to produce the semicoherent precipitates that provide attractive combinations of properties. In these cases, however, it is desirable to obtain the finest possible dispersion of semicoherent precipitates so that the strength decrease is not too large. This is accomplished by TMPs integrated into the aging steps, sometimes called final thermomechanical treatments or FTMTs [*105*,*106*].

These TMPs operate by using dislocations to influence precipitation. In the case of coherent precipitates, the lattice of the precipitate matches that of the matrix with perhaps a small misfit strain. There is little reason for coherent precipitates to form in the strain field of a dislocation because the precipitate and matrix lattices match. However, semicoherent precipitates nucleate much more easily on dislocations because some of the faces of semicoherent precipitates must contain dislocations.

It is apparent from the preceding discussion that precipitation of uniform distributions of semicoherent precipitates can be encouraged by introducing a uniform distribution of dislocations into the alloy prior to aging. These extra dislocations provide nucleation sites for semicoherent precipitates, yielding a more uniform distribution of these precipitates than would be achieved without the added dislocations. The supplementary dislocations are introduced by plastically straining the solution-treated alloy between 1.5 and 8% before aging. Rolling cannot be used to introduce the necessary dislocations because such small reductions do not provide a uniform distribution of dislocations through the thickness of the material. Instead, tensile deformation is used to introduce the extra dislocations—an operations called stretching. Note that stretching performed prior to solution-treatment does not have the desired effect because the dislocations introduced by stretching are removed by recovery at the solution-treatment temperature.

In addition to providing a uniform distribution of nucleation sites for semicoherent precipitates, plastic deformation prior to aging may serve other purposes. For example, quenching from the solution-treatment temperature usually introduces residual stresses into aluminum alloys. The residual stresses can be very detrimental for fatigue properties and corrosion resistance. Stretching after solution-treatment can relieve or redistribute the residual stresses and eliminate many of the associated problems. In addition, stretching may be used to straighten wrought products that have become warped during rolling or extrusion operations. Slight differences in friction between the workpiece and the two rolls or the sides of the extrusion die can cause pronounced warping of the workpiece. Stretching is often used to correct this situation.

V. Applications

A. *TMP for Imparting Superplastic Properties to 7475 Aluminum*

Superplasticity is characterized by the large elongations that are obtained when a material with fine grain size is slowly deformed at an elevated temperature [*107,108*]. The large, uniform elongations—which may exceed 1000%—are useful for fabricating sheet material into complex shapes. A light-weight superplastically formed component can replace ordinary components that are built up of several individual subcomponents assembled by many fasteners. (see Chapter 7)

Superplastic properties are obtained when deformation mechanisms such as grain-boundary sliding and diffusional creep make large contributions to deformation [*107,108*]. Microstructural requirements for superplasticity are a fine grain size that does not grow rapidly at the superplastic-deformation temperature and presence of predominantly high-angle boundaries. Since the mechanisms that confer superplastic properties require high-angle boundaries, a polygonized subgrain structure is not suitable for superplasticity because the subgrain boundaries are mostly low-angle boundaries.

Development of TMPs to produce the required grain structures has been one of the critical advances required for production of superplastically deformable structural aluminum alloys. Table I shows the strong effect of grain size on superplastic elongation in 7475 Al [*109*]. Owing in part to the potential benefits of superplasticity formed components, substantial research has been performed to develop and optimize TMPs capable of producing grain sizes near 10 μm in high-strength aluminum alloys. This

TABLE I

EFFECT OF GRAIN SIZE ON SUPERPLASTIC ELONGATION OF 7475 Al WITH STRAIN RATE = $2 \times 10^{-4}\ s^{-1}$ AND DEFORMATION TEMPERATURE = 516°C

Grain size[a] (μm)	Elongation (%)
13	1400
16	880
22	730
69	150
156	80

Source: After Hamilton et al. [*109*].
[a] Linear intercept grain size in longitudinal direction.

section will illustrate application of the principles outlined in Section III to the problem of designing a TMP to produce superplastically formable 7475 Al.

Recall from Section III that production of fine grain sizes by particle-stimulated nucleation of recrystallized grains requires an alloy with two types of particles: particles larger than η_c (typically 1 μm) to stimulate nucleation and small particles to restrict grain growth after recrystallization. First, consider whether 7475 Al meets these requirements.

1. Large Particles

Commercial alloys such as 7475 contain insoluble constituent particles formed by the impurities Fe and Si. Although most constituent particles have diameters larger than 1 μm, their average spacing is between 20 and 40 μm (depending on alloy purity), which is too large to yield the desired grain size. In addition, constituent particles are inhomogeneously distributed, which would lead to an inhomogeneous grain-size distribution.

Additional particles larger than η_c are needed to attain the desired grain size. The additional particles may be formed by precipitating the soluble alloying elements (Zn, Mg, and Cu in the case of 7475 Al) as particles larger than η_c. Since η_c is 50 times larger than the size of precipitate particles normally used to provide room-temperature strengthening, prolonged aging at elevated temperatures must be performed to provide sufficiently coarse particles to stimulate recrystallization. Experiments have shown that precipitate particles in most age-hardening aluminum alloys can be coarsened to provide a dispersion of 1-μm diameter particles with an average spacing of 5 to 10 μm.

2. Small Particles

Small additions of Cr are made to 7475 Al, leading to formation of small insoluble particles known as dispersoid particles. As illustrated in Section III, the dispersoid particles in 7475 Al give a Zener diameter of about 13 μm, a good compromise between the counterposed effects of allowing easy nucleation of recrystallizing grains and restricting grain growth.

From the considerations just described, it is clear that 7475 is quite well suited to TMPs for grain-size control. Now consider the stages required for the grain-size–control TMP. As discussed Section IIID, the first stage is formation of the necessary precipitate dispersions. In heat-treatable alloys, constituent particles and dispersoid particles are formed during solidification and homogenization treatments. These insoluble particles cannot be substantially altered by heat-treatment. However, the soluble alloying elements must be precipitated in the form of particles approximately 1 μm in diameter. Figure 11 shows a typical TMP for grain-size control in 7475 Al [8]. The first two steps in this process, solution-treatment and aging at an elevated temperature, are intended to produce the coarse precipitate particles.

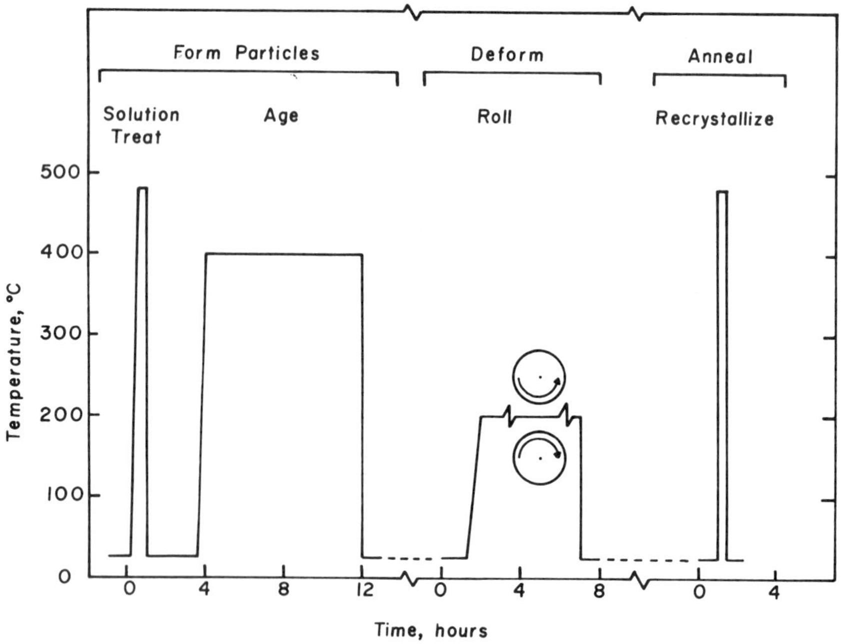

Fig. 11. Schematic diagram of TMP for grain-size control in 7475 Al alloy.

In the second stage, corresponding to the third step of the TMP illustrated in Fig. 11, the alloy is deformed to provide a driving pressure for recrystallization. Normally, 7475 Al is deformed by rolling to a total reduction of about 90% at temperatures near 200°C. Lower rolling temperatures usually lead to extensive cracking of the workpiece during rolling, while higher temperatures allow recovery processes to reduce the stored energy of deformation. Finally, the third TMP stage is an annealing step to allow recrystallization to a fine grain size. A rapid heating rate is needed to assure that many of the potential nucleation sites for recrystallization actually produce a recrystallized grain.

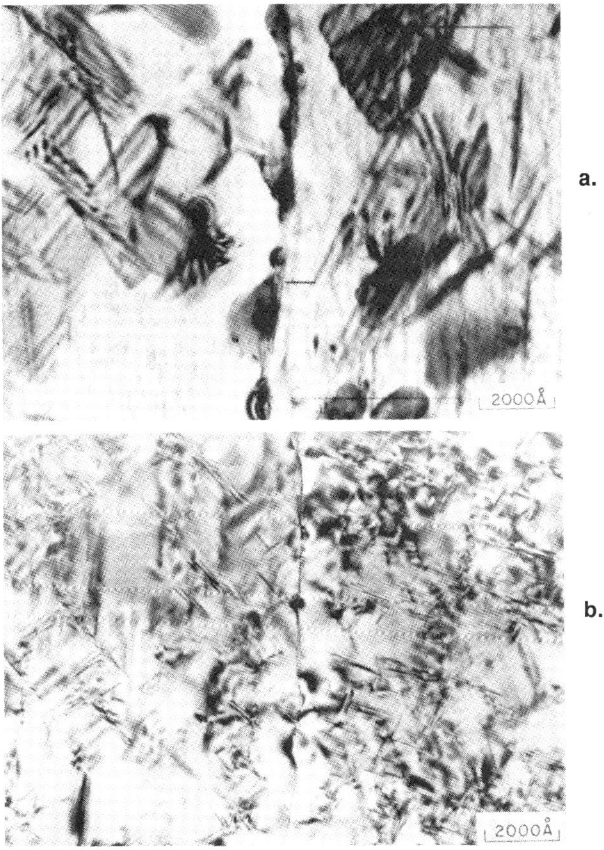

FIG. 12. TEM micrographs showing the effect of stretching on precipitate morphology after aging.
(a) Unstretched.
(b) Stretched 2% prior to aging. (Courtesy R.E. Crooks and E.A. Starke, Jr. [*110*])

B. TMP for Precipitate-Size and Distribution Control

Recall that the goal of this type of TMP is to introduce dislocations that can promote a uniform distribution of heterogeneously nucleated semicoherent precipitate particles. This example illustrates how such a TMP operates in an Al-Cu-Li-Mg-Zr alloy [*110*]. A simple precipitation-hardening treatment for this alloy consists of solution-treatment at 538°C followed by aging at 190°C. The semicoherent precipitates formed during the aging treatment are S′ phase (a metastable form of the equilibrium Al_2CuMg phase) and T_1 phase (Al_2CuLi). Figure 12a shows the microstructure produced by solution-treatment and aging. The S′ and T_1 particles precipitate on any available dislocations, but the scarcity and inhomogeneous distribution of dislocations in the solution-treated microstructure yeild an inhomogeneous distribution of relatively large precipitates.

Figure 12b shows the effects of a 2% stretch carried out after solution-treatment but before aging. As described earlier, stretching introduces a moderate number of homogeneously distributed dislocations which then act as nucleation sites for the semicoherent precipitates. The change in size and uniformity of S′ and T_1 precipitates is quite remarkable in the case and leads to an improvement in strength from 475 to 555 MPa. For this reason, stretching is a standard practice for most of the Al-Li family of alloys under development for aerospace applications.

VI. Summary

The chapter attempts to establish that controlled thermomechanical processing influences most components of the microstructure of aluminum alloys. It has been shown that a broad range of microstructures (and properties) can be tailored by varying process parameters. Furthermore, wide variations in process parameters can be employed to obtain the same desired microstructure, thereby providing processing flexibility. The principles described are generic and should be applicable to commercially established aluminum alloys as well as new materials such as PM alloys, metal-matrix composites (e.g., Al-SiC), and low-density–high-modulus and high-strength Al-Li–based alloys. Extensive use of TMP, however, has been hampered due to a lack of clear understanding of the influence of thermomechanical history and the process parameters on microstructure and properties of aluminum alloys. Substantial work is needed to understand these relationships by careful experimental simulation of industrial processes and development of quantitative, predictive models. Verification of such models applied to industrial-scale processing should provide the important link between research and applications.

Acknowledgments

The authors express their appreciation to Dr. R.E. Sanders and Professors T. Sheppard and E. Nes for discussions on many of the topics covered in this chapter. The authors would also like to thank Dr. V.M. Sample for providing constant *Z* test samples. The authors are grateful to Alcoa for providing support during the preparation of this chapter.

References

1. Hungtington, A.K. (1915). *J. Inst. Met.*, **3**, 23.
2. Andrade, E.N. daC. (1914). *Proc. Roy. Soc. A*, **90**, 329.
3. Compton, A.H. and Allison, S.K. (1935). In *X-Rays in Theory and Experiment*. D. Van Nostrand, New York.
4. Brown, A.F. (1950). In *Metallurgical Applications of the Electron Microscope*. Inst. Met. Monograph and Rep. series, no. 8, p. 103.
5. Castaing, R. (1960). In *Advances in Electronics and Electron Physics*, **Vol. 13**. Academic Press, New York and London, p. 317.
6. Anderson, R.J. (1918). *Metallurgical and Chem. Engg.*, **18**, no. 10, 523.
7. Wilson, D.V., and Chapman, J.A. (1963). *Phil. Mag.*, **8**, 1543.
8. Wert, J.A., Paton, N.E. Hamilton, C.H., and Mahoney, M.W. (1981). *Met. Trans.*, **12A**, 1267.
9. Jonas, J.J., Sellars, C.M., and McG.Tegart, W.J. (1969). *Int. Met. Rev.*, **14**, 1.
10 Farag, M.M., Sellars, C.M., and McG.Tegart, W.J. (1968). In *Deformation under Hot Working Conditions*. Special report no. 108). Iron and Steel Inst., London, p. 60
11. Osterman, F. (1971). *Met. Trans.*, **2**, 2897.
12. DiRusso, E., Conserva, M., Gatto, F., and Markus, H. (1973). *Met. Trans.*, **4A**, 1133.
13. Sanders, R.E., Jr., and Starke, E.A., Jr. (1978). *Met. Trans.*, **9A**, 1087.
14. McEvily, A.J. Jr., Clark, J.B., and Bond, A.P. (1967). *Trans. ASM*, **60**, 661.
15. Pugh, E.N., and Jones, W.R.D. (1961). *Metallurgia*, **63**, 3.
16. Hornbogen, E. (1967). *Aluminium*, **43**, 41, 115, 163.
17. Starke, E.A. Jr., and Lin, F.S. (1982). *Met Trans.*, **13A**, 2259.
18. Starke, E.A. Jr. (1977). *Mater. Sci. Engg.*, **29**, 99.
19. Santner, J.S. (1981). *Met. Trans.*, **12A**, 1823.
20. Urcola, J.J., and Sellars, C.M. (1987). *Acta. Met.*, **35**, 2637.
21. Christodoulou, N., Jonas, J.J., and Canova, G.R. (1982). In *Mechanical Testing for Deformation Model Development*, R.W. Rohde and J.C. Swearengen, eds. ASTM STP 765. ASTM, Pennsylvania.
22. Sample, V.M., and Lalli, L.A. (1987). *Mat. Sci. Tech.*, **3**, 28.
23. Urcola, J.J., and Sellars, C.M. (1987). *Acta Met.*, **35**, 2649.
24. Urcola, J.J., and Sellars, C.M. (1987). *Acta Met.*, **35**, 2659.

25. Sellars, C.M. (1986). In *Annealing Processes—Recovery, Recrystallization and Grain Growth*, N. Hansen, D. Juul Jensen, T. Leffers, and B. Ralph, eds. Riso National Laboratory, Roskilde, Denmark, pp. 167–187.
26. Wong, W.A., McQueen, H.J., and Jonas, J.J. (1967). *J. Inst. Met.*, **95**, 129.
27. McQueen, H.J., and Conrod, K. (1985). In *Microstructure Control in Aluminum Alloys: Deformation, Recovery and Recrystallization*, (E.H. Chia and H.J. McQueen, eds. Metallurgical Society, Pennsylvania, p. 197.
28. Dillamore, I.L., Morris, P.L., Smith, C.J.E., and Hutchinson, W.B. (1972). *Proc. Roy. Soc.*, **329A**, 405.
29. Ahlborn, H. (1965). In *Recrystallization, Grain Growth and Textures*. ASM, Metals Park, Ohio, pp. 374–381.
30. Walter, J.L., and Koch, E.F. (1963). *Acta Met.*, **11**, 923.
31. Doherty, R.D., and Bellier, S. (1977). *Acta Met.*, **25**, 528.
32. Faivre, P., and Doherty, R.D. (1979). *J. Mat. Sci.*, **14**, 897.
33. Noda, T., and Huber, J. (1978). *Z. Metallkde.*, **69**, 570.
34. Ridha, A.A., and Hutchinson, W.B. (1982). *Acta Met.*, **30**, 1929.
35. Dillamore, I.L., and Katoh, H. (1974). *Metal Sci.*, **8**, 73.
36. Pierce, D., Asaro, R.J., and Needleman, A. (1982). *Acta Met.*, **30**, 1087.
37. Rauch, E., Canova, G.R., Jonas, J.J., and Semiatin, S.L. (1985). *Acta Met.*, **33**, 465.
38. Korbel, A., Embury, J.D., Hatherly, M., Martin, P.L., and Erbsloh, H.W. (1986). *Acta Met.*, **34**, 1999.
39. Korbel, A., Raghunathan, V.S., Tierlinck, D., Spitzig, W., Richmond, O., and Embury, J.D. (1984). *Acta. Met.*, **32**, 511.
40. Hahn, G.T., and Rosenfield, A.R. (1975). *Met. Trans.*, **6A**, 653.
41. Van Stone, R.H., Merchant, R.H., and Low, J.R. (1974). In *ASTM STP 556*. ASTM, Philadelphia, pp. 191–215.
42. Humphreys, F.J., and Kalu, P.N. (1987). *Acta Met.*, **35**, 2815.
43. Koeller, R.C., and Raj, R. (1978). *Acta Met.*, **26**, 1551.
44. Humphreys, F.J. (1979). *Acta Met.*, **27**, 1801.
45. Brown, L.M., and Stobbs, W.M. (1976). *Phil. Mag.*, **34**, 351.
46. Ashby, M.F. (1970). *Phil. Mag.*, **21**, 399.
47. Humphreys, F.J. (1977). *Acta Met.*, **25**, 1323.
48. Dermarkar, S., Guyot, P., and Pelissier, J. (1983). *Acta Met.*, **31**, 1315.
49. Guyot, P., and Wintenberger, M. (1974). *J. Mat. Sci.*, **9**, 614.
50. Dahmen, U., and Westmacott, K.H. (1983). *Scr. Met.*, **17**, 1241.
51. Hardwick, D., and McG. Tegart, W.G. (1961). *Mem. Sci. Rev. Met.*, **58**, 869.
52. Sample, V.M., Fitzsimmons, G.L., and DeArdo, A.J. (1987). *Acta Met.*, **35**, 367.
53. Rossard, C. (1968). *Rev. Met.*, **65**, 181.
54. Luton, M.J., and Sellars, C.M. (1969). *Acta Met.*, **17**, 1033.
55. Sheppard, T., and Tutcher, M.G. (1980). *Metal Sci.*, **14**, 579.
56. Sheppard, T., Parsons, N., and Zaidi, M.A. (1983). *Metal Sci.*, **17**, 481.
57. McQueen, H.J., Evangelista, E., Bowles, J., and Crawford, G. (1984). *Metal Sci.*, **18**, 395.
58. Dillamore, I.L., and Roberts, W.T. (1965). *Met. Rev.*, **10**, 271.
59. Hirsch, J., and Lucke, K. (1986). In *Aluminum Technology '86*, T. Sheppard, ed. Inst. Met., London, p. 267.
60. Hollinshead, P.A. (1986). Ph.D. thesis. Univ. of London.
61. Sheppard, T., Zaidi, M.A., Hollinshead, P.A., and Ragunathan, N. (1985). In Microstructure Control in Aluminum Alloys: Deformation, Recovery and Recrystallization, E.H. Chia and H.J. McQueen, eds. Metallurgical Society, Pennsylvania, p. 19.

62. Panchanadeswaran S. (1986). Unpublished research, Alcoa Laboratories.
63. Naess, S.E., and Andersson, B. (1984). In *Textures in Non-ferrous Metals and Alloys*, H.D. Merchant and J.G. Morris, eds. Metallurgical Society, Pennsylvania.
64. Zaidi, M.A., and Sheppard, T. (1985). *Mat. Sci. Tech.*, **1**, 593.
65. Sellars, C.M., and McG.Tegart, W.G. (1972). *Int. Met. Rev.*, **17**, 1.
66. Zaidi, M.A., and Sheppard, T. (1982). *Metal Sci.*, **16**, 229.
67. McQueen, H.J., Wong, W.A., and Jonas, J.J. (1967). *Can. J. Phys.*, **45**, 1225.
68. McQueen, H.J., and Hockett, J.E. (1970). *Met. Trans.*, **1A**, 2997.
69. McQueen, H.J. (1977). *Met. Trans.*, **8A**, 807.
70. Abson, D.J., and Jonas, J.J. (1970). *Metal Sci.*, **4**, 24.
71. McElroy, R.J., and Szkopiak, Z.C. (1972). *Int. Met. Rev.*, **17**, 175.
72. Sellars, C.M., and McG.Tegart, W.J. (1966). *Mem. Sci. Rev. Met.*, **63**, 731.
73. Semiatin, S.L., Frey, N., Walker, N.D., and Jonas, J.J. (1986). *Acta Met.*, **34**, 167.
74. Sheppard, T., Zaidi, M.A., and Tan, G.H. (1983). *Metal Sci.*, **17**, 563.
75. Sheppard, T., Zaidi, M.A., Tutcher, M.G., and Parsons, N. (1985). In *Microstructure Control in Aluminum Alloys: Deformation, Recovery and Recrystallization*, E.H. Chia and H.J. McQueen, Metallurgical Society, Pennsylvania, p. 155.
76. Sah, J.P., Richardson, G.J., and Sellars, C.M. (1974). *Metal Sci.*, **8**, 325.
77. Sakai, T., and Jonas, J.J. (1984). *Acta Met.*, **32**, 189.
78. Zaidi, M.A. and Sheppard, T., (1984). *Metals Tech.* **11**, 313.
79. Zaidi, M.A. (1980). Ph.D. thesis. Univ. of London.
80. Chia, E.H., and Starke, E.A., Jr. (1977). *Met. Trans.*, **8A**, 825.
81. Tutcher, M.G. (1979). Ph.D. thesis. Univ. of London.
82. Cotterill, P., and Mould, P.R. (1976). In *Recrystallization and Grain Growth in Metals*. John Wiley & Sons, New York, pp. 180–265.
83. Kelly, A., and Nicholson, R.B. (1963). *Prog. Mat'l. Sci.*, **10**, 149.
84. Lorimer, G.W. (1978). In *Precipitation Processes in Solids*, K.C. Russel and H.I. Aaronson, eds. Metallurgical Society, Pennsylvania, pp. 87–119.
85. Sellars, C.M., and Whiteman, J.A. (1979). *Metal Sci.*, **13**, 187.
86. Raghunathan, N., Zaidi, M.A., and Sheppard, T. (1986). *Mat. Sci. Tech.*, **2**, 938.
87. Holl, H.A. (1967). *Metal Sci.*, **1**, 111.
88. Staley, J.T. (1976). In *Properties Related to Fracture Toughness*. ASTM STP 605. ASTM, Pennsylvania, pp. 71–103.
89. Sandstrom, R. (1980). *Z. Metallkde*, **71**, 681.
90. Sandstrom, R. (1980). In *First Riso Symposium on Metallurgy and Materials Science*, N. Hansen and T. Leffers, eds. Riso National Laboratory, Roskilde, Denmark, pp. 45–49.
91. Humphreys, F.J. (1979). *Metal Sci.*, **13**, 136.
92. Zener, C. (1984). Private communication to C.S. Smith, *Trans. AIME*, **175**, 15.
93. Nes, E., Ryum, N., and Hunderi, O. (1985). *Acta Met.*, **33**, 11.
94. Furrer, P., and Warlimont, H. (1978). *Aluminium*, **54**, 135.
95. Hornbogen, E. (1977). In *Fundamental Aspects of Structural Alloy Design*, R.I. Jaffee and B.A. Wilcox, eds. Plenum Press, New York, pp. 389–410.
96. Martin, J.W., and Doherty, R.D. (1976). In *Stability of Microstructure in Metallic Systems*. Cambridge Univ. Press, Cambridge, England, pp. 221–243.
97. Staley, J.T. (1976). *Metals Engineering Quarterly*, **16**, no. 2, 52.
98. Nes, E. (1976). *Acta Met.*, **24**, 391.
99. Nes, E. (1980). In *First Riso Symposium on Metallurgy and Materials Science*, N. Hansen and T. Leffers, eds. Riso National Laboratory, Roskilde, Denmark, pp. 85–95.

100. Nes, E., and Slevolden, S. (1979). *Aluminium*, **55**, 319.
101. Wert, J.A., and Austin, L.K. (1988). *Met. Trans.*, **19A**, 617.
102. Chan, H.M., and Humphreys, F.J. (1984). *Acta Met.*, **32**, 235.
103. Hornbogen, E., and Zum Gahr, K.H. (1975). *Metallography*, **8**, 181.
104. Spiedel, M.O. (1975). *Met. Trans.*, **6A**, 631.
105. Williams, J.C., and Starke, E.A., Jr. (1984). In *Deformation Processing and Structure*, G. Krauss, ed. ASM, Metals Park, Ohio, pp. 279–354.
106. Paton, N.E., and Sommer, A.W. (1973). In *Proc. Third Int. Conf. on Strength of Metals and Alloys*. Cambridge, England, pp. 101–108.
107. Ghosh, A.K., and Hamilton, C.H. (1982). *Met. Trans.*, **13A**, 733.
108. Langdon, T.G. (1982). In *Superplastic Forming of Structural Alloys*, N.E. Paton and C.H. Hamilton, eds. Metallurgical Society, Pennsylvania, pp. 27–40.
109. Hamilton, C.H., Bampton, C.C., and Paton, N.E. (1982). In *Superplastic Forming of Structural Alloys*, N.E. Paton and C.H. Hamilton, eds. Metallurgical Society, Pennsylvania, pp. 173–189.
110. Crooks, R.E., and Starke, E.A. Jr. (1984). *Met. Trans.*, **15A**, 1367.

6

Welding of Aluminum Alloys

C.E. CROSS

IMAC Research & Consulting
Idaho Springs, Colorado

and

G.R. EDWARDS

Center for Welding and Joining Research
Colorado School of Mines
Golden, Colorado

I. Introduction

Aluminum alloys may be readily joined using conventional welding processes, but not without incurring defects that can limit joint efficiency. The degree to which these inherent defects can be avoided, through weld-process control and alloy selection, will determine how closely the weldment properties approximate the base-metal properties. When joining modern aluminum alloys having exceptional mechanical properties, maintaining a low weld defect density is even more important if maximum joint

ISBN 0-12-341831-3

efficiency is to be achieved. In the truest sense, a weldment is a discontinuity in the base plate differing considerably from the base metal in both composition and microstructure. Therefore, it is unreasonable to expect a welded joint to be 100% efficient, even if defect-free.

Defects in aluminum weldments tend to fall into one of four categories: (1) hot tears in the weld metal or heat-affected zone, (2) porosity or inclusions in the weld metal, (3) lack of fusion or incomplete penetration of weld metal, and (4) softening in the heat-affected zone. The formation of any one of these defect types is influenced by a combination of process and metallurgical factors. For example, hot tears are formed when grains behind the weld pool are torn apart by tensile stresses which are largely influenced by process factors such as joint design, fixturing, and weld travel speed. However, hot-tearing is also influenced by grain size and the presence of liquid films at grain boundaries, both of which are metallurgical factors. Weld-metal porosity is influenced not only by the solution and partitioning of atomic hydrogen, but also by the time available for pore coalescence, as determined by weld travel speed. Weld penetration is determined not only by heat input, but by the solidus temperature and thermal properties of the alloy in question. The degree of softening and size of the heat-affected zone are determined not only by the time-temperature profile in the base metal, but also by the kinetics of the precipitation reactions in question. These factors will be considered here. In addition, certain special problems that complicate the welding of modern aluminum alloys will be examined.

II. Conventional Welding Practice

The two welding processes most often used to join aluminum are gas-metal-arc (GMA) and gas-tungsten-arc (GTA). Both processes utilize an inert gas (helium or argon) to protect the molten weld pool from oxidation and to provide ions for the arc plasma. In both arc processes, the weld heat input (per unit length) is defined by the arc power (voltage × current) multiplied by an arc efficiency factor (usually 50–80%, depending upon the welding process), and divided by the weld travel speed. GMA welding of aluminum is commonly done with direct current and a positive electrode (DCEP), using either a constant-voltage or constant-energy power supply. The effect of different gas mixtures on weld bead shape and penetration is discussed by Reichelt et al. [*1*], who have suggested argon to be the optimal shielding gas. In the GMA welding process, the filler wire serves as a consumable electrode, melting off as a spray of droplets at high welding currents or as discrete drops at lower welding currents. Sufficiently low welding current results in short circuit transfer, which is sometimes beneficial for welding out of position or in cases where dilution must be minimized. GMA welding permits high heat inputs and deposition rates, as evidenced by the joining of two-inch–thick plate in only two weld passes using 5/32-inch–diameter wire [*2*]. The GMA process has also become a popular method for making spot welds [*3*].

GTA welding of aluminum is done with a constant-current power supply, using either direct or alternating current and a nonconsumable tungsten electrode. Filler metal can be added to the weld pool either manually or automatically as cold wire feed, allowing multipass welding of thick-section aluminum plate. The DCEP mode results in low penetration and gives good cleaning action due to ionization at the weld pool surface, which helps remove surface contamination. The use of direct current with a negative electrode (DCEN) results in deeper penetration, but lacks the cleaning action of DCEP. Alternating current has the combined advantage of good cleaning and moderate penetration, but usually requires a superimposed high-frequency signal to assist arc reinitiation when passing from the "electrode-negative" (EN) to the "electrode-positive" (EP) portion of each cycle. Effect of gas mixture and electrode geometry on GTA weld bead morphology was studied by Key [*4*], who found that the use of helium in conjunction with a blunt electrode (60–90° included angle) will maximize penetration. In comparison, the use of a sharp-pointed electrode should increase the power input density, thereby increasing the magnitude of fluid flow driven by surface tension-temperature gradients, which should result in a lower depth-to-width ratio [*5*]. Fluid flow in aluminum weld pools has been modelled by Kou and Sun [*6*], taking into account surface tension, buoyancy, and electromagnetic forces.

Arc processes do not always lend themselves to the welding of thin-gage material, in which case resistance welding often becomes the preferred joining method. Resistance welds are made by passing current through overlapping sheets of aluminum positioned between copper electrodes. The interface between sheets, where the electrical resistance is highest, is where the greatest amount of heat is generated and a molten nugget is formed. The effect of welding parameters (current, electrode pressure, and weld duration) on surface cracking tendency and shear strength of Alloy 2036-T4 resistance spot welds was studied by Roest and Rager [*7*]; a linear correlation was found to exist between welding current and shear strength, independent of weld duration or electrode pressure. Large variations in strength are often encountered when analyzing resistance spot welds which, according to Chihoski [*8*], can be related to variations in the degree of diffusion bonding surrounding the weld nugget.

Other means of welding aluminum involve less conventional heat sources such as electron beam or laser. Electron beam welding can produce high-quality, through-thickness welds in six-inch aluminum plate with depth-to-width ratios as high as 20 to 1 [*9*]. However, problems with spiking may be encountered [*10*] and the high vacuum associated with this process tends to aggravate porosity and loss of volatile alloy constituents such as magnesium, which can lead to hot-tearing. Laser welding of thin-section aluminum can be readily accomplished with proper surface preparation and joint configuration to maximize power absorption [*11*]. Lewis and Dixon [*12*] have shown that, while the reflectivity of aluminum is high, the plasma formed during welding provides enhanced coupling and improved energy transfer to the base metal.

III. Metallurgical Concerns

A. *Filler-Metal Selection*

When selecting a filler metal for the welding of a particular aluminum alloy, an alloy is chosen that is as similar as possible to the base-metal composition. This is particularly important from the standpoint of corrosion-resistance, where large compositional differences between weld metal and base metal can lead to galvanic attack. The filler alloy must also be sufficiently resistant to hot-tearing for the given conditions of weld constraint and must provide adequate joint strength and ductility. The use of filler alloys of higher alloy content will result in the weld metal having higher strength but lower ductility.

There are four basic alloys groups from which to choose a commercial aluminum filler metal: 1XXX, 2XXX, 4XXX, and 5XXX. The more common of these filler metals are listed in Table I along with their nominal compositions. In the low-alloy 1XXX series, Alloy 1100 is used almost exclusively to join other low-alloy material. Similarly, in the aluminum-copper 2XXX series, Alloy 2319 is universally used to join other 2XXX-series alloys. Three aluminum-silicon 4XXX-series filler metals are commonly used: Alloys 4043, 4145, and 4047 (listed in order of increasing silicon content). While all 4XXX filler metals can be characterized as having moderate strength and low ductility, their exceptional compatibility and resistance to hot-tearing has made them valuable for general use in welding a wide variety of alloy combinations. They are not recommended, however, for welding base-metal alloys high in magnesium content (Mg

TABLE I

COMMON ALUMINUM FILLER-METAL ALLOYS WITH CORRESPONDING MAJOR-ALLOY CONSTITUENT

Alloy	Major alloy constituent	Wt%
1100	Fe + Si	1.0
2319	Cu	6.3
4043	Si	5.0
4145		10.0 (+4Cu)
4047		12.0
5554	Mg	2.8
5654		3.5
5183		4.5
5356		5.0
5556		5.2

Source: *Kaiser Aluminum Handbook* [*69*].

above 3 wt %). Five aluminum-magnesium 5XXX-series filler metals are commonly used: Alloys 5554, 5654, 5183, 5356, and 5556 (listed in order of increasing alloy content, strength, and resistance to hot-tearing). These filler metals are used to weld 5XXX, 6XXX, and 7XXX base-metal alloys.

Mechanical properties typical of aluminum-alloy weldments are compared in Table II for commonly used alloys. It should be noted that cold-worked alloys such as Alloy 3003-H18 will experience appreciable degradation in strength when welded, while a solid-solution strengthened alloy such as Alloy 5083-H321 will retain higher as-welded properties. When precipitation-hardened alloys are welded (e.g., Alloys 2219-T81 and 6061-T6), strength of the weldment can be partially recovered through postweld heat-treatment, although joint ductility is compromised. A more extensive compilation of mechanical properties for similar as well as dissimilar aluminum-alloy weldments is given by Liptak and Baysinger [*13*]. While aluminum-silicon filler metals (e.g., Alloy 4043) may not provide aluminum weldments with optimal mechanical properties, their use is often dictated by weldability requirements (i.e., the need to avoid hot-tears). A comparison of the weldability of different filler-metal alloys by Dudas and Collins [*14*] shows that filler alloys rich in silicon content have the greatest resistance to hot-tearing.

TABLE II

TRANSVERSE TENSILE PROPERTIES OF GAS-METAL-ARC WELDED ALUMINUM ALLOYS

Base-metal properties			Weldment properties		
Base metal	YS (ksi)	Elongation (% in 2 inch)	Filler metal	YS (ksi)	Elongation (% in 2 inch)
1100-0	5	35	1100	6	23
3003-H18	27	4	1100	9	15
2219-T81	51	10	2319	26	3
			2319 (T87)[a]	34	2
2014-T4	42	22	4043	20	4
2090-T8E41			2319	30	3
			2319(T4)[a]	37	nil
5083-H321	33	16	5356	21	16
6061-T6	40	12	4043	18	8
			4043 (T6)[a]	40	5
7005-T63	46	12	5556	30	10

Source: *Kaiser Aluminum Handbook* [*69*]. Data on 2090-T8E41 courtesy of R.P. Martukanitz, Alcoa Laboratories

[a] Post-weld heat treatment.

B. Weld Metal and Heat-Affected–Zone Microstructure

Aluminum-alloy weld-metal microstructures typically consist of epitaxial, columnar grains (100–500 μm in diameter) that nucleate on base-metal grains and grow normal to the weld pool interface. For example, Fig. 1 shows these epitaxial grains in the fusion zone of an autogeneous GTA weld made on Alloy 2219 plate. Alloy additions of titanium and zirconium are known to act as grain refiners [*15*] by promoting heterogeneous nucleation of grains. As discussed in a review by Mondolfo [*16*], there are two main theories regarding nucleants: the carbide-boride theory and the peritectic theory. The carbide-boride theory assumes that TiB_2 and TiC serve as nucleation sites due to their compatible lattice structure. In the peritectic theory, alpha-aluminum is formed on $TiAl_3$ intermetallic particles in a peritectic reaction. Kou and Le [*17*] have provided evidence that $TiAl_3$ particles, retained in the weld pool after melting of the base metal, can serve as nucleation sites. Furthermore, He et al. [*18*] suggest that constitutional supercooling related to titanium partitioning (partition coefficient above 1) enhances the likelihood for weld-metal nucleation (see p. 121). High-frequency oscillation of the welding arc has also been found to refine grain structure [*19*].

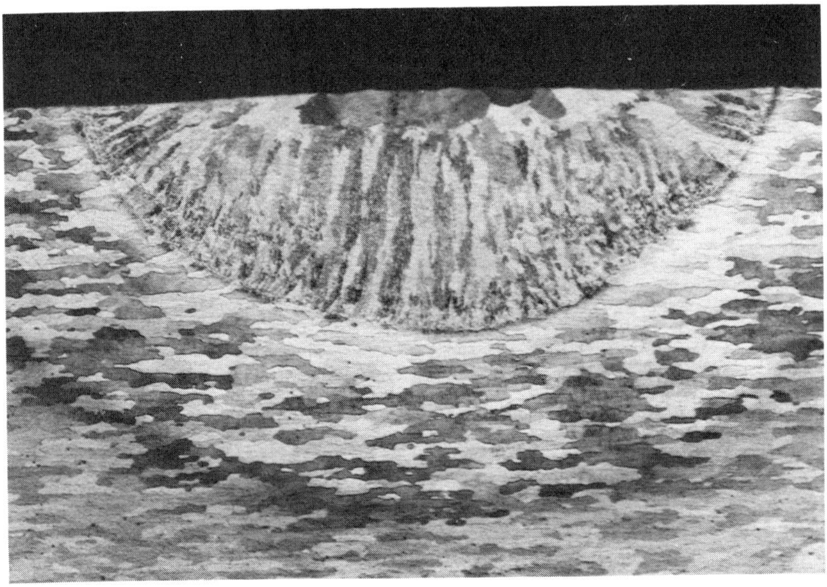

FIG. 1 Macrograph of a transverse cross section taken from an autogeneous GTA weld made on Alloy 2219 plate. Note the epitaxial (A) and columnar (B) nature of weld-metal grains. Also note how the HAZ etches differently from the base metal. (Courtesy of J.F. Capes, Colorado School of Mines)

Within each grain there exists a substructure consisting of a parallel array of cells or dendrites (1–10 μm in diameter) together with interdendritic second-phase constituents formed as the result of solute redistribution during solidification. An example of this is given in Fig. 2, which shows an enlarged view of one of the weld-metal grains of Fig. 1. The second phase usually forms from a eutectic reaction and has the following stoichiometries for Al-Cu, Al-Si, and Al-Mg alloys, respectively: $CuAl_2$, Si and, Mg_2Al_3. High-temperature phases such as $Fe_2Si_2Al_9$ can also form, as can primary crystals of silicon. Predictions for the quantity of eutectic second-phase as well as dendrite spacing can be made utilizing solidification theory [*20*].

Weld-process parameters directly influence weld-metal microstructure, as determined by the growth rate and temperature gradient at the weld pool interface. Growth rates will vary from grain to grain, depending upon their orientation relative to the welding direction and upon the orientation of the dendrites within the grain [*21*]. The upper limit for growth rate is simply the weld travel speed. The temperature gradient remains approximately constant around the liquid–solid interface and varies inversely with

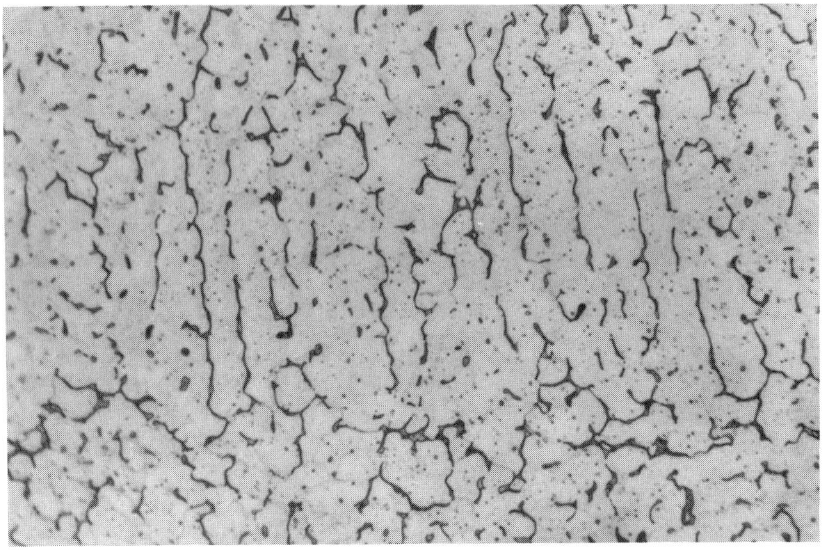

FIG. 2 Micrograph taken from the Alloy 2219 autogeneous GTA weld metal of Fig. 1. Note parallel array of columnar α-aluminum dendrites (light phase) and interdendritic θ-$CuAl_2$ (dark phase). (Courtesy of J.F. Capes, Colorado School of Mines)

heat input when welding thick plate [22]. It has been shown that dendrite spacing varies directly with the square root of heat input for thick-section welding [23] and varies directly with heat input for thin-section welding [24]. A comprehensive review of weld-metal solidification behavior is given by Davies and Garland [25].

The heat-affected zone (HAZ) is that portion of base metal so near the weld metal that its microstructure has been altered by the heat of welding. The HAZ microstructure in aluminum-alloy weldments may differ significantly from that of the base metal, particularly for heat-treatable alloys. As the distance from the fusion line increases, the base metal will experience a continuously decreasing peak temperature during the welding process. Thus, the transformations that may affect the precipitates (dissolution, reprecipitation, and coarsening) will vary with distance from the fusion line, as governed by peak temperature and cooling rate. The HAZ in both heat-treatable and non–heat-treatable alloys can also be altered by recovery, recrystallization, and grain growth. Reducing weld-heat input will constrict peak-temperature isotherms and minimize the volume of metal in the HAZ.

Dumolt et al. [26] have thoroughly studied precipitation reactions occurring in the HAZ of Alloy 2219 autogeneous GTA weldments. By welding on base metal containing only θ'' precipitates, Dumolt et al. found the HAZ to consist of regions of full or partial reversion and regions of θ' phase formation, both of which could be correlated with the thermal experience. Thermal profiles can be measured experimentally, calculated analytically [*27,28*], or simulated using finite element analysis [*29,30*].

C. *Hot-Tearing Susceptibility*

Hot-tears form when grains are torn apart in the weld metal immediately behind the weld pool. The orientation and size of the weld-metal grains, as influenced by low- and high-frequency weld pool oscillation, respectively, have been shown to significantly affect hot-tearing susceptibility [*31*]. The stress state in the region behind the weld pool is influenced by solidification shrinkage as well as by thermal contraction. Both weld parameters and the degree of constraint, as determined by joint configuration and weld fixturing, have been shown to affect the severity of this stress [*32*]. In particular, Chihoski [*32*] suggest that low welding speeds will intensify the tensile stresses trailing the weld pool. Hot-tears have also been observed in the HAZ and are thought to be a problem of liquation at base-metal grain boundaries [*33*].

Resistance to hot-tearing is known to vary with alloy composition. Pure aluminum and alloys of eutectic composition show the greatest resistance to hot-tearing. Numerous investigators [*34–36*] have reported that the hot-tearing susceptibility of binary alloys increases with solute content up to a peak value (between 1 and 3 wt %) and then decreases as the solute content is further increased. This observation was verified [*37*] using a varestraint test whereby binary alloys of variable composition were strained during welding and the accumulative length of induced cracks was measured. Data representative of this study are shown in Fig. 3. Aluminum-copper alloys were found to be much more susceptible to hot-tearing than either aluminum-magnesium, aluminum-lithium, or aluminum-silicon alloys (in descending order of susceptibility). The varestraint test operates by bending a weld coupon around a curved mandrel during the welding operation [*38*]. The axis of curvature of the mandrel is oriented normal to welding direction and its radius determines the degree of augmented strain applied in the vicinity of the weld pool. A variation of this test, the transvarestraint test, involves bending the weld coupon over a mandrel with its axis of curvature oriented parallel to the welding direction [*39*]. Numerous other types of tests are useful in evaluating weldability, including the Houldcroft test [*40*], the patch test [*41*], the spot varestraint test

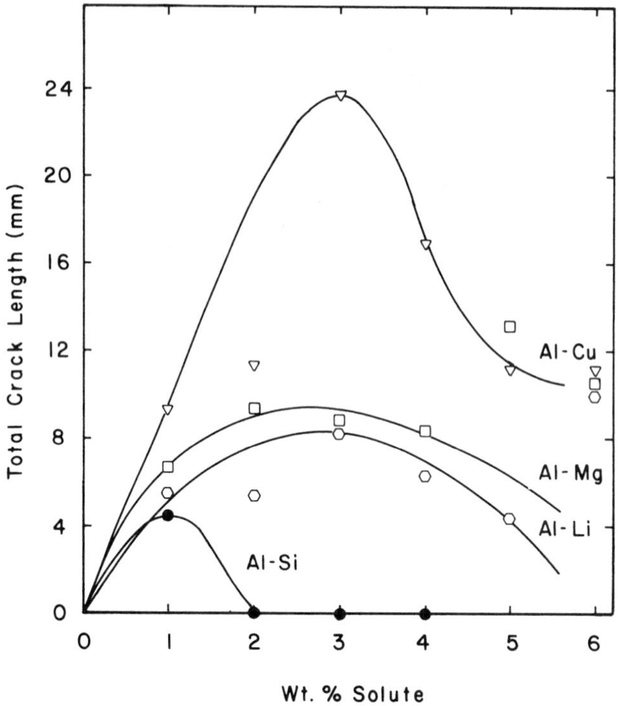

FIG. 3 Comparison of hot-tearing susceptibility for aluminum binary alloys (Al-Cu, Al-Li, Al-Mg, and Al-Si) as a function of alloy content. Hot-tearing susceptibility was determined by measuring the total accumulative length of cracks generated in a varestraint weldability tests run at 4% augmented strain [*37*].

[*42*], the variable tensile strain test [*43*], the variable deformation rate test [*44*], and the Sigmajig [*45*].

Two of the more recent theories on aluminum weld metal hot-tearing include the "generalized theory" of Borland [*46*] and the "critical deformation rate theory" of Matsuda et al. [*44*]. Borland suggests that hot-tears form during a critical stage of solidification, after dendrites have begun to interlock. It is argued that the continuity of liquid at this stage is important because it allows back-filling of any tears that might form. If the last liquid to solidify wets the dendrites, then a continuous liquid film will be available for healing of tears. If, on the other hand, this liquid does not wet dendrites but tends instead to agglomerate, then bridging betwen dendrites will be encouraged, and tears will not form in the first place. According to Rogerson and Borland [*47*], it is at intermediate values of wettability where problems arise.

Matsuda and coworkers examined hot-tearing from a thermal-mechanical standpoint using the brittle-temperature-range (BTR) concept of Senda et al. [*48*]. The BTR is defined graphically on a plot of temperature versus strain. The upper bound of the BTR is defined as the liquidus temperature of the alloy in question. The lower bound of the BTR is measured experimentally as the temperature at the very end of a crack generated in a transvarestraint test. As the augmented strain in the transvarestraint test is increased, the resulting crack will increase in length toward some maximum value relating to minimum crack-tip temperature and is thought to correspond to the lowest melting eutectic liquid in the alloy system. With the BTR boundaries thus determined as a function of augmented strain, a criterion for tearing is met whenever the weld experiences a temperature-strain cycle that intersects the critical BTR region. Cross and Olson [*37*] have shown that additions of eutectic-forming impurities to aluminum binary alloys do in fact extend the maximum crack length and increase hot-tearing susceptibility.

D. Porosity

Porosity in aluminum weld metal, as discussed in a comprehensive review by Devletian and Wood [*49*], forms during weld-solidification and is related to a large drop in atomic hydrogen solubility when aluminum transforms from liquid to solid. Pores are thought to nucleate at the liquid–solid interface, to which hydrogen atoms have been interdendritically partitioned. Nucleation of small pores requires that internal pore pressures meet and exceed ambient pressure by an amount necessary to overcome surface tension. Once nucleated, they either can grow interdendritically or can detach and float into the weld pool where coalescence is likely to occur. Weld travel speed can have a marked influence on porosity. At fast growth rates, the time available for hydrogen diffusion and pore coalescence is limited. At slow growth rates, time may be sufficient for escape of bubbles and outgassing. Primary sources of hydrogen include hydrated oxides and hydrocarbon films present on the surfaces of base metal or filler metal. Martukanitz and Michnuk [*50*] have found the surface condition of the filler metal (i.e., hydration of surface oxides) to be a controlling factor in pore formation during welding of aluminum alloys.

Among a variety of means to reduce porosity, besides degreasing of the joint and proper care of the filler wire, is welding at pressures above ambient [*51*]. This effectively increases the critical concentration of hydrogen needed to nucleate a pore. The use of freon mixtures in the shielding gas has also been found beneficial [*52*]. Freon decomposes to form chlorine and fluorine in the arc, which can then react with hydrogen. A special technique used for joint preparation, cathodic etching [53], also lessens the incidence of porosity. In this technique, a low-current GTA-DCEP arc is run along the joint prior to welding, where it believed that a cleaning action is achieved through cation bombardment.

IV. Welding of Modern Alloys

A. *High-Strength Wrought Alloys*

Alloys such as 2024 and 7075 have limited weldability because they are overly susceptible to hot-tearing. Experience has shown that the dual copper-magnesium content in these alloys is the apparent cause of limited weldability. However, 7XXX-series alloys that are low in copper (e.g., Alloys 7039 and 7005) are weldable, particularly when specially developed filler metals are used. Robinson and Baysinger [54] and Gibbs [55] discuss the advantages of using Filler Alloy 5039 (containing zinc) in welding Alloy 7039. Their results indicate both improved strength from natural aging and improved resistance to hot-tearing. Nakata et al. [56] have developed a high magnesium (7 wt % Mg) filler alloy showing exceptional hot-tearing resistance when used in welding a 7005-type alloy. Also, there is evidence that the new lithium-based alloys have good weldability as reviewed by Pickens [57]. Cross et al. [58] found that aluminum-lithium binary alloys exhibit high resistance to hot-tearing, superior to that of the commercially welded Alloy 5083. Aluminum-lithium-magnesium ternary alloys have been found prone to weld-metal porosity [59] although this problem was alleviated by chemically milling the weld joint prior to welding. Microhardness profiles in aluminum-lithium-copper (Alloy 2090) weldments have been characterized by Martukanitz et al. [60], indicating a large drop in HAZ strength. Corrosion tests performed on weldments of this alloy have not shown any tendency for stress-corrosion cracking [61].

B. *Rapidly Solidified Powder Alloys*

The primary problem encountered when welding any powder aluminum alloy, particularly one that is not 100% dense, is that of weld-metal porosity attributed to hydrated oxides at particle surfaces. This problem can be minimized by welding fully dense alloys that have been vacuum degassed at elevated temperatures (500°C), although this heat-treatment may result in coarsening of the base-metal microstructure. A patented process for degassing at lower temperatures (370°C) is described by Pickens [*62*] whereby a canister of powder is repeatedly evacuated and back-filled with dry nitrogen. Heat input should be minimized when welding powder alloys, not only to restrict coarsening in the HAZ, but also to get a fine microstructure in the weld metal approaching that of the rapid solidified powder. Baeslack [*63*] was able to successfully make autogeneous electron beam welds on RSP Al-8Fe-1.7Ni forged plate following a 24-hour degassing treatmment at 550°C. While the rapid cooling associated with the electron beam welding was shown to have the potential for generating weld-metal microstructures comparable to the base metal, a unique problem was found to exist in Al-8Fe-2Mo alloys due to the formation of coarse intermetallics near the fusion line in the weld metal. (See Fig. 4.)

C. Metal-Matrix Composite Alloys

Considerable attention is now being focused on means of joining aluminum-composite materials. A study to examine the arc weldability of a silicon-carbide–reinforced 6061 aluminum alloy (powder processed) by Ahearn, et al. [*64*] was successful, but resulted in some unique problems. First of all, the material had to be vacuum degassed to bring porosity levels

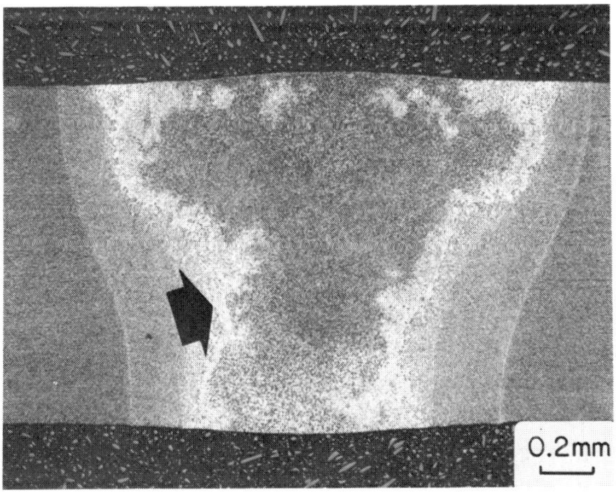

FIG. 4 Macrograph of transverse section taken from electron-beam weldment on RSP Al-8Fe-2Mo alloy. Note presence of coarse Al_3Fe-type intermetallics near fusion-zone boundary (arrow). (Courtesy W.A. Baeslack, Ohio State University)

to an acceptable level—a problem attributed to the powder. Second, the molten weld pool was found to be extremely viscous and difficult to control. However, strength of the resulting weldments was comparable to that of conventional wrought 6061 weldments. Another problem that reportedly can occur is delamination in the HAZ, although Ahearn and coworkers did not encounter this in their study. Resistance spot welding also shows promise for joining of composites. Hersh [*65*] showed the feasibility of resistance spot welding an aluminum-boron composite alloy (50 vol % boron-fiber–reinforced Alloy 6061). Soviet investigators [*66*] also have reported success in resistance spot welding of a boron-reinforced aluminum alloy, but found that the volume percent of boron fibers must be less than 33% to achieve good bonding between the weld nugget and the base metal. Devletian [*67*] has proven the capacitor-discharge process useful in spot welding aluminum/silicon carbide composites. Nondestructive evaluation is another major concern complicating the development of welding techniques for composite materials. Standard techniques such as ultrasonics have to be modified to account for background signals generated at fiber–matrix interfaces [*68*].

V. Summary

Acceptable welds can be made on most conventional aluminum alloys (excluding high-strength alloys bearing both copper and magnesium) provided that a proper filler alloy and welding technique are employed. Weldment efficiency is dependent upon the strength of the deposited filler alloy, the amount of microstructural damage in the HAZ, and the degree to which hot-tearing and porosity have been avoided. Hot-tearing susceptibility can be reduced by using a filler metal of high alloy concentration and by minimizing weld constraint through joint design and fixturing. Although the interaction between the many variables affecting hot-tearing is not clearly understood, it is known that grain refinement improves resistance to tearing and that aluminum-silicon alloys are least susceptible. Porosity can be minimized through proper joint preparation and through use of filler wire that has been protected from humid environments.

Special problems are associated with welding of new aluminum-base materials such as the RSP alloys and metal-matrix composites. For example, a preweld vacuum heat-treatment is necessary to avoid porosity when preparing RSP materials for welding. Porosity and delamination are recognized problems associated with the welding of composite materials, and these and other concerns make this area a very challenging one for aluminum-welding metallurgists.

Topics for future aluminum-welding research will, out of necessity, involve process and filler-metal development for new high-strength and composite alloys. Specialty filler-metal alloys will be made with compositions fine-tuned to the needs at hand: ultralow iron content for improved ductility, optimum titanium-boron-zirconium additions for grain refinement, and avoidance of low-melting, eutectic-forming elements for improved hot-tearing resistance. While the trend will be toward use of low heat input, high-energy-density welding processes to minimize HAZ damage, new base-metal chemistries and tempers must also be designed with HAZ damage in mind.

References

1. Reichelt, W.R., Evancho, J.W., and Hoy, M.G. (1980). *Welding J.*, *59*, 147–155s.
2. Evancho, J.W., Reichelt, W.R., and Hoy, M.G. (1981). *Welding J.*, **60**, 24–29.
3. Alcoa (1972). *Welding Alcoa Aluminum.* Handbook, 91–99. Pittsburgh, Penn.
4. Key, J.F. (1980). *Welding J.*, **59**, 364s–2370s.
5. Burgardt, P., and Heiple, C.R. (1986). *Welding J.*, **65**, 150s–155s.
6. Kou, S., and Sun, D.K. (1985). *Met. Trans.*, **16A**, 203–213.
7. Roest, C.A., and Rager, D.D. (1974). *Welding J.*, **53**, 529s–536s.
8. Chihoski, R.A. (1970). *Welding J.*, **49**, 567s–578s.
9. ASM (1983). *ASM Metals Handbook*, **6**, 641–642. American Society for Metals.
10. Mara, G.L., Funk, E.R., McMaster, R.C., and Pence, P.E. (1974). *Welding J.*, **53**, 246s–251s.
11. Huntington, C.A., and Eagar, T.W. (1983). *Welding J.*, **62**, 105s–107s.
12. Lewis, G.K., and Dixon, R.D. (1985). *Welding J.*, **64**, 49s–54s.
13. Liptak, J.A., and Baysinger, F.R. (1968). *Welding J.*, **47**, 173s–180s.
14. Dudas, J.H., and Collins, F.R. (1966). *Welding J.*, **45**, 241s–249s.
15. Ganaha, T., Pearce, B.P., and Kerr, H.W. (1980). *Met. Trans.*, **11A**, 1351–1359.
16. Mondolfo, L.F. (1982). In *Grain Refinement in Castings and Welds*, G.J. Abbaschian and S.A. David, eds. TMS-AIME, Warrendale, Penn., pp. 3–50.
17. Kou, S., and Le, Y. (1986). *Welding J.*, **65**, 305s–313s.
18. He, Y., Olson, D.L., Edwards, G.R., and Frost, R.H. (1986). *CSM-CWR Internal Report 086–033* Colorado School of Mines, Center for Welding Research, Golden, Colorado.
19. Pearce, B.P., and Kerr, H.W. (1981). *Met. Trans.*, **12B**, 479–486.
20. Cross, C.E., Capes, J.F., and Olson, D.L. (1987). In *Welding, Failure Analysis, and Metallography*, M.R. Louthan, I. LeMay, and G.F. VanderVoort, eds. ASM, Metals Park, Ohio, pp. 3–16.
21. Savage, W.F., and Hrubec, R.J. (1972). *Welding J.*, **51**, 260s–271s.
22. Adams, C.M. (1958). *Welding J.*, **37**, 210s–215s.
23. Brown, P.E., and Adams, C.M. (1960). *Welding J.*, **39**, 520s–524s.

24. Jarman, R.A., and Jordan, M.F. (1970). *J. Institute Metals*, **98**, 55–57.
25. Davies, G.J., and Garland, J.G. (1975). *Int. Met. Rev.*, **20**, 83–106.
26. Dumolt, S.D., Laughlin, D.E., and Williams, J.C. (1982). *Proc. 1st Int. Aluminum Weld. Conf., 1981.* Weld. Res. Council, New York, N.Y., pp. 115–134.
27. Rosenthal, D. (1941). *Welding J.*, **20**, 220s–234s.
28. Christensen, N., Davies, V. deL., and Gjermundsen, K. (1965). *British Welding J.*, **12**, 54–75.
29. Kou, S. (1981). *Met. Trans.*, **12A**, 2025–2030.
30. Kou, S., and Le, Y. (1983). *Met. Trans.*, **14A**, 2245–2253.
31. Kou, S., and Le, Y. (1985). *Met. Trans.*, **16A**, 1345–1352.
32. Chihoski, R.A. (1972). *Welding J.*, **51**, 9s–18s.
33. Gittos, N.F., and Scott, M.H. (1981). *Welding J.*, **60**, 95s–103s.
34. Singer, A.R.E., and Jennings, P.H. (1947). *J. Institute Metals*, **73**, 197–212.
35. Pumphrey, W.I., and Lyons, J.V. (1948). *J. Institute Metals*, **74**, 439–455.
36. Dowd, J.D. (1952). *Welding J.*, **31**, 448s–456s.
37. Cross, C.E., and Olson, D.L. (1986). *Proc. Conf: Aluminum Alloys—Their Physical and Mechanical Properties, Charlottesville 1986*, E.A. Starke and T.H. Sanders, eds. TMS-AIME, Warrendale, Penn., 1869–1875.
38. Savage, W.F., and Lundin, C.D. (1965). *Welding J.*, **44**, 433s–442s.
39. Bailey, N., and Regelous, G.R. (1976). *Welding Inst. Res. Bul.*, 201–204.
40. Houldcroft, P.T. (1955). *Brit. Weld. J.*, **2**, 471–475.
41. Borland, J.C., and Rogerson, J.H. (1962). *Brit. Weld. J.*, **9**, 494–498.
42. Turner, P.W., and Lundin, C.D. (1970). *Welding J.*, **49**, 579s–587s.
43. Tamura, H., Kato, N., Ochiai, S., and Katagiri, Y. (1977). *Trans. JWS*, **8**, 16–22.
44. Matsuda, F., Nakagawa, H., Nakata, K., and Okada, H. (1979). *Trans. Jap. Weld Res. Inst.*, **8**, 85–95.
45. Goodwin, G.M. (1987). *Welding J.*, **66**, 33s–38s.
46. Borland, J.C. (1960). *British Welding J.*, **7**, 508–512.
47. Rogerson, J.H., and Borland, J.C. (1963). *Trans. AIME.*, **227**, 2–7.
48. Senda, T., Matsuda, F., and Takano, G. (1973). *J. Jap. Weld. Soc.*, **42**, 48–56.
49. Devletian, J.H., and Wood, W.E. (1983). *Weld. Res. Council Bulletin.*, **290**, 1–18.
50. Martukanitz, R.P., and Michnuk, P.R. (1982). In *Trends in Welding Research*, S.A. David, ed. ASM, Metals Park, Ohio, pp. 315–330.
51. Brandon, E.D. (1970). *Welding J.*, **49**, 510s–520s.
52. Bicknell, A.C., and Patchett, B.M. (1985). *Welding J.*, **64**, 21–27.
53. Barhorst, S. 1985). *Welding J.*, **64**, no. 4, 28–31.
54. Robinson, I.B., and Baysinger, F.R. (1966). *Welding J.*, **45**, 433s–444s.
55. Gibbs, F.E. (1966). *Welding J.*, **45**, 445s–453s.
56. Nakata, K., Miyanaga, Y., Matsuda, F., Tsukamoto, K., and Arai, K. (1980). *Trans. Jap. Weld. Res. Inst.*, **9**, 205–216.
57. Pickens, J.R. (1985). *J. Mater. Sci.*, **20**, 4247–4258.
58. Cross, C.E., Olson, D.L., Edwards, G.R., and Capes, J.F. (1983). *AIME Proc. 2nd Int. Aluminum-Lithium Conf. Monterey 1983*, pp. 675–682.
59. Kainova, G.E., and Malinkina, T.I. (1969). *Metall. Termicheskaya Obrabotka Metallov*, 104–105.
60. Martukanitz, R.P., Stevens, R.H., and Johnston, L.R. (1987). In *Welding Failure Analysis, and Metallography*, M.R. Louthan, I. LeMay, and G.F. VanderVoort, eds. ASM, Metals Park, Ohio, pp. 53–64.
61. Symons, D.M., Natalie, C.A., Edwards, G.R., and Martukanitz, R.P. (1986). *Conf.*

Proc.: Aluminum-Alloys—Their Physical and Mechanical Properties Charlottesville 1986, E.A. Starke and T.H. Sanders, eds. EMAS, United Kingdom, pp. 1189–1195.
62. Pickens, J.R. (1981). *J. Mater. Sci.*, **16**, 1437–1457.
63. Baeslack, W.A. (1985). *Metallography*, **18**, 73–82.
64. Ahearn, J.S., Cooke, C., and Fishman, S.G. (1982). *Metal Construction*, **14**, 192–197.
65. Hersh, M.S. (1971). *Welding J.*, **50**, 515s–521s.
66. Ryazantsev, V.I., Fedoseev, V.A., and Yastina, N.V. (1981). *Svar. Proiz.*, 17–18.
67. Devletian, J.H. (1987). Welding J., **66**, no. 6, 33–39.
68. Rogovsky, A.J., and Holmquist, G.R. (1984). *Mater. Eval.*, **42**, no. 3, 318–324.
69. Kaiser (1978). *Welding Kaiser Aluminum.* Handbook Kaiser Aluminum Oakland, Calif.

7

Superplastic Aluminum Alloys

C.C. BAMPTON

Rockwell International
Science Center
Thousand Oaks, California

J. WADSWORTH

Lockheed Missiles
and Space Company
Palo Alto, California

and

A.K. GHOSH

University of Michigan
Ann Arbor, Michigan

I. Introduction

Superplasticity is the capability of polycrystalline materials to undergo extremely large neck-free tensile elongations. For many years, the phenomenon was considered to be only of scientific interest. In more recent

ISBN 0-12-341831-3

times, the usefulness of superplasticity in the manufacture of complex parts from metallic alloys has become evident, and the subject has been extensively reviewed from both fundamental and technological viewpoints [1–7]. As a result of these investigations, a number of key points regarding the manufacture, structural prerequisites, and properties of superplastic alloys have been proposed.

The practical significance of alloys that possess the property of superplasticity is that extremely complex shapes can be made using simple forming operations. Very low superplastic flow stresses, typically less than 7 MPa (1000 psi), allow gas pressure forming in single operations. This simplifies the necessary strain-rate control for complex shapes and greatly reduces tooling costs, since only one, single surface tool is needed. Conventional forming operations usually require multiple sets of matched dies, often with interstage anneals of the workpiece and allowances for part spring-back and residual stresses. The limitations of superplastic forming are the low production rates (0.5–2.0 hours per part), the need for relatively high processing temperatures (≥0.6 absolute melting temperature) and, in some cases, the development of intergranular cavities.

In the following section, a brief background on superplasticity is presented followed by a more detailed description of structurally useful superplastic aluminum alloys.

II. Fundamentals of Superplasticity

A. *Strain-Rate Sensitivity*

Most pure metals and many alloys will fail, by localized necking and rupture, with a ductility less than 100% if tested in tension, even at a high homologous temperature ($T \geqq 0.6T_M$). Superplastic alloys, on the other hand, frequently fail with ductility values of over 500% and can often achieve values of well over 1000%. The largest recorded elongation, shown in Fig. 1, was achieved with a Pb-62%Sn alloy and has the remarkable value of 4850% [8]. The property of superplasticity, as characterized by tensile elongations of greater than about 500%, has been successfully developed in many alloys, a number of which are of commercial importance [2].

From an applied-mechanics viewpoint, superplastic metallic alloys have one primary property in common. The strength (resistance to plastic flow) of such materials is highly strain-rate sensitive. That is, σ, the flow stress, increases rapidly with $\dot{\epsilon}$, the strain rate. A measure of this rate sensitivity of strength is given by the relation $\sigma = K\dot{\epsilon}^m$, where K is a material constant

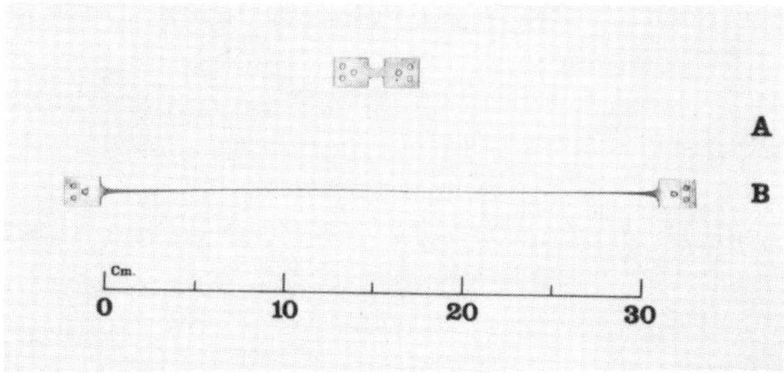

FIG. 1. Pb-62% Sn alloy tested by Ahmed and Langdon [8] to an elongation of 4850%.

and m is the strain-rate–sensitivity exponent. The value of m determines the rate at which the neck development progresses after localized plastic flow starts. The local strain rates in the neck region of a sample are high compared to the remainder of the gage length. If the value of m is high, the increase in flow strength of the material in the neck region will be large and further deformation will be strongly resisted. If the value of m is low, however, there will be little resistance to continued deformation in the neck region and localized flow, hence low elongation, will result.

It is an experimental fact that if m, the strain-rate–sensitivity exponent, increases, the elongation to failure increases [9]. When the relation between σ and $\dot{\epsilon}$ is linear ($m = 1$), a material is said to behave in a Newtonian-viscous manner and such materials can be classified as ideally superplastic and have tremendous elongations to failure. In the case of superplastic metals, values of m usually do not exceed 0.8. Alloys that are controlled by glide of dislocations, impeded by solute drag, have a value of $m = 0.33$. Most pure metals and alloys have a value of $m = 0.1$–0.2 and oxide-dispersion–strengthened (ODS) alloys have very low values of $m \cong 0.02$. Although a high value of m is a necessary condition for superplasticity, it is not a sufficient condition. Material-embrittling characteristics (such as grain-boundary separation, cavitation at interphase boundaries, and other premature failure modes) can lead to low elongations even when the material exhibits high strain-rate sensitivity at low strains.

Two microstructural mechanisms are available for achieving the high strain-rate sensitivity that is required for superplasticity. One is by internal stress superplasticity, such as by transformation cycling or by thermal cycling of thermally anisotropic polycrystalline metals [1], where values as high as $m = 1$ can be achieved. The other technique, leading to fine-structure superplasticity, is by grain-size refinement and is the most common method for aluminum alloys.

Fine-structure superplasticity thus refers to the high elongations that can be obtained in metallic alloys that have ultrafine grain sizes. Typically, the grain size, as measured by mean linear intercept, is less than about 20 microns (20 μm). Fine-structure superplastic alloys exhibit superplasticity only at high homologous temperatures ($T \geq 0.6T_M$). For a given temperature within this range, they exhibit a peak in m value at a certain strain rate. This peak–m-value strain rate increases with temperature. The peak–m-value strain rates are typically very slow compared with most metal-working processes. The mode of plastic deformation is believed to be primarily that of grain-boundary sliding accomodated by slip near the grain boundary [7]. A number of microstructural prerequisites have been established for alloys to possess superplasticity [*7*,*9*,*10*].

B. Cavitation

Cavitation has been found to occur in many superplastic alloys and in all superplastic aluminum alloys [*10*]. It is discussed here in the general context of high-temperature deformation. When polycrystalline materials are deformed above one-half of their homologous temperature, cavities often form on grain boundaries and can lead to premature failure. Cavitation has been reported for high-temperature deformation under creep [*11*] and fatigue [*12*] conditions as well as during superplastic flow.

During creep, the cavities tend to occur preferentially [*11*] on boundaries between 60 and 90° to the tensile axis, probably as a consequence of the near-normal stress across the boundary. Under fatigue conditions, cavities appear on the boundaries which are actively sliding during stress reversal (i.e., the maximum shear stress orientations) [*13*,*14*]. In superplasticity, there is no particular preferred location of the cavities with respect to the tensile axis, but this may well reflect the extensive grain rotation during such deformation. This randomization of superplastic cavity formation and growth has the important practical consequence that cavity interlinkage and failure may be significantly more delayed in superplastic strain than in creep or fatigue strain [*15*].

A general observation is that grain-boundary sliding (GBS) is a necessary but not sufficient criterion for cavitation under all deformation modes [*1*,*16–19*]. Evidence that sliding does occur is often obtained in the form of offsets at grain boundaries. In superplasticity, GBS occurs to such an extent that grain rotation reduces any initial texture [*1*].

Grain boundaries may contain ledges, run-in lattice dislocations, grain-boundary dislocations, and particles. Evidence suggests that it is these defects, together with grain-boundary triple points, that control the rate and extent of sliding. Initial sliding rates are often very fast until the stresses, built up at such defects, oppose sliding, which can then only continue if the stresses are relieved by a diffusion-controlled accommoda-tion process.

Nucleation of cavities appears to occur at any local stress raiser that is effective enough to cause local boundary separation or at any inhomogeneity, such as a particle, that can locally reduce the interfacial energy and decrease the critical nucleus size [*20*]. GBS causes cavities to develop at irregularities whenever the accomodating plastic flow or diffusion is not fast enough to prevent it. The role of the applied stress is to cause both the sliding and the consequent development of local stress concentration.

In the case of superplastic deformation, where the extent of GBS is so very much greater than in creep or fatigue, cavitation tends to occur preferentially along phase boundaries [*21*,*22*] or at particles [*21*,*23*] where diffusion or slip accomodation is more difficult. The diffusion accomodation of GBS in the presence of a second phase must be controlled by movement of one of the atom species. This causes shear-stress concentrations at points where diffusion is restricted and can lead to decohesion. In the dislocation model, particles or ledges can act as obstacles to dislocation motion, thus necessitating climb. If the climb rate is lower than the sliding rate, then a dislocation pile-up forms which acts as a stress concentrator leading to cavitation.

C. *Material Characterization and Analytical Modelling*

To utilize fully the superplastic potential of alloys in practical gas-forming operations, a thorough knowledge of the alloy's superplastic characteristics is required. These characteristics should be utilized in an analytical approach to predict and control local strains and strain rates in actual sheet-forming operations.

The relationships among flow stress, strain, strain rate, m value, and temperature for an alloy are complicated by significant dynamic changes in microstructure which generally occur during superplastic deformation. As we will describe in some examples later, the most significant of these changes include grain growth, cavitation, and recrystallization.

Several approaches have been proposed for measuring an alloy's superplastic characteristics so that some form of constitutive equation may be defined [*7*]. Similarly, analytical modelling approaches have been developed to utilize the material constitutive equations for realistic superplastic forming operations [*24–31*]. These are issues of fundamental importance to practical applications of superplastic forming and merit further study by those interested in the technology.

III. Aluminum Alloys

A. *Alloy Systems*

The development of superplasticity in Al alloys in general has been reviewed [*7,10*]. Specifically, work on superplasticity in aluminum alloys has included studies on Al-33Cu [*33–37*], Al-Zn-Mg-Zr [*38–40*], Al-Mg-Si [*41*], Al-Al_2O_3-Ga [*42,43*], Al-Ca [*44,45*], Al-Si [*46,47*], Al-Mg-Zr [*48*], Al-10%Mg [*49,50*]. and Al-Pd [*51*] alloys. The superplastic aluminum alloys of commercial interest, however, are considered to be the British Al-Cu-Zr alloys [*7,52–55*], the 7000-series alloys modified by thermomechanical treatments [*7,56–58*], the Al-Ca-Zn alloy [*7,59*], and Al-Li–based alloys [*60–64*].

The fine-grain, superplastic aluminum alloys may be categorized according to the nature of the grain-stabilizing secondary phases: (1) those with high volume fractions of relatively coarse secondary phase and (2) those with low volume fractions of small secondary-phase particles. In each case, the value of f/r, where f is the volume fraction of second phase and r is the second-phase particle radius, at the superplastic forming temperature, is as high as possible. This is to inhibit elevated-temperature grain growth of the primary phase (alpha-Al) by the classical Zener drag mechanism [*65–69*].

Aluminum binary eutectic alloys, such as Al-Cu, Al-Ca, and Al-Si, are all examples of the first category where f may be as high as 0.5. Although these alloys have been processed to yield very high superplastic ductilities, their room-temperature properties are generally unattractive as engineering materials, especially in structural applications. For this reason, the following discussion will be limited to the second category of aluminum alloys, although the reader is encouraged to review the extensive literature on the superplastic characteristics of the eutectic-type superplastic aluminum alloys.

The second category of superplastic aluminum alloys may be further categorized by the detailed nature of the fine-grain processing. All of the commercially useful superplastic aluminum alloys involve recrystallization of a deformed microstructure in the fine-grain processing. The recrystallization process may be either discontinuous, usually during a static heating stage, or continuous, usually during elevated-temperature deformation. Discontinuous recrystallization involves discrete nucleation and growth of entirely new grains at preferred sites in a deformed microstructure. Continuous recrystallization involves progressive recovery of a deformed microstructure such that the dislocation structure within the deformed grains rearranges to form subgrain boundaries (polygonization) which ultimately evolve to high-angle grain boundaries.

The mechanism of recrystallization has important practical significance for the superplastic forming process. During a superplastic forming operation, dynamic changes in microstructure and, hence, in superplastic characteristics are essentially limited to grain growth and cavitation in discontinuously recrystallized alloys. In continuously recrystallizing alloys, however, the recrystallization process itself is added to the grain growth and cavitation developments during the deformation. Maintaining optimum superplastic deformation conditions throughout a forming operation may be considerably more complex in the case of dynamically recrystallizing alloys.

The following discussions will describe in some detail the two main examples of superplastic aluminum-alloy systems developed to commercial status for engineering applications. This will be followed by a brief summary of other superplastic aluminum alloy systems that have shown potential for future commercialization.

B. 2004 Aluminum Alloy

The 2004 aluminum alloy, previously known as Supral 100 [*52–55*], has the distinction of being the first commercially successful superplastic aluminum alloy. This alloy was designed from first principles as an ultrafine-grain superplastic aluminum alloy which retains attractive room-temperature properties. A fine example of pragmatic research and development led to commercial production of the 2004, Al-6%Cu-0.4%Zr alloy with associated, specially developed superplastic-forming processes and equipment [*7*].

The alloy relies on a special combination of coarse $CuAl_2$ and very fine Al_3Zr particles to (1) inhibit static recrystallization of the cold-worked structure, (2) encourage dynamic (continuous) recrystallization during the early stages of superplastic deformation, and (3) inhibit grain growth after completion of recrystallization. Figure 2 shows the effect of Zr on the development of a fine, dynamically recrystallized grain size in a 2004-type alloy. The 2004-type superplastic alloys require a relatively large amount (0.3–0.5 wt %) of Zr. Such quantities are in excess of those typically found in commercial practice (0.1–0.15 wt %) where Zr is often added for grain refinement of castings or to prevent recrystallization at temperatures well below those used in superplastic forming. The additional amount of Zr can be problematical, since higher than normal casting temperatures are required. Furthermore, coarse undissolved Al_3Zr particles may form and these can promote cavitation during superplastic flow. The 2004 alloy is, therefore, processed by somewhat nonstandard ingot casting and thermomechanical mill practices.

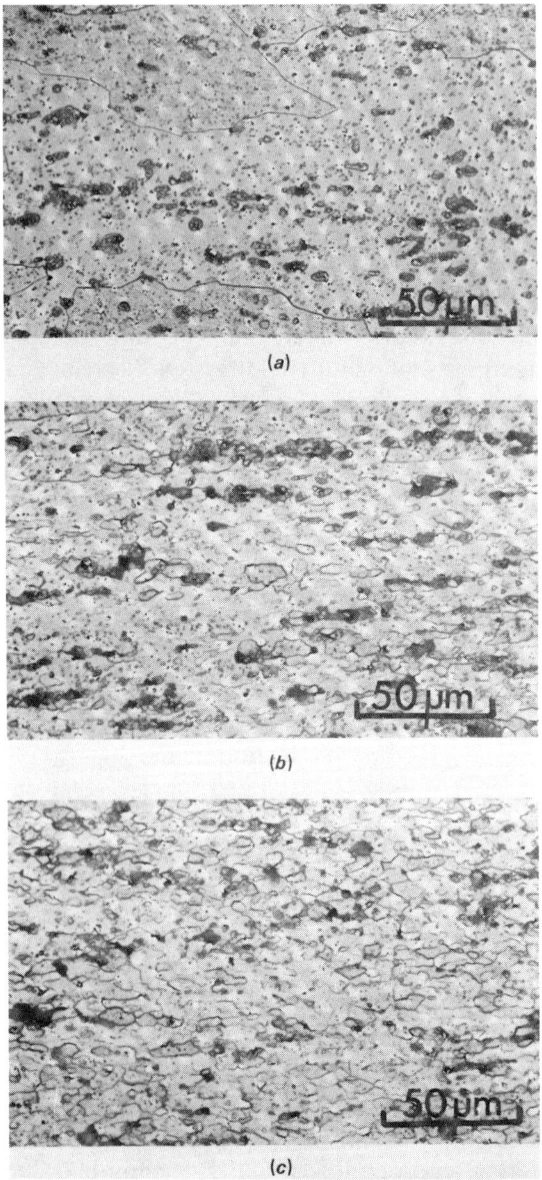

FIG. 2. Effect of Zr on grain structure of rolled Al-6Cu-Zr alloys tested at 425°C. (a) No Zr. (b) 0.25Zr. (c) 0.5Zr.

The 2004 alloy has shown superplastic tensile elongations above >1000% and is commercially attractive since the superplastic forming is carried out at a moderately high strain rate (about 10^{-3} S^{-1}) and a moderately low temperature (about 460°C). This typically involves processing times of 15 to 45 minutes per part and allows the use of low-cost, cast aluminum tooling.

The dynamically recrystallizing alloys, such as 2004, tend to exhibit lower m values (0.3–0.5) than statically recrystallized alloys, such as 7475-02 (to be discussed later). It appears that the dislocation creep process is dominant in these dynamically recrystallized 2004 types, whereas the 7475-02, statically recrystallized types derive their superplasticity from diffusional creep processes.

The main limitations of the 2004 alloy are its moderate room-temperature strength and the inability to control cavitation in a practical and economic manner. This latter effect is due to the superplastic flow stress being generally too high (above 7 MPa) to effectively utilize hydrostatic pressure techniques (to be described later) for cavitation control. Nevertheless, a very large number and variety of production applications have been produced in this alloy.

C. 7475-02 Aluminum Alloy

The advantage of high strength combined with superior fracture-toughness has been the main driving force for examining the possibility of SPF of this material. It is fortuitous that the compositional and processing modifications made to the basic 7075 alloy, in order to develop improved toughness in the 7475 alloy, also generally improve the superplastic capability.

A number of researchers have developed methods for grain refinement of the 7000-series alloy, of which at least one method [57] has been found to be relatively easy to implement. Starting from coarse-grain rolled plate microstructure, this process consists of the following steps: (1) overaging 8 h at 370 to 400°C after solutionizing to produce coarse precipitates, (2) rolling to 90% reduction at 200°C to develop strain fields around the precipitates, and (3) recrystallizing at or above 480°C to nucleate the fine grains from these high-strain areas to develop 10–14-μm overall grain size in the sheet material. A schematic of the process is illustrated in Fig. 3, which leads to a significantly small aspect ratio for the fine-grain–size sheet product. A critical aspect of the process is the heating rate in the recrystallization step, which must be extremely rapid to simultaneously activate as many nuclei as possible [58]. The chromium-containing dispersoids in these

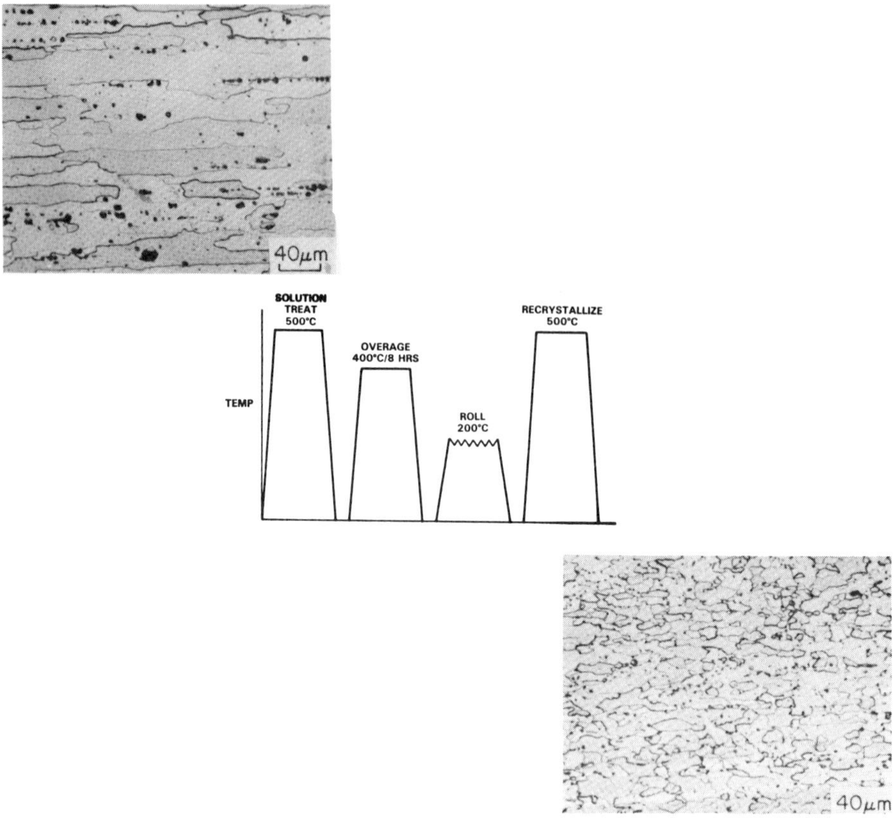

FIG. 3. Steps involved in thermomechanical processing of 7475 Al plates to fine-grain sheet [*13*].

alloys reduce grain-boundary mobility and help achieve and maintain a fine-grain structure. Thermomechanical processes of this type have been successfully used by the major aluminum producers to develop a superplastic 7475 alloy, known as 7475-02. The beneficial effect on superplastic ductility of decreasing grain size in 7475 is dramatically illustrated in Fig. 4.

Mechanical behavior and microstructural changes in this material during superplastic deformation have been examined in a number of investigations [*7,70–73*]. It is not possible to cover this subject in detail here; however, a few key points are made. The stress–versus–strain-rate plots and the corresponding *m* values are shown in Fig. 5. The effects of increasing grain size on raising the flow stress and corresponding decrease in *m* values are shown. Flow stresses are very low (200 psi) for 10–14-μm grain

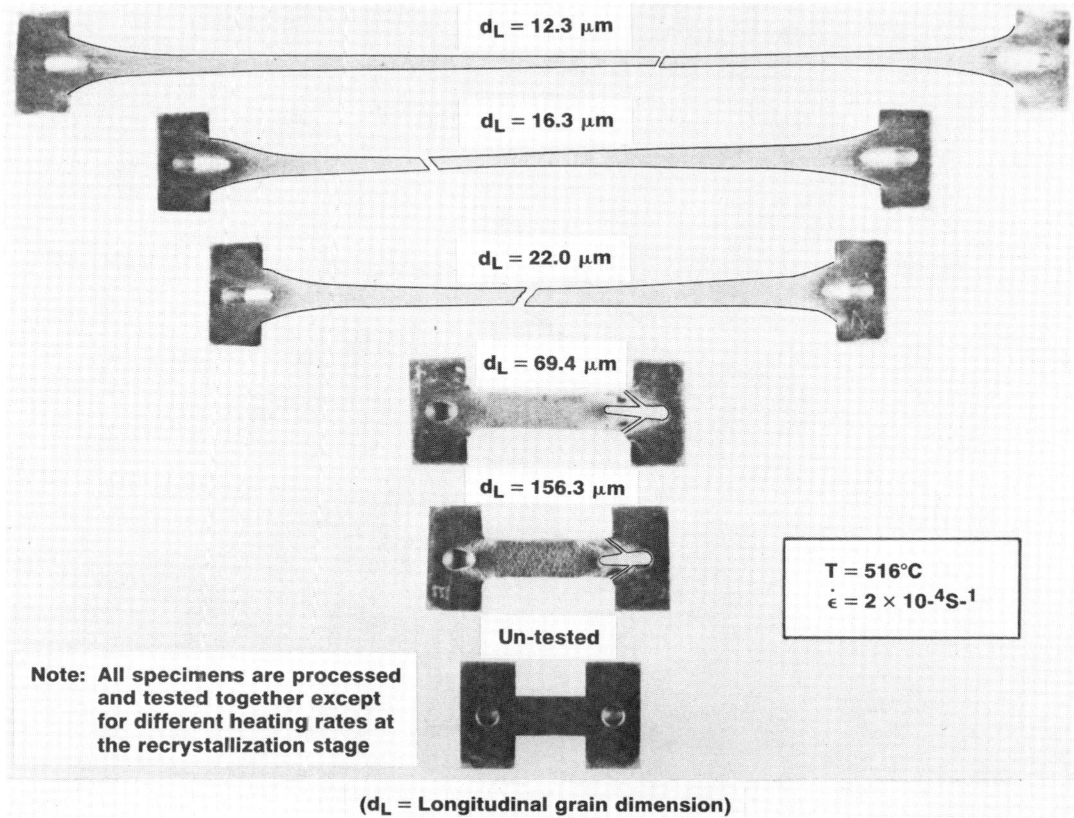

FIG. 4. Effect of grain size on superplastic ductility of 7475 alloy. T = 516°C. $\dot{\epsilon} = 2 \times 10^{-4}\ s^{-1}$. d_L = longitudinal grain dimension.

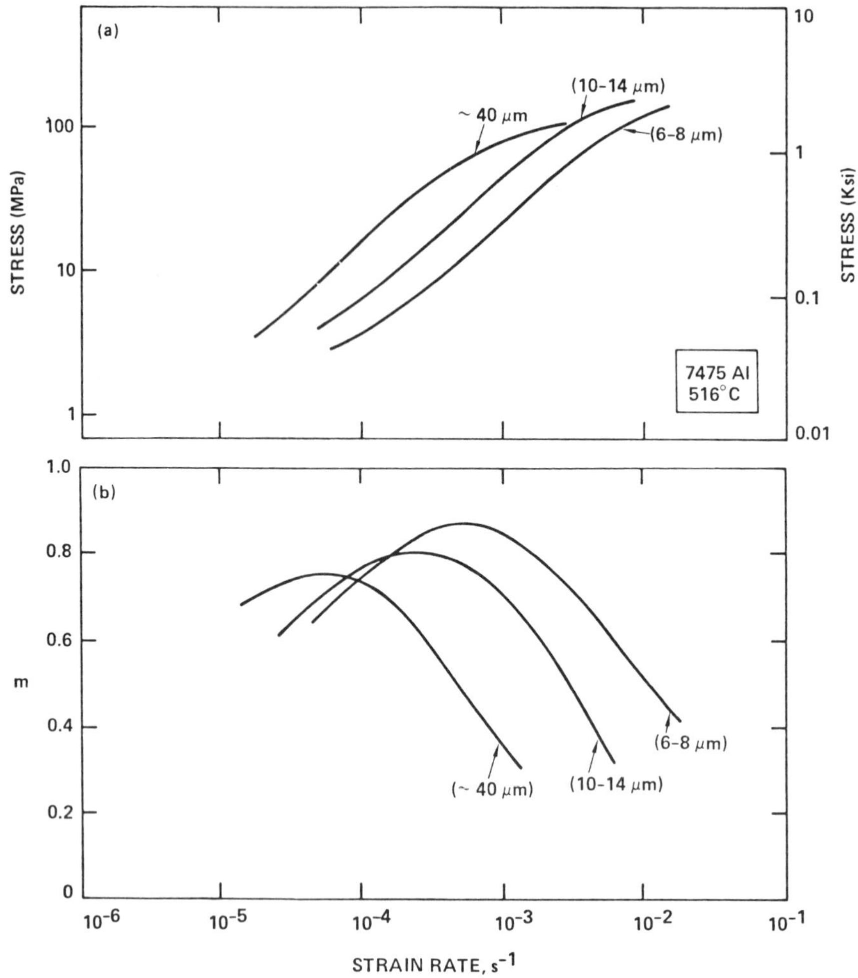

FIG. 5. Stress–versus–strain-rate plot and corresponding strain-rate sensitivity, m, values for 7475 Al for three different grain sizes.

sizes at the optimum strain rate of 2×10^{-4} s^{-1} and 515°C, thus making gas-pressure forming extremely convenient. Peak m values are very high (0.8–0.9), indicating that diffusional creep processes may be operational. These results are from step strain-rate tests carried out with small accumulated strain levels. With increasing levels of superplastic strain, considerable strain-hardening occurs in this material, which has been attributed primarily to dynamic grain growth. In a previous work [*71*], it was shown that the extent of strain-hardening increased with increase in strain rate, an effect not expected on the basis of grain growth alone.

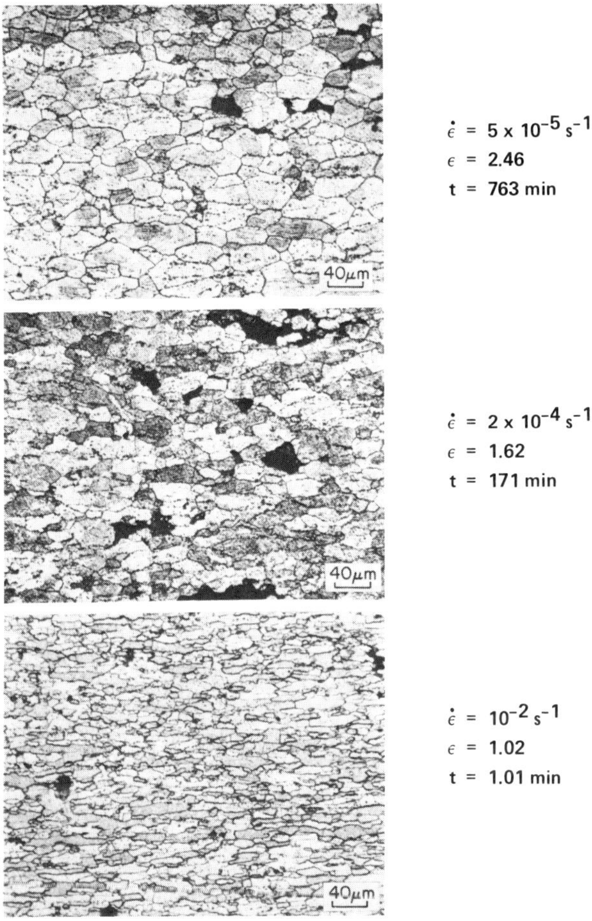

FIG. 6. Grain structure of 7475 Al under three different superplastic test conditions (all tested at 516°C). (a) $\dot{\epsilon} = 5 \times 10^{-5}$ s^{-1}, $\epsilon = 2.45$, $t = 763$ min (b) $\dot{\epsilon} = 2 \times 10^{-4}$ s^{-1}, $\epsilon = 1.62$, $t = 171$ min, and (c) $\dot{\epsilon} = 10^{-2}$ s^{-1}, $\epsilon = 1.02$, $t = 1.01$ min.

Microstructures are shown in Fig. 6 from near the fracture area of superplastically tested 7475-02 under three different test conditions. The slow strain rates of 2×10^{-4} s^{-1} and 5×10^{-5} s^{-1} promote a large extent of grain growth; however, the more rapid rate of 10^{-2} s^{-1} actually produced some degree of grain refinement. Although not shown here, extremely elongated grains with slip markings are visible at 10^{-1} s^{-1}, thus suggesting that dislocation activity is important at temperatures as high as 515°C at high strain rates. Grain-growth plots are shown in Fig. 7 for this material for longitudinal and transverse grain dimensions. Static grain

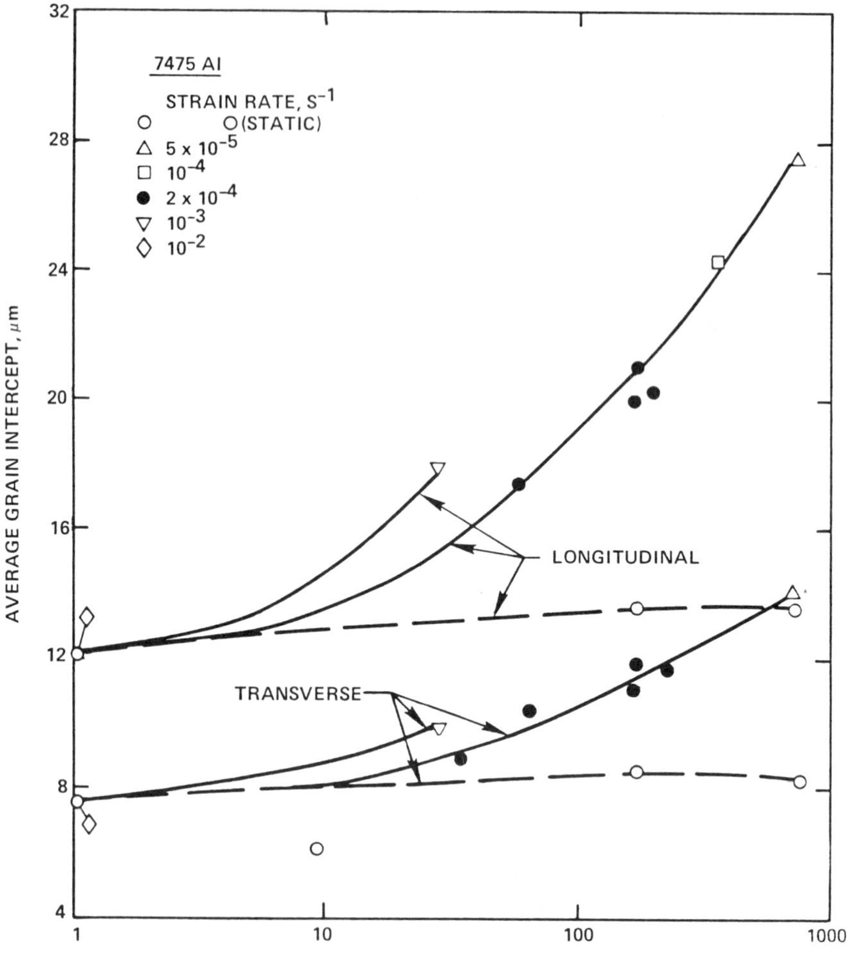

FIG. 7. Static and dynamic grain growth in 7475 Al at 516°C, tested at a variety of stain rates. Dashed lines show static grain growth. Solid lines indicate dynamic grain growth.

growth is indicated by dashed lines and dynamic grain growth for a variety of test strain rates by solid lines. As with other materials [*74,75*], dynamic grain growth in this material is significantly greater than static grain growth; however, unlike Ti-6Al-4V and Si-1%Bi alloys, the strain-rate dependence of dynamic grain growth is either absent or very small. At very high strain rates such as 10^{-2} s^{-1}, an increase in the longitudinal grain dimension with simultaneous decrease in the transverse grain dimension suggests grain elongation a manifestation of dislocation creep. While grain

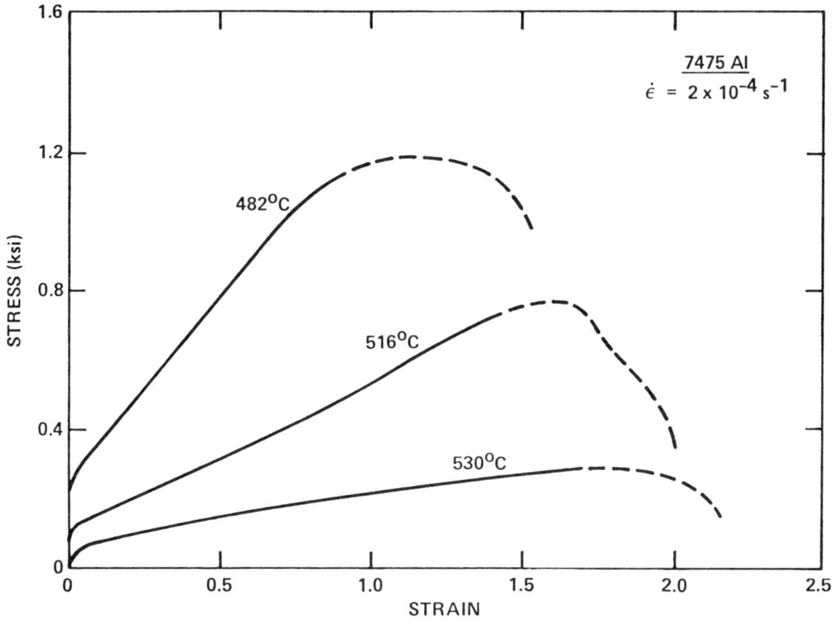

FIG. 8. Stress–strain curves for 7475 Al alloy at three different temperatures for a constant strain rate of 2×10^{-4} s^{-1}.

growth during SPF is significant and is primarily responsible for strain-hardening in this material, much evidence suggests that dislocation structure evolution processes also occur and may have important contributions to the hardening behavior.

The fact that dislocation structure evolution might be quite important in the strain-hardening process during superplastic flow is also suggested by the results of Fig. 8, which show the stress-strain curves determined at a constant strain rate (2×10^{-4} s^{-1}) for three different temperatures within the diffusional creep range. (Solution-treatment temperature for this alloy is 482°C.) The dashed portions of the curves are due to necking and strain-localization effects. While the rate of grain growth is expected to be greater at the higher temperatures, the extent of strain-hardening measured simply by $d\sigma/d\epsilon$ is smaller at the higher temperatures. However, these strain-hardening values, when normalized by the flow stress for each temperature, are more nearly equal. Similar effects have also been observed as a function of strain rate. On the other hand, hardening from dislocation processes is likely to explain this phenomenon directly. The reason why these hardening transients extend up to true strain levels of 1.0 or greater may be related to the strain partitioning that occurs between dislocation and diffusional processes.

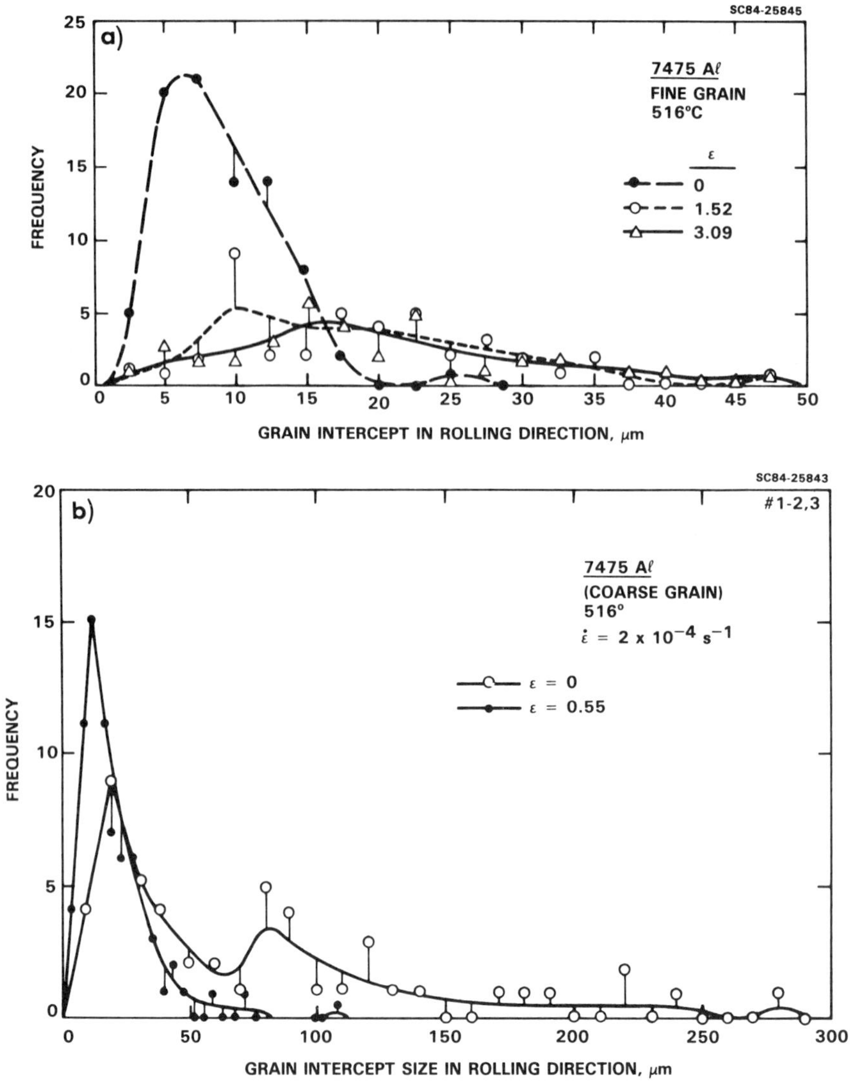

FIG. 9. Grain size distribution as a function of superplastic strain for (a) fine-grain and (b) coarse-grain 7475 Al alloys, both for $T = 516°C$, $\dot{\epsilon} = 2 \times 10^{-4}\ s^{-1}$.

At higher strain rates or for coarser-grain materials, the rate of dislocation accumulation within the grains could be rapid enough to initiate dynamic recrystallization (continuous) processes. Microstructural evidence of this has been obtained in a number of cases [*70,76*]. Figure 9 shows that grain-size distribution changes in an opposite manner if the starting grain size is fine (Fig. 9a) compared to the case in which it is coarse (Fig. 9b). An initial log normal distribution of a fine-grain material tends to flatten and spread out to larger grain sizes with increasing superplastic strain—a direct result of dynamic grain growth. On the other hand, a bimodal coarse grain distribution (Fig. 9b) sharpens to a large peak of fine-grain sizes as newly nucleated grains form—a direct result of dislocation accumulation leading to a dynamic recrystallization. Similar effects have also been observed for a fixed grain size, but as a function of strain rate. This suggests that somewhere within the superplastic strain-rate range, a combination of the two processes might occur in such a manner so as not to alter the overall grain size in a significant way. It has been suggested [*77*], however, that true grain refinement by dynamic recrystallization does not occur during high–strain-rate superplastic deformation of 7475. Rather subgrain-boundary formation could account for an apparent grain refinement.

Related to the dynamic grain-growth and recrystallization processes during superplastic deformation of 7475 is the observed bunching and coarsening of the $Al_{18}Mg_3Cr_2$ dispersoid particles [*77,78*]. The detailed mechanism by which the resulting dispersoid-free zones form adjacent to grain boundaries is another subject of conjecture.

While the superplastic flow behavior and the microstructural changes in this material continue to offer a challenge in terms of their complete understanding, large tensile elongations (800–1400%) are possible under the optimum superplastic conditions (516°C and $2 \times 10^{-4}\ s^{-1}$). In spite of this, internal microvoid formation or cavitation is a problem in this material as it is with almost all other superplastic Al alloys. Cavities are sparse and inhomogeneously distributed, but because the matrix is extremely superplastic (high m) and the strain rate is so low (long time for diffusional contribution), the cavities can grow to large sizes as large strain levels accumulate [*7,71–73,78*]. This is different from the 2004 alloy where numerous small cavities are homogeneously distributed throughout the matrix, yielding a large overall void volume. For the level of strain ($\epsilon \leqq 0.8$) encountered in most practical forming operations, cavitation in the 7475-02 alloy is imperceptible (less than 0.3 v/o).

Many factors can influence cavitation in 7475-02 [*78*], of which the important ones are alloy cleanliness, grain size, flow stress, strain rate, forming temperature, and hydrostatic pressure. Large Fe, Si, Cr-bearing, and Mg_2Si particles have been observed to initiate cavities in this alloy, although in many cases cavities could not be associated with preexisting particles. It is seen that separation at grain-boundary triple points can often be caused, triggered perhaps, by chemical segregation (or low bond strength) effects. Any factor that increases the flow stress of the alloy increases the tendency for cavitation. In this category are large grain size

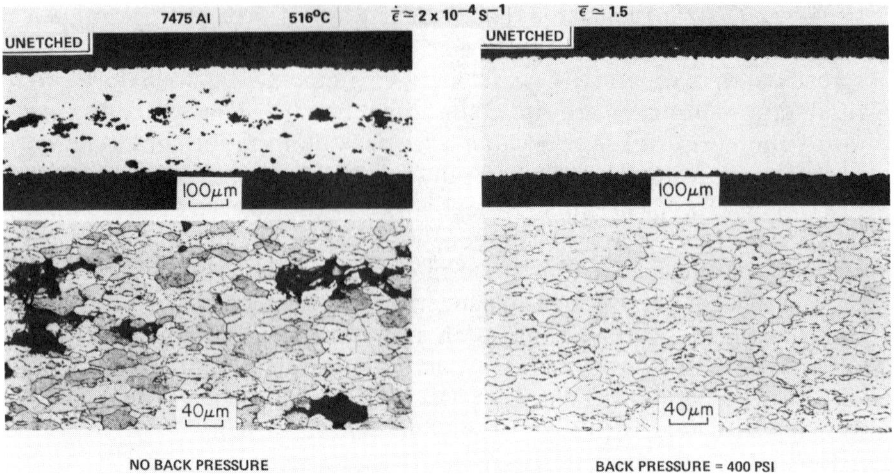

FIG. 10. Suppression of cavitation by the application of superimposed hydrostatic pressure (400 psi). Test conditions are T = 516°C, $\dot{\epsilon} = 2 \times 10^{-4}\ s^{-1}$ and $\epsilon = 1.5$.

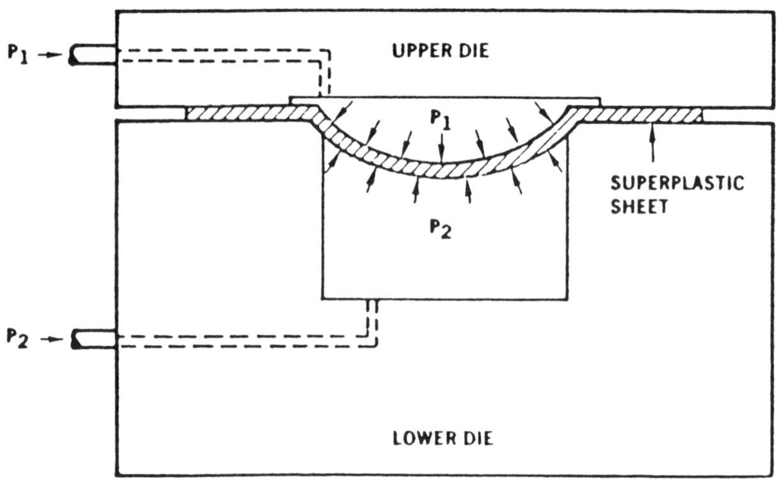

FIG. 11. Die apparatus for providing back pressure during SPF to suppress cavitation.

(or excessive rate of grain growth), higher applied strain rate, and lower forming temperature.

The effect of hydrostatic pressure is also an interesting and important one. Since both cavity nucleation and growth are influenced by hydrostatic pressure, it has been found that a superimposed pressure of the order of the flow stress during forming can suppress all cavitation (see Fig. 10.) It is important to remember that since strain-hardening raises the flow stress of the alloy in the course of forming, the level of hydrostatic pressure should be at least this increased level. A method and die apparatus for achieving this during forming is illustrated in Fig. 11. Using a common die arrangement, gas-pressure capability is also provided in the cavity into which the sheet forms. Against this back pressure, p_2, the sheet is formed by a forward pressure, $p_1 (p_1 > p_2)$ [*79,80*]. The differential pressure produces membrane stresses in the sheet which causes the sheet to form. The average pressure, $(p_1 + p_2)/2$, is the hydrostatic pressure that can suppress cavitation. This method is easy to implement in the existing SPF dies and is finding application in the aerospace industry.

D. Aluminum-Lithium Alloys

Superplasticity has been developed in a number of Al-Li based alloys [*60–64*]. These alloys are of considerable interest in the aerospace industry because lithium has the unusual effect of simultaneously increasing elastic modulus, decreasing density, and providing potent age-hardenability. (See Quist and Narayanan, p. 219).

The Al-Li–based alloy system provides an interesting case study for development of superplasticity in aluminum alloys. The alloys are all of the low particle volume fraction (f), low particle radius (r) category, as discussed earlier, and have been shown to be responsive to fine-grain processing by both static (discontinuous) and dynamic (continuous) recrystallization. However, since most of the useful Al-Li alloys contain Al_3Zr as the main dispersoid phase, they tend to be more closely related to the 2004-type superplastic behavior than the 7475-02 type. Thus, even in cases where fine-grain processing is carried out by static (discontinuous) recrystallization, optimum superplasticity is usually achieved at a relatively fast strain rate, compared with 7475-02. This is in some cases because of a strong tendency for dynamic (continuous) recrystallization in the Al_3Zr-containing alloys during superplastic deformation at strain rates toward the upper end of the superplastic range.

The Al-Li–based superplastic alloys also show similarity to the 2004 type in their cavitation behaviors [*81*]. The relatively high optimum superplastic strain rates, with the associated relatively high flow stresses, tend to cause early and copious nucleation of intergranular cavities. This generally occurs to such an extent that the apparent superplastic flow stress shows strain-softening rather than strain-hardening expected from dynamic grain growth. Fortunately, the Al–Li-based alloys have shown useful superplasticity in strain-rate/flow-stress regimes where the imposed hydrostatic pressure method for cavitation control is still practical and effective, as with the 7475-02 type discussed earlier.

Interestingly, the Al-Li–based alloys do not require the very high Zr contents nor the coarse, primary ($CuAl_2$) phase particles of the 2004-type alloys for the dynamic recrystallization–superplastic behavior. This seems to be related to the presence of Li, but has not been adequately explained as of mid-1988.

Another important aspect of the superplastic Al-Li alloys is that, even when cavitation is prevented, their post-superplastic properties are generally more adversely affected by the superplastic deformation than those of other superplastic aluminum-alloy systems. This is due to two effects: (1) The useful Al-Li based alloys developed to date have multiphase precipitation-hardening systems which provide most effective strengthening when a cold-deformation stage is included in the aging process. (See Quist and Narayanan, p. 219). This cold-deformation stage is not generally practical with a complex, net-shaped superplastically formed part. (2) The Al-Li–based alloys tend to develop strong crystallographic textures as mill-processed, unrecrystallized sheet. The resulting mechanical properties are strongly anisotropic with generally higher heat-treated strengths than in a fully recrystallized, weaker-textured microstructure, as found in post-superplastic-formed condition.

E. Powder-Metallurgy–Processed Alloys

The use of powder- or particulate-metallurgy (PM) methods to manufacture Al-based alloys is an area of technological interest. (See Hildeman and Koczak, p. 324.) This is especially true for the cases where the powder particles are manufactured using rapid solidification processing (RSP). It is generally accepted that RSP results in a number of microstructural benefits. Grain-size refinement is often considered to be one such benefit. It might therefore be anticipated that alloys produced by RSP could behave in a superplastic manner in the as-consolidated condition. Although superplasticity can be found in some consolidated RSP alloys (for example, in certain nickel-based alloys and high-carbon, Fe-based alloys [7]), studies have shown that in aluminum alloys manufactured by RSP, superplasticity is not necessarily an inherent property [*82*].

The transition from nonsuperplastic to superplastic behavior in a PM Al-3Cu-2Li-1Mg-0.2Zr alloy can be understood by examination of the microstructure before and after the thermomechanical treatment. In Fig. 12a, a transmission electron photomicrograph is shown of the alloy in

FIG. 12. (a) Transmission electron photomicrograph of a PM Al-3Cu-2Li-1Mg-0.22r alloy in the as-extruded condition after heating to 500°C for 0.5 hr and water quenching. The microstructure consists predominantly of medium-angle subgrains.
(b) Transmission electron photomicrograph of the same alloy after thermomechanical processing. The microstructure consists predominantly of large-angle, disordered boundaries.

the as-extruded condition (after heating to 500°C for 1/2 hr). Upon close inspection of this fine microstructure, it can be determined that the apparently fine grains are in fact medium-angle subgrains. The misorientation angles are in the 3–10° range. Furthermore, these boundaries are often quite complex, having a mixed twist–tilt character. Such boundaries are not considered to be particularly mobile and would not be expected to slide under an applied stress. After the thermomechanical-processing treatment (Fig. 12b), the microstructure contains very fine grains (2–3 μm) having an equiaxed configuration and the grain-boundary misorientations now generally fall into the classification of large-angle, disordered boundaries (i.e., >10° misorientation). In Fig. 12b, examples of such boundaries are shown. One particular disorientation was measured to be [0.603, 0.518, −0.256]/49.61°, which is approximately 7° from a $\Sigma 29_b$ boundary ($[22\bar{1}]$/46.39°); i.e., it is a large-angle, disordered grain boundary. Such boundaries would be expected to be highly mobile under stress and to be able to slide. It is noted that in some cases, (in Fig. 13), a low-angle boundary is also present within the grain. These subboundaries found in the recrystallized material, however, have misorientations of <3° and are of a relatively simple configuration, often of the type [100]/0.25–0.5°. Such low-angle subboundaries are relatively mobile by comparison with those having medium-angle misorientations and may even participate in superplastic flow either by accommodating some of the applied stress or by acting as sources and sinks for dislocations for adjacent, sliding, large-angle boundaries.

The finer-scale insoluble particles and included oxide dispersion found in PM-aluminum alloys, compared with IM alloys, have implications for both the recrystallization behavior and the subsequent cavitation behavior during superplastic forming. Significant increases in the strain rate, at which maximum strain-rate sensitivity occurs, have been observed in PM-aluminum alloys compared with equivalent ingot-aluminum alloys [*83*].

Superplastic-like behavior at high strain rates has been demonstrated in some PM-aluminum alloys [*84*]. These include the IN-9021 alloy (which is prepared by mechanical alloying) and the PM-2024 alloy with relatively large additions of Zr (0.6 wt %). These alloys showed about 300% ductility at strain rates of nearly 1 s^{-1}. The detailed microstructural mechanisms leading to this behavior are not well understood as of mid-1988.

F. Aluminum–Silicon-Carbide Metal-Matrix Composites

Silicon carbide–particulate–(or whisker–) reinforced aluminum composites (SiC_p/Al) have become commercially available in large quantities. The materials exhibit a unique combination of high specific modulus and

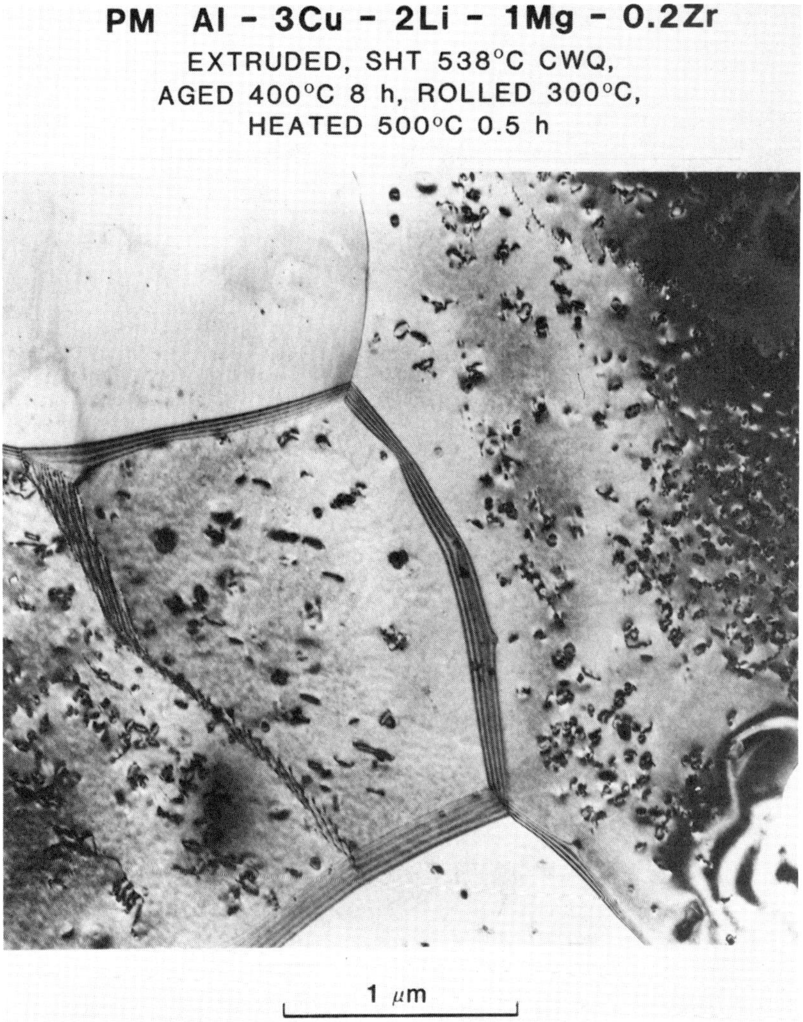

FIG. 13. A single large-angle grain from thermomechanically processed PM Al-3Cu-2Li-1Mg-0.2Zr alloy that contains a low-angle boundary.

strength at room temperature as well as good thermal stability. The advantages of these engineering materials over the continuous-fiber–reinforced composites are improved ductility and reduced anisotropy in mechanical properties. Most importantly, they are considerably less expensive. These properties make SiC_p/Al attractive for many structural applications.

The development of a manufacturing technology for SiC_p/Al and, in particular, development of forming operations have received only very

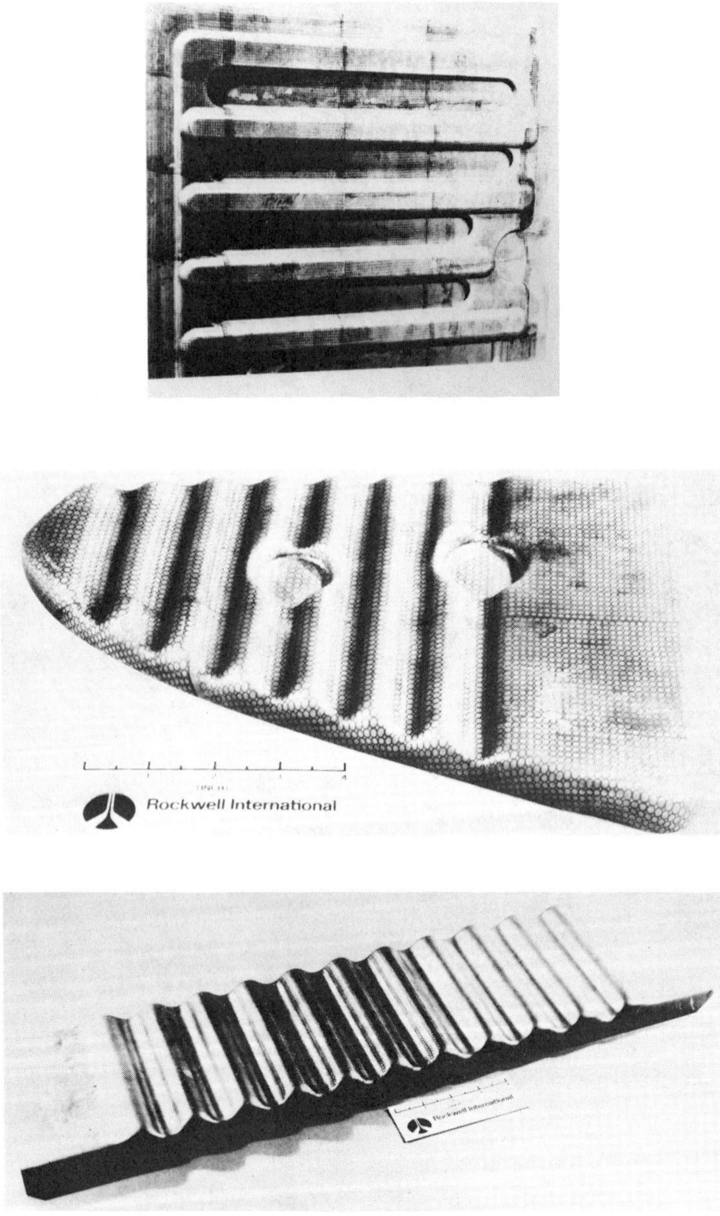

FIG. 14. Aircraft parts fabricated from SiC-reinforced Al
(a) foil-based composite fuselage-access door.
(b) PM-based composite wing rib
(c) PM-based sine-wave beam.

limited attention. The main difficulties encountered are due to the relatively low fracture-toughness and ductility of the composites, even at elevated temperature. The development of superplasticity in SiC/Al composite materials, however, has been investigated [*85,87*].

Special SiC_p/Al composites have been fabricated [*85*] by distributing SiC whiskers between thin foils of fine-grained, superplastic 7475 alloy that had been specially prepared. An optimum elongation-to-failure of 350% was obtained at 525°C with a slow strain rate of $2 \times 10^{-4}\ s^{-1}$.

Superplasticity has been demonstrated [*86*] in a commercial $SiC_p/2024$ Al using internal stress superplasticity. This procedure involved longtime, thermal cycling and resulted in an optimum elongation-to-failure of approximately 300% at a strain rate of about $4 \times 10^{-5}\ s^{-1}$.

Superplasticity has also been developed [*87,88*] in a composite consisting of the high-strength PM-7064 alloy matrix reinforced with 10–20 vol % silicon-carbide particulates (SiC_p). Elongations as high as 450% have been observed with processing conditions very similar to those for 7475-02. This approach may offer one of the most promising approaches to a high-strength, high-stiffness, superplastic aluminum alloy for highly loaded airframe structures, such as the examples in Fig. 14.

It has been demonstrated [*84*] that commercial $SiC_p/2124$ Al, following appropriate thermomechanical treatments, can exhibit superplastic-like behavior (300%) during isothermal deformation at relatively high strain rates of $3.3 \times 10^{-1}\ s^{-1}$. (See earlier PM-aluminum section.) Further development of one or more of these approaches seems to be justified due to the synergistic natures of the high-performance, high-cost, and difficult-to-fabricate materials with the low-cost, complex, net shape capabilities of the superplastic-forming process.

IV. Summary

Superplasticity has been developed in a wide range of aluminum-based materials. The superplastic-forming process is becoming widely recognized as an attractive manufacturing process, especially for high-performance materials that are otherwise expensive or sometimes impossible to fabricate to complex-shaped sheet-metal components.

The critical prerequisite of a thermally stable, fine grain size for superplasticity has been satisfied by a number of thermomechanical processes, with the overwhelming majority of useful aluminum alloys employing discontinuous or continuous recrystallization.

Improved knowledge and understanding of the dynamic microstructural changes during superplastic deformation (such as grain growth, recrystallization, dispersoid coarsening and bunching, and cavitation) would greatly facilitate development of enhanced superplastic alloys and processing methods.

References

1. Edington J.W., Melton, K.N., and Cutler C.P. (1976). *Progr. Mater. Sci.*, **21**, 61–58.
2. Edington, J.W. (1976). *Metals Technol.*, **3**, 138–153.
3. Taplin, D.M., Dunlop, G.L., and Langdon, T.G. (1979). *Ann. Rev. Mater. Sci.*, **9**, 151–189.
4. Mukherjee, A.M. (1979). Ibid, **9**, 191–217.
5. Padmanabhan, K.A., and Davies, G.J. (1980). *Superplasticity, Materials Research and Engineering*, **vol. 2**, B. Ilschner, ed. Springer-Verlag, Berlin.
6. Wadsworth, J., Oyama, T., and Sherby, O.D. (1980). In *Advances in Materials Technology in the Americas 1980*, **vol. 1**, I. LeMay, ed. ASME, New York, pp. 29–41.
7. Paton, N.E. and Hamilton, C.H. eds. (1982). *Superplastic Forming of Structural Alloys* AIME, Warrendale, Pennsylvania.
8. Ahmed, M.M.I., and Langdon, T.G. (1977). *Metall. Trans. A*, **8A**, 1832–1833.
9. Sherby, O.D., and Wadsworth, J. (1984). In *Deformation Processing and Structure*, G. Krauss, ed. ASM, Metals Park, Ohio, pp. 355–389.
10. Pearce, R, and Kelly, L., eds. (July 1985). *Superplasticity in Aerospace-Aluminum*, Proc. Intl. Conf., Ashford Press, Southampton, U.K.
11. Greenwood, J.N, Miller, D.R., Suiter, J.W. (1954). *Acta Met.*, **2**, 250.
12. Brookes, P.E., Kirby, N., and Burke, W.T. (1960). *J. Inst. Met.*, **88**, 500.
13. Williams, H.D. (1965). CERL Report RD/L/R 1341.
14. Corti, C.W., Broom, T. (1965). CERL Report RD/L/R 1304.
15. Sagat, S., Blenkinsop, P., and Taplin, D.M.R. (1972). *J. Inst. Met.* **100**, 268.
16. May, M.J., and Honeycombe, R.W.K. (1963). *J. Inst. Met.*, **92**, 41.
17. Chen, C.W. and Machlin, E.S. (1956). *Acta Met.* **4**, 655.
18. Grant, N.J. (1959). *Fracture.* John Wiley, New York, p. 562.
19. Intrater J., Machlin, E.S. (1959). *Acta Met.* **7**, 140.
20. Perry, A.J. (1974). *J Mats. Sci.*, **9**, 1016.
21. Dunlop, G.L., Shapiro, E., Taplin, D.M.R., Crane, J. (1973). *Met. Trans.*, **4**, 2039.
22. Taplin, D.M.R. and Sagat, S. (1972). *Mat. Sci. Eng.* **9**, 53.
23. Matuki, K., and Yamada, M. (1973). *J. Japan Inst. Met.* **37**, 448.
24. Jovane, F. (1968). *International Journal of Mechanical Sciences*, **10**, 403–427.
25. Cornfield, G.C., and Johnson, R.H. (1970). *International Journal of Mechanical Sciences*, **12**,, 479–490.
26. Holt, D.L. (1970). *International Journal of Mechanical Sciences*, **12**, 491–497.
27. Ghosh, A.K., and Hamilton, C.H. (1978). *Proceedings of American Society for Metals on Process Modeling-Fundamentals and Applications to Metals*, pp. 303–331.
28. Yu-Quan, S. (1987). *Material Science and Engineering*, **86**, 179–189.
29. Chandra, N., Goforth, R.E., and Haisler, W.E. (1985). ASM Metals Congress, Toronto, Canada, ASM Paper no. 8511-004.

30. Bellet, M., and Chenot, J.L. (1986). Proceedings of International Conference on Titanium Products and Applications, San Francisco, USA, October 1986.
31. Shah, K.N., Story, J.M., and Chandra, N. (January 1988). Poc. TMS-AIME Symp. *Superplasticity in Aerospace*, T.R. McNelley and H.C. Heikkenen, eds. Phoenix, Arizona.
32. Holt, D.L., and Backofen, W.A. (1966). *Trans. Quart, ASM*, **59**, 755–768.
33. Stowell, M.J., Robertson, J.L., and Watts, B.M. (1969). *Metal Sci. Jour.*, **3**, 41–45.
34. Watts, B.M., and Stowell, M.J. (1971). *Jour. Mater. Sci.*, **6**, 228–237.
35. Cutler, C.P., Edington, J.W., Kallend, J.S., and Melton, K.N. (1974). *Acta Metall.*, **22**, 665–671.
36. Rai, G., and Grant, N.J. (1975). *Metall. Trans. Ser. A*, **6A**, 385–390.
37. Kobayashi, Y., Ishida, Y., and Kato, M. (1977). *Scripta Metall.*, **11**, 51–54.
38. Matsuki, K., and Yamada, M. (1973). *J. Japan Inst. Metals*, **37**, 448–454.
39. Matsuki, K., Ueno, Y., and Yamada, M. (1974). *J. Japan Inst. Metals*, **38**, 219–226.
40. Matsuki, K., Morita, H., Yamada, M., and Murakami, Y. (1977). *Metal Sci.*, **11**, 156–163.
41. Otsuka, M., Mivra, Y., and Horichi, R. (1974). *Scripta Metall.*, **8**, 1405–1408.
42. Marya, S.K., and Wyon, G. (1976). In Proc. Fourth Int. Conf. Strength of Metals and Alloys, Nancy, France, 1976, **vol. 1**, pp. 438–442.
43. Weill, F., and Wyon, G. (1979). Proc. Fifth Int. Conf. Strength of Metals and Alloys Aachen, West Germany, 1979, pp. 387–392.
44. Piatti, G., Pellegrini, G., amd Trippodo, R. (1976). *J. Mater. Sci.*, **11**, 186–190.
45. Piatti, G. (1983). *Jour. Mater. Sci.*, **18**, 2471–2483.
46. Moles, M.D.C., and Davies, G.J. (1976). *Scripta Metall.*, **10**, 455–458.
47. Chung, D.W., and Cahoon, J.R. (1979). *Metal Sci.*, **13**, 635–640.
48. Matsuki, K., Yetani, Y., Yamada, M., and Murakami, Y. (1979). *Metal Sci.*, **10**, 235–242.
49. Lee, E.W., McNelley, T.R., and Stengel, A.F. (1986). *Met. Trans.*, **17A**, 1043.
50. McNelley, T.R., Lee, E.W., and Garg, A. (1986). *Aluminum Alloys—Their Physical and Mechanical Properties*, **vol. 11**, E.A. Starke and T.H. Sanders, eds. Chameleon Press, London, p. 1269.
51. Piatti, G., and Bardy, M. (1981). *Metal Sci.*, **15**, 55–62.
52. Grimes, R., Baker, C., Stowell, M.J., and Watts, B.M. (1975). *Aluminum*, **51**, 70–72.
53. Grimes, R., Stowell, M.J., and Watts, B.M. (1976). *Metals Technol.*, **3**, 154–160.
54. Watts, B.M., Stowell, M.J., Baikie, B.L., and Owen, D.G.E. (1976). *Metal Sci.*, **10**, 189–197.
55. Watts, B.M., Stowell, M.J., Baikie, B.L., and Owen, D.G.E. (1976). *Metal Sci.*, **10**, 198–206.
56. Wert, J.A., Paton, N.E., Hamilton, C.H., and Mahoney, M.W. (1981). *Metall. Trans. Ser. A*, **12A**, 1267–1276.
57. Paton, N.E., Hamilton, C.H., Wert, J.A., and Mahoney, M.W. (1982). *J. Metals*, **34**, 21–27.
58. Bampton, C.C., Wert, J.A., and Mahoney, M.W. (1982). *Metall. Trans.*, **13A**, 193–198.
59. Moore, D.M., and Morris, L.R. (1980). *Mater. Sci. Eng.*, **43**, 85–92.
60. Wadsworth, J., Henshall, C.A., and Nieh, T.G. (1986). *Aluminum-Lithium Alloys III*, C. Baker, P.J. Gregson, S.J. Harris and Peel, C.J. eds. Inst. Metals, London, pp. 199–212.
61. Grimes, R., and Miller, W.S. Ibid, pp. 153–168.
62. Lederich, R.J., and Sastry, S.M.L. (1984). *Aluminum-Lithium Alloys II*, E.A. Starke, Jr., and T.H. Sanders, eds. TMS-AIME, Warrendale, Pennsylvania, pp. 137–149.

63. Ghosh, A.K., and Gandhi, C. (August 1985). Proc. ICSMA7, Montreal, Canada, August 1985, pp. 2065–2072.
64. Bampton, C.C., Cheney, B.A., Cho, A., Ghosh, A.K., and Gandhi, C. (January 1988). Proc. TMS-AIME Symp. on *Superplasticity in Aerospace*, T.L. McNelley and H.C. Heikkenen, eds. Phoenix, Arizona, January 1988.
65. Zener, C. (1949). Private communication to C.S. Smith, *Trans. AIME*, **175**, 15.
66. Gladman, T. (1966). *Proc. Roy. Sci.*, **294**, 298–309.
67. Hillert, M. (1965). *Acta Met.*, **13**, 227.
68. Wold, K.G., and Chambers, F.M. (1968). *Australian Institute Metals*, **13**, 79–87.
69. Haroun, N.W., and Budworth, D.W. (1968). *J. Mat. Sci.*, **3**, 326–328.
70. Ghosh, A.K., and Hamilton, C.H. (May 1982). *Met. Trans. A*, **13A**, 733–743.
71. Ghosh, A.K. (1981). "Deformation of Polycrystals—Mechanisms and Microstructures." In *Proc. 2nd Riso Int. Symp. on Met. and Mat. Sci.*, N. Hanson et al., eds. Riso National Laboratory Roskilde, Denmark p. 277.
72. Bampton, C.C., and Edington, J.W. (1982). *Metal. Trans. A*, **13A**, 1721–1727.
73. Bampton, C.C., and Edington, J.W. (1983). *J. Engr. Mats. and Tech.*, **105**, 55–60.
74. Ghosh, A.K., and Hamilton, C.H. (1979). *Met. Trans.*, **10A**, 699.
75. Clark, M.A., and Alden, T.H. (1973). *Acta. Met*, **21**, 1195.
76. Ghosh, A.K. (February 1984). AirForce office of Scientific Research Interim Report, contract no. F49620-83-C-0055.
77. Wert, J.A., and Varloteaux, A. (1986). *Aluminum Alloys—Physical and Mechanical Properties*, **vol. II**, E.A. Starke, Jr., and T.H. Sanders, Jr., eds. EMAS, Warley, U.J., pp. 1255–1267.
78. Bampton, C.C., Ghosh, A.K., and Mahoney, M.W. in Reference 10, pp. 1–35.
79. Bampton, C.C. and Raj, R. (1982). *Acta Metall.*, **30**, 2043–2053.
80. Bampton, C.C., Mahoney, M.W., Hamilton, C.H., Ghosh, A.K., and Raj, R. (1983). *Metal. Trans. A*, **14A**, 1583–1591.
81. Pandey, M.C., Wadsworth, J., and Mukherjee, A.K. (1986). *Mat. Sci. and Engr.*, **78**, 115–125.
82. Wadsworth, J., and Pelton, A.R. (1984). *Scripta Met.*, **18**, 387–392.
83. Petit, J.I., Lege, D.J., and Story, J.M. (1983). *Materials and Processes-Continuing Innovations*, 26th National SAMPE Symposium, p. 1186.
84. Wadsworth, J., Henshall, C.A., Nieh, T.G., Pelton, A.R., and Gilman, P.S. (1985). *Aluminum Powder Metallurgy*, Proc. TMS-AIME Symp., Toronto, Canada, October 1985, pp. 137–154.
85. Mahoney, M.W., and Ghosh, A.K. (1982). *Silicon Carbide Reinforced Aluminum Structure*. AFWAL-TR-82-3051.
86. Wu, Mu-Yeh, and Sherby, O.D. (1984). *Scripta Metall.*, 773–776.
87. Mahoney, M.W., and Ghosh, A.K. (1987). *Metall. Trans.*, **18A**, 653–661.
88. Mahoney, M.W., Ghosh, A.K., and Bampton, C.C. (1987). *Composite Materials*, Proc. Intl. Conf., ICCM-6, Imperial College, London, July 1987.

Part IV

Current Aluminum Alloys

8

Aluminum-Lithium Alloys

W.E. QUIST and G.H. NARAYANAN

Boeing Commercial Airplanes
Seattle, Washington

I. Introduction

The single largest alloy-development effort in the history of aluminum metallurgy is taking place in an effort to bring to commercialization a special group of aluminum alloys containing lithium as one of the primary alloying elements. Lithium has the unique capability of increasing the modulus of elasticity while substantially lowering the density of aluminum. Studies of various types of aerospace vehicles have consistently shown that a reduction in density has a greater effect than any other property change

ISBN 0-12-341831-3

in reducing the structural weight of the vehicle. In response to that need, most major aluminum companies and aluminum research laboratories have put forth an unprecedented effort to develop this new class of materials. To date (summer 1988), seven new aluminum-lithium alloys have been registered as a result of this effort, with several more to come in the near future. The complexity of the various factors contributing to the strength and toughness of alloys in this system has presented a continuing metallurgical challenge. Aluminum-lithium–based alloys under development possess good combinations of strength, damage-tolerance, and durability and in general are quite weldable. In addition, several have demonstrated excellent cryogenic and elevated-temperature properties. Both ingot-metallurgy (IM) and powder-metallurgy (PM) appproaches have been explored for these materials, but major advances have been made by the former technique. Ingot-produced materials promise lower material costs and the properties achieved have been very competitive with those offered by conventional aluminum alloys.

II. History

Small amounts of lithium were added to Al-Zn-Cu alloys at least as early as 1924 in Germany, ostensibly to help improve properties such as strength [*1*,*2*,*3*], but it was the discovery by LeBaron in 1942 that lithium could be a major strengthening element in aluminum-copper alloys when accompanied by small amounts of various other elements, principally cadmium, that marks the true beginning of aluminum-lithium alloy metallurgy as we know it [*4*]. LeBaron's 1945 patent eventually led to the development of the Al-1.1Li-4.5Cu-0.5Mn-0.2Cd alloy X2020 by Alcoa in 1957, which was subsequently used by the RA-5C Vigilante aircraft for wing covers and horizontal stabilizers [*5*]. The emerging discipline of fracture mechanics soon discovered, however, that the fracture-toughness of alloy 2020 was the poorest of all commercially used aerospace alloys [*6*]. Several other aircraft programs considered the use of this strong, high-modulus, and light alloy, but its poor fracture characteristics soon brought the alloy into disfavor, and it was never used again for primary aircraft structure [*7*]. Alloy 2020 was subsequently withdrawn from registration in 1974. The Soviets, however, became interested in aluminum-lithium technology and soon developed a 2020-type alloy, VAD-23 [*8*], with a nominal composition of Al-1.1Li-5.3Cu-0.6Mn-.17Cd. They developed a lower-strength temper for their alloy, however, and it displayed somewhat improved fracture characteristics as compared to the alloy 2020-T6. Later, but still in the 1960s, the Soviets introduced another alloy, 01420, having the nominal composition of Al-2.0Li-5.3Mg-0.5Mn [*9,10*]. Alloy 01420 is the first aluminum-lithium alloy having characteristics similar to modern alloys of this system in that it displayed a density that was significantly lower than other aerospace aluminum alloys. The Soviets have claimed wide use of these alloys on their aircraft structures. Although a number of variations of alloy 01420 have been discussed in the literature, this alloy was nevertheless the last aluminum-lithium alloy publicly announced by the Soviets—that being in the mid-1960's.

Research on the aluminum-lithium system was resurrected in Britain and the United States in the early 1970s. Work started at first in Great Britain at the Fulmer Research Institute about 1970 [*11*] and in the United States (under Naval Air Systems Command sponsorship) at the Alcoa laboratories in 1973 [*12*]. These R&D efforts continued throughout the 1970s with a first by Balmuth and Schmidt [5] and the second by Staley, in Chapter 1 of this book.

In the late 1970s, however, several events combined to greatly accelerate developments and interest in the aluminum-lithium system. Perhaps the most important item was the growing belief that resin-matrix composites might soon replace aluminum as the primary aerospace material of construction, if significant advancements in aluminum metallurgy did not take place [*15*]. Another important factor was the recognition that density had a much greater influence in reducing the structural weight of an aerospace vehicle than did strength, fracture-toughness, fatigue, or elastic modulus [*16,17*]. These factors combined to stimulate and motivate major aluminum companies, aluminum metallurgists, aluminum users in the aerospace industries, universities, and government and independent laboratories to undertake an unprecedented development effort in the aluminum-lithium system. This high level of activity has persisted with the introduction of several new commercially viable AL-Li–based alloys by major aluminum companies in the United States, Britain, and Continental Europe. It is believed that these alloys have the distinct possibility of eventually supplanting most aluminum alloys being used in the aerospace industry as of the late 1980s. A depiction of many of the important events that led to the development of Al-Li alloys is shown in Fig. 1.

Al-Li–type alloys under development in the late 1980s rely on further additions of copper, zirconium, and/or magnesium as essential alloying elements to obtain the desired mix of engineering properties. Apart from density, the principal properties of interest include strength, fracture-toughness, modulus of elasticity, and fatigue, corrosion, and stress corrosion cracking resistance.

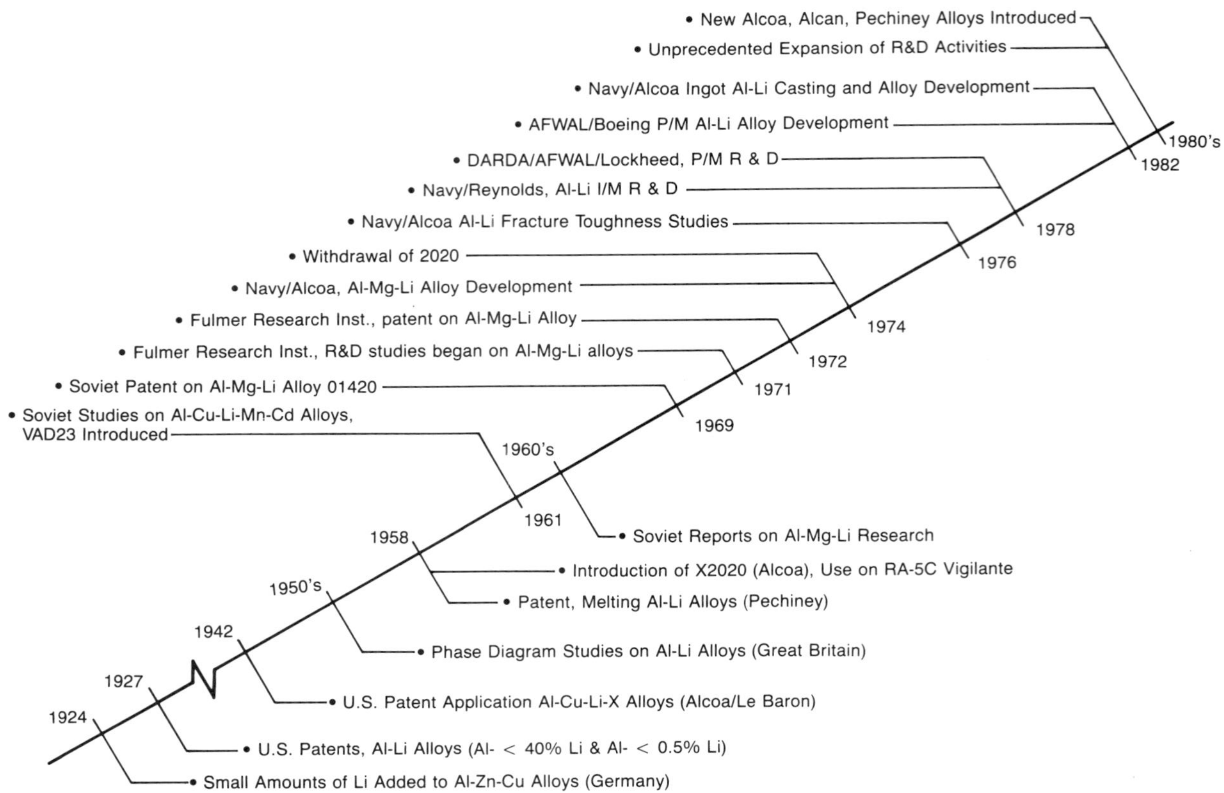

FIG. 1 A summary of important events in the developmental evolution of Al-Li alloys.

As of mid-1988, a total of eight aluminum-lithium type alloys have been registered as commercial alloys by the Aluminum Association. The first of these was alloy 2020, which was subsequently withdrawn in 1974. Two additional commercial alloys that have not been included in the Aluminum Association registrations list are the Soviet alloys VAD-23 and 01420. Of the seven alloys registered with the Aluminum Association in 1988, four have received foremost attention by the aerospace industry to this point:

- Alloy 2090, registered by Alcoa in August 1984
- Alloy 2091, registered by Pechiney in April 1985
- Alloy 8090, registered by Alcan and Pechiney in May 1985
- Alloy 8091, registered by Alcan in March 1985

The compositions of these four alloys, as well as those of the remaining three registered alloys, are shown in Table I. In addition to these alloys, there are perhaps a dozen more that are under active development at various laboratories throughout the world, a number of which may show advantages over the initial group in specific areas and will likely be brought to commercialization.

III. Physical Metallurgy of Al-Li–Based Alloys

As is the case with the conventional 2000- and 7000-series aluminum alloys, the Al-Li–based alloys under development are also precipitation-hardened. However, the phase equilibria and precipitation reactions that control the microstructure and hence the properties are significantly different from those of conventional alloys. A detailed discussion of the microstructures of the Al-Li alloys is presented by Williams and Howell in Chapter 11 of this book and so will not be included here.

As is evident from Table I, all commercial Al-Li–based alloys, as well as those under development as of the late 1980s, incorporate copper, Zirconium and usually magnesium as essential alloying elements to develop acceptable combinations of strength and toughness. The effect of these ternary, quaternary, and quinary alloy additions to Al-Li binary alloys is to significantly alter their phase equilibria and modify the precipitation sequence. When binary Al-Li alloys containing lithium in excess of about 1.3 wt % are quenched from above the solvus temperature, the breakdown of the solid solution occurs by the precipitation of the metastable ordered phase δ' (Al_3Li), which nucleates homogeneously throughout the matrix [*18–22*]. The δ' precipitate has an Ll_2 superlattice structure and is fully coherent with the solid-solution matrix. Owing to the small misfit strain, the precipitates exhibit a spherical morphology. Upon artificial aging, these

TABLE I

THE CHEMICAL COMPOSITION OF ALUMINUM-LITHIUM–TYPE ALLOYS REGISTERED WITH THE ALUMINUM ASSOCIATION AS OF MARCH 1988

	Alloy						
Element[a]	2090 Aloca August 6, 1984	2091 C. Pechiney April 8, 1985	8090 Alcan and C. Pechiney May 1985	8090A Alcoa Late 1985	8091 Alcan March 29, 1985	X8092 Alcoa May 1985	X8192 Alcoa August 1985
Si	0.10	0.20	0.20	0.10	0.30	0.10	0.10
Fe	0.12	0.30	0.30	0.15	0.50	0.15	0.15
Cu	2.4–3.0	1.8–2.5	1.0–1.6	1.1–1.6	1.8–2.2	0.5–0.8	0.4–0.7
Mn	0.05	0.10	0.10	0.05	0.10	0.05	0.05
Mg	0.25	1.1–1.9	0.6–1.3	0.8–1.4	0.5–1.2	0.9–1.4	0.9–1.4
Cr	0.05	0.10	0.10	0.05	0.10	0.05	0.05
Ni	—	—	—	—	—	—	—
Zn	0.10	0.25	0.25	0.10	0.25	0.10	0.10
Ti	0.15	0.10	0.10	0.15	0.10	0.15	0.15
Li	1.9–2.6	1.7–2.3	2.2–2.7	2.1–2.7	2.4–2.8	2.1–2.7	2.3–2.9
Zr	0.08–0.15	0.04–0.16	0.04–0.16	0.08–0.15	0.08–0.16	0.08–1.5	0.08–1.5
Other: each	0.05	0.05	0.05	0.05	0.05	0.05	0.05
total	0.15	0.15	0.15	0.15	0.15	0.15	0.15

[a] Numbers shown are either maximums or ranges.

precipitates coarsen, but they retain their coherency and spherical morphology to relatively large sizes ($\lesssim$50 nm). δ' contributes significantly to the precipitation-hardening of these alloys. The addition of a third, fourth, and perhaps a fifth element to the binary alloy modifies the precipitation sequence by forming new metastable and equilibrium phases with aluminum which may or may not incorporate lithium and which coprecipitate with δ'. The manner in which copper, magnesium, and zirconium impact the phase transformations and precipitation processes is central to the development of alloys of commercial interest.

Small amounts of zirconium (0.04–0.15 wt %) are added to all commercial Al-Li alloys as an essential alloying element for the control of grain structure. The addition of Zr results in the formation of a fine dispersion ($\lesssim$50 nm in size) of fully coherent, ordered Al_3Zr particles during ingot solidification and subsequent homogenization treatments. These dispersoids are very effective in retarding grain-boundary migration during the various subsequent thermal–mechanical treatments and thereby inhibit recrystallization [*23*]. A lower volume fraction of Al_3Zr dispersoids and special fabrication procedures are generally required if recrystallized microstructures are desired [*24*]. Zirconium also has a pronounced effect on the size, distribution, and coarsening kinetics of δ'. The Al_3Zr dispersoids, which are unaffected by the solution-heat-treatment, act as preferential sites for the nucleation of δ' (Al_3Li) during the quench and subsequent artificial aging because Al_3Zr is isostructural with Al_3Li. The enveloping of the preexisting Al_3Zr by δ' results in a distribution of composite precipitates coexisting with the homogeneously nucleated δ'. An example is shown in Fig. 2. The resulting bimodal distribution of δ' appears to have a significant favorable impact on the strength–toughness combination that can be achieved through the addition of Zr to Al-Li alloys.

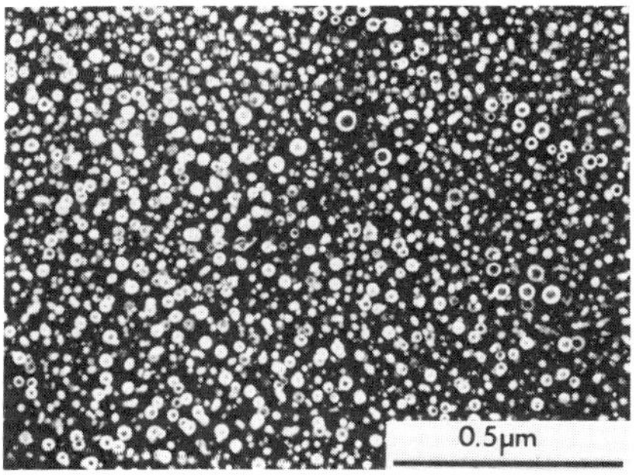

FIG. 2 Dark-field electron micrograph of δ' precipitates resulting from the preferential nucleation of δ' on preexisting Al_3Zr dispersoids.

The use of zirconium as a critical alloying element in modern Al-Li alloys is an important departure from early U. S. and Soviet alloys of this type that required a minor addition of cadmium (or silver, indium, zinc, mercury, or combinations thereof) to develop usable engineering properties [*4,8*]. It has now been established that the specific effects of these types of minor element additions are quite different. Zirconium modifies the grain structure, texture, and δ' precipitation mode, whereas cadmium promotes the homogeneous precipitation of copper-bearing phases such as θ' (in high copper alloys).

The addition of copper decreases the maximum solid solubility of lithium in aluminum at all temperatures [*21*]. Copper in solid solution does not influence the basic character of the δ' precipitation reaction [*25*], but it does introduce additional beneficial precipitation reactions that occur independently of the δ' reaction. The nature of the metastable and equilibrium phases that form within this alloy system has been shown to be dependent on the Cu:Li ratio and the aging temperatures [*21*]. For example, during aging at temperatures typical of commercial aging practices (~120–200°C), the decomposition of solid solution in a 2090-type alloy (Al-2.2Li-2.7Cu-0.12Zr) results in the formation of the metastable θ' ($CuAl_2$—the primary strengthening phase in Al-Cu alloys) and the equilibrium T_1-phase (Al_2CuLi) [*21,26–29*]. θ' has not been observed when the Cu:Li ratio is less than approximately 1:3. θ' has a tetragonal crystal structure and the precipitates form as platelets parallel to the {100} planes of the matrix. (See Fig. 3.) T_1 has a hexagonal crystal structure with $a = 0.497$ and $C = 0.934$ nm and also forms as thin platelets but with a {111} habit plane. (See Fig. 3.) In commercial Al-Li-Cu alloys such as 2090 (Al-2.2Li-2.7Cu-0.12Zr), T_1 is the predominant copper-bearing strengthening phase present after artificial aging to the near-peak–strength condition, though θ' may be present in a smaller amount. T_1 exhibits a strong tendency for heterogeneous nucleation on dislocations, low-angle grain boundaries, and other substructural features. (See Fig. 4.) Consequently, deformation prior to aging has a pronounced effect on their precipitation kinetics and on the size and distribution of T_1 precipitates within the grains [*30,31*]. In fact, the commercial production practices for alloys of this type involve a certain minimum degree of cold-work after solution-treatment and prior to artificial aging in order to promote more uniform matrix precipitation of T_1 and to achieve an improved strength-toughness combination over what is possible by aging alone.

Apart from the metastable θ' and equilibrium T_1 precipitates the addition of copper also introduces another copper-containing equilibrium phase, T_2, having the stoichiometry of Al_6CuLi_3 [*21*], which nucleates predomi-predominantly on high-angle boundaries. Electron [*32–35*] and X-ray

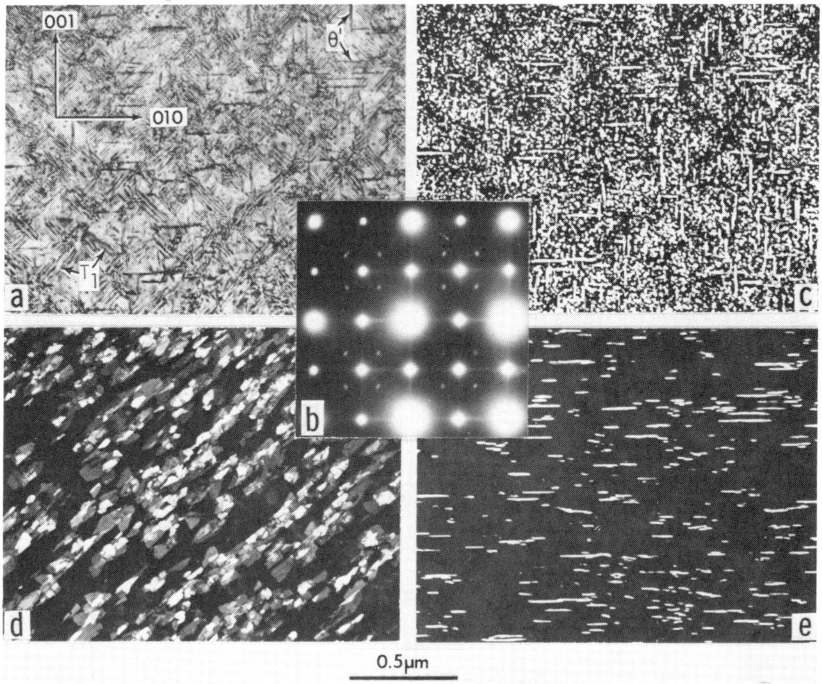

FIG. 3 Microstructure of nearly peak aged 2090 (Al-2.2Li-2.7Cu-0.12Zr) alloy stretched 2% prior to aging.
(a) Brightfield electron micrograph showing overall structure.
(b) Selected area diffraction pattern, ⟨100⟩ orientation.
(c) δ' precipitates, dark-field image.
(d) One variant of T_1 precipitates, dark-field image.
(e) One variantt of θ' precipitates, dark-field image.

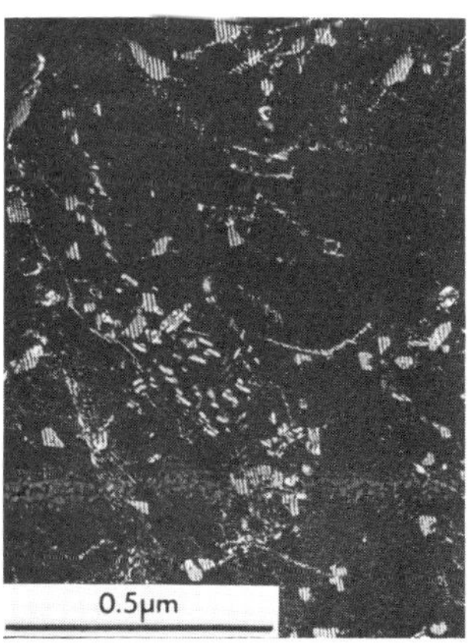

FIG. 4 Weak-beam dark-field image showing the preferential nucleation of T_1 on matrix dislocations.

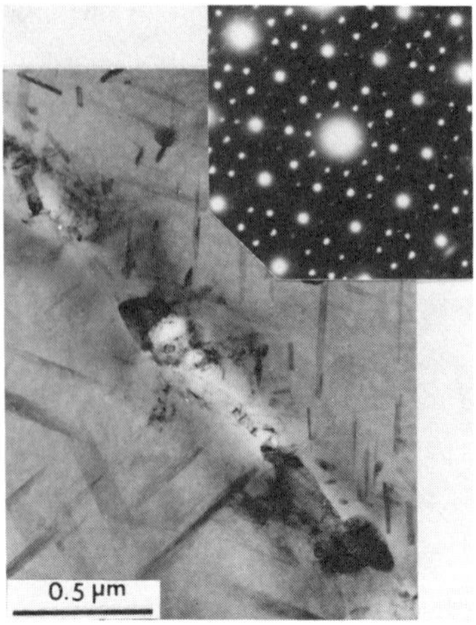

FIG. 5 Bright-field electron micrograph showing the grain-boundary precipitation of the equilibrium phase, T_2. Note the icosohedral symmetry of T_2 in the diffraction pattern (insert).

diffraction [34] analyses of this phase provide strong evidence that T_2 displays five fold icosohedral symmetry. (See Fig. 5.) The formation of the T_2 phase particles (which are richer in lithium than δ') along the grain boundaries leads, to the development of δ' precipitate-free zones adjacent to these boundaries (see Fig. 6) with concomitant reduction in ductility and

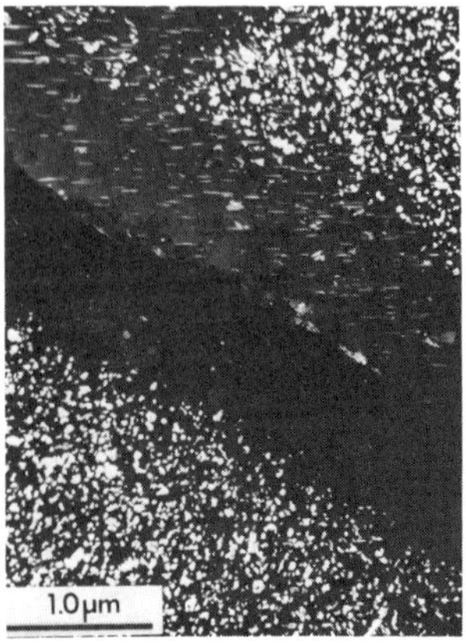

FIG. 6 Dark-field micrograph showing the presence of a relatively wide δ'-precipitate-free zone along the grain boundary of an overaged 2090-type alloy.

fracture-toughness. Cassada et al. [*34*] suggest that in 2090-type Al-Li alloys, T_2-phase is stable over the temperature range 170–520°C. Consequently, grain-boundary precipitation of this phase should be anticipated when thick products such as heavy forgings and plates are quenched from the solution-treatment temperature and/or when using slower quenching media to minimize quenched-in residual stresses. The combination of platic deformation prior to aging and lower-temperature ($\lesssim$150°C) artificial aging can be used as a means to promote copious matrix precipitation, inhibit grain-boundary precipitation of T_2 [*6, 37*], and thereby achieve improved ductility and fracture-toughness without sacrificing strength.

Apart from the T_2 phase, δ (AlLi)—which is the equilibrium precipitate phase in binary Al-Li alloys—has also been reported as a possible grain-boundary phase in 2090-type Al-Li alloys [*38*]. However, no positive identification of this phase has been made by using analytical electron microscopy techniques, possibly due to the highly reactive nature of this phase.

The effect of magnesium addition on the phase equilibria of Al-Li alloys has been reported by a number of investigators [*13,39,–41*]. Magnesium decreases the solubility of lithium in aluminum at all temperatures below about 425°C [*41*]. Similarly, the solubility of magnesium in aluminum is drastically reduced by the presence of lithium. Beyond the region of solid solution in Al-Li-Mg ternary alloys with compositions close to the Al-Li binary, the decomposition of the supersaturated solid solution occurs via the $\delta' \rightarrow \delta$ precipitation sequence. However, as the composition of the ternary alloy is varied from the Al-Li binary toward the Al-Mg binary by increasing the magnesium: lithium ratio, the phase that is in equilibrium with the α solid-solution changes from δ to Al_2MgLi. Alloys of commercial interest are those that lie in the $\alpha + Al_2MgLi$ phase field [*13,39*]. When supersaturated Al-Li-Mg alloys having composition in that region are aged at temperatures below about 240°C, the precipitation sequence occurs as follows:

$$\alpha_{ss} \rightarrow \alpha_m + \delta' \rightarrow \alpha_m + \delta' + Al_2MgLi \rightarrow \alpha_e + Al_2MgLi$$

where α_{ss}, α_m, and α_e refer to the supersaturated, the metastable, and the equilibrium α solid solutions, respectively.

The δ' phase that forms in Al-Mg-Li alloys is similar to that which occurs in the Al-Li binary. Al_2MgLi has a cubic α-Mn crystal structure and forms as rods along the $\langle 110 \rangle$ directions of the α matrix [*40*]. This phase is incoherent with the matrix and nucleates predominantly along the grain boundaries, dislocations, and other structural inhomogeneities. (See Fig. 7.) Consequently, it does not contribute to precipitation-hardening. δ' is the only phase responsible for the strength of Al-Li-Mg alloys, with the

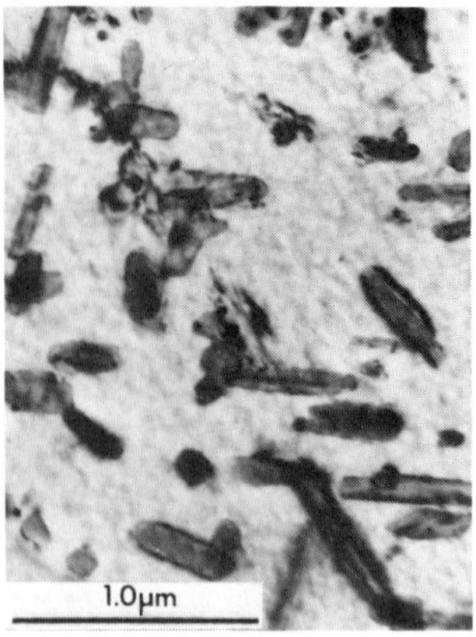

FIG. 7 Bright-field micrograph showing the preferential nucleation of rodlike Al_2MgLi phase precipitates in a peak-aged Al-2Li-4Mg-0.12Zr alloy.

magnesium contributing to some solid-solution–hardening. Consequently, Al-Li-Mg-Zr alloys do not develop strengths comparable to those of Al-Li-Cu-Zr or Al-Li-Cu-Mg-Zr alloys. The formation of the lithium-rich Al_2MgLi phase along the grain boundaries and subgrain boundaries leads to the development of δ'-precipitate-free zones during prolonged artificial aging (as is the case with other Al-Li alloys discussed previously), and contributes to the detrimental effect to alloy ductility, fracture-toughness, and possibly corrosion-resistance.

When magnesium is added to Al-Li alloys containing copper, matrix precipitation of S′ (Al_2CuMg) occurs. The exact nature of the phase equilibria of the quaternary Al-Li-Cu-Mg alloys depends on the relative concentrations of all three alloying elements. For example, the addition of small amounts (0.5–1.0 wt %) of magnesium to a high copper alloy such as 2090 (Al-2.2Li-2.7Cu-0.12Zr) suppresses the formation of Θ' and introduces S′ phase [*42*]. δ' precipitation is not markedly influenced by the magnesium addition and T_1 remains as the dominant secondary phase in such an alloy. (See Fig. 8.) In a higher-lithium–lower-copper alloy such as 8090 (Al-2.5Li-1.2Cu-0.7Mg-0.12Zr), S′, which coprecipitates with δ' and a small amount of T_1, becomes the dominant Cu-bearing phase [*43,44*]. (See Fig. 9.) As the magnesium: copper ratio is progressively increased, the precipitation of T_1 phase is fully suppressed and S′ phase becomes the primary Cu-bearing precipitation-hardening phase. This is the case with the 2091 alloy, which has a nominal composition of Al-2Li-2.2Cu-1.5Mg-0.08Zr [*45,46*]. Finally, when the concentration of magnesium exceeds that of copper, as in the case with an Al-2.5Li-1.5Cu-2Mg-0.12Zr alloy, during

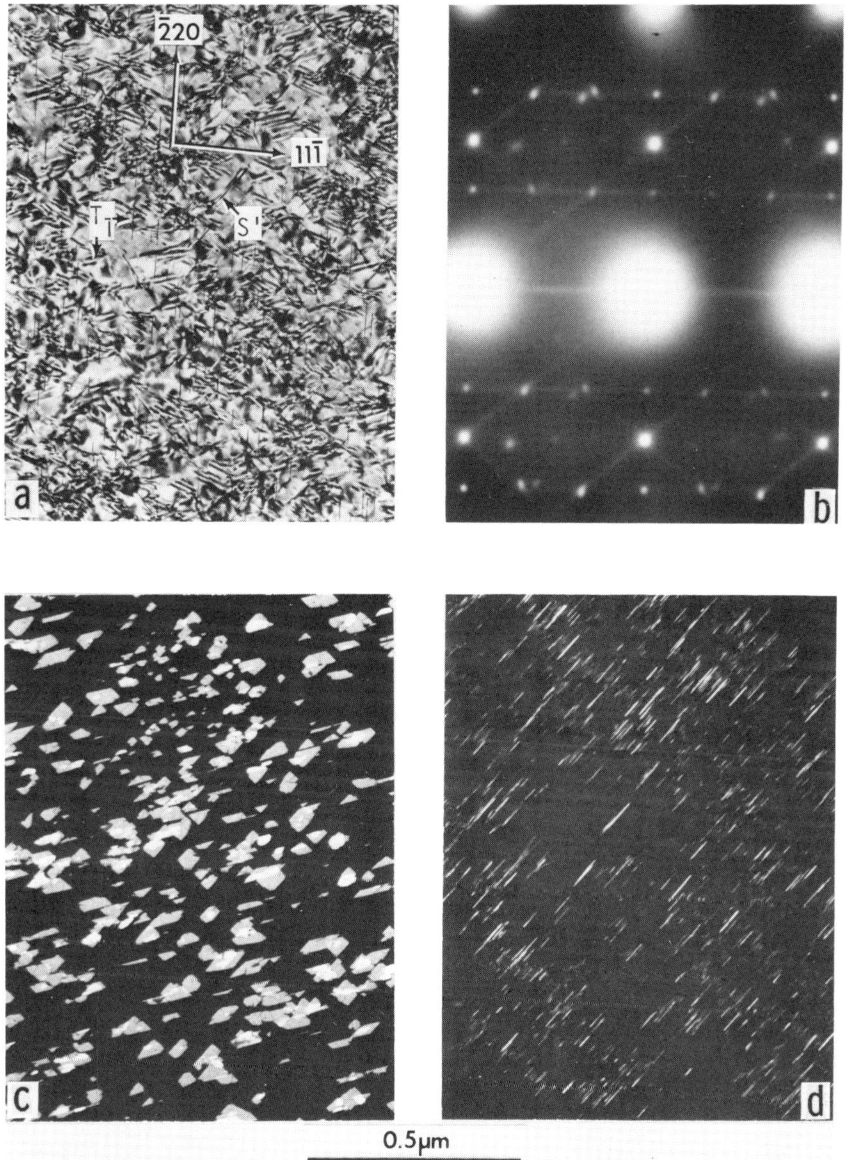

FIG. 8 Microstructure of nearly peak aged Al-2.2Li-2.5Cu-0.5Mg-0.12Zr alloy stretched 2% prior to aging.
(a) Bright-field micrograph showing overall features.
(b) Selected area diffraction pattern, ⟨112⟩ orientation.
(c) One variant of T_1 precipitates dark-field image.
(d) One variant of S′ precipitates, dark-field image.

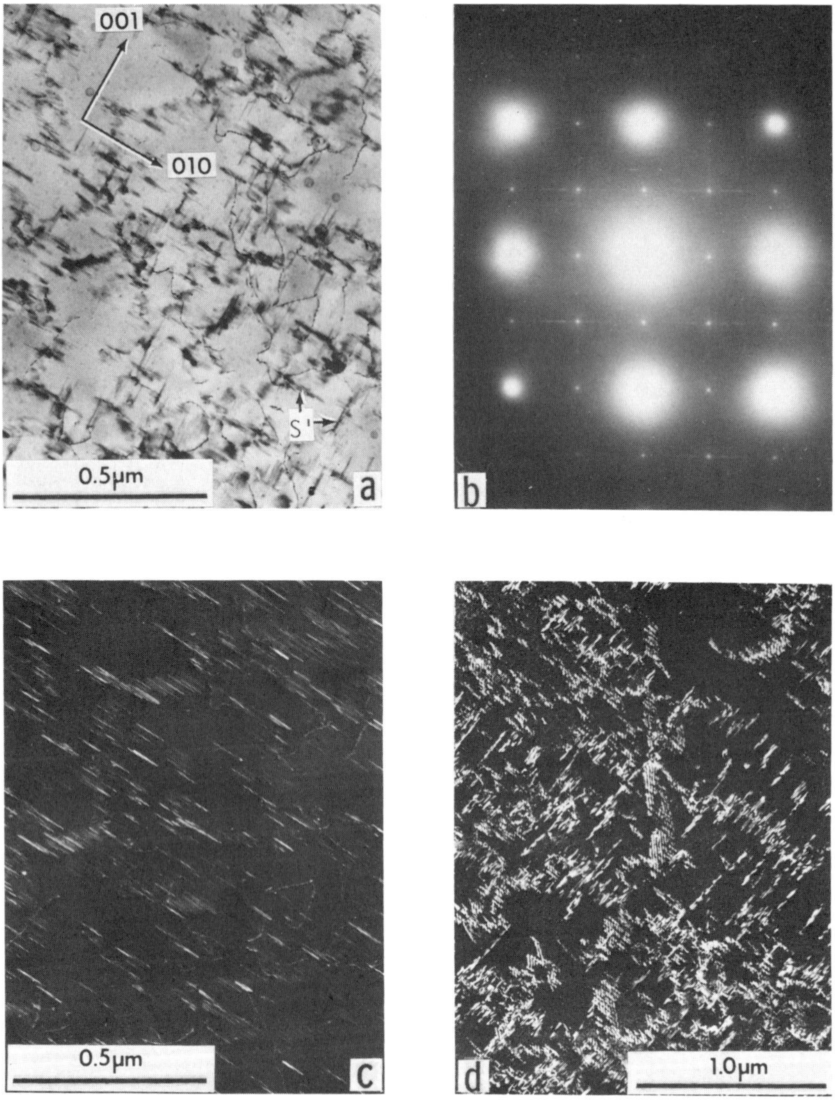

FIG. 9 Microstructure of a slightly underaged Al-2.4Li-1.3Cu-1.1Mg-0.14Zr (8090-type) alloy stretched 2% prior to aging.
(a) Bright-field micrography showing the overall features.
(b) Selected area diffraction pattern, ⟨100⟩ diffraction.
(c) One variant of S′-precipitates, dark-field micrograph.
(d) Weak-beam dark-field micrograph showing S′ and small amount of T_1.

prolonged aging the precipitation of the Al_2MgLi phase occurs along with the S′ [*47*].

The *S′* phase, which has an orthorhomibic crystal structure (a = 0.405 nm; b = 0.905 nm; c = 0.72 nm) forms initially as rods or needles aligned along the ⟨100⟩ directions in the matrix. The precipitates are partially coherent with the matrix and contribute significantly to the strength of the alloy. As in the case with the T_1 phase, S′ shows a strong tendency for heterogeneous nucleation along matrix dislocations, low-angle grain boundaries, and other structural inhomogeneities. Consequently, deformation prior to artificial aging has a pronounced influence on S′ precipitation, both in terms of precipitation kinetics and the size and distribution of the precipitates. Unlike the case of T_1 precipitates, the heterogeneous precipitation of S′ on low-angle boundaries or high-angle boundaries does not lead to the development of PFZs. However, several investigators have reported the occurrence of a T_2-type $Al_6Cu(LiMg)_3$ phase on the grain boundaries of 8090-type Al-Li-Cu-Mg-Zr alloys [*48–50*]. The formation of this phase on the grain boundaries also leads to the development of δ'-free precipitate-free zones (PFZs) in their vicinity and contributes to low ductility, fracture-toughness, and corrosion resistance in these alloys.

A schematic representation of the various phases found in basic as well as the more complex Al-Li systems that encompass modern Al-Li alloys is shown in Fig. 10.

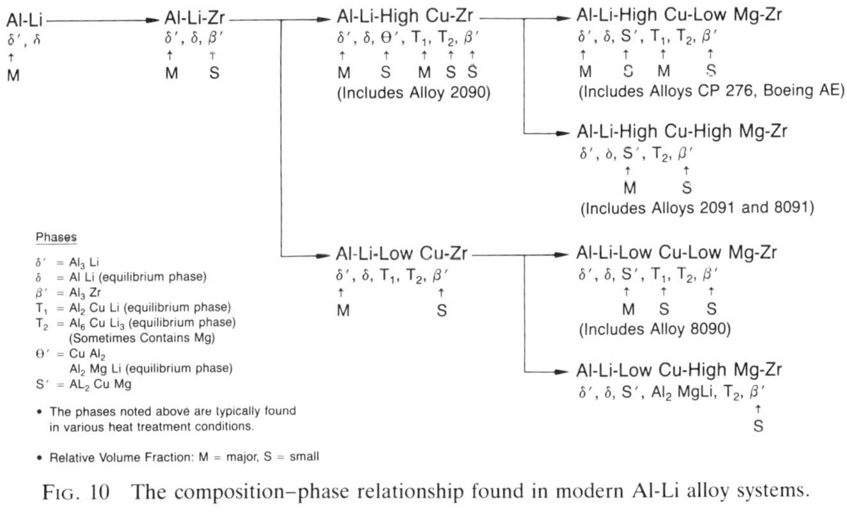

FIG. 10 The composition–phase relationship found in modern Al-Li alloy systems.

IV. Engineering Properties

The mechanical and physical properties of aluminum-lithium alloys differ from those of common aerospace aluminum alloys in many ways, and they present important challenges to optimizing their use on aircraft structure. A short description of the property characteristics of these alloys follows.

A. *Density*

The most significant property change associated with lithium additions to aluminum is a reduction in alloy density. Each wt % of lithium that is added causes approximately 3% reduction in the density of aluminum [*51*]. Although the presence of heavier elements such as copper will partially offset the density advantage afforded by lithium, the overall reduction in density achieved will still be dominated by the amount of lithium added [*52*]. (See Figs. 11 and 12.) Several investigators have developed impirical equations to predict alloy densities from known composition. One such

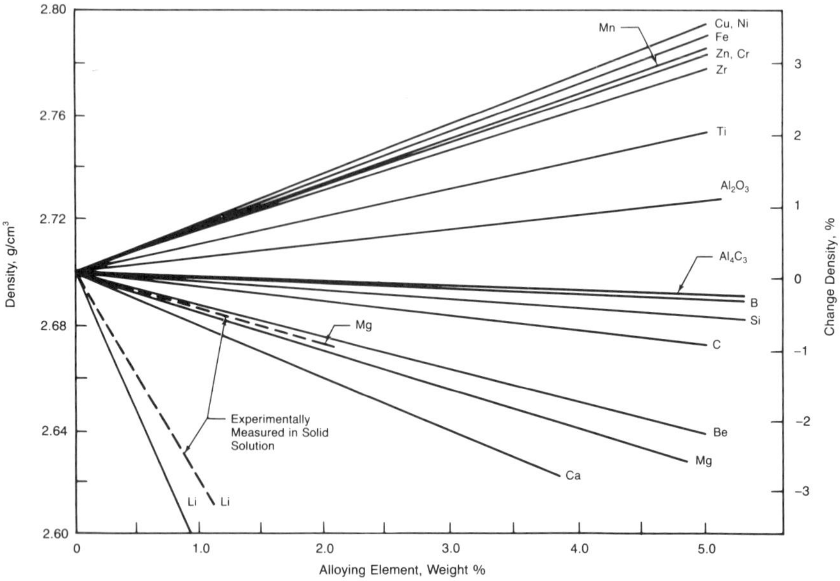

FIG. 11 Influence of alloying additions on the density of aluminum (calculated) [*52*].

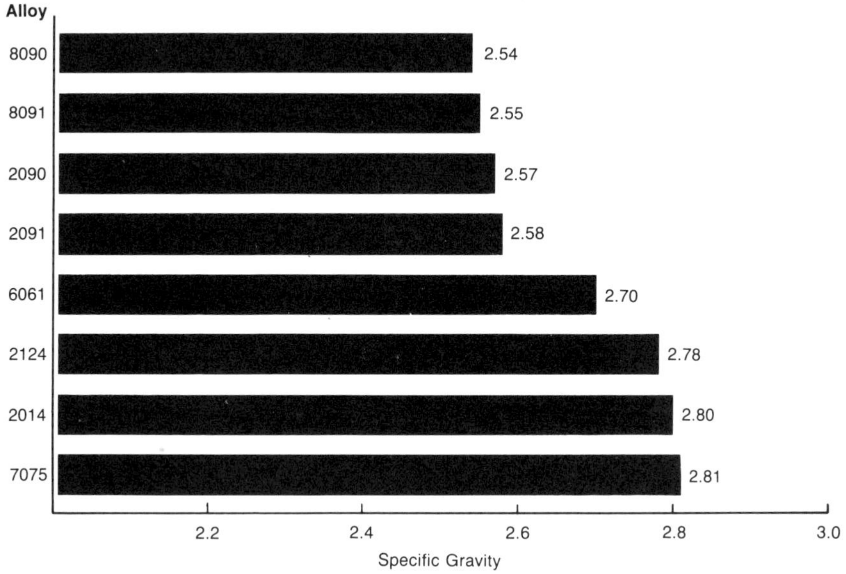

FIG. 12 A comparison of densities for several Al-Li and conventional alloys [*53*].

formula by Peel et al. [*43*] is as follows:

$$\text{Density (g/cc)} = 2.71 + 0.024\%\text{Cu} + 0.018\%\text{Zn} + 0.022\%\text{Mn} - 0.079\%\text{Li} - 0.01\%\text{Mg} - 0.004\%\text{Si}.$$

In this equation, the elemental concentrations are expressed in wt %.

Numerous trade studies on various types of aerospace vehicles have verified that a reduction in density is the most important material property for achieving a reduction in the structural weight of the vehicle [*14–16*]. (See Fig. 13 and Table II.) Density is often three to five times more effective in reducing structural weight than are commensurate changes in strength, fracture-toughness, modulus of elasticity, or fatigue properties.

B. Modulus of Elasticity

Lithium additions also significantly increase the modulus of elasticity, each wt % addition increasing the tension modulus by approximately 6% [*51,54*]. (See Fig. 14.) The modulus enhancement is due to solid-solution effects as well as the precipitation of lithium-bearing compounds, and it will change relative to both the amount of lithium added and the prior

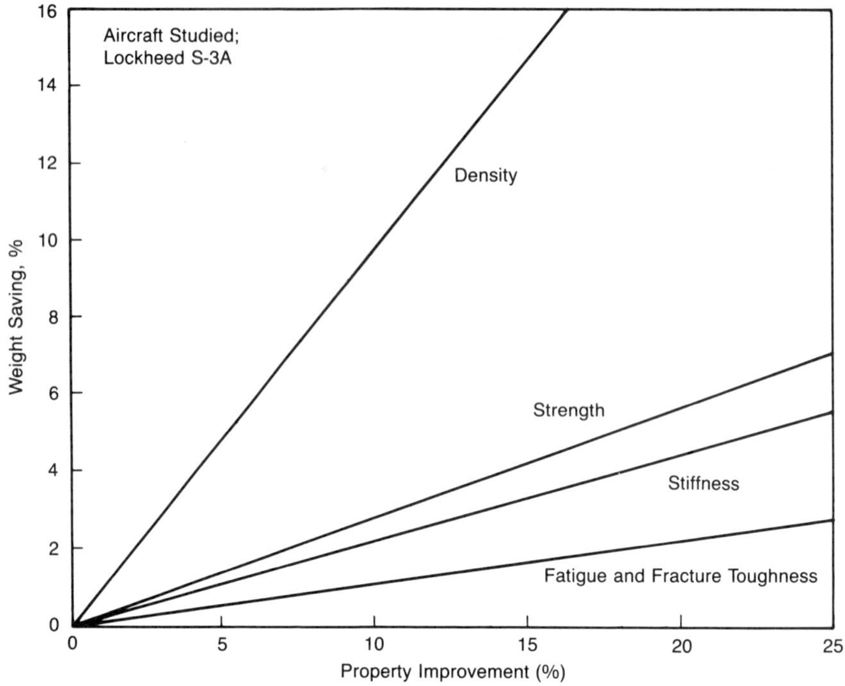

FIG. 13 The relative effect of various material properties on reducing structural weight in a Lockheed S-3A aircraft.

thermal treatment [*55*,*56*]. Copper has a slight beneficial effect and magnesium a slightly negative influence on the modulus of aluminum [*54*,*57*]. Investigations [*58*] have shown that a maximum in the modulus occurs when the alloys have been artificially aged to a state below the peak strength condition. Figure 15 shows a comparison of the modulus of elasticity of several Al-Li and baseline aluminum alloys [*53*].

C. Influence of Microstructure on Mechanical Properties

In binary Al-Li alloys, significant contributions to strengthening come from the precipitation of the δ' phase, the actual degree of strengthening achieved being a function of the volume fraction and the size distribution of the δ' precipitates [*59*,*61*]. In the underaged–to–peak-age condition, the coherent δ' precipitates are sheared by the slip dislocations during deformation and strength increases are related to the resistance to

TABLE II

A Summary of Published Trade Studies Showing the Effect That Low-Density Al-Li Alloys Would Have on Lowering the Structural Weight of Aircraft Structure

	Alloy(s)	Savings (on affected structure)		
Aircraft studied	Baseline/new	lbs	%	Date of study
RA-5C	7075-T6/X2020	168	(6)	1958
B-58	2024-T86/X2020	266	(13)	Early 1960s
F-111	2024-T86/X2020	130	(15)	Early 1960s
L-1011	7075-T76/(+20)(−10)[a]	17,030	(12.4)	1978
S-3A	7075-T76/(+13)(−7)[a]	1,022	(11.5)	1978
F-27 (mod)	7075, 2024/5 alloy types	836	(7.8)	1982
737 (mod)	7075, 2024/5 alloy types	3,223	(9.8)	1982
L-1011	7075, 2024/5 alloy types	20,279	(10.6)	1982
Advanced fighter (concept)	7075-T76/(+8 to +30)(−7)[a]	1,599	(11.7)	1982
F-20A	7075, 7050/several low-ρ types	570	(12.6)	1982
747F	7075/3 low-ρ types	519-686	(10–12.9)	1981
757-200	5 standard alloys/12 low-ρ types	2,023	(11.2)	1983
747	7075, 2024/2 low-ρ types	11,000–14,000	(~7.0)	1983–1984

Source: Quist et al. [15].

[a] (+20)(−10) means (engineering properties + 20%) (density − 10%).

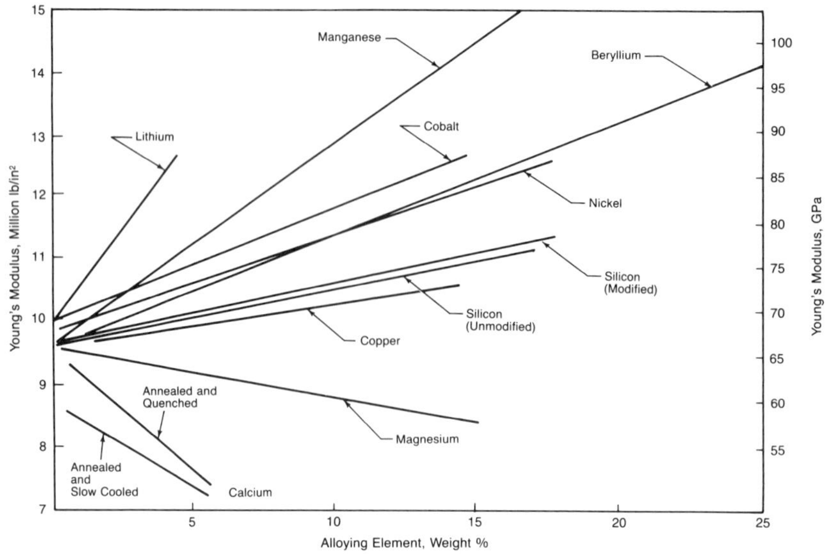

FIG. 14 Influence of alloying additions on Young's modulus of aluminum [57].

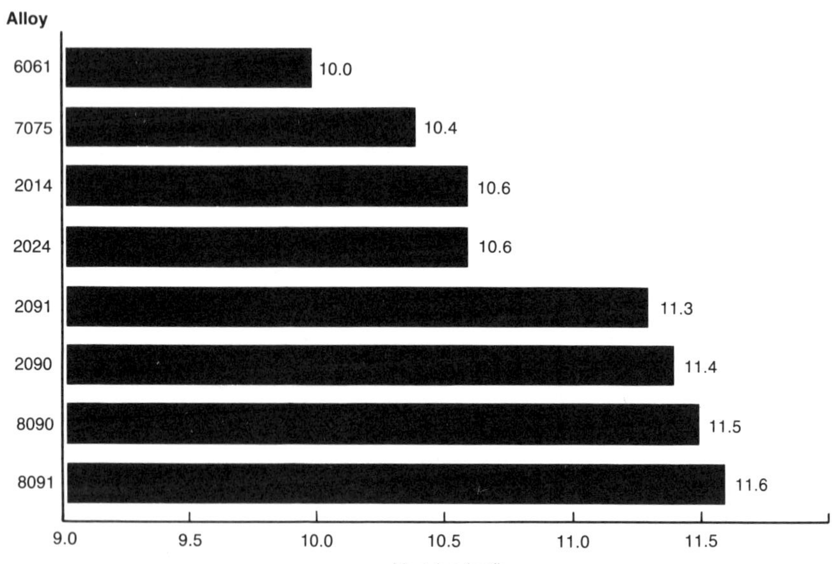

FIG. 15 A comparison of Young's modulus for several Al-Li and conventional alloys [53].

dislocation shearing of δ'. The dominant strengthening mechanism has been determined to be the order effect associated with δ' [*60,61*]. Owing to the $L1_2$ superlattice structure of δ', dislocations in δ'-strengthened alloys tend to move in pairs (i.e., as superdislocations). An important consequence of this mode of deformation is the tendency for slip localization. Once slip has occurred on a given plane, continued slip on that plane becomes energetically favorable over activating slip on another slip system. The occurrence of superlattice dislocations and planar slip are thus a characteristic feature of Al-Li alloys.

During artificial aging of binary Al-Li alloys, the maximum strength condition corresponds to the transition from dislocation shearing to the dislocation bypassing of δ' by the Orowan-bypassing mechanism, which occurs when the precipitates attain a certain critical diameter due to particle-coarsening. Despite the more homogeneous deformation due to the bypassing process, drastic reductions in ductility occur because of the precipitation of equilibrium δ phase on the grain boundaries and the simultaneous formation of precipitate-free zones adjacent to these boundaries, both of which contribute to intergranular fracture [*60*].

In more complex alloys of commercial interest, the coprecipitation of a second and perhaps a third age-hardening phase (θ', T, and/or S'), apart from the δ' phase, introduces additional contributions to the strength and modifies the deformation characteristics. For example, in 2090-type Al-Li-Cu-Zr alloys, the precipitation of θ' and T_1 occurs independently of δ' during artificial aging and provides significant additional contributions to the strength of the alloy. It has been also reported [*60*] that copper, although it does not directly participate in the δ' precipitation, tends to shift the critical diameter for transition from dislocation shearing to the bypassing mechanism to a smaller δ' size. This effect may be associated with the reduction in stacking-fault energy of the solid solution caused by the copper addition. Bypassing of δ', while rendering the deformation mode more homogeneous, also make δ' strengthening less efficient. The strength increases in this case are governed largely by the progressive precipitation of T_1. The T_1 platelets, especially the out-of-plane variants of T_1, act as unshearable obstacles to dislocation motion and contribute further to slip homogenization. Cold-work prior to artificial aging induces a more homogeneous nucleation of T_1 throughout the grain and provides an improved strength–fracture-toughness blend.

In alloys such as 8090, 8091, and 2091 that contain magnesium in addition to copper and lithium, the presence of semicoherent S' phase contributes to the strength of the alloys. In the medium-strength alloys 8090 and 2091, despite the presence of S', δ' remains as the dominant strengthening precipitate [*46*]. In the higher-strength alloy 8091, a more significant contribution to the strength arises from S′. The presence of S′, which is not easily sheared by dislocations, promotes cross-slip and hence more homogeneous deformation. In the underaged condition where S′ precipitates have not grown to sufficient size to induce slip dispersal, planar slip may persist.

D. *Strength*

The strength of most modern Al-Li alloys has been targeted to range from that of the moderate-strength 2014 alloy to that of the high-strength 7075-T6 alloy and this objective has been readily achieved. However, the interrelationship between the key elements of the microstructure (i.e, grain structure, matrix precipitate, and grain-boundary precipitates) and the engineering properties of this new family of alloys are somewhat different from that for the lithium-free conventional aluminum alloys.

As noted previously, all ingot-cast Al-Li alloys commercially available as of the late 1980s contain zirconium as a retardant to recrystallization. Zirconium, when present in sufficient amounts (0.08–0.15 wt %), in combination with lithium, makes these alloys extremely resistant to recrystallization. Consequently, even as finished products these alloys remain virtually uncrystallized, unless specifically processed to yield a recrystallized grain structure. In the unrecrystallized condition, all Al-Li–alloy products display pronounced crystallographic texture and substantial anisotropy in mechanical properties, with tensile and shear strengths displaying minima along directions 40–60 degrees from the major working direction (i.e., longitudinal direction) [*43,44,54,62,63*]. These property reductions have been found to be typically 20% or more as compared to the longitudinal properties. (See Fig. 16.) In order to reduce the degree of isotropy and also to improve fracture-toughness, sheet products of two alloys, 2091 and 8090, are being marketed in the recrystallized condition.

E. *Elongation*

In the high-strength condition, uncrystallized aluminum-lithium alloys often display relatively low elongations in the longitudinal direction, typically in the 3–7% range, but these values are substantially improved in the off-angle and long transverse grain directions [*65*]. This condition is also a result of the highly developed deformation texture in the unrecrystallized

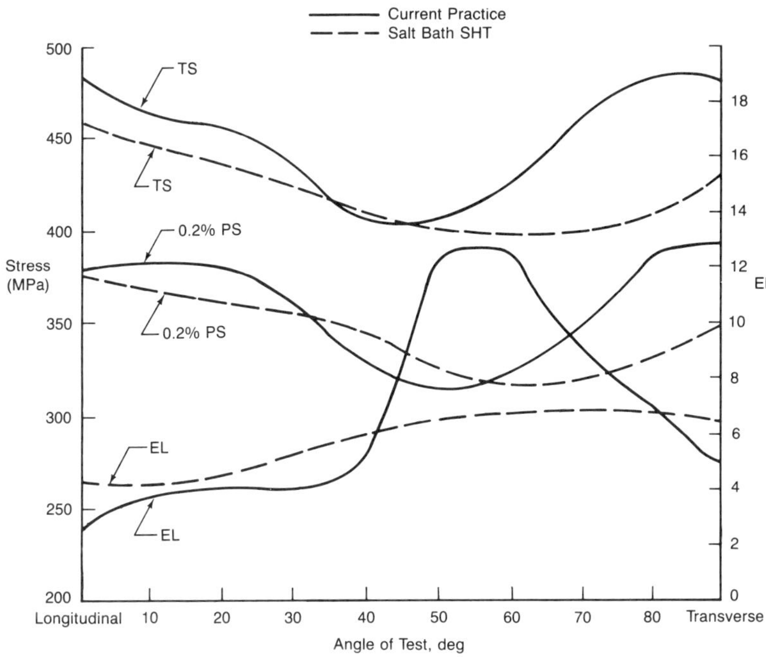

FIG. 16 The mechanical properties of Al-Li alloy 8090 as a function of solution-heat-treatment practice and angle to the rolling direction [*64*].

products. Again, improved isotropy can be obtained if the alloy is caused to recrystallize under controlled conditions. The low elongation values in the longitudinal grain direction for unrecrystallized Al-Li alloys adversely affects formability in that grain direction.

F. Fracture-Toughness

The fracture-toughness of Al-Li alloys has been a matter of concern since the first commercial alloy, X2020, was found to possess the poorest plane-stress fracture-toughness of any aluminum alloy developed for structural applications in aerospace vehicles. Modern commercial Al-Li alloys such as 2090, 2091, 8090, and 8091, when tested in the artificially aged condition in the L-T test orientation, fracture by a combination of transgranular shearing and intergranular delamination. The lowest ductility and fracture-toughness are observed in the S-T direction, in which case the grain boundaries parallel to the rolling plane are normal the loading axis. In this orientation, failure is purely intergranular, reflecting the inherent weakness of the grain boundaries of Al-Li alloys. An understanding of the fracture-toughness issue in these alloys has required extensive study, but significant progress has indeed been made. The following mechanisms have now been identified as controlling the fracture process in Al-Li alloys:

1. Planar slip, which results from the shearable nature of the ordered δ' phase (Al_3Li) [*66,67*]. Planar slip leads to the localization of deformation strains into intense slip bands. Failure occurs by either slip-band fracture or by grain-boundary fracture at the slip-band–grain-boundary intersections due to the stress concentration caused by the dislocation pileup.
2. Strain localization at relatively soft PFZs that form along the grain boundaries during artificial aging [*30,68*]. These strain localizations lead to premature ductile failure along the PFZs.
3. The hydrogen content in Al-Li alloys is normally higher than that for conventional aluminum alloys. It has been suggested this higher hydrogen content may cause embrittlement problems [*69*] although this has not been fully substantiated by experimental observations.
4. Owing to the very nature of the unrecrystallized structure possessed by most Al-Li alloys in their final product form, impurities in the melt that may have segregated during solidification tend to remain concentrated at or near grain boundaries and are never redistributed through a recrystallization process.
5. Grain-boundary embrittlement caused by the precipitation of equilibrium phases such as T_2 and δ along grain boundaries and subboundaries either during quenching or during subsequent artificial aging [*34,70,71*].
6. The unrecrystallized structure found in most Al-Li alloys becomes very highly textured in thin products. The resulting low-angle grain boundaries allow propagating cracks to proceed relatively unimpeded across these grain boundaries.
7. Grain-boundary embrittlement caused by the segregation of alkali-metal impurities such as Na and K [*72–74*] and perhaps lithium itself [*75*]. One apparent additional manifestation of sodium and potassium contamination in Al-Li alloys is the appearance of cleveage cracks on {100} and {110} planes during slow strain-rate deformation—a very unusual occurrence in aluminum alloys [*76*].
8. Certain combinations of alkali-element impurities can form very low melting-point constituents in aluminum. It has been suggested that

concentrations of these elements can be sufficient in Al-Li alloys to form these constituents and embrittle grain boundaries at room temperature [*77,78*]. (Most believe purity standards of the late 1980s prohibit such an occurrence, however.)

In addition, there are several other factors that govern the fracture-toughness of commercial Al-Li alloys:

1. The detailed chemical composition of the alloy (i.e., lithium content and the nature and concentration of other alloying elements). In particular, the relative volume fractions and types of precipitates (viz. δ' versus T_1) in which lithium is tied up in the heat-treated alloy have a strong influence on fracture-toughness. Other things being equal, an alloy with a higher lithium content generally exhibits lower fracture-toughness at a given strength level.
2. Heat-treatment. Natural aging, underaging, and low-temperature aging [*79*] give the best blend of strength and toughness for Al-Li alloys. Overaging, in general, tends to degrade fracture-toughness considerably.
3. Thermomechanical treatments (TMT). The fracture-toughness of Al-Li alloys has been found to be very sensitive to deformation at room temperature following solution treating and quenching but prior to artificial aging. Strains of up to at least 7% often progressively improve both strength and fracture-toughness [*36,80,81*].

Improvements in the in-plane and short-transverse fracture-toughness of Al-Li alloys have been sought by understanding, resolving, and/or circumventing one or more of the preceding factors. This has included alloying with copper and magnesium to induce coprecipitation of hardening phases such as T_1 and/or S', along with δ', taking all steps possible to cleanse the melt of unwanted impurities (iron, silicon, and alkali metals, for example), specially processing the alloys to obtain a fine-grained recrystallized microstructure with elongated grains [*24*] and developing compositions and heat-treatment procedures that minimize the formation of unwanted phases. As noted, the fracture properties of Al-Li alloys are also particularly sensitive to aging practice and thermomechanical treatments. Substantial fracture-toughness improvements have been realized by age-hardening at relatively low temperatures [*79*] and by cold-working the alloy following solution-treatment and quenching prior to low-temperature aging [*80,81*]. The fracture-toughness improvements achieved in modern Al-Li alloys have been gratifying and, in general, the toughness of these alloys is nearly equivalent to or better than those of baseline alloys, particularly at intermediate strength levels.

G. *Fatigue Crack Growth Rate*

The fatigue crack growth resistance of most Al-Li–type alloys under constant-amplitude and variable-amplitude loading is substantially better than those of the best aerospace structural aluminum alloys in use, particularly at moderate to high ΔK levels and for cracks that have grown beyond the initiation stage [*82–85*]. This behavior has been the object of many intensive investigations. It has been determined that this high resistance to fatigue crack propagation is due to the crystallographic nature of the crack path resulting from a combination of planar slip and the pronounced texture of the unrecrystallized material. The torturous crack path reduces the "crack-driving force" due to extrinsic effects such as crack deflection (or branching) and roughness-induced crack closure [*85–87*]. With respect to crack-growth resistance for very short fatigue cracks, Al-Li alloys have not demonstrated the same significant advantages over baseline alloys. For example, small fatigue cracks in high-strength 2090-type alloys grow significantly faster than expected from long-crack data. This is partially the result of reduced crack-tip shielding effects [*87,88*]. Still, Al-Li alloys show very respectable and competitive crack-growth behavior in this regimen as compared to baseline alloys. In addition, under programmed and spectrum-loading conditions, Al-Li alloys often demonstrate an even greater relative advantage over baseline alloys than for constant-amplitude sinusoidal loading conditions [*84*]. The fatigue-crack–initiation resistance of Al-Li alloys may be somewhat poorer than that of conventional alloys, however [*89,90*], an effect that has been attributed to the combined influence of planar slip and texture. However, overall fatigue performance under S-N-type (stress versus number of cycles to failure) test conditions is generally equivalent to, or better than, that of standard aluminum alloys.

H. *Cryogenic Properties*

It has been established since 1986 that Al-Li alloys such as 2090 and 8090 possess outstanding combinations of strength and fracture-toughness at cryogenic temperatures down to 4°K, which is superior to even those of the late 1980s industry standard, alloy 2219 [*91–95*]. For example, the commercial alloy 2090 in the high-strength T8E41 temper exhibits a 45%

increase in strength, 12% increase in modulus, and 90% increase in K_{IC} when cooled from room temperature to 4°K (L-T grain direction). The short transverse grain direction shows a slight decrease, however [*91–95*]. The origin of this behavior is not completely understood but is thought to result from the specific deformation characteristics of these alloys at low temperatures. In particular, the increased strain-hardening rates exhibited by this alloy at low temperatures may increase the resistance to strain-controlled fracture by suppressing strain localization [*95*]. It has also been suggested that the solidification and progressive strengthening of low—melting-point constituents present along the grain boundaries is a partial explanation for improvements in low-temperature fracture-toughness properties [*78*]. The magnitude of the observed fracture-property gains has been quite unexpected, however.

I. Other Engineering Properties

In addition to the preceding, there are certain other engineering properties of Al-Li alloys that are attractive in comparison to conventional baseline aluminum alloys. For example, most Al-Li alloys are amenable to superplastic forming [*96,97*], display moderate to good weldability [*98,99*], and are reasonably corrosion-resistant [*100*]. In addition, these alloys can be chemically milled, bonded, anodized, alclad, painted, and formed by normal and/or somewhat modified fabrication techniques [*101*].

On the debit side, as noted previously, aluminum-lithium alloys often display considerable anisotropy in strength and ductility with a minimum in strength properties occurring at angles of 40 to 60° to the longitudinal direction. This includes tensile, compression, shear, and bearing properties. In addition, they are somewhat more susceptible to surface oxidation at moderate temperatures and are prone to warping during quenching (mostly attributable to the relatively high quenching temperature). Another area of some concern is that during elevated-temperature heat-treatments, diffusion of lithium occurs out to the surface and this out-diffusion produces a lithium-depleted layer and sometimes a region of subsurface pores [*102,103*]. The depth of the lithium depletion zone can be as much as 200–300 μm and in thin sheet products this can lead to a reduction in strength [*102*]. In extruded shapes, larger than normal variations in mechanical properties have been noted between different legs of the extrusion, especially when they vary in thickness [*104,105*]. In addition, none of the first-generation Al-Li alloys have proven to be as applicable to forged parts as other product forms. Although these alloys are very forgeable, it has proven difficult to develop the high level of resistance to stress corrosion cracking normally desired in this product form, in conjunction with adequate strength and fracture-toughness. Relatively low toughness and ductility in the ST grain direction of thick forgings, extrusions, and plates are also of concern.

V. General Characteristics of Current Commercial Al-Li Alloys

All commercial and near-commercial Al-Li alloys have been produced by ingot-casting technology, albeit modified to meet the special requirements imposed by lithium additions to aluminum. The reactive nature of the molten Al-Li alloy bath and its substantially altered heat-transfer characteristics have made it necessary to employ modified refractories and degassing procedures, protective inert cover gasses and/or surface fluxes during melting and casting, and special ingot-cooling techniques (some employing organic coolants in place of water). The goal of producing sound Al-Li ingots, with acceptable recovery rates, has presented a significant technical and economic challenge to the aluminum industry—and has required extensively modified or new facilities (see p. 134). Powder-metallurgy technology has also been applied for the production of Al-Li alloys in an effort to develop alloys with improved combinations of properties, as compared to their ingot-metallurgy counterparts. It has been hoped that PM techniques could also circumvent some of the problem areas associated with IM technology such as anisotropy and poor short-transverse ductility and fracture-toughness. Among the various powder-production techniques that have been explored, the standard atomization techniques have not demonstrated a significant overall potential for Al-Li PM alloy production [*106*]. Significant progress has been made through other approaches, however, such as melt spinning (or planar casting) [*107,108*] and mechanical alloying techniques [*106,109*]. It would appear that alloys with improved combinations of strength, toughness, and stress corrosion cracking resistance and isotropic mechanical properties, suitable for forging applications, may be developed and marketed using these approaches.

Four Al-Li type alloys are receiving industry's principal attention and are being most actively developed. All have a relatively low density and high modulus of elasticity. A summary of their other characteristics are as follows.

A. Alloy 2090

This Al-2.2Li-2.7Cu-0.12Zr alloy is being developed for sheet, plate, and extrusion applications and is offered in a variety of medium- and high-strength temper conditions [*110,111*]. Thermomechanical processing is crucial to achieve improved combinations of strength and toughness in this alloy. Often this takes the form of a stretch of 4 to 6% following solution-treatment and quenching [*80,110*]. Alloy 2090 is relatively easy to cast and shows good weldability and corrosion-resistance characteristics. It is normally processed to have an unrecrystallized grain structure and therefore displays a relatively high degree of anisotropy with considerably reduced strength properties at 45 to 60° from the extrusion or rolling direction. The shear strength of 2090 is also below those of the baseline alloys. However, the alloy possesses very high plane-strain fracture-toughness (except in the short transverse direction) and fatigue crack growth resistance and exhibits excellent strength and toughness at cryogenic temperatures. In the T8E41 temper condition, this alloy also displays relatively good stress corrosion resistance. In addition, its elevated-temperature properties are attractive up to 300 to 350°F [*112*]. The room-temperature plane-strain fracture-toughness of alloy 2090 is somewhat below that of alloys 2024-T3 and 7075-T6 at equivalent strength levels.

B. Alloy 2091

This Al-2.0Li-2.2Cu-1.Mg-0.08Zr alloy is being developed for sheet, plate, extrusion, and forging applications [*113*]. The first applications are expected to be for relatively thin sheet where the alloy is processed to develop a recrystallized fine-grain structure. In this condition, and with either a natural or underaged heat-treatment, the alloy displays strength characteristics slightly below those of 2024-T3, but has outstanding plane-stress fracture-toughness characteristics, perhaps even superior to those of 2024-T3. In addition, its shear and off-angle strength properties are somewhat improved over Al-Li alloys having an unrecrystallized structure, but still do not match the levels of baseline aluminum alloys. The various fatigue and corrosion-resistance properties of alloy 2091 thin sheet range from adequate to excellent and generally match those of 2024-T3. As of 1988, however, the recrystallized grain structure for alloy 2091 can only be retained up to a thickness of about 0.15 inch. In addition, the alloy cannot be heat-treated to strengths much above those for 2024-T3 [*113*].

C. Alloy 8090

This Al-2.4Li-1.3Cu-0.9Mg-0.12Zr alloy is being developed in a large variety of moderate-to high-strength–level tempers having both recrystallized and unrecrystallized grain structures [*65,113*]. In addition, some tempers employ a 4–6% stretch following quenching prior to artificial aging. The alloy is offered as sheet, plate, extruded, and forged products and has found early industry applications in each of these areas. In common with other Al-Li alloys, alloy 8090 displays excellent resistance to fatigue crack growth and good plane-stress fracture-toughness [*114*]. Alloy 8090 retains its attractive engineering characteristics up to thickness of 4 to 6 inches, thicker than for other existing Al-Li alloys [*115*]. When processed to develop a recrystallized grain structure, the alloy is reported to have outstanding plane-stress fracture-toughness, comparable to that of alloy 2091 [*65*]. Alloy 8090, like alloy 2090, has been shown to be weldable. It also demonstrates excellent cryogenic properties as well as elevated-temperature properties up to approximately 350°F [*116*]. Alloy 8090 has been used in more product forms and for more diverse applications than any other Al-Li alloy developed to date.

D. Alloy 8091

This promising Al-2.6Li-2.0Cu-0.8Mg-0.12Zr alloy is being developed in a variety of sheet, thin-plate, and extrusion applications, although its development has lagged far behind that of alloy 8090 [*65,116*]. In general, the chemical constituents of this alloy are similar to that of alloy 8090, but it displays somewhat higher strength for a similar heat-treat condition. The engineering properties of alloy 8091 are again similar to those of alloy 8090, except that the strength–fracture-toughness blend is felt to be improved. Unfortunately, casting of this alloy has been found to be somewhat difficult and it is not available in products having more than moderate section thicknesses. Alloy 8091 has been found to display attractive super plastic-forming characteristics. Several studies have been undertaken to further develop this area.

VI. Applications

The use of Al-Li alloys for aerospace applications is moving forward toward a variety of well-defined areas of application [*117,119*]. General areas of most interest include the following:

- Primary fuselage structure for aircraft and helicopters including skins, frames, stringers, bulkheads, and doors.
- Secondary structure for aircraft and helicopters as stowage bins, clips, and brackets.
- Wing covers and associate structure for aircraft including ribs, stringers, spars, and chords.
- Cryogenic tank bodies and associate structure for missiles.
- Primary and secondary structure for various types of hardware to be used in satellite and space applications.

The product forms being produced for these various applications include sheet, plate, and extrusions and some forgings for specific applications. In addition, both thick and thin wall tubing have also been considered for commercial development.

Looking to the future, the application of Al-Li alloys is expected to broaden into most areas where aluminum is presently used in the aerospace industry. As of the late 1980s, the cost of Al-Li alloys is approximately two to three times that of conventional alloys due to the price of lithium itself; the special alloy melting, casting, and processing requirements; and the recouping of development expenditures by the producers. The type and extent of Al-Li–alloy usage is very sensitive to cost-benefit trade-offs and the price situation will call for selective applications of these alloys. Military uses, (including high-performance aircraft, missiles, and spacecraft) will be easiest to justify, with the commercial applications being carefully chosen to maximize cost-effectiveness. The economics of Al-Li use are expected to steadily improve as time progresses. Market penetration is expected to cover all types of applications including nonstructural, secondary, and primary structure. A second generation of Al-Li alloys having improved strength–fracture-toughness blends, weldability, elevated temperature properties, applicability to forging, and other needed engineering property attributes will likely emerge by the mid-1990s. Al-Li alloys are also expected to broaden their applications to metal-matrix composites and forgings—both possibly through the use of powder-metallurgy approaches. The need for greater cost-effectiveness in Al-Li–alloy applications will encourage increased activity in the area of superplastic forming of these alloys.

The use of Al-Li alloys for naval and land vehicles may also emerge. For example, the aluminum superstructures, navy ships are very sensitive to weight. Consequently, low-density Al-Li alloys may become an attractive alternative to current alloys as the material of construction. Likewise, both specialty automotive and military land vehicles may turn to Al-Li alloys for specific applications, such as for the bodies, frames, and wheels of racing cars and for armor plate on tanks and similar vehicles.

Of course, the future has always proved to be more unpredictable than predictable. The authors must admonish the reader to accept the preceding comments with that clear understanding.

References

1. Reuleaux, O. (1925). *J. Inst. Met.*, **33**, 346. Cf. *Z Metallkunde* (1924), **16**, 436.
2. Scheuer, E. (1927). *Z. Metallkunde*, **19**, 16 (in German).
3. "German Alloy Sold Commercially" (1929). *Iron Age*, **123** 718.
4. LeBaron, I.M. (1942). U.S. Patent no. 2,381,219 application date 1942, granted 1945.
5. Balmuth, E.S., and Schmidt, R. (1981). In *Aluminum-Lithium Alloys*, T.H. Sanders, Jr., and E.A. Starke, Jr., eds. TMS-AIME, Warrendale, Pennsylvania, p. 69.
6. Kaufman, J.G., and Holt, M. (1965). In *"Fracture Characteristics of Aluminum Alloys.* Tech. Paper no. 18. Alcoa Research Laboratories, Pittsburgh.
7. Spuhler, E.H. (1983). Formerly of The Aluminum Company of America, Pittsburg, private communication.
8. Archakova, Z.N., Romanova, O.A., and Fridlyander, I.N. (1961). *J. Inst. Met. Met. Abs.*, **28**, 945.
9. Fridlyander, I.N., Shamray, V.F., and Shiryayera, N.V. (1965). *Russian Metallurgy*, no. 2, pp. 83–90. Translated from *Izvestiya Akademii nauk SSSR Metally*, **2**, 153.
10. Fridlyander, I.N., Shiryaeva, N.V., Ambartsumyan, S.M., Gorokhova T.A., Gabidullin, R.M., Sidorov, N.G., Sorokin, N.A., and Kuznetsov, A.N.U.K. Patent no. 1,172,736, application date 1967, granted 1969; and French Patent no. 1,519,021, application date 1967, granted 1968.
11. "Low Density, High Strength Al-Mg Alloys Containing Lithium." Fulmer Research Institute, Flyer, Metal Show, Detroit, October 18–21, 1971, p. 2.
12. Evancho, J.W. (1974). "Development of An Al-Mg-Li Alloy." Naval Air Development Center Report, Contract no. N62269-73-C-0219.
13. Sanders, T.H., Jr. (1976). "Development of an Al-Mg-Li Alloy." Final Report, Naval Air Developmental Center Contract no. N622269-74-C-0328.
14. Lewis, R.D., Webster, D., Palmer, I.G. (1978). "A Feasibility Study for Development of Structural Aluminum Alloys from Rapidly Solidified Powders for Aerospace Structural Applications." Technical Report, AFML-TR-78-102, DARPA Order no. 3417. Lockhead Missiles & Space Company, Palo Alto, California.
15. Quist, W.E., Narayanan, G.H., and Wingert, A.L. (1983). In *Aluminum-Lithium Alloys II*, E.A. Starke, Jr., and T.H. Sanders, Jr., eds. TMS-AIME, Warrendale, Pennsylvania, p. 313.
16. Sakata, I.F. (1982). *Systems Study of Transport Aircraft Incorporating Advanced Aluminum Alloys—Final Report.* NASA CR-165820, NASA Contract NAS1-164634.
17. Ekvall, J.C., Rhodes, J.E., and Wald, G.G (1982). In *Design of Fatigue and Fracture Resistant Structures.* ASTM Special Tech. Publ. 761, p. 328.
18. Noble, B., and Thompson, G.E. (1971). *Met. Sci. J.*, **5**, 114.
19. Williams, D.B., and Edington, J.W. (1975). *Met. Sci*, **9**, 529.
20. Tamura, M., Mori, T., and Nakamura, T. (1970). *J. Japan Inst. of Met.*, **34**, 919.

21. Silcock, J.M. (1959–1960). *J. Inst. Met.*, **88**, 357.
22. Williams, D.B. (1981). In *Aluminum-Lithium Alloys*, T.H. Sanders, Jr., and E.A. Starke, Jr., eds. TMS-AIME, Warrendale, Pennsylvania, p. 89.
23. Ryum, N. (1969). *Acta Met.*, **17**, 269.
24. Palmer, I.G., Miller, W.S., Lloyd, D.J., and Bull, M.J. (1985). In *Aluminum-Lithium Alloys III*, C. Baker, P.J. Gregson, S.J. Harris, and C.J. Peel, eds. Inst. of Metals, London, p. 565.
25. Bauman, S.F., and Williams, D.B. (1983). In *Aluminum-Lithium Alloys II*, E.A. Starke, Jr., and T.H. Sanders, Jr., eds. TMS-AIME, Warrendale, Pennsylvania
26. Nobel, B., and Thompson, G.E. (1972). *Met Sci. J.*, **6**, 167.
27. Huang, J.C., and Ardell, A.J. (1985). In *Aluminum-Lithium Alloys III*, C. Baker, P.J. Gregson, S.J. Harris, and C.J. Peel eds. Inst. of Metals London, p. 455.
28. Rioja, R.J., and Ludwiczak, E.A. (1985). Ibid, p. 471.
29. Tosten, M.H., Vasudevan, A.K., and Howell, P.R. (1985). Ibid, p. 483.
30. Lin F.S., Chakrabortty, S.B., and Starke, E.A., Jr. (1982). *Met. Trans. A*, **13A**, 401.
31. Rioja, R.J., Bretz, P.E., Sawtell, R.R., Hunt, W.H., and Ludwiczak, E.A. (1986). In *Aluminum Alloys, Their Physical and Mechanical Properties*, **Vol. III**, E.A. Starke, Jr. and T.H. Sanders, Jr., eds. Conference Proceedings, Engineering Materials Advisory Services Ltd., p. 1781.
32. Cassada, W.A., Shiflet, G.J., and Starke, E.A., Jr. (1986). *Scripta Met.* **20**, 751.
33. Cassada, W.A., Shiflet, G.J., and Poon, S.J. (1986). *Phy. Rev. Let.*, **56**, 2276.
34. Cassada, W.A., Shiflet, G.J., and Starke, E.A., Jr. (1986). In *Aluminum Alloys, Their Physical and Mechanical Properties,* **Vol. II**, E.A. Starke, Jr., and T.H. Sanders, Jr., eds. Conference Proceedings, Engineering Materials Advisory Services Ltd., p. 695.
35. Barges, C., Tosten, M.H., Howell, P.R., and Ryba, E.R. (1987). *J. Mat. Sci.*, **22**, 1663.
36. Ashton, R.F., Thompson, D.S., Starke, E.A., Jr., and Lin, F.S. (1986). In *Aluminum-Lithium Alloys III*, C. Baker, P.J. Gregson, S.J. Harris, and C.J. Peel, eds. Inst. of Metals, p. 66.
37. White, J., and Miller, W.S. (1987). In *4th International Aluminum-Lithium Conference*, G. Champier, B. Dubost, D. Miannay, and L. Sabetay, eds. *Journal de Physique* **48**, Colloque C3, 425.
38. Tosten, M.H., Vasudevan, A.K., and Howell, P.R. (1985). In *Aluminum-Lithium Alloys III*, C. Baker, P.J. Gregson, S.J. Harris, and C.J. Peel, eds. Inst. of Metals, London, p. 490.
39. Fridlyander, I.N. Shamray, V.F., and Shiryaeva N.V. (1985). Translated from *Izvestia Akademii nauk SSSR Metally*, **2**, 153.
40. Thompson, G.E., and Noble, B. (1973). *J. Inst. of Met.*, **101**, 111.
41. Mondolfo, L.F. (1976). *Aluminum Alloys—Structure and Properties*, Butterworth, London, p. 554.
42. Narayanan, G.H. (1983). Unpublished research. The Boeing Company, Seattle, Washington.
43. Peel, C.J., Evans, B., Baker, C.A., Bennett, D.A., Gregson, P.J., and Flower, H.M. (1984). In *Aluminum-Lithium Alloys II*, T.H. Sanders, Jr., and E.A. Starke, Jr., eds. TMS-AIME, Warrendale, Pennsylvania 363.
44. Gregson, P.J., and Flower, H.M. (1985). *Acta Met*, **33**, 537.
45. Sainfort, P., and Dubost, B. (1987). In *4th International Aluminum-Lithium Conference*, G. Champier, B. Dubost, D. Miannay, and L. Sabetay, eds. *Journal de Physique*, **48**, Colloque C3 407.
46. Hautefeuille, I., Rahouadj. R., Barbaux, Y., and Clavel, M. (1987). Ibid, p. 669.
47. Narayanan, G.H., Wilson, B.L., Quist, W.E., and Wingert, A.L. (1983). *Low Density*

Aluminum Alloy Development. Third Interim Technical Report, AFWAL Contract no. F33615-81-C-5053, p. 84.

48. Ball, M.P., and Lloyd, D.J. (1985). *Scripta Met.*, **19**, 1065.
49. Sainfort, P., and Guyot, P. (1985). *Phil. Mag.*, **A51**, 575.
50. Sainfort, P., Dubost, B., and Dabus, A. (1985). *Comptes Rendus Acad. Sc. II*, **301**, 689.
51. Sankaran, K.K., and Grant, N.J. (1981). In *Aluminum-Lithium Alloys*, T.H. Sanders, Jr., and E.A. Starke, Jr., eds. TMS-AIME, Warrendale, Pennsylvania, p. 205.
52. Dean, W.A. (1984). In *Aluminum: Properties and Physical Metallurgy*, J.L. Hatch, ed. ASM, Metals Park, Ohio, pp. 200–241.
53. Wakeling, P.O. (1987). "ALCAN Developments in Al-Li Technology." NASA Workshop, Langley, Virginia.
54. Narayanan, G.H., Quist, W.E., Wilson, B.L., and Wingert, A.L. (1982). *Low Density Aluminum Alloy Development*. First Interim Tech. Report. AFWAL Contract no. F33615-81-C-5053, p. 95.
55. Muller, W., Bubeck, E., and Gerold, V. (1986). In *Aluminum-Lithium Alloys, III*, C. Baker, P.J. Gregson, S.J. Harris, and C.J. Peel, eds. Inst. of Metals, London, p. 435.
56. Agyekum, E., Ruch, W., Starke, E.A., Jr., Jha, S.C., and Sanders, T.H., Jr. (1986). Ibid, p. 448.
57. Dudzinski, N. (1952–1953). *J. Inst. Met.*, **81**, 49.
58. O'Dowd, M.E., Ruch, W., and Starke, E.A., Jr. (1987). In 4th International Aluminum-Lithium Conference, G. Champier, B. Dubost, D. Miannay, and L. Sabetay, eds. *J. de. Physique*, **48**, Colloque C3, B. 565.
59. Jensrud, O. (1985). In *Aluminum-Lithium Alloys III*, C. Baker, P.J. Gregson, S.J., Harris, and C.J. Peel, eds. Inst. of Metals, London, p. 441.
60. Sainfort, P., and Guyot, P. (1985). Ibid., p. 420.
61. Miura, Y., Matsui, A., Farukawa, M., and Nemoto, M. (1985). Ibid, p. 427.
62. Palmer, I.G., Miller, W.S., Lloyd, D.J., and Bull, M.J. (1985). Ibid, p. 565.
63. Welpmann, K., Buhl, H., Braun, R., and Peters, M. (1987). In *4th International Aluminum Lithium Conference Proceedings*, G. Champier, B. Dubost, D. Miannay, and L. Sabetay, eds. *J. de Physique*, **48**, Colloque C3, 677.
64. Reynolds, M.A., Gray, A., Creed, E., Jordan, R.M., and Titchener, A.P. (1986). In *Aluminum-Lithium Alloys III*, C. Baker, P.J. Gregson, S.J. Harris, and C.J. Peel, eds. Inst. of Metals, London.
65. Grimes, R., Miller, W.S., Reynolds, M.A., and Gray, A. *Aluminum-Lithium Development, Applications and Superplastic Forming*, Proceedings, WESTEC' 86, S.P. Agrawal, and J.R. Kar, eds. ASM, Metals Park, Ohio, p. 38.
66. Starke, E.A., Jr., Sanders, T.H., Jr., and Palmer, I.G. (1981). *J. of Met.*, **33**, 924.
67. Sanders, T.H., Jr. and Starke, E.A., Jr (1982). *Acta Met.*, **30**, 927.
68. Gysler, A., Crooks, R., and Starke, E.A., Jr. (1981). In *Aluminum-Lithium Alloys*, T.H. Sanders, Jr., and E.A., Starke, Jr., eds. TMS-AIME, Warrendale, Pennsylvania, p. 263.
69. Hill, D.P., and Williams, D.N. (1984). In *Aluminum-Lithium Alloys II*, T.H. Sanders, Jr., and E.A. Starke, Jr., eds. TMS-AIME, Warrendale, Pennsylvania, p. 201.
70. Suresh, S., Vasudevan, A.K., Tosten, M., and Howell, P.R. (1987). *Acta Met.*, **35**, 25.
71. Vasudevan, A.K., and Dohorty, R.D. (1987). *Acta Met.*, **35**, 1193.
72. Vanyblat, Y.M., Kopeliovich, D.A., and Golder, Yu. G. (1976). *Fiz. Metal. Metalloved*, **42**, 1021.
73. Vasudevan, A.K., Miller, A.C., Kersher, M.M. (1983). In *Aluminum-Lithium Alloys II*, Starke, Jr., and T.H. Sanders, Jr., eds. TMS-AIME, Warrendale, Pennsylvania, p. 181.

74. Wert, J.A., and Lumsden, J.B. (1985). *Scripta Met.*, **19**, 205.
75. Miller, W.S., Thomas, M.P., Lloyd, D.J., and Creber, D. (1986). In *Aluminum-Lithium Alloys III*, C. Baker, P.J. Gregson, S.J. Harris, and C.J. Peel, eds. Inst. of Metals, London, p. 584.
76. Fager, D.N., Hyatt, M.V., and Diep, H.T. (1986). *Scripta Met.*, **20**, 1159.
77. Webster, D. (1984). *Metal Progress*, **125**, 33.
78. Webster, D. (1987). *Met. Trans. A.* **18A**, 2181.
79. Curtis, R.E., Narayanan, G.H., and Quist, W.E. (1983). European Patent no. EP 0150456 A2. Published 1986.
80. Hunt, W.H., Jr., Sawtell, R.R., and Bretz, P.E. (1987). "Aluminum-Lithium Alloys and Method". U.S. Patent no. 4,648,913.
81. Cassada, W.A., Shiflet, G. J., and Starke, E.A. Jr. (1986). In *Aluminum-Lithium Development, Applications and Superplastic Forming*, Proceedings, WESTEC '86, Los Angeles, S.P. Agrawal and J.R. Kar, eds. ASM, Metals Park, Ohio, p. 2.
82. Coyne, E.J., Sanders, T.H., Jr., and Starke, E.A., Jr. (1981). In *Aluminum Lithium Alloys*, T.H. Sanders, Jr., and E.A. Starke, Jr., eds., TMS-AIME, Warrendale, Pennsylvania p. 293.
83. Vasudevan, A.K., Bretz, P.E., Miller, A.C., and Suresh, S. (1984). *Mat. Sci. and Engr.*, **64**, 113.
84. Scarich, G.V., Breshahan, K.M., and Bretz, P.E. (1985). In *Fatigue Crack Growth Resistance of Aluminum Alloys under Spectrum Loading*, **Vol. II**. *Aluminum-Lithium Alloys*. Tech. Rept. no. NOR-85-141, NASC. Northrop Corporation, Hawthorne, California.
85. Jata, K.V., and Starke, E.A., Jr. (1986). In *Aluminum-Lithium Alloys III*, C. Baker, P.J. Gregson, S.J. Harris, and C.J. Peel, eds. Inst. of Metals, London, p. 247.
86. Petit, J., Suresh, S., Vasudevan, A.K., and Malcolm, R.C. (1986). Ibid, p. 257.
87. Venkateswara Rao, K.T., Yu, W., and Ritchie, R.O. (1986). *Scripta Met.*, **20**, 1459.
88. James, M.R. (1987). *Scripta Met.*, **21**, 783.
89. Peters, M., Welpmann, K., Zink, W., and Sanders, T.H., Jr. (1986). In *Aluminum-Lithium Alloys III*, C. Baker, P.J. Gregson, S.J. Harris, and C.J. Peel, eds. Inst. of Metals, London, p. 239.
90. Farcy, L., Carre, C., Clavel, M., Barbaux, Y., and Aliaga, D. (1987). In *4th International Aluminum-Lithium Conference*, G. Champier, B. Debost, D. Miannay, and L. Sabetoy, eds. *J. de Physique*, **48**, Colloque C3, 565.
91. Glazer, J., Verasconi, S.L., Dalder, E.N.C., Yu, W., Emigh, R.A., Ritchie, R.O., and Morris, J.W. (1986). *Adv. Cryogenic Eng.*, **32**, 397.
92. Glazer, J., Morris, J.W., Kim, S.A., Aushi, M.W., and Ledbetter, H.M. (1987). *AIAA Journal*, **25**, 1271.
93. Dorward, C. (1986). *Scripta Met.*, **20**, 1379.
94. Glazer, J., Verzasconi, S.L., Sawtell, R.R., and Morris, J.W. (1987). *Met. Trans*, **18A**, 1695.
95. Venkateswara Rao, K.T., Hayashigatni, F.H., Yu, W., and Ritchie, R.O. (1988). *Scripta Met.*, **22**, 93.
96. Agrawal, S.P., and Kar, R.J., eds. (1986). *Aluminum-Lithium Development, Application and Superplastic Forming*, Proceedings, WESTEC '86. ASM, Metals Park, Ohio.
97. Wadsworth, J., Hensahall, C.A., Nieh, T.E. (1986). In *Aluminum-Lithium Alloys III*, C. Baker, P.J. Gregson, S.J. Harris, and C.J. Peel, eds. Inst. of Metals, London, p. 199.
98. Pickens, J.R. (1984). *A Review of the Weldability of Lithium-Containing Aluminum Alloys*. Martin-Marietta Report MML TR 84-22.
99. Edwards, M.R., and Stoneham, V.E. (1987). In *4th International Aluminum-Lithium*

Conference, G. Champier, B. Debost, D. Miannay, and L. Sabetoy, eds. *J. de Physique*, **48**, Colloque C3, 293.

100. Lin, F.S., Moji, Y., Quist, W.E., and Badger, D.V. (1987). Ibid, p. 905.

101. Quist, W.E., Hyman, M., Gane, D.H., and Badger, D.V. (1986). In *Aluminum-Lithium Alloys, Proceedings*, WESTEC '87, Los Angeles, R.J. Kar, W.E. Quist, and S.P. Agrawal, eds. ASM International, Metals Park, Ohio.

102. Papazian, J.M., Bott, G.G., and Shaw, P. (1985). In *Proc. 17th National SAMPE Technical Conference*, p. 688.

103. Papazian, J.M., Wagner, J.P.W., and Rooney, W.D. (1987). In *4th International Aluminum-Lithium Conference Proceedings*, G. Champier, B. Debost, D. Miannay, and L. Sabetoy, eds. *J. de Physique*, **48**, Colloque C3, 513.

104. Narayanan, G.H., and Axter, S.E. (1986). Unpublished research. The Boeing Company; Seattle, Washington.

105. Skillingberg, M.H., and Ashton, R.H. (1987). *In 4th International Aluminum-Lithium Conference*, G. Champier, B. Debost, D. Miannay, and L. Sabetoy, eds. *J. de Physique*, **48**, Colloque C3, 179.

106. Quist, W.E., Narayanan, G.H., Wingert, A.L., and Ronald, T.M.F. (1986). In *Aluminum-Lithium Alloys III*, C. Baker, P.J. Gregson, S.J. Harris, and C.J. Peel, eds. Inst. of Metals, London, p. 625.

107. Kim, N.J., Bye, R.L., Das, S.K. (1987). In *4th International Aluminum-Lithium Conference*, C. Champier, B. Dubost, D. Miannay, and L. Sabetoy, eds. *Journal de Physique*, **48**, Colloque C3, 309.

108. Narayanan, G.H., Graham, W.H., Quist, W.E., Wingert, A.L., and Ronald, T.M.F. (1987). In *Processing of Structural Metals by Rapid Solidification*, F.H. Froes, and S.L. Savage, eds. ASM International, Metals Park, Ohio, p. 321.

109. Schelling, R.D., Gillman, P.S., and Donachie, S.J. (1987). In *Aluminum-Lithium Development, Applications and Superplastic Forming*, WESTEC '86, Los Angeles, S.P. Agrawal and R.J. Kar, eds. ASM, Metals Park, Ohio.

110. Bretz, P.E., and Sawtell, R.R., (1986). In *Aluminum-Lithium Alloys III*, C. Baker, P.J. Gregson, S.J. Harris, and C.J. Peel, eds. Inst. of Metals, London, p. 47.

111. Bretz, P.E., and Gilliland, R.G. (1987). *Light Metal Age*, **45**, no. 314, p. 7.

112. Malcom, R.C., Cordier, F.J., and Bucci, R.J. (1986). *The Effects of Various Thermal Conditions on the Tensile Properties of Aluminum Alloy 2090–T8E41 Plate (0.5–1.0 Inch Thick)*. ALCOA Tech Report no. 56–86-AH429. ALCOA Laboratories Alcoa Center, Pennsylvania.

113. LeRoy, G., and Meyer, P. (1987). In *Aluminum-Lithium Development, Applications and Superplastic Forming*. WESTEC '86, Los Angeles, S.P. Agrawal and R.J. Kar, eds. ASM, Metals Park, Ohio, p. 103.

114. Miller, W.S., White, J., Reynolds, M.A., McDarmaid, D.S., and Starr, G.M. (1987). In *4th International Aluminum-Lithium Conference* G. Champier, B. Debost, D. Miannay, and L. Sabetoy. eds. *J. de Physique*, **48**, Colloque C3, p. 179.

115. Grimes, R., Davis, T., Saxty, H.J., and Fearon, J.E. (1987). Ibid, p. 11.

116. Peel, C.J., McDarmaid, D., and Evans, B. (1988). *Aluminum-Lithium Alloys, Proceedings*, WESTEC '87, Los Angeles, R.J. Kar, W.E. Quist, and S.P. Agrawal, eds. ASM International, Metals Park, Ohio.

117. *Aluminum-Lithium Alloys* (1987). WESTEC'87, R.J. Kar, W.E. Quist, and S.P. Agrawal, eds. ASM International, Metals Park, Ohio.

118. Little, D. (1986). In *Aluminum-Lithium Alloys III*. C. Baker, P.J. Gregson, S.J. Harris, and C.J. Peel, eds. Inst. of Metals, London, p. 15.

119. Roeder, J.P. (1987). In *4th International Aluminum-Lithium Conference*, G. Champier, B. Debost, D. Miannay, and L. Sabetoy, eds. *J. de Physique*, **48**, Colloque C3, 1.

9

Fiber-Reinforced Aluminum Alloys

S.J. HARRIS

Department of Metallurgy and Materials Science
The University of Nottingham
Nottingham, England

I. Introduction

Since 1960 attempts have been made to manufacture and use fiber reinforced aluminum-alloy composites. The main driving force for these developments has been the aerospace industries, where improved mechanical and physical properties are being sought in association with a reduction in the weight of finished components. Attempts to produce composites were initially centered upon larger-diameter fibers such as boron (100-μm–diameter single filaments). More recently, small-diameter (<20-μm) fibers (alumina, silicon carbide, and carbon) have come along which have attractions for metal-matrix composites (MMCs). These fibers may be produced in continuous or discontinuous forms. Continuous fibers tend to be used for uniaxial reinforcement of aluminum matrices, and only a small number of attempts have been made at producing laminated products (e.g., with 0°/90° and 0°/±45° fibers). Discontinuous fibers or whiskers are undergoing active consideration for more

ISBN 0-12-341831-3

random reinforcement of many heat-treatable aluminum alloys. The costs of such composites may not be excessive and their use in the automobile industry for engine parts has become a possibility.

Metal-matrix composites have to compete with developments in conventional alloys as well as with other new materials. The main competition comes on the mechanical-property front from lower-density, high-modulus, aluminum-lithium alloys and from carbon-fiber–reinforced plastics. This is particularly the case with materials that in-service operate close to ambient temperature. The addition of high-modulus ceramic fibers (e.g., silicon carbide, alumina, and boron) promotes improvements in specific modulus and strength values of aluminum alloys. The largest improvement in mechanical properties occurs parallel to uniaxially aligned continuous fibers. This type of reinforcement produces anisotropic materials with lower moduli and strengths in directions other than the fiber axis. A lower level of anisotropy in MMCs reduces the need for plied or woven-fiber constructions which are frequently employed in polymer matrices. The addition of fibers to metal and polymer matrices can promote damage-resistance in the resultant composites when operating under conditions of static and dynamic loading. Attempts have been made to combine polymer, metal, and fiber into a composite material. An example of this is ARALL, which is made up of aluminum alloy sheets laminated with an aramid-fiber–reinforced polymer. Good compressive strength properties and fatigue-resistance result from this low-density composite (see p. 295).

Short-fiber alumina or silicon-carbide whisker reinforcements also promote increases in elastic properties as well as providing better wear-resistance in aluminum alloys. The alignment of fibers or whiskers is limited and, hence, more isotropic properties are achieved than with continuous-fiber composites. Parallel to the short-fiber or whisker route, developments are taking place in MMCs with particle reinforcement (size range 1–10 μm). The more equiaxed particles of α-SiC are potentially cheaper materials than β-SiC whiskers and, therefore, the question of whether there are any technical advantages of whiskers over particles needs a clear answer.

For service life above 200°C, MMCs become more attractive materials than polymer-based composites and conventional aluminum alloys. Significant stiffness and strength levels are maintained at temperatures up to 400°C in reinforced aluminum alloys. Electrical and heat conduction are important properties that are associated with metals. Where particular combinations of mechanical and physical properties are required, MMCs can often carry out functions that are not possible with polymer-based materials.

The sections that follow first examine composite-fabrication processes and then appraise the physical and mechanical properties of these materials.

II. Composite Fabrication Techniques

The principal aim in devising a manufacturing process for MMCs is to produce a material that has fibers appropriately distributed in the matrix at the nominated volume fraction and at the same time to achieve minimal porosity. The bond strength between fiber and matrix should be suited to the application. Such bonding should be obtained without gross reaction between the two constituents; otherwise, in the case of ceramic fibers, significant weakening may result. The techniques can be considered under two headings: those relating to continuous fibers and those to discontinuous fibers. (See Fig. 1 and 2.) Under both headings, the processes use:

1. liquids to infiltrate fiber bundles or performs or
2. solid-state methods which may involve a preform stage

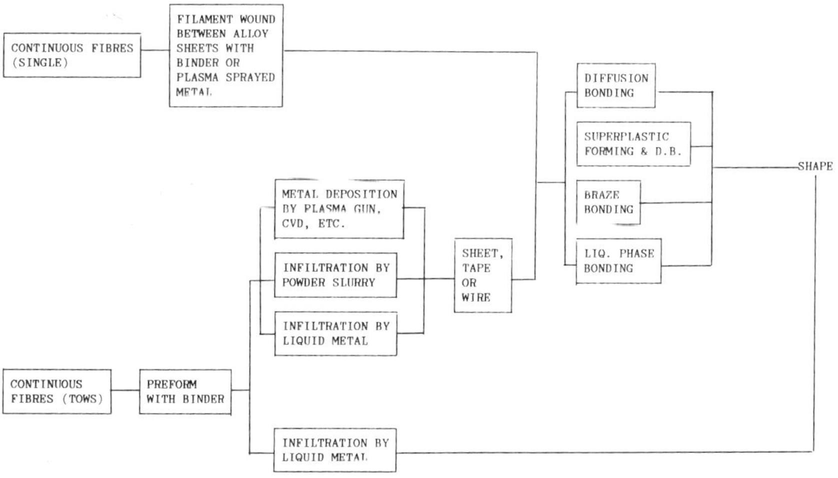

FIG. 1. Production methods and routes for continuous-fiber–reinforced MMCs. (after Harris [*34*])

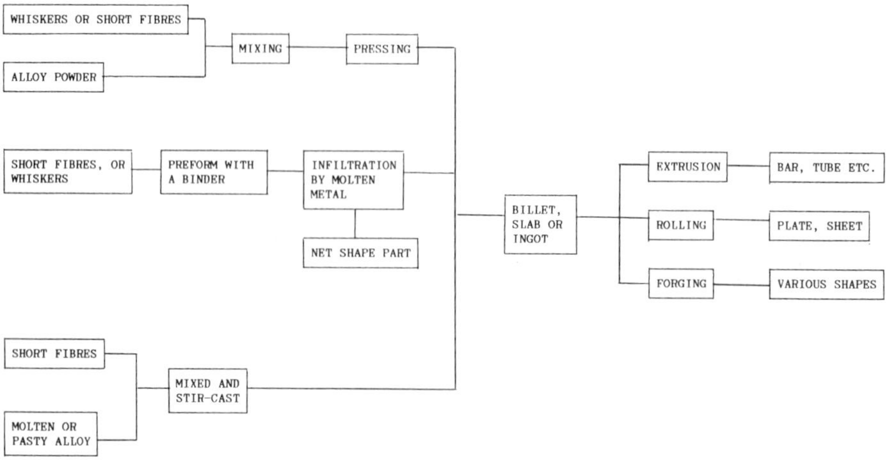

FIG. 2. Production methods and routes for discontinuous-fiber or whisker MMCs.

A. *Liquid Infiltration*

Both short-length and continuous fibers have been successfully infiltrated to produce low-porosity MMCs. Initial attempts to gravity feed fiber-preforms that occupied greater than 10% by volume of the cavity were not successful. This happened either because of excessive fiber–liquid-metal chemical interactions (e.g., aluminum-carbon where Al_4C_3 was produced) or lack of wetting, in which case porosity and poor fiber–matrix bonding resulted. Ways of tackling these problems are now available. These include precoating the fibers to promote good wetting, modifying the casting process so that the mold space may be evacuated, or applying pressure to the infiltrating metal. These methods produce much faster filling of the mold, which permits much reduced contact times between liquid metal and fibers, thus restricting the degree of possible chemical attack. Organic (polyvinyl acetate) or silicate binders can be used to hold the fibers in position in the mold cavity. Organic binders must be destroyed prior to the infiltration process.

Major interest has focused on squeeze casting, which requires a heated die and ram capable of withstanding and applying pressure to a superheated melt. (See Fig. 3.) The melt is introduced immediately prior to the ram coming into action. Within the die space, a fiber preform some 10–40 mm in thickness is already in place. Pressures applied to the system are in the range of 3 to 100 MPa. These are necessary to overcome

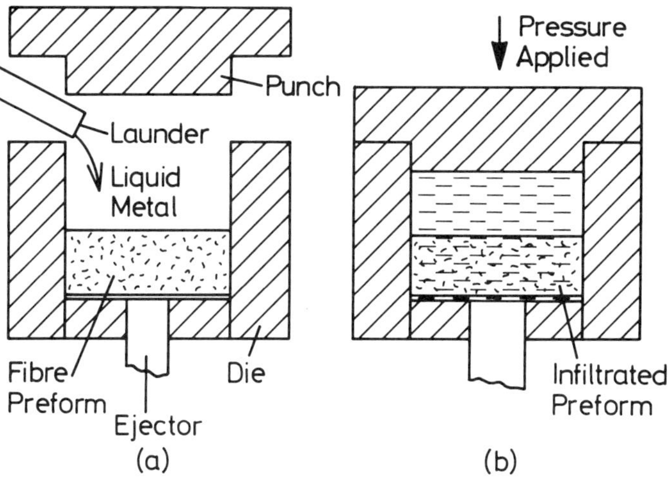

FIG. 3. Squeeze casting of a composite material using a short-fiber perform.
(a) Before casting.
(b) After casting and while applying pressure. (after Harris [*34*])

capillary effects and frictional forces experienced by liquid flowing through interstitial channels between fibers [*1*,*2*]. Once initiated, infiltration of the preform takes place very quickly with the melt front passing through a 20-mm–thick preform in a fraction of a second (i.e., at a velocity of ≈30 mm/s). Mold and preform temperatures, as well as the level of the metal superheat, need to be optimized [*3*]. Solidification does not appear to be nucleated by the fibers, and the rate of solidification is dependent on mold and preform preheat as well as metal superheat. In an alloy matrix (e.g., Al-4.5%Cu), it has been shown that the degree of microsegregation decreases with reduced interfiber spacings [*4*].

Evidence exists of strong bonding between fiber–matrix interfaces with magnesium-containing alloys and δ alumina fibers [*5*]. Nevertheless, in the short time of infiltration and solidfication, no evidence of gross reaction has been detected. (See Fig. 4a.) The absence of deleterious interfacial reactions between HM carbon fiber and aluminum-magnesium and aluminum-silicon alloys gives further evidence of the effectiveness of squeeze casting with its short time period of contact of molten metal and fibers [*6*] when compared with the earlier experiments [*7*,*8*].

Several variants of the vacuum-infiltration process have been used with continuous fibers to produce finished components or perform wires or tapes [*9*,*10*]. Fiber preforms with organic binders were employed and

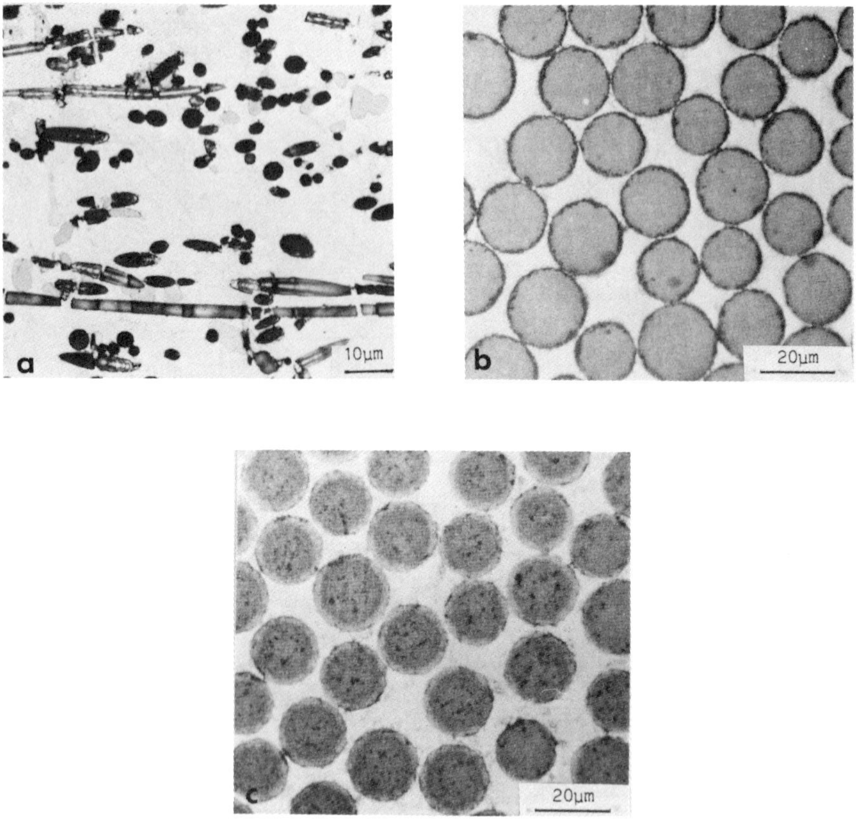

FIG. 4. Microstructures of infiltrated MMCs.
(a) Random Saffil δ-alumina short fibers in a squeeze cast Al-Si-Mg (LM25) alloy matrix. (after Harris [*34*])
(b) Uniaxial α-alumina continuous fibers after a short time exposure in the liquid state to an aluminum-lithium alloy during vacuum infiltration.
(c) As (b) but a longer exposure to the molten alloy. (after Champion et al. [*9*])

50–60% volume fraction composites have been achieved. It is important for the alloy to wet the fibers in this process; otherwise, porosity results. With carbon fibers, pretreatment with sodium [*11*] or titanium diboride [*12*] was found to be necessary to produce low-porosity composites. With α-alumina (20-μm–diameter), additions of lithium, calcium, or magnesium to aluminum have been of considerable assistance in improving the wetting behavior. (See Fig. 4b.) It has been shown that 2–3 wt % lithium alloy can be used with contact times at 700°C for a period of 10 minutes without significant fiber degradation.

Modifications of the squeeze-forming and vacuum-infiltration processes do exist. One such example is the process adopted by Yajima et al. [*13,14*] in which continuous silicon-carbide fibers (500-filament yarns, 13-μm fiber diameter) are performed around a drum and bonded with polystyrene. On release from the drum, 300 × 300-mm sheets could be cut and stacked alternately with an aluminum-alloy sheet into a stainless-steel mold. (See Fig. 5.) The mold was evacuated to 1 mm Hg and heated in a furnace to 720°C for 10 minutes. The mold was then immediately subjected to a pressure of 20 MPa for 5 minutes as the final stage of MMC production.

Mention should be made of two additional techniques for producing composites from the liquid phase. Stir-casting or compocasting requires the fibers to be stirred into an aluminum alloy which is held at a temperature between liquidus and solidus. After stirring, the mold is allowed to cool

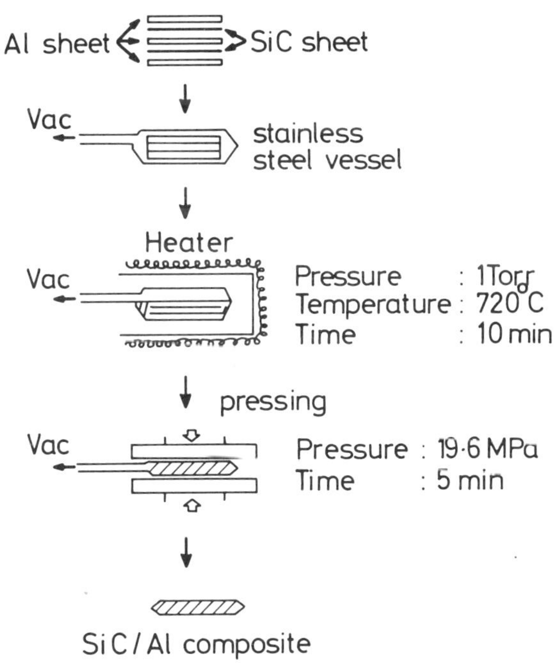

FIG. 5. Alternative technique of forming an MMC using vacuum evacuation of mold followed by hot-pressing with a molten matrix. (after Yajima et al. [*14*])

and fully solidify [*15–17*]. Properties obtained on composites produced by this technique have often been disappointing, with the exception of wear-resistance. *In-situ* composites can be prepared by directional solidification of eutectic alloy [*18,19*]. Problems can arise at changes in component section where the growth velocity of solidifying eutectic changes. This promotes local fiber misalignment and reduced tensile properties.

B. Solid-State Methods

This form of processing may involve the solid-state bonding of either

1. preinfiltrated preforms (i.e., wires or tapes) or
2. powders

1. Bonding of Fibers and Matrices or Infiltrated Preforms

The chief example of the use of solid-state bonding has been in association with the fabrication of boron or Borsic fiber MMCs. (Borsic is boron fiber with a SiC coating.) The single-filament continuous fibers (typically ≃150-μm diameter) were laid up using filament-winding techniques between either 1100 or 6061 alloy sheet; they were then subjected to hot-diffusion bonding. Fiber volume fractions of 60% were achieved using pressures of 14 to 70 MPa and temperatures of 485°C. Composites with Borsic fibers were pressed at higher temperatures (500–600°C). Lower pressures were used with alloys, which permitted some liquid phase to be present during consolidation [*20,21*]. Variations of this practice have been adapted using superplastic alloys which may reduce pressures and temperatures of operation.

Preforms can be wires, tapes, or sheet that have had the metal introduced by the vacuum infiltration by liquid metal, electrodeposition, chemical vapor deposition, thermal spraying by plasma gun, or by a powder slurry. Figure 6 shows a transverse section through a wire preform where Thornel carbon fibers have been infiltrated by a molten aluminum alloy. The nonliquid metal techniques can all experience difficulty with the penetration of dense layers containing a large number of filaments. As previously indicated, the preforms can be hot-diffusion-bonded to form the composite. Such a process can operate on preforms which may be placed in a mold at various fiber orientations to give composites with strengthening in more than one direction (e.g., 0°/90°, 0°/±45°). Besides hot-diffusion-

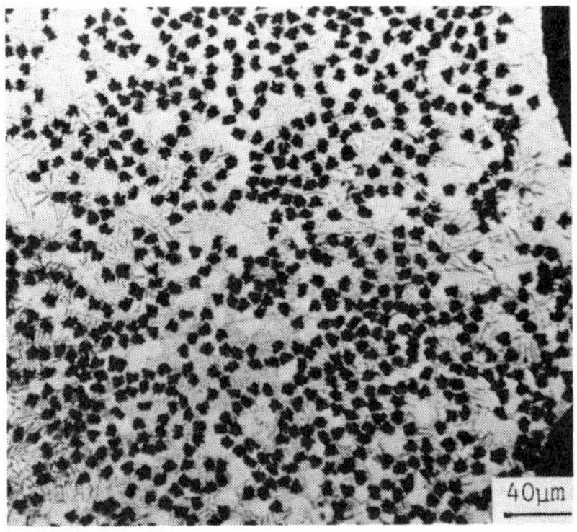

FIG. 6. A transverse microstructure through a strand of continuous Thornel 50 graphite fibers that have been infiltrated by an aluminum alloy to provide a wire preform. (after Kendall [*27*])

bonding in the solid state—which may require the application of higher pressure (up to 110 MPa) for long periods of time (30 minutes)—braze bonding (1 MPa, 10 minutes) or partial melting techniques (1 MPa, 5 minutes) may be applied. (See Fig. 7 and 8.) Increases in time and temperature tend to promote metal–fiber interactions and, thereby, fiber degradation, while high pressure can promote filament breakage.

2. Powder Forming

Developments since 1978 include the incorporation of silicon-carbide whiskers (50 μm long, under 1 μm in diameter, V_f between 0.05 and 0.30) into aluminum matrices by mixing them with −325 mesh inert gas atomized alloy powders. Initially the blended powders and whiskers are cold-compacted and then hot-pressed either above or below the solidus of the matrix alloy. This compacted billet (up to 300-mm diameter, 85 kg) is then given a further mechanical processing in the form of either rolling, extrusion, or forging [*22,23*]. Typical product forms are back-extruded tube of 300-mm diameter and 20-mm wall thickness, as well as straight and cross-rolled plate 6.35 mm thick.

Examination of the material at various stages in the processing operation reveals that the SiC whiskers contained a number of sharp-edged equiaxed impurity particles less than 5 μm in size. The initial sample contained a significant fraction of whiskers with length-to-diameter ratios of >50:1. After cold-compacting and hot-pressing, the majority of fibers had aspect ratios of between 5 and 10. (See Fig. 9.) Further working by extrusion,

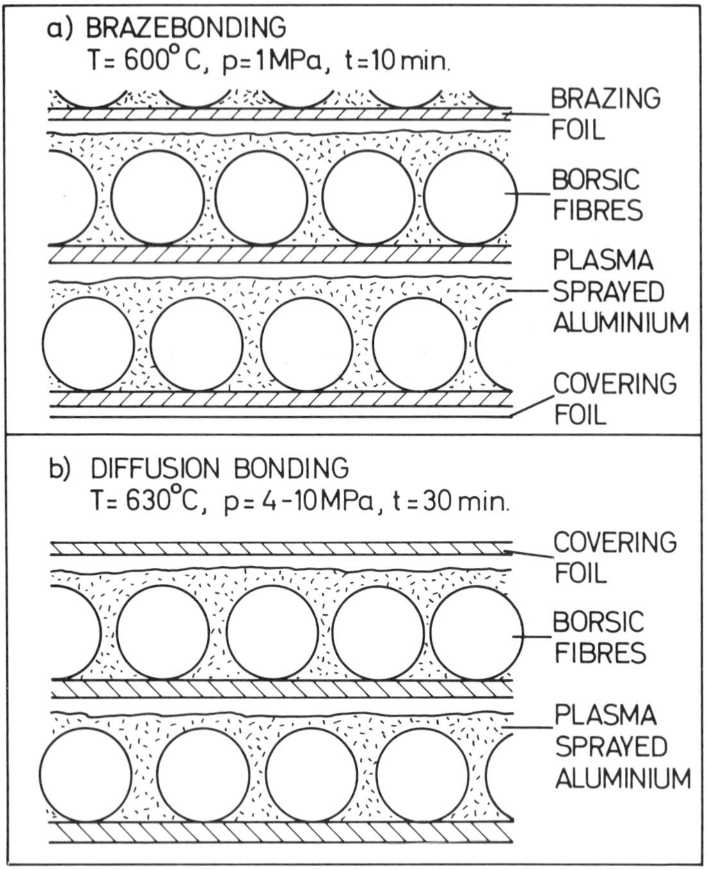

FIG. 7. Methods of bonding continuous-fiber preforms. (a) Braze bonding. (b) Diffusion bonding. (after Fitzer and Jacobson [*74*])

FIG. 8. Diffusion-bonded boron-aluminum alloy composite produced from plasma-sprayed 6061 tapes and continuous boron fibers by diffusion bonding at 550°C. (after Kreider and Prewo [*25*])

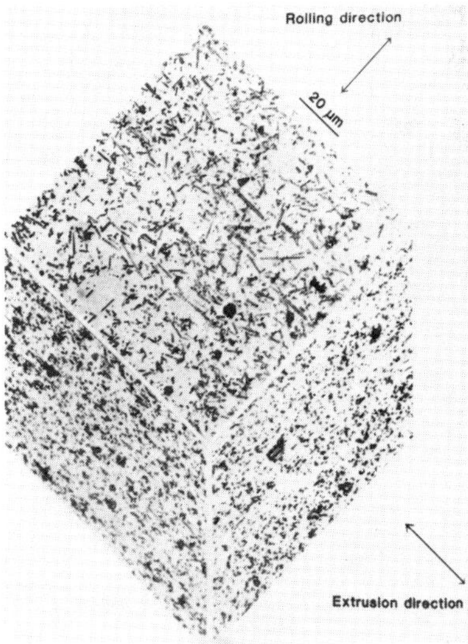

FIG. 9. Microstructure of a rolled sheet of a silicon-carbide–whisker (V_f = 0.20) aluminum composite, initially produced from blended powder–whisker mixture. (after Nieh et al. [*24*])

rolling, and so on reduced the ratio only to a small extent. Besides impurities introduced with the whiskers, intermetallics usually 10 μm in size and smooth in outline occurred from the alloy processing [*24*]. Comparisons between whisker-reinforced and particulate silicon-carbide–reinforced alloys processed down similar routes showed that the whisker-containing material was more prone to porosity.

III. Properties of Fiber-Reinforced Aluminum Alloys

Major attention has been placed on the determination of the mechanical properties of both discontinuous and continuous reinforced materials (i.e., strength, modulus, fatigue, and wear-resistance at ambient and elevated temperatures). Other properties that have been examined in detail are thermal conductivity and expansion, while only limited electrical data are available. In the chemical-property area, investigations have focused on internal reactions between fiber and matrix and on the effects of environmental conditions on MMCs.

A. *Mechanical Properties*

1. Elastic Moduli

a. *Continuous-Fiber–Reinforced Systems*

Longitudinal (E_{11}) and transverse (E_{22}) moduli have been determined on unidirectionally reinforced alloys. Figure 10 shows data plotted on the basis of moduli versus fiber volume fraction for boron (100-μm diameter),

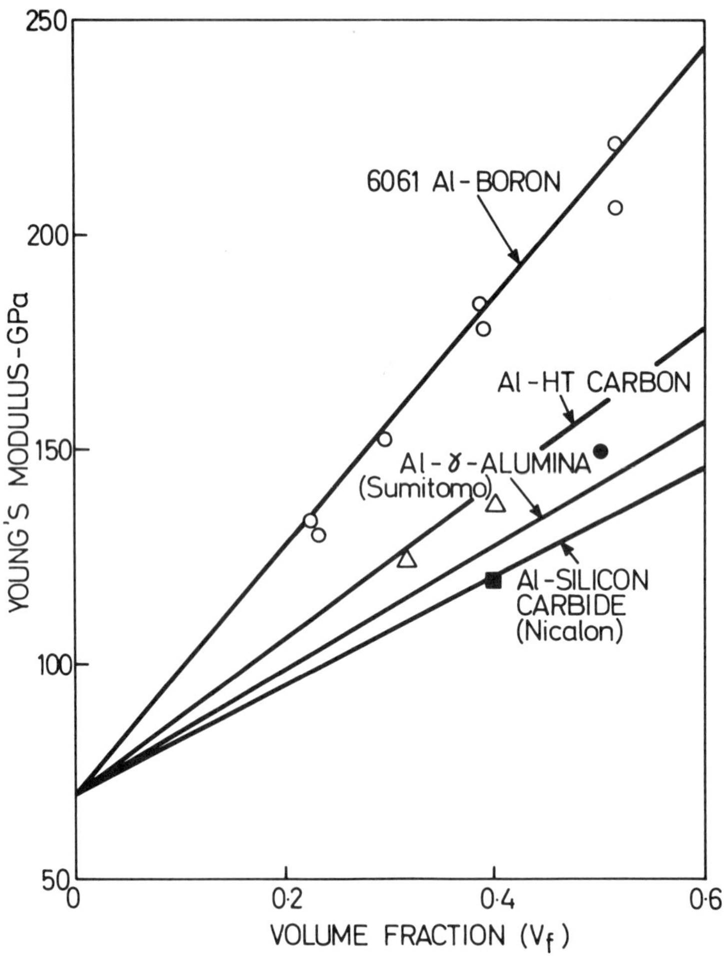

Fig. 10. Young's modulus values for the composite determined parallel to the fiber axis plotted against the volume fraction of continuous fibers. The lines are drawn on the basis of calculations using Eq. 1. (after Kreider and Prewo [*25*], Abe et al. [*26*], Tanaka et al. [*28*])

TABLE I

MECHANICAL PROPERTIES OF CONTINUOUS-FIBER COMPOSITES ($V_f = 0.5$) MEASURED PARALLEL TO THE UNIAXIAL REINFORCING DIRECTION

Fiber type	Matrix	Density (mg/m^3)	Modulus (GPa)	Specific modulus (GNm mg^{-1})	Breaking strength (MPa)	Specific strength (MNm·mg^{-1})
Boron/Borsic	6061 (T6)	2.70	225	83.3	1400	518.5
α-alumina (Dupont)	Al-Li(T6)	3.20	230	71.9	600	187.5
γ-Alumina (Sumitomo)	Al-5%Cu(T6)	2.98	150	50.3	860	288.6
Carbon (HM)	Al	2.33	235	100.9	550	236.1
Carbon (HT)	Al	2.23	160	71.7	600	269.1
Silicon carbide (Nicalon)	Al	2.63	135	51.3	780	296.6
Silicon carbide (CVD)	Al	3.10	235	75.8	1200	387.1
Carbon (HM)	Epoxy resin	1.60	215	134.0	975	609.4

HM ≡ high-modulus
HT ≡ high-tensile strength
CVD ≡ chemical vapor deposition

taken from References [*9*,*14*, and 25–28]. The influence of aluminum-alloy composition produces only small changes in modulus. The most useful change from this source is associated with the aluminum-lithium alloys which are used in conjunction with α-alumina fibers. Table I shows a range of data published on various reinforced composites that contain uniaxially aligned fibers at $V_f = 0.5$. Note that a number of systems (containing boron, α-alumina, HM carbon, and CVD silicon-carbide fibers) are capable at this volume fraction of achieving longitudinal moduli in excess of 200 GPa (i.e., three times the stiffness of aluminum). When the density of these composites is brought into consideration, the resultant specific modulus values indicate advantages for HM carbon and boron/Borsic over the higher-density α-alumina and CVD silicon-carbide fiber reinforcement. Nevertheless, the introduction of small-diameter γ-alumina and silicon carbide in multifilament tows has provided the opportunity for composites ($V_f = 0.5$) to have modulus values close to twice that of aluminum.

Transverse modulus (E_{22}) values determined perpendicular to the fiber axis are somewhat reduced from those obtained longitudinally. (See Table II.) In many of the MMCs, values in excess of 100 GPa were measured; these are at least 55% greater than the modulus of the unreinforced aluminum matrix. In this respect, carbon-fiber–reinforced plastics show

TABLE II

MECHANICAL PROPERTIES OF CONTINUOUS-FIBER COMPOSITES ($V_f = 0.5$) MEASURED TRANSVERSE TO THE UNIAXIAL REINFORCING DIRECTION

Fiber Type	Matrix	Modulus (GPa)	Specific modulus (GNm mg^{-1})	Breaking strength (MPa)	Specific strength (MNm mg^{-1})
Boron/Borsic (100-μm diameter)	6061 (T6)	130	48.1	100	37.0
Boron/Borsic (145-μm diameter)	6061 (T6)	130	48.1	240	88.9
α-alumina (Dupont)	Al-Li(T6)	150	46.9	196	61.3
γ-alumina (Sumitomo)	Al-5%Cu(T6)	110	36.9	230	77.2
Carbon (HM)	Al	—	—	70	30.0
Silicon carbide (Nicalon)	Al	77	29.3	140	53.2
Carbon (HM)	Epoxy resin	6.5	4.1	37.5	23.4

much greater differences between E_{11} and E_{22} values. The ratio $E_{22}:E_{11}$ for the polymer composite is 0.03; for almost all MMCs it is at least 0.55. Even allowing for favorable density values, the specific transverse modulus for the polymer-based materials remains low.

Raising the temperature of the metal-matrix system does not produce a great change in longitudinal modulus until temperatures exceed 450°C. For example, 0.5 V_f α-alumina in aluminum has $E_{11} = 140$ GPa at 450°C.

b. *Discontinuous-Fiber–Reinforced Systems*

The introduction of either SiC whiskers [*22*] or short δ-alumina fibers ($V_f = 0.2$) into aluminum alloys improves the modulus to values around 100 GPa in all directions in the plane of sheet material. This represents increases of 40% and 25% over conventional aluminum alloys and lower-density aluminum-lithium alloys respectively. (See Table III.) Whiskers and fiber additions keep the density of composites above that of unreinforced aluminum-lithium alloys. Hence, specific modulus values are only marginally better. A plot of modulus versus short-fiber or whisker V_f is shown in Fig. 11. As the temperature increases, the modulus does begin to fall. For example, in a discontinuous δ-Al_2O_3 6061 matrix, reductions of 9,15, and 28% below the 20°C value occur at test temperatures of 200, 300, and 400°C, respectively [*29*].

TABLE III

Mechanical Properties of Discontinuous-Fiber–Reinforced Composites ($V_f = 0.2$)

Fiber Type	Matrix	Density (mg/m^3)	Modulus (GPa)	Specific modulus ($GNm\ mg^{-1}$)	Yield stress (MPa)	Breaking strength (MPa)	Specific breaking strength ($MNm\ mg^{-1}$)	% strain to failure
Silicon carbide (whisker)	2124 (T6)	2.67	103.4	38.7	451	638	238.9	3.0
Silicon carbide (whisker)	6061 (T6)(L)	2.65	108.0	40.4	443	584	218.7	1.8
Silicon carbide (whisker)	6061 (T6) (T)	2.65	—	—	407	480	179.8	3.8
δ-alumina (Saffil)	1100	2.80	83.0	29.6	58	142	50.7	1.1
δ-alumina (Saffil)	6061 (T6)	2.80	92.0	32.9	—	362	129.3	1.1
δ-alumina (Saffil)	A1–4%Cu (T6)	2.81	86.0	30.6	353	436	155.2	1.5
δ-alumina (Saffil)	2618A (T6)	2.81	102.0	36.4	265	383	136.8	0.8
Unreinforced	Al-Li-Cu-Mg (8090)	2.54	80.0	31.5	400	500	196.9	6.0

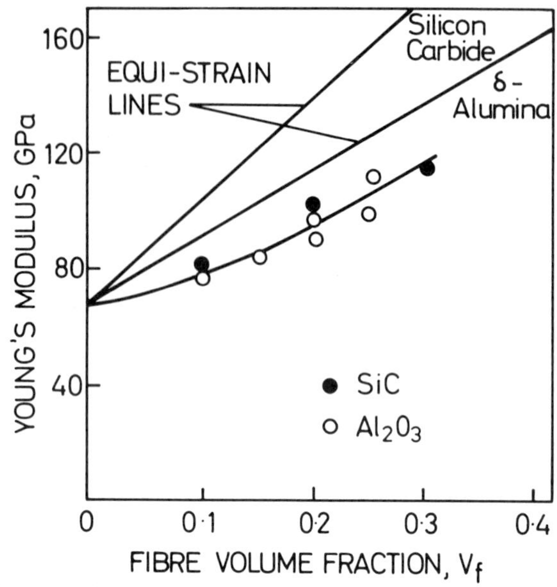

FIG. 11. Young's modulus values determined on aluminum alloys reinforced with δ-Al_2O_3 short fibers and SiC whiskers. Comparisons are made with plots obtained on the basis of Eq. 1. (after McDanels [*72*], Harris [*34*])

2. Stress–Strain Relationships

a. Continuous-Fiber–Reinforced Systems in Axial Tension

The stress–strain plots determined longitudinally on all composites of this type possess an initial elastic region (Stage I). A rule of mixtures relationship

$$E_{11} = E_f V_f + E_m V_m \tag{1}$$

predicts E_{11} with increasing fiber V_f at this stage [*30,31*]. This is followed by a nonlinear region in which the effective modulus decreases. The onset of plastic deformation in the matrix is responsible for the beginning of this Stage-II behavior. The tensile stress and strain at which the nonlinearity begins have been found to depend on the matrix yield strain and the level of residual stress which remains in the composite from the fabrication process. Once the matrix has yielded, the actual form and effective modulus of the composite become dependent on a number of factors including fiber modulus and the work-hardening characteristics of the matrix. Kreider and Prewo [*25*] claim that 2024 aluminum reinforced with 150-μm boron fibers has a Stage-I region which is separated by a nonlinear region before a linear Stage II occurs with an almost constant rate of work-hardening ($d\sigma_m/d\epsilon_m$), which can be represented by

$$E_{11} = E_f V_f + (d\sigma_m/d\epsilon_m)\, V_m \tag{2}$$

A final short portion of the plot beyond Stage II usually depends upon the pattern of fiber failures. In most ceramic-fiber–reinforced systems, the fiber has a significantly smaller coefficient of thermal expansion than aluminum and its alloys, (See Section IIIB.) This leads to the situation where after fabrication and cooling from elevated temperature, the matrix is in tension and fibers are in compression. The major effect of this is to bring about an effective increase in strain to failure of the fibers.

b. *Continuous-Fiber–Reinforced Systems in Transverse Tension*

The yield stress in transverse direction usually depends on the matrix material. Strain to failure of the composite is affected by the process that initiates fractures. In the large-diameter (150-μm) Borsic-fiber composites, failure occurs almost entirely through matrix rupture. In the smaller-diameter (100-μm) Borsic-fiber composites, failure (at 0.15% strain) involves longitudinal fiber splitting. With α-alumina fibers in an Al-Li matrix, the plot shows a higher strain to failure (0.5%) as was the case with the large-diameter 150-μm Borsic fiber [*9,25*].

c. *Discontinuous-Fiber–Reinforced Systems*

The stress–strain curves obtained on SiC-whisker–reinforced 6061 and δ-Al_2O_3-reinforced aluminum alloys show similar forms. (See Fig. 12.) After the initial elastic region, there is a pronounced knee in the curve at which point it would seem that plastic deformation of the matrix begins. The plot continues in the form of a distinct curve, the exact shape of which depends in part on the matrix work-hardening characteristics. Failure occurs abruptly without any necking down of the specimen, when a number of processes, including transverse fiber–matrix debonding and fiber cracking, occur at a significant number of sites in the composite cross section.

SiC-whisker reinforcement in 6061 alloy (annealed condition) increases the yield stress from 50 to 105 MPa ($V_f = 0.05$) and 145 MPa ($V_f = 0.2$) [*32*]. At a higher volume fraction ($V_f = 0.2$), silicon-carbide whiskers increase the yield stress of 6061 rolled plate in the T6 condition from 276 MPa (unreinforced) to 443 MPa (parallel to the rolling direction) and 409 MPa (transverse to the rolling direction) [*24*]. Saffil δ-Al_2O_3 fibers ($V_f = 0.2$) produce an increase in 0.2% proof stress of 99.85% aluminum (LMO) from 27 to 57 MPa in the as-cast condition [*33*]. In an Al-4%

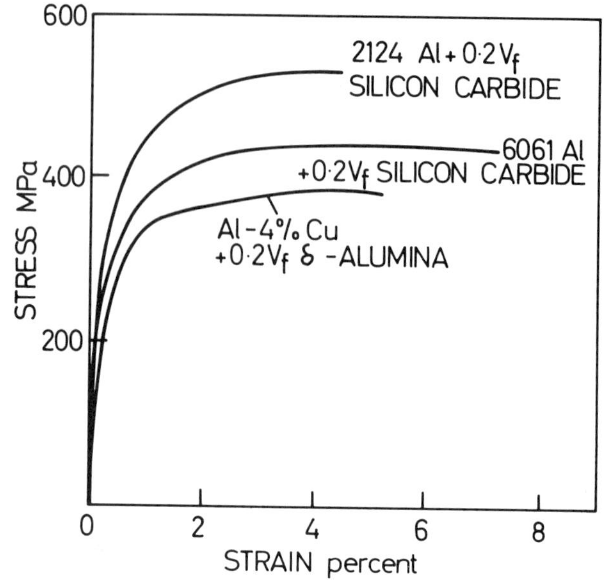

FIG. 12. Stress–strain plots obtained on two forms of discontinuous composite. (a) Silicon-carbide whisker ($V_f = 0.2$) 6061 alloy (T6) (b) Silicon-carbide whisker ($V_f = 0.2$) 2124 alloy (T6) (after McDanels [*72*]) (c) Saffil δ-alumina ($V_f = 0.2$) aluminum-4% copper alloy (T6).

copper alloy (T6 condition) with the same fiber content, the 0.2% proof stress increased from 157 to 196 MPa [*34*]. Al-Cu-Mg 2618A (T6 condition) and Al-Si-Mg LM25 (overaged) alloys have their 0.2% proof stresses reduced by ≃100 MPa by the introduction of these fibers.

Stress–strain plots in tension and compression have been determined by Arsenault and Taya [*32*] in SiC-whisker–reinforced 6061 alloy composites. The tensile yield stress, σ_y^T, was shown to be lower than the compressive yield stress, σ_y^c. This was explained in terms of residual stress patterns set up while cooling from fabrication or heat-treatment temperatures. The yield stresses are related by the expression

$$\sigma_y^T/\sigma_y^c = C_0 + C_1\alpha_m \Delta T$$

where C_0 and C_1 are functions of mechanical properties of matrix and whisker. α_m is the coefficient of thermal expansion of the matrix. ΔT is given by

$$\Delta T = T_1 - T_0 \quad (4)$$

where T_1 is the temperature below which dislocation generation becomes minimal during cooling and T_0 is ambient. Plotting $\sigma_y^c - \sigma_y^T$ versus V_f gives a linear relationship. At $V_f = 0.2$, the difference between the compressive and tensile yield stress was about ≃70 MPa.

To examine this situation in more detail, Withers et al. [*35*] used a neutron diffraction technique to measure separately the mean matrix and fiber strains, while an aluminum–silicon-carbide fiber composite was thermally cycled in the temperature range of 20 to 500°C. Figure 13 shows how the strain measurements derived from lattice plane spacings are influenced by rising and falling temperatures. *In situ* high-voltage electron microscope studies have been made on thermally cycled thin films of 6061-alloy–SiC-whisker ($V_f = 0.2$) composites [*36*]. It was revealed that on heating to 200°C, the dislocations had disappeared from the immediate vicinity of the whiskers (i.e., the temperature near to what gives the maximum change in matrix strain) (Fig. 13). On cooling down from 500°C, dislocations were observed to reappear at the whisker interface at 200°C and in much greater numbers at 100°C. Dislocation densities were nonuniform in the sample and were greater near large SiC particles and particularly near the ends of whiskers.

The variation in effective yield stress at ambient temperature (see Fig. 14) when δ-Al_2O_3 short fibers reinforce different alloys may be explained by different levels of residual stress residing in the matrix as well as by differences in precipitation distributions in age-hardening matrices. Composites with weak fiber-matrix bonding may allow more relaxation to occur in cooling, thus reducing matrix tensile stress. Strong bonding may inhibit such relaxation and encourage the removal of elements locally from the matrix—e.g., magnesium in the case of an Al-Cu-Mg alloy (2618A). The removal of magnesium due to fiber-matrix reaction may prevent the precipitation of S(Al_2CuMg) phase in the vicinity of the fiber and permit early yield to take place in a softened region where high tensile stresses reside [*34*].

Beyond the yield, the stress–strain plot depends, to a large degree, on matrix work-hardening behavior and transverse fiber matrix debonding. The degree to which these processes will take place depends upon the variation in work-hardening characteristics of the individual matrix alloy and, therefore, its heat-treatment, the fiber volume fraction and spacing, and the fiber–matrix bond strength. Attempts have been made by Tanaka and Mori [*37*] and Daimaru and Taya [*38*] to use Eshelby's [*39*] equivalent inclusion model to predict yield stress and work-hardening rates in SiC-whisker–reinforced composites. Arsenault and Taya [*32*] have demonstrated that in these models there is a need to take into account the residual stresses.

In the final stages of the stress–strain plot, the exact form of the curve can depend on damage processes taking place within the composite and also upon the ductility of the matrix. In the less well bonded systems with soft ductile matrices, larger elongations to failure can occur. This happens because a weak bonded interface is less capable of transferring stress into

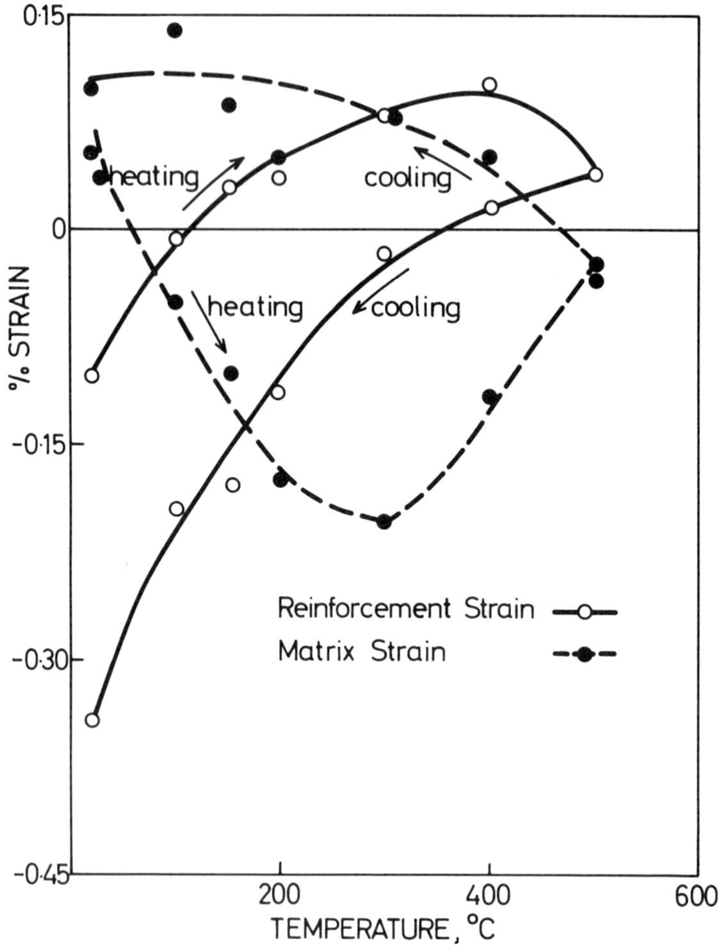

FIG. 13. Mean longitudinal strain values obtained on (a) Aluminum matrix and (b) The silicon-carbide–whisker reinforcement during a heating and cooling cycle (20–500°C). The strain values were calculated from lattice spacings measured by neutron diffraction analysis. (after Withers et al. [35])

the fiber and any premature cracks in the fiber or at transverse fiber interfaces can be contained within a ductile matrix. Radial compressive stresses acting on weak interfaces may also reduce their tendency to break down at small strains. In strongly bonded composites, fiber failure may occur at lower strains because of more effective stress transfer at the interface. The failure of the fibers then becomes a dominant factor and causes low composite failure strain.

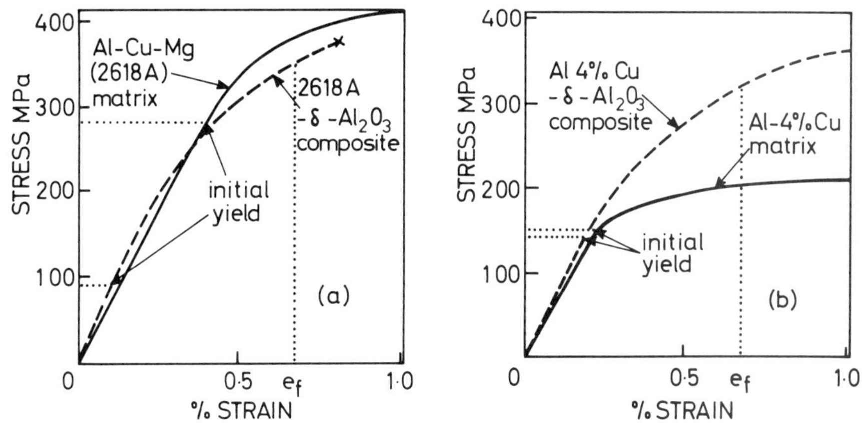

FIG. 14. Stress–strain plots for (a) An Al-Cu-Mg (2618) alloy reinforced with δ-Al_2O_3 ($V_f = 0.2$) short fibers and without reinforcement. (b) An Al-4% copper alloy reinforced with δ-Al_2O_3 ($V_f = 0.2$) and without reinforcement. The plots demonstrate the difference in the stresses and strains where the matrix starts to yield and how this may in part influence the level of fiber strengthening. (after Harris [*34*])

3. Tensile Strength

a. Continuous Fibers—Axial Tensile

Table I provides test data on a number of MMCs with $V_f = 0.5$. The most attractive breaking strengths measured come from the larger-diameter (145μm) Borsic fibers [*25*]. In the V_f range, 0.4–0.7, strengths in the 1200–1900 MPa range have been measured. These strength levels are dependent to a large degree on the fiber strength and volume fractions. The rule of mixture relationship closely predicts the breaking strength (σ_c^F) values:

$$\sigma_c^F = \sigma_f V_f + \sigma_m V_m . \tag{5}$$

Small-diameter (<20 μm) fibers (i.e., carbon, silicon carbide, and alumina), do not produce strength levels in the V_f range 0.4–0.6 that are as attractive as boron-based composites. (See Table I). Individual values have not exceeded 900 MPa on many occasions. High-tensile and high-modulus carbon [*40,41*] and α-alumina [*9*] reinforced alloys scarcely produce a value in excess of 700 MPa. Such strength levels can be approached by lower volume fraction composites using the same fiber (i.e., $V_f \simeq 0.3$). As a group, these small-diameter fibers do experience difficulty in achieving the kind of strengths forecast by Eq. 5 for $V_f \geq 0.4$. Using rule of mixtures for

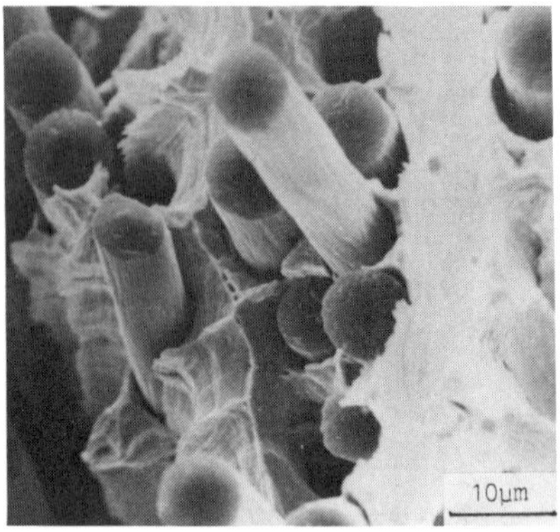

FIG. 15. Fiber pullout and matrix ductility as demonstrated in an aluminum–continuous-carbon-fiber composite.

densities, specific tensile strengths can be calculated for composites containing $V_f = 0.5$ fibers. (See Table I.)

Fracture surface characteristics of broken specimens vary according to the level of fiber–matrix bonding that exists in a given composite. In most composites, some fiber pullout occurs and there is usually evidence of local ductility in the matrix. (See Fig. 15.) Various theories for failure have been put forward to cover continuous-fiber–reinforced polymers and metals. These polarize into the cumulative weakening model [*42*] and the fiber break propagation model [*43*]. The weakening model predicts that progressive fiber fracture produces cumulative weakening and redistribution of load until at some section of the composite, the load cannot be sustained. The crack-propagation model assumes that in the neighborhood of a broken fiber, fracture occurs in the immediately adjacent fibers, because of the redistribution of load.

b. Continuous Fibers—Transverse Tensile

In most composites, reasonable levels of transverse strength have been achieved. (See Table II.) With the larger-diameter (145-μm) Borsic fibers in 6061 alloy, strengths equal to the breaking strength of the matrix have been attained (i.e., $\simeq$240 MPa). The 100-μm Borsic fibers gave strength values that are only about one half of this figure because of longitudinal fiber fractures [*25*]. Composites with small-diameter (20-μm) multifilament reinforcement are also capable of producing adequate transverse strength levels (e.g., α-alumina Al-Li alloy: 196 MPa; γ-alumina in Al-5%Cu: 230 MPa; and SiC in 6061 alloy: 140 MPa). More disappointing transverse properties are associated with carbon-reinforced aluminum where the maximum strength may not exceed 70 MPa [*27*]. All of the transverse strengths exceed those obtained on uniaxially reinforced graphite–epoxy-resin composites which usually give values below 40 MPa [*44*]. The ratio of the transverse to longitudinal breaking stress is in the 0.12–0.33 range for most metallic-matrix composites, whereas for polymer-matrix composites, the ratios are in the neighborhood of 0.04.

c. *Discontinuous Fibers*

The breaking strength of the discontinuous systems is dependent upon the type, aspect ratio, volume fraction, and distribution of the reinforcement, the alloy and its heat-treated conditions, and the fiber-matrix bond. If short fibers or whiskers are aligned parallel to the loading direction, then their strength levels can be represented in terms of a modified rule of mixture equation:

$$\sigma_c^F = \sigma_f^F \left(1 - \frac{1_c}{21}\right) V_f + \sigma'_m V_m \qquad (6)$$

where σ'_m is the stress carried by the matrix at a failure, 1 is the half-length of fiber, and 1_c is the minimum fiber length that will be capable of carrying the fracture stress of the fibers, σ_f^F in the composite, it is given by

$$1_c = \frac{\sigma_f^F d}{\tau} \qquad (7)$$

where d is the fiber diameter and τ is the shear stress acting parallel to the fiber axis. When fibers' or whiskers' 1_c values are 10 times d, then the strength of the composite is over 90% of that of a continuous-fiber–reinforced material assuming similar fiber strengths, and so on [*45*]. Attempts have been made to align and classify whiskers in water streams and sieving screens to provide precursors for composite fabrication [*46,47*] and the potential for high strengths has been demonstrated. Serious commercial attempts to reinforce aluminum alloys with short fibers and whiskers have not usually involved any prior alignment treatment. Fibers have remained random in the plane of sheet or bar material unless extrusion processes often used in powder forming produce a degree of whisker alignment. (See Table III.) With random reinforcement, it is to be expected that strength levels will be reduced below those forecast by Eq. 6.

Comparison made between silicon-carbide whiskers and δ-alumina fibers at $V_f = 0.2$ in 6061 alloy (T6) condition (see Table III) shows the superiority of the whisker reinforcement with an increase in strength of ≃60% over the δ-alumina reinforced composites [*29,32*]. Both types of composite appear to be dependent upon the matrix properties. For example, with δ-alumina fibers ($V_f = 0.2$), an increase in breaking strength from 142 to 436 MPa occurs as the matrix changes from 99.5% aluminum to Al-4%Cu (T6 condition) [*34*]. In the case of the silicon-carbide–whisker composites, fiber strengthening of virtually all aluminum alloys is usually found, producing an increase in breaking strength 30–40% above the values achieved on the wrought matrix alloy [*24,48*]. With δ-alumina, effective fiber reinforcement occurs where fiber–matrix bonding is relatively weak. Intermediate or strong bonding of δ-alumina in certain alloys produces either similar (6061) or reduced strength levels (−10% in 2618A) in the composite. As indicated in Section IIIA 2c, the major factor in initiating failure appears to be fiber cracking. This may be encouraged by strong interfacial bonding on two accounts:

1. More efficient transfer of load to the fibers whereby short fibers break at small composite strains.
2. Higher tensile residual stresses in the matrix which cause the composite to yield at lower stresses and, thus, reach fiber breaking strains also at lower stresses. (See Fig. 14.)

Elongation to failure values are reduced quite markedly in many cases to below 3.0% by the introduction of whiskers or fibers. Fiber cracking, fiber–matrix debonding to varying degrees are influential here in initiating matrix failure [*49*]. If matrix materials are ductile and fiber–matrix bonds are less strong, then elongation values can approach 3%. SiC-whisker reinforcement again appears to be less damaging in this respect than δ-alumina fibers.

4. Laminate Strength

Because of the low transverse strengths of the uniaxially reinforced polymers, many engineering components have to be fabricated with a laminate construction using fibers at ±45° to the main axis of the component or in 0°/90° arrangements. The improved transverse properties of metal-matrix composites make them superior to these laminates. Nevertheless, work has been done on ±45° and 0°/90° plied composites. The fiber volume fraction of the composite cannot be substantially increased above the $V_f = 0.5$–0.6 range so that in any direction, only about half of the fibers contribute to the modulus and strength in a given direction. Tensile tests on ±45° laminated composites containing α-alumina fibers (total $V_f = 0.5$) in an Al-Li alloy matrix gave moduli of 124 GPa and tensile strengths in the 138–172 MPa range in the 0° direction [*9*]. Besides the use of individual laminates, woven-fiber cloths have been experimented with in the small-diameter alumina and silicon-carbide fibers, but little test data are available.

5. Compression Properties

The axial compressive strength of Borsic-reinforced 6061 alloy was measured to be in the 1750–2140 MPa range [*50*]. Values over 3 GPa which are at least five times greater than the tensile strength have been found for α-alumina fiber ($V_f = 0.55$) reinforced Al-Li alloys [*9*]. In the same composite, transverse compressive strengths of 327 MPa were found. Axial compressive strengths at 2200 MPa of Sumitomo γ-alumina was three times the tensile strength [*26*]. It is generally accepted that metal-matrix composites tend to have much better compressive strengths than polymer-based materials. For glass-reinforced polyester, the tensile and compressive strengths are similar while the compressive strengths of carbon and Kevlar reinforced epoxy resins are 18% and 72%, respectively, of the values obtained in tension. Transverse compressive strengths in most polymer-matrix composites are less than 150 MPa [*44*]. Metals have much greater shear and compressive moduli than polymers and this, together with a lack of sensitivity to defects (e.g., porosity, fiber misalignment), appears to provide some explanation for this significant property improvement.

6. Fracture-Toughness

MMCs with continuous fibers show limited strain capability in tension; nevertheless, they do not behave in an entirely brittle fashion. This is because two processes can take place at the tip of a sharp notch:

1. local plastic deformation in the matrix
2. debonding of the fiber–matrix interface

These may render continuous-fiber composites notch-insensitive when they are subjected to axial loading. Tests carried out on Borsic–aluminum-alloy composites by several investigators [*51,52*] have demonstrated this situation. Tests on 0°/90° and ±45° laminates did not show any evidence of notch-sensitivity [*53*].

With discontinuous-fiber reinforcement, the situation can be quite different. It is claimed by Nieh et al. [*24*] and Hasson et al. [*54*] that silicon-carbide-whisker–reinforced aluminum alloys are notch-sensitive.

The fracture-toughness can depend upon the following considerations:

1. strength of whisker–matrix interface
2. defective whiskers (in the original material or damaged during composite fabrication)
3. presence of intermetallic or impurity particles (under 10 μm in size) in the matrix
4. heterogeneous dislocation distributions
5. whisker-free zones

Once a suitable or critical crack has initiated in the material, the crack propagation may be rapid. Silicon-carbide ($V_f = 0.2$) 6061 (T6) composites have given measured K_{IC} values of 23.4 MPa $\sqrt{m}$ in the extruded form (L-T) direction and values in the 6–7 MPa $\sqrt{m}$ range in rolled plate [55]. This fracture-toughness data should be compared with values in the 30–39 MPa $\sqrt{m}$ range which have been reported on unreinforced 6061 (T6) alloys. The higher toughness found in the reinforced extrusion in the L-T orientation may result from the larger mean free path between the SiC reinforcement in these circumstances. In plate materials, changes in the notch root radius (25–600 μm) increases the K_{IC} values from 6 to 25 MPa $\sqrt{m}$. From these tests, a linear relationship has been obtained between crack opening displacement at fracture initiation and root radius. This indicates that initiation is a strain-controlled process.

Scanning electron microscope examination of fracture surfaces produced on compact tension samples revealed that the crack path was irregular, following in part regions of high SiC-whisker reinforcement and in part matrix areas where a process of void growth had occurred. Such void growth could be related to the ends of SiC whiskers. Evidence was also found of the involvement of nonbonded interfaces between whisker and matrix and noninfiltrated areas in whisker-rich regions [55]. Nieh et al. [25] have found that intermetallic inclusions ($\simeq$50 μm) in the matrix and ceramic particles associated with the whiskers are often the prime initiators of failure in unnotched samples.

It is believed that notch blunting at crack tips occurs by the formation of a small-scale plastic zone in the matrix. The developing plastic zone interacts with SiC whiskers which leads to the opening of voids at whisker ends. As the crack surfaces move apart and crack tip blunting ceases, stretching of the voids in the matrix occurs. This leads to void coalescence and crack propagation. The toughness is controlled by size, shape, and distribution of silicon-carbide whiskers as well as the yield stress of the matrix.

7. Impact Behavior

Composites reinforced uniaxially with continuous fibers show large variations in Charpy test results, depending upon the relative orientation of the crack plane and fiber axis. When the notch is perpendicular to the reinforcement, the impact energy increases linearly from 3 to 6 J/cm^2 with increasing volume fraction of Borsic fibers in the range $V_f = 0.3$–0.5. By comparison, the energy is reduced threefold when the crack can run between the fibers or longitudinally through the fibers [56]. Higher impact values have also been measured on Sumitomo γ-alumina fibers ($V_f = 0.55$) in high-purity aluminum where a perpendicular cracked sample gave values of 9.4 J/cm^2. More highly alloyed matrices (e.g., Al-5%Cu) reduced this to 2.2 J/cm^2 [57]. Heat-treatment of alloys to higher yield strengths usually reduced the impact properties.

Impact tests on SiC-whisker–reinforced 6061 alloys gave values of 1.6 J/cm^2 as fabricated and 0.6 J/cm^2 when the alloy was heat-treated in the T6 condition [55].

8. Fatigue

a. Continuous-Fiber Systems

Uniaxially reinforced alloys usually possess very good fatigue properties when loaded parallel to the major fiber axis. In Borsic-reinforced alloys, the fatigue strength increased with fiber volume fraction. Kreider et al. [56] and Menke and Toth [58] have presented axial fatigue data in the form of Goodman diagrams, and these again showed that the material has excellent fatigue resistance over a range of alternating stress–to–mean-stress ratios. Figure 16 plots mean stress versus number of cycles for boron-6061 alloy ($V_f = 0.4$) composite using an alternating-stress: mean-stress ratio of one. It is noteworthy that the slope of the plot with increasing cycles is extremely shallow and that the ratio of fatigue strength at 10^7 cycles to tensile strength is greater than 0.55. Tension-tension fatigue tests on α-alumina–aluminum-lithium ($V_f = 0.5$) again gave a shallow S-N plot with a fatigue: tensile-strength ratio (at 10^7 cycles) of 0.77 [9]. This represents a doubling of the fatigue performance of unreinforced aluminum alloys.

Hancock [59] has shown that softer, lower-yield matrices are sometimes better in fatigue, while Dvorak and Johnson [60] have demonstrated that composite fatigue behavior can be dependent upon the fatigue behavior of the matrix. If fatigue cracks are initiated by random fiber breaks and the cyclic stress in the matrix does not exceed its 10^7 endurance limit, cracks

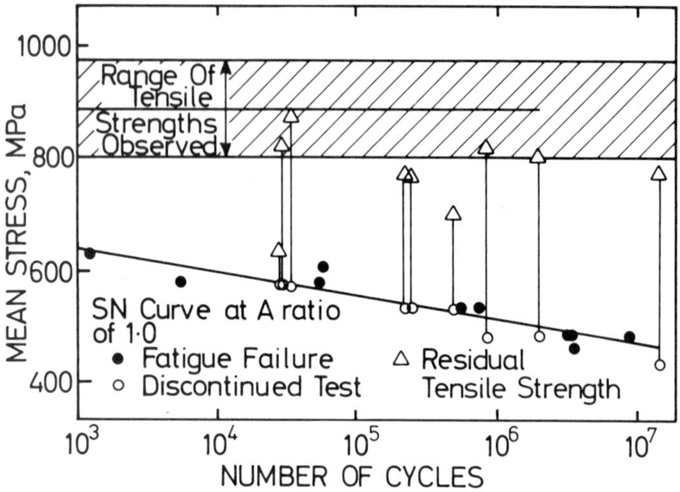

FIG. 16. Mean stress versus number of stress cycles plotted for a 6061 aluminum alloy reinforced uniaxially with continuous 100 μm boron fibers. The alternating to mean stress ratio was maintained at unity and the loading was parallel to the fiber axis. A number of tests were interrupted before failure under fatique conditions and the specimens were tensile-tested to give the residual strength values indicated. (after Kreider and Prewo [*25*])

will not propagate. This is more likely in annealed softer matrices rather than in heat-treatable alloys where the endurance limit is below the cyclic yield stress. Fiber fracture can take place during fatigue (as shown by X-radiography), but the extent to which it occurs prior to final failure appears to be small [*58*]. The residual tensile strengths of composites still approached the as-fabricated value even after 10^6 fatigue cycles. (See Fig. 16.) Failure occurs after many separately initiated fatigue cracks have propagated through a significant section of the matrix; then extensive fiber failure and final specimen separation occurs.

b. Discontinuous-Fiber Systems

S-N data obtained on 2014 (T6) and 6061 (T6) alloys reinforced with silicon-carbide whiskers (V_f = 0.15–0.20) have demonstrated that these composites have distinctly improved fatigue-resistance above that of the unreinforced alloys [*61,62*]. Similar behavior was found by Harris and Wilks [*33,63*] with Saffil δ-alumina-fiber–reinforced aluminum (99.85% Al, LMO). The improvement in each case amounts to a doubling of fatigue strength at 10^6 cycles. Harris and Wilks [*49,63*] have also shown that the addition of δ-alumina fibers to 2618A (Al-Cu-Mg) and LM25 (Al-Si-Mg) alloys reduces the fatigue strength under low (10^3) and high (10^6) stress-cycle conditions. Comparison of tensile and fatigue results indicates that for an improvement in fatigue-resistance, it is necessary for the fibers to reinforce the particular matrix alloy in a tensile test.

Williams and Fine [*64*] have carried out a study of microcrack initiation and growth on a silicon-carbide–whisker- ($V_f = 0.20$) reinforced 2124 alloy under fatigue conditions. Crack initiation occurred at the poles of silicon-carbide whiskers after a number of cycles had elapsed. Propagation then proceeded down the fiber–matrix interface provided that it was favorably oriented to the stress direction. The density of cracks increased quite rapidly (to $10^9/m^2$) after a few hundred stress cycles in appropriate stress-cycle ranges. The crack lengths then increased with the number of cycles, while the number of cracks decreased. Comparisons of the whisker-reinforced 2124 T6 ($V_f = 0.20$) with the unreinforced material revealed that the silicon-carbide composite had crack-growth rates that were slower than the matrix alone. However, the unreinforced material was able to support cracks that were 500 μm long at a peak cycle stress close to the yield point while the composite could only maintain a crack of 30-μm length at a stress which was $\simeq$60% of its yield strength. Some improvement in long crack-growth–resistance occurred at low ΔK values and when the crack was propagating normal to fibers aligned by the extrusion process during fabrication [*65*]. At high ΔK values, the crack-growth rates in the composite exceeded those in the matrix, thus supporting the fracture-toughness observations.

Crack-initiation and growth-rate studies on δ-Al_2O_3 fibers in 99.85% Al matrix have indicated that multiple-crack initiation occurs at fiber–matrix interfaces of relatively low bond strength. Thereafter, as in the 2124-SiC whisker experiments, crack growth and coalescence of the multitude of cracks occur. In the more well bonded systems (i.e., with 2618A and LM25 matrices), large ceramic particles (25 μm in size) appear to initiate cracks at much lower cyclic strains. Such cracks propagate across the strongly bonded interfaces and promote crack growth in the matrix and adjacent δ-Al_2O_3 fibers [*63*].

9. Elevated-Temperature Behavior

a. Tensile Properties—Continuous Fibers

The properties of unidirectional boron-aluminum composites parallel to the fibers are controlled at a given temperature to a large degree by decreasing fiber strengths and fiber–matrix interaction. The interaction degrades the fibers within relatively short periods of exposure to test temperatures in excess of 400°C. Strength levels with Borsic fibers (V_f = 0.4) in 6061 alloy at 300°C exceed 1000 MPa (i.e., 83% of measured values at 20°C [*66*]). Properties of α-alumina fiber (V_f = 0.55) in an Al-Li alloy matrix virtually remain constant up to test temperatures of 320°C. Silicon-carbide (Nicalon) fibers in 1100 Al show only minor reductions in strength up to 400°C [*14*]. The test data quoted so far relate to short-time (≃1 h) holding a temperature before test. Exposure of boron–aluminum composites for 10,000 h at temperature reduces its strength to 750 MPa at 230°C and to 600 MPa at 370°C. Transverse to the fibers, properties are more sensitive to test temperatures. At 200°C, a reduction in ambient transverse strength of 35% was observed in a number of alloys. For example, in boron-reinforced 6061, the strength was 170 MPa at 200°C [*25*].

b. Tensile Properties—Discontinuous Fibers

The properties for δ-alumina-6061 alloy composites are shown in Fig. 17. A strength reduction from 350 to 150 MPa occurred as the test temperature increased from 20 to 300°C. In comparison, at 300°C the unreinforced 6061 alloy had a strength of 80 MPa. Other alloy matrices (e.g., Al-9%Si-3%Cu) (see Fig. 17) are capable of maintaining strengths in excess of 200 MPa at 300°C when reinforced with δ-alumina fibers in the V_f = 0.12–0.24 range. It has been shown that the breaking strengths of a 6061 matrix reinforced with V_f = 0.25 silicon-carbide whiskers maintained a level of 325 MPa when tested at 200°C (See Fig. 17). At 300°C, the strength was down to 140 MPa, which is very similar to that achieved in the same matrix with δ-alumina fibers present. Strongly bonded matrices with short δ-Al_2O_3 (V_f = 0.2), which show virtually no reinforcement at ambient, begin to reveal improvements in tensile strength above those obtained on the unreinforced matrix as the test temperature exceeds 200°C [*29*]. At least part of this increase must result from the reduction in residual tensile stress shown to take place as the temperature rises. (See Section IIIA2c.)

c. Creep and Stress Rupture Properties

Uniaxially reinforced Borsic-6061 Al composites loaded parallel to the continuous fibers show excellent creep resistance up to 500°C. Times to rupture of 1000 h are achieved at 500°C under applied stress levels of 500 MPa, which is three times the load achieved by Ti-6Al-4V alloys [*25*]. Creep tests at these temperatures did not reveal any significant composite deformation. The fiber behavior controls these properties, and it is necessary for the boron fibers to be coated with silicon carbide. Otherwise, the

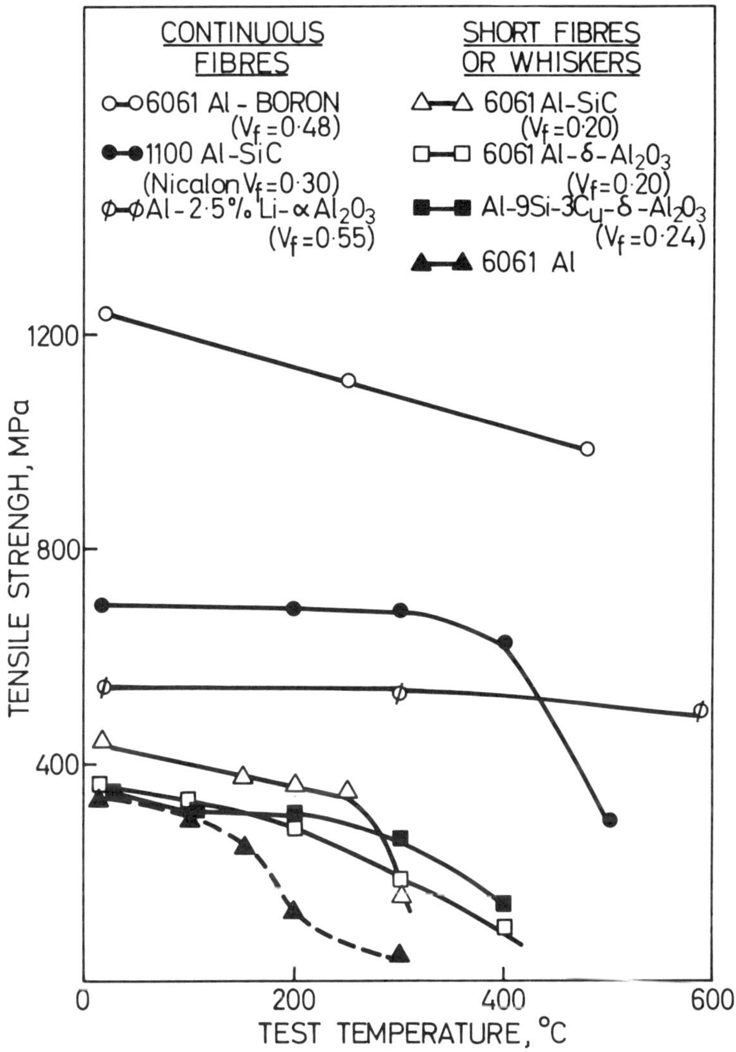

FIG. 17. Tensile strength values for continuous fiber (parallel to uniaxial aligned fibers) and whisker and discontinuous-fiber–reinforced aluminum alloys plotted against the test temperature. (data after Yajima et al. [*14*], Kreider and Prewo [*25*], Dinwoodie et al. [*29*], Champion et al. [*9*], McDanels [*72*])

composites begin to show inferior stress-rupture properties after 100-h exposure at 400°C.

Silicon-carbide–whisker- ($V_f = 0.2$) reinforced 2024 alloys have been stress-rupture–tested at 200 and 300°C. At the higher temperatures, these materials show that they can carry the same applied stress for a hundred times longer than the unreinforced 2024 alloy [*22*]. Tests carried out on silicon-carbide–whisker- ($V_f = 0.2$) reinforced 6061 alloys in the 232–371°C temperature range provided results that produce a creep curve having short primary, long steady-state, and negligible tertiary stages [*67*]. An equation for steady-state creep rage, ϵ_{ss} (s^{-1}) was obtained:

$$\epsilon_{ss} = 1.5 \times 10^{-9} \sigma^{20.5} \exp Q_a/RT \quad (8)$$

where σ is the applied stress in MPa and Q_a is the activation energy (390 kJ/mole). The large value of the exponent ($n = 20.5$) and the activation energy are similar to the kind of values found in oxide-dispersed aluminum alloys. The activation energy is almost three times that associated with the diffusion of aluminum. The difference in steady-state creep rate of 6061 alloy with and without silicon-carbide whiskers is shown in Fig. 18. The creep rate is controlled by the glide of dislocations in the alloy case, but in the reinforced composite, the explanation of the creep-rate equation (Eq. 8) is not understood. Lilholt and Taya [*68*] have confirmed that the stress-sensitivity of such composites is larger than that associated with the matrix alone. However, the creep strength is significantly smaller than is predicted by mid-1980's models for aligned short-fiber composites. Creep failure occurs in the ductile manner with little evidence of fiber failure.

10. Wear Properties

Work has been concentrated on composites involving δ-alumina and silicon-carbide short fibers. Wear tests have demonstrated a very large reduction in wear rates ($\simeq$30 times) with the introduction of δ-alumina fibers in the $V_f = 0.2$–0.3 range if the test is carried out against a hardened chromium-steel mating surface [*29*]. A larger survey of alumina ($V_f =$ 0.05–0.08) fiber-reinforced composites was carried out in connection with the development of new pistons for diesel engines [*69*]. Samples showed little evidence of wear when they were subjected to sliding against cast iron, chromium plate, and steel. The test results demonstrated that SG iron or chromium-plated countersurfaces wear more rapidly than hardened steel.

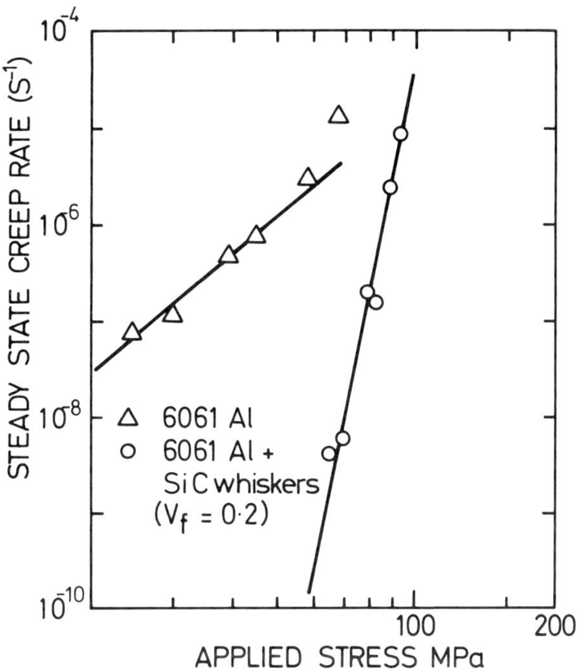

FIG. 18. Comparison of the steady-state creep rate versus stress for 6061 alloy and 6061–silicon-carbide whisker ($V_f = 0.2$) composite at 288°C. The whisker reinforcement increases the creep-resistance appreciably. (after Nieh, [*67*])

B. Thermal Properties

With the majority of ceramic-fiber–reinforced materials, there is a significant difference in thermal-expansion coefficients between the fiber and the aluminum alloy. For example, in the 0–100°C temperature range, the expansion coefficient for aluminum is 24×10^{-6} °C^{-1}, while that of α-alumina is 7×10^{-6} °C^{-1}. Measurements made on unidirectional composites prepared from continuous α-alumina fibers ($V_f = 0.55$) and aluminum show very different values for directions parallel to and perpendicular to the fiber axis. In the parallel direction, the values are close to those of the α-alumina (7×10^{-6} °C^{-1}) and give an intermediate value (16×10^{-6} °C^{-1}) in the transverse direction. This kind of behavior is maintained at least up to 400°C [*9*]. Similar behavior has been found with Sumitomo γ-gamma fibers in aluminum [*26*].

The coefficient of expansion of discontinuous-fiber composites has been shown to vary with fiber volume fraction and direction in the composite. δ-alumina fibers ($V_f = 0.24$) in an Al-9%Si-3%Cu alloy reduce the coefficient from 20.3×10^{-6} to 15.5×10^{-6}°C^{-1} in the plane of material and to 15.7×10^{-6} °C^{-1} normal to the plane. Changes in the value of the coefficient begin to take place at low values of V_f; however, as V_f increases above 0.12, changes in coefficient become smaller [*29,69*].

Thermal cycling can promote changes in the mechanical properties of fiber-reinforced aluminum alloys. It is usually the strength properties (breaking stress and yield stress) rather than modulus that change. This has been demonstrated in continuous-boron-fiber–reinforced aluminum and in silicon-carbide-whisker–reinforced 2124 alloy. After 100 cycles between 20 and 400°C, the whisker-reinforced alloy increases in length and decreases a small amount in thickness, while the width remains essentially constant. A reduction in breaking strength of 30% and yield-stress reduction of 57% occur after 1000 cycles up to 400°C. It is claimed that matrix plastic strain during each cycle helps bring about the property change. Cycling to higher temperatures of 500°C reduces the change in breaking strength (8%) and yield stress (38%) because of matrix creep processes and possible recrystallization [*70,71*]. Thermal fatigue cracks were found in an alumina-reinforced alloy after 240 cycles between 20 and 400°C [*69*].

The thermal conductivities of continuously reinforced composites differ in the longitudinal and transverse directions. Each depends upon the volume fraction of fibers present in the composite. Conductivity of these materials will be much better than those achieved in polymer-matrix materials. For example, transverse thermal conductivity of α-alumina–reinforced epoxy resin ($V_f = 0.6$) has been measured as 0.85 Wm^{-1}K^{-1} while that of α-alumina–reinforced aluminum ($V_f = 0.5$) is 34.8 Wm^{-1}K^{-1} [*9*]. Discontinuous alumina fibers ($V_f = 0.06$–0.08) reduce the conductivity of the matrix by 12% [*69*].

C. Electrical Conductivity

The electrical conductivity behaves in a similar fashion to the thermal conductivity. As the volume fraction of the nonconducting filaments increases, then the conductivity begins to go down. For unidirectional continuous-fiber–reinforced aluminum, this can be predicted parallel to the fiber axis by a rule of mixture equation.

D. Corrosion

With most ceramic fibers (e.g., Al_2O_3, SiC, B), there are no major aqueous corrosion problems other than those that already relate to the aluminum-alloy matrix behavior. Carbon fibers can introduce galvanic action because of their conducting nature, etc. In these circumstances, attempts to prevent carbon fibers coming to the surface (in particular, the fiber end regions) in unidirectional reinforced systems would be important. Many of the fibers should be able to improve the stress-corrosion properties of aluminum alloys.

As elevated temperatures, oxidation behavior of the aluminum alloy is not usually impaired by the presence of alumina or SiC. In the case of carbon fibers, care has to be taken to prevent fibers emerging at exposed surfaces. Major problems do arise internally in some systems as the temperature approaches the melting point of the alloy (e.g., in Al-graphite and Al–silicon-carbide systems).

IV. Summary

The development of fiber-reinforced metals has followed two distinct paths: those relating to continuous fibers and those relating to discontinuous fibers or whiskers.

Continuous-fiber composites usually contain a high volume fraction of fibers (V_f = 0.3–0.6). The fibers are usually expensive and require careful handling in fabrication. Use of uniaxially aligned fibers in composites produces significant increases in mechanical properties parallel to the fiber axis. Without exception, high elastic modulus values above 150 GPa are promoted and in some cases—with B, C, α-Al_2O_3, and SiC (CVD version fibers)—values similar to those of steel (200 GPa) are produced. The majority of fibers do not reduce the density of aluminum, the major exception being carbon. The specific modulus properties of continuous-fiber–reinforced aluminum do approach those of uniaxially polymer based composites reinforced with carbon fibers (CFRP).

Strength properties parallel to the fiber are more variable for continuous-fibers composites. Large-diameter fibers (100–150 μm)—e.g., B or SiC (CVD)—produce high tensile strengths (e.g., above 1200 MPa). These materials also behave in a notch-insensitive manner and have good fatigue properties. Smaller-diameter fibers do not promote such good strengths (600–900 MPa). Reasons for this behavior are not well understood, but increasing fiber volume fraction (above 0.4), fiber breakage, and reaction with the matrix in fabrication are seen as contributing factors. On specific strength properties, only boron-fiber composites can seriously compete with CFRP. There is a factor that acts in favor of MMCs, and this is where components are subjected to loads acting in directions other than parallel to the fibers. Modulus and strength values obtained transverse to the fibers and in shear are much superior in metal rather than polymer-matrix composites. At elevated temperatures (above 150°C), metal matrices retain stiffness, strength, and creep-resistance, whereas thermosetting polymers have limitations. Development with thermoplastic-polymer matrices (PEEK, etc.) may give increased competition in this area to metals.

Continuous-fiber MMCs have been used (particularly Borsic-Al) for panels, long beams, and stiffeners, particularly in aerospace applications. In addition to good stiffness and fatigue properties, these composite parts also have compressive properties superior to those of CFRP. While environmental factors such as corrosion cause problems to aluminum-alloy matrices, thermosetting polymers also have difficulties with moisture uptake. It is difficult to see application of continuous-fiber MMCs outside the aerospace field on the grounds of cost unless it is for very specialized purposes. Whether MMCs will successfully be used in the jet engine is difficult to ascertain. Higher temperature requirements may rule out aluminum as a matrix, and fiber-reinforced titanium may be more successful.

Discontinuous-fiber composites have received much intensive scrutiny during the 1980s with the advent of silicon-carbide whiskers and a variety of alumina fibers. The fiber volume fractions in these composites are smaller (i.e., $V_f = 0.05$–0.25), with occasional excursions up to $V_f = 0.40$. Fiber or whisker additions in the V_f range 0.15–0.20 improve the modulus of aluminum to values between 90 and 110 GPa, representing a ≥30% increase over conventional aluminum alloys. Both Al_2O_3 and SiC increase the density of aluminum to a small degree. The improvement in elastic modulus is enough to compete on a specific modulus basis with aluminum-lithium alloys. In this sense, the use of an aluminum-lithium matrix would make these MMCs more attractive.

Tensile strength can be increased by the addition of short fibers or whiskers. The degree of strengthening depends upon many factors. Strength increases with V_f although the improvements gained above $V_f = 0.2$ appear to be minimal [*72*]. Limitations on composite strength do arise from a matrix standpoint. SiC whiskers produce a 100% increase in strength with a matrix UTS of 150 MPa and 2% increase in strength with a matrix UTS of 550 MPa [*73*]. Fiber–matrix bonding and fiber-breaking strain are interrelated factors. The stronger the bond, the greater the opportunity is for efficient stress transfer to the fiber, which may then reach its failure strain at a lower composite stress. In these circumstances, residual matrix stresses arising in the composite fabrication may lower the matrix yield stress and allow the composite to reach its failure strain at lower stress values. Reduced strain to failure, low fracture-toughness, and scatter in tensile strength are interrelated problems in short-fiber– or whisker-reinforced composites. These related to void or crack initiation and propagation. Initiation make take place in the matrix, at the matrix–fiber interface, or in the fibers themselves. The more weakly bonded systems have elongations to fracture in the 1.5–6.0% range. The scatter depends on fiber content and the presence of impurity particles or matrix precipitates. Discontinuous-fiber composites are notch-sensitive, and there is a need to understand fracture processes in them in more detail so that attempts may be made to improve toughness. Whether fracture-toughness values can be increased significantly without raising the costs of the materials or their processing remains to be answered.

Improvements in fatigue-resistance can be obtained by the addition of short fibers or whiskers in composites when effective reinforcement occurs in tension. The differences between fatigue and fracture-toughness performance of SiC-whisker–reinforced alloys can be explained by the slower fatigue-crack growth rate at low ΔK values when compared with the unreinforced matrix at low ΔK values and an inferior performance at high ΔK where conditions approach those for fracture-toughness. One of the differences between SiC-particle and SiC-whisker–reinforced composites has been found with fatigue-resistance where the whisker reinforcement has been shown to be superior [*62*]. Further investigations of fatigue-crack initiation and propagation in whisker and particulate composites are needed.

Discontinuous reinforcement improves the tensile strength of most aluminum alloys significantly at temperatures up to 350°C and creep strengths are also maintained at 300°C. Accurate models of the creep behavior of short-fiber–reinforced composites have yet to be devised. Wear-resistance at ambient and elevated temperatures was one of the major reasons for introducing this form of composite strategically into the pistons of the diesel engine. High-temperature applications involve thermal cycling and this brings into question the thermal-fatigue–resistance and dimensional stability of these materials. Further investigations of these aspects are needed with composites that contain ceramics and metals with very different thermal expansion coefficients.

MMCs with discontinuous-fiber reinforcement have become of special interest to the automobile industry, particularly in respect to components for highly rated diesel engines. Whether MMCs can extend their applications beyond selective regions of the piston depends very much on a mixture of commercial and technical considerations. The dominant issue will always be the cost of materials and their fabrication.

References

1. Clyne, T.W., Bader, M.G., Cappleman, G.R., and Hubert, P.A. (1985). *J. Mater. Sci.*, **20**, 85.
2. Clyne, T.W., and Bader, M.G. (1985). In *Fifth Intl. Conf. on Composite Materials*, W.C. Harrigan, Jr., J. Strife, and A.K. Dhingra, eds. AIME, Philadelphia, pp. 755–771.
3. Fukunaga, H., and Ohde, T. (1982). In *Fourth Intl. Conf. on Composite Materials*, T. Hayashi, K. Kawata, and S. Umekawa, eds. Japanese Soc. Comp. Materials, Tokyo, pp. 1443–1455.
4. Cornie, J.A., Mortensen, A., Gungor, M., and Flemings, M.C. (1985). In *Fifth Intl. Conf. on Composite Materials*, W.C. Harrigan, Jr., J. Strife, and A.K. Dhingra, eds. AIME, Philadelphia, pp. 809–823.
5. Cappleman, G.R., Watts, J.F., and Clyne, T.W. (1985). *J. Mater. Sci.*, **20**, 2159–2168.
6. Sawada, Y., and Bader, M.G. (1985). In *Fifth Intl. Conf. on Composite Materials*, W.C. Harrigan, Jr., J. Strife, and A.K. Dhingra, eds. AIME, Philadelphia, pp. 785–793.
7. Baker, A.A., Shipman, C., and Jackson, P.W. (1972). *Fibre Sci. and Technology*, **5**, 213.
8. Khan, I.H. (1976). *Met. Trans.*, **7A**, 1281–1289.
9. Champion, A.R., Krueger, W.H., Hartmann, H.S., and Dhingra, A.K. (1978). In *Second Intl. Conf. on Composite Materials*, B. Noton, R. Signorelli, K. Street, and L. Phillips, eds. AIME, Philadelphia, pp. 882–904.
10. Harrigan, W.C., Jr. (1978). *Metall. Trans. A.*, **9A**, 503–507.
11. Goddard, D.M. (1977). In Report no. ATR-77 (8162)-3. Aerospace Corporation, El Segundo, California.
12. Amateau, M.F. (1976). *J. Composite Materials*, **10**, 279.
13. Yajima, S., Okamura, K., Tanaka, J., and Hayase, T. (1980). *J. Mater. Sci.*, **15**, 2130.
14. Yajima, S., Okamura, K., Tanaka, J., and Hayase, T. (1981). *J. Mater. Sci.*, **16**, 3033.
15. Mehrabian, R., Riek, R.G., and Flemings, M.C. (1974). *Met. Trans.*, **5A**, 1899.
16. Levy, C.G., Abbashian, G.J., and Mehrabian, R. (1977). In *Failure Modes in Composites IV*, Cornie J.A. and Crossman, eds. AIME, Philadelphia.
17. Quigley, B.F., Abbashian, G.J., Wunderlin, R., and Mehrabian, R. (1982). *Met Trans.*, **13A**, 93.
18. Chadwick, G.A. (1975). *Metal Sci. J.*, **9**, 300.
19. Jabczynski, F.S.J., and Cantor, B. (1981). *J. Mater. Sci.*, **16**, 2269.
20. Mangiapane, J.A., Gray, D.F., Saltor, S.A., and Timoshenko, J.A. (1968). AIAA Paper 68–1037.
21. Miller, M., and Schaeffer, W. (1971). In *Proc. 1971 SAMPE Conf. Space Shuttle Mater.*, Huntsville.
22. Rack, H.J., Baruch, T.R., and Cook, J.L. (1982). In *Fourth Intl. Conf. on Composite Materials*, T. Hayashi, K. Kawata, and S. Umekawa, eds. Japanese Soc. Comp. Materials, Tokyo, pp. 1465–1471.
23. Sakamoto, A., Hasegawa, H., and Minoda, Y. (1985). In *Fifth Intl. Conf. on Composite Materials*, W.C. Harrigan, Jr., J. Strife, and A.K. Dhingra, eds. AIME, Philadelphia, pp. 699–707.
24. Nieh, T.G., Rainen, R.A., and Chellman, D.J. (1985). In *Fifth Intl. Conf. on Composite Materials*, W.C. Harrigan, Jr., J. Strife, and A.K. Dhingra, eds. AIME, Philadelphia, pp. 825–842.
25. Kreider, K.G., and Prewo, K.M. (1974). In *Composite Materials*, **vol. 4**, L.J. Broutman and R.H. Krock, eds. Academic Press, New York, pp. 400–480.
26. Abe, Y., Horkiri, S., Fujimura, K., and Ichiki, E. (1982). In *Fourth Intl. Conf. on*

Composite Materials, T. Hayashi, K. Kawata, and S. Umekawa, eds. Japanese Soc. Comp. Materials, Tokyo, pp. 142–1484.

27. Kendall, E.G. (1974). In *Composite Materials*, **vol. 4**, L.J. Broutman and R.H. Krock, eds. Academic Press, New York, pp. 319–399.
28. Tanaka, J., Ichikawa, H., Hayase, T., Okamura, K., and Matsuzawa, T. (1982). In *Fourth Intl. Conf. on Composite Materials*. T. Hayashi, K. Kawata, and S. Umekawa, eds. Japanese Soc. Comp. Materials, Tokyo, pp. 1407–1413.
29. Dinwoodie, J., Moore, E., Langman, C.A.J., and Symes, W.R. (1985). In *Fifth Intl. Conf. on Composite Materials*, W.C. Harrigan, Jr., J. Strife, and A.K. Dhingra, eds. AIME, Philadelphia, pp. 671–685.
30. Kelly, A. and Davies, G.J. (1965) *Met. Rev.* **10**, 37.
31. Ashton, J.E., Halpin, J.C., and Petit, P.H. (1969). In *Primer on Composite Materials: Analysis*, Technomic, Stamford, Connecticut.
32. Arsenault, R.J., and Taya, M. (1985). In *Fifth Intl. Conf. on Composite Materials*, W.C. Harrigan, Jr., J. Strife, and A.K. Dhingra, eds. AIME, Philadelphia, pp. 21–36.
33. Harris, S.J., and Wilks, T.E. (1986). In *Fibre Reinforced Composites*. Institute of Mechanical Engineers, London, pp. 19–28.
34. Harris, S.J. (1988). *Materials Science & Technology*, **4**, 231.
35. Withers, P.J., Tensen, D.J., Lilholt, H., and Stobbs, W.M. (1987). In *Sixth Intl. Conf. on Composite Materials*, **vol. 2**, L. Matthews, N.C.R. Buskell, J.M. Hodgkinson, and J. Morton, eds. Elsevier, London, pp. 255–264.
36. Voselsang, M., Arsenault, R.J., and Fisher, R.M. (1986). *Met. Trans.*, **17A**, 379–389.
37. Tanaka, K., and Mori, T. (1970). *Acta. Metall.*, **10**, 131.
38. Daimaru, A., and Taya, M. (1982). In *Fourth Intl. Conf. Composite Materials*, T. Hayashi, K. Kawata, and S. Umekawa, eds. Japanese Soc. Comp. Materials, Tokyo, p. 199.
39. Eshelby, J.D. (1957). *Proc. Roy. Soc.*, **A241**, 376.
40. Jackson, P.W., Braddick, M., and Walker, P. (1972). *Fibre Sci. Technol.*, **5**, 219–236.
41. Harris, S.J., and Marsden, A.L. (1974). In *Proc. of Conf. on Practical Metallic Composites*. Institution of Metallurgists, London, pp. B35–B42.
42. Rosen, B.W. (1965). In *Fibre Composite Materials*. A.S.M., Metals Park, Ohio, ch. 3.
43. Zweben, C., and Rosen, B.W. (1970). *J. Mech. Phys. Solids*, **18**, 189–206.
44. Hull, D. (1981). In *An Introduction to Composite Materials*. Cambridge University Press, Cambridge, p. 126.
45. Kelly, A., and Tyson, W.R. (1965). In *High Strength Materials*, V.F. Zackay, ed. John Wiley & Sons, New York, p. 578.
46. Parratt, N.J. (1966). *Chem. Eng. Progr.*, **62**, 61.
47. Sutton, W.H. (1970). In *Whisker Technology*, A.P. Levitt, ed. Wiley-Interscience, New York, p. 273.
48. Webster, D. (1982). *Met Trans.*, **12A**, 1495.
49. Harris, S.J., and Wilks. T.E. (1985). In *First European Conf. on Composite Materials*, A.R. Bunsell, P. Lamicq, and A. Massiah, eds. European Assoc. for Composite Materials, Bordeaux, France pp. 595–603.
50. Abrams, E.F., and Davies, L.G. (1971). In *Proc. 1971 SAMPE Conf., Space Shuttle Mater.*, Huntsville.
51. Hancock, J.R., and Swanson, G.D. (1971). In *Composite Materials: Testing and Design*. ASTM-STP-497, p. 299.
52. Kreider, K.G., and Dardi, L. (1972). In *Failure Modes in Composites*, I. Toth, ed. AIME, Philadelphia.

53. Wright, M.A., Welsh, D., and Jollay, J. (1979). *J. Mater. Sci.*, **14**, 1218.
54. Hasson, D.F., Hoover, S.M., and Crowe, C.R. (1985). *J. Mater. Sci.*, **20**, 4147.
55. Crowe, C.R., Grey, R.A., and Hasson, D.F. (1985). In *Fifth Intl. Conf. on Composite Materials*, E.C. Harrigan, Jr., J. Strife, and A. Dhingra, eds. AIME, Philadelphia, pp. 843–866.
56. Kreider, K.G., Dardi, L., and Prewo, K. (1971). In *Metal Matrix Composite Technology*. AFML-TR-71-204.
57. Abe, Y., Nakatani, M., Yamatasuta, K., and Horikiri, S. (1985). In *ECCM* A.R. Bunsell, P. Lamicq, and A. Massiah, eds. E.A.C.M., Bordeaux, pp. 604–609.
58. Menke, G.D., and Toth, I.J. (1971). In *Time Dependent Mechanical Behavior of Composite Materials*. AFML-TR-70-174.
59. Hancock, J.R. (1975). In *Composite Materials*, **vol. 5**, L.J. Broutman and R.H. Krock, eds. Academic Press, New York, pp. 371–414.
60. Dvorak, G.J., and Johnson, W.S. (1980). *Int. J. of Fracture*, **16**, 585–607.
61. Hermann, R.A., Lave, P.J., Ordway, F., and Hahn, H. (reported in Nair et al. [73]).
62. Crowe, C.R., and Hasson, D.F. (1983). In *Strengths of Metals and Alloy*, R.C. Gifkins, ed. Pergamon Press, New York, p. 859.
63. Harris, S.J., and Wilks, T.E. (1987). In *Sixth Intl. Conf. on Composite Materials*, **vol. 2**, F.L. Matthews, N.C.R. Buskell, J.M. Hodgkinson, and J. Morton, eds. Elsevier, London, pp. 113–127.
64. Williams, D.R., and Fine, M.E. (1985). In *Fifth Intl. Conf. on Composite Materials*, W.C. Harrigan, Jr., J. Strife, and A.K. Dhingra, eds. AIME, Philadelphia, pp. 639–670.
65. Yau, S.S., and Mayer, G. (1986). *Materials Science and Engineering*, **82**, 45–57.
66. Schaeffer, W.H. and Christian, J.L. (1969). In AFML-TR-69-36.
67. Nieh, T.G. (1984). *Met. Trans.*, **15A**, 139–146.
68. Lilholt, H., and Taya, M. (1987). In *Sixth Intl. Conf. on Composite Materials*, **vol. 2**, F.L. Matthews, N.C.R. Buskell, J.M. Hodgkinson, and J. Morton, eds. Elsevier, London, pp. 234–244.
69. Donomoto, T., Funatani, K., Miura, N., and Miyake, N. (1983). *Ceramic fiber reinforced piston for high-performance diesel engines*, Society of Automotive Engineers, Warrendale, Pennsylvania. Paper no. 830067.
70. Chawla, K.K. (1974). In *Grain Boundaries in Engineering Materials*. Claitor's, Baton Rouge, Louisiana, pp. 435–444.
71. Patterson, W.G., and Taya, M. (1985). In *Fifth Intl. Conf. on Composite Materials*, W.C. Harrigan, Jr., J. Strife, and A.K. Dhingra, eds. AIME, Philadelphia, pp. 53–66.
72. McDanels, D.L. (1985). *Met. Trans.*, **16A**, 1105–1115.
73. Nair, S.V., Tien, J.K., and Bates, R.C. (1985). *Int. Met. Rev.*, **30**, 275–290.
74. Fitzer, E. and Jacobsen, G. (1982). In *Fourth Intl. Conf. on Composite Materials*, T. Hayashi, K. Kawata, and S. Umekawa, eds. Japanese Soc. Comp. Materials, Tokyo, pp. 1315–1322.

Suggested Reading

1. Hull, D. (1981). In *An Introduction to Composite Materials*. Cambridge University Press, Cambridge, UK.
2. Chawla, K.K. (1987). In *Composite Materials: Science & Engineering*. Springer-Verlag.
3. Kelly, A., and Davies, G.J. (1965). *Met. Rev.*, **10**, 37.

10
ARALL® Laminates

R.J. BUCCI and L.N. MUELLER

Alcoa Laboratories
Alcoa Center, Pennsylvania

L.B. VOGELESANG and J.W. GUNNINK

Delft University of Technology
The Netherlands

I. Introduction

ARALL® Laminates are a new family of structural materials developed for fatigue critical applications requiring light-gage sheet [*1–9*].* These materials are bonded arrangements of thin aluminum alloy sheets and alternating plies of epoxy-adhesive impregnated with unidirectional aramid fibers (Fig. 1). The principal benefit of the resulting hybrid composite is

* The term ARALL originally was an acronym (Aramid Aluminum Laminate) derived by its inventors at Delft University. ARALL Laminate is now a registered trademark of Alcoa.

ISBN 0-12-341831-3

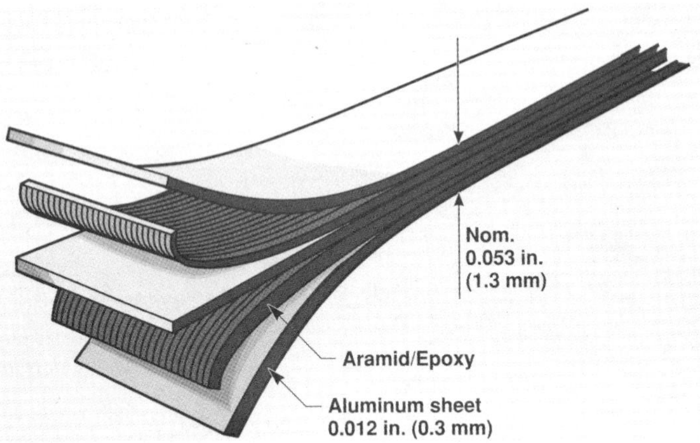

FIG. 1. Alcoa ARALL Laminates. Schematic of standard 3/2 lay-up.

ability to impede and self-arrest crack growth attributed to the component of cyclic loading aligned with the fibers (generally also the direction of greatest tensile stress). Once a through-thickness fatigue crack develops in the aluminum layer, controlled delamination between the metal, epoxy, and fiber interfaces accommodates stress redistribution from the metal to unbroken fibers in the crack wake. The bridging provided by the strong aramid fibers constrains crack opening, thereby reducing the driving force for metal crack advance [*1*,*2*,*10*].

ARALL Laminates, pioneered at Delft University in the Netherlands in the early 1980s [*1–6*], are now commercially produced by Alcoa [*7–9*]. Though originally developed for fatigue resistance, ARALL Laminates display a range of impressive, albeit directional, property improvements over those of monolithic high-strength aluminum, and they feature performance traits that compete with those of advanced composites [*1–9*]. These attractive characteristics include 15–20% lower density than aluminum; up to 60% higher strength than 7075 and 2024 aluminum at comparable stiffness; fabricability comparable to metal (e.g., it can be cut, sawed, drilled, joined, and inspected by conventional metal practices); ability of the outer metal layers to protect against fiber–resin system damage by moisture, thermal attacks, lightning, and impacts; and damping ability superior to monolithic aluminum. Envisioned aircraft usages include tension-dominated fatigue and fracture critical structure (e.g., lower wing and fuselage skins), damping critical structure, lightning-strike areas, and structure requiring resistance to impacts, where attendant trade studies have identified potential for significant (15–40%) weight savings over current designs [*3–6,11–16*]. Since October 1987, ARALL has been flying in the lower wing of a Fokker-50 prototype commercial transport aircraft [*14*], and Fokker and other airframers are evaluating the material for broader aircraft use.

II. Commercial ARALL Laminate Product Forms

ARALL Laminate property combinations can be modified by varying fiber-resin systems, aluminum alloys, sheet gages, stacking sequences, fiber orientations (such as uniaxial and cross-ply), surface preparation techniques, and the degree of postcure stretching or forming. Although several variations of these systems have been successfully fabricated, the four product variants listed in Table I have been standardized for commercial availability. All are fatigue-resistant with good static strength. ARALL-1 is 7075-T6–based, has nominal 0.4% postcure stretch, and is intended for high fatigue loaded structures together with high compressive

TABLE I

ARALL LAMINATE COMMERCIAL PRODUCT FORMS[a]

Description	Characteristics
ARALL-1 Laminate	Superior fatigue
Alloy 7075-T6[b]	High strength
250°F (120°C) cure prepreg[c]	
0.4% permanent stretch	
ARALL-2 Laminate	Excellent fatigure
Alloy 2024-T3[b]	Increased formability
250°F (120°C) cure prepreg[c]	Damage-tolerance
with or without 0.4% stretch	
ARALL-3 Laminate	Superior fatigue
Alloy 7475-T76[b]	Controlled toughness
250°F (120°C) cure prepreg[c]	Good exfoliation
0.4% permanent stretch	High strength
ARALL-4 Laminate	Excellent fatigue
Alloy 2024-T8[b]	Elevated temperature
350°F (175°C) cure prepreg[c]	
with or without 0.4% stretch	

[a] Also produced in a one-side or two-side clad condition for added corrosion protection.

[b] Bonding surfaces anodized and primed.

[c] With 50% by volume unidirectional aramid fibers.

loadings. As-cured ARALL-2 is intended for fatigue applications with low compressive loadings. The good ductility and toughness of alloy 2024 makes ARALL-2 better suited to applications requiring damage tolerance and formability. ARALL-3 is an ARALL-1 upgrade that in lieu of 7075-T6 uses controlled toughness alloy 7475 in a T76 exfoliation-resistant temper. ARALL-3 is recommended for ARALL-1 substitution when a guaranteed fracture-toughness capability is required for both the longitudinal and long transverse test directions. ARALL-4 is 2024-T8 based and employs a special prepreg system to achieve higher temperature capability. The prepreg system used in all four standard products consists of thermoset adhesive impregnated with unidirectional aramid fibers in a 50:50 fiber–resin weight ratio. The fibers are oriented parallel to the aluminum sheet rolling direction. ARALL versions 1 through 3 use a 250°F (120°C) cure resin prepreg with a recommended service temperature of 200°F (90°C). The ARALL-4 prepreg cures at 350°F (175°C) and is intended for use up to 325°F (160°C). Nominal thickness of the individual aluminum sheet layers is 0.012 in. (0.3 mm), and the prepreg nominal thickness is 0.0085 in. (0.2 mm). Standard laminate configurations vary from a 2/1 (two layers of aluminum surrounding one layer of prepreg), which is 0.033 in. (0.83 mm) thick, to a 5/4 (five layers of aluminum with four prepreg layers interspaced), which is 0.094 in. (2.4 mm) thick. Thicker laminates have also been made. Due to the low density of the aramid prepreg layers, the net density of ARALL Laminate is lower than that of monolithic aluminum. The plot of Fig. 2 shows that increasing laminate thickness by increasing the number of plies reduces density to an asymptotic limit of about 80% that of solid aluminum.

Prior to lay-up and cure, the aluminum surfaces are anodized and primed to provide good bond integrity and to inhibit aluminum bondline corrosion in the event of moisture intrusion. A chromic acid or phosphoric anodizing treatment is used. Early lab-scale panels were cured in a heated platten press, whereas the current commercial-scale product is autoclave-cured. Tests performed on lab-scale panels manufactured by both curing processes show no difference in mechanical properties. Curing results in a residual stress state that places the metal layers in tension and fibers in compression. ARALL-1 and -3 Laminates are given a nominal 0.4% permanent stretch to reverse the residual stress state to compression in the aluminum layers and tension in the higher-strength aramid fibers. Figure 3 shows how stresses in the aluminum layers are affected by the amount of postcure stretch. ARALL Laminates in the as-cured state have excellent fatigue properties, but the stretching operation makes ARALL Laminate practically insensitive to fatigue crack growth. For example, Fig. 4 shows that under constant cyclic loading, going from an as-cured to a 0.5%

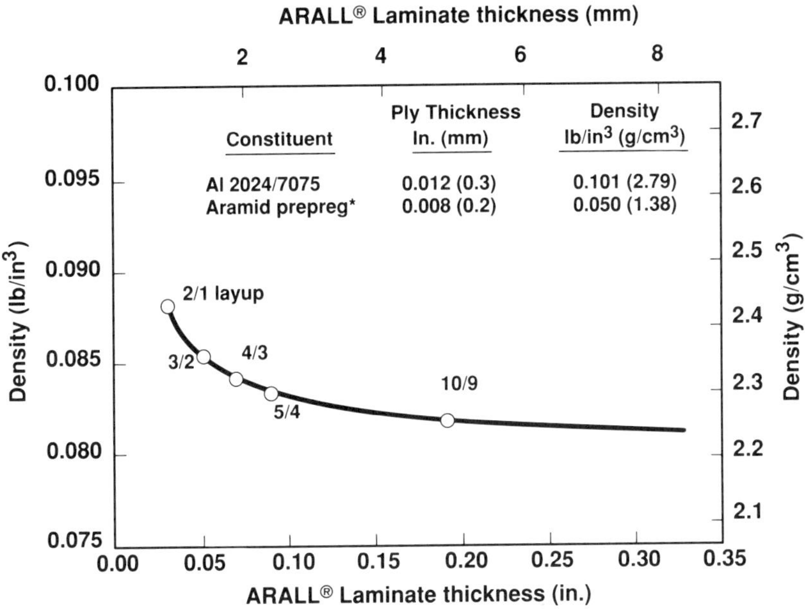

FIG. 2. ARALL Laminate calculated density versus product thickness.

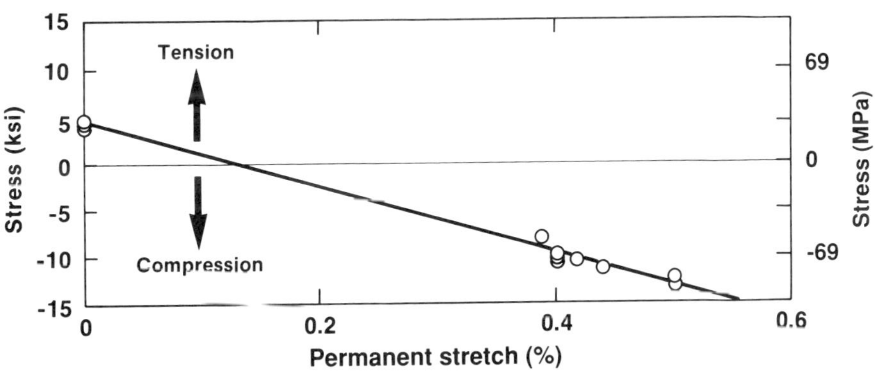

FIG. 3. Aluminum layer residual stress in 3/2 ARALL-1 Laminate, 0.053 inch (1.3 mm).

stretch condition increases the ARALL Laminate center crack panel fatigue life from tenfold to a thousandfold over that of the monolithic 7075-T6 baseline.

For added corrosion protection, cladding can be applied to the outer aluminum faces of the ARALL sandwich. Also, one-side cladding can be supplied to accommodate further bonding to the opposite bare face.

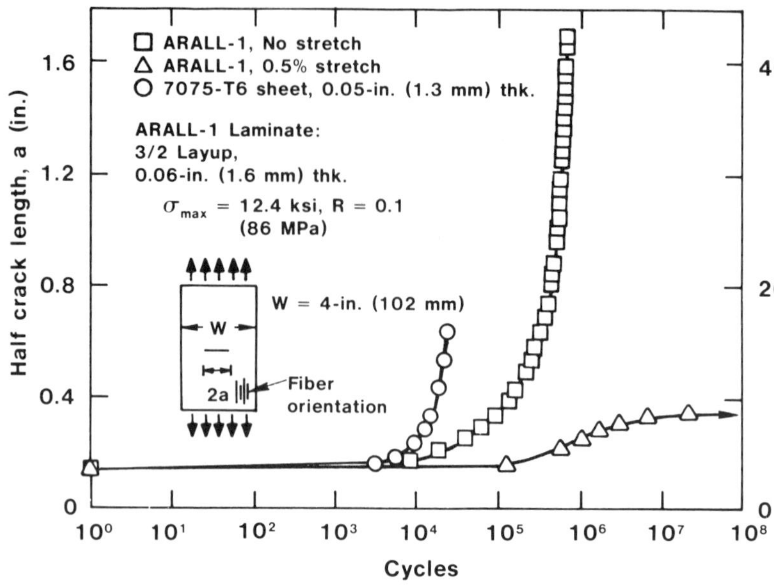

FIG. 4. Fatigue crack growth of stretched and unstretched ARALL-1 Laminate and 7075-T6 sheet. ARALL-1 Laminate: 3/2 lay-up; 0.06 in. (1.6 mm) thick. 7075-T6 sheet: 0.05 in. (1.3 mm) thick. σ_{max} = 12.4 ksi, R = 0.1.

III. Mechanical Properties

Table II compares ARALL Laminate room-temperature typical mechanical properties to those of monolithic aluminum-sheet alloys and two common carbon-fiber–composite layouts. As with most fibrous composite materials, ARALL Laminate properties are directional as dictated by fiber orientation. Due to space limitations, the ARALL properties are only given for the 3/2 configuration (i.e., three layers of aluminum and two layers of prepreg). Alternate lay-up properties are available from Alcoa data sheets on each of the standard product designations. Like density (Fig. 2), many ARALL Laminate properties can be estimated from rule of mixtures and the appropriate aluminum and directional prepreg properties. Typical tensile and compressive stress-strain behaviors of 3/2 ARALL-1 Laminate and its constituents are given in Fig. 5.

ARALL Laminate tensile ultimate strengths in the reinforcement (longitudinal) direction are significantly better than those of the respective aluminum counterparts; also, they are competitive with those of graphite–epoxy composite. The fibers contribute little to the ARALL transverse

TABLE II

ARALL LAMINATE, ALUMINUM SHEET, AND CARBON-FIBER-COMPOSITE ROOM-TEMPERATURE TYPICAL MECHANICAL PROPERTIES

Property	Test direct.	3/2 ARALL Laminate, 0.053 in. (1.3 mm) thick				Aluminum-alloy sheet, 0.063 in. (1.6 mm) thick				Carbon-fiber composite[a]	
		ARALL-1	ARALL-2	ARALL-3[b]	ARALL-4[b]	7075-T6	2024-T3	7475-T76	2024-T8	Unidirectional[c]	Typical structure[d]
Tens. ultimate	L	116(800)	104(717)	120(828)	106(731)	83(572)	66(455)	79(545)	72(497)	180(1241)	95(655)
strength, ksi (MPa)	LT	56(386)	46(317)	54(373)	49(338)	83(572)	65(448)	80(552)	72(497)	8(55)	40(276)
0.2% off, tens, yld.	L	93(641)	52(359)	85(587)	54(373)	74(510)	52(359)	69(476)	64(442)	—	—
strength, ksi (Mpa)	LT	48(331)	33(228)	46(317)	46(317)	72(496)	47(324)	69(476)	64(442)	—	—
Tens. elastic	L	9.8(68)	9.3(64)	9.8(68)	8.9(61)	10.3(71)	10.5(72)	10.0(69)	10.5(72)	21.0(145)	11.0(76)
modulus, msi (GPa)	LT	7.0(48)	7.1(49)	7.4(51)	7.1(49)	10.3(71)	10.5(72)	10.0(69)	10.5(72)	1.7(12)	5.0(35)
Tens. elong., %	L	0.7[e]	1.4[e]	1.0[e]	1.4[e]	11	18	12	8	—	—
	LT	7.1[e]	12.0[e]	—	3.9[e]	11	18	12	8	—	—
Tens. tot. strain to failure, %	L	1.9	2.5	2.2	2.6	12	19	13	9	0.9	0.5
	LT	7.9	12.7	—	4.6	12	19	13	9	0.5	0.5
Poisson ratio	L	0.33	0.32	—	—	0.33	0.33	0.33	0.33	0.31	0.48
(tension)	LT	0.25	0.26	—	—	0.33	0.33	0.33	0.33	0.017	0.23
0.2% off. compr. yld.	L	54(372)	38(262)	—	—	73(503)	44(304)	64(442)	63(435)	180(1241)[f]	95(655)[f]
strength, ksi (MPa)	LT	57(393)	34(234)	—	—	76(524)	50(345)	68(469)	64(442)	30(207)[f]	42(290)[f]
Compr. elastic	L	10.2(70)	9.7(67)	—	—	10.5(72)	10.7(74)	10.5(72)	10.7(74)	21(145)	11.0(76)
modulus, msi (GPa)	LT	7.5(52)	7.6(52)	—	—	10.5(72)	10.7(74)	10.5(72)	10.7(74)	1.7(12)	5.0(35)
0.2% off. shear yld.	L-LT	—	17(117)[g]	—	—	—	30(207)[g]	—	—	12(83)[h]	37(255)[h]
strength, ksi (MPa)	LT-L	—	16.5(114)[g]	—	—	—	30(207)[g]	—	—	—	—
Shear modulus,	L-LT	2.4(17)[g]	2.5(17)[g]	—	—	3.9(27)	4.0(28)	3.8(26)	4.0(28)	0.65(4.5)	3.1(21)
msi (Gpa)	LT-L	—	2.3(16)[g]	—	—	3.9(27)	4.0(28)	3.8(26)	4.0(28)	—	—

(*continues*)

TABLE II (*Continued*)

Property	Test direct.	3/2 ARALL Laminate, 0.053 in. (1.3 mm) thick				Aluminum-alloy sheet, 0.063 in. (1.6 mm) thick				Carbon-fiber composite[a]	
		ARALL-1	ARALL-2	ARALL-3[b]	ARALL-4[b]	7075-T6	2024-T3	7475-T76	2024-T8	Unidirectional[c]	Typical structure[d]
Bearing ult. strength	L	95(655)	77(531)	—	—	129(890)	110(759)	120(828)	107(738)	—	—
e/D = 1.5, ksi (Mpa)	LT	102(703)	79(545)	—	—	129(890)	110(759)	120(828)	107(738)	—	—
e/D = 2.0, ksi (Mpa)	L	107(738)	82(565)	—	—	166(1145)	137(945)	153(1055)	136(938)	—	—
	LT	105(724)	79(545)	—	—	166(1145)	137(945)	153(1055)	136(938)	—	—
Bearing yld. strength	L	85(586)	56(386)	—	—	110(759)	78(538)	100(690)	88(607)	—	—
e/D = 1.5, ksi (MPs)	LT	88(607)	56(386)	—	—	110(759)	78(538)	100(690)	88(607)	—	—
e/D = 2.0, ksi (Mpa)	L	102(703)	66(455)	—	—	128(883)	94(649)	113(780)	100(690)	—	—
	LT	97(669)	64(441)	—	—	128(883)	94(649)	113(780)	100(690)	—	—

[a] *DoD/NASA Advanced Composites Design Guide*, **Vol. 1A**, generic F180 graphite/epoxy composite, 60% fiber content, 0.056 lb./cu. in. (1.55g/cu. cm); e.g., AS/3501-6, T300/5208.

[b] Single laboratory lot.

[c] 100% fibers in L direction.

[d] Fiber orientations: 42%, 0 degree; 50% ± 45 degrees; 8% 90 degree.

[e] Plastic strain determined from test record by back extrapolating elastic slope from point of fracture.

[f] Compressive ultimate strength.

[g] Iosipescu shear test method; L-LT means fibers parallel to long specimen axis, LT-L means fibers normal to long specimen axis.

[h] In-plane shear ultimate.

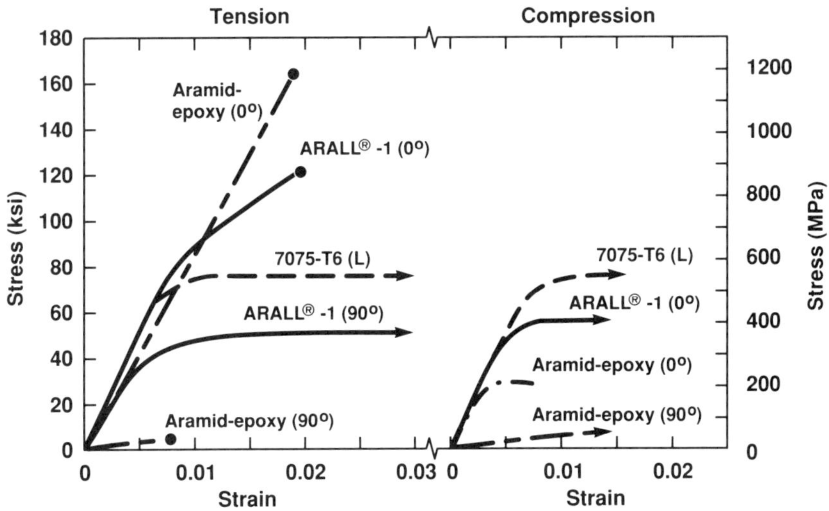

FIG. 5. Tensile and compressive stress-strain behaviors for ARALL-1 Laminate (3/2 lay-up, 0.053 inch thick), aramid-epoxy prepreg and 7075-T6 sheet.

properties; however, the isotropic aluminum is advantageous when loading is multidirectional. ARALL Laminate yield properties are linked to plastic strength of the metal layers. The higher ARALL-1 and -3 longitudinal tensile yield strengths over monolithic material are, in part, attributed to aluminum hardening during stretching and, in part, due to the poststretch residual compression in the metal layers. In contrast, ARALL variants 2 and 4, without stretch, show no yield-strength advantage over their monolithic aluminum counterparts. The aramid prepreg and aluminum have similar elastic moduli, so it is not surprising that ARALL Laminate stiffness in the reinforcement direction is compatible to that of aluminum. The slight tension modulus increase of stretched ARALL-1 and -3 over the as-cured ARALL-2 and -4 versions is attributed to fiber straightening from the postcure stretch. Tensile data shown in Fig. 6 illustrate the increased strength capability of ARALL-4 at elevated temperatures beyond 200°F (90°C).

The ARALL Laminate longitudinal and transverse compressive yield strengths are comparable since the fibers contribute little in compression. The compressive strength of ARALL-2 is slightly lower than that of 2024-T3, while ARALL-1 has a higher compressive yield than 2024-T3, but it is lower than that of 7075-T6. The ARALL Laminate longitudinal and transverse elastic moduli in compression are similar to the tension elastic moduli. Like aluminum, ARALL Laminate compressive failures in short

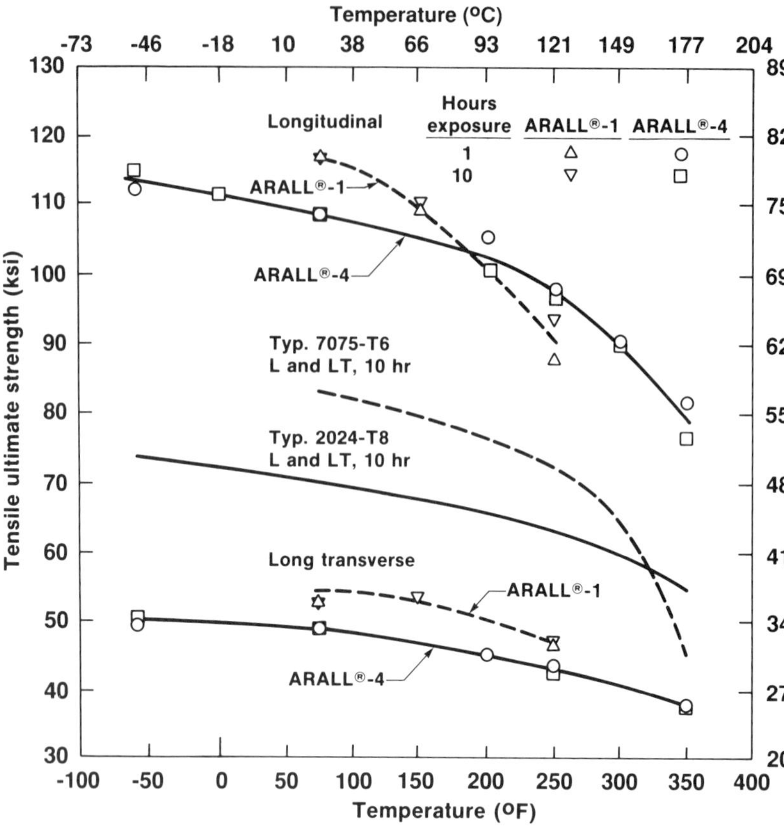

FIG. 6. Elevated-temperature tensile strength of ARALL-1 and ARALL-4 Laminates, 3/2 lay-up, 0.053 inch (1.3 mm) thick.

coupon specimen tests occur by plastic buckling. Compressive strength of stiffened structures is related to Euler buckling (elastic stiffness) and compressive yield strength as shown in Fig. 7. Because of its lower density, ARALL-stiffened structure can still be designed to match the compressive load-carrying ability and weight of an all-aluminum structure [*5*,*6*,*13*].

ARALL Laminates are prone to notch sensitivity like most fibrous composite materials. Due to ineffectiveness of the fibers in bearing, the ratio of bearing strength to tensile strength is lower than that experienced by aluminum (Fig. 8). ARALL Laminate bearing strength is reasonably estimated by the summation of the bearing strengths of the aluminum layers, and this approach can be used to determine riveted-joint static strength from existing rivet tables [*17*]. With proper care, ARALL joints

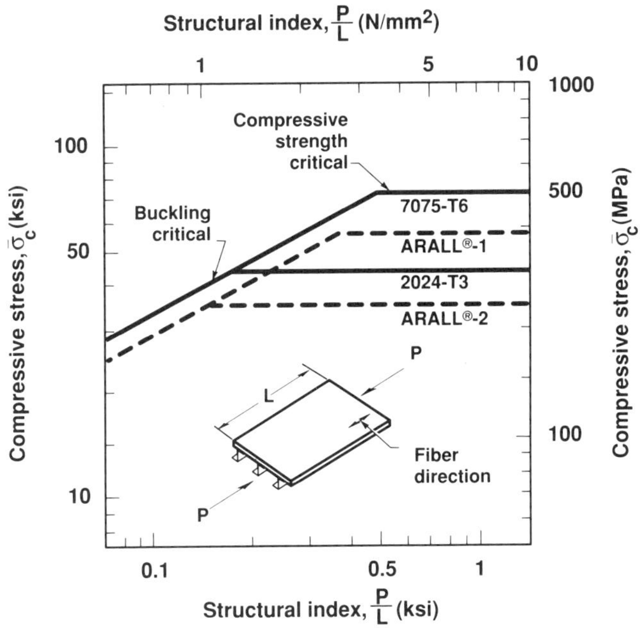

FIG. 7. ARALL Laminate versus aluminum stiffened panel compressive strengths.

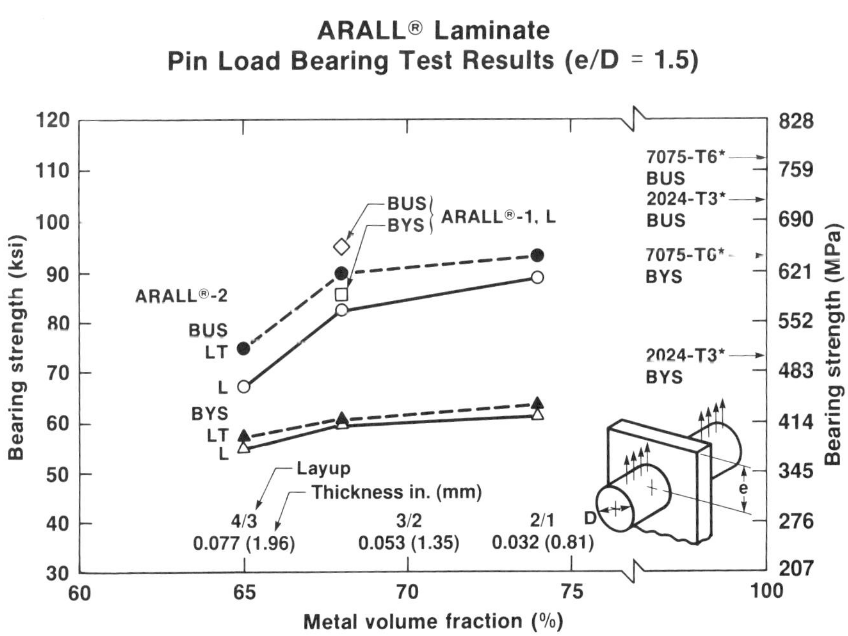

FIG. 8. ARALL Laminate pin-load–bearing test results (e/D = 1.5).

can be designed to equivalent aluminum-joint static strength. However, to optimize ARALL joint integrity, it may prove advantageous to employ combinations of adhesive bonding and mechanical fastening.

There is presently no standard test practice for shear of aluminum sheet. Historically, much of the thin-aluminum-sheet data appearing in Mil-Hdbk-5 [*18*] was developed by the blanking (punch) shear test method. When tested by the same practice, 3/2 ARALL-1 Laminate has a blanking shear strength of 39 ksi (269 MPa), which is about 20% less than the 47 ksi (324 MPa) minimum reported for 7075-T6 at comparable gage. This decrease is expected since the aramid epoxy layers constitute about 30% of the 3/2 laminate sheared area. The Iosipescu shear test [*19*] is better suited for evaluating in-plane shear properties of composite materials, and ARALL properties developed by this method are given in Table II. Ultimate shear values are not reported since plastic buckling always preceded failure in shear.

Results of short-beam interlaminar shear (ILS) tests on 3/2 ARALL-1 Laminate are given in [*7*]. Assuming initiation of resin failure at the point of slope change on the load versus head-deflection test record and using the corresponding load, the ILS strength of 3/2 ARALL-1 Laminate was calculated at 4.7 ksi (32 MPa), which is comparable to that of epoxy-based composite. Fiber orientation was found to have no effect on ILS strength, and the specimens failed by plastic collapse.

IV. Fatigue

ARALL Laminates were developed primarily to obtain material with good fatigue resistance. Their primary attribute is ability to slow and self-arrest fatigue crack growth, particularly when tested in the stretched condition. (See Fig. 4). Extensive fatigue testing on ARALL Laminate versions 1 and 2 has been done under both constant load amplitude and realistic flight simulation loading conditions [*1–15*]. These tests include smooth and notched coupons, cracked specimens, lugs, and mechanically fastened and bonded joints in combination with a variety of environmental exposure conditions. Results from several of these studies are shown in Figs. 9 through 13. Under all of the preceding conditions, ARALL Laminates showed significant fatigue-life increase over solid aluminum in all cases where principal fatigue loading was tension-dominated and in the direction of reinforcement. Depending on test conditions, naturally initiated cracks grew very slowly or arrested after a small amount of crack advance. Crack arrest was observed, even in cases where ARALL

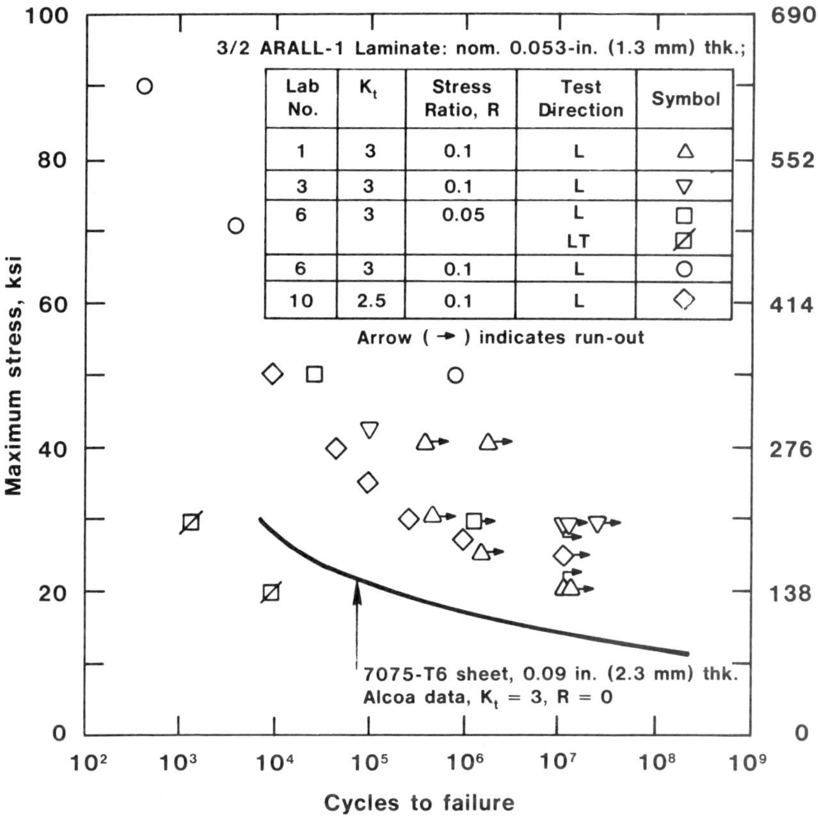

FIG. 9. Notch fatigue behavior of 3/2 ARALL-1 Laminate, nom. 0.053 in. (1.3 mm) thick (various laboratories)

specimens were subjected to higher stress than the aluminum specimens. The results of Fig. 10 show that ARALL fatigue resistance, as expected, depends on fiber orientation [*15*]. The 0° and 15° specimens are shown to be nearly fatigue-insensitive. As the fibers rotate away from the loading direction, the rate of fatigue cracking increases, until at 45° ARALL-1 Laminate exhibits similar behavior to 2024-T3 (considered one of the best high-strength aluminum performers in fatigue). Preliminary design studies done for aircraft lower-wing and fuselage components have shown that, in general, ARALL Laminate fatigue properties are so good that governing design criteria shift away from fatigue to other factors, such as compression and static notch strength [*6,11–13*].

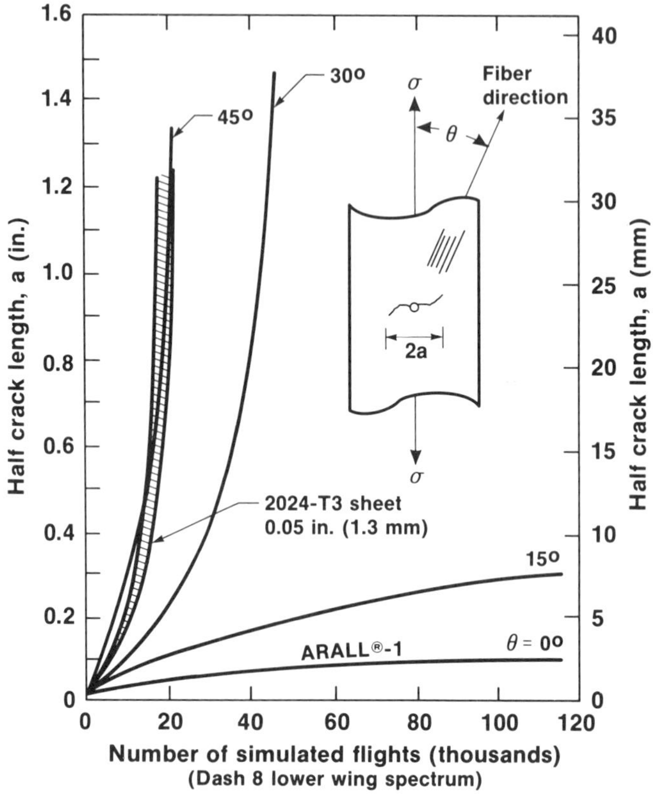

FIG. 10. Effect of fiber orientation on 3/2 ARALL-1 Laminate (0.053 inch) fatigue crack growth. (Courtesy of deHavilland Aircraft)

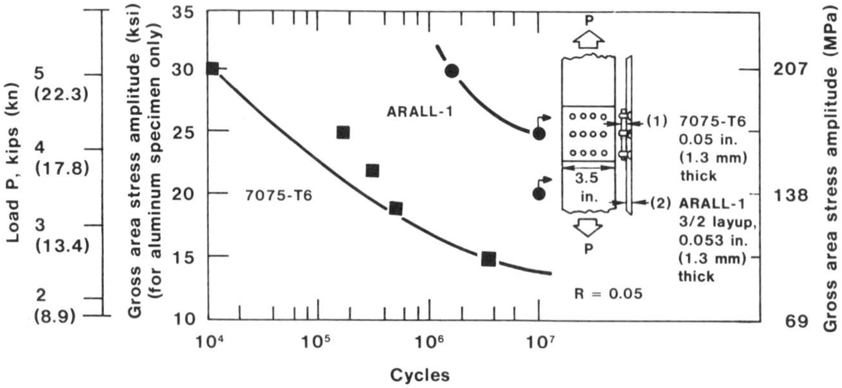

FIG. 11. Joint fatigue strength of ARALL-1 Laminate and 7075-T6 aluminum sheet. (Courtesy of McDonnell Douglas, Long Beach, California)

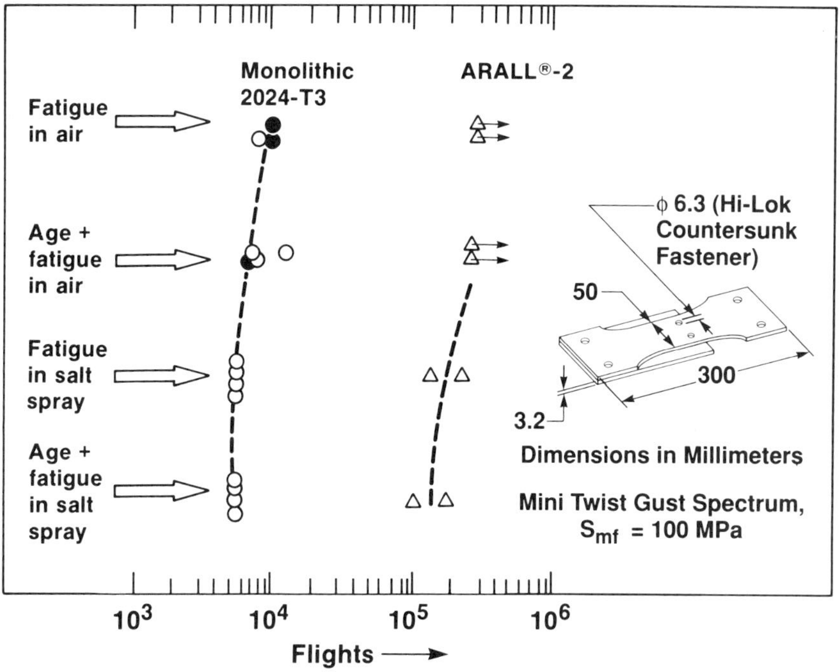

FIG. 12. ARALL-2 Laminate and 2024-T3 bolted joint specimen fatigue lifetimes under various exposure conditions.

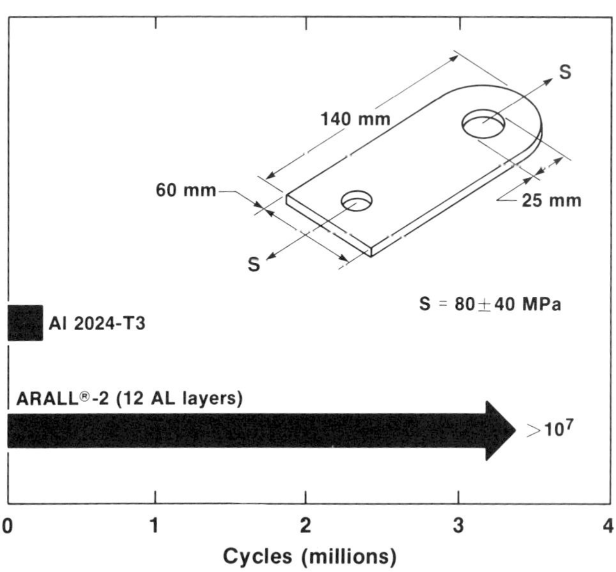

FIG. 13. ARALL-2 and 2024-T3 fatigue test results on lugs.

V. Fracture Toughness and Static Notch Strength

Fracture toughness defines the ability of a material to resist rapid fracture in the presence of fatigue cracks or other cracklike damage. ARALL Laminate is a hybrid material with its own characteristic fracture toughness behavior. Its static notch strength is dependent upon a number of factors including the metal-to-fiber volume fraction, the fiber direction, the base metal tensile strength, the fiber failure strain, the state of residual stress, and the type of stress concentration. When loaded in the fiber direction, ARALL panels containing fatigue-cracked metal layers show residual strengths significantly higher than those of monolithic aluminum panels containing equivalent length through cracks [*4–9,20*]; see also Fig. 14. This superior ARALL Laminate toughness response is attributed to unbroken fibers bridging the metal layers in the fatigue crack wake. Prior cyclic loading, in this case, produces local fiber–matrix debonding which enhances strain redistribution along fiber lengths both ahead and behind the crack tip. If instead cracks are introduced by saw cut rather than

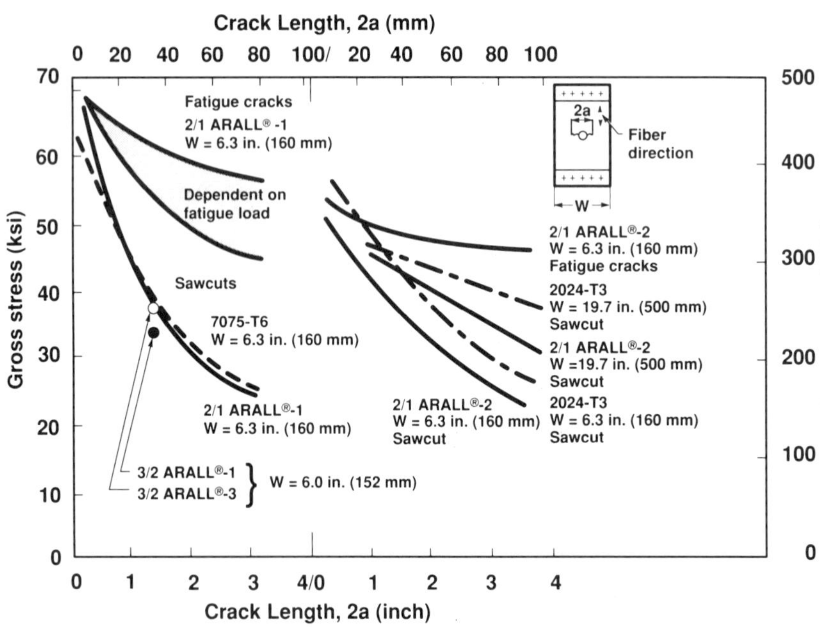

FIG. 14. Residual strength of ARALL Laminate and aluminum sheet center-slotted panels.

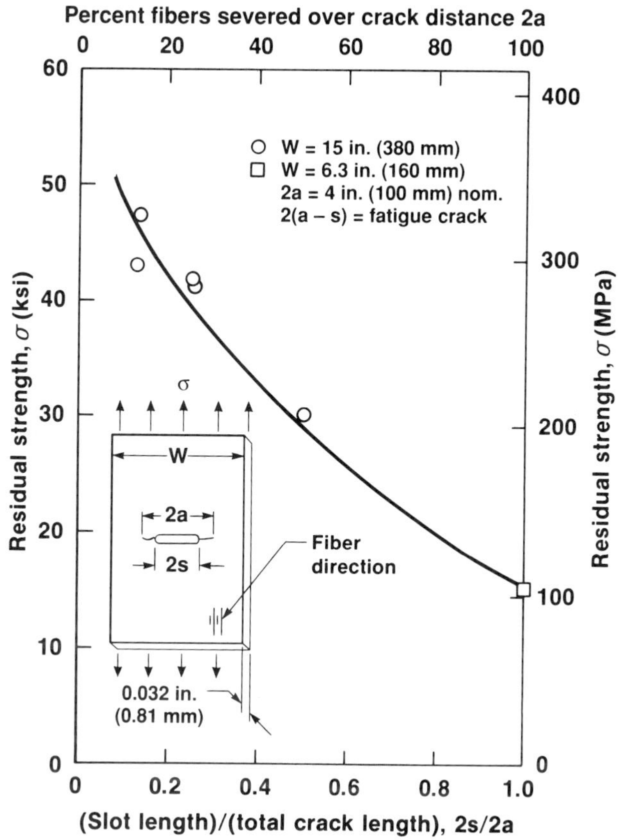

FIG. 15 Residual strength of fatigue-damaged 2/1 ARALL-1 Laminate center-crack panels.

by fatigue, then there are no bridging fibers, and fiber strains ahead of the crack are more localized. In this case, the ARALL panel residual strengths approach those of their aluminum counterparts as shown in Fig. 14. The center crack panel results of Fig. 15 [*20*] show that for fixed crack length, ARALL residual strength increases with the fraction of total crack length extended by fatigue. This is in direct contrast to monolithic metal behavior, where fatigue cracks represent the most severe form of stress concentration.

Ductile aircraft metals tend to exhibit net section failure stresses approaching yield, and development of fracture toughness parameters for such alloys generally requires large test panels to validate complex structural designs. Figure 14 shows that there is an appreciable size dependence on

TABLE III

TYPICAL ARALL LAMINATE AND ALUMINUM SHEET CENTER-SLOT PANEL FRACTURE TOUGHNESS VALUES[a]

Product	Nominal gage in. (mm)	Nominal panel width in in. (mm)	Typical Kc ksi$\sqrt{\text{in.}}$ (MPa$\sqrt{\text{m}}$) L-T	T-L
3/2 ARALL-1	0.053 (1.3)	6 (150)	59 (65)	54 (59)
2/1 ARALL-2[b]	0.041 (1.0)	20 (500)	110 (121)	[c]
3/2 ARALL-3	0.053 (1.3)	6 (150)	61 (67)	[c]
7075-T6	0.06 (1.5)	16 (400)	68 (75)	62 (68)
2024-T3	0.06 (1.5)	>20 (510)	123 (135)	120 (132)

[a] All test panels contained a centrally located 0.010-in.–(25-mm–) wide machined slot of length approximately one-fourth the total panel width. ARALL Laminate Kc values determined in accordance with metal test procedure.
[b] Substituted 0.016-in–(0.4-mm–) thick 2024-T3 sheet for 0.012-in. (0.3-mm) 2024 sheet in standard 2/1 configuration.
[c] No data available (Sept. 1988).

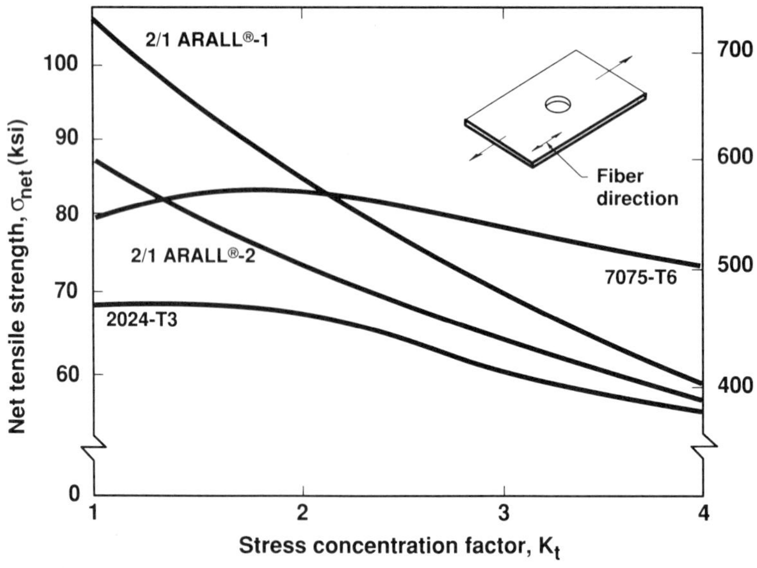

FIG. 16. Open-hole–specimen notch strength: ARALL Laminate versus aluminum for different K_t factors.

the residual strength behavior of ARALL-2 panels. Because of higher strength, the size dependence on ARALL-1 and -3 panel toughness response is considerably less than that exhibited by the ARALL-2 panel results [*6,21*]. Though more work needs to be done, ARALL Laminate Kc values can be estimated from the available center-slot panel data. These results are compared to typical 7075-T6 and 2024-T3 sheet fracture toughness values in Table III.

Figure 16 compares ARALL Laminate and aluminum-sheet static-strength sensitivity to blunt notch stress concentrations, such as open holes. When contrasted to their metal counterparts, ARALL Laminates 1 and 2 show greater notch strength reductions from the unnotched condition. The increased notch sensitivity is caused by the combination of linear elastic response and small failure strains of the aramid-prepreg layers, which is common to all fiber-reinforced composites. As with composites, it is emphasized that although ARALL Laminate static-strength design allowables are degraded by notch factors (more so than metals), this does not preclude the weight efficiency and good fatigue resistance attainable in actual structures [*6,11–13*]. It may imply, however, that ARALL structural qualification criteria are different from those of metal and traditional composite designs.

VI. Environmental Durability

Extensive testing is in progress to address the effects of various aircraft environments on ARALL Laminate mechanical and bondline properties. The available results are encouraging [*6,7,22,23*]. Table IV shows that exposure to high humidity, thermal-cycle testing, salt fog, and seacoast atmosphere has little effect on the tensile and compressive strengths of ARALL Laminates 1 and 2. In cases where strength loss is shown, surface corrosion of the outer aluminum faces is in large part responsible. The base-metal properties given in Table IV show that alloy 2024-T3 is slightly more prone to environmental strength degradation than alloy 7075-T6, particularly in the salt fog accelerated test environment. Cladding the outer aluminum faces offers a possible solution to this problem.

As with pure fibrous composites, the ARALL Laminate prepreg layer is susceptible to temperature and humidity controlled moisture absorption. In spite of comparable diffusion coefficients, the ARALL Laminate moisture gain is negligible compared to that of pure composite because of the excellent aluminum barrier properties [*22,23*]. Under lab-accelerated exposures, the fiber–adhesive interface has been shown to be the weakest

TABLE IV

EFFECT OF ENVIRONMENTAL PRESOAK ON ARALL LAMINATE[a] AND ALUMINUM SHEET[b] TENSILE AND COMPRESSIVE PROPERTIES

	Percent property retention after exposure[c]															
	Tensile ultimate strength[d]								Compressive yield strength[e]							
	ARALL-1		ARALL-2		7075-T6		2024-T3		ARALL-1		ARALL-2		7075-T6		2024-T3	
Conditions	L	LT	L	LT	L	LT	L	LT	L	LT	L	LT	L	LT	L	LT
100% RH at 140°F (60°C)																
1 mo.	101	99	102	99	100	100	101	100	107	107	101	99	100	102	101	100
3 mo.	101	99	100	94	101	101	101	100	103	104	98	98	99	101	102	101
6 mo.	99	101	99	90	102	102	103	97	104	100	99	101	103	102	102	99
Hot-wet/cold/hot-dry, cycled daily[f]																
1 mo.	101	99	102	91	100	100	100	98	104	100	97	98	101	101	100	99
3 mo.	98	97	99	81	99	100	96	93	105	101	97	98	104	105	101	98
6 mo.	101	97	98	96	98	99	96	90	105	100	97	99	104	104	106	102
5% salt fog at 95°F (35°C)																
500 hrs.	101	97	99	88	99	99	98	92	100	98	98	98	101	102	100	100
1000 hrs.	99	89	98	83	98	97	92	85	104	105	99	99	99	100	100	100
Seacoast atmosphere, Pt. Judith, RI																
1 yr.	99	98	95	84	100	99	96	90	98	97	96	96	100	100	99	98

[a] 3/2 lay-up, 0.053 in. (1.3 mm) thick.
[b] 0.05-in. (1.3-mm) sheet gage.
[c] Average of triplicate tests.
[d] 8.5 × 0.75 in. (21.6 × 1.9 cm) tensile blank; edges unprotected.
[e] 2.6 × 0.63 in. (6.6 × 1.6 cm) compression blank; edges unprotected.
[f] Daily cycle of 18 hrs. at 100% RH, 5 hrs. at −10°F (−23°C), 1 hr. at 180°F (82°C).

link in the ARALL Laminate bondline, particularly when loading normal to the fiber direction is present [*22*]. Insufficient adhesion strength between the aramid fiber and adhesive results in low peel strengths (Bell peel test) and energy-release rates (thick adherend shear and wedge peel tests) of the prepreg layer. Experience, however, suggests that this feature will not impair ARALL Laminate usage in actual structures [*6,11,13*]. Though more work is necessary in the area of environmental durability, it has been suggested that the best use of bondline exposure testing is for setting material environmental limits, rather than for purposes of design [*22*]. Since moisture can only gain bondline access at free edges, it is reasonable to expect that sealants could be relied upon for protection in actual structure. Additionally, real structure will have a much smaller free-edge–volume fraction than present in the small laboratory coupon simulations.

VII. Auxiliary Characteristics

A. *Impact Resistance*

Though not a major concern for monolithic aluminum, sensitivity to impact is inherent to fibrous composite materials [*24*]. However, unlike pure composites, which can delaminate internally without showing external damage after low-velocity impact, the aluminum surface of ARALL Laminates will dent before internal delamination, allowing visual inspection [*6,7,25,26*]. ARALL Laminate postimpact residual strength and fatigue performances are shown in Figs. 17 and 18, respectively. When subjected to the same level of impact, ARALL Laminates consistently outperform pure composites on a percent-strength-loss basis, as shown in Fig. 17 [*6,25*], and cracks emanating from impacted areas grow much slower than those in solid aluminum sheets, as shown in Fig. 18 [*6,27*]. In both examples, 2024-T3–based ARALL-2 outperforms 7075-T6–based ARALL-1. The exhibited ARALL-2 Laminate superiority is attributed to increased toughness of the 2024-T3 aluminum layers. As with pure composites, potential degradation by impact will require attention in damage-tolerant certification of ARALL Laminate structures.

B. *Vibrational Damping*

Adhesively bonded metal laminates typically show improved vibrational-damping ability over their monolithic counterparts [*28–30*]. The

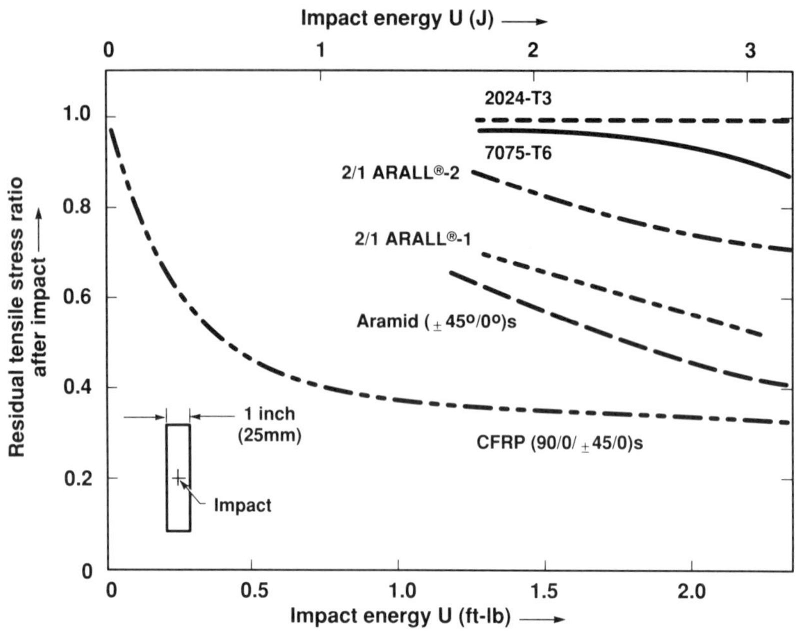

FIG. 17. Residual tensile strength of different materials after impact.

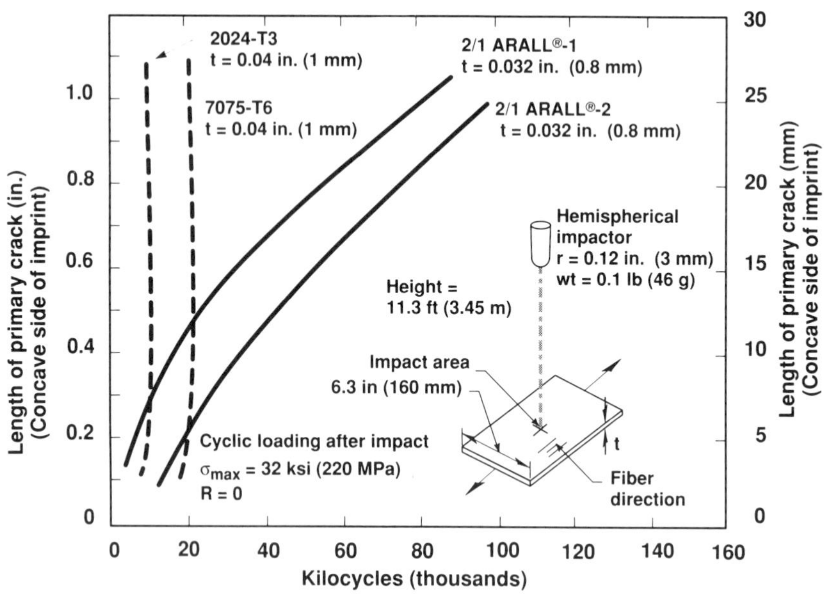

FIG. 18. ARALL Laminates fatigue crack growth after impact.

same good damping qualities have been verified for ARALL Laminates (two to five fold improvement over monolithic aluminum sheet), but with the added advantage that reinforcing fibers add stiffening and strength so that damping augmentation is not at the expense of other properties [*7*,*31*]. Several airframers are presently conducting evaluations to exploit ARALL Laminates as potential solutions to acoustic fatigue and buffeting problems.

C. Lightning Strikes

Sheet-gage selection in commuter aircraft, particularly in areas around fuel tanks, is chosen so that burn-through or molten metal cannot be detected on the opposite side of a sheet struck by lightning. In tests conducted by deHavilland Aircraft [*15*], ARALL-2 Laminate was shown to offer a 42% improvement over 2024-T3 sheet in the ability to cope with lightning strikes; that is, 0.053-inch-(1.3-mm-) thick ARALL-2 Laminate offered the same resistance as 0.090-inch (2.3-mm) 2024-T3 sheet. Coupling the improved lightning resistance with the reduced density of ARALL Laminates implies potential for considerable weight savings on sheet parts sized by lightning strike.

VIII. Design and Evaluation of Full-Scale Structure

Any new structural material considered for aircraft must be evaluated for airworthiness in accordance with today's design criteria. To do this, it is important to recognize inherent differences between ARALL Laminates, metals, and pure composites so that design can be tailored to emphasize advantages of the material of choice.

The structural-integrity issues of greatest concern are static strength, durability, and damage tolerance. Static-strength allowables for composites are determined by notch-sensitivity factors (e.g., holes and cutouts) and environmental knockdowns, whereas aluminum strength allowables are determined from the base-material properties. ARALL Laminates are more notch-sensitive than solid aluminum, but not to the extent of pure composites. Also, environmental effects, which are decisive in pure composites are virtually absent in tests on ARALL Laminate specimens, even in aggressive environments [*6*,*13*,*22*]. Thus, blunt notch static-strength results, of the type shown in Fig. 16, are generally key to establishing ultimate design limits for ARALL Laminates.

Structural durability is most often related to time-dependent damage evolution under the combination of cyclic loading and environmental attack. Pure composite materials generally exhibit fatigue response vastly superior to aluminum, even in the presence of aggressive environments [*24,32*]. For aluminum structures, durability relates to crack nucleation and early-stage growth. Though in ARALL, subcritical cracks may develop in the aluminum plies, these cracks grow very slowly and tend toward self-arrest. Consequently, crack initiation is not expected to be the decisive factor controlling ARALL Laminate structural durability.

To be damage-tolerant requires that a structure retain adequate strength after some damage has occurred by fatigue, corrosion, or accidents (e.g., impacts) until detected through inspection and repaired in such a way that the structure is able to sustain the design ultimate load. The absence of damage or slow further deterioration after damage onset implies good damage tolerance. To obtain a damage-tolerance assessment, it is important to know the nature and shape of (critical) defects, the damage- (crack-) propagation rates, and residual strength of the degraded structure. Damage events resulting in delamination and loss of reinforcement (e.g., impacts and penetrations) are decisive in determining the residual strength capability of fibrous composite materials [*24*]. In this regard, Fig. 16 shows that postimpact strength retention of ARALL Laminates is superior to that of true composites. Tracking subcritical damage in pure composites is difficult, and these materials mainly rely on their strength advantage to assure adequate damage tolerance. In all-aluminum structures, damage onset is generally followed by crack growth at an accelerating rate, which requires diagnostic attention to avoid functional impairment, and possibly catastrophic failure. ARALL Laminate's inspectability and unique tolerance to cracking gives it a decided advantage over aluminum. These features provide opportunity for greater weight efficiency, increased safety margin, and sizable cost savings by reducing the need for frequent inspections and fatigue-related repairs.

During the development stage, it became clear that design, manufacture, and testing of full-size aircraft parts were necessary in order to prove the advantages of ARALL Laminates. In 1987, Fokker Aircraft completed full-scale structural fatigue testing of a F-27 commuter-aircraft lower-wing section [*11*], redesigned to take full advantage of ARALL Laminate properties [*12–13*]. The test verified that both significant life extension and a 33% weight saving over the existing all-aluminum design could be attained. Inspection of the ARALL wing panel after testing to three times the economic design lifetime revealed some small cracking in an area isolated to a mounting flange. Most of these cracks had only penetrated one layer of metal (out of six), and it was apparent that after initiation of the cracks, their growth had decelerated and halted altogether. No other cracking problems were noted, and 20,000 additional simulated flights were applied without incident. At this point, the fatigue testing was discontinued and the ARALL panel tested for residual strength. The residual strength attained exceded 1.4 times the design limit load (highest load expected in service). RLD (the Dutch certification agency) requires a residual strength of 1.1 × limit load for damaged structures, so the ARALL-2 panel surpassed this requirement by 30%. The success of the Fokker experiment was so convincing that Fokker and several other airframers are actively pursuing introduction of flying parts. Already Fokker is flying ARALL parts on its F-50 prototype aircraft [*14*].

IX. Summary

ARALL Laminates were developed to fill a materials need in the aerospace industry—that need being a material to couple the higher specific strength and fatigue performance capabilities of advanced composites with the classical durability, fabricability, and supportability characteristics of aluminum. In addition, the material's enhanced vibrational damping and lightning-protection abilities provide new options for innovative aircraft usage. Since the early 1980s, ARALL Laminates have shown themselves to be a most exciting new material for aerospace structure. The unique set of ARALL Laminate properties match up well with existing structural-integrity requirements and the more demanding performance needs of next-generation aircraft. Though intended primarily for application in fatigue-prone parts, the cyclic properties, in general, are so good that the criterion for design transitions from fatigue to one of the following: blunt notch static strength, bearing strength, compression strength, or transverse properties. Through use of good design practices, goals of 20 to 40% weight savings and no-repair structure have been demonstrated to be within reach.

ARALL Laminates are now shifting from the research–development and test–evaluation stages to the commercialization stage with the advent of flying articles. Almost every major aircraft company in the Free World has an ongoing internal ARALL Laminates program, and industry specifications and certification issues are being brought to the forefront. Given the success experienced to date, it is not unreasonable to expect that ARALL Laminates can compete seriously for a significant share (10–15%) of the airframe structure in next-generation aircraft.

Acknowledgments

The work presented represents more than six years of accumulated characterization effort by many individuals, and any list of names is likely to miss valued contributors. Particular acknowledgment is owed to the following Alcoans: K.E. Luyk and A. Govada (ARALL fabrication) and R.W. Schultz, H.F. Wu, J.L. Teply, and S. Lee (mechanical and durability behaviors). Also acknowledged are the many excellent contributions by faculty and students of the Delft University Department of Aerospace Engineering, in particular J. Schijve. Gratitude is also expressed to Fokker Aircraft, 3M Company, and AKZO for their contributions to the total ARALL development effort.

References

1. Marissen, R., and Vogelesang, L.B. (January 1981). "Development of a New Hybrid Material: ARALL." Intercon. SAMPE Mtg., Cannes, France.

2. Vogelesang, L.B., Marissen, R., and Schijve, J. (May 1981). "A New Fatigue Resistant Material: Aramid Reinforced Aluminum Laminate (ARALL)." 11th ICAF Symp., Noordwijkerhout, The Netherlands.

3. Gunnink, J.W., Vogelesang, L.B., and Schijve, J. (1982). "Application of a New Hybrid Material (ARALL) in Aircraft Structures." ICAS-82-2.6.1, 13th Cong. International Coun. Aeronautical Sci., Seattle, Washington, p. 990.

4. Vogelesang, L.B., and Gunnink, J.W. (1983). "ARALL, A Material for the Next Generation of Aircraft. A State of the Art." Report LR-400, Dept. Aerospace Eng., Delft Univ. of Tech., The Netherlands.

5. Gunnink, J.W., Verbruggen, M.L.C.E., and Vogelesang, L.B. (1986). "ARALL, A Light Weight Structural Material for Impact and Fatigue Sensitive Structures," *Vertica*, **10**, no. 2, 241.

6. Vogelesang, L.B., and Gunnink, J.W. (1986). "ARALL: A Materials Challenge for the Next Generation of Aircraft," *Materials and Design*, **7**, no. 2.

7. Bucci, R.J., Mueller, L.N., Schultz, R.W., and Prohaska, J.L. (1987). "ARALL-Laminates—Results from a Cooperative Test Program," *Advanced Materials Technology 87*, **vol. 32**. Sci. of Adv. Matls. & Proc. Eng. (SAMPE), p. 902.

8. Bucci, R.J., and Mueller, L.N. (October 25–28, 1987). "ARALL Laminate Performance Characteristics." ARALL Laminates Technical Conf., Champion, Pennsylvania.

9. Bucci, R.J., Mueller, L.N., Vogelesang, L.B., and Gunnink, J.W. (1988) "ARALL Laminates: Properties and Design Update," *Materials—Pathway to the Future*, **vol. 33**. Sci. of Adv. Matls. & Proc. Eng. (SAMPE), p. 1237, Alcoa Laboratories, Alcoa Center, PA.

10. Ritchie, R.O., Yu, Weikang, and Bucci, R.J. (1987). "Fatigue Crack Propagation in ARALL Laminates: Measurement of the Effect of Crack-Tip Shielding from Crack Bridging." University of California, Berkeley Report No. UCB/R/87/A1049 (submitted to *Engineering Fracture Mechanics*).
11. van Veggel, L.H., Jongebreur, A.A., and Gunnink, J.W. (June 10–12, 1987). "Damage Tolerance Aspects of an Experimental ARALL F27 Lower Wing Skin Panel," 14th ICAF Symp., Ottawa, Canada.
12. Gunnink, J.W., and Schee, P.A. v. d. (July 1987). "Design of the ARALL F-27 Lower Wing Fatigue Panel." 4th Internat. Conf. on Composite Structures (ICCS4), Paisley, Scotland.
13. Gunnink, J.W. (July 20–24, 1987). "Design Studies of Primary Aircraft Structures in ARALL Laminates." ICCM 6/ECCM 2, Imperial Coll., London.
14. van Veggel, L. (October 25–28, 1987). "ARALL Applications in Fokker Aircraft Wing Structures." ARALL Laminates Technical Conf., Seven Springs Resort, Champion, Pennsylvania.
15. Ioannou, M., Kok, L.J., Fielding, T.M., and McNeill, N.J. (June 10–12, 1987). "Evaluation of New Materials in the Design of Aircraft Structure." 14th ICAF Symp., Ottawa, Canada.
16. Procedings ARALL Laminates Technical Conf., Champion, Pennsylvania (October 25–28, 1987), Alcoa Laboratories, Alcoa Center, PA.
17. Gunnink, J.W., and Rothwell, A. (1985). "An Assessment of the Static Strength of Flush Head Aluminum Alloy Rivets in Two Grades of ARALL Sheet." Report LR-391, Faculty Aero. Eng., Delft Univ., The Netherlands.
18. Military Std. Handbook, Mil Hdbk 5D, *Metallic Materials and Elements for Aerospace Vehicle Structures*. U.S. Dept. of Defence, rev. May 1, 1985.
19. Lee, S., and Munro, M. (1986). "Evaluation of In-plane Shear Test Method for Advanced Composite Materials by the Decision Analysis Technique." *Composites*, **Vol. 17**, p. 13.
20. Teply, J.L., DiPaolo, B., and Bucci, R.J. (1987). "Residual Strength of ARALL Laminate Panels." *The Nations Future Materials Needs*, Vol. 19, Intl. SAMPE Tech. Conf. Ser., p. 353.
21. Macheret, Y., and Teply, J.L. (1988). "Residual Strength of ARALL Laminates." *Mechanics of Composite Materials—1988*, AMD—Vol.92, American Society of Mechanical Engineers, p. 53.
22. Verbruggen, M.L.C.E. (1987). Aramid Reinforced Aluminum Laminates: ARALL Adhesion Problems and Environmental Effects." Report LR 504 and LR 505. Ph.D. thesis, Faculty of Aero. Eng., Delft University, The Netherlands.
23. Lee, S., and Kipp, T.R. (1988). "The Environmental Resistance of ARALL Laminates," *Materials—Processes: The Intercept Point*, Vol. 20, Intl. SAMPE Tech. Conf. Ser., p. 617.
24. Whitehead, R.S. (December 1–3, 1987). "Lessons Learned for Composite Aircraft Structures Qualification, 1987 Aircraft/Engine (ASIP/ENSIP) Conf., San Antonio, Texas.
25. Chen, D. (1986). "Drop Weight Impact Test on Some Sheet Materials." Doc. B2-86-03, Dept. Aero. Eng., Delft Univ., The Netherlands.
26. Johnson, W.S. (1986). "Impact and Residual Fatigue Behavior of ARALL and AS6/5245 Composite Material." NASA Langley TM 89013.
27. Roebroeks, G. (1985). "Fatigue Behavior of ARALL after Impact." Doc. B2-85-01, Dept. Aero. Eng., Delft Univ., The Netherlands.

28. Hyer, M.W., Anderson, W.J., and Scott, R.A. (1976). "Nonlinear Vibrations of Three-Layer Beams with Viscoelastic Cores, I: Theory." *Journal of Sound and Vibration*, **46**, no. 1, 121.
29. Hyer, M.W., Anderson, W.J., and Scott, R.A. (1978). "Nonlinear Vibrations of Three-Layer Beams with Viscoelastic Cores, II: Experiment." *Journal of Sound and Vibration*, **61**, no. 1, 25.
30. Ely, R.A. (1984). "Laminated Damped Fuselage Structures." In *Vibration Damping 1984 Workshop Proceedings*, AFWAL TR-84-3064, p. 00-1.
31. Nutter, N.R., Gray, R.A., Crowe, C.R., and Kasper, R.G. (October 4–9, 1986). "Damping Characteristics of SiC/Al Metal Matrix Composites and Aramid Reinforced Aluminum Laminates (ARALL-1)." ASM Matls. Week 86, Lake Buena Vista, Florida.
32. Whitehead, R.S., Ritchie, G.L., and Mullineaux, J.L. (October 1983). "Durability of Composites." 9th Mechanics of Composites Rev., Dayton, Ohio.

11

Powder-Metallurgy Aluminum Alloys

GREGORY J. HILDEMAN

Alcoa Laboratories, Alcoa Center
Alcoa Center, Pennsylvania

and

MICHAEL J. KOCZAK

Department of Materials Engineering
Drexel University
Philadelphia, Pennsylvania

ISBN 0-12-341831-3

I. Introduction

A. *Historic Development of Aluminum Powder-Metallurgy Alloys*

Over one hundred years ago, Hall and Heroult independently discovered a process to produce aluminum by electrolysis. This refining technology opened the door for economical production of the light metal and initiated a search for new applications. Use of aluminum as powders began when flake-type aluminum powder was manufactured in the early 1900s by milling or attriting thin sheets into tiny flakes used as an additive pigment in paint formulations. Introduction of the atomization process in the 1920s provided for a more efficient production process and marked a new era in aluminum-powder production. This process, coupled with the beginning of World War II and the development of aluminized explosives, created a major market for atomized aluminum powder. Subsequently, atomized aluminum powder has found commercial applications in diverse commercial and defense-related areas of technology. A partial listing includes pigments for paints and roofing, commercial blasting materials, reducing agents for the manufacture of metals and alloys, pharmaceuticals, alumiweld conductors, cellular concrete, pyrotechnics, chemicals, and rocket propellants. Furthermore, the commercial viability of low-cost pressed and sintered aluminum powder-metallurgy parts emerged in the 1960s, resulting in a further expansion of the powder-metal market into near-net shape components. The chronological development of dispersion and precipitation powder-metallurgy alloys is charted in Tables I and II. Table I can be divided into three classes: high-temperature, low-density, and high-modulus alloys, which can include powder-metallurgy aluminum metal-matrix composites. In contrast, the precipitation-strengthened PM alloys (Table II) can be conveniently categorized by conventional wrought-alloy designations (e.g., 2000 and 7000 series). Exception however, is noted for the pressed and sintered alloys, which are designated by 201, 601, and so on.

Considering dispersion alloys during the 1940s and 1950s, aluminum powders were investigated for use in high-temperature applications. Sintered aluminum products (SAP) or (XAP) utilized Al_2O_3 dispersions to improve the high-temperature strength and creep response of parts consolidated from milled aluminum powder [*1–4*]. Shortly after this period, experimental work [*5–7*] on atomized aluminum powders containing high-solute contents was initiated and provided the genesis of research that has evolved into development of the current aluminum PM alloys. These alloys utilized the benefits derived from rapid solidification, which include

TABLE I

Development of Aluminum Powder-Metallurgy Alloys—Dispersion Alloys

Alloy type	1950	1960	1970	1980	Reference[a]
Elevated Temperature:	(SAP)-XAP-Ae				
	Al-Al_2O_3				1–3
		Al-X[a]			7
		Al-Fe, Al-X			5,6
		Al-Si			(Dixon and Skelly, 1965)
			Al-Al_2O_3(MA)[b]		51
			Al-Mn-Si[c]		37
			Al-X[d]		(Read et al., 1975)
				Al-X[e](MA)	(Erich, 1980)
				Al-Fe-Ce	68,69
				Al-Fe-Mo	73
				Al-Cr-Zr	(Miller, 1984)
				Al-Fe-V-Si	(Skinner et al., 1984)
High modulus:[f]					
	Al-TiC, Al-Mn, Al-Fe[g]				(MacDonald and Ransley, 1954)
	Al-Mn, Fe, Cr				5
				Al-Fe-NiCo	(Vidoz, Lewis et al., 1986)
				Al-SiC (MMC)	(Divecha et al., 1981)
			Al-SiC (MA)		(Donachie and Gilman, 1983)
				Al-Cu-Li-Zr	(Meschter et al., 1986)
Low Density					
	Al-Li, Al-Be				(Lyle and Towner, 1960)
				2024 + Li	63
				Al-Li-Cu-Mg	(Palmer et al., 1982)
				Al-Mg-Li(MA)	(Donachie and Gilman, 1983)
				Al-Li-Be	(Lewis and Starke, 1986)

Notes:
[a] Reference numbers in right-hand column refer to references at end of chapter.
[b] X = Mn, Ni, Co, Cr, Ti, Zr, V
[c] MA = mechanically alloyed.
[d] Also has low thermal expansion.
[e] X = Ti, V, Cr, Mn, Fe, Co, Ni, Zr, Nb, Mo, Si.
[f] X = Fe, Co, Ti, Mg, Mn, Zn, Ni, Cr, Cu.
[g] A number of elevated temperature alloys also have a high elastic modulus.
[h] Blended Al and elemental Mn, Ni, Fe powders.

TABLE II

Development of Aluminum Powder-Metallurgy Alloys Precipitation Strengthened

Alloy type	1950	1960	1970	1980	Reference[1]
Pressed and sintered (Al-Cu-Mg-Si):					
	Al-Cu				53
	Al-Cu-Mg-Mn				54
		Al-Cu			55
		201AB, 601AB			56
Prealloyed/wrought					
2XXX (Al-Cu):		Al-Mn-Cu			61
		2024			22
			2124		47
			IN9021		(Erich and Donachie, 1982)
					50
				2024, 2024 + Zr	40
Corrosion-resistant					
5XXX (Al-Mg):[a]				IN9051, IN9052	57
					6
High-strength					
7XXX (Al-Zn-Mg-Cu):					
	7XXX[b]				22, (Lyle and Towner, 1962)
		7XXX			24
			7XXX + Co, Fe + Ni		19,28 (Grant, 1982)
			7XXX + (Co, Zr, Cr)		30
				7090/7091	(Alcoa, 1980)
				7064	(Kaiser, 1985)
				7XXX + Zr + Ni	(Hildeman et al., 1985)
				7XXX + Mn	(Koczak, 1982)
				7XXX + Cr, Zr	(Alcan, 1986)

Notes

[1] Reference numbers in right-hand column refer to references at end of chapter.

[a] Al-Mg alloys exhibit excellent corrosion resistance.

[b] Investigated elemental additions up to 15% Zr, 15% Mg, 20% Cu plus transition elements (Mn, Fe, Ni, Cr, Ti, V, Zr, Co, Mo, W) additions up to 4%

alloying flexibility, increased solid-solubility limits, fine-scale microstructures, and compositional homogeneity.

From early research efforts emerged several types of alloy systems that can be classified into dispersion-strengthened alloys, precipitation-strengthened alloys and hybrid systems (e.g., mechanically alloyed or composite-reinforced systems). The dispersion-strengthened aluminum PM alloys contain fine dispersions of relatively insoluble intermetallic phases. These alloys have good elevated-temperature strength, thermal stability, and wear resistance and they exhibit a high elastic modulus. An example of this type of alloy system is the Al-Fe-X alloys where X represents a variety of elements that form a high volume fraction (typically ranging from 5 to 40% of intermetallic aluminide phases in an aluminum matrix). In addition, Al-Li and Al-Li-Be alloys for lower-density and blended powder-metallurgy composites (e.g., 2000 series with SiC whiskers and particulate reinforcement) for greater stiffness are being explored as of the late 1980s. Given the advantages of rapid solidification and the flexibility inherent in blending different powders, an opportunity exists to develop novel alloys with tailored or designed physical and mechanical properties for a wide variety of high-performance applications.

A second major alloy class (see Table II) developed along the lines of traditional heat-treatable 2XXX and 7XXX ingot-metallurgy compositions. In this sector, rapid solidification resulted in the ability to produce a fine grain size and refined constituent microstructure in powders. After consolidation, subsequent PM alloys have demonstrated combinations of strength, fracture toughness, fatigue resistance, and corrosion resistance superior to ingot alloys. Alloys 7064, 7090, and 7091 were introduced in the 1980s and are examples of advanced PM materials developed for commercial aerospace applications.

Techniques for processing powder-metallurgy alloys have also evolved. Rapidly solidified particulate, in the form of powder, ribbon, or attrited flake, must be consolidated to full density to maximize mechanical properties. Processing procedures must, therefore, be carefully designed to maintain the refined structure and to obtain the benefits inherent in powder alloys. Advances in powder processing such as vacuum degassing, consolidation, hot isostatic pressing, direct spray deposition, extrusion, forging, and rolling have aided the development of aluminum powder-metallurgy alloys for structural applications. To this end, rapidly solidified particulate—produced by atomization, mechanical alloying, or attriting melt spun ribbon—has been consolidated to full density with excellent mechanical properties. Additional information on the development of aluminum powder-metallurgy alloys and processing techniques is contained in reviews by Jones [*8*], Pickens [*9*] Palmer and Tietz [*10*], Singer [11], and Froes et al. [12].

B. Development of Rapid-Solidification Concepts

Since the 1950s, rapid solidification of aluminum alloys has been based on gas-atomization techniques. This processing technique allowed researchers to explore formation of aluminum-base alloys containing high nonequilibrium levels of a wide variety of solute elements. Since the early work of Roberts [*7*] Towner [*5*], and Lyle and Towner [*6*] greater understanding of the effect of cooling rate on refinement of dendrite arm spacing (DAS) has evolved.

Based on the work of Spear and Gardner [*13*], Dean and Spear [*14*], and Flemings [*15*], a linear relationship between dendrite arm spacing and cooling rate was established from measurements on Al-Cu alloy castings, atomized powder, and a splat-cooled sample. Subsequent research by Matyja et al. [*16*] confirmed extrapolation between DAS and cooling rate to cooling rates of over 10^8 K/sec. Using Al-Fe and Al-Si splat-quenching techniques pioneered by Pol Duwez [*17*]. The general relationship between cell size and cooling rate is illustrated in Fig. 1. The graph shows that rapid solidification inherent in atomization and melt spinning can refine the cell size by over two orders of magnitude compared to direct-chill (DC) cast ingot. Today, we know from the work of many researchers, which has been summarized by Mehrabian [*18*], that a microstructural characteristic such as dendrite arm spacing is a function of the liquid–solid interface

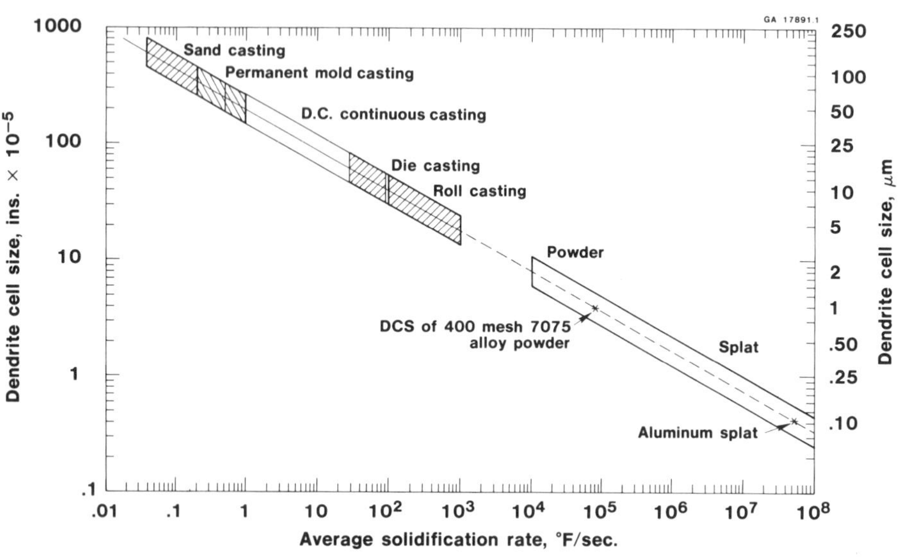

FIG. 1. Cooling rate and cell pacings of various casting processes.

velocity, which is controlled by variables (e.g. the amount of undercooling developed before nucleation, the rate of heat flow, as well as thermodynamic and kinetic factors which are dependent upon the alloy system considered). Figure 1 is, however, useful in that it provides a perspective on the ability of rapid-solidification processes such as atomization, melt spinning, planar flow casting and splat quenching to refine the microstructure of an alloy. A wide variety of rapid-solidification processes have been applied to aluminum alloys [*19*]. These include gas atomization, centrifugal atomization, rotating cup, rotating electrode process, plasma rotating electrode process, single- and double-roll quenching, ultrasonic gas atomization, the Osprey process, and the soluble-gas-atomization process.

In addition to refinement of the dendrite arm spacing, extension of solid solubility is of importance in maximizing the volume fraction of second phases which can be precipitated from an extended solid solution. Table III lists the maximum extended solid solubility reported by various investigators. Following production of alloy powders, secondary processes (i.e., mechanical alloying combining carbide or oxide reinforcement) provide opportunities for high-temperature strength and modulus improvements. The refinement of microstructure via rapid solidification and the extension of solid solubility are important physical features of the process. The resulting benefits (i.e., the mechanical-property improvement versus weight

TABLE III

Extension of Solid Solubility for Binary Aluminum Alloys

Elements	At % equilibrium	At % reported maximum
Ce	0.01	1.9
Co	0.01	0.5–5
Cr	0.44	5–6
Cu	2.5	17–8
Fe	0.025	4–6
Mg	18.9	37–40
Mn	0.7	6–9
Mo	0.07	>1.0
Ni	0.023	1.2–7.7
Si	1.59	10–16
Ti	0.15	.22–2
V	0.2	1.4–2
Zn	66.5	38
Zr	0.083	1.15–1.5

Source: After Jones [*8*].

saving) demonstrate the future utility of aluminum powder-metallurgy alloys in aerospace applications (Fig. 2). For example, a 30% increase in strength or a 50% increase in elastic modulus in a PM alloy or composite can translate into a 10% savings in weight in aerospace applications. This goal for weight savings is equivalent to that of Al-Li ingot alloys.

C. *Advantages and Barriers to Aluminum Powder-Metallurgy Alloys*

The ability to achieve a refined microstructure combined with economical powder-processing techniques creates the potential for aluminum PM alloys to be applied where high strength and stiffness, elevated-temperature performance, corrosion, or wear resistance are required. In addition, powder-metallurgy alloys can offer combinations of mechanical (strength, modulus, fracture toughness) and physical properties (e.g., thermal and electrical conductivity, tailored coefficients of thermal expansion, density) that are superior to cast alloys'. The benefits of rapid solidification are highlighted in Table IV. Advances in PM aluminum have been achieved in the areas of alloy design, atomization, and consolidation processes. The future challenges involve the development of even greater levels of properties coupled with reliability and reproducibility in powder properties and processing, so that design confidence and cost-effective applications can be developed in commercial and defense sectors. Significant achievements in weight savings are possible as illustrated in Fig. 2, when strength, modulus, and fatigue improvements are developed in alu-

TABLE IV

ADVANTAGES AND BARRIERS TO GROWTH OF WROUGHT ALUMINUM PM ALLOYS

Advantages:
Unique compositions
Fine-grain, constituent, and disperoid sizes
Homogeneous distribution of alloy elements
Superior combinations of strength, toughness, and resistance to corrosion and fatigue
Potential for improved modulus, thermal stability, and low density
Barriers:
High cost
Product availability and size limitation
Quality-control issues

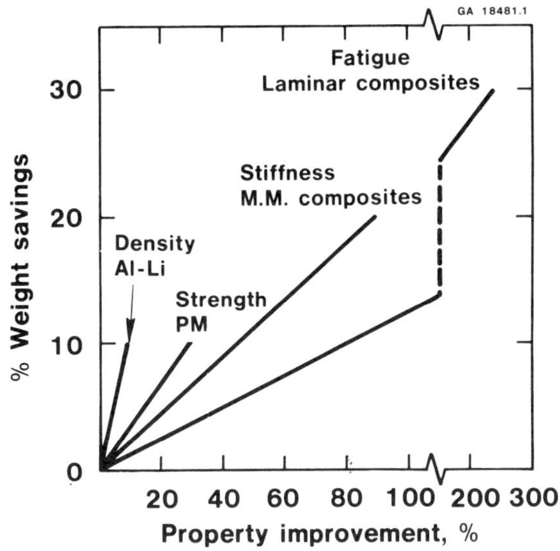

FIG. 2. The relationship between weight savings and property improvements.

minum powder-metallurgy alloys. Barriers to the future development of aluminum PM alloys include the necessary processing quality control and reliability of design properties required for structural applications coupled with a required cost reduction necessary for wider acceptance in commercial automotive and aerospace applications. Initial commercial applications in four areas include (1) high-strength alloys for aircraft structural components, (2) wear-resistant alloys and composites for automotive engines and appliances, (3) high-modulus composite alloys for stiffness, and (4) high-temperature alloys for airframe and engine applications.

Although aluminum PM alloys offer significant potential in terms of mechanical properties, there are several barriers that limit their application and acceptance in the marketplace. These are listed as barriers to growth in Table IV. As with most new materials, the cost of substitution is usually a major hurdle. In addition, the cost-effectiveness must be clearly established in the total product life cycle. Excluding pressed and sintered parts, the cost of wrought PM aluminum alloys are greater than alloys produced from ingot metallurgy because of the additional processing steps as shown in Fig. 3. As of the late 1980s, estimates indicate that PM alloys will cost 1.5–5 times as much as ingot alloys. Therefore, applications for PM alloys will initially be limited to components where superior performance and effective materials utilization justify their use. It is expected, however, that the cost of PM alloys should be reduced as the volume of powder-metallurgy components increases. This is because the capital cost in terms

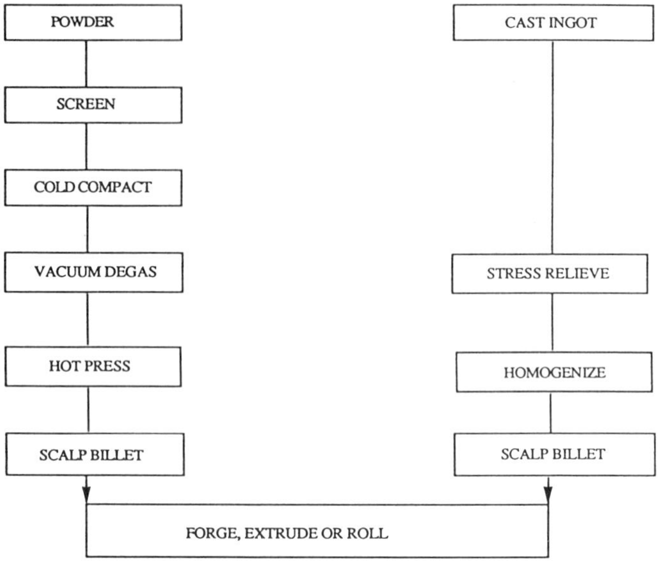

FIG. 3. Comparison of powder-metallurgy and ingot-metallurgy processing steps.

of equipment will be spread over a larger quantity of material produced. In addition, producers of rapidly solidified components will seek more efficient methods to produce atomized powders and attrited melt spun ribbons for consolidated components.

The growth of PM alloys has also been constrained by the size and availability of billets that can be processed into wrought products. In the 1970s and 1980s, applications were limited to forgings, extruded shapes, and sheet products that could be fabricated from billets of less than 160 kg. As of the late 1980s, a number of material producers (including Alcoa, Allied, Kaiser, Novamet, and Pratt and Whitney) are in the process of increasing the size of the consolidated billet up to 550 kg [*20–21*], which will also be a step to reduce the cost as well as increase the availability of advanced powder-metallurgy materials.

Quality is another issue that must be addressed. Mechanical properties such as fracture toughness and fatigue are extremely sensitive to the size, number, and distribution of inclusions. At least part of the improvement of PM alloys in these areas can be attributed to a refined microstructure that results in intrinsic metallurgical benefits compared to ingot-metallurgy alloys. However, care must be taken during processing of the powders to avoid contamination by particles that could degrade inclusion-sensitive mechanical properties.

Another factor that may be important to the growth of higher-cost materials is a greater attention to the life-cycle cost of a component. For example, over the 20-year life of an aerospace component, the cost of a PM part may be lower due to savings in fuel because of lower weight. Reduced frequency of inspection as well as fewer maintenance or part-replacement requirements for fatigue or corrosion damage are also factors in evaluating life-cycle costs. Consideration of these factors will aid acceptance of wrought powder-metallurgy as cost-effective structural materials. There is therefore a need for joint effort combining the skills of design and materials engineers to clearly assess performance–cost–life-cycle trade-offs and to determine future directions of powder processed aluminum alloys for commercial, defense and aerospace markets.

II. Precipitation–and Solid-Solution–Strengthened Powder-Metallurgy Alloys

A. *7000 Series PM Alloys*

1. Exploratory Investigations

High-strength 7XXX PM alloys are strengthened by precipitates (e.g. Guinier-Preston Zones) composed of elements such as Zn, Mg, and Cu which are soluble in aluminum. These alloys offer combinations of strength and damage tolerance as determined by fracture-toughness, fatigue, crack propagation, and corrosion-resistance evaluations, which are superior to those of their ingot-metallurgy counterparts. Atomized powder compositions in the 7XXX series (Al-Zn-Mg-Cu) were among the earliest systems to be explored. In the 1950s and 1960s, a number of researchers were active in investigating combinations of Zn, Mg, and Cu for high strength with a variety of dispersoid-forming elements to control recrystallization and maintain a fine grain size. Initial efforts were directed at achieving high strength levels. For example, Roberts [*22*] obtained yield strengths as high as 810 MPa. Towner and Lyle [*23*] and Haar and Lyle [*24*] produced Al-Zn-Mg-Cu extrusions with yield strengths of 858 MPa. Unfortunately, the ductility of extrusions at these strength levels was low, typically less than 1% elongation. From this early work, there were indications that high-strength made by powder metallurgy had corrosion-resistance and fatigue-endurance limits that were superior to ingot-metallurgy alloys'. Subsequent work by Cebulak and Truax [*25*] demonstrated that good combinations of strength (654 MPa), elongation (8%), exfoliation resistance (immune), fatigue limit (207 MPa), and fracture toughness (21 MPa-m^{12}) were obtainable in 7 XXX PM alloys.

2. Process Development of Al-Zn-Mg-Cu Alloys

During the 1960s and 1970s, processing of powder-metallurgy alloys advanced specifically in the understanding of powder handling prior to consolidation. Cebulak [*26*] and Roberts [*27*] independently reported that vacuum degassing greatly improves toughness in high-strength PM alloys. Figure 4 from Lyle and Cebulak [*28*] illustrates improvements in the ratio of notch tensile strength to yield strength for a die-forged Al-8.0% Zn-2.5% Mg-1.0% Cu-1.5% Co alloy obtained by vacuum degassing compared to forgings fabriated from billets that were heated in argon. Discovery of the benefit of vacuum degassing was a significant step following a series of powder-metallurgy processing patents [*29,71*] discussing the importance of heating the powder in a flowing protective atmosphere in order to degas the powders. Later in the 1970s and 1980s, degassing procedures became more sophisticated by incorporating a depurative gas treatment [*30*] progressive evacuation cycles [*31*], and vacuum hot compaction without a canister [*32,33*]. The nature of complex decomposition processes of the aluminum hydroxides during degassing has also been investigated [*34–36*].

In addition, effort has been directed at scaling up the powder billet size to allow fabrication of plate and sheet products combined with cost reduc-

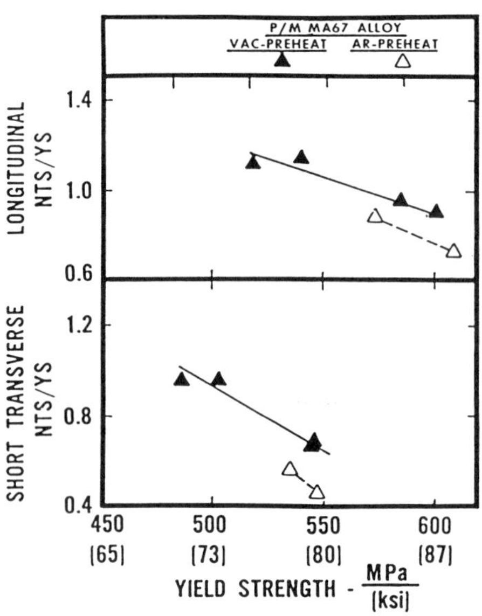

Fig. 4. Effects of preheat method on toughness and strength of PM MA 67 die forgings.

tion. One of the first efforts to scale up billet size was conducted by Cebulak [*37*]. In that program, billets weighing 1500 kg composed of a high-strength Al-Zn-Mg-Cu PM alloy were produced that were subsequently rolled into plate. More recently, Billman and Jatkar [*21*] developed processing techniques to produce 550 kg billets for both heat-treatable and dispersion-strengthened PM alloys. In addition to producing large billets that could be rolled to sheet gage, alternative powder-to-sheet techniques have been examined. For example, Daugherty [*38*] discussed a method where by plus 200 mesh aluminum powders are fed continuously into a hot rolling mill to yield sheet. More recently, Ro [*39*] examined the concept of blending Al-Zn-Mg-Cu alloy powders with Al powders to develop a low-cost method to produce high-strength 7XXX sheet.

3. Commercial and Second-Generation 7XXX Alloys

The decade of the 1980s has been a milestone in the development of commercial high-strength alloys. In 1980, the Aluminum Company of America (Alcoa) registered powder-metallurgy alloys 7090 and 7091 and Kaiser Aluminum and Chemical Corporation (Kaiser) registered X7064 in 1985. The compositions of these and other alloys are listed in Table V. Comparison of the compositions between 7091 and X7064 shows these alloys contain similar amounts of Zn, Mg, and Cu but have different dispersoid-forming elements. As shown in Table VI, properties of X7064-TX73 and 7091-T7E69 are similar and offer higher strength, fracture toughness, and resistance to stress corrosion cracking compared to ingot-metallurgy 7075-T73.

TABLE V

Nominal Compositions of 7XXX Aluminum PM Alloys

Alloy	Zn	Mg	Cu	Co	Cr	Zr	Ni	O	Al
7064	7.1	2.3	2.0	0.2	0.1	0.2	—	.2	Bal.
PM61	8.5	2.5	1.5	0.6	—	0.2	—	.2	Bal.
7090	8.0	2.5	1.0	1.5	—	—	—	.35	Bal.
7091	6.5	2.5	1.5	0.4	—	—	—	.35	Bal.
CW67	9.0	2.5	1.5	—	—	0.14	0.1	.35	Bal.

Source: For 7064 and PM61, Kaiser Aluminum and Chemical Corp., Data Sheet, April 1985. For 7091 and 7092, Aluminum Company of America, Technical Brief, March 1985. For CW67, Alcoa Internal Research and Development (IRAD), April 1983.

Note: Elements in wt %.

TABLE VI

TYPICAL PROPERTIES OF 7XXX-T7 ALUMINUM PM ALLOY DIE FORGINGS

Alloy temper	Test direction	YS (MPa)	UTS (MPa)	Elong. (%)	K_{Ic}[a] (MPa · $m^{1/2}$)	SCC (MPa)
X7064-TX76	L	572	607	15	30	—
	LT	551	600	12	—	—
	ST	545	593	6	—	241
X7064-TX73	L	517	572	15	33	—
	LT	503	551	12	—	—
	ST	496	551	9	—	310
7090-T7E71	L	550	600	13	22	—
	T	540	574	7	23	241
7091-T7E69	L	514	566	14	30	—
	T	496	549	9	28	290
7075-T73	L	467	525	12	22	—
(IM)	T	414	474	7	20	290

Source: Kaiser Aluminum and Chemical Corp., Data Sheet, April 1985, for X7064-TX76, X7064-TX73, and 7075-T73. Aluminum Company of America, November 1984, for 7090-T7E71 and 7091-T7E69.

Notes:

[a] L-T orientation per ASTM E399.

L = Parallel to grain flow.

T = Nonparallel to grain flow.

ST = Short transverse

LT = Long transverse

The compositions of two other 7XXX alloys, CW67 (developed by Alcoa) and PM61 (a Kaiser-developed composition), are also listed in Table V. Ping [*40*] reported that PM61 was developed for peak-strength T6 applications with yield strengths of 752 MPa and has been evaluated for seat tracks in aircraft. Research at Alcoa by Hildeman et al. [*41*], was directed at achieving high strengths similar to 7090's and a fracture toughness equivalent to 7091's in the T7 stress-corrosion-resistant temper. Figures 5 and 6 illustrate that the second-generation alloys i.e. CW67, X7090, and X7091 has met the desired goals by demonstrating the highest combination of strength and toughness in a commercial die forging.

Fatigue properties are vital to many aerospace applications. Alcoa studies reported by Hart [*42*] demonstrated that S-N fatigue resistance of powder-metallurgy alloys 7090 and 7091 was superior to that of 7075 and 2024. Evaluation of fatigue response addressed three major areas: S-N fatigue resistance, fatigue crack growth (da/dN) versus ΔK, and spectrum fatigue response. Early studies in the 1970s showed constant amplitude fatigue crack growth (FCG) rates of the PM alloys were greater than

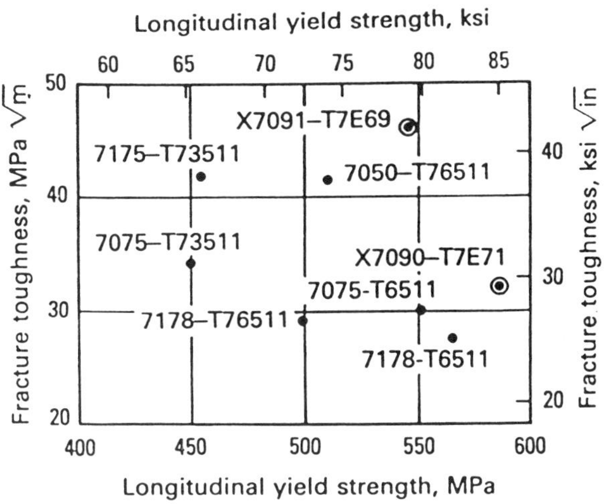

FIG. 5. Typical fracture toughness versus yield strength for high-strength aluminum extruded shapes. Specimens were 6.4–38 mm (0.25–1.5 in.) thick.

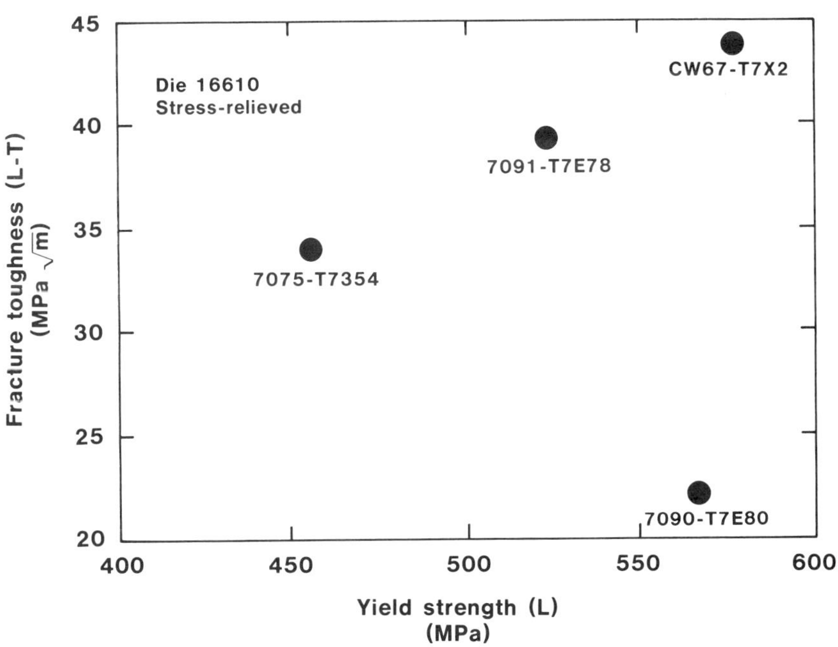

FIG. 6. Fracture toughness and yield strength for 7XXX PM versus 7075 IM Forgings.

conventional alloys'. Subsequent studies [*43–45*] have shown that although constant-amplitude FCG rates of PM alloys may be higher than those of ingot alloys, under spectrum fatigue conditions simulating actual load conditions, crack growth rates of powder-metallurgy alloys can be lower than growth rates of ingot alloys. It is, therefore important to conduct screening tests for fatigue that simulate the load and notch-stress fatigue conditions that are expected in the actual application.

B. 2000 Series PM Alloys

Powder metallurgy alloys containing Cu and Mg as the major alloy additions fall into the 2XXX alloy-composition classification. As in the case of 7000 series compositions, 2000 series PM alloys exhibit improved combinations of strength and toughness compared to ingot-metallurgy alloys which have moderately high strengths (typically 300 to 450 MPa yield strength) with good damage-tolerant properties, including fatigue and corrosion resistance.

While most of the effort since the 1950s in heat-treatable alloys has been directed at 7XXX compositions, a number of investigators have examined application of powder-metallurgy processing to Al-Cu-Mg compositions. Roberts [*22*], and later Lebo and Towner and Lyle [−], Grant [*47*] and Voss [*31*] examined alloy 2024. Compared to alloy 2024, a gain in tensile strength and fatigue life of PM 2024 fabricated from the splat-quenched droplets was obtained by Lebo and Grant. However, to maximize the

TABLE VII

NOMINAL COMPOSITIONS OF 2XXX AND 5XXX PM ALLOYS

Alloy	Cu	Mg	Mn	Zr	Cr	O	C	A
PM63	4.5	1.8	0.8	0.25	—	N.D.	—	Bal.
PM042	3.7	1.8	0.2	0.6	—	N.D.	—	Bal.
IN9021	4.0	1.5	—	—	—	0.8	1.1	Bal.
2024 IM	4.4	1.5	0.4	—	—	—	—	Bal.
IN9051	—	4.0	—	—	—	0.6	0.75	Bal.
IN9052	—	4.0	—	—	—	0.8	1.1	Bal.
5083 IM	—	4.4	0.7	—	0.15	—	—	Bal.

Source: Kaiser Aluminum and Chemical Corp., April 1985, for PM63. Alcoa/Lockheed California, August 1981, for PM042. Novamet Aluminum, Data Sheet, August 1984, for IN9021, IN9051, and IN9052.

Note: Elements in wt %.

N.D. = not determined.

benefit that powder-metallurgy processing can provide, alternative compositions and process techniques appear to be necessary.

Addition of Zr in order to form stable dispersoids was one route pursued by Kaiser and Alcoa/Lockheed. Table VII includes two 2024 powder-metallurgy alloys: PM63 under development by Kaiser [*48*] and PM042 developed by Alcoa [*49,50*]. Both of these alloys contain levels Zr that are close to or exceed the solid-state solubility limit of 0.28% Zr in aluminum. These additions of Zr exceed that which is possible for ingot metallurgy alloys and results in a fine grain size PM alloy with excellent combinations of strength and toughness. Figure 7 demonstrates that powder-metallurgy variations of 2000 series alloys containing Zr as a dispersoid result in

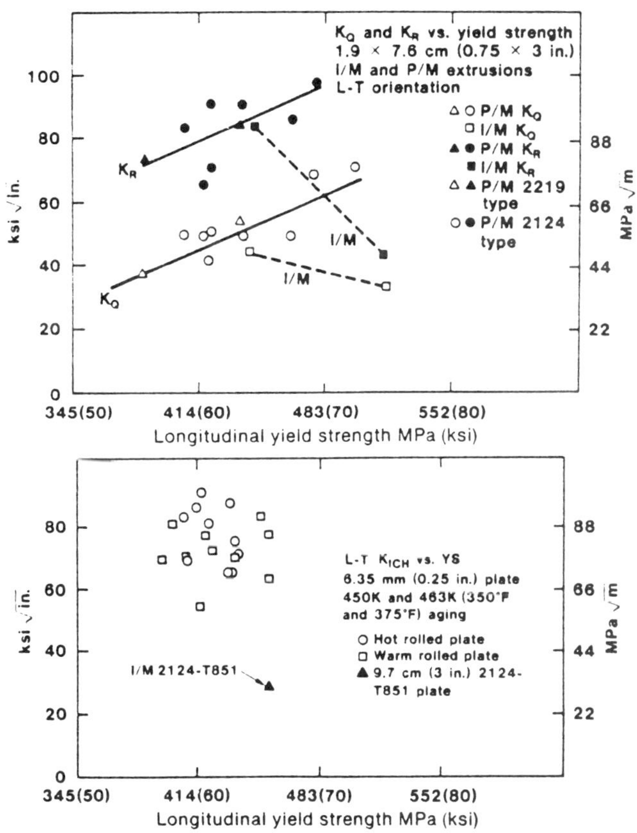

FIG. 7. Toughness versus yield-strength relationships in extrusion and plate. Top: Dependency of K_Q and K_R on the yield strength of the extrusions. Bottom: Dependency of charpy toughness on the yield strength of the 6.35-mm plate.

superior combinations of strength and fracture toughness compared to conventional 2XXX alloys.

Another route to improve 2000 series alloys by powder-metallurgy processing is by mechanical alloying. Using the process technology developed by Benjamin and Bomford [*51*] for Al-Al_2O_3, Novamet has developed alloy IN9021, which has a composition shown in Table VII to be similar 2024 except for the oxygen and carbon contents, which are a result of the process. These alloys are produced by attriting mixtures of Al, Cu, and Mg powders with organic additives that control welding and fracturing during milling. During this process, fine dispersions of oxides and carbides are formed from the reaction of the organic additives with the milled powder. After consolidation and heat treatment, the alloy is strengthened by precipitates and a fine grain size typically less than 0.5 μm. As shown in Table VIII, IN9021-T4 forgings [*20,52*] exhibit greater strength, toughness, and stress corrosion cracking resistance compared to 2024-T4's.

C. *Pressed and Sintered PM*

An economical processing method to produce aluminum PM parts is by pressing loose powders in a die and sintering the compact in a low–dew–point atmosphere. Early research [*53,54*] demonstrated the feasibility of producing specimens with tensile strengths of 275 MPa for Al-Cu sintered

TABLE VIII

Typical Properties of IN-9021 and IN-9052 Forgings

	IN-9021-T4	2024-T4	IN-9052	5086-H34
Ultimate tensile strength, MPa (-Ksi)	538 (78)	469 (68)	448 (65)	324 (47)
Yield strength 0.2% MPa (-Ksi)	469 (68)	324 (47)	379 (55)	255 (37)
Elongation, %	13	19	13	10
Fracture toughness, Klc MPa $\sqrt{m}$ (-Ksi $\sqrt{in.}$)	40 (36)	29 (26)	44 (40)	—
Elastic modulus, GPa (-Msi)	74.4 (10.8)	(10.6)	71.0 (10.3)	71.0 (10.3)
Density, g/cm^3 (-lb/in.^{3})	2.80 (0.101)	2.80 (0.101)	2.66 (0.096)	2.66 (0.096)
Threshold for SCC, [a] MPa (-Ksi)	>413 (>60)[b]	296 (43)[b]	>379 (>55)[c]	Susceptible if heated at 150° for long times[c]

Source: 2024-T4 and 5086-H34 data from *Aluminum Standards and Data* (Al. Assoc. Inc.) and *Military Standardization Handbook-5C* (U.S. Department of Defense).

[a] 90-day alternate immersion on 3% NaCl.

[b] Long transverse.

[c] Sensitized at 200°F for 7 days; short transverse.

compacts. Subsequent work by Storchheim [55] and Dudas and Dean [56] have made pressed and sintered aluminum PM a commercial success. There are primarily two types of alloys that are extensively utilized for pressed and sintered PM. Alloy 201AB contains 4.4% Cu, 0.8% Si, and 0.4% Mg, while 601 AB contains 0.25% Cu, 0.6% Si, and 1.0% Mg. Both alloys offer combinations of strength, ductility, and finishing characteristics to provide a low-cost alternative for replacement of die-cast or machined parts.

D. 5000 Series PM Alloys

In the 1960s, only a limited amount of research on Al-Mg alloys was reported. Towner and Lyle [6] explored aluminum atomized Al-Mg powder alloys in extruded form. Roberts [46] evaluated the mechanical properties of cast and powder extrusions of alloy 5083. A twofold increase in yield strength was obtained for the PM alloy compared to the cast alloy. In a comparison of 5005 sheet with roll-compacted 5005 particles, Daugherty [38] also demonstrated increases in strength for the powder-metallurgy alloy, which was considerably more resistant than conventional alloys to recrystallization.

In the early 1980s, International Nickel Company (INCO) developed a mechanically alloyed Al-4% Mg designed IN9051 [57]. This alloy utilized a combination of strengthening mechanisms including solid-solution strengthening from Mg, substructure strengthening from a fine grain size of less than 0.5 μm, and dispersion strengthening from oxides and carbides. Alloy IN9051 was subsequently replaced by IN9052 (Table VII), which has an equivalent level of Mg but contains a higher dispersoid content. As shown in Table VIII, IN9052 alloy forgings [52] exhibit excellent combinations of strength, toughness, and SCC resistance compared to ingot-metallurgy alloy 5083.

E. Low-Density Aluminum Alloys

Based upon a calculated rule of mixtures, [58,59] additions of Li, Mg, or Be to aluminum have a significant effect on reducing density. Lithium is the most potent since every 1 wt % addition reduces the density of aluminum by 3% coupled with a modulus improvement. Additions of almost 6 and 7 wt % of Mg and Be, respectively, are required to reach equivalent density reductions. Since the maximum solid-solubility limits of Li, Mg, and Be are 4.0, 18.9, and 0.063 wt %, respectively, density reductions are limited to approximately 12% for Li, 8% for Mg, and 0.03% for Be solely based upon solid-solubility considerations. It is, therefore, clear that power-metallurgy processing offers alternatives in providing the ability to reduce density through supersaturation of alloy additions beyond the equilibrium solid-solubility limits and increase strength via dispersion strengthening from fine second-phase particles produced by rapid solidifications.

In the 1950s and early 1960s, Lyle et al. [*40*] and Towner and Lyle [*6,46*] explored Al-Li and Al-Be atomized powder alloys. Since these elements simultaneously increase the elastic modulus while decreasing density, they are attractive additions for alloys that require high stiffness which is controlled by the ratio of modulus to density. Unfortunately, the reactivity of Li and Be as well as the toxicity of Be have hindered development of low-density alloys containing these elements.

Driven by the economics of aircraft operational costs (such as the increases in the cost of fuel in the late 1970s and early 1980s [*62*]), development of aluminum alloys containing lithium was initiated. While most of the efforts by the major aluminum companies focused on Al-Li compositions—such as 2090, 2091, 8090 and 8091 (Table IX), which could be formed into cast and wrought products—a number of studies utilized the potential powder-metallurgy advantages of a fine homogeneous microstructure to improve the low ductility and fracture toughness associated with alloy 2020, an Al-Cu-Li alloy developed by Alcoa in the 1950s. Initial research on PM Al-Li alloys investigated compositions similar to 2024 with additions of Li. Sankaran and Grant [*63*] showed that additions of 1 to 3% Li increased the specific strength of 2024.

TABLE IX

COMPOSITION OF LOW-DENSITY ALUMINUM ALLOYS

	Li	Cu	Mg	Zr	O	C	Mn	Cd
2020	1.0	4.5	—	—	—	—	0.8	0.15
2024	—	4.4	1.5	—	—	—	0.6	—
01420 (USSR)	2.2	2.7	—	0.12	—	—	—	—
Alcoa (CW56)	2.2	2.7	—	0.12	—	—	—	—
Lockheed Alloy 1	3.0	1.5	1.0	0.2	—	—	—	—
Lockheed Alloy 2	2.0	3.0	1.0	0.2	—	—	—	—
Novamet (INCO) Alloy	1.4	—	3.8	—	—	—	—	—
Allied 643	34.–3.6	0.8–0.1	0.4–0.6	0.4–0.6	0.6	1.1	—	—
Allied 644	3.0–3.2	0.8–1.1	0.4–0.6	0.4–0.6	—	—	—	—
Alcan	2.5–1.9	1.2	0.7	—	—	—	—	—
2090	1.9–2.6	2.4–3.0	0.25	0.10	—	—	0.05	—
2091	1.7–2.3	1.8–2.5	1.1–1.9	0.25	—	—	0.10	—
8090	2.2–2.7	1.0–1.6	0.6–1.3	0.25	—	—	0.10	—
8091	2.4–2.8	1.8–2.2	0.5–1.2	0.25	—	—	0.10	—

TABLE X

COMPARISON OF MECHANICAL PROPERTIES OF 8090 ALUMINUM ALLOYS

		Conventional direct chill cast and forged	Spray-cast and hot isostatic pressed
^{s}y	Yield strength MPa (ksi)	310 (45)	310 (45)
^{s}UTS	Ultimate strength MPa (ksi)	345–380 (50–55)	449 (65)
$^{\epsilon}$f		0.25–0.55	9–10
K_{ISB}	$MPa(m)^{1/2}[ksi(m)^{1/2}]$	11–20 (10–17)	31–35 (28–32)
As-cast grain size	(microns)	200–500	15–30
Grain-boundary film thickness	(microns)	2–3	0.1–0.2

Source: Lewis, R.E. [Private Communication]

Additional studies by Gysler et al. [*64*] compared the properties of PM and IM versions of 2020. In 1981, Palmer and Tietz [*10*] reviewed the properties of PM and IM alloys that were being investigated at that time. They concluded that powder-metallurgy Al-Li alloys exhibited better combinations of a specific strength and elongation compared to ingot metallurgy alloys. Subsequent studies [*65*] have shown that similar combinations of strength and elongation were obtained for PM and IM Al-3% Li-1.5% Cu-1.0% Mg-0.15% to 0.2% Zr alloys.

It is, therefore, apparent that if the mechanical properties of low-density PM alloys are to greatly exceed those of ingot-metallurgy compositions, alloy developers will need to focus on chemistries that provide weight savings greater than the 8–10% offered by IM alloys such as 2090, 2091, or 8091 (Table IX). A comparison of a spray-deposited Al-Li alloy (e.g. 8090) versus the cast and wrought counterparts, Table X depicts the improvement in strength and ductility. A number of companies are exploring Al-Li PM alloys in order to take advantage of higher specific modulus, e.g. Allied, Lockheed, INCO (Novamet), Table X.

III. Dispersion Strengthened Powder Metallurgy Alloys

A. *Elevated Temperature Alloys—Pioneering Studies*

The concept of using alluminium alloys for elevated-temperature applications dates back to the late 1940s, when oxide-dispersion–strengthened aluminum was developed. At that time in Switzerland, Irmann discovered that sintered aluminum had mechanical properties similar to those of aluminum alloys; he attributed this enhancement in strength to the presence of oxide films surrounding the individual powder particles in the sintered matrix. On subsequent cold compaction and hot-working, the oxide was broken up and distributed as small dispersoids, thereby giving rise strengthening of the aluminum matrix. These materials are known as the sintered aluminum products (SAP).

Subsequently in the United States, Alcoa developed a range of oxide dispersion-strengthened aluminum materials similar to SAP or XAP. The first detailed study of SAP-type materials in the United States was conducted by Lyle [*3*], who investigated SAP-type alloys containing 7.8–16.5% oxide by weight. It was found that these materials had moderate room-temperature strength, compared to 2000 series aluminum alloys. However, SAP alloys were stronger than ingot-metallurgy alloys at temperatures above 200°C and had the ability to retain strength after prolonged exposure at temperatures as high as 482°C. Moreover, when temperature was increased, strength decreased at a lower rate than for conventional aluminum alloys with comparable room-temperature strength.

The pioneering work on SAP-type alloys led to the development of other types of dispersion-strengthened aluminum alloys based on rapid solidification in which intermetallic compounds generally 5–40% by volume constituted the dispersed phase. These fine dispersoid reinforcements, typically less than 0.5 microns, are formed in situ during atomization of the alloy melt. Such powders are termed prealloyed powders and are subsequently compacted and hot consolidated to produce a wrought alloy.

Roberts [*7*] and Towner and Lyle [*5*,*6*] independently studied several alloys in which the dispersoid was an aluminum intermetallic. The rationale for selecting the alloying elements was to achieve a fine dispersion of an insoluble constituent in the aluminum matrix. Hence, alloying elements chosen were those such as Fe, Cr and other transition elements having low solid solubilities, high liquid solubilities, and low diffusion rates in solid aluminum. In addition, many of the transition elements form thermally stable intermetallic compounds of relatively low density, so that appreciable amounts of alloying elements may be added without a prohibitive increase in the density of the final alloy.

Lyle and Towner [5,6,2a] found that binary Al-Fe alloys containing 7.5 wt %, 10 wt %, and 12.8 wt % Fe gave the highest tensile strengths at 315 and 427°C. More complex alloys of composition 7.8 wt % Fe and 0.2 wt % each of Cr, Ti, V, and Zr and a quarternary alloy of aluminum with 2.5 wt % Fe, 0.4 wt % Cr, and 3.4 wt % V also gave excellent high-temperature strength up to temperatures of 427°C. The complex alloy of aluminum containing 7.8 wt % Fe and 0.2 wt % each of Cr, Ti, V, and Zr had the best stress-rupture properties.

These alloys exhibit excellent high-temperature strength due to microstructural stability at elevated temperatures relative to precipitation strengthened alloys. This is because aluminides are relatively thermodynamically stable and resistant to coarsening, which is a consequence of the extremely low equilibrium solid solubilities and small diffusion rates of the alloying elements in the aluminum matrix. To achieve sufficient strength at ambient and elevated temperatures, it is necessary to have a high volume fraction of the dispersoids on a fine scale that are uniformly distributed throughout the matrix. This can easily be achieved by rapid solidification and PM processing. It is impossible to produce these alloys by conventional ingot solidification since attendant slower cooling rates result in considerable macrosegregation. In addition, the intermetallic dispersoids formed at cooling rates of one degree per second in ingots are relatively coarse and would not make a significant strength contribution.

In addition to research directed at exploring alloy compositions, Microstructural transitions in rapidly solidified binary alloys of Al with Fe, Ni, Cr, Mn, Co, Ti, and V were studied by Jones [66] among others. It was observed that metallographically two types of distinct zones were present, they were designated zone A and zone B. Zone A consisted of a very fine dendritic or cellular aluminum structure and was optically featureless. The microhardness of this zone was approximately twice that of zone B and did not change significantly with annealing up to temperatures of 200°C, after which it decreased. Zone B consisted of a much coarser structure. The transition from the zone-A to the zone-B structure was explained in terms of the cooling rates experiences by the two regions.

The microstructural stability of strengthening dispersoids is a very important criterion in the selection of a new class of aluminum alloys for high-temperature applications. Pontikakas and Jones [67] studied the coarsening behavior of intermetallics in several Al-transition metal alloys produced by rapid solidification. Al-Fe, Al-Ni, and Al-Cr alloys were heat treated at 500 to 625°C for time periods up to 1000 hours and the size and distribution of the corresponding dispersoids were analyzed with respect to current particle-coarsening theories. At lower temperatures, coarsening was enhanced due to residence of the dispersoids on the grain boundaries. The observed increases in the average particle size of $FeAl_3$ were found to be in good agreement with predictions based on grain boundary diffusion control at 500°C and lattice diffusion control at 600°C. The highest resistance to coarsening was exhibited by $CrAl_7$ at 500°C while $FeAl_3$ was more resistant than $NiAl_3$ for all conditions.

As a result of these preliminary investigations, Al-8 wt % Fe was considered an excellent host binary candidate system. There were further developments during the late 1970s. In that time period, significant research efforts sponsored by the Air Force as well as companies (e.g., Alcoa, Alcan, Allied, Sumitomo Light Metals, Pechiney, Pratt and Whitney, Lockheed, and Kaiser) spawned the development of alloy systems such as in Al-Fe-Ci, Al-Fe-V, Al-Fe-Ni, and Al-Fe-Mo ternaries as well as quarternary variants of these systems which exceeded the mechanical properties of 2000 series ingot alloys. The objective of these programs was to determine optimum production processes and alloy compositions for dispersion-strengthened alloys. Major factors include powder-solidification rate, powder cleanliness, oxygen content, outgassing conditions (time, temperature), and powder-consolidation conditions (e.g., extrusion temperature and reduction ratio). Each of these factors can contribute or detract from mechanical properties by altering the second-phase aluminide size, effecting bonding between powder particles and modification of microstructural texture.

An example of the advantage of a dispersion-strengthened alloy evolved from these studies is shown in Fig. 8. The yield strength of Al-Fe-Ce PM alloy [*68*] is plotted against typical properties of 2219, 2024, and 7178. It is

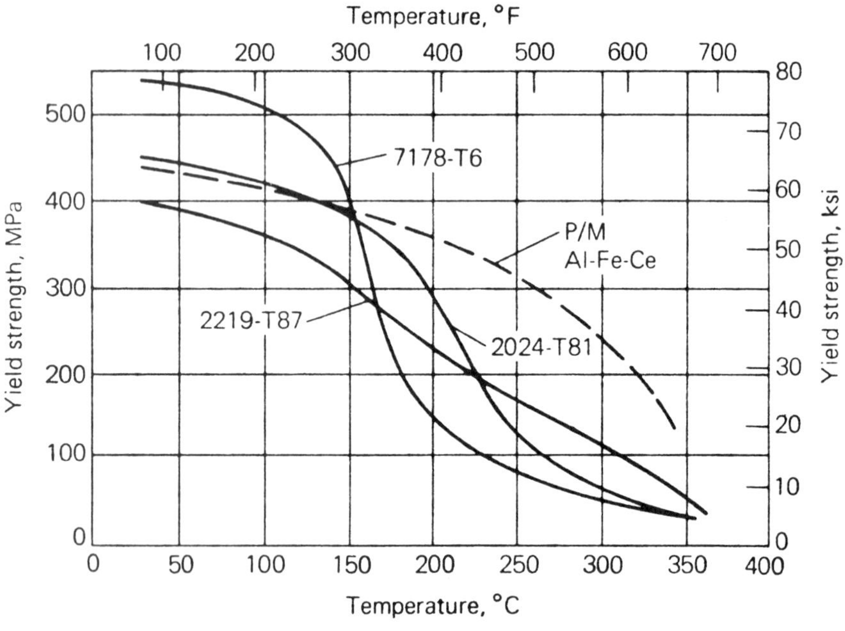

FIG. 8. Yield strength versus temperature for PM Al-Fe-Ce compared to ingot-metallurgy alloys 2219-T87, 2024-T81, and 7178-T6 following 100 hours exposure at the temperature indicated.

apparent that at temperatures above 150°C, the disepersion-strengthened PM alloy exhibits better strength. In addition, the room-temperature properties of the dispersion-strengthened Al-Fe-Ce alloy are retained after exposure to elevated temperatures whereas conventional precipitation strengthened alloys will lose much of their strength after exposure to elevated temperatures. In addition to the Al-Fe-Ce alloy by Alcoa, several other alloys were under development (e.g., Al-Fe-Mo by Pratt & Whitney, Al-Fe-V-Si by Allied Signal, and Al-Cr-Mn-Zr by Alcan).

B. Elevated-Temperature Alloys—Development Efforts

Based on the earlier composition and powder-processing studies, the development of high-temperature aluminum alloys narrowed and matured to a number of candidate alloy systems: Al-Fe-Ce, Al-Fe-Mo-V, Al-Fe-Zr-V, Al-Fe-V-Si, and Al-Cr-Zr-Mn. Table XI compares strength elongation and modulus for the Al-8Fe, Al-8Fe-4Ce, Al-8Fe-2Mo, and two Allied alloys and an Alcan high-temperature alloy. Alternative approaches for production of rapidly solidified alloys consider atomization, melt spinning, splat quenching, and rotating-disc techniques. The eventual successful

TABLE XI
PROPERTIES OF DISPERSION-STRENGTHENED PM ALLOYS

		YS	UTS		E	
Alloy	Temp. °C	MPa (ksi)	MPa (ksi)	EL %	GPa	(msi)
Al-8Fe	RT	504 (73.1)	572 (83.0)	5.0	81.4	(11.8)
	232	260 (37.7)	280 (40.6)	9.6	55.2	(8.0)
	343	148 (21.4)	172 (24.9)	12.3	40.0	(5.8)
Al-8Fe-4Ce	RT	445 (64.5)	550 (79.8)	9.0	78.1	(11.3)
	232	367 (53.3)	395 (57.3)	6.0	—	—
	343	152 (22.0)	176 (25.5)	7.0	—	—
Al-8Fe-2Mo	RT	472 (68.4)	492 (71.3)	6.7	86.2	(12.5)
	204	427 (62.0)	452 (65.5)	7.0	79.3	(11.5)
	343	200 (29.0)	222 (32.2)	10.0	75.9	(11.0)
Allied 1	RT	596 (86.4)	621 (88.9)	6.3	—	—
	232	425 (61.6)	433 (62.8)	6.8	—	—
	343	253 (36.7)	270 (39.2)	11.4	—	—
Allied 2	RT	632 (91.7)	645 (93.6)	8.7	—	—
	232	414 (60.0)	420 (60.9)	7.3	—	—
	343	240 (34.8)	252 (36.6)	10.2	—	—
Alcan	RT	540 (78.3)	580 (84.1)	12	82	(11.9)
	250	415 (60.2)	437 (63.4)	14	—	—
	300	270 (39.2)	300 (43.5)	15	—	—

Note: RT = Room Temperature (20°C)

alloy system must combine high performance with cost-effective processing. In each of these alloy systems, the ability of the dispersoids and intermetallic compounds to achieve a uniform and fine intermetallic particle distribution is essential for high-temperature strength of the candidate systems. Consequently, the development of a low-angle grain boundary structure that is pinned by fine dispersions can provide enhanced elevated-temperature stability.

As of the late 1980s, dispersion-strengthened aluminum PM alloys are being examined for use in aircraft and jet engines as a replacement for 2000 series aluminum or titanium alloys. Rainen and Ekvall [*70*] determined that Al-Fe-Ce and Al-Fe-V-Si sheet show potential for significant weight and cost savings. Figure 9 plots the mechanical properties of Al-7.1Fe-6.0Ce (Alcoa designation CZ42) and Allied's Al-8Fe-1V-2Si (e.g., FVS0812) on a specific property basis as a function of test temperature and exposure. In addition, if a comparison is made of retained strength of the aforementioned alloys, the stability of FVS0812 and CZ42 are shown to be superior to that of 2090, 2024, and 2219 ingot-metallurgy alloys (Fig. 10). Furthermore, the dispersion-strengthened PM alloys exhibit secondary properties (e.g., fracture toughness and crack propagation rates) that are equivalent to those of 7075-T6 as shown in Figs. 11 and 12.

In an evaluation of aluminum PM alloys, it is useful to compare the performance with not only ingot-metallurgy alloys but with other isotropic metals, metal and polymer-matrix composites. On this basis, Allied's FVS1212 (Al-12.4 wt % Fe-1.2 wt % V and 2.3 wt % Si) is compared

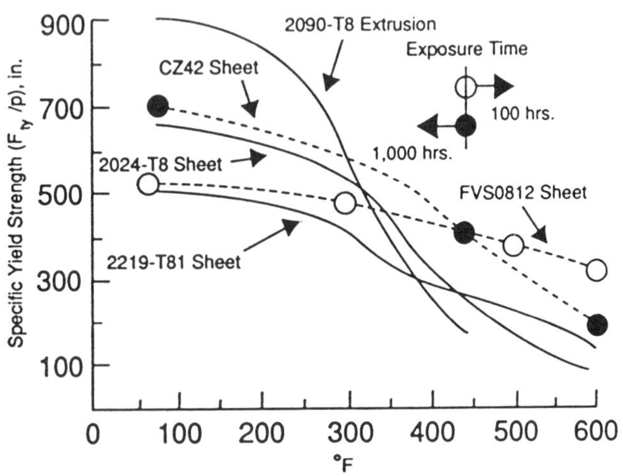

FIG. 9. Comparison of elevated-temperature PM aluminum alloy (FVS0812 and CZ42 and 2090-T8) as a function of thermal exposure test temperature.

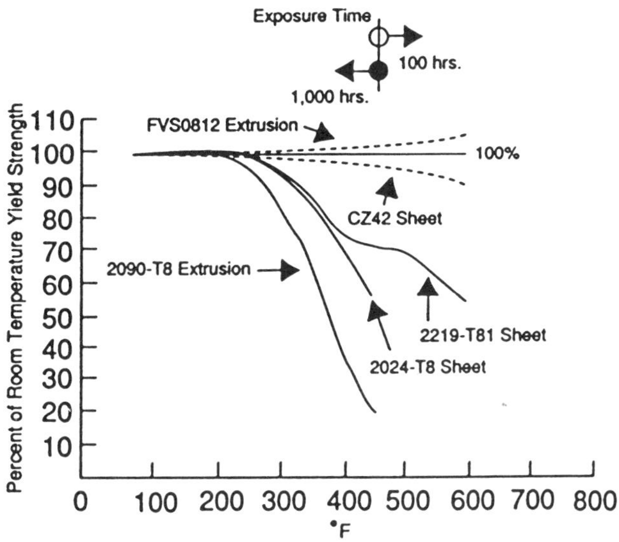

FIG. 10. Elevated-temperature stability of CZ42 and FVS0812 as a percentage of room-temperature strength.

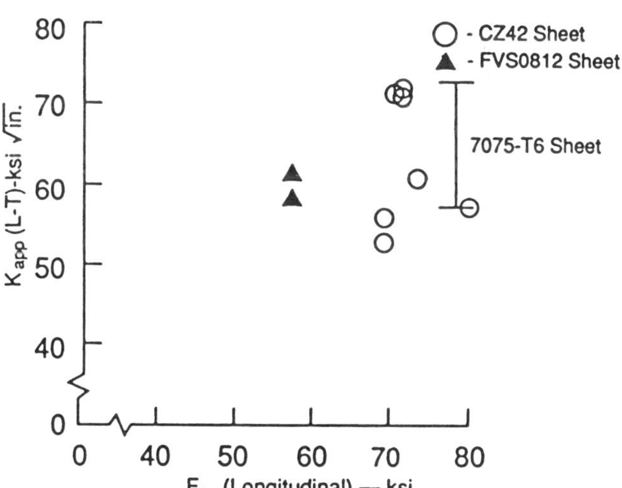

FIG. 11. A comparison of fracture toughness versus longitudinal yield strength for FVS0812, CZ42, and 7075 sheet. [73]

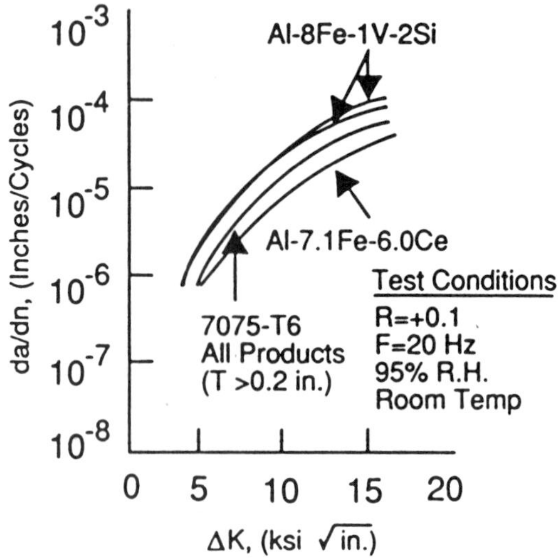

FIG. 12. Fatigue crack propagation da/dN versus ΔK of thin extrusions. [73]

against a particulate SiC-aluminum metal-matrix composite, 2618-T61, and a graphite-polymide-resin–matrix composite. The ulimate tensile strength and specific modulus versus temperature are shown in Figs. 13 and 14, respectively. Unidirectional polymer-matrix composites and some quasi-isotropic composites generally excel with regard to specific strength and

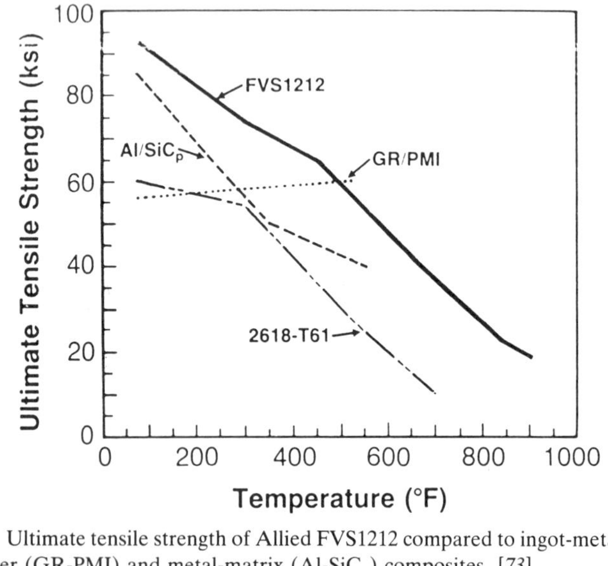

FIG. 13. Ultimate tensile strength of Allied FVS1212 compared to ingot-metallurgy (2618-T61) polymer (GR-PMI) and metal-matrix (Al-SiC_p) composites. [73]

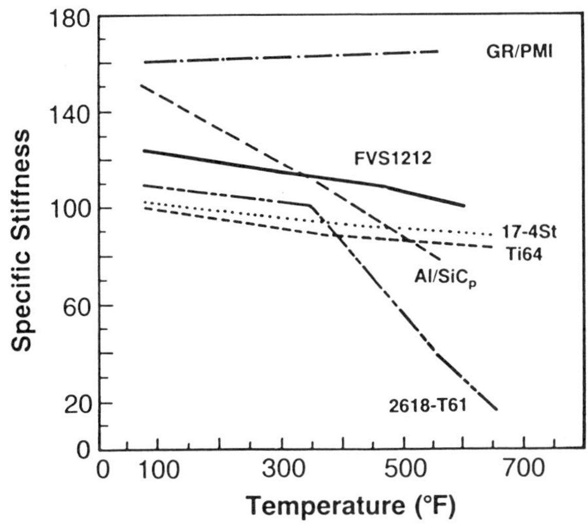

FIG. 14. Specific stiffness versus temperature of Allied FVS1212 compared to ingot metallurgy (2618-T61), polymer (GR-PMI), and metal-matrix composites (Al-SiC_p). [73]

specific modulus. However, the properties must be considered on a cost-effective basis and in terms of the life-cycle cost of the product.

In summary, the elevated-temperature aluminum alloys utilize a dispersion of aluminides and silicides ranging in volume fraction from 5 to 40% in an aluminum matrix for strengthening. The size, distribution, and coarsening response for the intermetallic phases provide for excellent elevated-temperature strength with a significant increase in modulus. Important issues to resolve are the processing and mechanical-property reliability, economy of scale of billet production coupled with the ability to produce complex structural shapes via forging, extrusion, or rolling superplastic forming or forging approaches. Issues that also pose challenges to these high-temperature structural-alloy systems are the elevated-temperature coarsening response, fatigue crack initiation, and propagation response coupled with the oxidation and stress corrosion performance.

IV. Hybrid Powder-Metallurgy Alloy Systems

A. *Mechanically Alloyed Aluminum Powder Alloys*

With the development of high-strength 7000 series alloys and dispersion-strengthened elevated-temperature alloys, it would be desirable to further strengthen these alloys by superimposing an additional constitutent, (e.g.

an oxide dispersoid) to improve elevated-temperature response, or a high-volume fraction of a whisker or particulate carbide to increase the strength and modulus. To this end, several hybrid aluminum powder-metallurgy alloys have been developed. These include mechanically alloyed elemental and pre-alloyed aluminum alloy powder systems; the mechanical alloying of aluminum powders and carbides; aluminum metal-matrix composites developed by mixing and consolidating aluminum powder and high-modulus, low-density micron-and submicron-sized carbides (e.g. SiC, TiC, B_4C); and direct-spray deposition of aluminum powders with refractory reinforcements (e.g., Osprey). In each of these processes, the additional constituents provide for desirable and detrimental changes to the alloy's mechanical and physical properties. With the addition of refractory SiC particulates and/or whiskers, the strength, modulus, and wear resistance increase with a reduction of ductility and fracture toughness. In regard to physical and electrical properties, a reduction is experienced with respect to the coefficient of thermal expansion, electrical and thermal conductivity. Nevertheless, significant increases in the modulus can be achieved (e.g., 50%) with the addition of high-volume fraction of carbides in an aluminum matrix. The nominal upper limit of carbide reinforcements is a 20 volume percent addition in order to achieve a reasonable balance of strength, modulus, fracture toughness, and ductility.

The addition of oxide and carbide dispersions via mechanical alloying (MA) has been developed for 2000 and 5000 series alloys, and Novamet has introduced N9021 (Table VII). In the 5000 series alloys, mechanical alloying has produced Al-4 wt % Mg alloys designated as IN9051 and IN9052. The strengthening effects are hierarchical, namely, combining multiple strengthening mechanisms. (i.e. a microstructure that has a Mg solid solution; a submicron grain size, dispersions of MgO, Al_2O_3, and Al_4C_3; and the low-angle textured microstructure) developed from the attrition and extrusion processes. These alloys were developed from the attrition of mixed elemental powders with the additions of process control agents during the mechanical alloying process to provide for the additions of carbon and facilitate the controlled oxidation of the alloy constituents. The benefits include improved ambient strength and good levels of toughness, a stable microstructure, ductility, and stress corrosion resistance (Figs. 15 and 16). In addition, an improvement in the elevated-temperature creep response is achieved (Figs. 17 and 18).

Efforts have also been undertaken to improve the elevated-temperature properties of powder-metallurgy aluminum alloys by combining rapid solidification and mechanical alloying in high-strength, corrosion-resistant, and high-temperature aluminum alloy systems, e.g., Al-Zn-Mg, Al-Cu, Al-Mg, Al-Fe-Ni, Al-Fe-Ce, and Al-Ti. By superimposing the dispersion

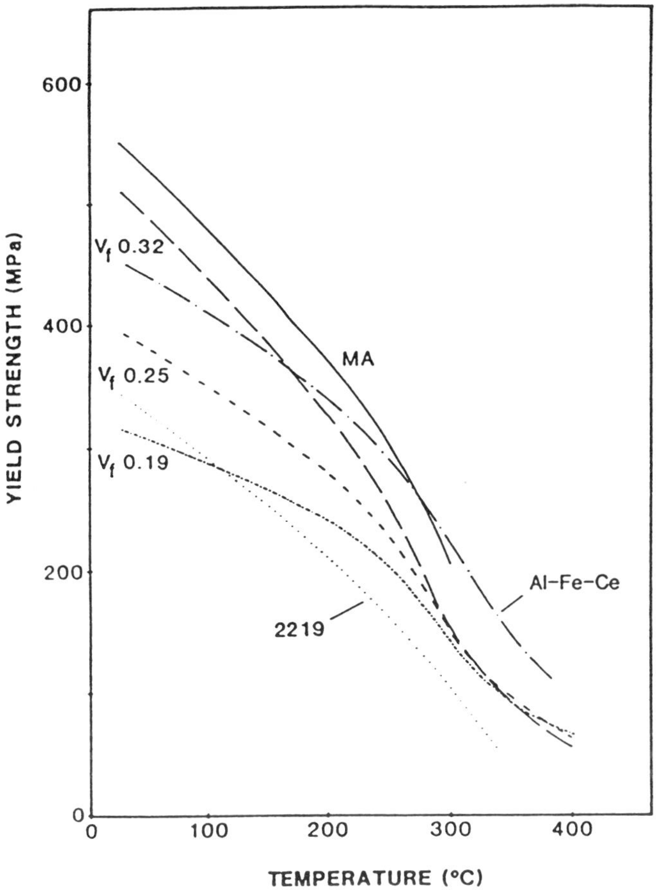

FIG. 15. Elevated-temperature yield-strength response for Al-Fe-Ni @ V_f = 0.19, 0.25, and 0.32 of $FeNiAl_9$, 2219, Al-Fe-Ce and Al-Fe-Ni @ V_f = 0.19 which has been mechanically alloyed (MA). From (Ezz et al, 1986)

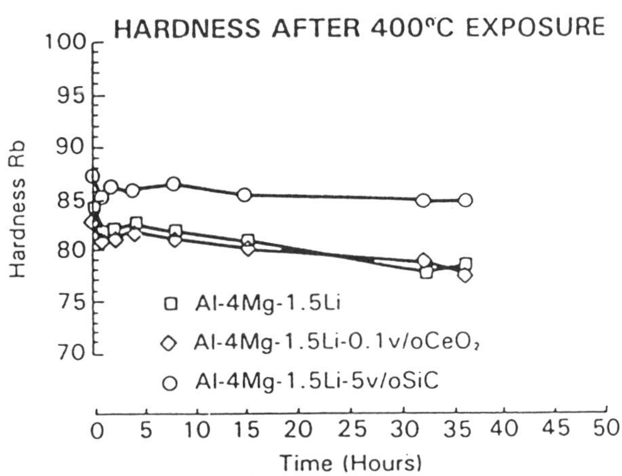

FIG. 16. Hardness of various Al-Mg alloys following exposures at 400☆ C. [*72,73*]

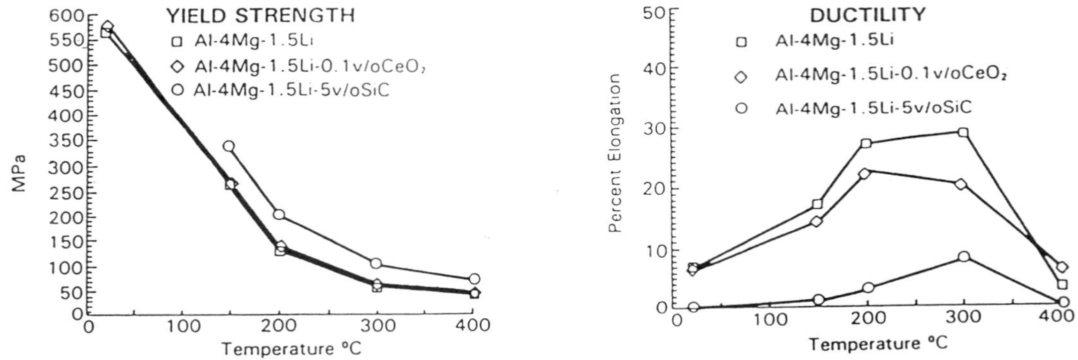

FIG. 17. Yield strength and ductility response versus test temperature for various mechanically alloyed Al-Mg alloys. [*72*,*73*]

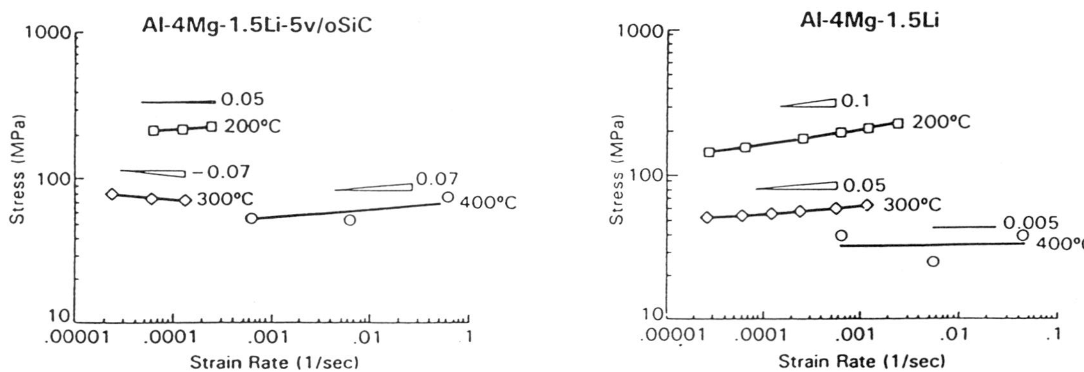

FIG. 18. Creep response at various temperatures for SiC-reinforced and unreinforced mechanically alloyed Al-Mg-Li Alloys. [*72*,*73*]

strengthening provided by the fine, stable oxides and carbides introduced by mechanical alloying and the intermetallic dispersoids provided in the rapidly solidified aluminum powder, the alloy design provides for an improved ambient and elevated-temperature strength and room-temperature strength after elevated-temperature exposure. Furthermore, mechanical alloying alters the texture by inducing the formation of stabilized low-angle boundaries and grain boundaries pinned by oxides and carbides. Static recrystallization, grain growth, and intermetallic coarsening are, therefore, inhibited. A significant increase in microstructural stability at elevated temperature, compared to nonmechanically alloyed material, has been reported in pre-alloyed and mechanically alloyed Al-Fe-Ce and Al-Fe-Ni alloys (Fig. 15). In addition, a large improvement in strength in the 25–400°C temperature range as well as enhancement in creep resistance compared to nonmechanically alloyed material has been reported (Fig. 18).

In Al-Fe-Ni, yield strength increases of 77% at 25°C and 35% at 400°C were achieved by the mechanical alloying. The hardness levels remained unchanged after one hour of exposure at up to 500°C. At temperatures over 500°C, the sharp drop in hardness indicated considerable coarsening of $Al_9(FeNi)$. Furthermore, exposing the MA material at 455°C for 624 hours resulted in a hardness drop of only 10%, compared to 70% for the non-MA alloy. A key question is the mechanism by which fine dispersion inhibits the coarsening of the intermetallic particles at elevated temperatures, e.g., 450°C, as well as the elevated-temperature deformation mechanism at temperatures of 200 to 300°C. The initial studies were extended to pre-alloyed Al-Fe-Ce and Al-Ti PM powders to demonstrate that the stability due to MA can be generalized to increase high-temperature strength and creep resistance in other aluminum systems. In addition, Dispal (i.e., dispersion aluminum) alloys have been developed based on mechanical alloying with carbon contents ranging from 0 to 3.0 wt % C and 0.4 to 2.5 wt % 0. Based upon the level of oxygen and carbon, a range of properties can be achieved (Figs. 19 and 20). The stress-rupture response of the Dispal alloys are compared with various titanium alloys in Fig. 20.

For structural applications, it is anticipated that the high-temperature strength and creep resistance must be balanced by the alloy toughness and ductility. The achievement of acceptable properties requires the appropriate alloy chemistry and powder-processing history, which, in turn, dictates the aluminide size and dispersoid volume fraction. These results support the assertion that more thermally stable aluminide phases and microstructures with oxide and carbide dispersions will lead to promising high-temperature aluminum-base alloys.

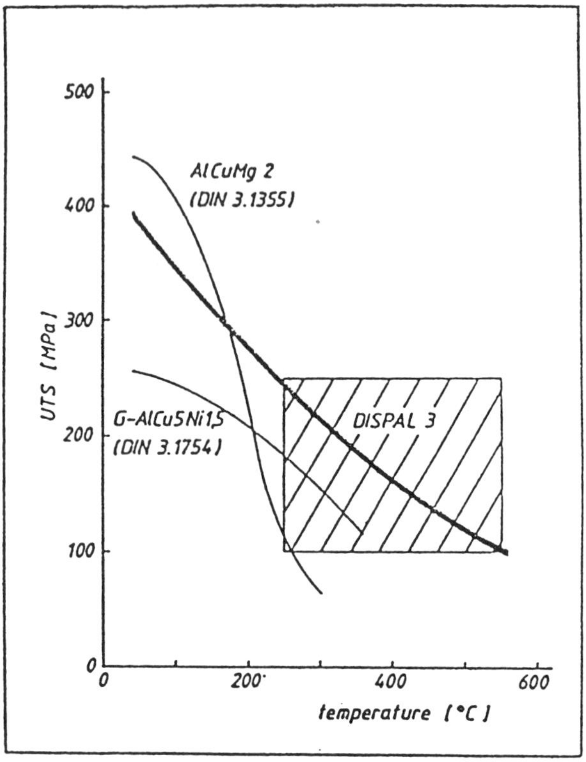

Fig. 19. Elevated-temperature response of Al-Cu alloys versus Dispal 3.

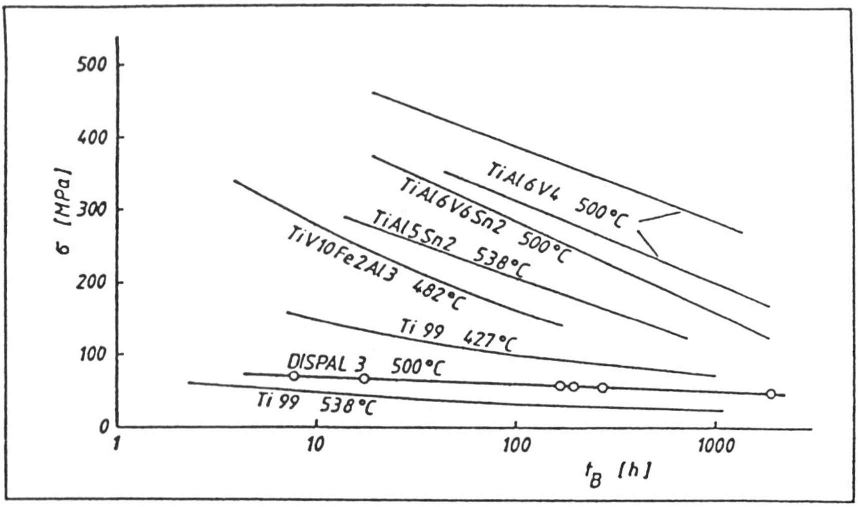

Fig. 20. Elevated-temperature stress-rupture response for various alloys.

B. Aluminum PM Composite Alloys

Advances in fiber, whisker, and particulate chemistry and processing have led to programs to reinforce aluminum-base materials. Specifically, the addition of fiber, whiskers, or particulate SiC produces a composite material with excellent specific mechanical properties. Techniques can include mechanical alloying, powder blending, and consolidation or spray deposition (Figs. 21, 22, and 23). The strength of fiber-, whisker-, and particulate-reinforced aluminum clearly depends upon the fiber and matrix properties, whisker volume fraction, aspect ratio, and orientation. In addition, the excellent fiber- or whisker-reinforcement properties must be maintained during the consolidation and extrusion process in order to achieve the full potential of metal-matrix composites. If a reduction of fiber length or surface damage occurs, the subsequent metal-matrix composite will suffer a natural strength reduction. These factors have been examined analytically and experimentally to define structurally competitive properties of reinforced aluminum-base material systems. A balance of properties must be achieved in these reinforced alloys. Specifically, reinforcements, (e.g., SiC) will enhance modulus, strength, and creep resistance

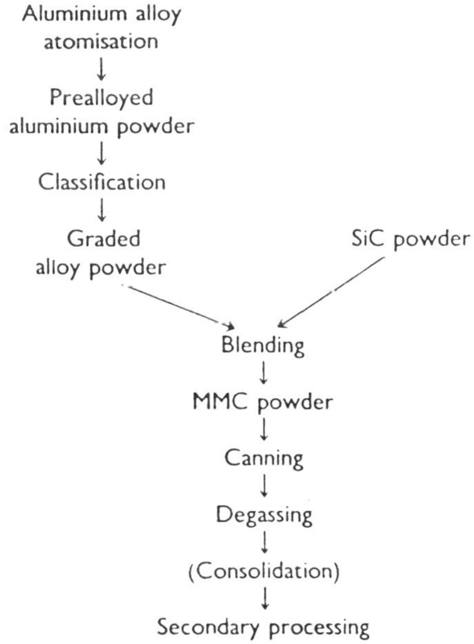

FIG. 21. Powder-blending route for metal-matrix composites.

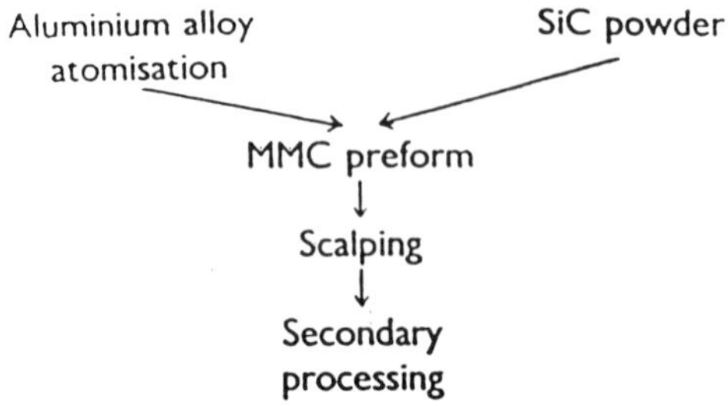

FIG. 22. Spray-deposition route for metal-matrix composites.

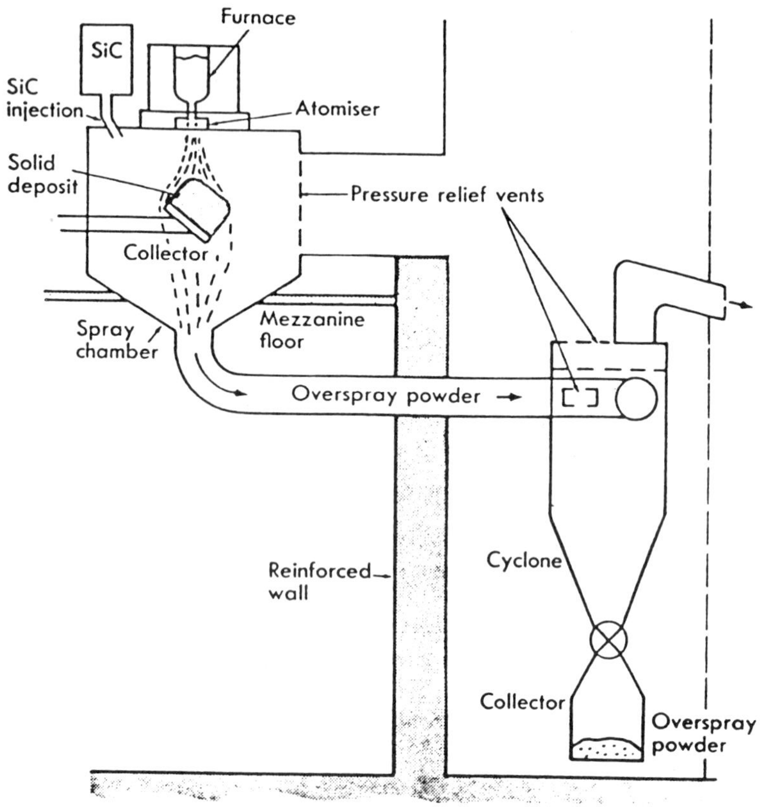

FIG. 23. Schematic diagram of Osprey fabrication of metal-matrix composites.

while reducing ductility, fracture toughness, and notched tensile properties. A compromise in the volume fraction, size, and distribution of the reinforcement phase is required in order to achieve a useful structural alloy. Figure 24 compare 8090 alloys and 8090 with SiC additions formed by spray deposition. In general, increases in the modulus and yield strength

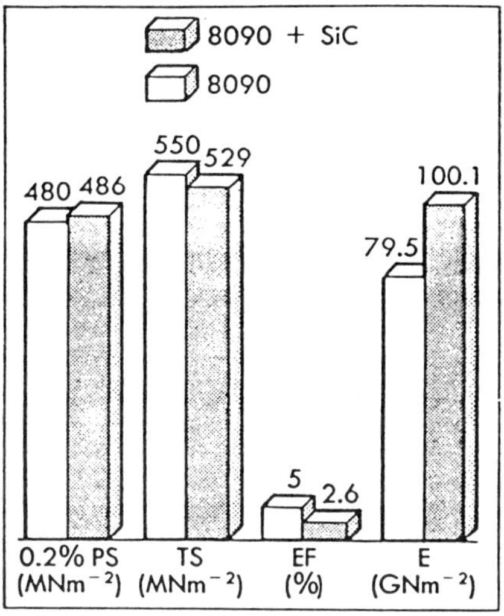

FIG. 24. Longitudinal yield strength, tensile strength, percent elongation, and modulus for Al-Li (8090) alloy and Al-Li (8090) with SiC additions.

occur with a marked reduction in failure strain. Furthermore, when carbide or oxide additions are combined by mechanical alloying, similar increases in strength and a reduction in ductility are noted (Figs. 17 and 18). The application of aluminum-metal matrix composites via the powder-metallurgy route or combined by mechanical alloying or spray deposition offers a high-modulus, high-temperature material. However it is not clear whether the additional processing costs will limit acceptance in the commercial marketplace.

V. Summary

Aluminum powder metallurgy offers a processing technology with great potential. Progress to date includes the development of dispersion-strengthened, precipitation-hardened, and metal-matrix composite PM alloys. These alloys have been shown to exhibit superior combinations of mechanical properties (such as strength, fracture toughness, elastic modulus, and fatigue and corrosion resistance) compared to ingot-metallurgy alloys. As of the late 1980s, the high cost of wrought PM components appears to be the major barrier hindering acceptance and growth of aluminum powder-metallurgy alloys as structural materials. Nevertheless, the technology has developed new alloys and processing concepts for aluminum alloys with superior ambient- and elevated-temperature, which can be utilized in future structural applications to take advantage of their higher performance.

References

1. Irmann, R. (1949). *Technische Rundschau,* **41**, 36, p. 19.
2. Irmann, R. (1952). *Metallurgia*, **46**, p. 125.
3. Lyle, J.P. Jr. (December 1952). *Metal Progress*. p. 109.
4. Gregory, E., and Grant, N.J. (1954). *Trans. AIME*, **200**, p. 247.
5. Towner, R.J. (1958) *Metal Progress*, **73**, *no. 5*, p. 70.
6. Towner, R.J., and Lyle, J.P. Jr. (1960). Alcoa Internal Report.
7. Roberts, S.G. (1957). WADC Report no. 56–481.
8. Jones, H. (1978). *Aluminum*, **54**, p. 274.
9. Pickens, J.R. (1981). *J. Mater. Sci.*, **16**, p. 1437.
10. Palmer, I.G., and Tietz, T.E. (1981). In *Advances in Powder Technology*, ed. G.Y. Chin. American Society for Metals, Metals Park, Ohio, pp. 189–224.
11. Singer, R.F. (1986). *Powder Met. Intl.*, **18**, no. 2, pp. 79–83.
12. Froes, F.H., Kim, Y-W., and Hehmann, F. (1987). *J. of Metals*, **39**, no. 8, pp. 14–21.
13. Spear, R.E., and Gardner, E.R. (1963). *Trans. Am. Foundrymens Soc.*, **71**, pp. 209–215.
14. Dean, W.A., and Spear, R.E. (1966). "*Proceedings of the 12th Army Materials Research Conference*". Syracuse University Press, Syracuse, N.Y., p. 268.
15. Flemings, M.C. (1966). "In *Proceedings of the 12th Army Materials Research Conference.*" Syracuse University Press, Syracuse, N.Y., p. 235.
16. Matyja H., Giessen, B.G., and Grant, N.J. (1968). *Journal Institute of Metals*, **96**, pp. 30–32.
17. Duwez, P. (1967). *Trans. ASM*, **60**, p. 607.
18. Mehrabian, R. (1982). *International Metals Reviews*. **27**, no. 4, pp. 185–208.
19. Grant, N.J. (1982). In *High Strength Powder Metallurgy Aluminum Alloys*, eds. M.J. Koczak and G.J. Hildeman. TMS-AIME, Warrendale, Penn., p. 3.
20. Schelling, R.D., and Donachie, S.J. (1983). *Metal Powder Report*, **38**, no. 10.
21. Billman, F.R., and Jatkar, A.D. (1986). AFWAL Contract F33615-82-C-5063.
22. Roberts, S.G. (1961A). "In *Powder Metallurgy*," ed. W. Leszynski. Interscience, p. 799.
23. Towner, R.J., and Lyle, J.P. Jr (1962). ASTIA Report AD 289526.
24. Haar, A.P., and Lyle, J.P. Jr. (1966). Contract no. DA-36-034-ORD-3559RD. Report no. AD 487,764.
25. Cebulak, W.S., and Traux, D.J. (1971). Contract DAAA 25-70-C-0358.
26. Cebulak, W.S. (1973). Contract DAAA 25-72-C-0593.
27. Roberts, S.G. (1973). Contract DAAG 46-73-C-0040.
28. Lyle, J.P. Jr., and Cebulak, W.S. (1975). *Met. Trans.*, **6A**, p. 685.
29. Lyle J.P. Jr and Towner, R.J., (1960) U.S. Patent 2,963,780.
30. Roberts, S.G. (1978). U.S. Patent 4, 104,061.
31. Voss, D.P. (1979). AFOSR Grant 77-3440, Report no. DFVLR-FB79-34.
32. Kuli, J.C., and Billman, F.R. (1982). AFWAL TR-82-4071.
33. Singleton, O.R. (1984). AFWAL TR-83-4142.
34. Morgan, J.T., Gegel, H.L., Doraivelu, S.M., Matson, L.E., Martorell, I.A., and Thomas, J.F. Jr. (1982). In *High Strength Powder Metallurgy Aluminum Alloys*, eds. M.J. Koczak and G.J. Hildeman. AIME, pp. 193–206. Warrendale, Penn.
35. Kim, Y-W., Griffith, W.M., and Froes, F.H. (1983). *ASM Metals/Materials Technology Series*, no. 8305-048. American Society for Metals, Metals Park, Ohio.
36. Ackerman, L., Guillemin, I., Lalauze, R., and Pijolat, C. (1985). In *High Strength Powder Metallurgy Aluminum Alloys*, II, eds. G.J. Hildeman and M.J. Koczak. AIME, pp. 175–191.

37. Cebulak, W.S. (1977). Contract DAA 25-72-C-0593. Report FA-TR-76067.
38. Daugherty, T.S. (1964). *Journal of Metals*, **16**, pp. 827–830.
39. Ro, D.H. (1981). AFWAL Contract no. F33615-80-C-5161.
40. Ping, S.W. (1986). In *Rapidly Solidified Powder Aluminum Alloys*, eds. M.E. Fine and E.A. Starke, Jr. ASTM STP890. ASTM, Philadelphia, pp. 369–380.
41. Hildeman, G.J., Labarre, L.C., Hafeez, A. and Angers, L.M. (1985). In *High Strength Powder Metallurgy Aluminum Alloys*, II. TMS-AIME, Warrendale, Penn., pp. 25–43.
42. Hart, R.M. (1981). Alcoa Green Letter.
43. Bretz, P.E. (1984). Alcoa Technical Brief.
44. Langenbeck, S.L. (1982). In *High Strength Powder Metallurgy Aluminum Alloys*, eds. M.J. Koczak and G.J. Hildeman. TMS-AIME, Warrendale, Penn. pp. 87–105.
45. Scarich, G.V. (1984). IRAD Spectrum Fatigue Data. Northrop Corporation, Hawthorne, California.
46. Lyle, J.P., Jr., and Towner, R.J. (1970). U.S. Patent 3,544,932.
47. Lebo, N., and Grant, N.J. (1974). *Met. Trans.*, **5**, p. 1547.
48. Ping, S.W. (April 1985). Kaiser Aluminum and Chemical Corporation Data Sheet.
49. Paris, H.G. (1986). US Patent, 4,629,505.
50. Paris, H.G., and Chellman, D.J. (1986). In *Rapidly Solidified Powder Aluminum Alloys*, eds. M.E. Fine and E.A. Starke, Jr. ASTM STP890. ASTM, Philadelphia, pp. 527–544.
51. Bomford, M.J., and Benjamin, J.S. (1974). U.S. Patent 3,816,080.
52. Novamet Aluminum (1984). Data Sheet.
53. Hardy, J. (1940). Unpublished Alcoa Research Laboratory data.
54. Cremer, G.D., and Cordiano, J.J. (1943). *Trans. AIME*, **152**, pp. 152–1162.
55. Storchheim, S. (1962). *Progress in Powder Metallurgy*, **18**, pp. 124–130.
56. Dudas, J.H., and Dean, W.A. (April 1969). *International Journal of Powder Metallurgy*, 5.
57. Pickens, J.R., Schelleng, R.D., Donachie, S.J., and Nichol, T.J. (1981). U.S. Patent 4,292,079.
58. Dudzinski, N., Murray, J.R., Mott, B.W., and Chalmers, B. (1947). *Journal Inst. of Metals*, **74**, pp. 291–314.
59. Dean, W.A. (1967). In *Aluminum, Vol. I: Properties, Physical Metallurgy and Phase Diagrams*, ed. K.R. Van Horn. American Society for Metals, Metals Park, Ohio, p. 168.
60. Lyle, J.P., Jr., Towner, R.J., and Vruggink, J.E. (1960). Alcoa Internal Report.
61. Towner R.J., and Lyle J.P. Jr. (1961). U.S. Patent 2,966,736.
62. Quist, W.E., and Lewis, R.E. (1986). In *Rapidly Solidified Powder Metal Aluminum Alloys*, eds. M.E. Fine and E.A. Starke, Jr. ASTM STP890. ASTM, Philadelphia, pp. 7–38.
63. Sankaran, K.K., and Grant, N.J. (1981). In *Aluminum-Lithium Alloys*, eds. T.H. Sanders, Jr., and E.A. Starke, Jr. TMS-AIME, Warrendale, Penn. p. 101.
64. Gysler, A., Crooks, R., and Starke, Jr. E.A. (1981). In *Aluminum-Lithium Alloys*, eds. T.H. Sanders, Jr., and E.A. Starke, Jr. TMS-AIME, Warrendale, Penn. p. 263.
65. Palmer, I.G., Lewis, R.E., Crooks, D.D., Starke, E.A. Jr., and Crooks, R.E. In *Aluminum-Lithium Alloys*, **II**, eds. T.H. Sanders, Jr., and E.A. Starke, Jr. TMS-AIME, Warrendale, Penn. p. 91.
66. Jones, H. (1969). *Mater. Sci. Eng.*, **5**, p. 1.
67. Pontikakas, I., and Jones, H. (1982). *Metal Science*, **16**, 27.
68. Sanders, R.E. Jr., and Hildeman, G.J. (1981). AFWAL-TR-81-4076.
69. Hildeman, G.J., and Sanders R.E. Jr. (1983). U.S. Patent 4,379,719.
70. Rainen, R.A., and Ekvall, J.C. (1988). *J. of Metals*, **40**, no. 5, p. 18.
71. Roberts, S.G. (1969). U.S. Patent 3,462,248.
72. Gilman, P.S., Brooks, J.W., and Bridges, P.J. (1985). "Proceedings of the Third International Conference on Aluminum Lithium," Oxford, England.

73. Allied Technical Data Sheet, Allied Signal Corporation (1988) Morristown, N.J.
74. Lyle, J.R., and Granger, D.A. (1985) *Ullman's Encyclopedia of Industrial Chemistry*, vol. 1A, VCH Publishers, pp. 481–528.

Further Reading

Adam, C.M. (1981). U.K. Patent Application GB 2088409A.
Adam, C.M., Bourdeau, R.G., Broch, J.W., and Cox, A.R. (1981). Final Report, AFWAL Contract no. F33615-76-C-5136.
Ansell, G.S., and Weertman, J. (1959). AIME Trans., **215**, p. 838.
Ansell, G.S. (1966). In *Oxide Dispersion Strengthening*, eds. G.S. Ansell, T.C. Cooper, and F.V. Lenel. AIME, p. 61.
Ardell, A.J., and Nicholson, R.B. (1966). *J. Phys. Chem. Solids*, **27**, p. 1793.
Ardell, A.J. (1972A). *Acta. Met.*, **20**, p. 61.
Ardell, A.J. (1972B). *Acta. Met.*, **20**, p. 601.
Arnold, V. (1984). In *PM Aerospace Alloys*. MPR Publishing **1**, p. 31–1.
Benjamin, J.S., and Bamford, M.J. (1977). *Met. Trans.*, **8A**, p. 1301.
Bloch, E.A. (1961). *Met. Rev.*, **6**, no. 22, p. 193.
Brailsford, A.D., and Wynblatt, P. (1979). *Acta. Met.*, **27**, p. 489.
Bridenbaugh, P.R., Cebulak, W.S., Billman, F.R., and Hildeman, G.J. (1985). *Light Metal Age*, **43**, nos 9, 10, p. 18.
Bridges, P.J. In *PM Aerospace Alloys*. MPR Publishing **1**, p. 30–1.
Cebulak, W.S. (1979). U.S. Patent 4,135,922.
Cebulak, W.S., Johnson, E.W., and Markus, H. (1976). *Met. Eng. Quart.*, p. 37.
Chellman, D. J., and Ardell, A. J. (1974). *Acta. Met.*, **22**, p. 577.
Coble, R. L. (1963). *J. of Appl. Phys.*, **34**, p. 1679.
Courtney, T.H. (1982). In Solid-Solid Phase Transformations, eds. H.I. Aaronson, D.E. Laughlin, R.F. Sekerka, and C.M. Wayman. TMS AIME, Warrendale, Penn., p. 1057.
Davies, G.K.L., Nash, P., and Stevens, R.N. (1980). *Acta. Met.*, **28**, p.179.
Decker, R.F. (1973). *Met. Trans.*, **4**, p. 2495.
Divecha, A.P., and Fishman, S.G. (1979) *Proc. 3rd Int. Conf. of Composite Materials*, **3**, p. 351, AIME, Warrendale, Penn.
Divecha, A.P., Fishman, S.G. and Karmarkar S.D., (1981). *Journal of Metals* p. 12.
Dixon, C.F., and Skelly, H.M. (1965). *International Journal of Powder Met.*, **1**, p. 28.
Donachie, S.J., and Gilman, P.S., (1983). In *Aluminum-Lithium Alloys II*, eds. E.A. Starke, Jr., and T.H. Sandes, Jr. AIME, p. 507.
Doyle, W.M. (1969). *Aircraft Engineering*, **41**, no. 11, p. 11.
Duwez, P., and Willens, M. (1963). *TMS-AIME*, **227**, p. 362.
Erich, D.L. (1980). AFML-TR-79-4210.
Erich, D.L., and Donachie, S.J. (February 1982). *Metal Progress*, p. 22.
Ezz, S., Koczak, M.J., Lawley, A., and Premkumar, M.K. (1985). In *High Strength Powder Metallurgy Aluminum Alloys*, II, eds. G.J. Hildeman and M.J. Koczak. TMS-AIME, Warrendale, Penn., p. 287.
Frazier, W.F., and Koczak, M.J. (1985. In *High Strength Powder Metallurgy Aluminum Alloys* II, eds. G.J. Hildeman and M.J. Koczak. TMS-AIME, Warrendale, Penn., p. 353.
Froes, F. H., and Pickens, J.R. (1984). *J. of Metals*, **36**, no. 1, p. 14.

Gemmell, G.D., and Grant, N.J. (1957). *J. of Metals*, p. 417.
Gibbs, G.B. (1967). *Mater. Sci. Engr.*, **2**, p. 269.
Grant, N.J., and Preston, O. (1957). *Trans. AIME*, **209**, p. 349.
Greenwood, G.W. (1956). *Acta. Met.*, **4**, p. 234.
Griffith, W.M., Sanders, R.E. Jr., and Hildeman, G.J. (1982). In *High Strength Powder Metallurgy Aluminum Alloys*, II, eds. G.J. Hildeman and M.J. Koczak. TMS-AIME, Warrendale, Penn. p. 209.
Herring, C. (1950). *J. of Appl. Phys.*, **21**, p. 437.
Hirchner, H.O.K. (1971). *Met. Trans.*, **2**, p. 2861.
Hughes, I.R., Marshall, G.J., and Miller, W.S. (1985). In *Rapidly Quenched Metals*. eds. S. Steeb and H. Warlimont. p. 1743.
Humphreys, F.J. (1983). In *Deformation of Multi-Phase and Particle Containing Materials, Proceedings of 4th RISO International Symposium on Metallurgy and Materials Science*, p. 41. Elsevier Science Publishers, Holland.
Inco Alloys (1988). Inco Map Alloy-AL-905XL. International Publication no. IAI-50. Huntington, West Virginia.
Jatkar, A.D., Gilman, P.S., and Benn R.C. (1986). U.S. Patent 4,624,705.
Koczak, M.J., and Topolski, M.J. (1982). In *High Strength Powder Metallurgy Aluminum Alloys*, **I**, eds. M.J. Koczak and G.J. Hildeman. TMS AIME, Warrendale, Penn.,
Lagneborg, R. (1972). *Int. Met. Rev.*, **17**, p.130.
Langenbeck, S.L., Rainen, R.A., Simenz, R.F., Wald, G.G., Himmel, E.J., Cox, J.M., Gillette, T., Hildeman, G.J., Denzer, D.K., Rioja, R.J., Cho, C.W., Malcolm, R.C., Simon, J.W., Gorman, K., Rea, R.S., Hilliard, D.A., Adam, C.M., and Bjeletech, J.G. (1986). AFWAL-TR-86-4027.
Lewis, R.E., and Starke, E.A. Jr., (1985). In *Mechanical Behavior or Rapidly Solidified Materials*, eds. S.M.L. Sastry and B.A. McDonald. p. 131.
Lewis, R.E., Palmer, I.G., Crooks, D.D., Paris, H.G., Billman, F.R., Starke, E.A. Jr., Sanders, T.H., Wald, G., Himmel, E., and Ekvall, J.C. (1986). AFWAL-TR-86-4020.
Liddiard, P.D. (1984). In *PM Aerospace Alloys*. MPR Publishing 1, 26–1.
Lifshitz, I.M., and Sylozov, V.V. (1961). *J. Phys. Chem. Solids*, **19**, p. 35.
Lund, R. W., and Nix, W.D. (1976). *Acta. Met.*, **24**, p. 469.
Lyle, J.P. Jr., and Ennor, W.T. (1957). U.S. Patent 2,809,891.
Lyle, J.P. Jr., and Towner, R.J. (1970B). U.S. Patent 3,544,394.
MacDonald, N.F., and Ransley, C.E. (1954). In *Symposium on Powder Metallurgy*. Iron and Steel Institute, London, pp. 242–248.
Martin, J.W. (1980). In *Micromechanisms in Particle Hardened Alloys*, Cambridge University Press New York.
Martin, J.W., and Doherty, R.D. (1976). In *Stability of Microstructure in Metallic Systems*, Cambridge University Press New York.
McLean, D. (1962). Met. Rev., **7**, no. 28, p. 481.
Meschter, P.J., Lederich, R.J., and O'Neal, J.E. (1986). NASA Report no. 178145.
Milan, P.P. Jr. (1983). *J. of Metals*, **35**, p. 76.
Miller, W.S. (1984). European Patent Application no. 83304950.5.
Mukharjee, A.K. (1975). In *Treatise on Materials Science and Technology*, **Vol. 6**, ed. R.J. Arsenault, Academic Press, London, p. 164.
Mullendore, A.W., and Grant, N.J. (1984). *J. of Metals*, p. 973.
Murthy, K.L., Mohamed, F.A., and Dorn, J. E. (1972). *Acta Met.*, **20**, p. 1009.
Nabarro, F.R.N. (1948). "In *Proceedings of the Conference on the Strength of Solids", Bristol*. Physical Society, London, p. 75.
Oriani, R.A. (1964). *Acta. Met.*, **12**, p. 1399.
Ostwald, W. (1900). *Z. Phys. Chem.*, **34**, p. 495.

Ovecogulu, M.L., and Nix, W.D. (1982). In *High Strength Powder Metallurgy Aluminum Alloys*, eds. M.J. Koczak and G.J. Hildeman. TMS AIME, Warrendale, Penn.), p. 225.
Palmer, I.G., Lewis, R.E., and Crooks, D.D. (1982) In *High Strength Powder Metallurgy Aluminum Alloys*, eds. M.J. Koczak and G.J. Hildeman. TMS-AIME, Warrendale, Penn., pp. 369–390.
Perepezko, J.H., LeBeau, S.E., Mueller, B.A., and Hildeman, G.J. (1986). In *Rapidly Solidified Powder Aluminum Alloys*, eds. M.E. Fine and E.A. Starke, Jr. ASTM-STP 890. ASTM, Philadelphia, p. 118.
Premkumar, M.K. (1985). Ph. D. dissertation. Drexel University, Philadelphia.
Ray, R., Polk, D.E., and Giessen, B.C. (1982). U.S. Patent 4,347,076.
Read, P.J., Latimer, K.G., Reynolds, T.D.W., and Munson, D. (1975). U.S. Patent 3,899,820.
Roberts, S.G. (1961B). Contract no. DA-04-200-507-ORD-866, Report no. MSPR 61–69.
Ronald, T.M.F., and Griffith, W.M. (1984). In *PM Aerospace Alloys*. MPR Publishing 1, p. 25–1.
Scamans, G.M., and Rehal, A.S. (1979). *J. of Mater. Sci.*, **14**, p. 2459.
Sellars, C.M., and Petkovic-Luton, R.A. (1980). *Mater. Sci. Eng.*, **46**, p. 75.
Skinner, D.J., Chipko, P.A., and Okazaki, K. (1984). European Patent Application 84109894.0.
Skinner, D.J., Ker, R., Koczak, M.J., and Lawley, A. (1981). In *Modern Developments in Powder Metallurgy*, **Vol. 13**, eds. H.H. Hausner, H.W. Antes, and G.D. Smith. (MPIF, Princeton, N.J., p. 483.
Skinner, D.J., Okazaki, K., and Adam, C.M. (1984). In ASTM "*Symposium on Rapidly Solidified Powder Aluminum Alloys*, Philadelphia.
Staley, J.T. (May 1976). *Mat. Eng. Quart.*, p. 52.
Starke, E.A. Jr. (1977). *Mater. Sci. Eng.*, **29**, p. 99.
Starke, E.A. Jr., and Wert, J.A. (1985). *In High Strength Powder Metallurgy Aluminum Alloys*, **II**, eds. G.J. Hildeman and M.J. Koczak. TMS-AIME, Warrendale, Penn., p. 3.
Tietz, T.E., and Palmer, I.G. (1981). In "*Advances in Powder Technology*," ed. G.Y. Chin. ASM, Metals Park, Ohio.
Towner, R.J. (1961). *Met. Eng. Quart.*, **1**, no.1, p. 24.
Vidoz, A.E., Crooks, D.D., Lewis, R.E., Palmer, I.G., and Wadsworth, J. (1986). In *Rapidly Solidified Powder Aluminum Alloys*, eds. M.E. Fine and E.A. Starke, Jr. ASTM STP890. ASTM, Philadelphia, pp. 237–251.
Wagner, C. (1961). *Z. Electrochem*, **65**, p. 581.
Webster, D. (1979). *Met. Trans.*, **10A**, p. 1913.
Weertman, J. (1955). *J. of Appl. Phys.*, **21**, p. 1213.
Wilcox, B.A. and Clauer, A.H. (1966). *Trans. AIME*, **236**, p. 570.
Willis, T.C. (1988). *Metals and Materials*, **4**, p. 485.

12

The Microstructure of Aluminum-Lithium Base Alloys

D.B. WILLIAMS

Department of Materials Science and Engineering
Lehigh University
Bethlehem, Pennsylvania

and

P.R. HOWELL

Department of Materials Science and Engineering
The Pennsylvania State University
University Park, Pennsylvania

I. Introduction

Alloys based on the aluminum-lithium system are of considerable scientific and engineering interest. The addition of Li lowers the density of Al and increases both its strength and modulus. Thus, such alloys offer substantial (~10%) weight savings over conventional Al-base engineering alloys and are therefore strong candidates for the next generation of

ISBN-0-12-341831-3

aerospace alloys to succeed the 2000 and 7000 series. In addition, fatigue properties and corrosion resistance are generally improved. However, the addition of Li also reduces the ductility significantly, usually to unacceptable levels ($<\sim 9\%$). Hence, it is important to understand why these radical changes in behavior occur in order that the properties of the alloys may be optimized. The scientific understanding of the full role of Li in Al alloys is not aided by the great difficulty in detecting and quantifying the presence of Li on a scale of <100 nm, which is comparable to the dimensions of the microstructural features of interest.

In this chapter, the microstructures of a number of Al-Li base alloys are described. Due to space limitations, only a few alloys are discussed in any detail. These include the binary Al-Li, the Al-Li-Zr, and the Al-Li-Cu-Zr systems. In addition, the discussion is restricted to ingot-metallurgy alloys. For details on powder-metallurgy alloys, see, e.g., the International Conference Proceedings: *Aluminum-Lithium Alloys* [*1–3*].

II. The Binary Al-Li System

A. *The Al-Li Phase Diagram*

The primary phases of interest in binary Al-Li alloys are the face-centered cubic (fcc) Al solid solution α; the equilibrium AlLi phase δ (occasionally designated β), which has the NaTl structure; and the metastable, coherent Al_3Li phase with the ordered Ll_2 superlattice structure. This Al_3Li phase is usually termed δ', but is perhaps more accurately described as α'. The α/δ solvus line is well documented and there is good agreement between several investigators [*4*]. The $\alpha + \delta'$ solvus line and δ' phase field are not yet completely defined. There is reasonable agreement on the low Li solvus line, and the electron microscopy data of Williams and Edington [*5*] have been generally accepted in subsequent investigations. Cocco et al. [*6*] proposed a δ' coherent miscibility gap and a high Li solvus line for the $\alpha + \delta'$ field, based on their small-angle X-ray scattering (SAXS) work. However, this result has been criticized by Gayle and Vander Sande [7], as thermodynamically unreasonable. Gayle and Vander Sande proposed instead a monotectoid system (Fig. 1a) in which δ' exhibited a wide stoichiometry range. Sigli and Sanchez [*8*] calculated a slightly different phase diagram (Fig. 1b) in which the δ' is more stoichiometric around $\sim$25 at % Li, at least at the low Li side of the phase field. There is no monotectoid calculated, but a congruent point exists at the δ' critical temperature of $\sim$537°C [*9*]. Either Fig. 1a or 1b is reasonable, but both

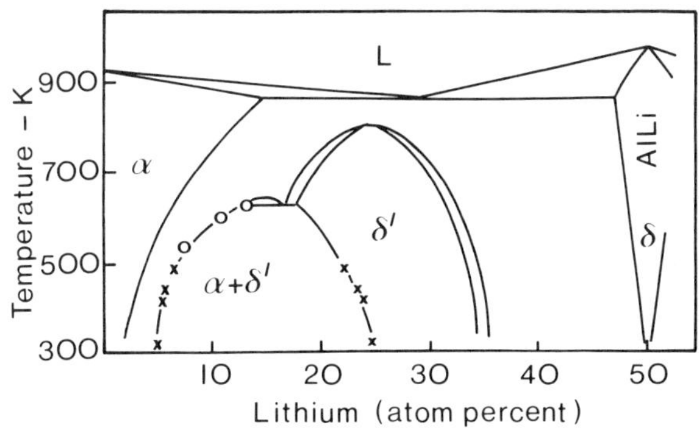

FIG. 1a. The Al-rich end of the Al-Li phase diagram proposed by Gayle and Vander Sande [7] showing the $\alpha + \delta'$ and δ' phase fields. The open circles are the TEM data of Williams and Edington [5] and the crosses represent the SAXS data of Cocco et al. [6]. (Reproduced by permission of Bull. Alloy Phase Diagrams)

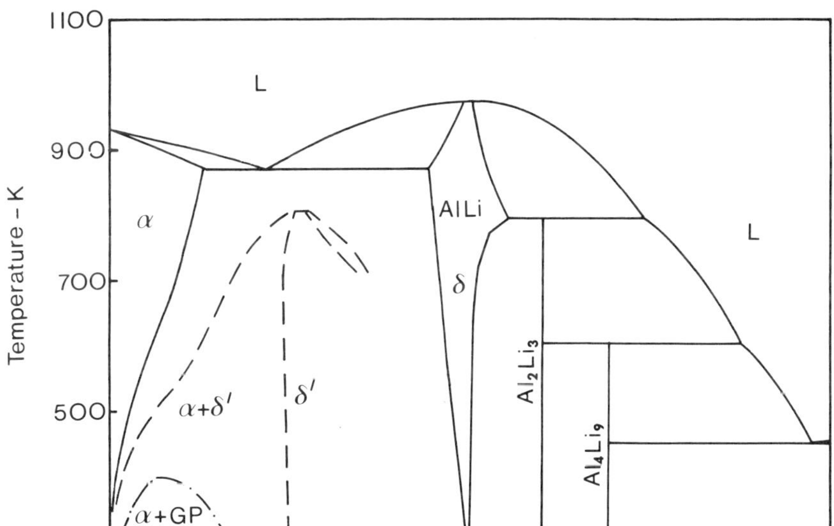

FIG. 1b. The calculated Al-Li phase diagram after Sigli and Sanchez [8]. The Al_3Li (δ') phase field shows that stoichiometry is maintained to approximately 537°C at the low Li side. A miscibility gap is also calculated to occur within the two-phase $\alpha + \delta'$ region indicating possible GP zone formation prior to metastable δ' precipitation. (Courtesy of J.M. Sanchez) (Reproduced by permission of Pergamon Press)

diagrams lack firm experimental support. The easiest way to distinguish the two diagrams may be to determine the variation of δ' composition with temperature. Preliminary electron-energy–loss spectrometry (EELS) data [*10*] indicate a reduction in Li content of δ' at 300°C in agreement with the Gayle and Vander Sande proposal [*7*] and the SAXS data of Cocco et al. [*6*]. However, the errors involved in quantification of the EELS data at this stage are too large to infer a clear deviation from stoichiometry [*11*]. A typical EELS spectrum showing the poor signal-to-noise ratio associated with the LiK edge in Al_3Li is shown in Fig. 2a. It is also possible that the composition of the δ' can be determined using atom-probe field-ion microscopy (APFIM). Figure 2b is a FIM image of an Al-7 at % Li alloy, aged at 150°C for 1000 minutes. The δ' precipitates are the small, discrete dark regions in the image (arrowed). Since the APFIM detects all elements with equal efficiencies, it may well prove to be a powerful tool for the analysis of Li.

It should also be noted that the δ' coherent solvus originally proposed by Williams and Edington [*5*] has since been reinterpreted in terms of a capillarity effect [*12*]. This line is therefore not included in Figs. 1a or 1b, either of which may be considered as the best available version of the Al-rich end of the binary diagram.

B. Precipitation in Binary Al-Li Alloys

From the point of view of structure–property relationships, the formation of δ' is the most important precipitation reaction since this confers precipitation-hardening capability on Al-Li base alloys. Many of the characteristics of δ' discussed shortly can be summarized in the structure image (Fig. 3), which shows a segment of an α–δ' interface. The interface is coherent (note the continuity of lattice planes) and there is little evidence of strain at the interface, which is generally smooth, with some indication of small ledges (arrowed). The coarser superlattice spacing in the δ' is obvious.

Precipitation of the highly reactive equilibrium δ phase results in the formation of Al_3Li precipitate-free zones (PFZs), a general loss of ductility, increased susceptibility to fatigue failure, and a low resistance to stress corrosion cracking.

1. δ' Precipitation

In alloys containing $>\sim 7$ at % Li, the δ' phase precipitates during the quench from the solution treatment temperature. Due to its low misfit

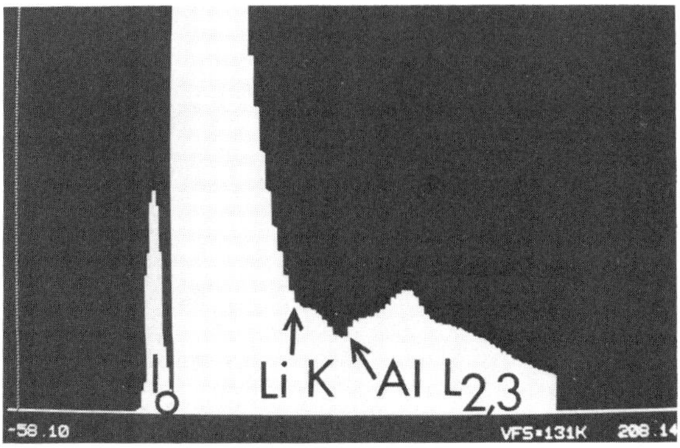

FIG. 2a. EELS spectrum from $\delta'(Al_3Li)$ particle taken at 120 kV showing the zero loss peak (I_0) followed by a display gain change. The rapidly decreasing background intensity is disturbed by the LiK edge at 55 eV and the $AlL_{2,3}$ edge at 73 eV. Despite the presence of 25 at % Li and the very thin foil, the LiK edge is very small. This makes detection of $<\sim 10$ at % Li impossible at 120 kV. (Courtesy of H.M. Chan)

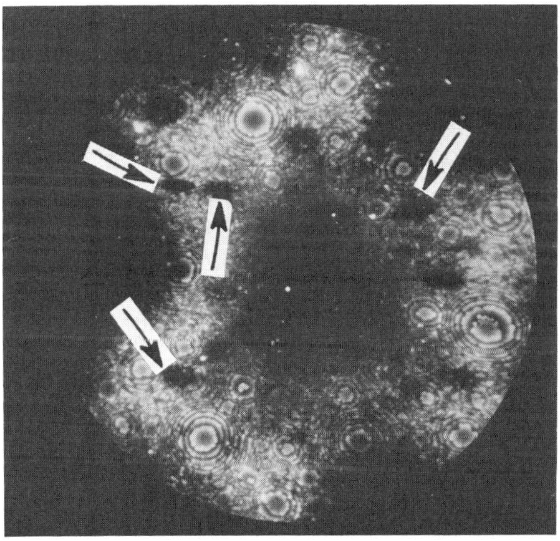

FIG. 2b. A field-ion micrograph of an Al-7 at % Li alloy that had been aged at 150°C for 1000 minutes. This micrograph was recorded at −253°C using a 50% He, 50% Ne imaging gas at 8×10^{-5} Torr pressure. Several δ' precipitates are arrowed. (Courtesy of K. Hono and H.W. Pickering)

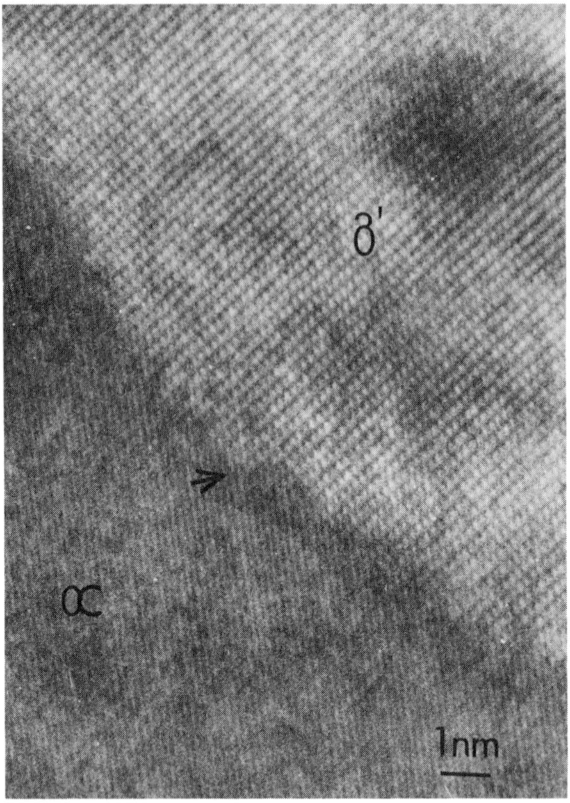

FIG. 3. A structure image showing crossed-lattice planes at a section of the α–δ' interface. The lattice planes are continuous across the interface and the coarser superlattice spacing in the δ' is visible. The arrow indicates possible evidence for a ledge at the otherwise smooth interface. (Courtesy of R. Gronsky)

parameter (−0.08%), which is easily determined from the strain contrast in Fig. 4a [*13*] and its low interfacial energy (to be discussed shortly), the δ' forms by homogeneous nucleation. This conclusion is further supported by the observed effect of vacancies on the δ' formation kinetics [*14*]. Spinodal decomposition at higher Li compositions (>∼13 at % Li) cannot be ruled out although experimental verification of this transformation mode is extremely difficult. Early calorimetric studies by Nozato and Nakai [*15*] indicated the existence of two precursor phases GP zones and δ''. There was little support for this idea until Sigli and Sanchez [*8*] calculated the possibility of a miscibility gap within the $\alpha + \delta'$ phase field (i.e., metastable with respect to δ') as shown in Fig. lb. Subsequent differential scan-

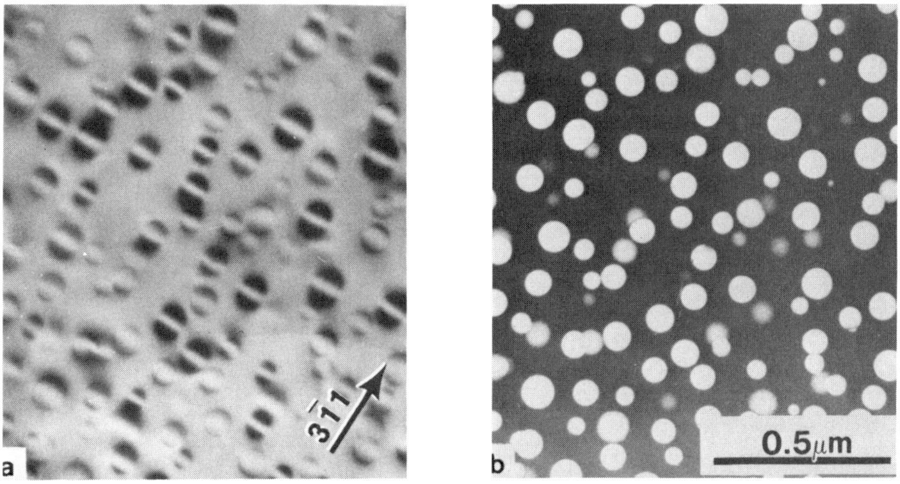

FIG. 4. A bright-field/δ′ centered dark-field (CDF) pair showing strain contrast in the matrix due to the coherent nature of the δ′ (a) and the spherical morphology of the δ′ (b) (Courtesy of S.F. Baumann)

ning calorimetry work by Papazian et al. [*16*] and SAXS work by Spooner et al. [*17*] have lent credence to the possibility of such precursors. Direct TEM imaging of the decomposition products expected within the miscibility gap has so far not been achieved.

The morphology of δ′ is almost invariably spherical and the distribution is homogeneous and insensitive to variations in pre-aging thermal history. (See Fig. 4b.) The exception to this generalization is when nucleation occurs at low undercoolings [*14*]. At $\Delta T < 40°C$, δ′ nucleates near to dislocations and low-angle grain boundaries. (See Fig. 5.) The δ′ shape is then disc-like and this is considered to arise because of rapid impingement of Li diffusion fields parallel to the dislocation line. Growth of δ′ is controlled by volume diffusion. At very small particle sizes and low undercoolings ($\Delta T < 40°C$), surface perturbations develop and lead to dendritic-shaped precipitates (Fig. 6).

The δ′ exhibits classical Lifshitz-Wagner (time $^{1/3}$) coarsening kinetics as observed by many investigators [*5,18–22*]. However, the δ′ particle size distributions are frequently normal in character rather than the log-normal or negatively skewed distributions predicted by various Ostwald ripening theories [*22,23*].

Coarsening kinetics can be affected by particle coalescence, a phenomenon documented by several workers (e.g., [*18*]). On a purely statistical basis, coalescence should lead to a translational antiphase boundary

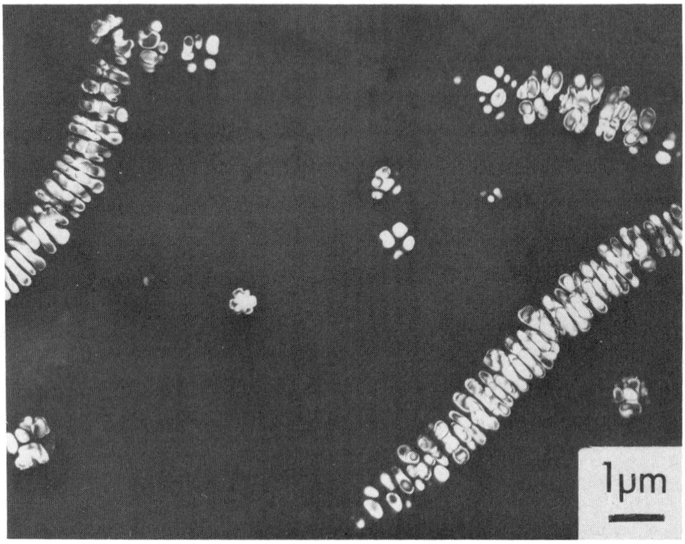

FIG. 5. Nucleation of δ′ on dislocations in an Al-6.7 at % Li alloy aged at 305°C, which corresponds to an undercooling (ΔT) of ≅20°C. (Courtesy of S.F. Baumann and D.B. Williams [*14*]) (Reproduced by permission of ASM)

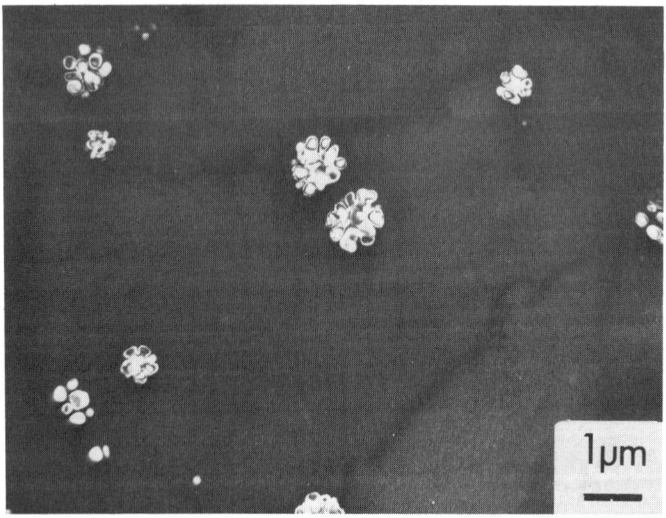

FIG. 6. The development of solid-state δ′ dendrites in an Al-6.7 at % Li alloy aged at 290°C (ΔT≅35°C). (Courtesy of S.F. Baumann and D.B. Williams [*14*]) (Reproduced by permission of ASM)

(APB) for three in four coalescence events. However, Kulwicki and Sanders [*23*] and Gu et al. [*22*] suggested that coalescence is restricted to those precipitates that are in phase. This is a reasonable hypothesis in that the APB energy has been calculated to be approximately 160 mJm^{-2} [*24*], which is an order of magnitude higher than the accepted value of the α–δ' interfacial energy (to be discussed shortly). Hence, coalescence of two δ' particles that are out of phase will replace two low-energy δ'–α interfaces with a high-energy translational APB. This may account for the absence of δ' APBs in the literature, but it is worth noting that the δ' extinction distance is usually >300 nm, implying that APB fringe contrast will only be visible between coalesced δ' particles of greater than this size, which are almost impossible to obtain. However, APB contrast in δ' particles that have been sheared by a single dislocation is occasionally visible [*25*].

Coarsening kinetics have been used to generate Li diffusion data as well as values of the α–δ' interfacial energy (γ). The values of γ determined in this way are usually overestimates (240 mJm^{-2} [*18*] and 90 mJm^{-2} [*20*]). For a particle that retains its spherical shape and remains coherent up to ~300 nm diameter, one expects both a low γ and a low misfit with the matrix. Perhaps a more reasonable value for γ is ~14 mJm^{-2} [*26*], which was determined through a novel method not involving kinetic measurements. SAXS work by Spooner [*27*] also indicates a value for γ of 15 mJm^{-2}.

In addition, δ' can coarsen discontinuously at grain boundaries [*28*] as shown in Fig. 7.

2. δ'–δ TRANSFORMATION

It is generally considered that δ nucleates heterogeneously due to the large misfit between the α and δ lattices. δ forms initially at grain boundaries where it generates δ' PFZs (Fig. 8) but also forms within the matrix.

The formation of δ' PFZs at grain boundaries has been studied by several investigators [*19,20,29*]. There is some uncertainty concerning the kinetics of PFZ formation. For example, Sanders et al. [*19*] found that the PFZ width grew as time $(t)^{1/3}$. Conversely, Jensrud and Ryum [*20*] analyzed PFZ formation theoretically and determined that PFZ growth should obey $t^{1/2}$ kinetics. Their experimental data substantiated their hypothesis. Figure 9 is a plot of PFZ half width versus $t^{1/2}$ for an Al-3% Li-0.5% Mn alloy.

Comparatively little attention has been paid to the development of the intergranular δ distribution as a function of aging time. This, in part, is a reflection of the reactive nature of the δ. However, preliminary work by

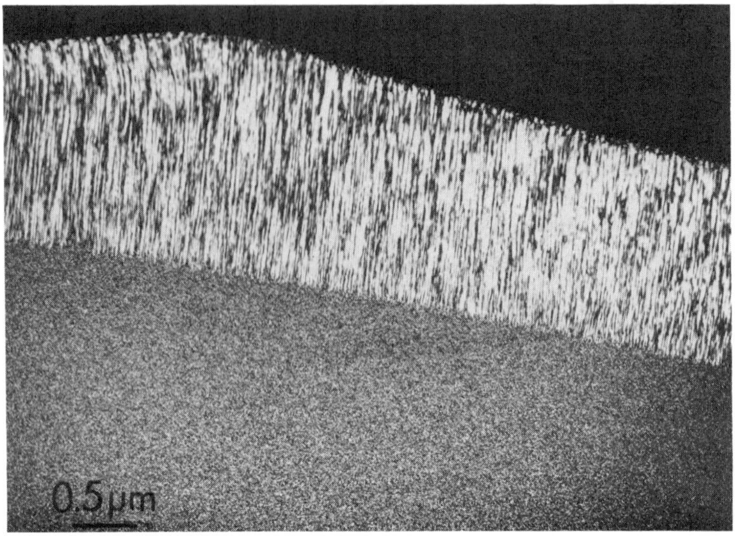

FIG. 7. Discontinuous coarsening in an Al-10.7 at % Li alloy aged at 140°C for two hours.

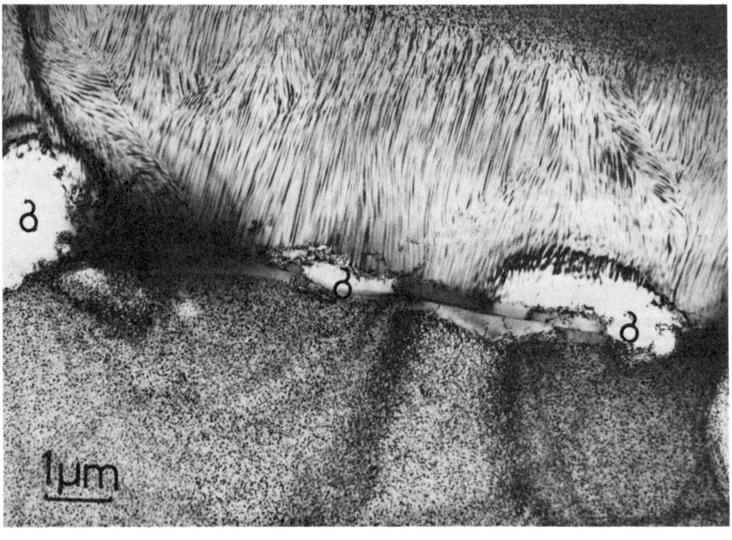

FIG. 8. Precipitation of δ on a high-angle grain boundary in an Al-13 at % Li alloy aged at 190°C. The precipitation of δ leads to the development of a δ' PFZ and termination of the discontinuous reaction.

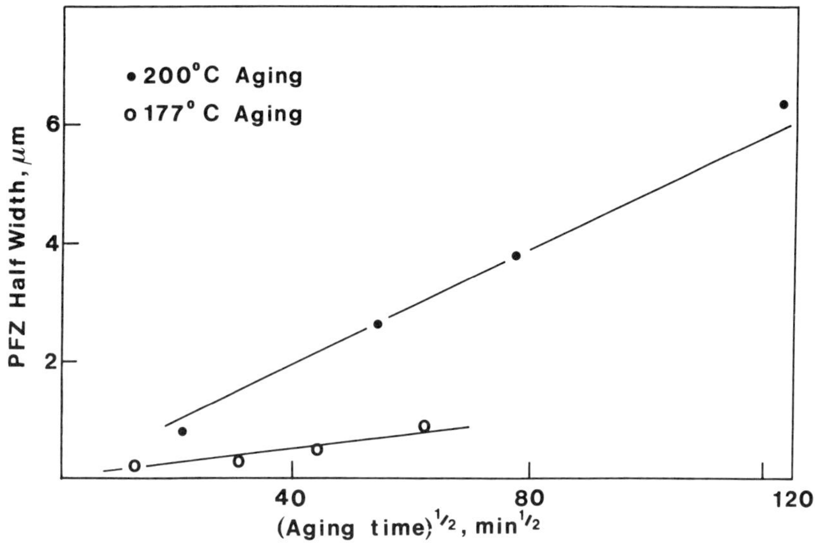

FIG. 9. The variation of PFZ width with (time)$^{1/2}$ for an Al-3% Li-0.5%Mn alloy. (Courtesy of A.K. Vasudevan).

Vasudevan et al. [*30*] indicates that the average δ diameter also increases as $t^{1/2}$ (Fig. 10).

The precipitation of δ can lead to the termination of the discontinuous reaction (see Fig. 8) since δ both pins the advancing grain boundary and generates δ' PFZs [*28*]. It has been proposed that δ' transforms to δ via a shear mechanism [*31*] and also that δ nucleates directly on the α–δ' interface [*32*]. Given the low interfacial energy and the very low misfit between α and δ', it would appear that a matrix dislocation or an array of dislocations (sub-grain boundary) would be much more potent sites ($\gamma_{SGB} \sim 200$–500 mJm^{-1}) with more than 10 times the available energy to lower ΔG^* for nucleation of δ.

III. The Al-Li-Zr System

Zirconium is a frequent addition to high-strength Al alloys since it supresses recrystallization and improves toughness and stress corrosion resistance [*33,34*]. In all the Zr-containing alloys described in this chapter, the Zr forms the metastable Al_3Zr (β') phase. In alloys containing <0.18% Zr, the β' is approximately spherical (Fig. 11) and can be coherent or

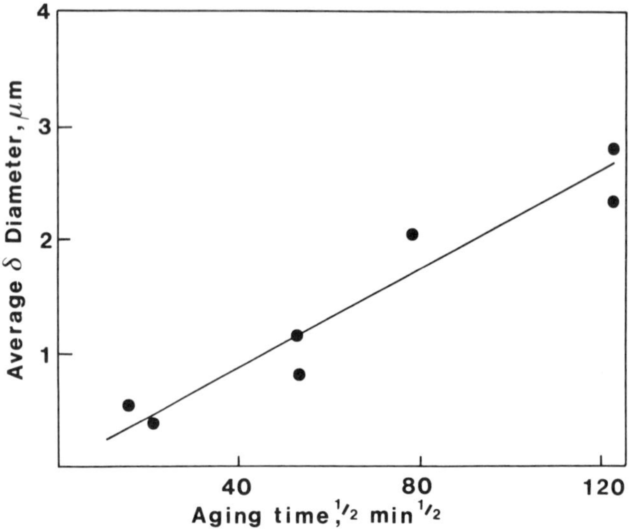

FIG. 10. The variation of δ diameter as a function of aging time for an Al-3% Li-0.5% Mn alloy. (Courtesy of A.K. Vasudevan).

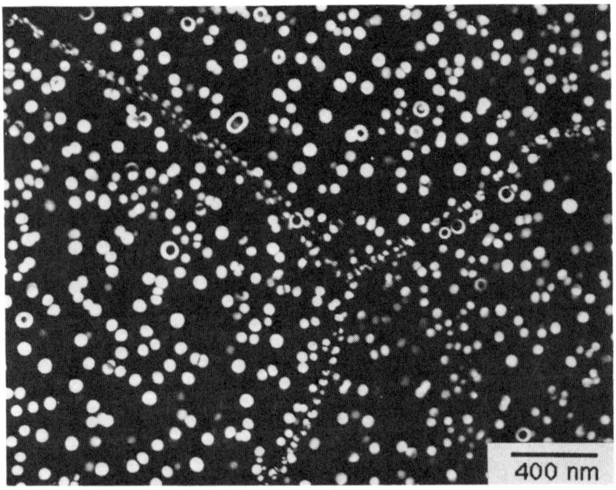

FIG. 11. The bimodal distribution of δ' developed in an Al-2% Li-0.12% Zr alloy aged at 190°C for four hours. The duplex β'/δ' precipitates contain a dark-imaging β' core. (Courtesy of M.H. Tosten)

semicoherent depending on processing history [*35*]. β' has a marked effect on the δ' distribution, and aging below the δ' solvus leads to the development of duplex δ'/β' precipitates and a bimodal precipitate distribution [*21*,*35–38*]. Figure 11 is an example of the bimodal δ' distribution. The development of the duplex precipitates had been ascribed to nucleation of δ' on the β–α interface [*21*,*36*,*37*] as well as impingement of preexisting δ' and β', followed by coarsening and eventual coalescence [*35*].

Reference to Fig. 11 shows that the β' images as a dark center in the δ' centered dark-field image. Several theories have been advanced for this observation [*21*,*35*,*38*,*41*] and the reader is referred to these papers for further discussion.

The presence of β' increases the overall coarsening rate of the δ' [*21*,*37*,*42*] when compared with the equivalent Al-Li or Al-Li-Mn alloys.

In common with the binary Al-Li alloys, the equilibrium δ phase precipitates at high-angle grain boundaries and eventually on sub-boundaries and within the matrix [*40*].

Finally, a discontinuous reaction is observed in Al-Li-Zr alloys [*37*] for aging temperatures in the range 160–200°C.

IV. The Al-Li-Cu-Zr System

A. *Phases Present*

The precipitation sequence in Al-Li-Cu alloys [*43–46*] has been suggested to occur as follows:

$$\alpha_{\varepsilon\varepsilon} \rightarrow \text{GP zones} \rightarrow \theta'' \rightarrow \theta' \rightarrow \theta\ (\mathrm{Al_2Cu})$$

$$\alpha_{ss} \rightarrow \mathrm{T_1}\ (\mathrm{Al_2CuLi})$$

$$\alpha_{ss} \rightarrow \delta' \rightarrow \delta\ (\mathrm{AlLi})$$

and in the presence of Zr:

$$\alpha_{ss} \rightarrow \beta' \rightarrow \beta\ (\mathrm{Al_3Zr})$$

As noted by Sanders [*45*], the relative size and volume fraction of the various precipitates are functions of both composition and aging temperature.

In addition to the preceding, several other phases have been proposed to form in Al-Li, Al-Li-Cu, and Al-Li-Cu-Zr alloys. A full listing of precipitate phases is given in Table I (after Suresh et al. [*47*]). The equilibrium phases (Table I) present in Al-rich alloys are δ, $\mathrm{T_1}$ ($\mathrm{Al_2CuLi}$), $\mathrm{T_2}$ ($\mathrm{Al_6CuLi_3}$), $\mathrm{T_B}$ ($\mathrm{Al_{7.5}Cu_4Li}$), and θ ($\mathrm{Al_2Cu}$).

TABLE I

PHASES REPORTED IN Al-Li, Al-Li-Zr AND Al-Li-Cu-Zr ALLOYS

Phase	Reference
δ'	Williams and Edington [5]
δ	Williams and Edington [5]
GP zones	Schneider and von Heimendahl [59]
θ''	Schneider and von Heimendahl [59]
T_1	Noble and Thompson [44]
T_1' (1)	Rioja and Ludwiczak [71]
θ'	Schneider and von Heimendahl [58]
T' (2)	Sainfort and Guyot [11]
θ_B' (2)	Silcock [43]
T_B' (2)	Rioja and Ludwiczak [71]
T_B	Schneider and von Heimendahl [59]
T_2	Hardy and Silcock [72]
T_2' (2)	Rioja and Ludwiczak [71]
β'	Rioja and Ludwiczak [71]

(1) The T_1' phase is proposed to be a metastable precursor to the T_1 phase. It is suggested that T_1' is orthorhombic with the Pt_2Mo structure.

(2) These phases are based on the θ' structure and, for example, the θ_B' and T_B' phases are proposed to be intermediate in nature between θ' and the equilibrium T_B phase.

Since this chapter is concerned primarily with the microstructures developed in Al-Li base alloys, the reader is referred to the original publications cited in Table I concerning phase identification.

B. Precipitation in Al-Li-Cu-Zr Alloys

1. ALLOYS WITH A HIGH LI:CU RATIO

Noble and Thompson [*44*] presented the first transmission electron microscope study of the aging behavior in Al-Li-Cu alloys and documented the presence of T_1, GP zones, θ', and δ'. Noble and Thompson showed that the T_1 phase formed as thin, hexagonal plates on {111} matrix planes with the same orientation relationship as that determined by Silcock [*43*]. It was also proposed that in alloys of low Cu supersaturation, the T_1 phase formed by the dissociation of a/2 ⟨110⟩ dislocations into a/6 ⟨112⟩ Shockley partials, followed by Cu and Li enrichment of the intrinsic stacking faults.

In alloys of high Cu supersaturation, it was suggested that T_1 formed by nucleation on GP zones.

Lin et al. [*48*] investigated two Al-3% Li-2% Cu-0.2% Zr-*X*% Cd alloys using both transmission electron microscopy and Guinier X-ray patterns and found that T_2, δ, δ', T_1, and β', together with the constituent phase Al_7Cu_2Fe, were present after aging for 48 hours at 160°C. Lin et al. also demonstrated that stretching prior to aging increased the nucleation rate of T_1.

Sankaran and O'Neal [*49*] examined an Al-2% Li-4% Cu-0.2% Zr alloy at a series of aging temperatures. Below 100°C, the only phases present were GP zones and δ', while above 165°C, the major phases encountered were T_1 and δ'. For an Al-2.5% Li-2.5% Cu-0.2% Zr alloy, T_1 was the only precipitate for aging temperatures in excess of 225°C. They attributed the lack of δ' to the lowering of the δ' solvus by the Cu. However, Baumann, and Williams [*13*] found that Cu did not affect the δ' solvus. Sankaran and O'Neal also concluded that T_1 can nucleate on preexisting GP zones or on defects such as helical dislocations and stacking faults.

Figures 12a and 12b are typical selected area diffraction patterns (SADP) (⟨001⟩ and ⟨110⟩ beam directions, respectively) from an Al-2% Li-3% Cu-0.12% Zr alloy aged at 190°C for 18 hours (peak age). Superlattice spots from the δ' are observed (arrowed) and the streaks in the ⟨111⟩ and ⟨001⟩ directions arise from the T_1 and θ' precipitates, respectively. (Although some uncertainties still remain concerning the exact structures of the T_1 and θ' phases, the nomenclature originally employed by Noble and Thompson [*18*] will be used in this section.) Figures 13a–c are θ', T_1, and δ' centered dark-field micrographs from the same area. θ' precipitates as thin plates on the {001} matrix planes while T_1 forms on the {111} matrix planes [*44*,*50*]. Reference to Fig. 13c shows that a trimodal δ' distribution develops in these alloys [*50*]. In addition to the small spherical δ' which is characteristic of binary Al-Li alloys (Fig. 4) and the duplex δ'/β' precipitates that form in Al-Li-Zr alloys (Fig. 11), the δ' coats the broad coherent faces of the θ'. This may be shown by noting that there is a one-to-one correspondence between the θ' plates of Fig. 13a and the "plate-like" δ' in Fig. 13c (e.g., arrowed B). After long-term anneals at 190°C, T_1 becomes the dominant matrix phase and little θ' remains.

The equilibrium phases that precipitate on the high-angle grain boundaries depend on the Li : Cu ratio. According to the isothermal section of the Al-Li-Cu ternary phase diagram of Suzuki et al. [*51*], for an Al-2% Li-3% Cu alloy, T_2 and not δ should precipitate. This has been confirmed by Tosten and Howell [*50*]. Figure 14a is a bright-field micrograph of a T_2 precipitate and Fig. 14b is a corresponding convergent-beam electron-diffraction pattern from this precipitate. As can be seen, T_2 displays

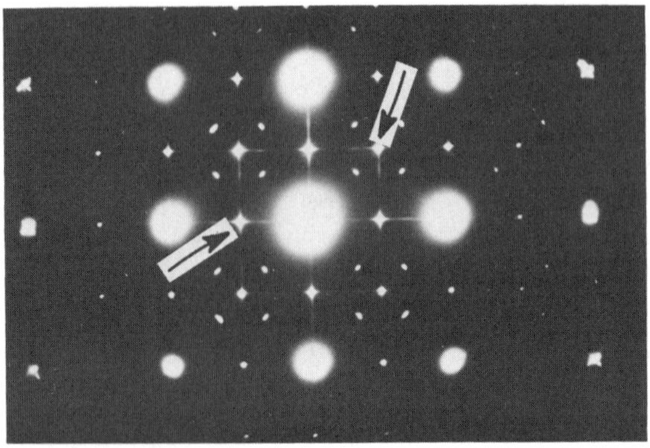

(a)

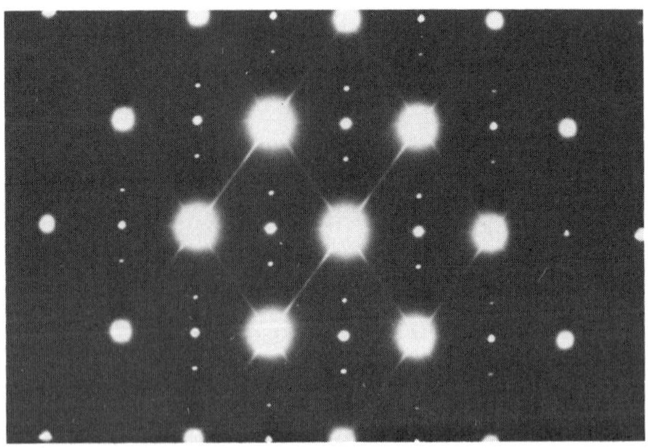

(b)

FIG. 12. Typical SADPs from an Al-2% Li-3% Cu-0.12% Zr alloy. (a) ⟨001⟩ beam direction. (b) ⟨110⟩ beam direction.

apparent icosahedral (five-fold) symmetry. This is consistent with the electron-diffraction work of Sainfort et al. [*52*] and the combined X-ray and electron-diffraction studies of Bartges et al. [*53*]. It is also interesting to note that Crooks and Starke [*54*] also found T_2 on high-angle grain boundaries and presented a selected-area diffraction pattern that displayed fivefold symmetry. Subsequent work [*55*] has questioned the assumption of apparent icosahedral symmetry, proposing instead that the T_2 phase is microcrystalline and microtwinned.

In common with binary Al-Li alloys, δ' PFZs are associated with precipitation on high-angle boundaries [*48*,*56*]. In addition, δ' PFZs also form

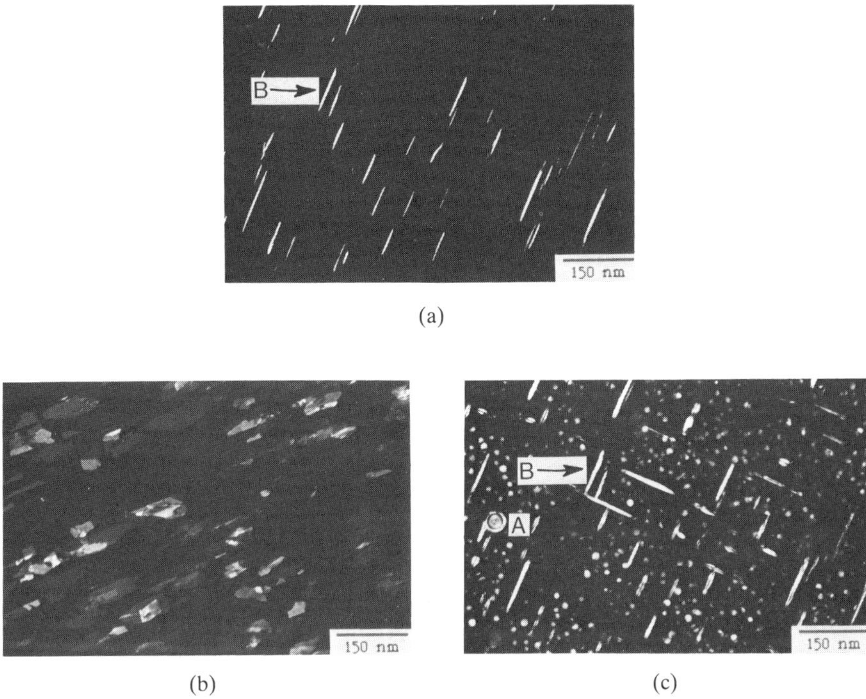

FIG. 13. Precipitate distribution in an Al-2% Li-3% Cu-0.12% Zr alloy that had been strained by 2% prior to aging at 190°C for 18 hours. (a) θ' CDF. (b) T_1 CDF from the same region. (c) δ' CDF from the same region. In addition to the small spherical δ' and the δ' that has encapsulated the β' (at A), the δ' has coated the broad faces of the θ' (arrowed B). (Courtesy of M.H. Tosten)

in the vicinity of low-angle grain boundaries both at the early stages of aging (due to copious precipitation of T_1) and after extended annealing (due to the precipitation of both T_1 and T_2 and/or δ).

Increasing the Li:Cu ratio leads to a change in the relative volume fraction of the various precipitate phases. Figures 15a–d show δ', θ' and two T_1 centered dark-field micrographs, respectively (⟨110⟩ beam direction) of an Al-3% Li-1% Cu-0.12% Zr alloy after aging at 190°C for 50 hours (peak age). A trimodal distribution of δ' is still observed (Fig. 15a). but the dominant δ' morphology is spherical, which reflects the lower volume fraction of θ' in this alloy. (Compare Fig. 15b with Fig. 13a.) In addition, a lower volume fraction of the T_1 phase is present (Figs. 15c and 15d). A high density of precipitates is observed on high-angle boundaries which leads to wide δ' PFZs. No δ' PFZs develop in the vicinity of

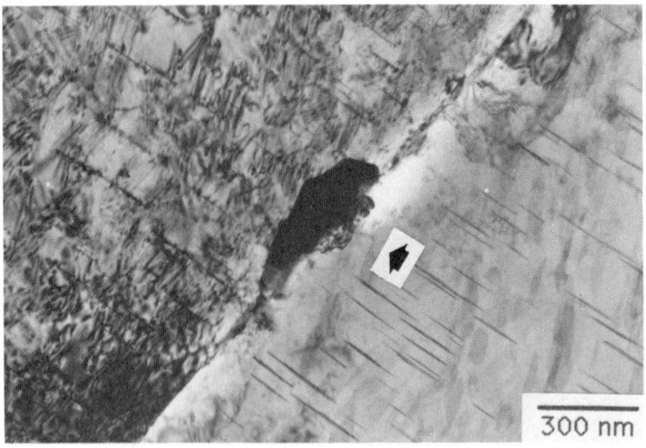

(a)

(b)

FIG. 14. T_2 precipitation in an Al-2% Li-3% Cu-0.12% Zr alloy. (a) A low magnification BF image. The T_2 precipitate is arrowed. (b) A convergent-beam electron-diffraction pattern from the T_2 precipitate. Note the icosahedral symmetry.

low-angle grain boundaries when T_1 and θ' are the only precipitate species on the boundary [56]. However, PFZs are formed after extended aging times when T_2 and or δ precipitate on the boundaries [57].

Gregson and Flower [39] investigated a similar alloy containing 3% Li, 1.5% Cu, and 0.12% Zr and found that, in addition to δ', T_1 precipitated after four hours aging at 190°C and after 15 hours at 170°C. Gregson and Flower did not detect any θ'. However, Fig. 6 of their paper does indicate

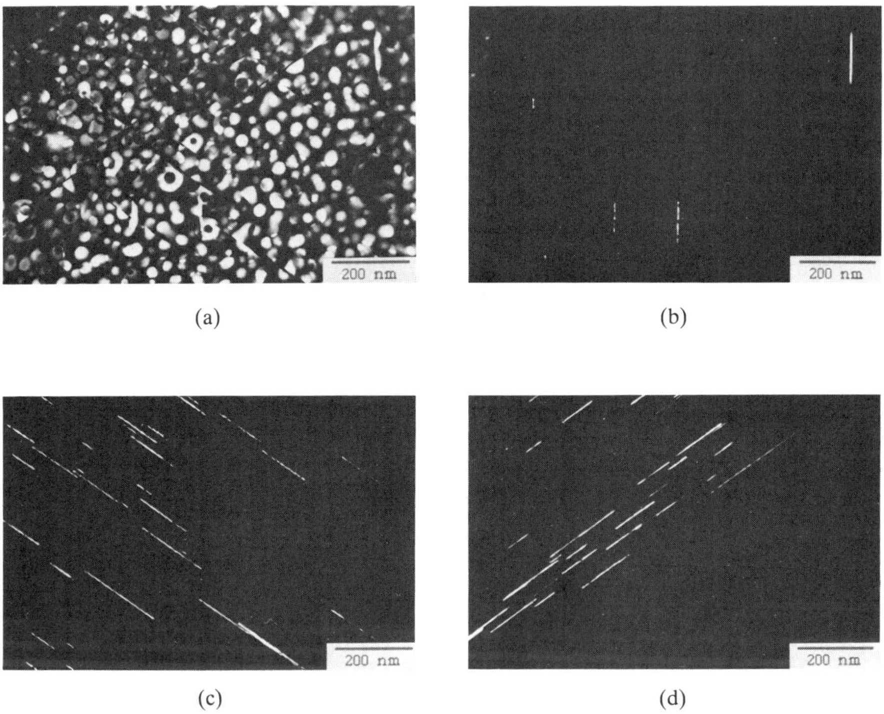

FIG. 15. Precipitate distributions in an Al-3% Li-1% Cu-0.12% Zr alloy that had been strained by 2% prior to aging at 190°C for 50 hours. (a) δ' CDF. The dominant δ' morphology is spherical. (b) θ' CDF. Although the number density of θ' precipitates is low, they are still associated with a δ' coating. (c) T_1 CDF. (d) T_1 CDF. (Courtesy of M.H. Tosten)

the presence of a low number density of plate-like features on the {001} matrix planes. In view of the results of Tosten et al. [*50*], these features are most probably θ'. Both Tosten et al. [*50*] and Gregson and Flower [*39*] showed that δ' PFZs do not develop in the vicinity of the T_1 precipitates. (See Fig. 15a.) Hence, for alloys containing Li and Cu in the approximate ration range of 3:1 to 2:1, it would appear that the T_1 growth is accomplished by the accretion of Li from the α solid solution. However, for lower Li:Cu ratios (e.g., 2:3), it is likely that T_1 can grow at the expense of the δ' [*50*].

Lin et al. [*48*] documented the occurrence of both T_2 and δ on high-angle grain boundaries in an alloy containing 3%Li and 2%Cu. Again, this is consistent with the phase diagram presented by Suzuki et al. [*51*].

2. Alloys with a Low Li:Cu Ratio

A number of investigators have examined alloys with a composition close to that of Alloy 2020 [*51,59-63*]. In general, for alloys containing ≃4.5% Cu and ≃1% Li, δ' formation is suppressed [*59,60*]. However, increasing the Li content to ≃1.2% does lead to a small volume fraction of δ'[*62,63*].

The major matrix phases encountered are T_1 and θ' and it has been demonstrated by both Rinker et al. [*61*] and Feng et al. [*62*] that stretching increases the nucleation rate of both θ' and T_1. These authors also found that additions of Cd increased the nucleation rate of θ', which is consistent with earlier findings on Al-Cu-Cd alloys [*64,65*].

Figures 16a and b are β'/θ' and T_1 centered dark-field micrographs respectively (⟨110⟩ beam direction) from an Al-1% Li-4.5% Cu-0.12% Zr alloy that had been aged at 190°C for 128 hours. The spherical precipitates in Fig. 16a are β' and the close association between the β' and the θ'

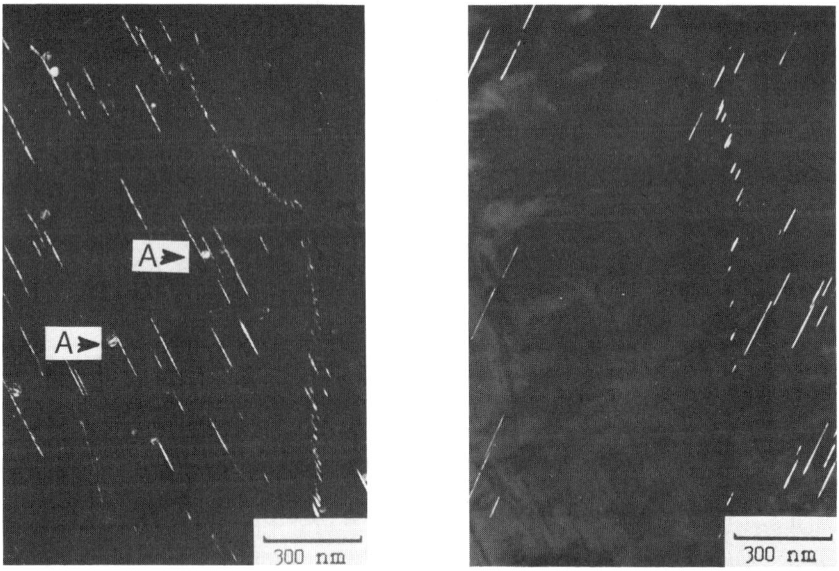

(a) (b)

Fig. 16. Precipitate distributions in an Al-1% Li-4.5% Cu-0.12% Zr alloy aged at 190°C for 128 hours. (a) β'/θ' CDF. Many of the θ' plates have their broad faces in contact with the β'. (b) T_1 CDF. Note the precipitation of T_1 and θ' on the low-angle grain boundary. (Courtesy of J.M. Galbraith) (arrowed A) suggests that β' can act as a nucleation substrate for the θ' [*60*]. Heavy precipitation of θ' and T_1 on a low-angle grain boundary is also illustrated in Figs. 16a and b.

V. The Al-Li-Cu-Mg System

Precipitation in Al-Li-Cu-Mg alloys is similar to that which occurs in the relevant ternary alloys [*46*] and δ', S′ (Al_2CuMg), and T_1 have been identified in the matrix. S′ is a metastable precursor to the equilibrium S phase and in ternary Al-Cu-Mg alloys, the following precipitation sequence has been documented [*66*].

$$\alpha_{ss} \rightarrow \text{GP zones} \rightarrow S'' \rightarrow S' \rightarrow S\ (Al_2CuMg).$$

S′ develops as narrow laths on {210} matrix planes and nucleates on dislocations and subboundaries [*67*]. In common with the Al-Li-Cu alloys, the equilibrium phases—S, T_2, and Al_2MgLi—form at both high- and low-angle grain boundaries.

Gregson and Flower [*36,39*] investigated a series of Al-Li-Mg-Cu-Zr alloys. In common with the Al-Li-Cu-Zr alloy, they found both β' and δ' in the as-quenched condition and aging at 170 or 190°C led to a bimodal distribution of δ' that is very similar to that found in the Al-3% Li-1% Cu-0.12% Zr alloy (Fig. 15a). Neither Cu nor Mg affected the δ' coarsening rates which still exhibited classical Lifshitz-Wagner coarsening kinetics.

The volume fraction of the phases is a sensitive function of alloy composition. For example, Gregson and Flower [*39*] found that for an Al-2.6% Li-1.2% Cu-0.5% Mg-0.13% Zr alloy, coprecipitation of S′, δ', and T_1 occurred, while for an Al-2.7% Li-1.2% Cu-0.7% Mg-0.09% Zr alloy, only S′ and δ' developed. Similarly, Chen and Starke [*68*] found that for an Al-2.7% Li-1.5% Cu-1% Mg-0.15% Zr aged at 190°C, the dominant matrix phase was δ' at the early stages of aging and S′ only precipitated after prolonged aging. They also noted that the laths of S′ eventually form corrugated sheets on the {210} planes. δ' PFZs were found at high-angle grain boundaries but not at low-angle boundaries. The amount of T_1 precipitation was very small. In common with the T_1 precipitates in the Al-Li-Cu alloys, Chen and Starke found that the S′ could grow through the δ' precipitates. Ahmad and Ericsson [*69*] examined an Al-2.5% Li-1.2% Cu-0.7% Mg-0.12% Zr alloy that had been aged at 190°C and found only δ' after one hour whereas after six hours of aging, both T_1 and S′ were present in addition to the δ'. Crooks and Starke [*54*] investigated an alloy of a lower Li : Cu ratio (Al-3% Cu-1.6% Li-0.8% Mg-0.2% Zr) and noted the absence of δ and δ'. Both S′ and T_1 were found in the matrix whereas on boundaries whose misorientation exceeded 10°, both the T_2 and S phases were present.

It is also worth noting that the now almost ubiquitous icosahdral symmetry phenomenon has been reported in these alloys when chill cast [*70*]. In light of the work presented by Sainfort et al. [*52*], Tosten et al. [*38*] and Bartges et al. [*53*], this phase is most probably T_2.

VI. Concluding Remarks

Precipitation reactions in binary Al-Li and ternary Al-Li-Zr alloys are well understood. The major question remaining is the exact nature of the Al-rich end of the Al-Li phase diagram.

The situation with regard to Al-Li-Cu-Zr and Al-Li-Cu-Mg-Zr alloys is less clear and this is a reflection of the increased complexity of these systems. The major problems associated with quantifying microstructures are the paucity of equilibrium diagrams and the lack of a full structural analysis of transition and equilibrium phases. However, much effort is being expended in a number of research groups concerning equilibrium in these complex alloys and it is hoped that the fundamental questions will be solved in the near future.

Acknowledgments

The authors are grateful to a number of their research students, both past and present, who have contributed significantly to the work described in this chapter. Dr. S.F. Baumann and Mr. M.H. Tosten are particularly acknowledged in this respect. Valuable discussions with Dr. A.K. Vasudevan and Dr. S. Suresh are also gratefully acknowledged. The authors are grateful to the National Science Foundation (Grant #DMR-84-00427 to DBW), the Alcoa Foundation, and the Alloy Technology Division of Alcoa Laboratories, Alcoa Center, Pennsylvania, for their financial support.

References

1. Sanders, T.H. Jr. and Starke, E.A. Jr. eds. (1981) *Aluminum-Lithium Alloys*, TMS-AIME, Warrendale, Pennsylvania.

2. Starke, E.A. Jr. and Sanders, T.H. Jr. eds. (1984) *Aluminum-Lithium Alloys II*, TMS-AIME, Warrendale, Pennsylvania.

3. Baker, C., Gregson, P.J., Harris, S.J., and Peel, C.J. eds. (1986) *Aluminium-Lithium Alloys III*, The Institute of Metals, London.
4. Williams, D.B. (1981). *Aluminum-Lithium Alloys*, T.H. Sanders Jr. and E.A. Starke Jr., eds. TMS-AIME, Warrendale, Pennsylvania p. 89.
5. Williams, D.B., and Edington, J.W. (1975). *Met. Sci*, **9**, 529.
6. Cocco, G., Fagherazzi, G., and Schiffini, L. (1979). *J. Appl. Cryst.*, **110**, 325.
7. Gayle, F.W., and Vander Sande, J.B. (1984). *Bull. Alloy Phase Diagrams*, **5**, 19.
8. Sigli, C., and sanchez, J.M. (1986). *Acta Met. 34*, 1021.
9. Tamura, M., Mori, T., and Nakamura, T. (1970). *J. Jap. Inst. Met.*, **34**, 919.
10. Chan, H.M., and Williams, D.B. (1985). *Phil. Mag. B. 52*, 1019.
11. Sainfort, P., and Guyot, P. (1985). *Phil. Mag. A*, **51**, 575.
12. Baumann, S.F., and Williams, D.B. (1985). *Acta Met.*, **33**, 1069.
13. Baumann, S.F., and Williams, D.B. (1984). *Aluminum Lithium Alloys II*, E.A. Starke Jr. and T.H. Sanders Jr., eds. TMS-AIME, Warrendale, Pennsylvania p. 17.
14. Baumann, S.F., and Williams, D.B. (1985). *Met. Trans.*, **16A**, 1203.
15. Nozato, R., and Nakai, G. (1977). *Trans. J.I.M.*, **18**, 679.
16. Papazian, J.M., Sigli, C., and Sanchez, J.M. (1986). *Scripta Met.* **20**, 201.
17. Spooner, S., Williams, D.B., and Sung, C.M. (1986). *Aluminium-Lithium Alloys III*. Institute of Metals, London, p. 329.
18. Noble, B., and Thompson G.E. (1971). *Met. Sci. J.*, **5**, 114.
19. Sanders, T.H., Ludwiczak, E.A., and Sawtell, R.R. (1980). *Mat. Sci. Eng.*, **43**, 247.
20. Jensrud, O., and Ryum, N. (1984). *Mat. Sci. Eng.*, **64**, 229.
21. Gu, B.P., Liedl, G.L., Sanders, T.H., and Welpmann K. (1985). *Mat. Sci. Eng.*, **76**, 147.
22. Gu, B.P., Liedl, L.G., Kulwicki, J.H., and Sanders, T.H. (1985). *Mat. Sci. Eng.*, **76**, 217.
23. Kulwicki, J.H., and Sanders, T.H. (1984). *Aluminum-Lithium Alloys II*, E.A. Starke Jr. and T.H. Sanders Jr., eds. TMS-AIME, Warrendale, Pennsylvania p. 31.
24. Noble, B., Harris, S.J., and Dinsdale, K. *Metal Science* (1982). **16**, 425.
25. Baumann, S.F., and Williams, D.B. (1986). Unpublished research.
26. Baumann, S.F., and Williams, D.B. (1984). *Scripta Met.*, **18**, 611.
27. Spooner S., (1984). *Proc. Mat. Res. Soc. Symp.*, **41**, 89.
28. Williams, D.B., and Edington, J.W. (1976). *Acta Met.*, **24**, 323.
29. Williams, D.B. and Edington, J.W. (1974). *Phil. Mag.*, **30**, 1147.
30. Vasudevan, A.K., Tosten, M.H., and Howell, P.R. (1986). Unpublished research.
31. Shchegoleva, T.V., and Rybalko, O.F. (1976). *Phys. Met. Metall.*, **52**, 120.
32. Venables, D., Christodoulou, L., and Pickens, J.R. (1983). *Scripta Met.*, **17**, 1263.
33. Nes, E., and Ryum, N. (1971). *Scripta Met.*, **5**, 987.
34. Nes, E. (1972). *Acta Met.*, **20**, 299.
35. Stimson, W., Tosten, M.H., Howell, P.R., and Williams, D.B. (1986). *Aluminium-Lithium Alloys III*. Institute of Metals, London, p. 386.
36. Gregson, P.J., and Flower, H.M. (1984). *J. Mat. Sci. Letters*, **3**, 829.
37. Makin, P.L., and Ralph, B. (1984). *J. Mat. Sci.*, **19**, 3835.
38. Gayle, F.W., and Vander Sande, J.B. (1984). *Scripta Met.*, **18**, 473.
39. Gregson, P.J., and Flower, H.M. (1985). *Acta Met.*, **33**, 527.
40. Makin, P.L., and Ralph, B. (1984). *Proc. 42nd Annual Meeting of the Electron Microscopy Society of America* G.W. Bailey, ed. San Francisco Press, San Francisco, California p. 500.
41. Makin, P.L., Lloyd, J., and Stobbs, W.M. (1985). *Phil. Mag A*, **51**, L41.
42. Starke, E.A., Sanders, T.H., and Palmer, I.G. (1981). *J. Met.*, **33**, 24.
43. Silcock, J.M. (1959–1960). *J. Inst. Met.*, **88**, 357.

44. Noble, B., and Thompson, G.E. (1972). *Met. Sci. J.*, **6**, 167.
45. Sanders, T.H. (1981). *Aluminum-Lithium Alloys*, T.H. Sanders Jr. and E.A. Starke Jr. eds. TMS-AIME, Warrendale Pennsylvania p. 63.
46. Sanders, T.H., and Starke, E.A. (1984). *Aluminum-Lithium Alloys* II, E.A. Starke Jr. and T.H. Sanders Jr. eds. TMS-AIME, Warrendale, Pa Pennyslvania p. 1.
47. Suresh, S., Vasudevan, A.K., Tosten, M.H., and Howell P.R. (1987), *Acta Met. 35*, 25.
48. Lin, F.S., Chakrabortty, S.B., and Starke, E.A. (1982). *Met. Trans. A*, **13A**, 401.
49. Sankaran, K., and O'Neal, J.E. (1984). *Aluminum Lithium Alloys II*, E.A. Starke Jr. and T.H. Sanders Jr., eds. TMS-AIME Warrendale, Pennsylvania p. 398.
50. Tosten, M.H., Vasudevan, A.K., and Howell, P.R. (1986). *Aluminium-Lithium Alloys III*. Institute of Metals, London, p. 483.
51. Suzuki, H., Kanno, M., and Hayashi, N. (1982). *J. Jap. Inst. Light. Met.*, **32**, 88.
52. Sainfort, P., Dubost, B., and Dubus, A. (1985). *C.R. Acad. Sc. Paris*, **301**, 52.
53. Bartges, C., Tosten, M.H., Howell, P.R., and Ryba, E.R. (1987). J. Mat. Sci., *22*, 1663.
54. Crooks, R.E., and Starke, E.A. (1984). *Met. Trans.*, **15A**, 1367.
55. Vecchio, K.S. and Williams, D.B. (1988). Phil. Mag., B *57*, 535.
56. Tosten, M.H., Vasudevan, A.K., and Howell, P.R. (1986). *Aluminium-Lithium Alloys III*. Institute of Metals, London, p. 490.
57. Tosten, M.H., and Howell, P.R. (1986). Unpublished research.
58. Tosten, M.H., and Howell, P.R. (1986), *Aluminum Alloys: Their Physical and Mechanical Properties*. E.A. Starke Jr. and T.H. Sanders, Jr., eds. Engineers Materials Advisory Services Ltd. Warley U.K. p. 727.
59. Schneider, K., and von Heimendahl, M. (1973). *Z. Metall.*, **64**, 342.
60. Galbraith, J.M., Tosten, M.H., and Howell, P.R. (1987), J. Mut. Sci., *72*, 27.
61. Rinker, J.G., Marek, M., and Sanders, T.H. (1984). *Mat. Sci. Eng.*, **64,** 203.
62. Feng, W.X., Lin, F.S., and Starke, E.A. (1984). *Met. Trans.*, **15A,** 1209.
63. Starke, E.A., and Lin, F.S. (1982). *Met. Trans.*, **13A,** 2259.
64. Silcock, J.M., Heal, T.J., and Hardy, H.K. (1955–1956). *J. Inst. Met.*, **84,** 429.
65. Hardy, H.K. (1955–1956), *J. Inst. Met.*, **84,** 429.
66. Bagaryatski, Y.A. (1952). *Doklady Akraud Kauk SSSR*, **87,** 397.
67. Wilson, R.N., and Partridge, P.G. (1965). *Acta Met.*, **13,** 1321.
68. Chen, R.T., and Starke, E.A. (1984). *Mat. Sci. Eng.*, **67,** 229.
69. Ahmad, M., and Ericsson, T. (1985). *Scripta Met.*, **19,** 457.
70. Ball, M.D., and Lloyd, D.J. (1985). *Scripta Met.*, **19,** 1065.
71. Rioja, R.J., and Ludwiczak, E.A. (1986). *Aluminium-Lithium Alloys III*. Institute of Metals, London, p. 471.
72. Hardy, H.K. and Silcock, J.M. (1955–1956). *J. Inst. Met.*, **84,** 423.

13

TEM Characterization of Aluminum-Base Composites

S.R. NUTT

Division of Engineering
Brown University
Providence, Rhode Island

I. Introduction

Microstructural characterization of aluminum-based composites is a necessary step in determining the effects of processing parameters on mechanical behavior and in understanding the fundamental mechanisms of composite failure. Transmission electron microscopy (TEM) is a powerful tool for microstructural characterization, enabling one to analyze local compositions and structures and to image the internal microstructure with high spatial resolution. These capabilities are particularly important for studying composite materials, where macroscopic behavior often depends critically on the fine structure at interfaces. The objective of the present chapter is to demonstrate the application of TEM methods to the characterization of aluminum-based composites, focusing particular attention on aluminum alloys reinforced with silicon carbide whiskers.

Aluminum alloys reinforced with short fibers, whiskers, or particles are generally classified as discontinuous reinforcement composites (DRCs). The advantages of DRCs include increases in modulus, yield strength, and ultimate tensile strength and in some cases, improved resistance to creep and fatigue as well [*1–3*]. Because of these improvements in mechanical

ISBN 0-12-341831-3

properties, there is potential for significant weight savings and cost savings where the materials can replace more expensive alloys in use as of the late 1980s [*4*]. In addition, DRCs have a distinct advantage over most continuous-fiber composites because they can be processed by conventional metallurgical methods [*2*].

In spite of the significant advantages of DRCs, there are limiting problems, the most serious of which is poor fracture toughness [*1,5*]. In addition, typical elongations to failure in tensile tests are 1–2% compared to 10–20% for unreinforced alloys with similar thermal histories [*6*]. The problem of low toughness prohibits the use of DRCs in most applications to which they otherwise would be ideally suited, such as aerospace structures and certain engine components. An additional problem characteristic of most composite materials is thermal-expansivity mismatch between matrix and reinforcement phases. The difference in thermal expansivity can be as much as an order of magnitude in some composites, and the resulting thermal stresses can exceed the matrix yield stress [*7,8*]. Finally, wide variations in measured mechanical properties are often observed, a phenomenon that reflects nonuniform microstructures in the composites [*9*]. These problems persist in most DRCs because of incomplete understanding of the microstructure, its relation to processing parameters, and its effect on failure mechanisms.

Applications of transmission microscopy to the study of composite materials have been limited, partly because of the difficulty in preparing suitable thin specimens. The difficulty stems from the different thinning rates of the constituent phases, making it hard to obtain a specimen of uniform thickness in which both phases are electron transparent. This difficulty can be partly overcome by polishing techniques that minimize surface relief, such as dimple polishing, and by prethinning specimens so that ion-milling times are kept under 2 to 4 hours.

Several investigators have successfully used TEM methods to study high-temperature interfacial reactions in continuous-fiber composites [*10,11*], although studies of short-fiber composites have been few. Arsenault and coworkers observed high dislocation densities in slowly cooled discontinuous Al-SiC composites using a high-voltage TEM [*8,12*]. The dislocations observed were generated by differential thermal expansion between the matrix and the reinforcements. Their work demonstrated the usefulness of *in situ* TEM observations for providing insight into the thermally induced inelastic response of the composite. Nutt and Carpenter applied high resolution and analytical methods of TEM to study the interface chemistry and phase distribution in a 2124 Al-SiC_W composite [*13*]. Their observations were used to explain the unusual mechanical behavior reported by others [*6,9*]. Finally, Christman and Suresh [*14*] investigated the aging kinetics in Al-SiC_W composites using TEM and microhardness measurements. The aging precipitates evolved more quickly in the composite than in the unreinforced alloy, a phenomenon that was attributed to the presence of dislocations generated by thermal stresses that developed during quenching.

The works just referenced illustrate most of the standard techniques used in the TEM approach to characterizing composite microstructures. One of the first objectives is often to identify the phases present, and this is typically accomplished by a set of complementary techniques for analyzing local structure and composition. Structural information is obtained by selected area diffraction (SAD), microdiffraction, and convergent beam diffraction (CBED), while compositional measurements are made by energy-dispersive spectroscopy of X-rays (EDS) and electron-energy-loss spectroscopy (ELS) [*15*]. The last two methods are also essential for detecting segregation (or depletion) at composite interfaces. The fine-scale structure of interfaces can be studied by phase contrast imaging at atomic-level resolution, a technique that has been made more accessible by the advent of intermediate-voltage TEMs that operate at 300 to 400 kV. High-resolution images provide valuable structural information about the interface morphology, the presence of reaction or transition layers, and other subnanometer-scale features that are difficult to resolve by diffraction-contrast imaging methods.

TEM studies can (and should) be extended to the study of deformation and failure mechanisms in composites. One relatively simple way to do this is to join two ends of a fractured tensile specimen with adhesive and section through the fracture surfaces [*16,17*]. Examination of deformed material beneath the fracture surface can reveal localization of plastic flow, sites of void initiation, and possible fracture nucleation sites. A second possible approach is to study the evolution of damage mechanisms with increasing plastic strain by preparing specimens from tests interrupted at various stages prior to failure. Experimental TEM observations can then be combined with model-based calculations and predictions to develop a quantitative framework for predicting the bulk mechanical response of composites and for developing design criteria for new composites with superior properties.

II. Composite Microstructures

The addition of reinforcements generally has a strong effect on the microstructure of the composite matrix. The presence of reinforcements within the matrix can cause a variety of phenomena including accelerated

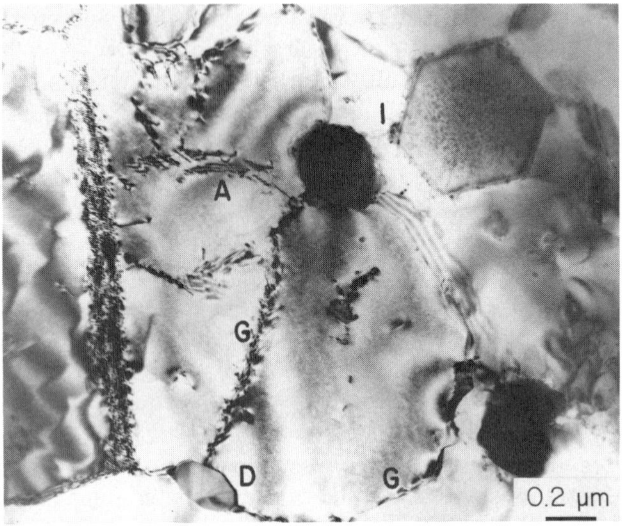

FIG. 1. Representative area of naturally aged 2124 Al reinforced with SiC whiskers, viewed in the extrusion direction: interface precipitates (I), grain-boundary precipitates (G), age-hardening precipitates (A), and dispersoid (D) [*13*].

aging, localized plastic flow, interface precipitation, grain-size refinement, and uneven solute distributions. Several of these effects are found in aluminum alloys reinforced with silicon carbide whiskers. These composites are fabricated by a powder metallurgy (P-M) method involving consolidation by hot-pressing (often *above* the alloy solidus temperature), followed by hot extrusion and conventional aging treatments. The composite microstructure is characterized by a slightly nonuniform distribution of well-aligned whiskers linked by low-angle and high-angle grain boundaries (Fig. 1). Whiskers invariably lie on the grain boundaries and apparently act as nucleation sites for matrix solidification during processing. The whiskers also pin the grain boundaries and restrict grain growth, resulting in a fine subgrain structure with subgrain sizes typically between 0.5 and 2.0 microns. The defect density within the grains can be low, as shown in Fig. 1, or extremely high, as reported by several authors [*8*,*12*]. High defect densities result from plastic flow that occurs during cooling, when the different thermal expansions of the components cause large thermal stresses to develop [*12*,*18*]. Subsequent heat treatment can in some cases allow the matrix to undergo recovery, and the defects generated during cooling are then stored in grain boundaries (Fig. 1). Thus, the defect densities observed in whisker composites will vary according to the thermal and mechanical history of the matrix alloy and the volume fraction of reinforcement.

The distribution of alloy phases is critical to the mechanical behavior of composites and is often different from the phase distribution in the unreinforced matrix processed in the same way. The grain size refinement and thermally induced plastic flow caused by the presence of composite reinforcements can strongly influence the evolution of matrix phases. Figure 1 shows the general types of matrix phases found in 2124 Al-SiC, including interface phases (I), grain-boundary precipitates (G), dispersoids (D), and constituent particles. Second-phase particles at the interface are particularly important with respect to the mechanical response of composites. Since load-transfer must occur at the interface, matrix stresses are often highest at this point. Furthermore, microstructural damage leading to composite failure often begins by interface debonding and second phases can influence the cohesive strength of the interface.

A. *Interface Regions*

Composite interfaces are often marked by the presence of microstructural irregularities such as second-phase particles, precipitate-free zones (PFZs), and solute enrichment or depletion. These features can be caused by interfacial reaction, heterogeneous nucleation, interfacial diffusion, or combinations of these. Nutt and Carpenter [*13*] studied whisker-reinforced 2124 Al composites fabricated by P-M methods and found small MgO crystallites along the borders of nearly all SiC whiskers. The crystallites were under 10 nm in size and often occurred in clusters along the interface (Fig. 2a). Phase identification was accomplished by the combined use of selected area diffraction, microdiffraction, EDS, and ELS. The energy-loss spectrum shown in Fig. 2b was acquired from the region shown in Fig. 2a and the presence of oxygen was indicated. To distinguish the presence of oxygen in MgO from oxygen present on the surfaces of the thin foil specimen, spectra were also acquired from the nearby matrix.

Oxide particles are normally incorporated into P-M products because of oxide surface films present on the powder particles from which the alloy is made. The surface films on aluminum-alloy powders generally consist of noncrystalline Al_2O_3. However, in P-M aluminum alloys containing Mg, the prevalent oxide phase is typically MgO [*19*]. The Al_2O_3 on the powders is usually reduced by Mg during high-temperature processing, a phenomenon that can alter the amount of Mg available for subsequent precipitation reactions.

The distribution of oxide phases can be of critical importance for

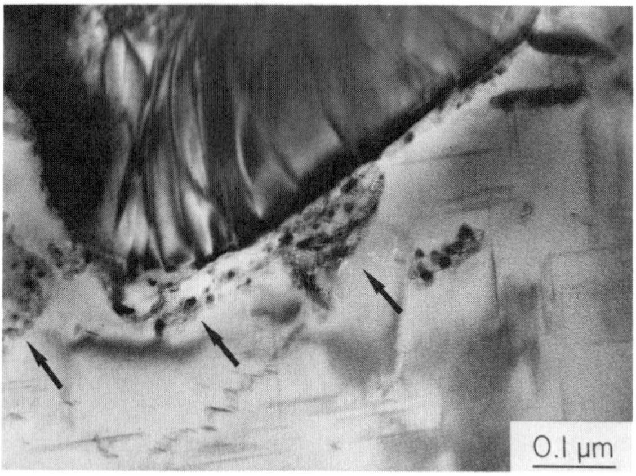

a

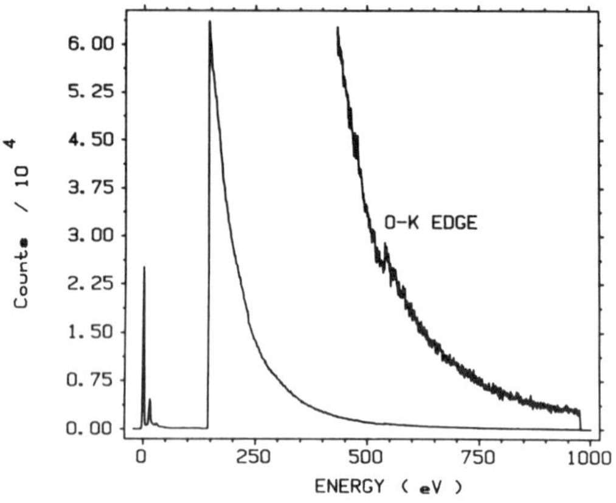

b

FIG. 2. (a) End-on view of SiC whisker in 2124 Al matrix showing small oxide particles clustered along the interface. (b) Energy-loss spectrum from interface particles. Enlarged region shows a characteristic oxygen edge.

optimizing and controlling the properties of P-M composites. In adverse cases, oxides (and SiC inclusions) can accelerate precipitation, reduce ductility, and promote intergranular fracture [*20*]. These effects are normally minimized by thermomechanical working, which breaks up the networks of prior particle boundaries (PPBs) that develop during hot-pressing. In P-M whisker composites, the oxides are normally distributed along whisker interfaces and occasionally agglomerated in large clusters near grain boundaries. The peculiar distribution of the oxide particles appears to be a consequence of the high-temperature processing of the composite (compression above the alloy solidus). During hot-pressing, oxides apparently evolve at interfaces and form clusters that are not completely dispersed by subsequent thermomechanical working. The presence of oxides along interfaces has a deleterious effect on the interface cohesive strength and promotes fiber debonding, as will be discussed later.

Fine structural details of composite interfaces can be resolved in phase-contrast images, *provided the interface is planar and can be tilted edge-on to the beam*. Images of inclined and/or stepped or curved interfaces present difficulties in interpretation because of contrast effects from overlapping features. Figure 3a shows a segment of SiC–Al interface in which the whisker is end-on and a {220} side face is parallel to the electron beam. Crossed lattice fringes in the whisker are 0.267 nm, while those in the aluminum are 0.2 nm. A thin film of noncrystalline material is evident between the whisker and the adjoining aluminum (arrow). The width of the interfacial layer varies irregularly along the interface, suggesting that there are local patches where the film is thicker than in other regions. Surface analysis of as-grown whiskers has revealed the presence of SiO_2 on whisker surfaces [*21*], and the interface film shown in Fig. 3a may contain SiO_2. This example illustrates one of the frequently encountered problems of phase-contrast imaging—namely, that while one can detect the presence of extremely small features, it is usually not possible to determine the structure and composition of those features from the image alone. For this reason, it is generally necessary to combine data from high-resolution images with data from analytical techniques such as CBED, ELS, and EDS.

One feature of some types of SiC whiskers that greatly facilitates the interpretation of interface images is the fact that they often have hexagonal cross sections and planar sides [*22*]. This feature allows one to tilt the specimen such that two of the six sides are parallel to the beam, and the interface is directly edge-on. Figure 3b shows a phase-contrast image of a whisker in this orientation in a 6061 Al matrix (an Al-Mg-Si alloy). Along the interface, there is a continuous polycrystalline layer 2–3 nm wide consisting of aluminum, magnesium, silicon, and oxygen. Microdiffraction patterns from the layer indicated the presence of crystallites of Mg_2Si, and it was concluded that the layer was composed of a mixture of Mg_2Si and noncrystalline Al_2O_3. the whisker orientation chosen for imaging the interface (⟨110⟩) also reveals the defect structure within the whisker. The banded contrast in the whisker is caused by microtwins on {111} planes oriented normal to the whisker growth axis [*22,23*] (Two sets of {111} planes are edge-on to the beam, resulting in crossed lattice fringes.) The

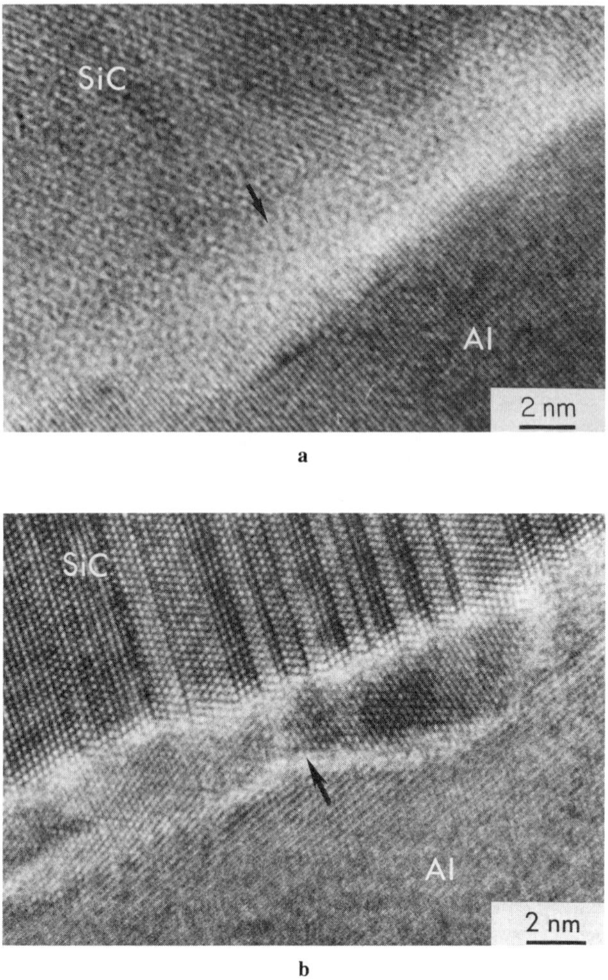

FIG. 3. High-resolution images of composite interfaces. (a) Narrow transition layer between SiC whisker and 2124 Al matrix in naturally aged composite. Whisker orientation is near [111] (end-on), and aluminum grain is [001]. The boundary plane of the whisker is (220). (b) Edge-on view of SiC whisker interface showing continuous layer of polycrystalline phases in 6061 Al-SiC_W composite.

whisker side exhibits a sawtooth profile, the cusps of which tend to coincide with the terminations of the planar defects. While the magnitude of surface roughness shown here is probably not significant, whiskers with much rougher surfaces are not uncommon, and larger notches in whisker sides may act as strength-limiting flaws, causing low stress-level fractures. In

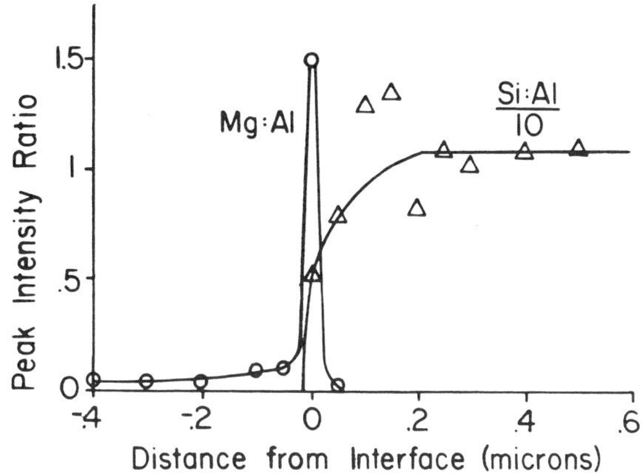

FIG. 4. EDS concentration profile of magnesium-to-aluminium ratio, measured along a trace normal to a precipitate-free Al–SiC interface in naturally aged 2124 Al-SiC_W [*13*].

composite matrices, such fractures reduce the mean aspect ratio of the whiskers and can contribute to low composite ductility [*24*]. Furthermore, surface roughness on whisker sides will tend to increase frictional stresses and inhibit fiber pullout, an important mechanism for increasing fracture toughness.

The matrix composition at composite interfaces often differs from the composition of the grain interiors because of solute segregation. Figure 4 is a concentration profile across a precipitate-free interface in naturally aged 2124 Al-SiC. The EDS profile shows enrichment of magnesium confined to a narrow region less than 50 nm wide at the whisker–matrix interface. The magnesium-to-aluminum ratio varies along the length of the interface, suggesting a possible correlation with the interface layer shown in Fig. 3a. Magnesium enrichment at the interface is not surprising in view of the studies of magnesium segregation to grain boundaries and surfaces in Al-Mg alloys [*25*,*26*]. A significant implication of this observation is that the solute distribution in the matrix is altered by the presence of the whiskers, and the solute depleted from grain interiors can no longer participate in the normal evolution of age-hardening precipitates.

The influence of composite reinforcements on the evolution of age-hardening precipitates can be seen in carbon-fiber–aluminum (CFAL) composites. CFAL composites are reinforced with *continuous* fibers and are often fabricated by continuous casting of composite wires followed by hot-pressing and conventional aging. Aging produces a high density of GP (Guinier-Preston) zones, as shown in Fig. 5, a transverse section of an A201 Al alloy reinforced with carbon fibers. In this material, solute segregation to the interface has occurred during heat treatment and has

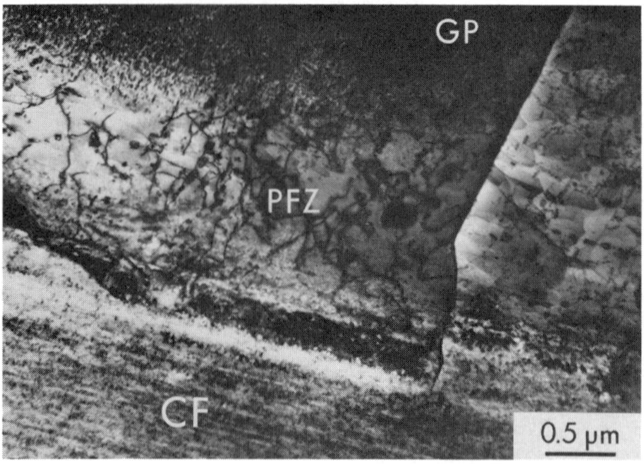

FIG. 5. Precipitate-free zone adjacent to carbon fiber (CF) in naturally aged A201 Al matrix (Al, 4.6-Cu, 0.7-Ag). GP zones and dispersoid (D) are indicated.

caused a precipitate-free zone to develop at the fiber interface (Fig. 5). The PFZ shown in Fig. 5 contains several dislocations caused by thermal stresses that develop during cooling as a result of the different thermal expansivities of the fiber and the matrix [7,27].

PFZs in CFAL composites are generally the result of the localized depletion of solute atoms and/or vacancies through their migration to the incoherent fiber interface. An excess vacancy concentration is normally frozen into the matrix when the material is quenched from the solutionizing temperature. These excess vacancies enhance the homogeneous nucleation of precipitate phases by facilitating the diffusion of solute during aging. However, the fiber–matrix interface is an effective vacancy sink, and excess vacancies from the adjoining matrix are often depleted during quenching. A reduction in excess vacancy concentration near the interface inhibits the homogeneous nucleation of GP zones and contributes to the formation of PFZs as shown in Fig. 5.

A second PFZ extends into the matrix along a low-angle boundary, which intersects the fiber interface (Fig. 5). Note that the subboundary is nearly edge-on to the beam, but the interface is inclined, as evidenced by the curvature of subboundary where it intersects the fiber. (The subboundary curvature reflects the curvature of the fiber surface.) The subboundary PFZ is narrow and tapered so that it vanishes a few microns from the fiber. The semicoherent subboundary is generally a less effective vacancy sink than the incoherent interface, although vacancies are channeled along the subboundary to the interface during quenching, resulting in the tapered PFZ shown in Fig. 5.

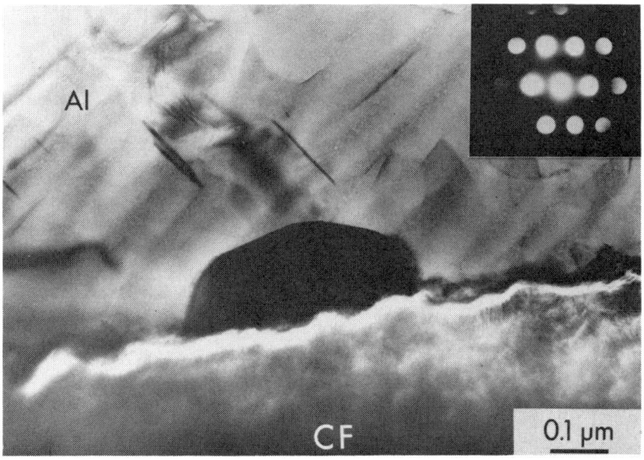

FIG. 6. Heterogeneous nucleation of $CuAl_2$ precipitate at carbon fiber–aluminum interface in A201 matrix. Inset microdiffraction pattern is $[0\bar{2}1]$ $CuAl_2$.

A second factor contributing to the formation of PFZs involves the depletion of solute atoms from the interface during heat treatment. Solute depletion can result from interface segregation and from the heterogeneous nucleation of precipitate phases on the fiber–matrix interface. Figure 6 shows a $CuAl_2$ precipitate nucleated on the fiber interface in an artificially aged CFAL (A201) composite. The specimen is a transverse section and provides an edge-on-view of the interface, revealing microscopic roughness on the fiber surface on a scale of 30 to 50 nm. The fiber is coated with polycrystalline titanium diboride to inhibit interfacial reaction and interdiffusion. During ion thinning of the specimen, the coating usually debonds from the fiber surface, leaving a narrow gap. The $CuAl_2$ precipitate nucleates on the TiB_2 coating and consumes excess copper from the matrix as it grows. Heterogeneous nucleation of precipitate phases at interfaces is also commonly observed in whisker composites [*13*]; in many composite systems, it appears to be an almost unavoidable consequence of the large areas of incoherent interface present.

Interfacial reactions are often unavoidable in composite materials, even when protective coatings are applied to the reinforcements. For example, some silicon carbide monofilaments grown by chemical-vapor deposition are coated with a carbon-rich surface layer designed to improve the fiber strength and inhibit interface reactions at elevated temperatures [*28*]. Aluminum-SiC_f composites are typically fabricated by arranging plies of filaments between aluminum foils, then hot-pressing below the solidus temperature of the alloy in an inert atmosphere. During the hot-pressing, aluminum reacts with the surface coating, forming small crystallites of

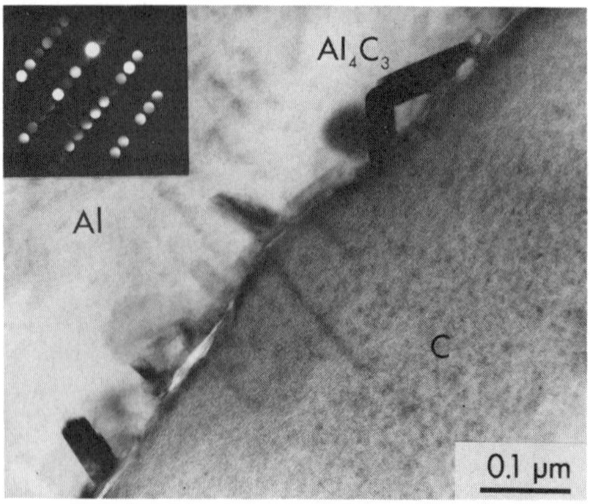

FIG. 7. Interface between carbon-rich coating on SiC filament and the aluminum matrix. Crystallites of Al_4C_3 have grown into the matrix during composite fabrication. Inset shows [$2\bar{1}\bar{1}0$] microdiffraction pattern from Al_4C_3 phase [*20*].

aluminum carbide, Al_4C_3 (Fig. 7). The crystallite assumes a platelet morphology and grows into the matrix by the out-diffusion of carbon from the filament coating. Although the strength of the SiC filaments is not affected by this small amount of interfacial reaction, the aluminum carbide is a water-soluble phase. Thus, the interface reaction will render the composite susceptible to environmental attack, particularly in applications that involve exposure to humid atmospheres.

B. Grain Boundaries

Grain-boundary precipitation is commonly observed in aluminum-base composites with continuous or discontinuous reinforcements. In Al-SiC_W composites, the grain-refining effect of SiC whiskers results in large grain-boundary areas and an abundance of heterogeneous nucleation sites. The ubiquitous precipitate phase in the 2124 Al matrix alloy is $CuMgAl_2$, an orthorhombic structure referred to as S phase. Figure 8a shows a typical example of a grain-boundary S precipitate in an artificially aged 2124 Al-SiC_W composite. The configuration of the boundary near the precipitate indicates that the precipitate has grown primarily into the upper grain and has pinned the boundary migrating toward the lower grain. The size and proximity of the precipitate to the SiC inclusion is also noteworthy. Grain-boundary precipitates near SiC inclusions are typically larger than other precipitates on the same boundary, a phenomenon that is related to

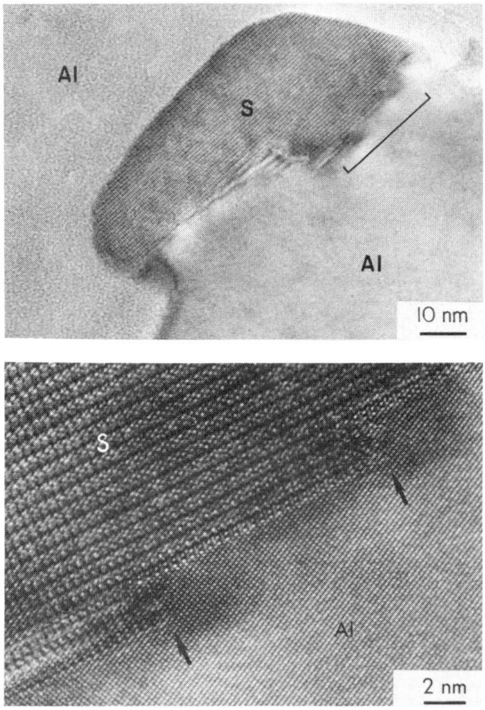

FIG. 8. Grain-boundary precipitation in artificially aged 2124 Al-SiC_w composite. (a) $CuMgAl_2$ (S-phase) particle on grain boundary pinned by SiC whisker. (b) Enlargement showing growth ledges (arrows) separated by coherent segments of boundary [*13*].

localized solute enrichment at SiC whiskers and enhanced diffusion along the interface and grain boundary.

The oblong shape of the precipitate particle in Fig. 8a derives from an earlier lath-like morphology that develops in the initial stages of precipitate growth. Coherent S phase normally precipitates on the grain boundary and grows into the grains, a process of *discontinuous precipitation* [*29,30*]. In this process, particle growth occurs only at the boundary and is therefore not continuous with time. Nucleation of precipitate laths is facilitated by the grain-boundary energy and by the low interface energy between specific lattice planes of the particle and the matrix [*13*]. As the laths thicken, they remain partially coherent on at least one side of the boundary and sometimes on both sides, depending on the relative misorientation between adjacent grains. In Fig. 8a, the lower interphase boundary in the image is viewed edge-on, and it exhibits several steps called growth ledges. Thickening of the precipitate can occur by the edge-wise propagation of ledges along this face. The distribution of the ledges indicates that they formed at the center of the face and migrated toward the ends, a widely reported mechanism of precipitate growth [*31*,*32*]. Structural details of the interphase boundary in the upper portion of the image are obscured because of the inclination with respect to the electron beam, although the curvature implies that it is less coherent.

Phase-contrast images reveal the atomic-level structure of the interphase boundary, as shown in Fig. 8b, an enlargement of Fig. 8a. Both phases are in ⟨100⟩ orientations, and the boundary planes are $(001)_S$ and $(01\bar{2})_{Al}$. Arrows mark the growth ledges, the height of which is equal to 2.1 nm, or three S-phase unit cells. The ledge structure is composed of dislocations that are not resolved in the orientation shown. The ledges are consistently one or three unit cells high, apparently because of an elastic interaction between the ledge dislocations [*31*]. The dark regions of contrast surrounding the ledges are caused by coherency strain associated with the ledge defects. Between the ledges is an atomically flat segment of boundary that is perfectly coherent. This segment exhibits a row of particularly bright spots at the interface, a common phenomenon normally attributed to Fresnel diffraction.

Most grain boundaries in continuous CFAL composites extend transversely between the fibers, since the fiber–matrix interface effectively replaces much of the grain-boundary area in longitudinal orientations. Many of these transverse grain boundaries in CFAL composites are composed of dislocation arrays, as shown in the weak-beam dark-field image in Fig. 9, taken from a 6061 Al-carbon fiber composite. The structure of the defect array is locally perturbed at points A and B by dislocations intruding from the matrix. The material has been artificially aged to increase the matrix yield strength and prevent plastic flow during cooling (caused by thermal stress [7]). The aging has resulted in a PFZ along the subboundary and several grain-boundary particles, generally Mg_2Si or Fe-Cr-Mn dispersoids. In the upper right, several parallel matrix dislocations of the same type and Burgers vector have emanated from a common source. Dislocation glide can occur more easily in the PFZ than in the precipitate-hardened matrix, and the defects that are parallel and straight in the PFZ become tangled away from the boundary where glide is impeded by $Mg_2Si(B')$ precipitates. Transverse grain boundaries in 6061 CFAL composites also provide a growth path for iron-bearing constituent particles that nucleate at the fiber interface during casting. The particles extend two to three microns along the transverse boundaries and sometimes span interfiber distances. Because these constituent particles are inherently brittle, their presence can cause premature crack initiation and facilitate crack growth between fibers.

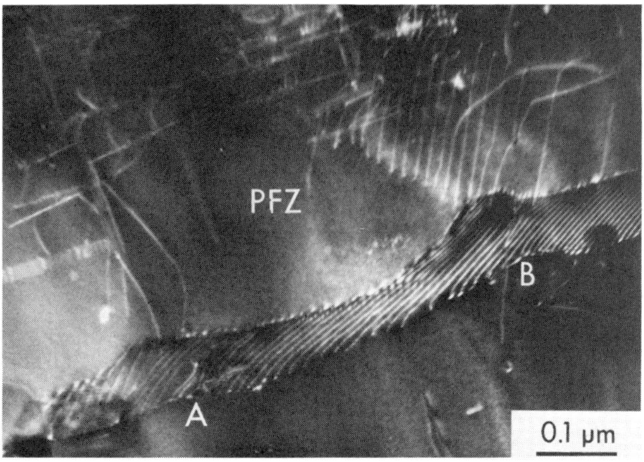

FIG. 9. Dislocation structure in low-angle boundary between carbon fibers in 6061 matrix, artificially aged. Intruding matrix dislocations perturb the dislocation array at A and B.

III. Deformation and Fracture

Failure of ductile metals generally involves the concentration of deformation into localized regions. When hard second-phase particles are dispersed in a soft matrix, ductile rupture is generally initiated by the nucleation of voids at particles [*24*]. The formation of internal voids can occur by particle–matrix decohesion or by fracture of particles. Particles with large aspect ratios will often fracture internally, while equiaxed particles tend to form voids by interface decohesion [*24*]. Additional plastic straining leads to void growth and coalescence by internal necking. These processes result in fracture surfaces that exhibit characteristic dimples. Failure modes in whisker-reinforced aluminum composites generally resemble those of certain particle-reinforced alloys in that the fracture surfaces are also typically dimpled, with whiskers often situated in the dimple centers [*1*,*2*,*24*]. Despite this resemblance to ductile fracture surfaces, the strain-to-failure of whisker composites is generally very low ($\leq 2\%$) [*1*,*6*,*33*], and the mechanisms of void formation are significantly different.

TEM observations of deformed composites can provide important insight into the localization of plastic strain and the process of damage accumulation that occurs during mechanical loading. Nutt and Duva [*16*] examined the deformed microstructures at different depths below fracture surfaces of whisker-reinforced aluminum composites tested in uniaxial

a

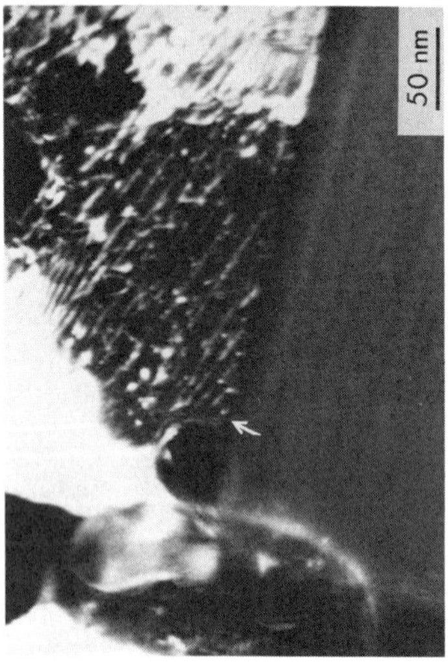

b

FIG. 10. Void initiation at whisker end near fracture surface in tensile fracture specimen of artificially aged 6061 Al-SiC_W composite. (a) Bright field image, **g** = 200. (b) Weak-beam dark field image of region shown in (a), **g** = 200, 3**g** [*16*].

tension. The formation of voids at fiber ends was a characteristic damage mechanism of near-surface regions. Figure 10 shows a whisker end where small voids (<50 nm) have formed at the corners. The intense plastic strain that has occurred around the fiber end causes strong diffraction contrast and makes it difficult to resolve individual dislocations. However, a weak-beam dark-field image of the same region shows a set of parallel matrix dislocations (Fig. 10b). The arrangement of the dislocations indicates a common source located at the base of the void and the fiber end (arrow). These fibers have aspect ratios of 4 to 10 and are approximately right circular cylinders with flat ends. The sharp corner profile formed by the fiber end acts as a stress concentrator and causes localized plastic flow and void formation [*16*].

Experimental observations made by transmission electron microscopy of deformation microstructures can be combined with model-based analyses to provide a description of the void nucleation process at fiber ends that is both quantitative and predictive. Nutt and Needleman [*34*] presented comparisons between predicted and observed modes of failure in SiC_W-reinforced aluminum composites. Their analyses were based on a continuum model for void nucleation by interface decohesion [*35*], and the TEM observations were made from regions immediately below tensile fracture surfaces. Figures 11a–c show the TEM observations of void

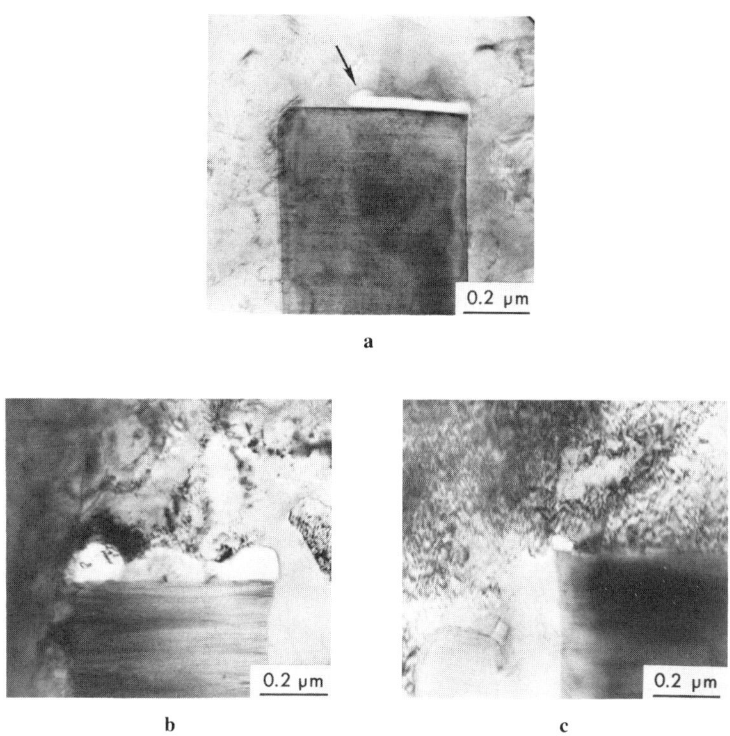

FIG. 11. Interface decohesion at fiber ends leading to different modes of void development in 6061 Al-SiC_W composite near tensile fracture surface [*34*].

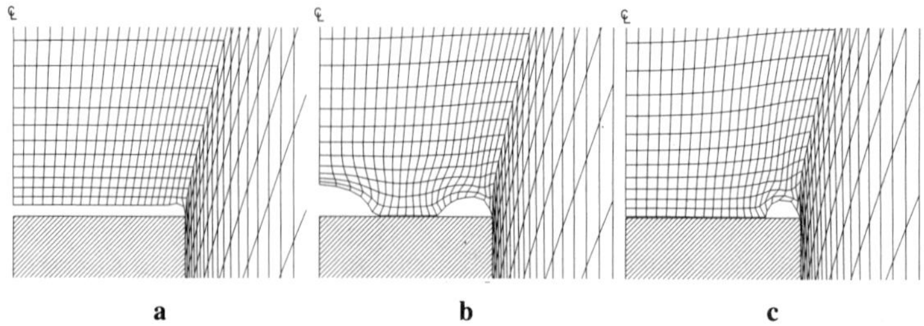

FIG. 12. Deformed finite element meshes in the vicinity of the fiber end for (a) $\sigma_{max}/\sigma_0 = 3.0$ at $\epsilon_a = 0.0151$; (b) $\sigma_{max}/\sigma_0 = 5.0$ at $\epsilon_a = 0.231$; and (c) $\sigma_{max}/\sigma_0 = \epsilon_a = 0.291$. The SiC inclusion is shaded and, for reference, the axis of symmetry is marked. The fiber axis is aligned with the tensile axis [*34*].

formation at fiber ends, while Fig. 12a–c show the predicted modes of void development. In Fig. 11a, a segment of the fiber end has debonded instantaneously, although debonding was preceded by the nucleation of a void at the site indicated by the arrow. In Fig. 11b, voids appear to have initiated separately at the fiber corners and at the fiber center and then coalesced. The most frequently encountered configuration however is shown in Fig. 11c, where a void has initiated at the fiber corner only. Note that decohesion along the whisker sides (after void formation on the whisker end) does *not* occur except in rare cases where a whisker is substantially misoriented with respect to the tensile axis.

Using the continuum model developed by Needleman [*35*], the process of void development can be predicted. In this model, the constitutive relations are specified independently for the matrix, the reinforcements, and the interface. The SiC inclusions are taken to be rigid and the Al matrix is modelled as a homogeneous, isotropic-hardening, elastic-viscoplastic solid [*34*]. Neither the residual stress distribution nor the nonhomogeneous matrix strength distribution that arises from prior thermomechanical processing are accounted for. The interface is characterized by three parameters: the strength, the work of separation, and a shear parameter. In the calculations, the properties of the aluminum matrix and the SiC reinforcements are chosen to be representative of the average properties reported elsewhere [*36*].

The predicted modes of void development are shown in Fig. 12a–c, where the only property that differs in the three calculations is the interface strength. Both the inclusion and the cell length-to-diameter ratios are 4 and the inclusion diameter is 0.585 times the cell diameter, corresponding to a fiber volume fraction of 20%. In all three calculations, debonding initiates near the fiber corner and the fibers debond from the top, but not from the sides. The ratios of interface strength-to-matrix strength (σ_{max}/σ_0) are 3.0, 5.0, and 6.0 for Fig. 12a, b, and c, respectively, and ϵ_a denotes the logarithmic tensile strain of the cell [*34*]. Thus, as the interface strength increases, the debonding behavior changes as shown in Figs. 11 and 12. Increased interfacial strength appears to inhibit void formation at the fiber center, as shown in Fig. 12c. Intense straining develops at the void tip, and at an applied strain level only slightly greater than shown, the resolution of the finite element mesh becomes inadequate to resolve the gradients that develop at the void tip. Assuming a matrix strength of $\sigma_0 = 332$ MPa for 6061 Al [*36*], the interface strength levels in Fig. 12a–c correspond to 996 MPa, 1660 MPa, and 1992 MPa, respectively. The calculations suggest that for a sufficiently high interface strength, complete decohesion of the fiber end can be substantially delayed, if not precluded altogether, even though void formation occurs at the fiber corner at relatively small strains.

The similarities between observed and predicted modes of void evolution suggest that it is possible to quantify interface properties in composites by comparisons between TEM observations and predictions based on model calculations. The calculations also indicate how qualitative features of void nucleation (such as void size, shape, and location) depend on parameters such as interface strength, matrix properties, and fiber geometry. Furthermore, the comparisons provide important insight into how voids nucleate and coalesce prior to failure and how the process of failure depends on controllable parameters of the microstructure. Detailed understanding of the process of void initiation and growth seems critical to explaining the low ductility of composites, which is currently the major limitation preventing their use. In view of these factors, it would appear that significant progress could be made by studies attempting to combine TEM observations of damage mechanisms in composites under various loading conditions with model-based analyses of the micromechanical processes.

References

1. Nair, S.V., Tien, J.K., and Bates, R.C. (1985). *Int. Met. Rev.* **30**, 275.
2. Divecha, A.P., Fishman, S.G., and Karmarkar, S.D. (1981). *J. Metals* **33**, 12–17.
3. Nieh, T.G. (1984). *Metall. Trans.* **15A**, 139–146.
4. McDanels, D.L. (1985). *Metall. Trans.* **16A**, 1105–1115.
5. Hasson, D.F., Hoover, S.M., and Crowe, C.R. (1985). *J. Mater. Sci.* **20**, 4147–4154.
6. Arsenault, R.J. (1984). *Mater. Sci. & Engr.* **64**, 171–181.
7. Kural, M.H., and Min, B.K. (1984). *J. Compos. Mater.* **18**, 519–535.
8. Arsenault, R.J., and Fisher, R.M. (1983). *Scripta Metall.* **17**, 67–71.
9. Nieh, T.G., and Chellman, D.J. (1984). *Scripta Metall.* **18**, 925–928.
10. Prewo, K., and McCarthy, G. (1972). *J. Mater. Sci.* **7**, 919–928.

11. Levi, C.G., Abbaschian, G.J., and Mehrabian, R. (1978). *Metall. Trans.* **9A**, 697–711.
12. Vogelsang, M., Arsenault, R.J., and Fisher, R.M. (1986). *Metall. Trans.* **17A**, 379–389.
13. Nutt, S.R., and Carpenter, R.W. (1985). *Mater. Sci. & Engr.* **76**, 169–177.
14. Christman, T., and Suresh, S. (1988) *Acta Metall.* **36**, 1691–1704.
15. Hren, J.J., Goldstein, J.I., and Joy, D.C., eds. (1979). *Introduction to Analytical Electron Microscopy*. Plenum, New York.
16. Nutt, S.R., and Duva, J.M. (1986). *Scripta Metall.* **20**, 1055–1058.
17. Bravman, J.C., and Sinclair, R. (1984). *J. Elect. Microsc. Tech.* **1**, 53–61.
18. Derby, B. (1985). *Scripta Metall.* **19**, 703–707.
19. Kim, Y.-W., Griffith, W.M., and Froes, F.H. (1983). *American Society for Metals Congress*, Paper 8305–048. American Society for Metals, Metals Park, Ohio.
20. Lyle, J.P., and Cebulak, W.S. (1975). *Metall. Trans.* **6A**, 685–699.
21. Rahaman, M.N., and De Jonghe, L.C. (1987). *Amer. Ceram. Soc. Bull.* **66**, 782–785.
22. Nutt, S.R. (1984). *J. Amer. Ceram. Soc.* **67**, 428–431.
23. Nutt, S.R. (1988). *J. Amer. Ceram. Soc.* **71**, 149–156.
24. Martin, J.W. (1980). *Micromechanisms in Particle-Hardened Alloys*. Cambridge University Press, Cambridge, England.
25. Malis, T., and Chatruvedi, M.C. (1982). *J. Mater. Sci.* **71**, 1479–1486.
26. Lea, C., and Molinari, C. (1984). *J. Mater. Sci.* **19**, 2336–2352.
27. Tsai, S.-D., Deepak, M., Marcus, H.L., Noyan, I.C., and Cohen, J.B. (1981). *Mater. Sci & Engr.* **47**, 145–149.
28. Nutt, S.R., and Wawner, F.E. (1985). *J. Mater. Sci.* **20**, 1953–1960.
29. Tu, K., and Turnbull, D. (1967). *Acta Metall.* **15**, 3619.
30. Meyrick, G. (1976). *Scripta Metall.* **10**, 649.
31. Howe, J.M., Aaronson, H.I., and Gronsky, R. (1985). *Acta Metall.* **33**, 649–658.
32. Laird, C., and Aaronson, H.I. (1969). *Acta Metall.* **17**, 505.
33. Nieh, T.G., and Karlak, R.F. (1983). *J. Mater. Sci. Lett.* **2**, 119–122.
34. Nutt, S.R., and Needleman, A. (1987). *Scripta Metall.* **21**, 705–710.
35. Needleman, A. (1987). *J. Appl. Mech.* **54**, 525.
36. Nardone, V.C., and Prewo, K.M. (1985). *Mater. Sci. & Engr.* **75**, 169.

14

Microstructure and Properties of Rapidly Solidified Aluminum-Transition Metal Alloys

J.M. SATER

Institute for Defense Analysis
Alexandria, Virginia

S.C. JHA

Marko Materials, Inc.
North Billerica, Massachusetts

and

T.H. SANDERS, JR.

School of Materials Engineering
Georgia Institute of Technology
Atlanta, Georgia

ISBN 0-12-341831-3

I. Introduction

In response to the demands for improved mechanical properties at elevated temperatures (500K), alloy and process developers have been exploring solute contents well beyond those used in conventional ingot metallurgy as well as nontraditional solute additions such as the transition metals. In order to utilize some of these compositions, techniques such as rapid solidification are necessary to ensure that damage-tolerant properties such as toughness and crack growth are not sacrificed at the expense of they contain elements that have limited solubility and low diffusivity, they strength principally from a dispersion-hardening mechanism, and, because they contain elements that have limited solubility and low diffusivity, they are more thermally stable than the conventional, high-strength, precipitation-hardening 2XXX and 7XXX alloys.

As early as 1947, investigators showed interest in the development of high-modulus, high-temperature aluminum alloys. The elements silicon, beryllium, cobalt, manganese, and nickel with compositions in excess of the limits of solid solubility in aluminum were found to offer superior combinations of modulus and strength. The Royal Aircraft Establishment suggested that yield strengths of 290 MPa, tensile strengths of 386 MPa with a modulus of 82.7 GPa, and ductilities of 4% might be possible. However, due to the restrictions on the solidification processes of the day, the size and distributions of the intermetallic phases containing these elements were sufficiently coarse to preclude having acceptable ductilities.

It was not until the late 1950s that alloys containing large quantities of insoluble elements had the potential to be used as engineering alloys. With the pioneering work at Alcoa Research Laboratories and at Kaiser Laboratories, binary Al-Fe, ternary Al-Fe-Ni and Al-Fe-Cr, and quaternary Al-Fe-Cr-V alloys were fabricated into extruded sections from pre-alloyed, atomized powders. Although the microstructures were coarse, they were significantly finer than those produced by ingot-casting methods.

Over the past 25 years, new nonequilibrium alloy phases and aluminum alloys with extended solid solubilities and refined microstructural features have been produced. These nonequilibrium structures require care in processing in order to maintain the fine, as-cast microstructure. Consequently, alloy developers must also be oriented toward process development.

The use of rapid solidification for aluminum alloys has led to the evolution of three main alloy classes: high-strength (room-temperature), corrosion-resistant alloys; low-density alloys with improved elastic modulus; and high-temperature alloys with acceptable properties up to approximately 340°C [*1*].

TABLE I

RELATIVE SIZES OF TYPICAL MICROSTRUCTURAL FEATURES IN 7XXX-ALLOY WROUGHT PRODUCTS

Feature	IM product	PM product
Constituent particles (Fe, Si types)	2–100 microns	0.5 micron
Nonmetallic inclusions (common limits of melt-filtration technology)	10	10
Dispersoid particles	0.01–.05 (Zr, Cr, Mn types)	0.01–0.5 (Zr, Cr, Mn, Fe, Ni, Co types)
Age-hardening precipitates	0.002–0.008	0.002–0.008
Oxide-film thickness on powder particle	—	~0.01
Individual oxide particle after hot-deformation	—	~0.01
Random oxide clusters after hot-deformation	—	~0.05

The high-strength, corrosion-resistant alloys are heat-treatable and include several variations of 7XXX and 2XXX alloys. The improved properties of these alloys in rapidly solidified form are brought about by fine grain sizes and fine, homogeneously distributed constituent particles and intermetallics, as well as by extended solid solubility. Table I shows the relative sizes of typical microstructure features in 7XXX alloys produced by conventional and rapid solidification techniques [*2*]. Wanhill and Schra [*3*] have compared properties of powder-metallurgy (PM) alloy 7091 and ingot-metallurgy (IM) alloys 7075, 7175, 7010, and 7050. From their study, they concluded (1) that the PM alloy possessed a better combination of strength and resistance to stress corrosion and exfoliation and (2) that it was more isotropic in strength and fracture toughness. Tension test results can be seen in Table II. Paris and Chellman [*4*] compared properties of a PM alloy based on a modified IM 2034 alloy and a PM alloy based on a modified IM 2219 alloy to IM 2034 and IM 2219, respectively. They also found that the PM alloys offered superior combinations of strength and toughness. Compositions and tensile properties are shown in Table III and IV. Bridenbaugh et al. [*2*] have noted similar results in other Alcoa work on PM and IM 7XXX and 2XXX series alloys. These alloys will have to compete with the conventional IM 7XXX and 2XXX alloys as well as the lighter weight Al-Li alloys.

The low-density lithium-containing alloys continue to be investigated extensively, in both ingot and rapidly solidified forms. These alloys are expected to replace the baseline alloys currently utilized by the aircraft industry: 2024-T3, 7075-T6, and 7075-T73. In one study, Meschter et al. [*5*]

TABLE II

MECHANICAL PROPERTIES OF 7XXX IM AND PM ALLOYS

	0.2% yield stress (MPa)[a]		Ultimate tensile strength (Mpa)	Elongation (%)
IM				
7075-T73	L	482.0	536.0	10.0
	T	436.0	500.0	7.0
7175-T736	L	500.0	555.0	11.0
	T	468.0	527.0	9.0
7075-T736	L	500.0	550.0	12.0
	T	455.0	536.0	8.0
7050-T736	L	473.0	527.0	11.8
	T	436.0	500.0	7.5
PM				
7091-T7E69	L	525.0	564.0	11.3
	T	510.0	555.0	10.0

[a] L denotes longitudinal, and T denotes transverse in relation to the metal flow directions.

TABLE III

CHEMISTRIES OF THE PM 2XXX ALLOYS EVALUATED IN THIS STUDY, IN WT %

Alloy	Composition type[a]	Cu	Mg	Si	Fe	Ni	Mn	Zr
Al-Cu-Mg								
2024-type PM								
514041	A	3.73	1.81	0.02	0.04	0.01	0.14	0.12
	T	3.70	1.85	—	—	—	0.20	0.14
IM								
503315	A	4.36	1.56	0.07	0.06	0.0	0.90	0.10
	T	4.30	1.50	—	—	—	0.90	0.12

[a] A denotes analyzed composition; T denotes target composition.

investigated the effects of copper, magnesium, and zirconium additions on rapidly solidified Al-3Li alloys. They concluded that, although ductility was often low, it was possible to match or exceed the tensile properties of the commercial 7075 and 2024 alloys currently in use by varying composition and heat treatment. For example, as shown in Table V, the strength–ductility combination of a peak-aged RSP Al-3Li-1Mg-0.2Zr alloy exceeds that for conventional 2024-T851. Again, RS allows production of fine-grain

TABLE IV

TENSILE PROPERTIES OF 2XXX ALLOY PM AND IM EXTRUSIONS

Alloy	Temper[a]	Yield strength, MPa (ksi)	Tensile strength, MPa (ksi)	Elongation, %	Reduction in area, %
2024 Type PM					
514041	NA	438 (63.5)	536 (77.6)	18	20
	PA	494 (71.6)	533 (77.3)	13	27
IM					
503315	NA	442 (64.1)	572 (82.8)	14	13
	PA[b]	525 (76.1)	570 (82.6)	11	25

[a] NA indicates naturally aged; PA indicates peak aged. All the alloys were stress-relieved by stretching 1.5–2.0%.
[b] Solution-heat-treated at 775 K (935°F), peak aged for 4 h at 464 K (375°F).

TABLE V

MECHANICAL PROPERTIES OF Al-3Li-Mg-Zr ALLOYS

	0.2% yield stress (MPa)	Ultimate tensile strength (MPa)	Elongation (%)
IM			
2024-T851	448	483	7.0
PM			
Al-3Li-1Mg-0.2Zr			
	L 453	542	9.0
	T 443	524	9.5
Al-3Li-2Mg-0.2Zr			
	L 454	550	7.9
	L 444	498	4.0

Note: Alloys were solution-treated at 530°C for one hour and aged at 177°C for 16 hours.

and particulate sizes in these alloys. The extension of solid solubility is important in investigations of RSP Al-Li-Be alloys by Vidoz et al. [*6*] since beryllium normally has low solid solubility in Al. The use of rapid solidification allows production of fine cast microstructures necessary for structural alloys. The additions of both beryllium and lithium lower the density and increase the elastic modulus of aluminum alloys. The tensile properties of a conventionally cast 2219-T87 (10 wt % beryllium) can be compared to an RSP 2219 alloy (8.2 wt % Be) in Table VI [*6*]. (For other

TABLE VI

MECHANICAL PROPERTIES OF RSP 2219 Al-8.2Be EXTRUSIONS

Condition	Young's modulus, GPA (Msi)	Yield strength, MPa (ksi)	Ultimate strength, MPa (ksi)	Elongation, %
As extruded	—	186.7 (27.1)	284.6 (41.3)	18.7
	—	195.0 (28.3)	288.7 (41.9)	18.2
	86.9 (12.6)	—	—	—
Solution-heat-treated and quenched	—	180.5 (26.1)	323.8 (47.0)	7.5[a]
Solution-heat-treated and quenched, strained, and, aged[b]	—	298.3 (43.3)	336.9 (48.9)	15.0
Aluminium 2219 T87[c]	(10.6)	(57.0)	(69.0)	10.0

[a] The test was discontinued at 7.5% elongation.
[b] The treatment included a 7.5% plastic strain before aging.
[c] This alloy is not RSP. It is included for comparison purpose only.

work, see [7].) The potential for successful application of the RSP Al-Li alloys depends on their ability to offer improved properties over the competitive IM Al-Li alloys.

Techniques for producing several rapidly solidified forms and typical methods of consolidating them will be discussed at some length. The effects of oxide-film formation on properties will also be included in this chapter. Theoretical discussions will include heat transfer in both powder and melt-spun ribbon as well as thermodynamic and kinetic parameters necessary for rapid solidification. Attention can then be focused on the influence of composition and casting parameters on the microstructure of rapidly solidified aluminum alloys. The chapter closes with a section devoted to the influence of the as-cast structure on mechanical properties and a discussion of the design of high-temperature, rapidly solidified aluminum alloys, particularly the Al-Fe-X and Al-Zr-X alloy systems.

II. Rationale for Utilization of Rapid Solidification

Solidification from the melt is an essential step in practically every sequence of metal processing. Examples include the production of ingots or continuous strands for working into more useful forms and the manufacture of net-shaped castings for direct application. The properties of the final product are determined to a great extent by the nature of the solidification process itself. For large castings that undergo conventional solidification processes, the amount of segregation is significant. An advantage, then, of completing solidification in a short period of time (seconds or less) and in thin sections (millimeters or less) is that there is little time or space within which to form a coarse, segregated microstructure.

Perhaps the most important consequence of the small dimensions necessary for rapid solidification is the achievement of cooling rates up to and possibly exceeding 10^6 K per second. These high cooling rates may give rise to undercoolings of 100°C or more before solidification occurs, in comparison to the fraction of a degree obtainable for conventional casting cooling rates of one degree per second. These high undercoolings, with the resulting rapid solidification, permit large deviations from equilibrium, which, in turn, offer the advantages of (1) extended solid solubility, (2) reduced grain size, (3) reduced number and/or size of segregated phases, and (4) new, nonequilibrium phases having interesting properties.

III. Production and Processing of Rapidly Solidified Materials

A. *Production of the Particulate*

The microstructures formed by rapid solidification in alloys under investigation in the late 1980s would be unattainable without the proper production techniques. There are three major processing techniques for producing rapidly solidified material, all involving a primary stage during which a small, cross-sectional dimension, in good contact with an effective heat sink, is rapidly generated. They are: spray methods, chill methods, and weld methods.

Heat-flow conditions prior to and during solidification have a pronounced effect on achievable interface velocities and, therefore, on the resulting microstructures. Solidification over small dimensions (a small thermal mass) can be accomplished by removing the heat either convectively or conductively. One rapid solidification processing (RSP) form, powder, is cooled primarily by convection. The heat is transferred away from the fine molten droplets by another impinging fluid. The RSP forms such as ribbons, fibers, and splat are cooled primarily by conduction. The heat is transferred away by a substrate.

Spray methods include atomatization processes, those processes producing a number of fine droplets from liquid metal, usually by direct impingement of the molten charge with a jet of another fluid. The various atomization processes can be subdivided depending on the mechanism by which the droplets form from the melt [*8*]. Twin fluid atomization processes are characterized by the use of a high-pressure/velocity working fluid which causes droplet formation in the material to be atomized. Gas and water atomization employ atomizing fluids of gas and water, respectively, that impinge on the melt to disrupt the melt-stream thereby forming droplets. Argon or nitrogen may be used in gas-atomization processes for easily oxidized alloys. Another more efficient variation of gas atomization is ultrasonic gas atomization (USGA), a process in which a high-velocity gas jet (up to Mach 2.5) is generated by a Hartman shock tube device [*9*] to shear the melt stream into fine particles. The drum-splat process is yet another variation of an atomization process. Rapidly solidified flakes are produced when gas-atomized metal droplets are sprayed onto the periphery of a rotating drum substrate. The individually solidified droplets, which are removed from the surface before other droplets contact them, experience higher cooling rates than conventional gas atomized powders.

There are several variations of the centrifugal atomization process. All of them involve disruption of molten metal at the rim of a rotating disc, cup, or bar, which may be concave or flat. Molten metal is supplied to the rotating object by pouring from a crucible—the rapid solidification rate (RSR) method—or by melting the end of a bar *in situ*—rotating electrode process (REP) method. The droplets, thrown off by centrifugal force, spheroidize and solidify in flight.

Mechanical atomization processes include a couple of techniques to disrupt a melt stream by purely mechanical means. In twin roll atomization, the main technique to lend itself to commercial production, powder is produced by directing a stream of molten metal between two high-speed, rotating rolls [*10*]. Particle size and shape are controlled by variations in the roll gap and the rotational speed.

Chill methods require stabilization of the melt source so that a relatively continuous product forms. For both free-flight melt-spinning (FFMS) and chill-block melt-spinning (CMBS), the molten charge is extruded through an opening in the bottom of the crucible by an overpressure of inert gas to form a continuous stream. In the case of FFMS, the cooling rate of the free-falling wire is enhanced by various gas or liquid quenchants. For CBMS, however, the extruded melt impinges on a rotating substrate to produce a flat ribbon. Cooling rate can be increased by decreasing metal flow rate and increasing substrate velocity. Planar flow casting (PFC) and melt drag (MD) processes are offshoots of CBMS that allow production of wider strips. Liquid metal is pressure-fed to the nozzle in PFC and gravity fed in MD. In crucible melt extıaction (CME) and pendent drop melt extraction (PDME) processes, metal solidifies on the edge of a disc brought into contact with a liquid metal. A short contact time is maintained before the metal solidifies and is released. The continuity of the casting can be disrupted by the introduction of notches at the edge of the disc, and staple fibers can be cast.

A primary limitation of both spray and chill methods is that heat transfer occurs across a boundary layer at the surface in contact with the heat sink, perfect contact occurring only at isolated points. An advantage of directed energy processes (weld methods) is that the liquid layer is in perfect thermal contact with the chill surface (the bulk material) such that very high cooling rates can be attained. These processes rely on an intense source of directed energy (laser, plasma, or electron beam) to melt a surface to a variety of depths.

An alternative technique for producing particulate is that of mechanical alloying (MA). This process has been used successfully to produce particulates in several ternary Al-Li-X PM alloys [*11–13*] and in an Al-Fe-Ni PM alloy [*14*]. The process of alloying and dispersing insoluble constituents, accomplished completely in the solid state, involves high-energy dry-ball milling in an attritor grinding mill of a blend of Al powders with alloy additions in the form of elemental and/or master alloy powders [*15–17*]. The powder particles become cold-welded as a result of the milling action and are subsequently fragmented. Submicron scale homogeneity is eventually achieved as the cold-weld-fracture process is repeated over some period of time. Diffusion also aids in homogenizing the microstructure as a result of temperature increases and high lattice defect densities. Fine oxide (Al_2O_3) and carbide (Al_4C_3) dispersions can also be introduced by MA, carbon being added to prevent excessive cold-welding. These dispersoids contribute significantly to increased strength over conventionally cast IM alloys of similar chemistry. The techniques for producing rapidly solidified particulate are summarized in Table VII [*8*].

B. Consolidation

The most common consolidation sequences for rapidly solidified aluminum alloys are shown in Table VIII [*2*]. The multistep approach is costly and requires knowledge and control of each separate step for a given sequence. Rapidly solidified particulate is normally cold-compacted to 70–80% theoretical density followed by placement in some type of container which is then vacuum degassed and hot-pressed. The billet, upon removal from the container, can be fabricated by conventional extrusion, forging, or rolling techniques.

A vacuum degassing step is necessary in each sequence to remove

TABLE VII

RAPID SOLIDIFICATION TECHNIQUES

Technique	Product type	Typical dimensions (μm)	Typical cooling rate (Ks^{-1})
Gas atomization	Smooth, spherical powder	50–100 dia.	10^2–10^3
Water atomization	Rough, irregular powder	75–200 dia.	10^2–10^4
USGA	Smooth, spherical powder	10–50 dia.	up to 10^6
Drum splat	Flake	1–3 mm dia. × 100 thick	10^4–10^5
REP	Smooth, spherical powder	125–200 dia.	10^2
RSR	Smooth, spherical powder	25–80 dia.	10^5
Twin roll atomization	Variable, powder to flake	200 thick flake	10^5–10^6
FFMS	Circular section filament	100–200 dia. wire	10^2–10^4
CBMS	Rectangular section tape	10–100 thick up to 3 mm wide	10^5–10^7
PFC/MD	Wide rectangular section tape	20–100 thick up to 150 mm wide	10^5–10^6
CME/PDME	Filament of fiber	20–100 thick	10^5–10^6
Directed energy	Surface layer	10–1,000	10^5–10^8

TABLE VIII

COMMON CONSOLIDATION SEQUENCES

Path	1	2	3	4
Cold compact	x	x		
Can	x		x	x
Vacuum degas	x	x	x	x
Hot-press[a]	x	x	x	x
Decan	x		x	x
PM billet	x	x	x	x
Mill product form[b]	x	x	x	x

Note: An x indicates that step is performed in a given sequence: 1, 2, 3, or 4.

[a] For path 1: hot-pressing
path 2: vacuum hot-pressing
path 3: hot isostatic pressing
path 4: hot extruding

[b] Types: extrusions, forgings, plate, sheet.

hydrogen and water vapor. Degassing temperatures should be as high as possible, but not so high that the initial, as-cast microstructural features are adversely affected by coarsening. Degassing temperatures must be more carefully controlled for the non-heat-treatable, dispersion-strengthened alloys in order to retain the desirable RS microstructural features while optimizing subsequent process requirements.

The surface of RS aluminum-alloy products consists of a thin, hydrous oxide layer formed during atomization which thickens during subsequent handling. Though the oxide content in the final, consolidated product is generally small, the redistributed oxides from stringers that may have detrimental effects on mechanical properties. These deleterious effects can be reduced by lowering the overall oxide level or by processing the material to obtain a more homogeneous oxide distribution. Production of RSP products having the thinnest possible surface oxide layer is one way to reduce the overall oxygen level. Formation of a continuous surface oxide film on aluminum is, however, thermodynamically favored, even at very low oxygen partial pressures in inert gases. The oxide-layer thickness is a function of the production techniques and parameters and (for a given set of conditions) alloy composition, with alloys containing Li, Mg, or Ce having the thickest layers [*18*]. The oxygen level for a given alloy system will also be a function of particulate size and shape, which, in turn, are functions of production technique and cooling rate, among others.

Surfaces of RSP aluminum alloys contain hydroxides, which form during the production process, and adsorbed moisture, which is picked up during subsequent handling and storage, in addition to the ever-present oxide film. When aluminum powders, for example, are heated, adsorbed moisture can evaporate, and hydroxides can decompose. The presence of stable oxides and temperature-sensitive hydroxides on RSP aluminum-alloy–product surfaces places a constraint on particulate bonding during the consolidation processes.

Though fracture of the oxide film is facilitated by the degassing step through film decomposition, it is during the hot-pressing and working steps that the oxide film is actually fragmented at prior particle boundaries (PPBs). These steps are necessary for sufficient breakup of the oxide film to occur so that the amount of metal–metal bonding can be increased. Oxide layers form aligned stringers with further processing. This subsequent deformation increases PPB area and interoxide spacing. Beyond some critical strain level, PPBs are not observed, and oxide stringers are not continuous [*19*].

Extreme thermal excursions evident in conventional consolidation processes can be avoided by dynamic powder compaction [*20,21*]. Powders are impacted by a high-velocity projectile in this process, and the passage of the resulting shock wave through the powders effects consolidation. Interparticulate bonding is believed to occur by localized melting at powder surfaces caused by frictional heat and locally high deformation rates. These small liquid regions solidify rapidly, and the main RSP microstructural features are retained.

In the direct powder-rolling process, alloy powder particles are directly roll-compacted to form thin metal strips of near-final thickness [22]. A sintering step, necessary to homogenize the strip, follows the compaction step. Final product form is obtained by multiple rolling and heat-treating cycles, thus eliminating several of the steps from the conventional consolidation sequence and reducing costs.

IV. Theoretical Discussion

A. *Heat Transfer in Atomized Powders*

Heat-transfer conditions inside a supercooled liquid droplet have been described by Levi and Mehrabian [23,24]. In the atomization process heat flow conditions involve solidification into a supercooled melt, where most of the heat of fusion is absorbed by the liquid through recalescence. Achievable interface velocities are limited by the amount of undercooling rather than by the rate of heat extraction or sample volume. Two distinct solidification regimes can be identified in the atomization process. The first is the rapid solidification stage: the heat of fusion is absorbed by the supercooled droplet. During this stage, heat lost to the environment is negligible. During the second stage, the external heat transfer limits the progress of solidification. This state, one of relatively slow growth, develops after the droplet undercooling has been relieved and is approximately isothermal.

From the results of an in-depth study of the thermal history of a solidifying droplet considering a nonuniform temperature distribution, Levi and Mehrabian [23,24] concluded the following:

1. Local recalescence allows the interface to change its velocity and its adjacent thermal field to accommodate the heat of fusion.
2. At a given undercooling with increasing particle size, the extent of the rapid solidification regime is reduced.
3. Decreasing the particle size and the nucleation temperature and increasing the heat-transfer coefficient (h) will increase the interface velocities and the amount of undercooling.
4. Production of a homogeneous solid requires an undercooling described by the following relationship:

$$\Delta H_{\mathrm{m}} \leq \int_{T_{\mathrm{N}}}^{T_{\mathrm{c}}} \mathrm{C}_1 dT \qquad (1)$$

where ΔH_m is the latent heat of fusion, C_1 is the specific heat of the liquid, T_N is the nucleation temperature, and T_c is some critical temperature.

As long as the interface temperature is below T_c, which is less than T_0 [25], the interface remains planar. To obtain a completely homogeneous solid, the droplet must be supercooled to T_N or "hypercooled" such that the heat of recalescence produced never increases the melt temperature above T_c.

B. *Heat Transfer in Splat and Melt-Spun Materials*

Both melt-spun and splat RSP materials experience heat loss by conduction. The solidification front moves into a superheated liquid, and the heat of fusion is withdrawn through the solid. The achievable interface velocities are limited by the rate of heat extraction to the substrate and the thickness of the material being processed. The undercooling obtained in melt-spinning is limited, at best, and, due to the positive temperature gradient ahead of the solid–liquid interface, the controlling factor in microstructural development is the high interfacial velocity. One of the problems associated with splat or melt-spinning processes is that a buildup of material on the substrate will reduce the heat transfer and, therefore, the interface velocity and, thus, very coarse microstructures consisting of equilibrium phases may form.

Numerous investigators have attempted to model the melt-spinning process by considering heat and mass flow during the casting operation [26–30]. The following observations are consistent with their analyses:

1. For large h (better thermal contact bettween substrate and melt), ribbon thickness decreases.
2. An increase in substrate velocity decreases ribbon thickness.
3. An increase in the initial temperature of the melt may increase ribbon thickness. Effects vary due to the interaction of thermal and flow transport mechanisms.

Clyne [31] found values of h ranging from 10^3 up to 10^6 Wm^{-2} K^{-1} for melt-spinning. Increased ribbon thickness raises both solidification time and temperature differences in the system such that it is possible to have initially undercooled liquid become superheated and, therefore, solidify more slowly through recalescence. These changes may be accompanied by structural degeneration such as microsegregation or phase separation. Additional research is required to determine the complex effects of h, ΔT, and other process variables to eliminate adiabatic recalescence (net latent heat evolved during recalescence much greater than net heat extracted through surface during recalescence), through improved processing. Considerations of melt-substrate contact time for continuous processes should also enter into such studies.

C. General Thermodynamic and Kinetic Considerations

The stability of the solidification interface associated with rapid growth following undercooling is important in determining the final microstructure. The classical theory of constitutional undercooling has generally been used as a fair approximation for determining growth conditions that result in unstable interfaces [*32*]. The temperature gradient in the liquid, G_L, preceding the moving front and its velocity, R, control solidification-front morphologies [*33*]. A planar interface, necessary for partitionless solidification, is stabilized by a steep temperature gradient and destabilized by a steep solute gradient resulting from microsegregation of solute and solvent atoms. This is the boundary between the planar and cellular regions. When the temperature gradient dominates, plane front growth is morphologically stable; when the solute gradient dominates, cellular and dendritic structures are more stable (constitutional supercooling regime).

The product of G and R is a measure of the average cooling rate during solidification. Increasing values of $G \times R$, therefore, correspond to the finer microstructures obtainable in rapid solidification. Since the dimensions of the as-cast RSP products are very small and casting speeds are very high, it is virtually impossible to directly measure cooling rate. Average cooling rates can be estimated by measuring some microstructural feature such as secondary dendrite arm spacing. This spacing is related to the time available for microstructural coarsening (local solidification time) and, therefore, to the heat-removal rate. An estimate of cooling rate can be made from plots as shown in Fig. 1 [*33*], where the measurements at low cooling rates are extrapolated to the rapid solidification regime. A reliable estimate of cooling rate can only be given by an order of magnitude, based on the relative uncertainties of the results.

For very high solidification velocities, the plot of G versus R is as shown in Fig. 2 for an Al-Si alloy [*34*]. Morphological instabilities resulting from microsegregation of solute and solvent atoms can no longer form. Stability is reestablished, and planar growth resumes beyond line EF in Fig. 2 for very large R. This behavior occurs because the perturbations are so small that surface tension becomes a major stabilizing force on the interface. Consideration of the stabilizing effect of solid–liquid surface energy led, in

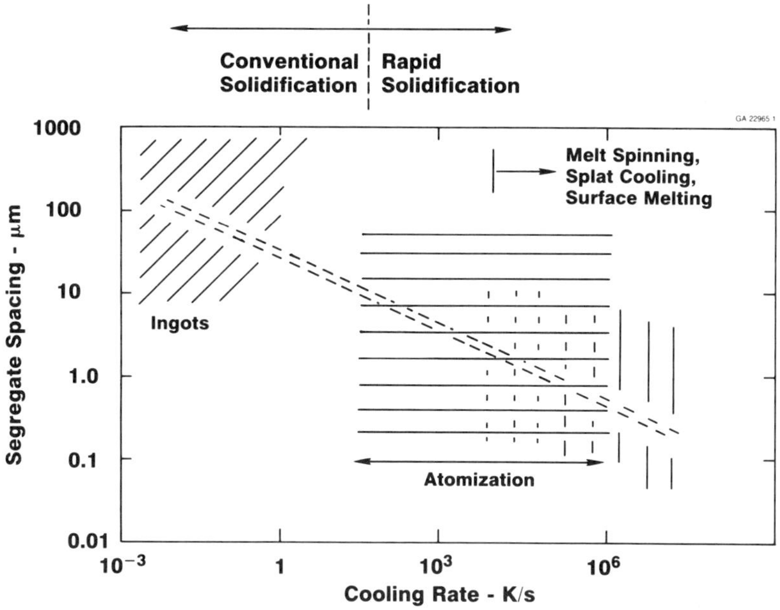

FIG. 1. Segregate spacing versus cooling rate for aluminum alloys solidifying in the dendritic mode.

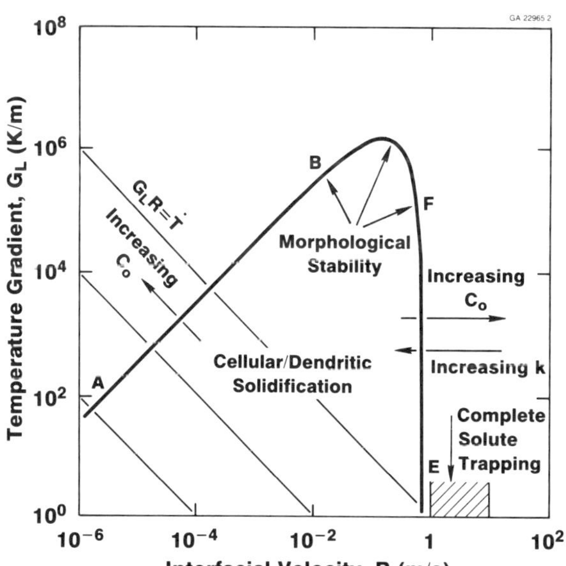

FIG. 2. Regimes of homogeneous plane front growth versus segregated cellular and dendritic growth as determined by the liquid temperature gradient (G_L) and the interface velocity (solidification rate, R) for Al-0.1% Si. The dome-shaped line separating the two regimes, the locus of morphological stability, is calculated assuming equilibrium partitioning of Si at the moving interface. The cross-hatched band indicates the estimated interfacial velocity, which is fast enough to achieve partitionless solidification.

part, to the theory of morphological stability by Mullins and Sekerka [*35,36*] and Sekerka [*37*]. At high growth rates, surface energy becomes more important than the destabilizing effect of the solute gradient. Figure 2 illustrates the fact that interface stability can be attained for sufficiently fast growth rates, which become insensitive to temperature gradient at the limit of absolute morphological stability [*28,30*], the locus of morphological stability (cellular–dendritic solidification) being a dome-shaped line. The locus is calculated assuming equilibrium partitioning at the moving interface. Increasing the cooling rate increases the number of gradient–rate combinations that can be used to avoid microsegregation. (See Coriell and Sekerka [*27*] and Baker [*38*] for additional information.) Morphological stability theory does, generally, predict far greater stability for rapid solidification than that predicted by constitutional supercooling. Solidification at the limit of morphological stability (absolute) does not necessarily imply partitionless solidification. Partitioning between the solid and liquid in the direction of growth may take place though no lateral partitioning can occur. The range of velocities for which the former will not occur is shown as the cross-hatched region in Fig. 2.

One of the most important factors in allowing rapid solidification to occur is the amount of undercooling at the onset of nucleation, particularly in powder materials. In numerous studies by Perpezko et al. [*39–42*], the amount of liquid undercooling has been found to be a key factor in controlling phase-selection kinetics, which in turn determines microstructural development. A range of possible nucleation sites showing different nucleating capabilities exist in any liquid sample. By continually isolating or removing nucleants from the sample, and undercooling the characteristic of the most potent remaining nucleant can be observed [*42*]. The use of small volumes via RSP basically allows the number of nucleant sites to be reduced so that large undercoolings can occur. If the amount of undercooling is large enough and the solidification rate is fast enough, partitionless solidification or complete solute trapping can occur.

From the preceding discussions of heat transfer and kinetic considerations, an important statement can be made: a homogeneous microstructure can be produced either by large undercoolings, as for atomized powders, or by high interface velocities, as for splat and melt-spun ribbon.

Some consideration of thermodynamic conditions is important in any discussion of nonequilibrium solidification microstructures, particularly in terms of partitionless solidification. Baker and Cahn [*25*] have discussed the thermodynamics of nonequilibrium solidification for binary alloys. The basis of their discussion rests on the assumptions that thermodynamics may be applied locally and that different processes at the interface are independent. From these stems the assumption of constrained equilibrium (i.e.,

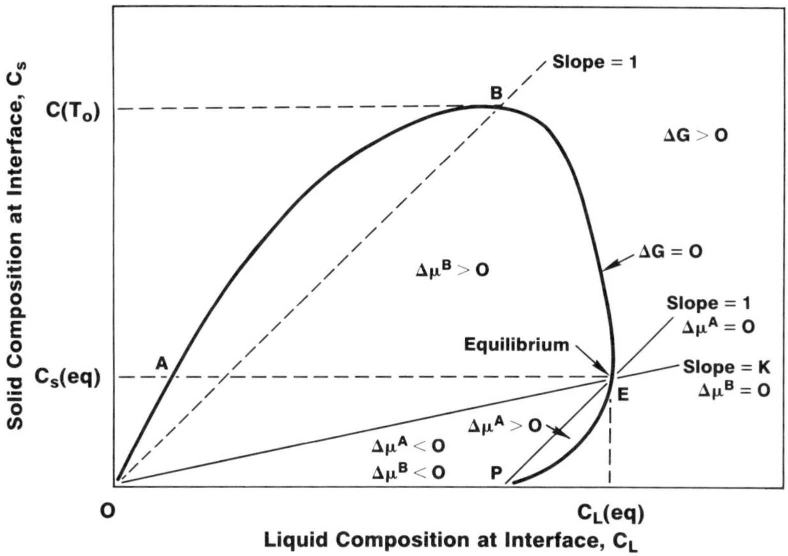

FIG. 3. The domain of all possible solid compositions that can form from various liquid compositions at a given temperature is enclosed within the curve OABEP.

certain processes are assumed to occur or not to occur so slowly that thermodynamics is applicable to the remaining processes). Examples include metastable phase equilibrium (constraint-absence of one or more stable phases) and partitionless solidification (constraint-diffusional mobility of one or more components).

The range of solid compositions, C_s, that can form from different liquid compositions, C_L, can be plotted as shown in Fig. 3 [*25*], the curve where the free-energy change is zero. Point E is the equilibrium composition. Tangent 1 in Fig. 4, the tangent to the liquid, touches the solid at a single point [*25*]. C_s increases and then decreases as C_L moves to the left (tangent 2). Tangent 2 (C(T_0) in Fig. 3) is the maximum solid composition and is the point where the two free-energy curves intersect. These diffusionless transformations are shown by line OB on Fig. 3. The locus of the compositions and temperatures where the liquid and solid free energies are equal is known as the T_0 curve. It represents the highest interface temperature at which the partition coefficient, $k = C_s/C_L$ (concentrations of solute in solid and liquid), can be unity. Partitionless transformations occur at some $T < T_0$ for a given composition, provided the temperature on recalescence is below T_0. The area within the curve above line OE corresponds to solute trapping; below line PE, it corresponds to solvent trapping. In these two regions, although the overall free energy decreases, one of the components

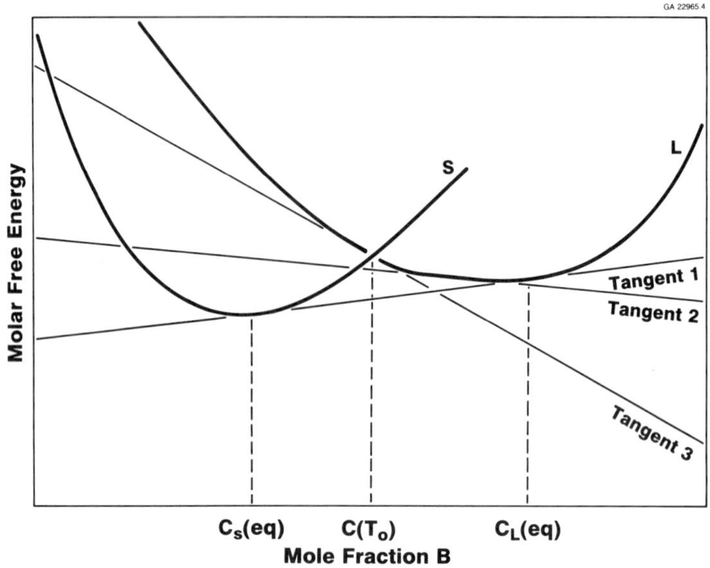

FIG. 4. Schematic free-energy curves to illustrate evolution of the curve OABEP in Fig. 3.

experiences an increase in chemical potential on crossing the interface, which, in turn, implies that the species are solidifying in a cooperative manner. A component experiences an increase in chemical potential on crossing the interface because (1) it is trapped by the advancing solidification front or (2) it is a necessary participant in an independent solidification reaction mechanism involving several species. Consideration of the reaction kinetics may further decrease the critical temperature below the T_0 value [*44*]. The major problem with classical thermodynamics in general is its inability to predict precisely what structures will appear. One must certainly question Baker and Cahn's [*25*] two basic assumptions.

V. Microstructural Effects

A. General Microstructural Effects: RSP Materials

Microstructures resulting from rapid solidification generally contain the features of conventional casting including grain, dendritic, and eutectic morphologies, though on a much finer scale. The sizes and quantities of microstructural features are strongly dependent on particle volume through undercooling and interface velocity, as well as composition.

Thin sections of gun splats may exhibit elongated or equiaxed grains with diameters of ~0.1 μm [*45*]. Thicker, multilayer regions may exhibit columnar growth extending from layer to layer [*46*]. Melt-spun ribbons may exhibit an initial chill zone on the substrate side leading to columnar growth with the possibility of an equiaxed zone adjacent to the air side [*47*]. Most powder particles having an average diameter much less than 100 μm are single grains, though some Al-Si powders have been observed to contain two or more grains and even twins [*24*].

Rapidly solidified materials exhibit extremely fine dendrites, often with no secondary arms for samples cooled at very high rates. In addition to dendritic and cellular growth, "predendritic" growth has been observed on the chill face of splat and melt-spun RSP products [*48*]. These regions are points of contact and nucleation on the chill face from which growth begins with a smooth, curved front retaining the original solute content of the parent melt and are surrounded by a region in which the dendritic arms are noncrystallographic. This interface breaks down into a cellular-dendritic mode with the associated solute microsegregation [*49*]. For a totally featureless, nonsegregated microstructure, partitionless solidification is necessary. Cellular-dendritic structures have also been observed in numerous powder studies.

Eutectic microstructures have also been observed in splat and melt-spun materials as well as powders. Kofler [*50*] found that a melt of noneutectic composition could solidify as a "quasi-eutectic" over a composition range that expands with increasing undercooling. These regions are referred to as coupled growth zones, whose shapes are determined by growth-rate variations with undercooling and composition [*51*]. However, because of the requirements of constant interface velocity and unidirectional heat flow, it is difficult in practice to see a ribbon or a variety of splats having a completely coupled eutectic structure.

B. Specific Microstructural Effects

The majority of the product microstructures in RSP aluminum-transition metal alloys consist of finely divided secondary intermetallic phases. These phases arise from breakdown of the dendritic-cellular, as-cast microstructure and from formation of a small fraction of precipitates by decomposition of the supersaturated, as-cast microstructure during degassing, preheating, and consolidation. This dendritic-cellular breakdown is illustrated in Fig. 5 for an Al-Fe-Ni-Co alloy. These secondary phases are primarily responsible for the strengthening. A small fraction of the microstructure may contain coarse, primary intermetallics which have formed in

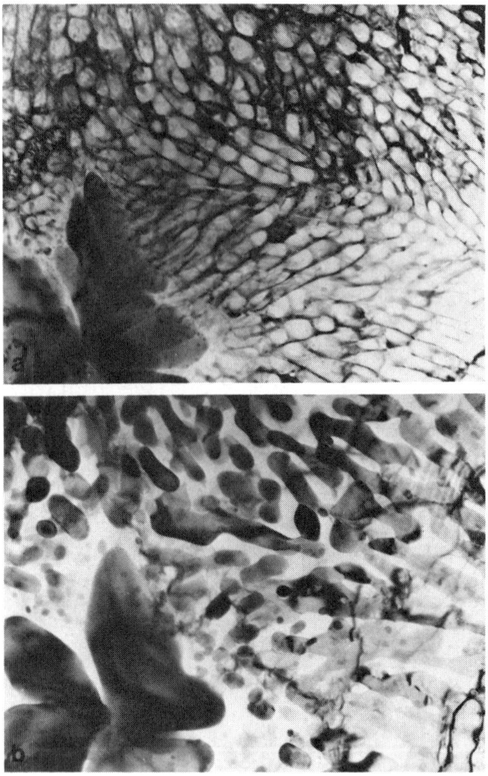

FIG. 5. Distribution of solute in a relatively slowly cooled Al-3.3 Fe-2.3 Ni-4.6 Co splat (a) before heat treating and (b) after heat treating one hour at 673K.

areas of relatively slow solidification. These coarse particles do not change size or shape significantly upon exposure to elevated temperature, though solute redistribution does occur in the surrounding cellular network. However, the size and spacing of the individual particles formed by solute redistribution are a function of the time at some temperature. The fractions of the various constituents are, thus, a function of the cooling rate. By increasing the cooling rate, the secondary intermetallics can be refined and the coarse, primary intermetallics can be eliminated. Optimum tensile properties will result from RSP particulate free of coarse, primary intermetallics but containing a high volume fraction of fine, secondary intermetallics.

The solidification structures found in splat and melt-spun aluminum-transition metal alloys have been studied and described in detail by Jones [*46*] and, Garrett and Sanders [*52*]. Zone A, consisting of fine, primary aluminum dendrites free of optically visible precipitates, is found on the substrate side of splat and ribbon, while zone B, consisting of primary intermetallic particles surrounded by primary aluminum, is generally found on the air side. Primary intermetallics on the air side appear to nucleate simultaneously with primary aluminum on the substrate side. Primary aluminum with solute segregated at dendrite-cell boundaries surrounds each intermetallic and grows radially outward until reaching other growth fronts. The kinetics of the solidification process suppress nucleation of the intermetallic so that aluminum nucleates and grows. Although solute is found in the intermetallics and segregated in the cellular structure in zone B, and it remains in the supersaturated matrix in zone A, average overall compositions of each zone would be similar. Primary intermetallic growth is controlled by external heat transfer conditions, hence the critical importance of ribbon thickness on microstructural features. The faster a material can be cooled and solidified (i.e., the thinner the ribbon or splat), the finer the cells and primary intermetallics will be. Eventually, at high enough cooling rates, the primary intermetallics are eliminated, and a featureless microstructure is formed. Examples of the coarse structures consisting of primary intermetallics and cells are shown in Fig. 6. Featureless microstructures of the thinnest ribbon (most rapidly solidified) are shown in Fig. 7. Through-thickness sections of melt-spun ribbon were examined to study the formation of zones A and B. As mentioned previously, a given ribbon may contain zone A, zone B, or both due to process variations. For a ribbon containing only zone B, the intermetallics may have existed prior to melt-spinning if, for example, the superheat necessary to dissolve the particles initially is inadequate. Conversely, for a ribbon containing only zone A, critical superheat has been attained. The appearance of a combined zone A-zone-B structure seems to discount the notion that zone B existence is only related to superheat and preexisting nuclei in the melt, although sufficient superheat and amount of time at temperature are necessary to ensure dissolution of all preexisting phases. In particular, the observation that zone A appears exclusively on the wheel side (in a ribbon containing both zones) precludes the possibility that the sole mechanism of zone B formation is growth of preexisting nuclei in an insufficiently superheated melt. A through-thickness section or ribbon containing both zone A and zone B can be seen in Fig. 8. The transition region between zones A and B is small to nonexistent. This is due to the rate-limiting, competitive growth of the numerous grains on the air side. Coarse intermetallics with small grains do appear near the air side while the columnar regions of zone A appear on the substrate side.

In the atomized powders, nucleation generally occurs at the droplet surface. A homogeneous region of zone A is produced as solidification

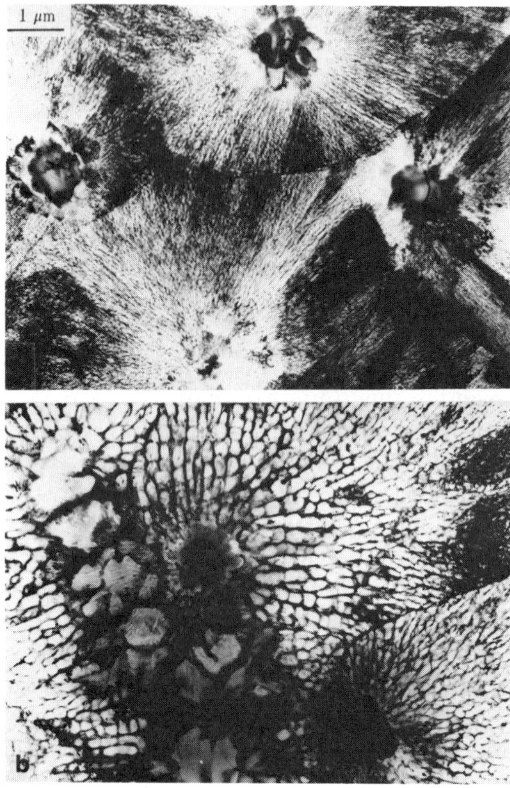

FIG. 6. Section of Al-10 wt % Co ribbon comparing slowly solidified microstructures. Section (a) solidified faster than section (b).

FIG. 7. Section of Al-10 wt % Co ribbon showing growth fronts that have undergone the fastest solidification.

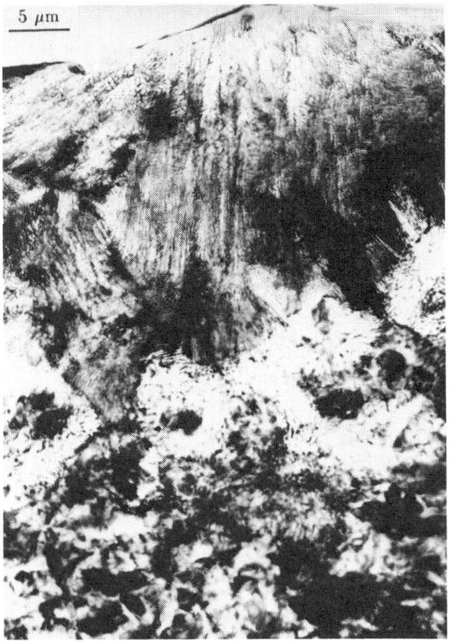

FIG. 8. Section of compacted (ROC) Al-10 wt % Co ribbon showing the edge of the columnar zone (zone A) to the transition region.

proceeds into a highly undercooled melt, provided the previously mentioned heat conditions are satisfied. Interface velocity and undercooling are reduced as the heat of fusion is absorbed into the liquid. When the interface temperature becomes greater than T_c, solute rejection into the liquid occurs. Eventually the interface becomes unstable, and dendrite–cell growth occurs. Different powder sizes will also contain a variety of solidification structures. These microstructures are affected by the powder size through the amount of undercooling. The largest powders or powders containing a high solute content generally show enhanced segregation including, for example, the presence of coarse primary intermetallics and cells in Al-Fe-Ni, Al-Fe-Ni-Co, and Al-Mn-Si alloys [*53*]. Smaller particles may exhibit a totally cellular structure or a combination of cells and zone A. A typical powder structure is shown in Fig. 9. The smallest particles will exhibit a complete solid solution, zone A.

Most microstructural work on consolidated-fabricated, rapidly solidified aluminum alloys has centered on forms produced from powder and splat [*54–56*]. Studies by Sanders et al. [*54*], utilizing extruded powder and splat materials, showed that final microstructures are affected by particulate morphologies and solidification rates as well as by consolidation and extrusion temperatures. Two general but distinct regions (one of large particle

FIG. 9. Typical powder microstructure of Al-Fe-X showing a possible combination of cellular structure and zone A.

distribution, the other of small) have been observed. The large, coarse particles formed during the initial solidification process. These two regions can be observed in Fig. 10. The fabricated powders reveal a more uniform particle size with fewer large particles than the splat product. This behavior is related to the buildup of splat on the substrate, which reduces the heat transfer, thus allowing coarse phases to form. Basic particle shapes present in these products include (1) large, irregularly shaped particles, (2) small, equiaxed particles, and (3) small, oriented, rod-like particles [*52,54*]. Increasing the consolidation temperature has the effect of creating in the product microstructure a narrow particle size distribution having fewer small particles.

The advantages of rapid solidification processing have opened several new avenues for novel alloy design. The extended solid solubility due to rapid solidification leads to solid-solution strengthening, and, by suitable thermal processing, a high volume fraction of precipitates can be generated from the supersaturated matrix for precipitation hardening. Rapid solidification also results in the formation of finely divided intermetallic particles, which are homogeneously dispersed throughout the microstructure. High levels of strength can therefore be obtained in rapidly solidified alloys by precipitation and dispersion-hardening mechanisms. The RS particu-

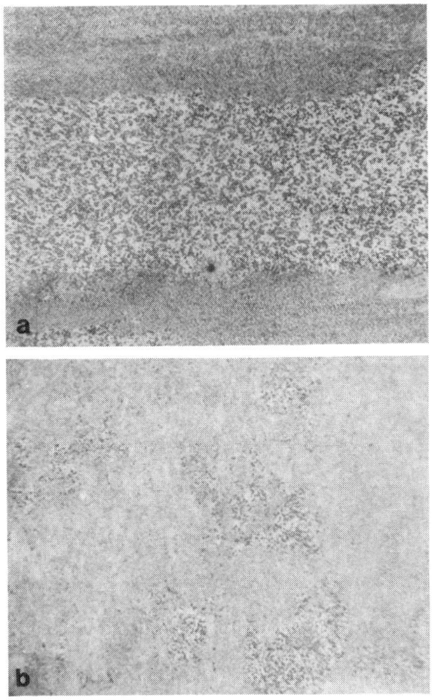

FIG. 10. Microstructures of sections from (a) air-splat and (b) air-atomized particulate consolidated and extruded at 675K.

lates are compacted and processed via the powder-metallurgical route to the final product form. Due to the small powder-particle size, an additional component due to grain-size strengthening may be realized.

VI. Design of High-Temperature, Rapidly Solidified Aluminum Alloys

Microstructural conditions necessary to attain maximum strength include a high volume fraction of intermetallics, a fine intermetallic particle size, a uniform particle distribution, and a thermally stable particle. Fabricated products having a high volume fraction of fine, uniformly sized particles should exhibit the highest yield and tensile strengths. This result has been confirmed by Sanders et al. [*54*] and Paris et al. [*57*] and is summarized in Table IX. The air-atomized powder found to have the smallest, most uniform particles exhibited the highest strengths. Paris et al.

TABLE IX

TENSILE PROPERTIES OF PM Al-Fe-Ni-Co AND Al-Mn ALLOYS

Particulate morphology	Yield strength (MPa)	Tensile strength (MPa)	Elongation (%)	NTS/YS[a]	Consolidation temperature (K)	Extrusion temperature (K)
Argon splat	236	314	17.8	1.94	675	675
Air splat	287	360	17.1	1.77	675	675
Air atomized	360	426	9.2	1.47	675	675
Air splat	295	360	11.2	1.50	675	666
Air splat	206	286	19.7	2.03	725	666

[a] NTS/YS = ratio or notched tensile strength to yield strength.

[57] have shown that increasing the solute level from 5.0 to 7.5 at % for the same processing conditions increases the volume fraction of particles. A simultaneous reduction in interparticle spacing occurs, a result that has the effect of increasing the yield and tensile strengths over those of the lower solute alloys.

Because of the possible applications of these rapidly solidified, wrought alloys in the aerospace industry, some indication of the toughness must be developed to ensure a usable product. The notched tensile strength/yield strength ratio (NTS/YS) is one such indicator. Generally speaking, a decrease in the NTS/YS ratio occurs with an increase in yield strength. The relationship of NTS/YS versus YS has been found to be similar for the materials studied by Paris et al. [57] and Sanders et al. [54] shown in Fig. 11 for alloys listed in Table X. Alloys containing up to 25 vol. % second phase show an increased strength-toughness relationship, while the reverse is observed for alloys containing more than 25 vol. % second phase.

The subject of precipitation and dispersion hardening has received extensive consideration in the literature and is qualitatively well understood. Orowan [58] proposed the model for hardening of an alloy containing a dispersion of strong obstacles that impede the motion of glide dislocations during deformation. However, the model described by Orowan considers only athermal glide of the dislocations. At high temperatures, the yield stress falls drastically, a behavior attributed to the local climb of dislocations around the particles [59]. Several theoretical treatments accounting for the decline of yield stress at elevated temperatures due to dislocation climb are available in the literature [59]. These theories predict that the limiting value of climb-controlled yield stress is approximately 0.4 times the Orowan stress at low temperatures. A good understanding of the high-temperature-deformation behavior, therefore, is essential for alloy development.

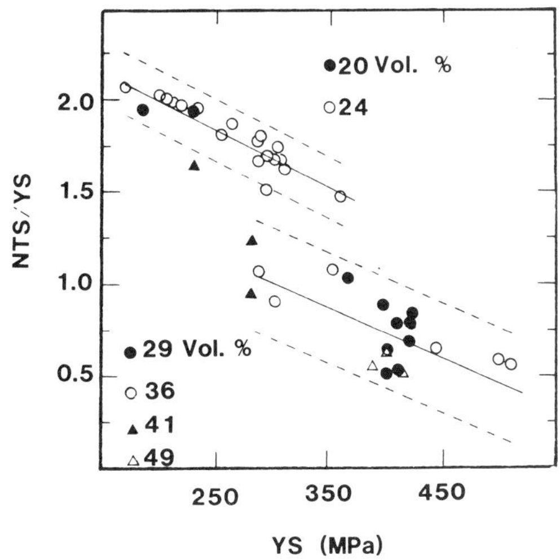

FIG. 11. Longitudinal notched tensile strength/yield strength NTS/YS versus YS plot for alloy extrusions noted in Table X (powder and splat particulate).

TABLE X

COMPOSITIONS OF PM ALLOYS (AT %)

Mn	Si	Fe	Ni	Co	Total solute	Volume fraction
5.00	—	—	—	—	5.00	0.41[a]
—	—	1.67	1.67	1.67	5.00	0.24[b]
—	—	1.67	1.11	2.22	5.00	0.24
—	—	1.67	2.22	1.11	5.00	0.24
6.00	—	—	—	—	6.00	0.49
—	—	2.00	2.00	2.00	6.00	0.29
—	—	3.00	3.00	—	6.00	0.29[c]
7.50	—	—	—	—	7.50	0.61
5.00	2.50	—	—	—	7.50	0.24[d]
—	—	2.50	2.50	2.50	7.50	0.36
6.00	3.00	—	—	—	9.00	0.29

[a] Calculated assuming 25% $Al_{12}Mn$, 75% Al_6Mn.
[b] Calculated for $Al_9(Fe,Ni,Co)_2$.
[c] Calculated assuming Al_9FeNi.
[d] Calculated assuming $Al_{12}Mn_3Si$.

The principal difference between the classic precipitation-hardenable 2XXX and 7XXX alloys and the PM high-strength aluminum alloys is that, although microstructure and, therefore, strength may be affected by processing (i.e., extrusion, rolling, or forging conditions), the majority of the microstructural elements derive their origins from the quenching and aging practices. Heat treatments for the IM 2XXX and 7XXX are reversible in that an aged alloy can be re-solution-heat-treated and aged to obtain similar mechanical properties as before. With PM aluminum-transition metal systems, however, the strength is controlled by the particle dispersion. Since the microstructure is supersaturated with respect to the solute elements, if inappropriate processing occurs, the structure cannot be returned to its initial state by any processing. Consequently, it is important to have a thorough understanding of processing and structure in order to control properties. The fact that PM-processed materials can be fabricated with properties equivalent to or superior to those of similar IM-processed material is encouraging. It suggests that the process itself does not adversely affect structure and properties. Research at Alcoa [*60*] has demonstrated that the PM-processed 7XXX aluminum alloys possess superior mechanical properties as compared to those of IM-processed 7XXX alloys. Aluminum alloys are also being designed for elevated temperature applications.

For elevated temperature applications, the alloy microstructure must exhibit a prolonged thermal stability, and the dispersed phase in the microstructure must be thermodynamically stable. Moreover, in order to maintain the alloy strength and, in addition, to have acceptable fracture properties, the dispersed phase should resist coarsening when the alloy is exposed to elevated temperatures. The driving force for coarsening is a reduction in overall interfacial energy of the system, and the mechanism for coarsening may be volume diffusion controlled, grain-boundary or substructure diffusion controlled, or interface controlled. An ideal high-temperature aluminum alloy, therefore, would have the following characteristics:

1. low diffusivity of solute in the aluminum matrix,
2. thermodynamically stable dispersed phase, and
3. low interfacial energy between the matrix and the dispersed phase.

Table XI contains the diffusivity values of several elements in aluminum at 350°C [*61*]. The transition-metal elements and the rare-earth elements have, in general, relatively low diffusivity. A number of alloys based on the aluminum-transition-metal systems have been produced, and their microstructure and properties have been studied extensively. Two major alloy

TABLE XI

INTERDIFFUSION COEFFICIENTS OF VARIOUS ELEMENTS IN ALUMINUM AT 350°C

Ce	8.6×10^{-16} cm^2/sec
Co	1.1×10^{-13} cm^2/sec
Cr	1.2×10^{-12} cm^2/sec
Cu	3.7×10^{-12} cm^2/sec
Fe	5.4×10^{-14} cm^2/sec
La	4.6×10^{-16} cm^2/sec
Li	9.0×10^{-12} cm^2/sec
Mn	1.7×10^{-16} cm^2/sec
Mo	2.6×10^{-4} cm^2/sec
Nb	6.7×10^{-14} cm^2/sec
Nd	7.9×10^{-16} cm^2/sec
Ni	8.8×10^{-14} cm^2/sec
Sm	3.4×10^{-15} cm^2/sec
V	7.9×10^{-1} cm^2/sec

systems have been identified, namely, rapidly solidified alloys based on the Al-Fe system and on the Al-Zr system. In the following sections, the properties of both systems will be described.

A. *Al-Fe Alloys*

Rapidly solidified Al-Fe alloys have been reported to contain three different metastable phases, depending on the processing parameters [*62*]. The as-cast-structure of a melt-spun Al-Fe alloy is shown in Fig. 12a. During compaction and hot extrusion of the powders, some of these metastable phases either dissolve or transform to the stable phases. Figure 13 shows the breakdown of the cellular structure in Al-Fe-X powders. The stable phase in the Al-Fe system is $Al_{13}Fe_4$, and the most commonly observed metastable phase is Al_6Fe. Janot and Lelay [*63*] studied the Al-Fe solid solution containing a very small amount of solute ($C < 240 \times 10^{-6}$ at %) using Mossbauer spectroscopy and concluded that a true random solid solution is not obtained in Al-Fe alloys. They also deduced the presence of solute-rich clusters in the matrix after annealing their samples at 300°C. Solid-state precipitation from the supersaturated matrix has been observed in melt-spun ribbons of Al-Fe alloys [*64*]. During the first stage of precipitation, ordered precipitates form, but their structural details are not known [*65*,*66*], through with continued aging, these

FIG. 12. (a) Section of Al-8 wt % Fe ribbon showing primary intermetallics. (b) Section of Al-8 wt % Fe-8 wt % Ce showing growth fronts that have undergone the fastest solidification (as-cast).

ordered precipitates transform to semicoherent needles of the $Al_{13}Fe_4$ phase. After prolonged aging at 350°C, the needles of $Al_{13}Fe_4$ phase coarsen, and their morphology becomes irregular. Attendant with the precipitation in the matrix is the formation of incoherent Al_6Fe particles at the grain boundaries.

Addition of Mo to an Al-8%Fe alloy increases the solid-solution strength of the alloy. The presence of Mo does not appear to bring about any drastic changes in the precipitation sequence and the microstructure of Al-Fe alloys.

Addition of Ce to an Al-8%Fe alloy brings about a significant change in the microstructure and the physical metallurgy of Al-Fe alloys. An as-cast microstructure of a melt-spun Al-Fe-Ce alloy can be seen in Fig. 12b.

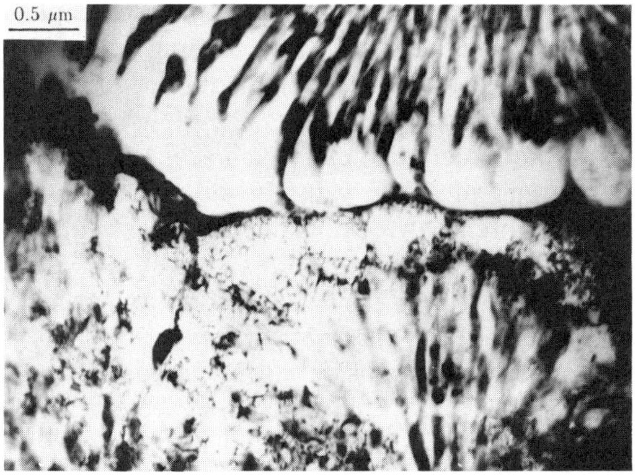

FIG. 13. Typical powder microstructure of Al-Fe-X showing the breakdown of the cellular structure and the formation of precipitates at powder boundaries in a consolidated material.

Several ternary metastable phases with undetermined crystal structures have been observed in the Al-Fe-Ce alloys. In melt-spun Al-Fe-Ce ribbon, solid-state precipitation of needles of $Al_{13}Fe_4$ phase is observed, with the possible incorporation of Ce into the structure [*65*], and precipitation of an unidentified ternary phase is observed at the grain boundaries.

A relatively small number of quantitative studies have been performed on RS Al-Fe-Ce alloys to understand the microstructure and mechanical properties of compressed and extruded alloys. Angers et al. [*67*] studied the nature of precipitates and their coarsening behavior in an Al-8%Fe-4%Ce alloy that had been cold compacted and hot isostatically pressed followed by hot extrusion. They identified the equilibrium $Al_{13}Fe_4$ and $Al_{10}Fe_2Ce$ phases and three metastable Al-Fe-Ce phases in the hot-extruded state. After aging the alloys at 425°C, only the presence of equilibrium phases was observed. Although a progressive decrease in the amount of metastable phases was observed after aging at 316°C, the microhardness measurements showed that the strength could be maintained for more than 2000 hours at this temperature. Certain quantitative measurements to evaluate the coarsening kinetics of stable particles in Al-8%Fe-4%Ce alloy at 425°C were attempted by Angers et al., but their coarsening data could not be explained by any of the available coarsening models.

An examination of the microstructure of extruded Al-Fe-Ce alloys aged at high temperatures shows a high volume fraction of dispersed phases coarsening at the subgrain boundaries and at the high-angle grain boundaries. Therefore, one would expect the coarsening behavior to be dominated by diffusion through grain boundaries and dislocation cores. Before good agreement between the experiments and theory can be found, better direct estimates of the basic parameters such as inter-diffusion coefficients and interfacial energies are essential. The diffusion data for Al-Fe and Al-Ce binary systems are known with some accuracy, but the actual diffusional interactions between Fe and Ce as solutes in Al have not been evaluated, and in the absence of such valuable information, it is difficult to analyze the experimental data systematically.

Other alloys under serious considerations based on the Al-Fe system contain additions of Co, Ni, Si, V, and other rare-earth elements. Prem Kumar et al. [*68*] have studied the mechanical behavior of RS Al-Fe-Ni alloys with different volume fractions of dispersed Al_9NiFe phase ranging from 19 to 32%. Their yield-stress data is shown in Fig. 14a and 14b. In Fig. 14a, it is seen that the room-temperature yield strength of Al-Fe-Ni alloy decreases monotonically with the average dispersoid size obeying Orowan's theory of dispersion hardening. In Fig. 14b, the yield stresses of the alloys are plotted against the temperatures of measurement for alloys containing different volume fractions of dispersoid. One alloy that has been aged to

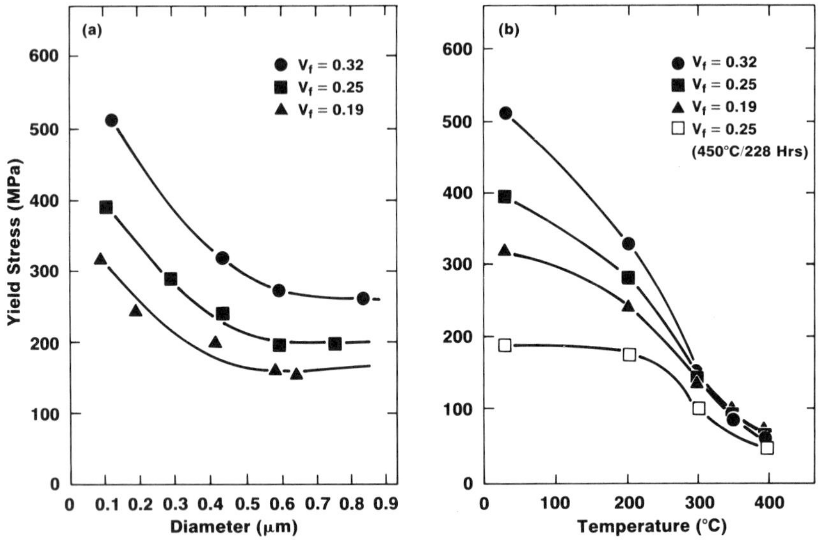

FIG. 14. (a) Room-temperature yield strength versus average dispersoid size. (b) Yield strength versus temperature (also included is one aged sample) for Al-Fe-Ni alloys containing different volume fractions of dispersoid.

coarsen the dispersoid phase has been included in the plot. All the stress values converage to one strength level at temperatures above 300°C, an observation clearly showing the effect of temperature on the deformation mode of the alloys. The high-temperature strength of Al-Fe-Ni alloys is relatively insensitive to the volume fraction and the size of the dispersoids. Similar observations have also been made in some Ni-based superalloys. The high-temperature strength of the RS alloys, therefore, is expected to be more sensitive to the subgrain structure, grain size, and other microstructural features. The nature and distribution of the dispersed particles, however, will dominate the fracture properties of the alloy.

B. Al-Zr Alloys

In order to explore systems containing thermally stable dispersed particles having low interfacial energy with the aluminum matrix, a significant amount of research has been done on RS Al-Zr alloys. With aging, the supersaturated solid solution in the RS Al-Zr alloys decomposes into the $L1_2$ Al_3Zr phase, which is either spherical or rod-like depending upon the mode of precipitation. Upon continued aging, the Ll_2 structure changes to the tetragonal plates of a DO_{23} structure. A number of aluminum alloys containing Zr, Hf, V, and Ti have been studied by Zedalis and Fine [69], since Al_3X (X = Zr, Hf, V or Ti) phases show considerable solid solubility in aluminum due to their isostructural lattices. They have determined lattice parameter variations of Al_3X phases having different amounts of Zr, Hf, V, and Ti in solid solution, and they have observed their total misfit with the aluminum lattice. In addition, Zedalis and Fine have shown that the coarsening rate of Al_3X precipitates decreases as the total misfit parameter decreases. A reduction in the interfacial energy of the precipitates should also lead to homogeneous precipitation within the matrix, avoiding preferential precipitation at the grain boundaries. Although some research has been done to develop a more complete understanding of the physical metallurgy of Al-Zr alloys, mechanical-property evaluations of these alloys have not been reported.

To summarize, the principles pertaining to elevated-temperature aluminum-alloy development are straightforward, though the proper selection of alloying elements to yield desirable properties appears to be a major concern as of the late 1980s. Most of the RS aluminum alloys exhibit attractive mechanical properties and thermal stability, but their fracture properties must be improved in order for these alloys to be used successfully in the aerospace industry. Fundamental studies elucidating proper strategies for designing PM aluminum alloys having good fracture properties continue to be an important avenue of investigation.

References

1. Quist, W.E., and Lewis, R.E. (1986). In *Rapidly Solidified Powder Aluminum Alloys ASTM STP 890*, M.E. Fine and E.A. Starke, Jr., eds. ASTM, Philadelphia, pp. 7–43.
2. Bridenbaugh, P.R., Billman, F.R., Cebulak, W.S., and Hildeman, G.J. (1985). *Light Metal Age*, **43** (9,10), pp. 18–26.
3. Wanhill, R.J.H., and Schra, L. (1986). In *Rapidly Solidified Powder Aluminum Alloys, ASTM STP 890*, M.E. Fine and E.A. Starke, Jr., eds. ASTM, Philadelphia, pp. 464–484.
4. Paris, H.G., and Chellman, D.J. (1986). In *Rapidly Solidified Powder Aluminum Alloys, ASTM STP 890*, M.E. Fine and E.A. Starke, Jr., eds. ASTM, Philadelphia, pp. 527–544.
5. Meschter, P.J., Rao, P.S., Lederich, R.J., and O'Neal, J.E. (1986). In *Rapidly Solidified Powder Aluminum Alloys, ASTM STP 890*, M.E. Fine and E.A. Starke, Jr., eds. ASTM, Philadelphia, pp. 512–526.
6. Vidoz, A.E., Crooks, D.D., Lewis, R.E., Palmer I.G., and Wadsworth, J. (1986). In *Rapidly Solidified Powder Aluminum Alloys, ASTM STP 890*, M.E. Fine and E.A. Starke, Jr., eds. ASTM, Philadelphia, pp. 237–251.
7. Sanders, T.H., Jr., and Starke, E.A., Jr., eds. (1984). *Aluminum-Lithium Alloys II*. TMS-AIME, Warrendaler, Pennsylvania.
8. Savage, S.J., and Froes F.H. (1984). *J. Metals* **36**, no. 4, pp. 20–33.
9. Anand, V., Kaufman, A.J., and Grant N.J. (1980). In *Rapid Solidification Processing: Principles and Technologies II*, R. Mehrabian, B.H. Kear, and M. Cohen, eds. Claitor's Baton Rouge, Louisiana, pp. 273–286.
10. Singer, A.R.E., and Roche, A.D. (1977). In *Modern Development in Powder Metallurgy*, H.H. Hausner and P.V. Taubenblat, eds. Metal Powder Industries Federation, American Powder Metallurgy Institute, Princeton, New Jersey, pp. 127–140.
11. Gilman, P.S. (1984). In *Aluminum-Lithium Alloys II*, T.H. Sanders, Jr., and E.A. Starke, Jr., eds. TMS-AIME, Warrendale, Pennsylvania, pp. 485–506.
12. Donachie, S.J., and Gilman P.S. (1984). In *Aluminum-Lithium Alloys II*, T.H. Sanders, Jr., and E.A. Starke, Jr., eds. TMS-AIME, Warrendale, Pennsylvania, pp. 507–515.
13. Narayanan, G.H., Wilson B.L., and Quist, W.E. (1984). In *Aluminum-Lithium Alloys II*, T.H. Sanders, Jr., and E.A. Starke, Jr., eds. TMS-AIME, Warrendale, Pennsylvania, pp. 517–541.
14. Ezz, S.S., Lawley, A., and Koczak, M.J. (1986). In *Aluminum Alloys Their Physical and Mechanical Properties* , **vol. II**, T.H. Sanders, Jr., and E.A. Starke, Jr., eds. Engineering Materials Advisory Services, London, England, pp. 1013–1028.
15. Benjamin J.S. (1970). *Met. Trans.*, **1**, pp. 2943–2951.
17. Benjamin, J.S., and Bomford, M.J. (1977). *Met. Trans. A*, **8A**, pp. 1301–1305.
18. Billman, F.R., Juli, J.C., Jr., Hildeman, G.J., Petit, J.I., and Walker J.A. (1982). In *Rapid Solidification Processing Principles and Technologies III*, R. Mehrabian, ed. NBS, Gaithersburg, Maryland, pp. 532–546.
19. Griffith, W.M., Kim, Y.-W., and Froes, F.H. (1986). In *Rapidly Solidified Powder Aluminum Alloys, ASTM STP 890*, M.E. Fine and E.A. Starke, Jr., eds. ASTM, Philadelphia, pp. 283–303.
20. Morris, D.G. (1986). In *Rapidly Solidified Materials*, P.W. Lee and R.S. Carbonara, eds. American Society for Metals, Metals Park, Ohio, pp. 67–75.
21. Sears, J.W., Miller, D.J., and Fraser, H.L. (1986). In *Rapidly Solidified Powder Aluminum Alloys, ASTM STP 890*, M.E. Fine and E.A. Starke, Jr., eds. ASTM, Philadelphia, pp. 304–318.

22. Ro, D.H., Toaz M.W., and Moxson, V.S. (1983). *J. Metals*, **35**, no. 1, pp. 34–39.
23. Levi, C.G., and Mehrabian R. (1982). *Met. Trans. A*. **13A**, pp. 221-234.
24. Levi, C.G., and Mehrabian, R. (1982). *Met. Trans. A*, **13A**, pp. 13–23.
25. Baker, J.C., and Cahn, J.W. (1971). *Solidification*, American Society for Metals, Metals Park, Ohio, pp. 23–58.
26. Jones H. (1978). In *Rapid Solidification Processing: Principles and Technologies*, R. Mehrabian, B.H. Kear, and M. Cohen, eds. Claitor's Baton Rouge, Louisiana, pp. 28–45.
27. Coriell, S.R., and Sekerka, R.J. (1978). In *Rapid Solidification Processing: Principles and Technologies*, R. Mehrabian, B.H. Kear, and M. Cohen, eds. Claitor's, Baton Rouge, Louisana.
28. Coriell, S.R., and Sekerka, R.F. (1980). In *Rapid Solidification Processing: Principles and Technologies II*, R. Mehrabian, B.H. Kear, and M. Cohen, eds. Claitor's, Baton Rouge, Louisiana, pp. 35–49.
29. Takeshita, K., and Shingu, T.H., (1983). Trans. Jap. Inst. Metals, **24**, pp. 529–536.
30. Cahn, J.W., Coriell, S.R., and Boettinger, W.J. (1980). In *Laser and Electron Beam Processing of Materials*, C.W. White and P.S. Peercy, eds. Academic Press, New York, pp. 89–103.
31. Clyne, T.W. (1984). *Met. Trans B.*, **15B**, pp. 369–381.
32. Tiller, W.A., Rutter, J.W., Jackson, K.A., and Chalmers, B. (1953). *Acta Met.*, **1**, pp. 428–437.
33. Cohen, M., Kear, B.H., and Mehrabian, R. (1980). In *Rapid Solidification Processing: Principles and Technologies II*, R. Mehrabian, B.H. Kear, and M. Cohen, eds. Claitor's, Baton Rouge, Louisiana, pp. 1–23.
34. Cohen M., and Mehrabian, R. (1982). In. *Rapid Solidification Processing: Principles and Technologies III*, R. Mehrabian, ed. NBS, Gaithersburg, Maryland, pp. 1–27.
35. Mullins, W.W., and Sekerka, R.J. (1963). *J. Appl. Phys*, **34**, pp. 323–329.
36. Mullins, W.W., and Sekerka, R.J. (1964). *J. Appl. Phys*, **35**, pp. 444–452.
37. Sekerka, R.J. (1973). In *Crystal Growth: An Introduction*, P. Hartman, ed. North Holland, Amsterdam, p. 403.
38. Baker J.C., (1970). "Interfacial Partitioning during Solidification," Ph.D. thesis, Massachusetts Institute of Technology, Cambridge, Massachusetts, ch. 5.
39. Perepezko, J.H. (1978). In *Rapid Solidification Processing: Principles and Technologies*, R. Mehrabian, B.H. Kear, and M. Cohen eds. Claitor's, Baton Rouge, Louisiana.
40. Perepezko, J.H., and Anderson I.E. (1980). In *Synthesis and Properties of Metastable Phases*, E.S. Machlin and T.J. Rowland, eds. TMS-AIME, Warrendale, Pennsylvania, pp. 31–63.
41. Perepezko, J.H., and Paik, J.S. (1982). In *Rapidly Solidified Amorphous and Crystalline Alloys*, B.H. Kear, B.G. Geissen, and M. Cohen, eds. Elsevier Science, New York, pp. 1–15.
42. Perepezko, J.H., and LeBeau, S.E. (1981). In *Aluminum Transformation Technology and Its Applications*. American Society for Metals, Metals Park, Ohio, pp. 309–346.
43. Baker, J.C., and Cahn, J.W. (1969). *Acta Met.*, **17**, pp. 575–578.
44. Hillert, M. and Sundman B. (1977). *Acta. Met.*, **25**, pp. 11–18.
45. Jones, H. (1973). *Rep. Progr. Phys.*, **36**, pp. 1425–1497.
46. Jones, H. (1969). *Mat. Sci. Eng.*, **5** pp. 1–18.
47. Davies, H.A., Shohoji N., and Warrington, D.H. (1980). In *Rapid Solidification Processing: Principles and Technologies II*, R. Mehrabian, B.H. Kear, and M. Cohen, eds. Claitor's, Baton Rouge, Louisiana pp. 153–164.
48. Biloni, H., and Chalmers, B. (1965). *Trans. AIME*, **233**, pp. 373–379.
49. Burden, M.H., and Jones, H. (1970). *Fizika*, **2**, suppl. 2, paper 17.

50. Kofler, A. (1965). *J. Austr. Inst. Met.*, **10**, pp. 132–139.
51. Gigliotti, M.F.X., Jr., Colligan, G.A., and Powell, G.L.F. (1970). *Met. Trans.*, **1**, pp. 891–897.
52. Garrett, R.K., Jr., and Sanders, T.H., Jr. (1983). *Mat. Sci. Engr.*, **60**, pp. 239–274.
53. Paris, H.G., Mullins, J.W., and Sanders, T.H., Jr. (1983). *Aluminum* **59**, pp. 163–168.
54. Sanders, T.H., Johnson, J.W., and Underwood, E.E. (1980). In *Rapid Solidification Processing: Principles and Technologies II*, R. Mehrabian, B.H. Kear, and M. Cohen, eds. Claitor's, Baton Rouge, Louisiana, pp. 141–152.
55. Sanders, T.H., Mullins, J.W., and Paris, H.G. (1982). In *Rapidly Solidified Amorphous and Crystalline Alloys*, B.H. Kear, B.C. Geissen, and M. Cohen, eds. Elsevier Science, New York, pp. 369–374.
56. Hildeman, G.J., Lege, D.J. and Vasudevan, A.K. (March 1982). "Fundamentals of Compaction Processes for Rapidly Quenched Prealloyed Metal Powders." Final Report, U.S. Air Force, AFWAL Contract F33615-79-C-5037.
57. Paris, H.G., Billman, F.R., Cebulak, W.S., and Petit, J.I. (1980). In *Rapid Solidification Processing: Principles and Technologies II*, R. Mehrabian, B.H. Kear, and M. Cohen, eds. Claitor's Baton Rouge, Louisiana, p. 331–341.
58. Orowan E. (1948). In *Symposium on Internal Stresses in Metals and Alloys*, Institute of Metals, London, England, p. 451.
59. Martin, J.W. (1980). *Micromechanisms in Particle Hardened Alloys*. Cambridge University Press, Cambridge, England p. 162.
60. Hildeman, G.J., Labarre, L.C., Hafeez, A., and Angers, L.M., (1986). In *High Strength Powder Metallurgy Aluminum Alloys II*, G.J. Hildeman and M.J. Koczac, eds. TMS-AIME, Warrendale, Pennsylvania, pp. 265–284.
61. Bishop, M., and Fletcher, K.E. (1972). *Int. Met. Rev.*, **17**, pp. 203-225.
62. Anantharaman, T.R., and Suryanarayana, C. (1972). *J. Mat. Sci.*, **6**, 1111–1135.
63. Janot C., and Lelay, G. (1969). *Compf. Rend.*, **269** no. 7, pp. 823–826.
64. Fürrer, V.P., and Warlimont, H. (1973). *Z. Metall.*, **64**, pp. 236–248.
65. Jha, S.C., and Sanders, T.H., Jr., (1986). In *High Strength Powder Metallurgy Aluminum Alloys II*, F.J. Hildeman and M. Koczak, eds. TMS-AIME, Warrendale, Pennsylvania, pp. 243–254.
66. Jacobs, M.H., Doggett, A.G., and Stowell, M.J. (1974). *J. Mat. Sci.*, **9**, pp. 1644–1660.
67. Angers, L. Chen, Y., Fine, M.E., Weertman, J.R., and Zedalis, M.S. (1986). In *Aluminum Alloys Their Physical and Mechanical Properties*, **vol 1**, E.A. Starke, Jr., and T.H. Sanders, Jr., eds. Engineering Materials Advisory Services, London, England, pp. 321–337.
68. Prem Kumar, M.K., Koczak, M.J., and Lawley, A. (1986). In *High Strength Powder Metallurgy Aluminum Alloys II*, G.J. Hildeman and M.J. Koczac, eds. TMS-AIME, Warrendale, Pennsylvania, pp. 265–284.
69. Zedalis, M.S., and Fine, M.E. (1986). *Met. Trans. A*, **17A**, pp. 2187–2198.

15

Fracture and Fatigue Characteristics in Aluminum Alloys

A.K. VASUDEVAN

Office of Naval Research
Arlington, Virginia

R.D. DOHERTY

Drexel University
Philadelphia, Pennsylvania

S. SURESH

Brown University
Providence, Rhode Island

The most important application of precipitation-hardened high-strength aluminum alloys is in aerospace structural components. From 1950 to the early 1970s, strength, corrosion resistance, and short transverse properties were the main requirements for aircraft applications, and the improvements in these properties have dramatically increased as newer, improved fabrication techniques became available. Since the mid-1970s, increasing demands from fail-safe designs and damage-tolerant constraints have made improved fracture toughness and fatigue crack growth resistance as critically important properties.

This chapter attempts to summarize the various microstructural effects on fracture toughness and fatigue crack growth resistance in precipitation-hardened aluminum alloys, with special reference to Al-Li alloys where

ISBN-0-12-341831-3

most recent interest has been concentrated. The article by Quist and Narayan (Chapter 4) gives critical information on processing and on the fracture and fatigue properties of commercial Al-Li alloys. Wherever relevant and possible, comparisons to some conventional aluminum alloys and powder metallurgy (PM) alloys will be cited. The chapter will not include S–N fatigue or elevated-temperature effects on fracture and fatigue. The main discussions pertain to identification of matrix and grain boundary microstructural effects on the overall failure characteristics of the alloys.

I. Fracture Toughness

Fracture-toughness values of alloys generally decrease as the strength level is increased by alloy composition, fabrication, and aging treatment. This phenomenological trend is illustrated in Fig. 1 for a variety of aluminum alloys. In all these cases, the loss in fracture toughness with increasing yield strength appears to be mainly due to the changes in precipitate microstructures caused by different aging treatments. The metallurgical

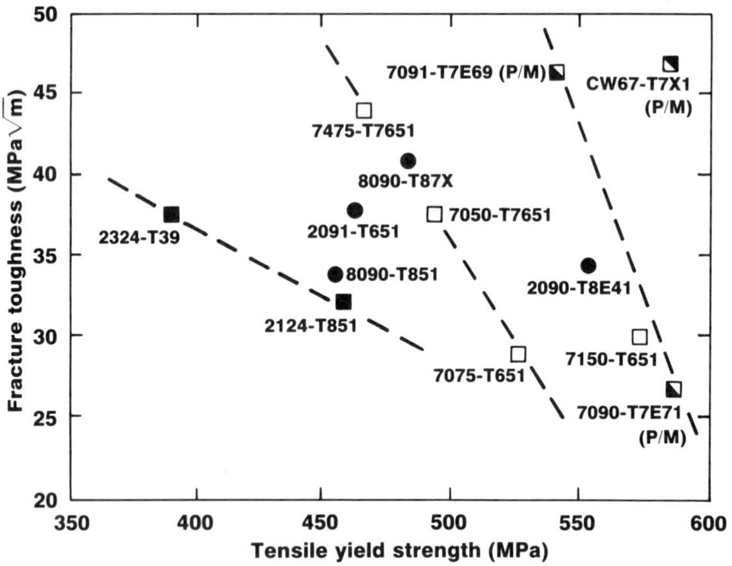

FIG. 1. Effect of composition, aging, and fabrication on fracture toughness and yield strength of various commercial aluminum alloys: ingot-metallurgy 2xxx, 7xxx, Al-Li alloys (2090, 2091, 8090), and powder-metallurgy 7xxx alloys.

and microstructural factors that control toughness and fatigue resistance of the aluminum alloys depend upon: (1) reduction of large Fe-, Si- and Cu-rich constituents from ingot casting, (2) addition of Zr, Mn, Cr, and Co to form fine dispersoids to control the post working grain shape and the degree of recrystallization, (3) control of "planar slip deformation"—the tendency to concentrate strain on particular {111} slip planes due to dislocation shear through coherent precipitates such as Guinier-Preston (GP) zones in general and the δ' (Al_3Li) precipitates in Al-Li alloys in particular, and (4) reduction of incoherent grain boundary (GB) precipitates (for example, by reducing the amount of GB precipitation formed specially during slow quenching). For conventional ingot metallurgy (IM) 2xxx (Al-Cu–based) and 7xxx (Al-Zn-Mg–based) alloys, the fracture effects arising from the role of constituents, dispersoids, and processing (such as quench-rate effects) have been well documented in several reviews [*1–3*]. As a result, we shall concentrate here on the fracture behavior of Al-Li alloys in relation to matrix and GB precipitation and only draw comparisons to IM and to powder-metallurgy (PM) alloys of the 7xxx and 2xxx series. We shall not dwell on PM alloys since the quality issues in fabrication procedures can sometimes mask intrinsic effects on toughness [*4–5*]. In addition, a detailed discussion of PM aluminum alloys is given in the Chapter 4 by Koczak and Hildeman. However, it migh be noted that the main effect of PM processing seems to be a significant reduction in the size of the cast constituents. This improves fracture toughness, and together with the very fine grain size of the PM alloys appears to give better S–N fatigue properties but may promote faster fatigue crack growth rates, particularly at low stress intensities.

Increasing the aging treatment of precipitation-hardened aluminum alloys from the underaged (UA) up to the peak-strength (PA) condition always leads to a monotonic decrease in fracture toughness (Fig. 1). Overaging (OA) beyond the PA condition can, however, lead either to a significant restoration of the toughness (in alloys in which matrix microstructural changes appear to dominate fracture, as in Fig. 2(a)) or to small recovery or no recovery of toughness (in alloys where grain boundary precipitate microstructures appear to dominate, as in Fig. 2(b)). The type of behavior seems to depend upon the alloy composition. In general, 2xxx and 7xxx alloys tend to show the behavior indicated by Fig. 2(a), while Al-Li alloys with high Li contents tend to show the behavior of Fig. 2(b). This is illustrated in Fig. 3, which shows the influence of aging treatment (expressed by the yield strength) on plane strain fracture toughness for three Al-Li-Cu-Zr alloys. The alloy compositions are given in the top right side of the graphs. In all three cases, the toughness gradually decreases with an increase in yield strength from severely UA to PA treatments.

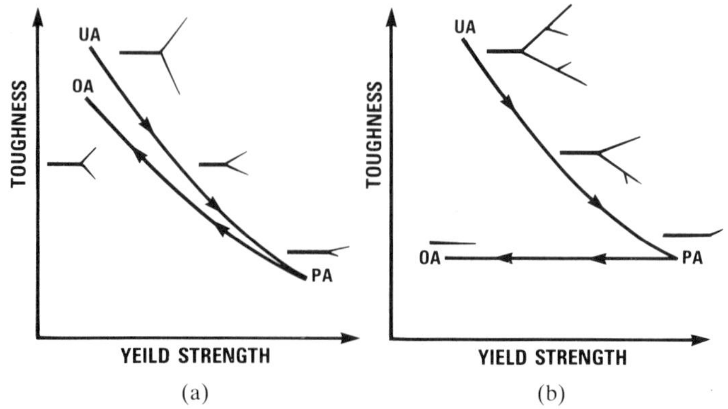

FIG. 2. Schematic representation of the fracture-toughness variation with yield strength for alloys showing: (a) less pronounced toughness variation from UA → PA → OA and (b) significant variation in toughness from UA → PA → OA condition. Also, changes in the crack geometry accompanied with changes in toughness are shown.

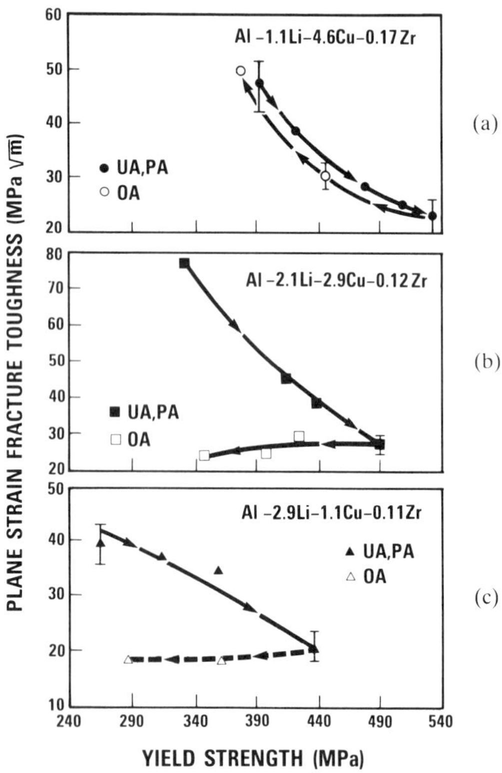

FIG. 3. Variation in plane strain fracture toughness with yield strength for three Al-Li-Cu-Zr alloys: (a) Al-1.1Li-4.6Cu-0.17Zr, (b) Al-2.1Li-2.9Cu-0.12Zr, and (c) Al-2.9Li-1.1Cu-0.11Zr.

Aging A1-1.1LI-4.6Cu-0.17Zr (Fig. 3(a)) beyond peak strength tends to increase its toughness. This observation is consistent with the fracture behavior of IM aluminum alloys of the 2xxx and 7xxx type [*1*,*2*,*7*], though it is less pronounced in PM alloys [4]. However, aging of alloys with high Li/Cu ratios beyond the PA temper does not improve fracture toughness of the alloys, as shown in Figs. 3(b) and 3(c) even though the yield stress falls significantly on overaging. Observations similar to that seem in the high-Li alloys (that is, limited recovery of toughness with OA) have been reported in a completely recrystallized coarse-grain 2024 alloy [8] and in Al-Mg-Si alloys [*9*]. In PA and OA alloys with high Li contents, GB fracture occurs from voids initiated around GB precipitates. In the UA alloys, where the amount of GB precipitate is limited, mixed-mode (transgranular and intergranular) fractures occur often with the types of crack branching illustrated schematically in Fig. 2.

The influence of composition and microstructure on crack growth toughness, characterized by the nondimensional tearing modulus T_R suggested by Paris et al. [*10*], is shown in Fig. 4, where the variation in T_R with fracture toughness K_{Ic} is plotted for three lithium-containing aluminum alloys. The aging treatments for the three alloys were between the severely UA and PA tempers. The most underaged alloys in each case were those with the highest values of T_R and K_{Ic}. Dramatic differences between these alloys are clearly seen. The important observations are:

(1) Compared to thc quarternary Al-2.9Li-1.1Cu-0.11Zr alloy with a nearly identical Li content; the ternary Al-3.1Li-0.12Zr alloy displays a much stronger resistance to crack advance (T_R), in the same range of K_{Ic} values.

(2) The differences between the ternary and low-Cu quarternary alloy are least in the most underaged condition.

(3) The higher Li/Cu ratio alloy (2.9Li–1.1Cu) exhibits significantly higher crack growth toughness (T_R) when compared to the other quaternary alloy with a lower Li/Cu ratio (2.1Li–2.9Cu).

These observations suggest that microstructural characteristics that are generally considered undesirable for crack initiation (K_{Ic}) are, in some cases, beneficial for tearing modulus (T_R). In the preceding example, the amounts of Cu-containing, platelike, T_1 precipitates relative to the spherical coherent δ' precipitates increases both with the Cu/Li ratio and with increased aging [*6*]. Microstructural observations showed that there was, in the Cu-containing alloys, a decrease in the amount of δ' between the severely UA and PA conditions with concomitant increase in T_1 matrix and T_2 GB precipitates. The Cu-containing plate-like T_1 precipitates in the quarternary alloys tend to suppress the planar slip deformation promoted

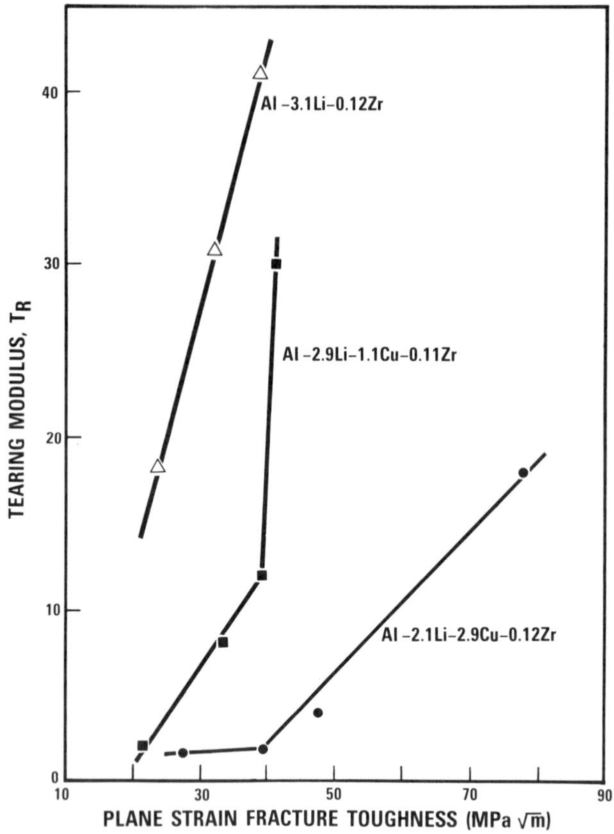

FIG. 4. Correlation between crack growth toughness (T_R) and crack initiation toughness (K_{Ic}) for three Al-Li alloys)

by the coherent δ' precipitates. It seems possible that increased planar slip may in some cases improve the crack growth toughness (T_R) at a constant K_{Ic}, by the crack-branching model described shortly.

The geometry of the growing crack during the outset of the fracture and subsequent growth dramatically influences the K_{Ic} and T_R values. Alloys with higher Li-content showed pronounced susceptibility to crack branching in the UA condition where planar slip is more important [6]. Less crack branching was observed in the PA tempers, though some branching was seen in the PA ternary alloy. That is, the UA micorstructures that contained large amounts of δ' showed a tendency, at the crack tip, to fail along slip bands where the shear strains had localized [6]. Precipitates sheared by dislocation offer a reduced resistance to subsequent dislocations on the same slip plane thus encouraging strain localization and possible slip plane cracking. Experiments also indicated that the tendency for crack branching decreased dramatically as the area fraction of the GB precipitates increased, for example in the PA and OA tempers.

Suresh and Shih [*11*] have outlined a procedure for determining the near-crack-tip stress and deformation fields for branched cracks in ductile materials under plane strain and small-scale yielding conditions. Since a branched crack has a lower effective crack-tip K-value than a linear crack of the same length under the same applied K_I, crack branching leads to an apparent enhancement in measured K_{Ic}. The analysis of Suresh and Shih [*11*] also shows that crack-tip plasticity further enhances the beneficial effects of crack deflection (i.e., further reduces the effective near-tip driving force), irrespective of whether the criterion for failure is the attainment of a critical stress or critical strain ahead of the advancing branched crack. In Figs. 5(a) and (b), the data from Figs. 3(b) and (c) have been replotted

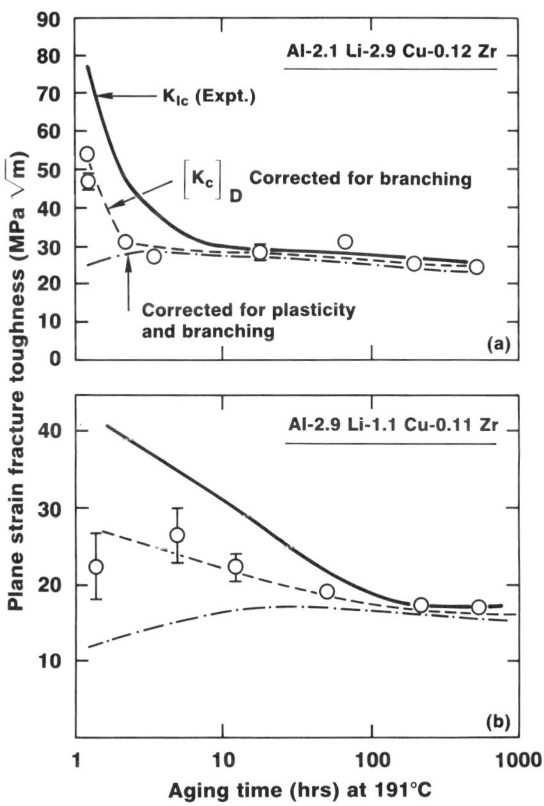

FIG. 5. Variation in nominal K_{Ic}, crack-deflection-modified effective toughness $[K_c]D$ and corrected for plasticity and crack branching as a function of aging time at 191°C for: (a) AL-2.1Li-2.9Cu-0.12Zr, (b) Al-2.9Li-1.1Cu-0.11Zr.

in terms of toughness versus aging time. The toughness values are shown as nominal K_{Ic} values without accounting for crack deflection and also after correction for crack branching (the true toughness values, $[K_c]_D$ using linear-elastic solutions, which do not account for crack-tip plasticity. It can be clearly seen that the variations of toughness as a function of aging are significantly reduced when corrected for these effects. In addition, if plasticity corrections [*11*] are also made on $[K_c]_D$, then the corrected toughness becomes nearly independent of aging time. Figure 5 suggests that the high nominal K_{Ic} values observed in UA microstructures are predominantly due to microstructurally induced changes in the crack geometry and to the effect of plasticity at lower yield strengths. In addition, all three toughnesses, K_{Ic}, $[K_c]_D$, and plasticity-corrected K values, approach a plateau at longer aging times where less crack branching is observed, suggesting that this is the "intrinsic" fracture toughness in a given alloy composition. The types of crack branching in different micorstructures have been shown schematically in Fig. 2.

The effect of GB precipitates on fracture was investigated with the method of microstructural reversion experiments in a high-purity recrystallized Al-3Li-0.5 Mn alloy [*6,12*]. This was chosen as a model alloy simply because it exhibited the same trend in the toughness versus yield-strength dependence as in the stronger quarternary alloys while giving very simple microstructures consisting just of matrix δ' and GB δ precipitates. In this Al-3Li alloy, simple reversion experiments were possible. These redissolved completely the matrix δ' while leaving the GB δ largely unchanged. These heat treatments gave independent control of the two phases allowing the relative contributions of the GB and matrix precipitates to the strength and toughness properties to be observed. Figure 6 shows the slow bend Charpy toughness variation with yield strength for various heat treatments for Al-3Li-0.Mn. After direct aging treatments at 177°C, from the as-quenched to the PA conditions, the usual significant toughness loss was seen with the increase in yield strength. This is due to matrix δ' precipitation and coarsening as well as to precipitation of GB δ. With overaging at 200°C, no increase in toughness was observed, even though yield stress was dramatically reduced. The microstructural changes on OA showed continued coarsening of δ' and additional growth of δ at the expense of the δ'. The property changes are similar to those previously discussed for the quarternary alloys (Figs. 3(b) and 3(c)), though observed here at lower yield-strength levels. The two curves in Fig. 6 that lie in between the UA and OA results correspond to the two δ'-reverted and-reaged materials. After reversion which redissolved all the matrix δ' while leaving the GB δ both before and after reaging treatments, the reverted PA and OA materials show significant loss in toughness compared to the initial as-

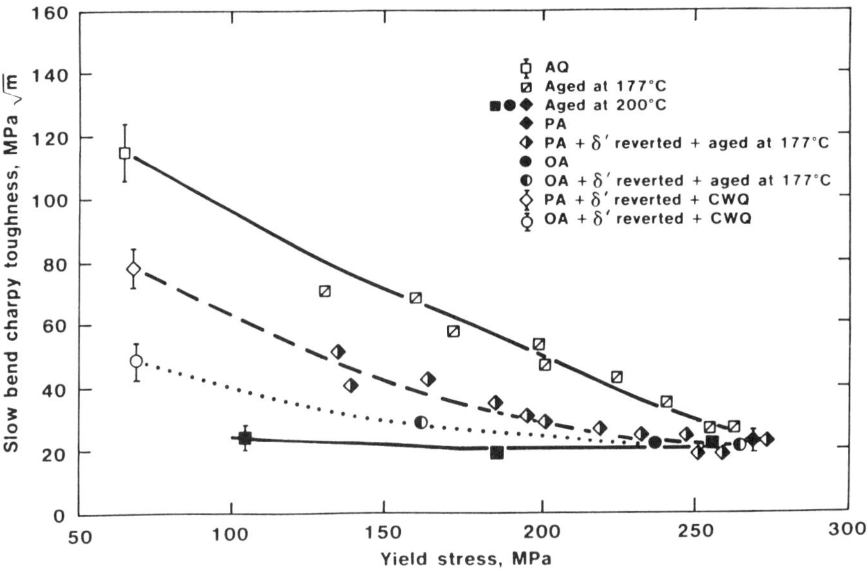

FIG. 6. Variation in slow bend Charpy toughness with yield strength for various heat treatments for the alloy Al-3Li-0.5 Mn.

quenched material, though at constant yield-strength levels. Qualitatively similar observations were made for reverted quarternary alloys. Full details of the reversion experiments and observations are given in reference [*12*].

In order to decouple further the role of matrix δ', in terms of its effects on yield strength and planar slip, from the influence of GB δ, the following critical experiment was conducted. The alloys were first aged to different times at 200°C, from UA (8 hrs) to severely OA (250 hrs), so that the size and area fractions of the GB δ could be varied significantly. These microstructures were then δ'-reverted and reaged to a fixed UA condition of 1 hr at 177°C. This produced microstructures that contained different amounts of GB δ but with nearly identical matrix δ', and that exhibited constant yield strengths [*12*]. The variation in plane strain fracture toughness of these microstructures was compared to the direct-aged (177°C/1 hr) material. This UA matrix structure was intentionally chosen to promote severe planar slip deformation. Figure 7 shows the result of such an experiment. For comparison, previous results of Unwin and Smith [*13*] from Al-6Zn-3Mg alloys are also included, where a different method of heat treatment was used to vary the amount of GB precipitates at nearly constant strength. The two alloys seen in Fig. 7 have very different

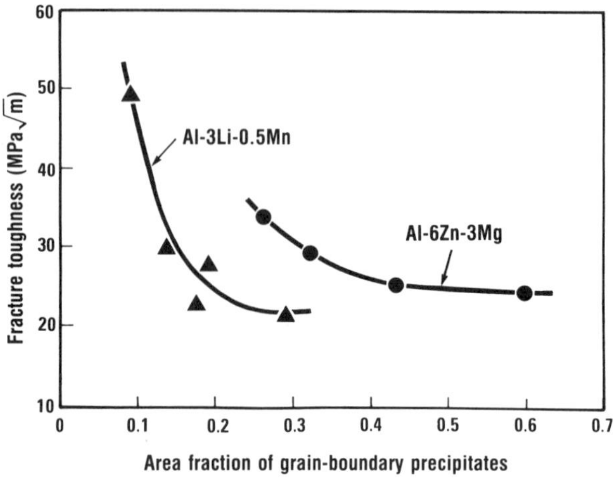

FIG. 7. Variation in plane strain fracture toughness with area fraction of grain boundary precipitates, at constant-matrix yield strength, for Al-3Li-0.5 Mn [12] and Al-6Zn-3 Mg [13] alloys.

compositions, yield strengths, and also sizes (morphology) of GB precipitates; but they exhibit similar trends. The alloys all show full intergranular fracture. It is clearly evident that there is a large loss in toughness as the area fraction of the GB precipitates increases. These observations demonstrate the major importance of GB precipitation to loss of toughness on aging. Earlier models of intergranular fracture in Al-Li alloys considered only the influence of matrix planar slip deformation applying stress concentrations at the GB, (see the references quoted in [*6*]). The experiments of Vasudevan & Doherty [*12*] show that a high concentration of embrittling GB particles leads to poor fracture toughness irrespective of the planarity of slip. Suresh et al. [*6*], in alloys with low levels of GB precipitates, show that planar slip can lead to enhanced toughness, in particular improved tearing modulus T_R, through the promotion of crack branching. A detailed inspection of Fig. 6 shows an interesting result. The alloy overaged for 250 hrs at 200°C (YS of 180 MPa) on reverting and underaging to a YS of 160 MPa had somewhat *higher* toughness than the various overaged alloys at all strength levels. Underaged alloys are more susceptible to planar slip[*6*]—so that the suggestion of *higher* toughness with planar slip, due to crack branching [*6*], seems to apply even with heavily embrittled grain boundaries. At the very least, this example suggests that planar slip does not seem significantly damaging in these alloys.

The significance of GB precipitation on toughness is also clearly seen in commercial alloys when inadequate heat treatments (either from incomplete solutionizing or by slow quenching after solutionizing) are applied.

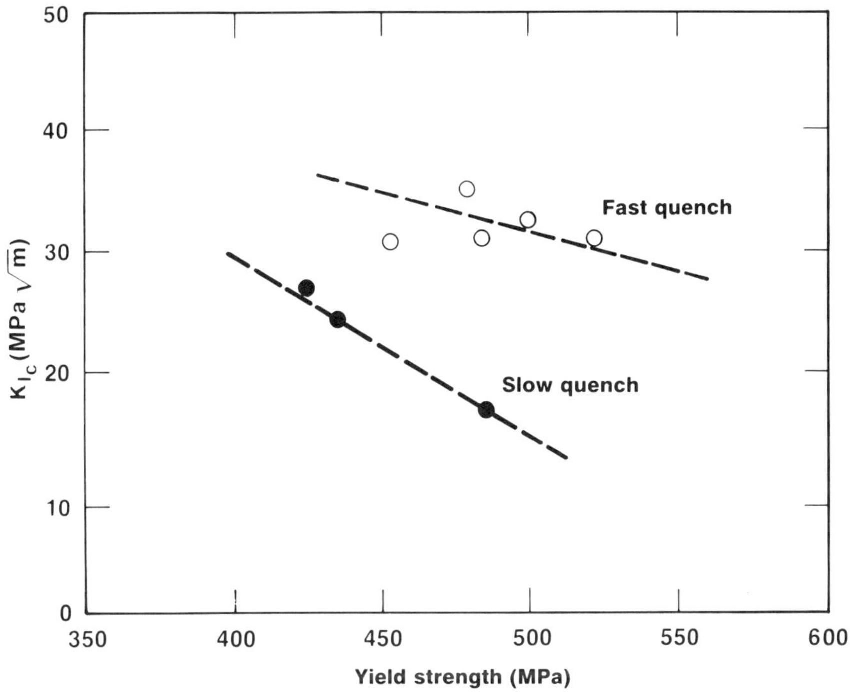

FIG. 8. Effect of quench rate on plane strain fracture toughness of 7075 alloy plate at different yield strengths.

This phenomenon is demonstrated, for example, in the quench-rate effects in 7075 alloy [2] (Fig. 8) and in 8090 alloys [*14*]. In both cases, more grain boundary precipitation occurred during slow quenching prior to the final aging treatment. This effect is particularly severe in thick section products where it is difficult to achieve fast quenching following solution heat treatments. For experimental results in commercial products, references can be made to various conference proceedings [*4,5,7,15–17*].

Glazer et al [*18*] have observed that at cryogenic temperatures, outstanding combinations of strength and toughness in a commercial Al-Li alloy, 2090-T8E41, are found. They observed a 45% increase in yield strenght and a 90% increase in K_{Ic} between 300 and 4°K. The origin of this behavior is not clearly understood, but was interpreted as being due to high work-hardening rates exhibited by the alloy at cryogenic temperatures [*18*]. Subsequent studies by Venkateswara Rao et al. [*19*] showed that the improved toughness results at low temperatures were restricted to certain loading directions. These directions were ones that gave increased crack branching, mainly along prior grain boundaries parallel to the rolling plane, in the unrecrystallized wrought material. Through thickness, toughness values were found to be reduced as the test temperatures were lowered. These low-temperature toughness results [*19*] again suggest a crucial role both for grain boundary fracture induced by grain boundary precipitates and for crack branching and crack deflection.

II. Fatigue Crack Growth

It has been well accepted that the fatigue crack growth (FCG) life of structural members is often dominated by the propagation of cracks at low-stress-intensity factor range (ΔK). Below a critical or "threshold" value of ΔK, ΔK_{th} (fatigue crack growth per cycle) tends to zero, (see, for example, the papers contained in ref. [*20*]). Extensive studies of constant-amplitude FCG behavior at near-threshold stress-intensity ranges in high-strength aluminum alloys have indicated that the FCG rates are strongly influenced by composition, microstructure, load-ratio (the R value), and the environment [*20*]. In particular, Al-Li alloys tend to exhibit much more pronounced effects of microstructure and environment on ΔK_{th} than is shown by conventional aluminum alloys [*21–23*]. This is illustrated in Fig. 9, where several Al-Li-Cu-Zr alloys with differing Li/Cu ratios are compared to conventional alloys.

Figure 9 shows the variation of ΔK_{th} as a function of the atomic Li/Cu ratio at a load ratio, R, of 0.33. (The load ratio is defined as the ratio of the minimum load to the maximum load of the fatigue cycle.) The results indicate that ΔK_{th} increases with increasing Li/Cu ratios [*22*]. The reasons for this trend seem to be (1) oxide-induced "closure" effects and (2) highly branched fatigue cracking. The closure phenomenon arises from debris on the fracture surface that prevents the crack from fully closing during the reduction of tensile load—with the effect of partially shielding the crack tip from the full applied ΔK. A highly branched crack will also induce closure effects as well as reducing the stress concentration at a branched or deflected crack [*11*]. Analysis of the fatigue fracture surfaces revealed somewhat thicker oxide layers in alloys with higher Li/Cu ratios. Experiments in vacuum, however, showed a decrease in ΔK_{th} with increase in R ratio in Al-Li-Cu-2r alloys, a trend that is different from the commonly observed effect of vacuum on conventional high-strength aluminum alloys where ΔK_{th} is then almost independent of R ratio [*24–26*]. Optical metallographic observations of crack profiles for the alloys indicated a linear crack for the low Li/Cu ratio of 2.2 alloy while the alloy with an Li/Cu ratio of 25 exhibited a highly branched and deflected crack profile [*22*]. Examples of deflected crack paths in fatigue are seen in Fig. 10.

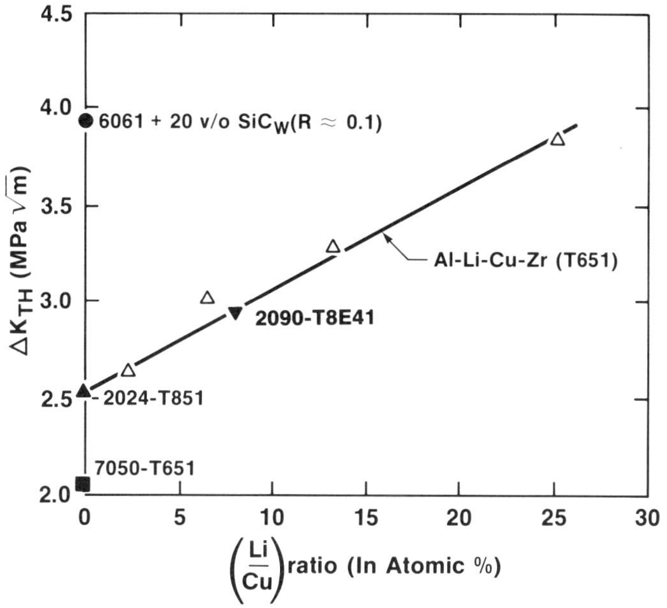

FIG. 9. Variation in threshold stress intensity ΔK_{TH} with atomic Li/Cu ratio. For comparison, data for 7050, 2090, 2024, and 6061 + SiC alloys are included.

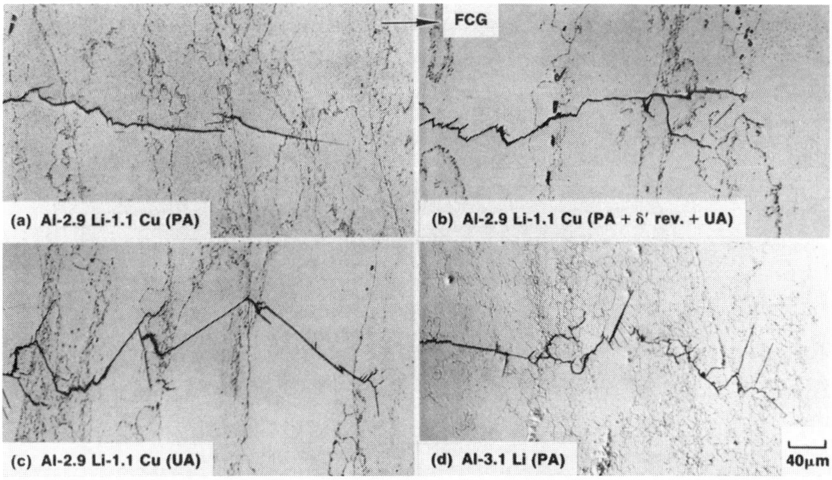

FIG. 10. Fatigue crack profiles of Al-3.1Li-0.12Zr (PA) and Al-2.9Li-1.1-Cu-0.11Zr alloy under PA, UA, and PA + δ' rev. + UA conditions, at near-threshold ΔK_{TH} FCG conditions.

These results indicate that the improved crack growth rates at low ΔK arise mainly from crack-deflection processes combined with increased closure effects due to fracture surface roughness and oxidation. The crystallographic FCG mechanism appears to be induced by coherent and shearable δ' participates in alloys with higher Li/Cu ratios. This seems to be a fatigue manifestation of the microstructural planar slip effect seen in developing a 'tortuous crack path in toughness testing of underaged alloys. At the highest lithium levels, some of the Al-Li-Cu-Zr alloys exhibit fatigue crack growth properties comparable to 6061-SiC composites [27]. (See Fig. 9.)

The preceding observations can be clearly demonstrated in Fig. 11, where the fatigue crack growth rates of alloys with various microstructures are shown. The PA, Al-3.1Li-0.12Zr ternary alloy, containing coarse δ' precipitates, is compared with the quarternary Al-2.9Li-1.1Cu-0.11Zr alloy data in the UA, the PA, and the PA + δ' reverted + UA conditions. The UA material contained large amounts of fine δ' and limited amounts of fine T_1; the PA material contained coarse T_1 and some coarse δ', while the PA–reverted–UA material contained coarse T_1 but fine δ'. The PA

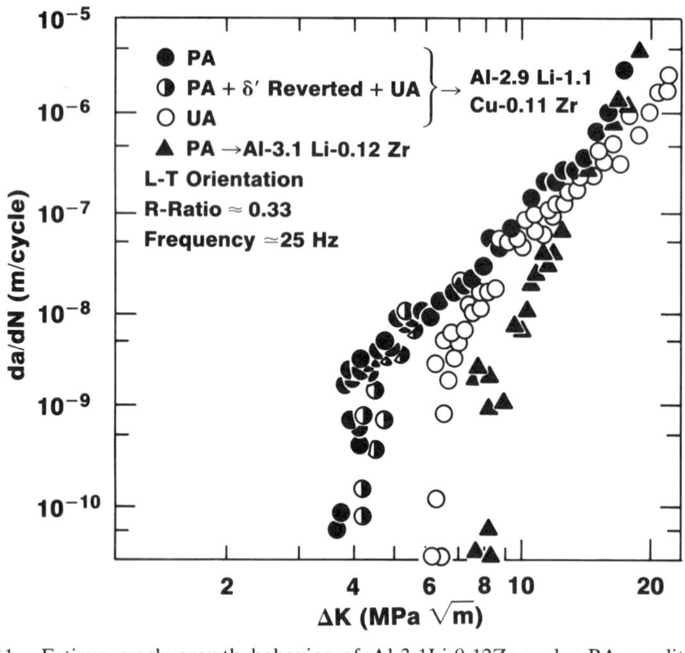

FIG. 11. Fatigue crack growth behavior of Al-3.1Li-0.12Zr under PA conditions and Al-2.9Li-1.1Cu-0.11Zr alloys under PA, UA, and PA + δ'-reverted + UA conditions.

ternary alloy showed the highest ΔK_{th} of 8.4 MPa$\sqrt{m}$ compared to the PA Cu-containing alloy whose ΔK_{th} was only 3.7 MPa$\sqrt{m}$. When the Cu-containing PA alloy was given a δ' reversion treatment and again UA, the crack growth rates differed little from the straight PA alloy. However, when the Cu-containing alloy was initially UA producing large amounts of very fine δ' and small amounts of T_1, the crack growth rates approached those of the binary PA alloy. This experiment indicates that the planar slip deformation caused by δ' results in high threshold values corresponding to crack branching, while coarse T_1 inhibits planar slip and the resultant crack branching, thus producing much poorer FCG properties at low ΔK values. The observed optical microscope crack profiles in these alloys at near ΔK_{th} are shown in Fig. 10. The Al-Li alloy in Fig. 10(d) had the most deflected crack profile compared to the Cu-containing PA alloy (Figs. 10(a) and (b)), while the UA Al-Li-Cu (Fig. 10(c)) showed intermediate amounts of crack branching. These results on fatigue crack growth are consistent with the previously discussed crack growth behavior (T_R) observed under monotonic loading conditions (Fig. 4). Similar observations have also been reported in conventional 7075 alloys [*26*] and in the commercial Al-Li 2090-T8E41 alloy [*28*].

Since Al-Li alloys tend to exhibit strong crystallographic texture, it is observed that FCG can vary significantly between various sample orientations. It was experimentally observed that high crack closure levels in TL,* LT, and TS orientations of a 2090-T8E41 alloys [*29*] lead to higher ΔK_{th}, since these orientations showed severe crack branching. However, the SL and ST orientations showed the lowest closure levels resulting in low ΔK_{th}, since a linear crack path developed due to intergranular fracture. Yoder et al. [*28*] have predicted the slip-band facet and fracture-surface asperity angles from texture in a 2090-T8E41 alloy and suggested that the tortuous back profile in the thickness direction is due to a consequence of texture.

When a single periodic overload spike (tension or compression mode) is applied, at a given ΔK, every x number of constant-amplitude cycles to a high-strength aluminum alloy, then a large FCG retardation occurs. Examination of the fracture-surface topography of the postoverloaded FCG in 7xxx alloys [*30–32*] and in PM CW67 alloys [*33*] appears to be highly crystallographic, similar to that in a constant-amplitude near ΔK_{th} fracture. Suresh [*32,34,35*] has proposed that this postoverload retardation is effectively governed by crack-branching micromechanisms to that of constant-amplitude near-threshold (ΔK_{th}) FCG.

* In TL, T stands for transverse loading direction, and L stands for longitudinal crack growth direction. LT, TS, and SL orientations are similarly defined, with s meaning short transfer direction.

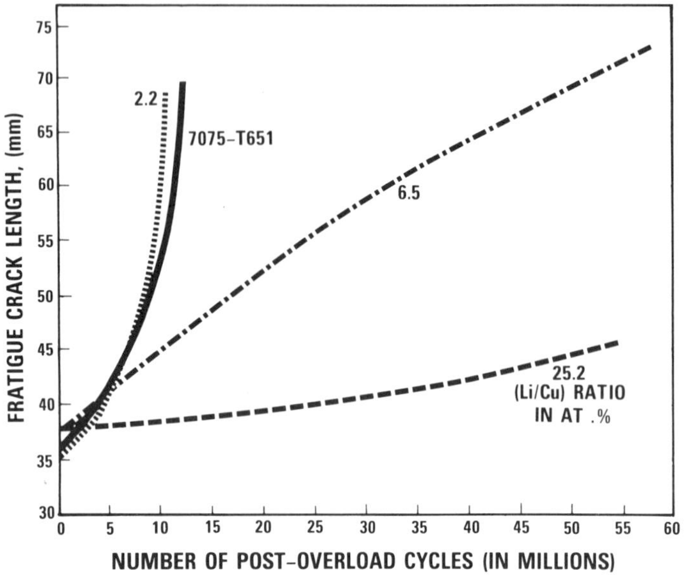

FIG. 12. Variation of fatigue crack length with the number of postoverload cycles for Al-Li-Cu-Zr alloys with low (A), medium (B), and high (C) Li/Cu ratios and for 7075-T651.

Under variable-amplitude loading, Al-Li-Cu-Zr alloys exhibit a large increase in retardation as the Li/Cu ratio increases [*23*] (Fig. 12). This marked retardation occurs in a higher Li/Cu ratio alloy where a strong propensity for crystallographic crack growth is observed, giving crack deflection, in both the near-threshold and postoverlodad regions. In general, Al-Li-Cu-Zr alloys tend to be superior to conventional 7xxx aluminum alloys (Fig. 12) [*23*,*30*] in tension-dominated overloads. However, in compression-dominated overloads, where compressive stresses tend to compact the fracture-surface asperities giving less closure, the Al-Li alloys tend to be less promising [*31*].

Doherty et al. [*36–38*] have discussed possible mechanisms of precipitate damage during fatigue of precipitation-strengthened alloys. Several examples have been produced showing localized fatigue-induced dissolution of shearable precipitates in persistent slip bands [*39*,*40*]. The different possible mechanisms all involve dislocation-induced damage to the precipitates in a slip band in underaged high-purity alloys. This damage leads to local weakening and thus to intense strain concentration and total precipitate destruction in the slip bands. In some cases [*36*,*37*], loss of solute from the slip band after fatigue fracture on {111} slip planes was demonstrated, but this was not universally found [*40*]. However, detailed studies of the influence of microstructural instability on the mechanical response of precipitation-hardened alloys have not yet been reported, though such studies would be of considerable value. It would be of interest, for example, to discover if commercial alloys containing dispersoid phases (which would be expected to help homogenize slip) show the same tendency to precipitate damage in persistent slip bands as is found in high-purity alloys.

III. Conclusion

Studies on fracture and fatigue in high-strength aluminum-based alloys have shown that there is a complex pattern of interaction between the mechanical behavior of the alloys and their microstructure. In the conventional concept of materials engineering, microstructural features directly influence properties (for example, the reduction of toughness directly produced by significant amounts of coarse as-cast constituent phases and by large fractions of GB precipitates produced by inefficient quenching and by aging). The concept of strain localization in planar slip bands appears to be significant in both monotonic testing and in fatigue with crack deflection and branching leading to important consequences for the mechanical behavior. This chapter suggests that much quantitative work remains to be done in this area, although the early results indicate that there may be considerable potential for improved toughening of both current and new aluminum alloys.

Acknowledgments

The authors are grateful to Alcoa for support of much of the experimental work described here, and to numerous colleagues for stimulating discussions of this topic.

References

1. Hahn, G.T., and Rosenfield, A.R. (1975). *Met. Trans.* 6A, 653.
2. Thompson, D.S. (1975). *Met. Trans.* 6A, 671.
3. Starke, E.A. Jr., and Lutjering, G. (1979). In "Fatigue and Microstructure," M. Meshii, ed. ASM, Metals Park, Ohio, p. 205.
4. Hildeman, G.J., and Koczak, M.J., eds. (1985). "High Strength Powder Metallurgy Aluminum Alloys," **II**. TMS-AIME.
5. Koczak, M.J., and Hildeman, G.J. eds. (1982). "High Strength Powder Metallurgy Aluminum Alloys", TMS-AIME.

6. Suresh, S., Vasudévan, A.K., Tosten, M., and Howell, P.R. (1987). *Acta Met.* **35**, 25.
7. Staley, J.T. (1975). *ASTM STP* **605**, 71.
8. Burghard, H.C. (1974). *Met. Trans.* 5, 2083.
9. Evensen, J.D., Ryum, N., and Emmbury, J.D. (1975). *Mater. Sci. Engr.* **18**, 221
10. Paris, P.C., Tada, H., Zahoor, A., and Ernst, H. (1979). *ASTM STP* **668**, p. 5.
11. Suresh, S., and Shih, C.F. (1986). *Int. J. Fract.* **30**, 237.
12. Vasudévan, A.K., and Doherty, R.D. (1987). *Acta. Met.* 35, 1193.
13. Unwin, P.N.T., and Smith, G.C. (1969) *J. Inst. Metals* **97**, 299.
14. Colvin, G.N., and Starke, E.A. Jr. (1988). *SAMPE Quart.* 19, 10.
15. Champier, G., Dubost, B., Miannay, D., and Sabetoy, L., eds. (1987). "4th International Al-Li Conference." *J. de Physique* **48**, colloque C3.
16. Starke, E.A. Jr., and Sanders, T.H. Jr., eds. (1986). "Aluminum Alloys: Their Physical and Mechanical Properties," **I, II, II**. EMAS.
17. Rioja, R.J., Bretz, P.E., Sawtell, R.R. Hunt, W.H., and Ludwiczak, E.A. (1986). In "Aluminum Alloys and Their Physical and Mechanical Properties", **III**, E.A., Starke, Jr., and T.H. Sanders Jr., eds., p. 1781.
18. Glazer, J., Verasconi, S.L., Sawtell, R.R., and Morris, J.W. (1987). *Met. Trans.* **18A**, 1695.
19. Venkateswara Rao, K.T., Hayashigatami, F.H., Tu, W., and Ritchie, R.O. (1988). *Scripta Met.* **22**, 93.
20. Suresh, S., and Davidson, D.L., eds. (1983). "Fatigue Crack Growth Threshold Concepts." TMS-AIME Publs.
21. Baker, C., Gregson, P.J., Harris, S.J., and Peel, C.J., eds. (1986). "Aluminum-Lithium Alloys," III, Inst. of Metals, London.
22. Vasudévan, A.K., and Suresh, S. (1985). *Met. Trans.* **16A**, 475.
23. Petit, J., Suresh, S., Vasudévan, A.K., and Malcolm, R.C. (1986). In "Aluminum-Lithium Alloys," **III**, C. Baiker, P.J. Gregson, S.J. Harris, and C.J. Peel, eds. Inst. of Metals, London, p. 257.
24. Kirby, B.R, and Beevers, C.J. (1979). *Fat. Eng. Mat. and Struct.* **1**, 203.
25. Lafarie-Frenot, M.C., and Gasc, C. (1984). *Fat. Eng. Mat. and Struct.* **6**, 329.
26. Suresh, S., Vasudévan, A.K., and Bretz, P.E. (1984). *Met. Trans.* **15A**, 1.
27. Nair, S.V., Tien, J.K., and Bates, R.C. (1985). *Int. Metals Reviews* **30**, 275.
28. Yoder, G.R., Pao, P.S., Imam, M.A., and Cooley, L.A. (1988). *Scripta Metall.* **22**, 1241.
29. Ritchie, R.O. (1988). *Mater Sci and Eng.* **A103**, 15.
30. Bretz, P.E., Vasudévan, A.K., Bucci, R.J., and Malcolm, R.C. (1984). *ASTM STP* **833**, 242.
31. Au, P., Topper, T.H., and El Haddad, M.H. (1983. "AGARD Conferencce, Toronto, Proceedings" **328**.
32. Suresh, S., and Vasudévan, A.K. (1983). In "Fatigue Crack Growth Threshold Concepts," D.L. Davidson and S. Suresh, eds. TMS-AIME Pubs., p. 561.
33. Kleek, J.J., Nethercott, R.B., and Kim, Y.W. (1986). In "Aluminum Alloys Their Physical and Mechanical Properties," E.A. Starke, and T.H. Sanders, Jr., EMAS, p. 1709.
34. Suresh, S. (1982). *Scripta Metall.* **16**, 995.
35. Suresh, S. (1983). *Eng. Fract. Mech.* **18**, 557.
36. Brett, S.J., and Doherty, R.D. (1978). *Mater. Sci and Eng.* **32**, 255.
37. Lea, C., Brett, S.J., and Doherty, R.D. (1979). *Scripta Metall.* **13**, 45.
38. Doherty, R.D. (1983). "*In* MicroMechanisms of Plasticity and Fracture," M.H. Lewis and D.M.R. Taplin, eds. Parsons Press, Trinity College, Dublin, p. 303.
39. Stoltz, R.E., and Pineau, A.G. (1978). *Mater. Sci and Eng.* **34**, 275.
40. Vogel, W., Wilhelm, M., and Gerold, V. (1982). *Acta Metall.* **30**, 21, 31.

16

Stress Corrosion of High-Strength Aluminum Alloys

N.J.H. HOLROYD

Alcan International Limited,
Banbury, Oxon, England

A.K. VASUDEVAN

Office of Naval Research,
Arlington, Virginia

and

L. CHRISTODOULOU

Martin Marietta Laboratories,
Baltimore, Maryland

I. Introduction

Stress corrosion cracking (SCC) of aluminum alloys is a complex phenomenon involving time-dependent interactions between alloy microstructure, mechanical deformation, and local environmental conditions. SCC is almost exclusively intergranular in nature other than when transgranular SCC is promoted by severe mechanical loading conditions and/or unfavorable local grain boundary orientations. As of 1988, despite extensive

ISBN 0-12-341831-3

research, the attribution of a SCC mechanism to a particular high-strength aluminum-alloy, environmental interaction remains controversial.

Continual development of new-generation aluminum alloys such as powder-metallurgy products (7090, 7091), aluminum-lithium–based alloys (2090, 8090, 8091) and new variants of existing alloys (7010, 7050, 7150) warrants a review of the current understanding of SCC mechanisms. Excellent review papers summarizing published information up until the mid-seventies [*1–2*] are updated by a more recent review [*3*] and the reader is directed to these together with a general text on aluminum alloys [*4*] for background on metallurgical, mechanical, and environmental aspects of conventional aluminum alloys. The aim of this chapter is to summarize the current mechanistic understanding of the SCC of aluminum alloys using the 1980's experimental data and, where possible, to focus discussion upon modern high-strength aluminum alloys.

A. *SCC Mechanisms: A Brief Review*

Although numerous detailed mechanisms abound, two basic theories have evolved. One envisages cracking as highly localized anodic dissolution of the alloy under the combined influence of stress and environment, while the other suggests that embrittlement and loss of ductility is promoted after the ingress of an aggressive species (usually atomic hydrogen) onto or into the alloy. Until the early 1980s, the "anodic-dissolution"–based models were almost universally accepted and the early work of Gruhl and coworkers in Germany [*5*] was ignored until these ideas were revived by the work of Gest and Troiano [*6*], which triggered other studies showing that hydrogen embrittlement generally has a role during the SCC of Al-Zn-Mg(-Cu) alloys.

1. Anodic Dissolution

Many variations of the basic theme of electrochemical dissolution originally proposed by E.H. Dix [*7*] have now been incorporated into anodic-dissolution–controlled SCC models to describe SCC behavior. (A summary of the factors considered critical in these models is given in Table I.) Initial interpretations involved preferential corrosion sites at grain boundaries (GBs) with precipitate-free zones (PFZs) or GB precipitates (PPTs) providing "active paths" for SCC. Later variations accepted that SCC is not generally a special case of intergranular corrosion and, therefore, included an additional "driving force" with preferential corrosion being stimulated by stress, strain, GB segregation, oxide film rupture, or coplanar slip.

Arguably the most elegant of these is a hybrid model known as the

TABLE I

CONTROLLING FACTORS IN SCC MODELS

Anodic dissolution	Hydrogen embrittlement
Grain boundary (GB) precipitate size, spacing, and/or volume fraction	Hydrogen absorption leading to:
GB precipitate-free-zone width, solute profile, or deformation characteristics	GB or transganular decohesion
Matrix precipitate size/distribution and deformation characteristics	Internal void formation via gas pressurization
Oxide rupture and repassivation kinetics	Enhanced plasticity (adsorption and absorption arguments exist)

critical-charge criterion [*8,64*] in which SCC propagation is deemed to require sufficient anodic dissolution to occur after film rupture and before film repair or passivation. This approach has been successfully used to describe transgranular SCC for several alloy systems other than aluminum but considerable modification is required before the intergranular SCC of aluminum alloys can be successfully and quantifiably modelled.

In all the dissolution-based models, the required end product is the localized removal of metal at the crack tip with minimal attack once the crack front moves forward and crack walls are created. If crack propagation is solely via anodic dissolution, metal removal must in general be on an atomic scale, i.e. 30–50 atomic layers, as detailed high-resolution fractographic examination of SCC of 7XXX alloys reveals that fracture faces perfectly match away from crack-arrest markings [*9*] (Fig. 1).

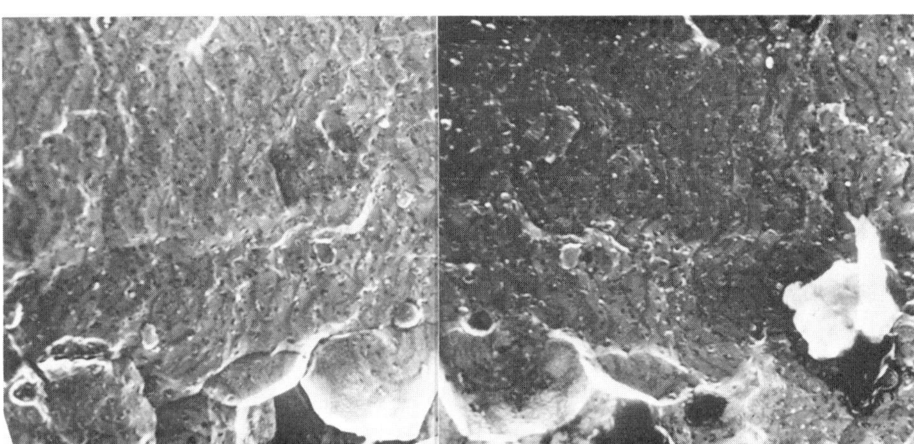

FIG.1. Matched stress corrosion fracture surfaces for a failed C-ring specimen of an Al-Zn-Mg alloy, 7017. (Courtesy of G.M. Scamans) (5000 × magnification).

2. Hydrogen Embrittlement

Embrittlement by the specific adsorption of ionic species reducing bond strengths at the crack tip [*10*] is no longer considered a likely SCC mechanism other than when atomic hydrogen in involved. Hydrogen-embrittlement (HE)–based theories have received much attention since the late 1970s and the majority of 1980's experimental evidence points to a strong hydrogen involvement. This evidence extends from direct observation of hydrogen entry into grain boundaries [*11*] and hydrogen-permeation–rate measurement [*6*] to the detection of reversible embrittlement [*5*] and near-perfect fracture-face matching after SCC in aqueous saline environments [*9*].

A basic assumption in all the HE models for intergranular SCC is that atomic hydrogen is produced by a cathodic electrochemical process, it enters alloys via grain boundaries, and it then promotes embrittlement. The mechanism of "embrittlement" is not understood, although various models have been proposed including reduction of cohesive strength [*5*], increased plasticity [*12*], brittle hydride formation [*13,14*], dislocation transport [*15*], and void pressurization [*16*]. For some newer observations in hydrogen embrittlement, refer to Chapter 18 by Watson and Meshii.

The most difficult mechanistic issue to resolve is the phenomenological similarity of SCC for various aluminum-alloys systems over a wide range of compositional, microstructural, and heat-treatment conditions. Minor differences exist, however, also requiring rationalization. To achieve this, well-designed experiments are needed to study the localized interactions between the chemistry, microstructure, and deformation characteristics of grain boundaries when exposed to the local environmental conditions developed with cracks.

In the following sections, the authors have attempted to summarize mechanistically important published data and to indicate gaps in our knowledge preventing the complete description of the stress corrosion cracking of aluminum alloys.

II. Experimental Observations

A. *Microstructural*

A fundamental observation often overlooked is that pure aluminum is immune to stress corrosion, whereas aluminum binary or ternary alloys containing copper, magnesium, zinc, or lithium can all be susceptible. The implication of this is that SCC of aluminum alloys is dependent on the presence of alloying elements and microstructural state. As a broad generalization, and alloy's potential SCC susceptibility tends to increase as the Zn, Mg, Cu, or Li total solute in solid solution increases. Microstructure/aging condition is also known to influence SCC. In general, for 2XXX (Al-Cu), 7XXX (Al-Zn-Mg[-Cu]) and Al-Li(-Cu-Mg) alloys may be summarized as:

Microstructure		Susceptibility
As solution-treated*	------------	Immune
Underaged	------------	Maximum susceptibility
Peak-aged	------------	Susceptible
Overaged	------------	Decreasing with aging

* with no significant room-temperature aging

On the microscopic grain-structure scale, wrought products of commercial alloys (e.g., plate and extrusions) tend to have elongated pancake grain structures and potential SCC susceptibility restricted to the short-transverse direction. High-purity analogue alloys, on the other hand, normally are completely recrystallized with isotopic SCC properties and susceptibilities increasing with grain size. Whether a recrystallized grain boundary structure is inherently more susceptible than an unrecrystallized short-transverse pancake grain structure is not yet established. Beneficial effects resulting from minor alloying have been attributed to several effects (for example, grain refining and/or influence upon deformation characteristics for chromium, manganese, or zirconium [*17*] and the provision of hydrogen trap sites for chromium and copper [*18*]).

Many attempts have been made to correlate the variation of SCC with metallurgical parameters known to be influenced by thermal heat-treatment. These include:

1. grain boundary precipitate size and spacing
2. precipitate-free zone width
3. preferential slip in the PFZ
4. solute profiles in the PFZ
5. matrix precipitate size/distribution
6. matrix-deformation characteristics
7. grain boundary segregation of elements (e.g., solutes or impurities)

In the late 1960s, it was hypothesized that dislocation–precipitate interactions play an important role during the SCC of the Al-Zn-Mg-Cu alloys. Initial proposals involved preferential flow in GB precipitate-free zones and localized corrosion in the regions subjected to plastic flow. These views were questioned when slip was shown to concentrate in bands across grains even in the most susceptible alloys and arguments involving the overall distribution and nature of precipitates within grains and their interaction with dislocations during plastic deformation became popular [*19*]. For instance, Speidel [*20,21*] suggested that the SCC susceptibility of Al-Zn-Mg-Cu alloys is enhanced by precipitates being sheared during plastic deformation, with the effectiveness of particle shearing being determined by particle size, volume fraction, and coherency stress field.

A limitation must be acknowledged for most the preceding data, as it is almost impossible to vary independently any of the parameters previously listed. For example, all the parameters could change during aging of a solution heat-treated and quenched alloy. Consequently, correlation of SCC susceptibility with one parameter need not exclude the possible dependence on other parameters and independence of the ones associated with the apparent correlation. The reader is referred to reviews by Bernstein and Thompson [*22,23*], which describe the possible influences of metallurgical variables upon environment-sensitive fracture and to the review by Holroyd and Scamans [*3*] for detailed information on aluminum alloys. A further limitation with most published mechanistic work is that it all involves, almost exclusively, either laboratory-cast and -processed alloys or commercial alloys subsequently subjected to laboratory-based thermal treatments. Complications arise because microstructural development during commercial thermal-mechanical practices almost inevitably differ from laboratory-based treatments and significantly modify SCC susceptibilities. Due care, therefore, must be exercised when generalizations are made for commercial alloys based upon data from noncommercial alloys and/or fabrication/thermal practices.

An exception to the preceding is the work of Sarker et al. [*24*], who used high-purity Al-Zn-Mg-Cu alloys fabricated and thermally treated via a commercial process route. These authors augmented the arguments of Speidel [*20,21*] with electrochemical ones to explain the influence of aging upon SCC of Al-6Zn-2Mg-XCu alloys containing various levels of copper. Homogeneity of slip was shown to depend upon copper level and the number of partially coherent and incoherent precipitates. Beneficial effects of overaging were attributed to a deformation mode transition from predominantly inhomogeneous, for peak-aged alloys, to a homogeneous mode generating reduced stress concentrations at GBs, for overaged alloys. In peak-aged alloys it was observed that copper levels up to 1 wt % had little effect upon SCC growth rates or deformation mode whereas for higher levels of copper slip homogeneity became more significant and SCC

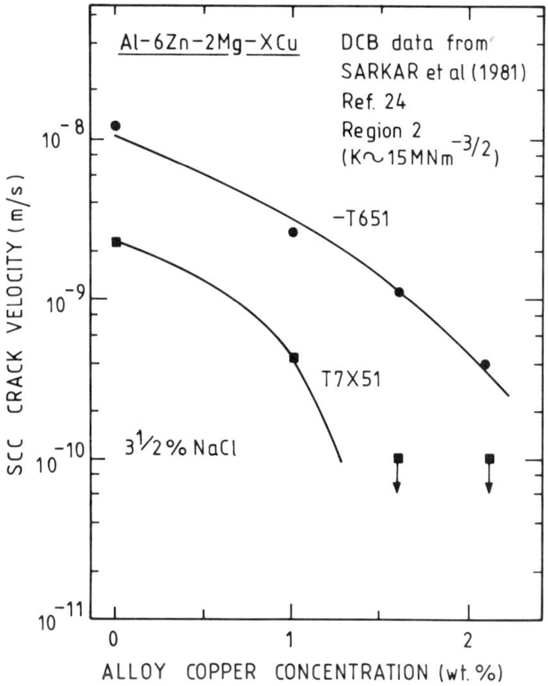

FIG. 2. Influence of copper content and alloy temper upon SCC crack growth rates for Al-6Zn-3Mg-XCu alloys in $3\frac{1}{2}\%$ sodium chloride (stress-intensity factor 15 $MNm^{-3/2}$). Double cantilevered beam (DCB) data from Sarker et al. [*24*].

resistance improved. An electrochemical argument was used to explain the progressively decreasing region 2 crack-growth rates as copper levels increased with copper being suggested to enter grain boundary precipitates and influencing local electrochemical potential differences in the crack-tip region (Fig. 2).

These arguments provide a link between the dislocation–precipitate interaction and the grain boundary anodic-dissolution models. In the latter, GB precipitates have been considered to be both detrimental and beneficial for SCC resistance (e.g., provision of active paths for corrosion [*25*] and sacrificial anodes retarding crack growth [*26*]. Concurrent with these studies, during the mid-1970s, other scientists, notably Doig and coworkers [*27*], focused attention upon solute concentrations within grain boundary precipitate-free zones and concluded that local copper levels controlled the SCC resistance of Al-Zn-Mg(-Cu) alloys by influencing the local cathodic electrochemical kinetics. These authors used high-resolution transmission electron microscopy (TEM) techniques, as have others in later studies on weldable aluminum alloys and weld heat-affected zones to detect grain boundary segregation in Al-Zn-Mg alloys (Mg and Zn segregation [*28*]) and Al-Mg-Si alloys (Mg segregation [*29*]). Experimental evidence from in-situ scanning auger microscopy (SAM) studies on Al-Zn-Mg(-Cu) alloys conducted by a research group at Martin Marietta laboratories, Baltimore (reviewed by Pickens et al. [*30*]), suggested that "free magnesium" segregates on grain boundaries which subsequently was proposed to play a crucial role during SCC by:

1. localizing hydrogen entry from aqueous electrolytes
2. accelerating hydrogen transport and/ or
3. providing sites for grain boundary embrittlement by hydrogen chemisorption or hydride formation.

Work published by Schmiedel and Gruhl [*31*] questions the preceding by suggesting that SCC in Al-Zn-Mg(-Cu) alloys is independent of magnesium segregation but is solely dependent upon grain boundary zinc concentration which is suggested to control the local hydrogen solubility and hence the SCC susceptibility. Initial findings from work in progress as of 1989 [*32*] show that copper segregation occurs on Al-Zn-Mg-Cu alloy grain boundaries (within a few nm's) susceptible to SCC but decreases dramatically as SCC resistance increases. This suggests that the earlier ideas of Doig et al. [*27*] are worthy of reconsideration, as does more recent evidence [*32,33*] suggesting that grain boundary magnesium levels increase and zinc levels only marginally decrease with improved SCC resistance.

The further work needed to clarify these differences of opinion should be forthcoming as electron microscopes providing 20-nm analytical capabilities are now generally available along with specialist equipment with 2-nm resolution. Experimentation using these techniques coupled with in-situ straining stages and quantitative image analysis should also resolve the issue of whether and when homogeneity of slip and grain boundary precipitation are controlling SCC behavior. Vasudevan et al. [*34*] have presented some convincing data for the SCC in Al-Li-Cu alloys being controlled by grain boundary precipitation and being insensitive to the matrix slip characteristics. By employing reversion and reaging heat treatments, alloy microstructures were developed such that peak-aged grain boundaries could be associated with an underaged matrix. The SCC growth behavior of the latter microstructure resembled that of peak-aged material rather than underaged material (Fig. 3), implying slip morphology is not rate-controlling and grain boundary precipitation is the dominant factor. Equivalent experiments for other aluminum systems are feasible and should yield important mechanistic information.

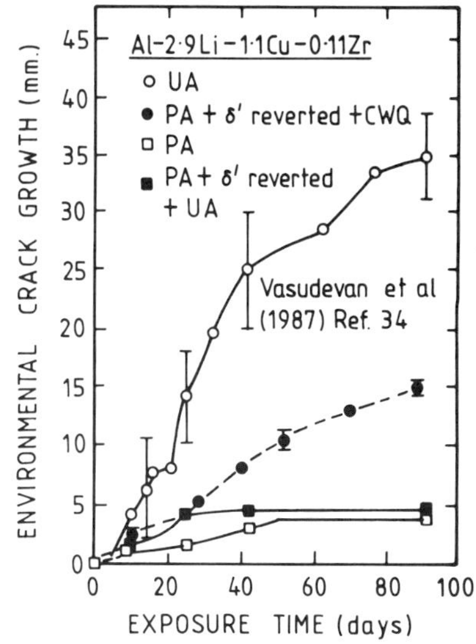

FIG. 3. SCC growth as a function of time for DCB precracked specimens for an Al-3Li-1Cu alloy in various tempers exposed to 3.5% sodium chloride. Vasudevan et al. [*34*].

For most applications of high-strength aluminum alloys, strength is of paramount importance. In general, however, the potential risk of environment-sensitive cracking prevents the exploitation of the maximum strength offered by 7XXX series, Al-Zn-Mg-Cu alloys. The commercial compromise adopted as of 1989 utilizes duplex aging practices developed during the 1960s [*1*] that minimize service problems [*21*] but introduce a 10–15% loss of peak strength (Fig. 4). The duplex aging practices, e.g. T73 temper are only available for Al-Zn-Mg-Cu alloys containing more than about 1 wt % Cu, where the aging kinetics are preferably accelerated at grain boundaries and near- peak–aged matrices can be combined with slightly overaged grain boundaries.

Naturally it would be advantageous if a high resistance to SCC (e.g., T73 temper) could be combined with peak strength (e.g. T6). Such an alloy temper has been claimed via a heat treatment known as retrogression and reaging (RRA) [*35*]. RRA consists of a retrogression stage, where peak or underaged material is heated in the temperature range 180–280°C followed by reaging at a lower temperature. Initial applications were restricted to thin sections [*36*] because it was not realized that retrogression temperatures below 220°C could be used to extend the regression times (Fig. 5) and allow successful RRA treatment of thick commercial plate [*37*].

Thompson et al. [*38*] have reviewed literature in this field up to 1986 and

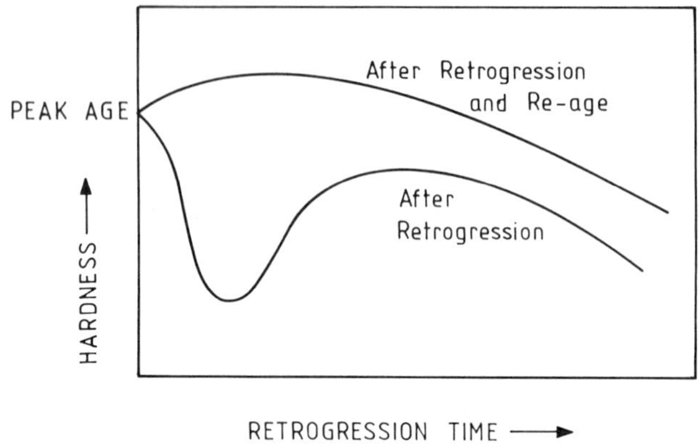

FIG. 4. A schematic diagram illustrating the effect of isothermal aging on the strength and resistance to SCC of 7XXX series aluminum alloys.

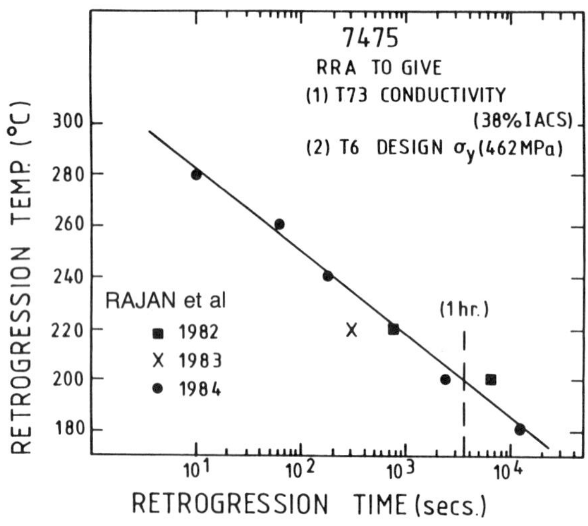

FIG. 5. Retrogression conditions needed to impart peak strength (T651) and a high resistance to environment-sensitive cracking (T73) after subsequent reaging 7475 [*32,39*].

concluded that there is no general understanding of how the RRA microstructure provides a high resistance to SCC and exfoliation. For instance, it has been proposed that the improved SCC resistance promoted by RRA heat treatments is due to grain boundary precipitates being above a critical size, 250–300 nm [*39,40*], as suggested previously for T73 duplex aging treatments [*11,18*]. In addition, Park [*41*] has reported that steady-state region 2 SCC plateau velocities for 7075-RRA tempers are directly related to the volume fraction of grain boundary precipitates, in qualitative agreement with the earlier work of Poulose et al. for conventional aging practices [*26*]. On the other hand, more recent work suggests that GB copper segregation is minimized by RRA and may well contribute to the improved SCC by modifying local cathodic electrochemical reaction kinetics [*32*] as suggested by Doig et al. [*27*].

B. Environmental

1. Subcritical Crack Growth in Gaseous Environments

The influence of gaseous environments upon the SCC of Al-Zn-Mg(-Cu) alloys is summarized by Speidel as follows: Stress corrosion cracking

1. neither initiates nor propagates in dry gases,
2. initiates almost immediately in wet gases when precracked specimens are subjected to stress-intensity factors approaching K_{Ic},
3. fracture morphology is intergranular with crack-growth kinetics controlled by the humidity level and independent of gas composition,
4. is observed at sufficiently low humidity levels, e.g. below 30% R.H., for condensation not to occur at propagating crack tips and therefore eliminate the possibility of anodic dissolution.

On the basis of the preceding, Speidel argues that a hydrogen-embrittlement mechanism must control the SCC of 7XXX alloys in moist gaseous atmospheres [*42*].

Work on 7090 and 7091 rapidly solidified powder alloys has failed to detect SCC in moist environments. However, the data available is limited, as is true in general for ingot-metallurgy (IM) aluminum-alloy systems, other than 7XXX series alloys. The data available implies that IM Al-Cu (2XXX), Al-Mg (5XXX), Al-Mg-Si (6XXX), and Al-Li (8XXX) alloys are resistant to SCC in moist gaseous atmospheres although some nonenvironment-sensitive slow crack growth may occur at high stress-intensity factors [*43*]. Further work is warranted to confirm this observation and establish whether 7XXX series alloys alone are susceptible to SCC in moist environments.

The inability of dry hydrogen to promote nucleation and growth of subcritical slow crack growth in high-strength aluminum alloys has been used in the past as an argument against hydrogen embrittlement being a potential SCC mechanism. More recent work has demonstrated that this observation results from hydrogen entry problems since embrittlement is promoted in dry hydrogen when ionization occurs or very high gas pressures (~300 atmosphere) are combined with dynamic loading close to overload.

2. Stress Corrosion in Aqueous Environments

Most SCC studies on aluminum alloys in aqueous environments involve chloride solutions. Some experimental data are available for 7XXX series alloys in solutions containing other halide ions, while that for nonhalide ions is limited, notable exceptions being Speidel's data [*21*,*42*] (for 7075-T651 in various aqueous solutions) and Foley and coworkers' data [*44*,*45*] (for 7075 in chloride, nitrate, and sulphate solutions).

Work by Holroyd et al. [*46*] has characterized the local solution chemistries developing within propagating stress corrosion cracks for 7475-T651 in sodium-chloride bulk solutions. These environmental conditions are insensitive to alloy temper and are acidic, as initially reported by Brown et al. [*47*], with the precise details of the solution chemistry development depending upon local geometry, bulk chloride concentration, electrochemical potential, and SCC growth rate (Fig. 6). Published data for powder-metallurgy (PM) alloys (limited to 7090- and 7091-type alloys [48] suggest little difference exists between PM and IM alloys in terms of local solution chemistry development within restricted geometries. Efforts to model local environmental conditions within propagating stress corrosion cracks are in progress as of the late 1980s. It is thought the acidic solutions involve $Al(OH)_2Cl$- and $Al(OH)Cl_2$-type gels on crack surfaces with ionic species such as $AlOH^{2+}$, $Al(OH)_2Cl_2^-$, and Al^{3+} dissolved in aqueous solution.

As of 1989, it remains questionable whether the SCC of 7XXX (or other aluminum-alloy systems) in aqueous environments occurs solely via a hydrogen-embrittlement mechanism. The indirect evidence for crack propagation being discontinuous with sudden rapid bursts is now very convincing [*3*] as too is data for subsurface crack initiation as described next.

C. *Mechanical*

1. SCC Initiation

SCC-initiation studies have been limited due to the experimental difficulties associated with detection of crack initiation. Consequently, nearly all mechanistic studies have involved at best both initiation and propagation but in general only crack propagation. Initial exceptions to this include the

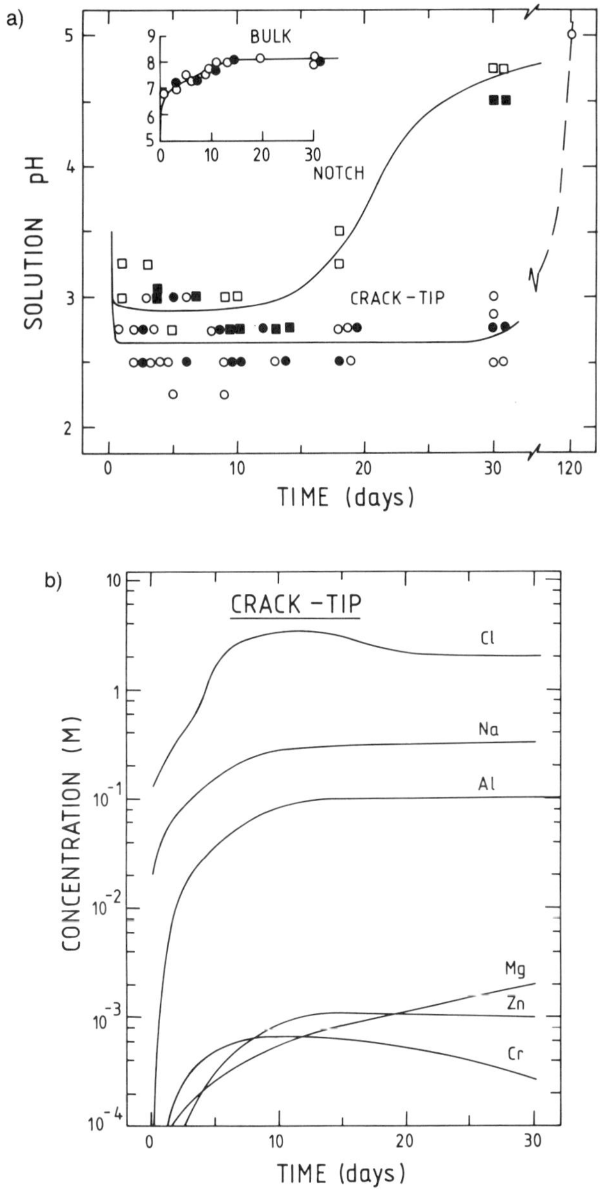

FIG. 6. Environmental conditions associated with the SCC of 7475 in 3% sodium chloride [58]: a) Long-term pH-time behavior (open symbols: T651; solid symbols: T73). b) Solution chemistry within a propagating SCC crack (T651).

work of:

1. Pathania and Tromans [*49*] reporting SCC-initiation data as a function of stress-intensity factor for notched cylindrical tensile specimens of laboratory Al-Mg, Al-Zn, and Al-Zn-Mg alloys in a saline environment and
2. Dorward and Hasse [*50*] evaluating the incubation effects introduced by different precracking methods prior to SCC propagation for 7075 in double cantilever beam (DCB) tests.

Pathania and Tromans observed that SCC-initiation times decrease with increasing stress-intensity factor, temperature, or hydrogen precharging. Their results were consistent with control by a thermally activated process (quoted activation energy ~110kJ/mole), which in view of the observed precharging effects is consistent with a hydrogen-embrittlement (HE) mechanism being involved. The latter comment needs qualification in view of the findings of Magnin [*51*] for a medium-strength Al-Zn-Mg alloy, 7020 in 3.5% sodium chloride, which suggests HE is only operative in the presence of a stress concentration and therefore does not influence SCC or corrosion fatigue initiation from smooth surfaces unless a stress raiser is established by another process (e.g., corrosion). This implies HE requires a triaxial stress state ahead of a crack tip and, if correct, validates the assumption made by Green and coworkers [*52*] and others [*53*] when using mode I/III loading experiments to determine whether HE is involved during SCC. These authors observed that SCC-initiation and hydrogen-charging effects in notched 7075-T651 samples in saline environments occurred far more readily under mode I loading (tensile), where triaxial stresses are available, compared to mode III loading (torsional), where they are limited (Fig. 7). On the basis of this and the assumption that HE requires a triaxial stress state, it was concluded that HE is involved during SCC in aqueous environments.

Perhaps the most convincing evidence for the involvement of HE during SCC of 7XXX series alloys in aqueous environments is provided by Ratke and Gruhl [*54*], who clearly demonstrated that SCC initiation for an Al-Zn-Mg alloy occurs at a site within an alloy and not in contact with the aqueous environment. To demonstrate this, these authors devised an elegant experiment involving an externally notched hollow tube that was loaded in tension with the external surfaces carefully protected to prevent moisture access and the inside of the tube filled with a saline environment (Fig. 8). Careful fractrographic studies revealed that crack initiation was internal and associated with a region ahead of the external notch. It was proposed the only possible explanation was that, during the SCC test, hydrogen was generated via a cathodic electrochemical process in the

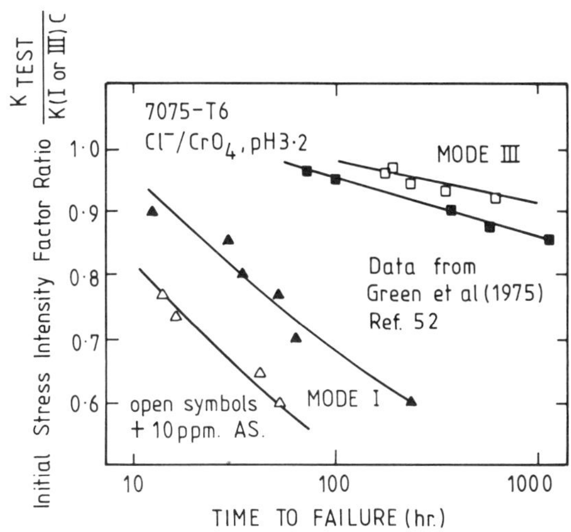

FIG. 7. Influence of loading mode upon SCC susceptibility for 7075-T6 in an acidified inhibited saline solution with and without an arsenic addition to poison hydrogen recombination and H_2 gas formation.

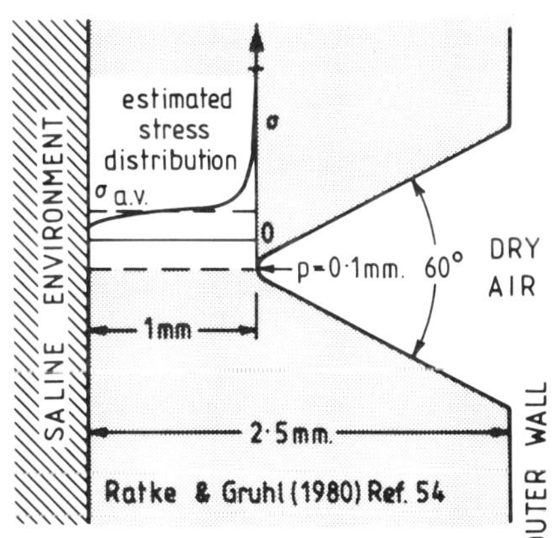

FIG. 8. Notched hollow tube specimen design used by Ratke and Gruhl [54].

aqueous electrolyte with some diffusing into the alloy via grain boundaries to accumulate at sites of high stress and promote internal crack initiation.

Minimal SCC initiation information has been published for rapidly solidified 7XXX series powder alloys. The data available are consistent with the view that SCC will be more difficult for PM alloys than IM alloys since for PM alloys the number of heterogeneous localized corrosion initiation

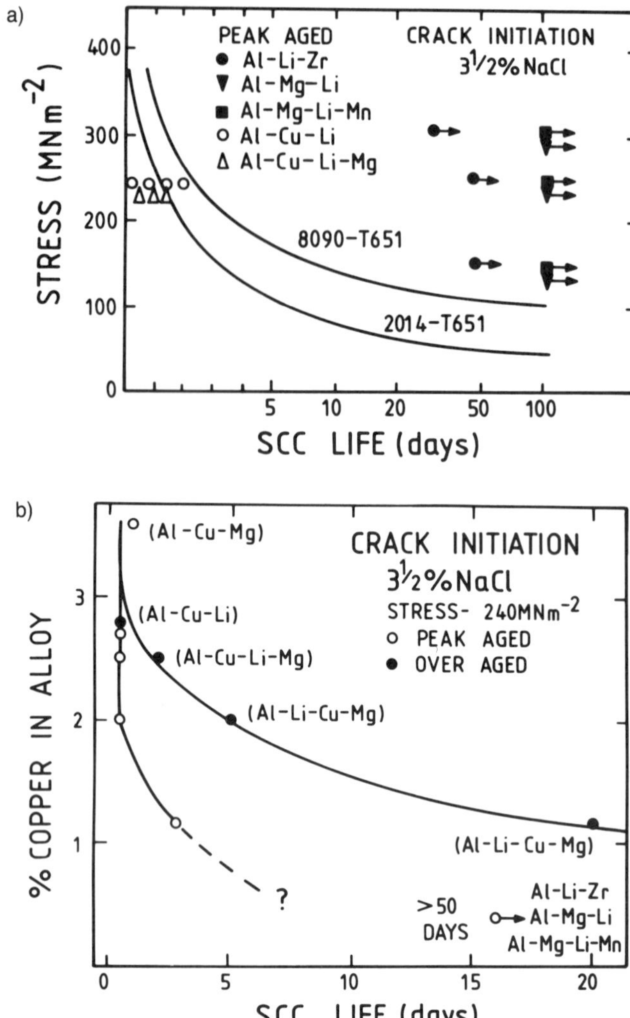

FIG. 9. Crack-initiation data from C-ring and tension-bar testing for various lithium-containing aluminum alloys tested in $3\frac{1}{2}$% sodium chloride under alternate immersion conditions: (a) influence of tensile stress and (b) influence of alloy copper level for a given stress level, 240 MNm^2 [55].

sites associated with large iron-rich constituents is reduced and corrosion attack in general is more uniform in nature. For aluminum-lithium–based alloys, a certain amount of SCC-initiation data is available from routine standard test data where test specimens have been diligently monitored and test conditions rigorously maintained during testing [55,56]. SCC initiation in copper-free aluminum-lithium–based alloys (e.g., Al-3Li(-Zr), Al-2.5Li-0.6Mg(-Zr)) rarely occurs in the absence of mechanically formed stress raises [55]. In the presence of copper, SCC initiation occurs more readily as copper content increases such that at concentrations above around 2%, SCC initiation occurs at low stresses even in overaged alloys [55,56] (Fig. 9).

2. SCC Propagation

SCC-propagation data for IM 7XXX series alloys in saline environments are available in abundance [2,21,42,57]. Testing generally involves using bolt-loaded double cantilever beam specimens with data being presented as crack growth rate (v) as a function of stress-intensity factor (K) along v-K curves. Idealized v-K curves (Fig. 10) have a K-independent crack-velocity plateau at intermediate Ks (region 2) along with K-dependent regimes at low Ks (region 1) and high Ks (region 3). SCC propagation in regions 1 and 2 is intergranular under either fixed-load or displacement-control conditions whereas SCC in region 3 is transgranular and normally

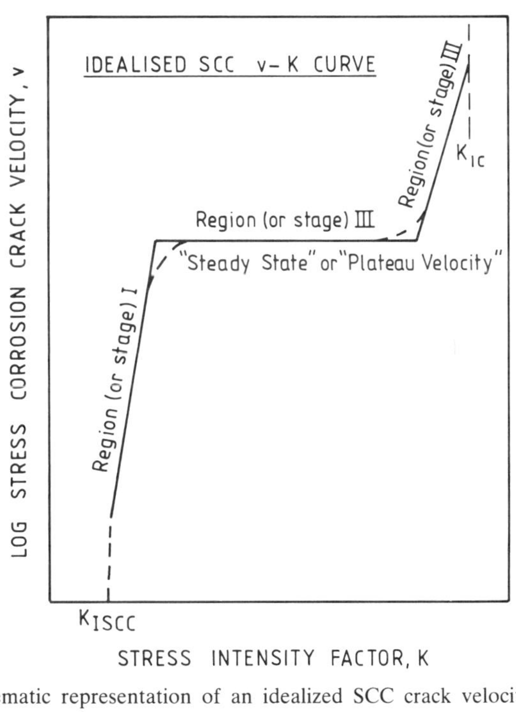

Fig. 10. Schematic representation of an idealized SCC crack velocity–stress-intensity factor (v-K) curve.

only occurs under rising K conditions at Ks just below K_{1c}. Now crack growth can be continuously monitored, using techniques such as potential-drop or compliance methods. It is evident that true "plateau velocities" rarely occur in region 2; rather, crack velocities fluctuate with an overall tendency for the mean velocity to increase with increasing K [58].

In the past, it has generally been assumed that v-K behavior provides a unique relationship independent of test procedure. Experimental [58] and theoretical evidence [59] now indicates that this assumption is generally invalid and significant difference can occur—e.g., between increasing and decreasing K experiments or if DCB specimens are preexposed to aqueous environments prior to testing [60]. Nevertheless, νv-K data rovide a useful measure of SCC susceptibility and remain one of the most accepted SCC test methods available.

Work reported during the early 1970s suggested that PM 7XXX series alloys offer better combinations of strength and resistance to SCC and exfoliation corrosion than conventional IM alloys [21,61]. While it is a fact that rapid solidification rates allow a wider range of alloy compositions, to be cast and hence the possibility of alloys with higher strength and resistance to SCC than IM alloys, it is not inevitable for a given composition that IM alloys have inferior SCC resistances to PM alloys.

Published ν-K data for PM 7XXX alloys 7090 and 7091 [62,63] suggest that crack-propagation characteristics can be matched by conventional IM

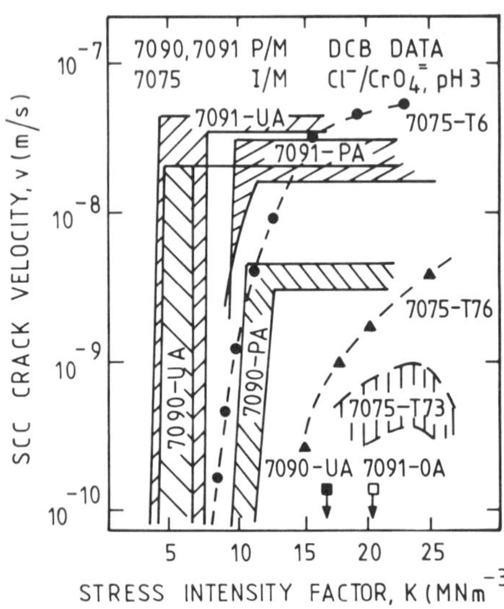

FIG. 11. SCC crack-propagation data for PM alloys 7090 and 7091 [62,63] and IM alloy 7075 [50] in an acidified inhibited saline environment.

alloys, since the latter can display a wide range of susceptibilities for a given strength level, depending upon the chosen thermal process route [3] (Fig. 11).

III. Summary

Since the 1970s, studies on SCC in high-strength aluminum alloys reveal complex interactions between an alloy's microstructure and its SCC behavior. In general, it has been observed that for a given alloy composition and fabrication route, the heat treatment resulting in different microstructures, such as T6 to T73 in a 7XXX-type alloy, has a pronounced effect on SCC behavior. While such effects are to some extent attributed to the matrix deformation, there is recent experimental evidence that grain boundary precipitation and local chemical composition in the GB precipitate-free-zone region can play an additional and more important role during SCC.

This chapter concludes that mechanistic interpretation of the SCC of high-strength aluminum alloys remains controversial and much systematic work remains to be done. On a more positive note, available experimental evidence indicates that there may be considerable potential for improvement in the SCC of both current and the newer emerging aluminum alloys using microstructure optimization via thermal-heat-treatment routes such as retrogression and reaging.

References

1. Sprowls, D.O., and Brown, R.H. (1969). *Proc. Conf. Fundamental Aspects of Stress Corrosion Cracking*, R.W. Staehle, A.J Forty, and D. van Rooyen, eds. NACE. pp. 466–506.
2. Speidel, M.O., and Hyatt, M.V. (1972). In *Advances in Corrosion Science and Technology*, **vol. 2**, M.G. Fontana and R.W. Staehle, eds. Plenum Press, pp. 115–335.
3. Holroyd, N.J.H., and Scamans, G.M. (1989). In *Environment-Induced Cracking of Metals*, R.P. Gangloff and M.B. Ives, eds. NACE, in press.
4. Hatch, J.E., ed. (1984). *Aluminum-Properties and Physical Metallurgy*. ASM.
5. Gruhl, W. (1984). *Z. Metallkunde*, **75**, 819–826.
6. Gest, R.J., and Troiano, A.R. (1974). *Corrosion*, **30**, 274–279.
7. Dix, E.M. (1940). *Trans. AIME*, **137**, 11–40.
8. Scully, J.C. (1975). *Corrosion Science*, **15**, 207–24.
9. Scamans, G.M. (1980). *Met. Trans.*, **11A**, 846–850.
10. Uhlig, H.H. (1969). *In* [*1*], pp. 86–97.
11. Scamans, G.M. (1978). *J. Materials Science*, **13**, pp. 27–36.

12. Lynch, S.P. (1977). *In* [*8*], pp 201–212.
13. Tuck, C.D.S. (1981). *Hydrogen Effects in Metals*. I.M. Bernstein and A.W. Thompson, eds. AIME, pp 503–511.
14. Pugh, E.N. (1981). In [*13*], pp 437–447.
15. Albrecht, J, Thompson, A.W., and Bernstein, I.M. (1982). *Met. Trans.*, **13A**, 811–820.
16. Van Leeuwen, H.P. (1973). *Corrosion*, **29**, 197–204.
17. Martin, J.W. (1980). In *Micromechanisms in Particle-Hardened Alloys. Cambridge University Press, p. 112.*
18. Christodoulou, L., and Flower, H.M. (1980). *Acta Met.*, **28**, 481–487.
19. Holl, H. (1967). *Corrosion*, **23**, 173–180.
20. Speidel, M.O. (1967). In [*1*], pp. 561–573.
21. Speidel, M.O. (1975). *Met. Trans.*, **6A**, 631–651.
22. Bernstein, I.M., and Thompson, A.W. (1978). In [*2*], vol. 7, pp. 53–175.
23. Bernstein, I.M., and Thompson, A.W. (1975). *Reviews on Coatings and Corrosion*, **2,** 3–44.
24. Sarkar, B., Marek, M., and Starke, E.A. (1981). *Met. Trans.*, **12A,** 1939–1943.
25. Middleton, W.R., and Parkins, R.N. (1972). *Corrosion*, **28,** 88–94.
26. Poulose, P.K., Morall, J.E., and McEvily, A.J. (1974). *Met. Trans.*, **5A,** 1393–1400.
27. Doig, P., Flewitt, P.E.J., and Edington, J.W. (1977). *Corrosion*, **33,** 217–221.
28. Rahman, M.S., and Polmear, I. (1982). *Z. Metallke*, **73,** 589–593.
29. Rahman, M.S., and Polmear, I. (1983). *Z. Metallke*, **74,** 733–736.
30. Pickens, J.R., Langan, T.J., and Green, J.A.S. (1987). In *Environment Sensitive Fracture of Metals and Alloys*, ONR, Washington, pp. 115–131.
31. Schmiedel, H., and Gruhl, W. (1983). *Z. Metallke*, **74,** 777–782.
32. Hepples, W., Jarrett, M.R., Crompton, J. S., and Holroyd, N.J.H. (1989). As ref *3*.
33. Rao, B.V.N. (1981). *Met. Trans.*, **12A**, 1356–1359.
34. Vasudevan, A.K., Liu, J., and Ricker, R.E., (1987). In *Environmental Degradation of Engineering Materials III*. Pennsylvania State University, pp. 321–327.
35. Cina,B. (1974). *U.S. Patent* 3,856,584.
36. Kaneko, R.S. (1980). *Met. Progress.*, **117,** 41–43.
37. Thompson, J.J., Tankins, E.S. and Agwarwala, V.S. (1987). In *Environmental Degradation of Engineering Materials III*. Pennsylvania State University, pp. 321–339.
38. Thompson, J.J., Tankins, E.S., and Agwarwala. V.S. (1987). *Materials Perform*, **26,** 45–52.
39. Rajan, K., Wallace, W., and Beddoes, J.C. (1982). *J. of Materials Science*, **17,** 2817–2824. (Note misprint acknowledged in (1985) *Met. Trans.*, **16A,** 2068.)
40. Park, J.K., and Ardell, A.J. (1984). *Met. Trans.*, **15A,** 1531–1543.
41. Park, J.K. (1988). *Mat. Sci. and Eng.*, **103A,** 223–231.
42. Speidel, M.O. (1971). In *The Theory of Stress Corrosion Cracking in Alloys*, J.C. Scully, Ed., NATO, pp. 289–344.
43. Lewandowski, J.J., Kohler, V., and Holroyd, N.J.H. (1987). *Materials Sci. and Eng.*, **96,** 185–195.
44. Lee, A.H., Brown, B.F., and Foley, R.T. (1980). *Corrosion*, **36,** 673–679.
45. Lee, A.H., and Foley, R.T. (1982). *Corrosion*, **39,** 379–383.
46. Holroyd, N.J.H., Scamans, G.M., and Hermann, R. (1987). In *Corrosion Chemistry within Pits, Crevices and Cracks*, A. Turnbull, ed. Her Majesty's Stationery Officer, London, pp. 495–510.
47. Brown, B.F, Fujii, C.T., and Dahlerg, E.P. (1969). *J. Electrochem. Soc.*, **116,** 218–219.
48. Langan, T.J., Christodoulou, L., Venables, D., Green, J.A.S., and Pickens, J.R. (1988). *Corrosion*, **44,** 165–169.

49. Pathania, R.S., and Tromans, D. (1981). *Met. Trans.*, **12A**, 607–612.
50. Dorward, R.C., and Hasse, K.R. (1979). *Corrosion Science*, **19,** 131–140.
51. Magnin, T. (1987). *Scripta Met.*, **21,** 907–911.
52. Green, J.A.S., Hayden, H.W., and Montague, W.G. (1975). In *Effect of Hydrogen on Behavior of Materials*, I.M. Bernstein and A.W. Thompson, eds. AIME, pp. 200–215.
53. Swanson, R.E, Thompson, A.W., Bernstein, I.M., and Maloney, J.L. (1981). As ref. **13,** pp. 459–466.
54. Ratke, L., and Gruhl, W. (1980). *Werkstoffe und Korrosion*, **31,** 768–773.
55. Holroyd, N.J.H., Gray, A., Scamans, G.M., and Hermann, R. (1985). In *Aluminium-Lithium III*. Institute of Metals, pp. 310–320.
56. Gray, A., Holroyd, N.J.H., and Miller, W.S. (1987). In *Conf. Proc. 1st Int. SAMPE Metals Conf.* SAMPE. pp. 339–351.
57. Hyatt, M.V. (1970). *Corrosion*, **26,** 487–503.
58. Holroyd, N.J.H., and Scamans, G.M., and Hermann, R. (1984). In *Embrittlement by the Localized Crack Environment*, R.P. Gangloff, ed. AIME, pp. 327–347.
59. Smith, E. (1983). *Res. Mechanica*, **9,** 237–247.
60. Hermann, R., and Holroyd, N.J.H. (1987). *Mat. Sci. and Technology*, **2,** 1238–1244.
61. Lyle, J.P., and Cebulak, W.S. (1974). *Metals Eng. Quart.*, **14,** 52–63.
62. Christodoulou, L., Gordon, J.R., and Pickens, J.R. (1985). *Met. Trans.*, **16A**, 945–951.
63. Pickens, J.R., and Christodoulou, L. (1987). *Met Trans.*, **18A,** 135–149.
64. Scully, J.C. (1977). In *Proc. Conf. on Mechanisms of Environment Sensitive Cracking of Materials*, P.R. Swann, F.P. Ford, and A.R.C. Westwood, eds. Metals Society, London, pp. 1–18.

17

A Surface-Engineering Approach to the Corrosion of Aluminum

G.M. SCAMANS, J.A. HUNTER, and N.J.H. HOLROYD

Alcan International Limited
Banbury, Oxon, England

Within the aluminum industry, corrosion has generally been considered from a rather negative and defensive standpoint and has been treated with a pragmatic approach relying heavily on service experience and the use of controlled exposure sites for long-term rate-of-degradation measurements. This experience and experimental data together with various methods of service-environment characterization have led to the extensive and successful use of many aluminum alloys in a wide range of potentially hazardous situations. The reasons for this heavy reliance on experience rather than a detailed understanding of the basic electrochemistry of aluminum and its alloys has been due to the difficulties of controlled and reproducible electrochemical experimentation on a highly reactive film-forming metal where breakdown of the protective film and subsequent corrosion is nearly always a localized phenomena and is very often controlled either by an external cathodic influence (e.g., a galvanic couple), by dispersed internal cathodic phases (e.g., intermetallic particles and/or precipitates) or by impurities that are either inherent in pot-line aluminum or are an environmental contamination.

Rather than continue to view corrosion phenomena of aluminum and its alloys as a negative property for the purpose of this chapter, we sought to examine the more positive aspects of aluminum corrosion and to identify

ISBN 0-12-341831-3

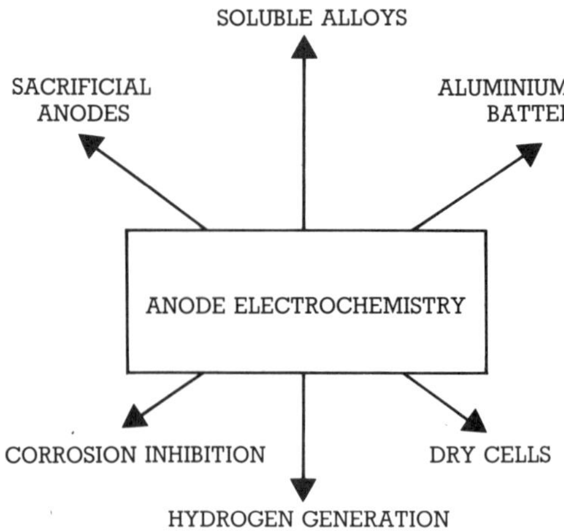

FIG. 1. Surface engineering of aluminum. The control and exploitation of the properties of oxidized aluminum surfaces.

opportunities for their exploitation. Activities in this area can be considered under the generic title of "surface engineering of aluminum" (as shown in Fig. 1), which was defined as the control and exploitation of the properties of oxidized aluminum surfaces. Significant progress has been made in most of these areas in the past five years. Specifically, improved bonding systems have been developed that, have made the aluminum structured car a more viable proposition [*1*] and exploitation of the anodic film in areas other than coloring or corrosion protection has led to its use as a membrane with excellent control of pore size [*2*]. Our own program was particularly concerned with the exploitation of electrochemical properties.

The areas of opportunity in anode electrochemistry can be subdivided as shown in Fig. 2. Of the available options, we chose to concentrate our activities on aluminum-anode development first for aluminum-air batteries and second for the sacrificial protection of high-strength weldable Al-Zn-Mg structural alloys. The present chapter will concentrate on outlining these particular positive aspects of aluminum corrosion.

I. Aluminum as an Electrochemical Fuel

Aluminum is an energy-dense material that has an electrochemical equivalent of 2.98 Ah/g (as compared to lithium at 3.86 Ah/g and zinc at 0.82 Ah/g), which means that if the protective oxide barrier on aluminum

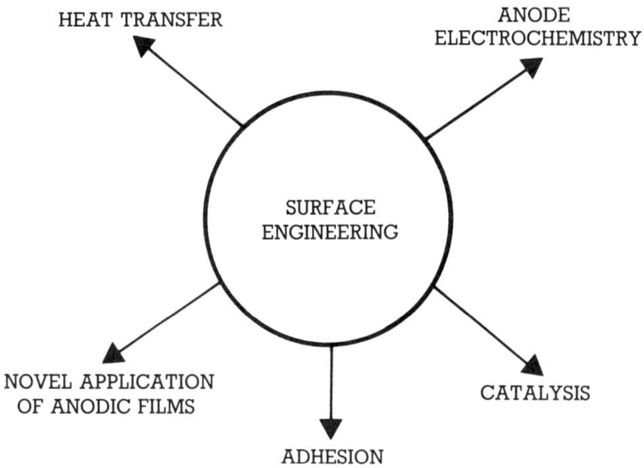

FIG. 2. Opportunities in surface engineering of the electrochemistry of aluminum anodes.

can be overcome, it should be possible to use aluminum anodes in a range of battery systems when coupled to a variety of cathodes (e.g., air electrodes, manganese dioxide, and silver oxide). Aluminum batteries are particularly suitable for reserve applications (e.g., emergency lighting) and power generation where a high portable energy density is required. Desirable properties of an aluminum anode are listed in Table I and it was from these considerations that targets were selected for our anode-development program.

The approach adopted was to examine anode activation using a specially constructed reaction microcell [*3*] as shown in Fig. 3. This enabled anode surfaces to be examined before and after discharge in the microcell in a scanning electron microscope. In this way, the dissolution morphology could be directly related to the anode-alloy microstructure and the composition of phases on the reaction surface could be readily identified using

TABLE I

IDEAL ANODE REQUIREMENTS

Most negative open-circuit voltage with low polarization under load.
Low corrosion rate at open circuit and under load.
Uniform dissolution morphology.
Rapid activation and reactivation.
Minimum additions of low-cost activations.
Tolerance of impurities in pot-line metal.
Compatibility with battery system.
Wide range of operating current densities from 1 mA/cm^2 to $1A/cm^2$.

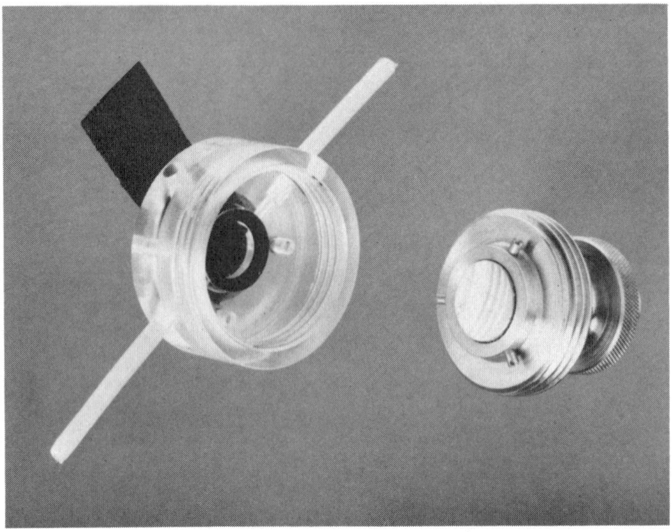

FIG. 3. Photograph of the reaction microcell disassembled to show the polished disc anode and the reaction chamber with the air cathode.

X-ray microanalytical techniques. Initial studies were conducted on a series of superpurity-based (99.995% aluminum) binary alloys containing controlled additions of gallium, indium, and tin. These elements were identified as activating elements from a study of the relevant scientific and patent literature [*4,5*]. Microcell discharge curves for a series of 200-ppm binary alloys in both 2M NaCl and 4M NaOH are shown in Fig. 4. In these experiments, an anode of 1-cm^2 surface area was coupled with a 1-cm^2 sample of air electrode of well-defined characteristics [*3,4*] and discharged through an external load controlled through a resistance box. The cell gap was 6 mm and the electrolyte was slowly pumped through the cell using a simple peristaltic pumping arrangement. The experimental technique has been described more fully in references [*3–5*]. For these experiments, the anode alloys had been given a high-temperature homogenization treatment to fully solutionize the activating element. In the solutionized condition, the alloying elements (e.g., Ga, Sn, and In) are taken fully into solid solution at high temperature (600°C) and are retained in solid solution by quenching to room temperature in cold water. Maximum solid solubilities for tin and indium are of the order of 0.12 wt % and 0.17 wt % by this particular solutionizing technique; gallium is highly soluble in aluminum. Under the conditions of discharge shown in Fig. 4, only the indium and tin additions have modified the behavior of aluminum compared to nonal-

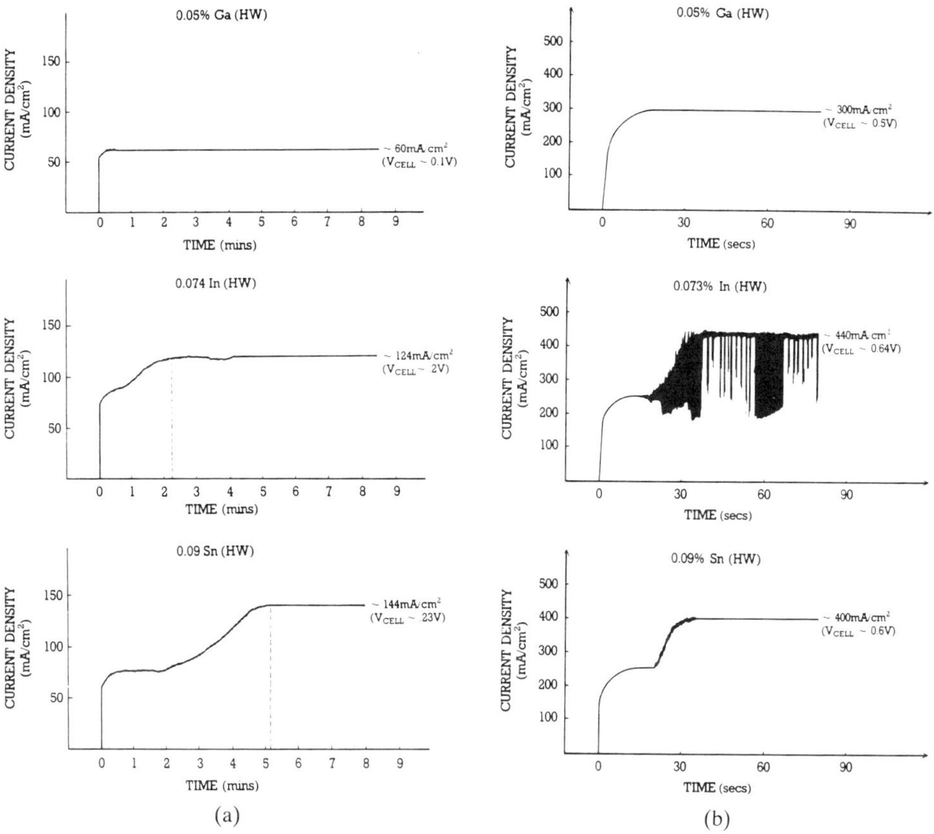

FIG. 4. Microcell discharge curves at 1 ohm for Al-Ga, Al-In, and Al-Sn fully solutionized (HW) binary alloys in (a) 2M NaCl at 25°C and (b) 4M NaOH at 60°C. Note only the In and Sn alloys have superactivated compared to pure aluminum (99.995%).

loyed pure aluminum metal (99.995%). The curve for the gallium binary alloy in both electrolytes is effectively the same as that for superpure aluminum in that the anode activates to a stable current density and voltage that remain constant with time. Additions of tin and indium do not effect this initial activation, but show a time-dependent exponential rise to a further activation after a period of activated discharge that we have termed superactivation. Once superactivation has occurred, the current and potential then stabilize again at high levels (increased power output). It is this superactivated state that provides the battery anodes with acceptable energy density and power-output characteristics.

Indium superactivation is characterized by rapid current and voltage oscillations as the anode rapidly switches between the superactive state and

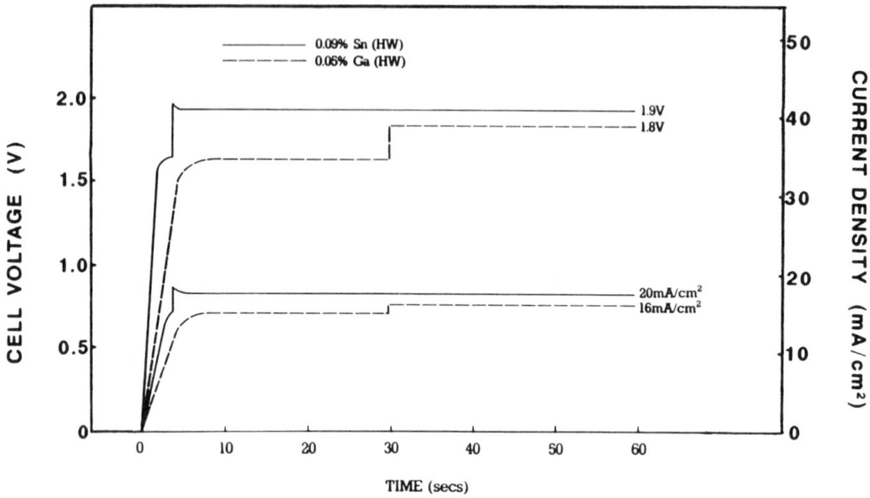

FIG. 5. Comparative microcell discharge of Al-Sn and Al-Ga binary alloys at 100 ohm in 4M NaOH at 60°C. Note that under this discharge condition, the gallium and tin alloys are both superactivated.

the active state of superpure aluminum. These oscillations eventually damp out and stable superactivation is obtained. More negative anode potentials are required for gallium-induced activation as is shown in Fig. 5, where the external circuit resistance has been increased to 100 ohms. After approximately 30 seconds of this high-resistance discharge, both the cell voltage and current surged as the anode suddenly superactivated. This gallium-induced superactivation is maintained when the external resistance is switched back to 1 ohm from 100 ohms. A particularly important feature of anode superactivation and the modification of the electrochemical properties of aluminum alloys is that the superactivating element must be in solid solution, as is shown graphically in Fig. 6. Precipitation of tin from a tin solid-solution alloy by a heat treatment of 24 hours at 400°C reverts the electrochemical behavior back to that of superpurity aluminum.

Modification of aluminum electrochemistry from elements in solid solution occurs because dissolution of the solid solution at appropriate anode potentials results in the selective dissolution of aluminum and the retention of the more noble alloying element on the dissolving anode surface. Once on the anode surface atoms of the superactivator can surface diffuse and agglomerate and grow to a critical size to form an active site for enhanced aluminum dissolution. These active sites can be readily identified for each of the superactivating elements, as shown in Fig. 7. Once the active site has developed, aluminum dissolution is then thought to be

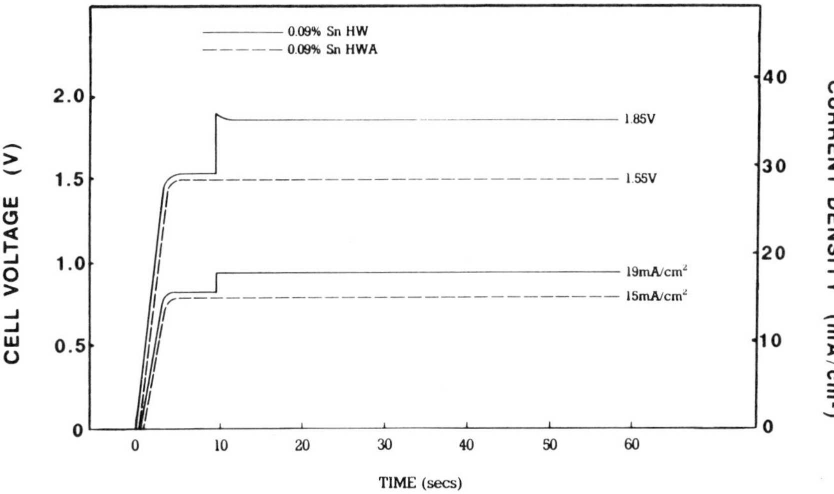

FIG. 6. Effect of tin-precipitation treatment, HWA (24 h at 400°C) on tin superactivation at 100 ohms in 4M NaOH at 60°C. Note that the reduction in tin solid-solution level completely suppresses superactivation.

achieved via the active site rather than by ionic transport through an oxide layer. This mechanistic discussion will be expanded in a future publication; transportation through surface films may still be important, but it is the film on the activator agglomerate that is important rather than a doped aluminum oxide. This is responsible for the significant depolarization of the anodic dissolution process and demonstrates why superactivation of aluminum can be similarly achieved by a surface alloying treatment (chemical, physical, or mechanical [*6*]) and by electrolyte additions of the superactivating element [*7*]. The dissolution and corrosion characteristics of the activated surface are then controlled by the specific characteristics of the active site in each case.

A second and very important principle of anode superactivation is that activation is a competitive process and that there is a sequence of activator dominance. For example, for superactivation in the alkaline electrolytes considered in this chapter, tin dominates over indium, which in turn dominates over gallium. This is illustrated in the discharge curves shown in Fig. 8. This means that the electrochemical characteristics of an anode of mixed-superactivator composition are controlled by the properties of the dominant superactivator, which is generally the most noble activating element in the anode. Dominance can occur both from solid solution and from the electrolyte. This means that additions of stannate to the electrolyte for corrosion control [*8*] provide the anode–electrolyte system with tin

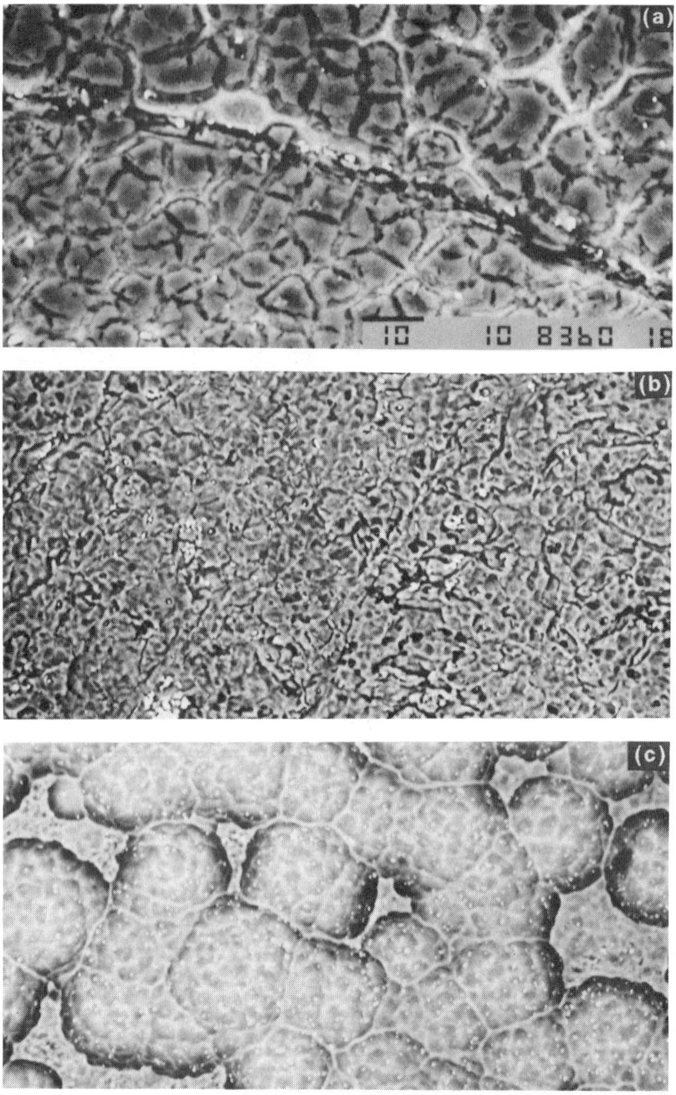

FIG. 7. Back-scattered scanning electron micrographs of superactivated surfaces of Al-Sn, Al-In, and Al-Ga binary alloys is 4M NaOH at 60°C. The bright particles on the surface were identified as (a) tin, (b) indium, and (c) gallium by X-ray microanalysis.

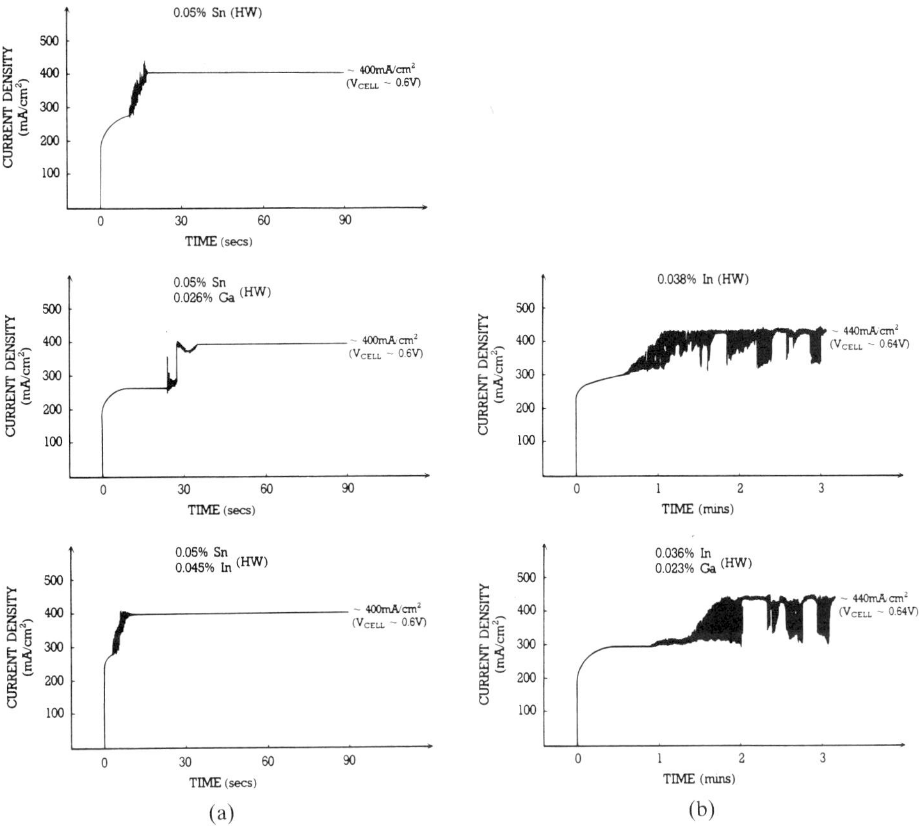

FIG. 8. Superactivation in 4M NaOH at 60°C of mixed-activator alloys in the microcell using a 1-ohm external resistance (a) showing tin dominance over gallium and indium and (b) showing indium dominance over gallium in the absence of tin.

as the dominant superactivator, which will completely mask other anode-compositional–modification effects.

Additional alloying elements to the dominant superactivator can have important secondary effects, particularly in terms of control of self-corrosion both at open circuit and under load. These additions can be nonactivating species—elements with a high hydrogen overpotential that moderate corrosion (e.g., manganese [9]—or can be the dominated addition. For example, gallium operates as a corrosion inhibitor for indium superactivated anodes in alkaline electrolytes and acts as superactivation stimulator for tin and indium superactivated anodes in saline electrolyte. This role of gallium is particularly important in practical anodes.

The combination of the simple principles of solid-solution or solution-phase activation together with the understanding of superactivator dominance greatly simplifies anode design.

Although details of the mechanism of superactivation and activator dominance are not understood, the establishment of simple anode-design principles has meant that our anode-development program in practical terms is almost complete and exploitation of aluminum as an electrochemical fuel is now reliant on the development of appropriate engineered battery systems that can manage the aluminum hydroxide reaction product effectively as has been discussed in a 1984 review article [*10*].

II. Sacrificial Protection of Aluminum Alloys

Weldable Al-Zn-Mg (7000-series) alloys are generally stronger than other weldable aluminum alloys and have the advantage that they recover their strength after welding by natural aging. However, their full commercial exploitation has been severely restricted by the occurrence of both exfoliation corrosion and weld-toe cracking of weldments in service. Weld-toe cracking, which is the more critical phenomenon,, initiates at the toes of welds and propagates by stress-corrosion cracking through the interfacial region between the fusion zone and the heat-affected zone. This region is commonly known as the white-zone due to its etching response in nitric acid [*11*]. Cracking through the white-zone can lead to further stress-corrosion cracking of the parent material and to extensive joint failure.

Our first approach to the prevention of weld-toe stress-corrosion cracking was not to elaborate on the previous amelioration techniques based on alloy or filler wire modification [*12*] or on joint design [*13*], but rather to identify the local environment in weld-toe stress-corrosion cracks and to determine the electrochemical conditions in that environment where cracking could be prevented. We then hoped to provide a surface coating by spray metallization that would hold the joint within an appropriate protection-potential range.

Figure 9 shows the "potential window" required to moderate stress-corrosion cracking in a slow strain rate test of welded specimens of an Al-Zn-Mg plate alloy (7017) in an environment that simulates crack-tip conditions. The potential must be held between −1130 and −1200 mV (SCE) to minimize stress-corrosion-cracking susceptibility and all localized corrosion phenomena.

Wires for spray metallization of welded joints to provide electrochemical control of potential were made from superpurity aluminum with controlled

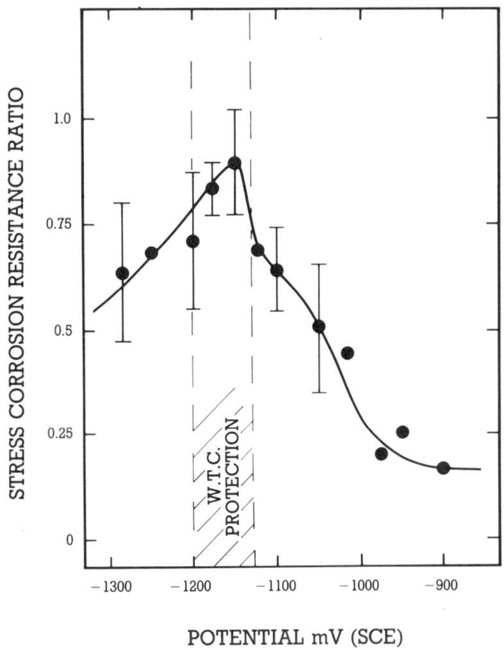

FIG. 9. Effect of applied potential on the weld-toe stress-corrosion cracking (W.T.C.) resistance of welded Al-Zn-Mg (7017) alloy plate in LN65666 (2% NaCl + Na_2CrO pH3.0) tested by the slow strain rate method. The safe potential window is between −1130 and −1200 mV (SCE).

additions of tin, indium, and gallium as with the battery anode development work. These were sprayed onto welded joints by both flame-spraying and by arc-spraying. Specimens for this operation were prepared by running a weld bead across the surface of a plate sample (referred to as a "bead-on-plate" sample), which was then cleaned and prepared for spray metallization by sand blasting. The complete heat-affected area and adjacent parent plate were covered by the deposited spray. For arc-spraying, it was necessary to increase the levels of elements added to compensate for volatilization losses experienced during the high-temperature spraying process. Typical volatilization losses are of the order of 60 to 75% for indium and 15 to 25% for gallium; tin conversely can be concentrated by up to 25%.

Significantly, since the spray-metallization process produces a rapidly solidified coating, elements in the spray wire are retained in solid solution in the as-sprayed coating to levels in excess of the maximum solid-solution level that can be obtained by solution heat treatment. This means that the

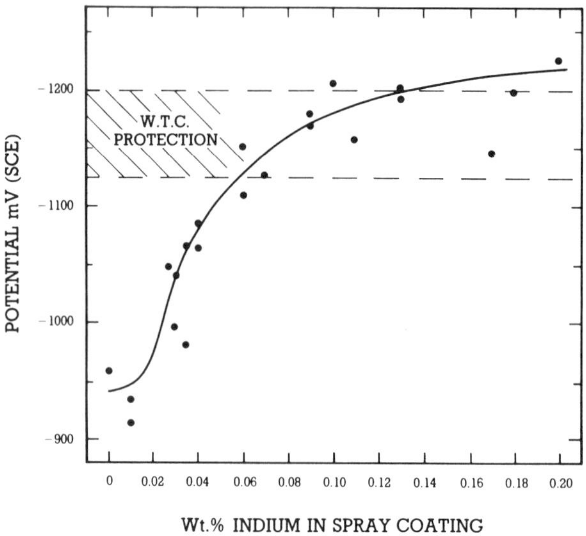

FIG. 10. Effect of measured indium solid-solution level in the sprayed coating after spraying on the corrosion potential after 70 hours immersion in LN65666. Note that at best 0.06 wt % indium is required to hold the potential in the required range for protection of welded joints.

activating elements are in the desired microstructural form to modify electrochemical behavior.

Figure 10 shows that the potential of mixed superactivator alloys is controlled by the indium level of the as-sprayed coating. Hence, in this type of environment (pH 3) indium is the dominant superactivator although the additional elements can have important secondary effects, particularly on the rate of self-corrosion after extended exposure times. These studies have led to the development of suitable spray-wire compositions for the protection of welded Al-Zn-Mg alloy joints when either flame-spray or arc-spray coating processes are employed. Suitable compositions are listed in Table II [*12,14*].

A second approach to the protection of welded joints of 7000-series alloys has been based on the observation that weld beads of alloys of the 2000, 5000, and 6000 series are electronegative with respect to the heat-affected zone and the parent alloy, whereas for 7000 series the weld bead is electropositive. For the majority of welded alloys, the weld bead itself acts sacrifically (i.e., it is preferentially corroded) and protects the surrounding parent metal from exfoliation and weld-toe cracking. We therefore sought to superactivate the weld beads of 7000-series alloys by modification

TABLE II

Composition Ranges of Arc- and Flame-Spray Wires

Element	Flame spray	Arc spray
Zinc	0–1.5 wt %	0–20 wt %
Indium	0.06–0.08	0.10–0.18
Tin	0–0.25	0–0.5
Gallium	≤0.05	0–0.26
Iron	≤0.04	≤0.08

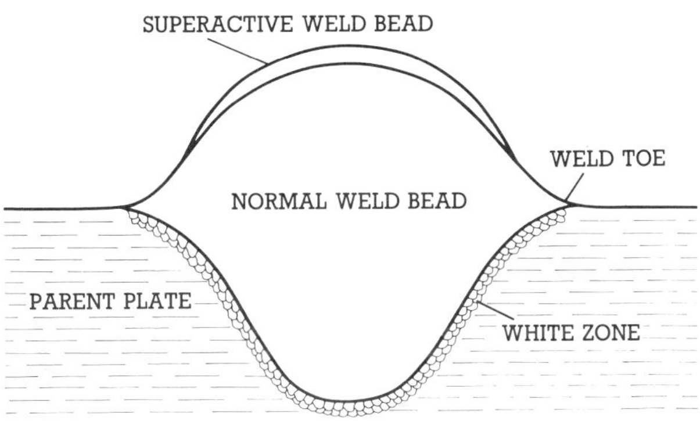

Fig. 11. Schematic representation of a weld bead superactivated by a capping pass with a filler wire containing additions of tin, indium, or gallium. Note that the capping pass must not cover the weld-toe.

of the filler-wire composition and thereby make the weld bead more electronegative than the parent alloy. To achieve this aim, a series of superpurity-based Al5Mg filler wire were produced containing various combinations of the superactivators tin, indium, and gallium. These were used as a capping pass after a conventional welding operation with conventional Al5Mg filler wire (NG61). This ensured that there was limited mixing of the superactivated filler material with the bulk of the weld pool enabling a high base purity to be maintained and prevented any diffusion of tin, indium, or gallium into the white-zone grain boundaries. The capping pass technique is shown schematically in Fig. 11.

Welded joints made by this technique did not suffer from either weld-toe corrosion (frequently a precursor to weld-toe cracking) or from exfoliation even after prolonged exposure times in 3% sodium chloride. Weld-toe stress-corrosion cracking was also prevented as shown in Fig. 12. Note that

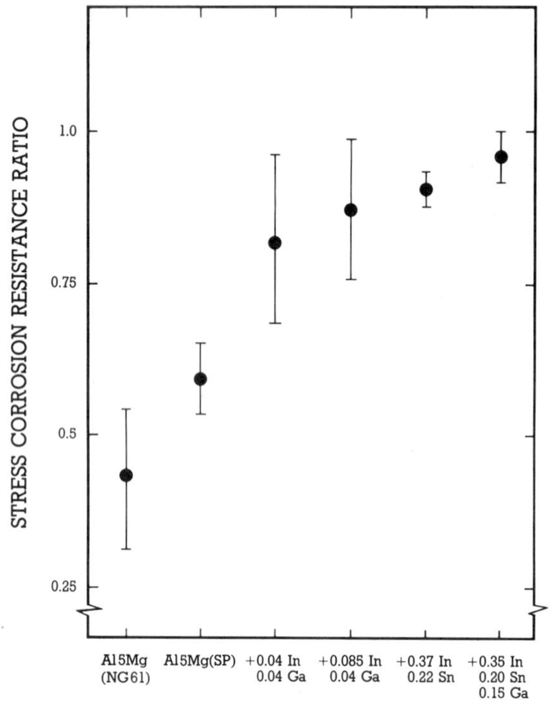

FIG. 12. Effect of additions of superactivators to the welding wire (A15Mg, NG61), used as a capping pass for welded joints of an Al-Zn-Mgt alloy 7017, on the resistance to weld-toe stress-corrosion cracking (W.T.C.) determined by slow strain rate stress-corrosion testing in LN65666.

this type of slow strain rate weld-toe cracking test is particularly severe due both to the low pH of the test environment (LN 65666) and to the mechanical lifting of the weld-toe that occurs by decohesion of the weld-toe–plate interface and effectively notches or precracks the white-zone. The potential of the welded specimen with the superactive capping pass was within the safe window as determined for the spray-metallized samples. Immunity was observed using both tin and indium as the dominant superactivating element. The gallium addition in each case accelerates the rate of indium or tin superactivation as was shown for the battery anodes. Further work is now required to optimize the composition of the capping-pass superactive filler wire and to determine the required control of the welding parameters during the capping pass to ensure that the welded joint is protected. This must include consideration of the effect of impurities (e.g., iron) and deliberate alloying additions (e.g., zinc and copper) accumulated by mixing of the superactive capping-pass material with the weld-pool material.

III. Discussion

The surface-engineering approach to the corrosion of aluminum has led to important advances in our understanding of the modification of the electrochemical behavior of aluminum by solid-solution and/or electroyte additions of certain superactivating elements. This understanding has been used to develop improved aluminum anodes for use in battery systems employing both alkaline and saline electrolytes and also to develop an improved sacrificial protection system for welded aluminum alloys. Work is now in progress in other areas of the diagram shown in Fig. 2 and in using the general principles of activation and dominance developed in the present work to studies of the wide range of localized corrosion phenomena shown by aluminum alloys. These electrochemical systems are generally much more complex than the simple superpurity-based model alloys we have used so far for superactive anode alloy development. To understand such phenomena (e.g., pitting, and exfoliation), the electrochemically dominant alloying additions must be determined, critical impurities must be identified both in the alloy and in the electrolyte, and the potency and hydrogen overpotential of local cathodes must be considered. Hopefully, in this way we can reach a better understanding of localized corrosion phenomena of aluminum alloys that presently impair service performance and thereby develop better methods of prevention and control.

Acknowledgments

The study of anode superactivation for battery anodes was carried out by J.A. Hunter as part of his D.Phil. program at Oxford University under the joint supervision of Dr. J.M. Sykes and Dr. G.M. Scamans. The development of sacrificial protection systems for welded 7000-series alloys was carried out by Dr. W. Hepples and Mr. C.R. Wiseman under the supervision of Dr. N.J.H. Holroyd.

Valuable discussions on various aspects of anode superactivations were held with our colleagues on the battery-development program, notably Dr. C.D.S. Tuck and our consultants, Prof. A.R. Despic and Dr. O.R. Brown.

References

1. Wheeler, M.J. Sheasby, P.G., and Kewley, D., SAE Technical Paper 870146. Selwood, P.G., Law, F.J., Sheasby, P.G. and Wheeler, M.J., SAE Technical Paper 870149. Kewley, D., Campbell, I.G., and Wheatley, J.E., SAE Technical Paper 870150. Marwick, W.F., and Sheasby, P.G., SAE Technical Paper 870151. Seeds, A., and Sheasby, P.G., SAE Technical Paper 870152. International Congress and Exposition, Detroit, Michigan, February 1987.
2. Furneaux, R.C., Rigby, W.R., and Davidson, A.P. (1986). European Patent 0,178,831.
3. Scamans, G.M. (1986). *Chemistry and Industry*, pp. 192–196.
4. Scamans, G.M., O'Callaghan, W.B., Fitzpatrick, N.P., and Hamlen, R.P. (1986). American Chemical Society Paper 86923, *Proceedings of 21st Intersociety Energy Conversion Engineering Conference, San Diego*, pp. 1057–1061.
5. Scamans, G.M., O'Callaghan, W.B., Fitzpatrick, N.P., and Hamlen, R.P. (1986). 15th International Power Sources Symposium, Brighton.
6. Despic, A.R., Stevanovic, R.M., and Vorkapic, A.M. (1984). ISE Meeting, Berkeley, paper A2-19, pp. 93–95.
7. Bohnstedt, W. (1980). *J. Power Sources* **5**, 245–253.
8. Katoh, M. (1971). U.S. Patent 3,563,803.
9. Jeffrey, P.W., and Halliop, W. (1987). ECS Meeting, Honolulu.
10. Fitzpatrick, N.P., and Scamans, G.M. (1986). *New Scientist* **111**, 1517, 34–37.
11. Cordier, H., Schippers, M., and Polmear, I.J. (1977). *Z. Metallkde* **68**, 280.
12. Holroyd, N.J.H., Hepples, W., and Scamans, G.M. (1987). *Environmental Degradation of Engineering Materials*. Penn State.
13. DiRusso, E., and Abis, S. (1985). Conference on Light Alloys in Transport, European Metals Society, Strasbourg, France.
14. Holroyd, N.J.H., and Scamans, G.M. (1987). U.K. Patent Application 8704251.

18

Hydrogen in Aluminum and Aluminum Alloys*

J.W. WATSON

Armco Research and Technology
Middletown, Ohio

M. MESHII

Department of Materials Science and Engineering
Northwestern University
Evanston, Illinois

I. Introduction

It has been generally thought that face-centered cubic metals and alloys are less sensitive to hydrogen embrittlement effects. The solubility of hydrogen in solid aluminum is extremely small. Hydrogen effects in aluminum and aluminum alloys, consequently, have not received adequate

* This work was supported by the Materials Research Center under Grant No. DMR 8520280 and the ALCOA Foundation under Grant No. TC 132996TC.

ISBN-0-12-341831-3

attention in the past. Since the mid-1970s, however, the extent of hydrogen damage in aluminum alloys has been recognized to be substantial. It has been accepted that hydrogen embrittlement plays a key role in stress corrosion cracking, the major concern in technological applications of high-strength aluminum alloys. In order to solve these problems, hydrogen-induced degradation of the microstructure and mechanical properties of aluminum and aluminum alloys must be understood. The deleterious effects of hydrogen, however, are multifaceted and complex. The fundamental knowledge of hydrogen in aluminum is essential to the understanding of various hydrogen effects in aluminum and aluminum alloys. Nevertheless, the hydrogen-induced degradation of microstructure and mechanical properties remains a serious problem in the application of aluminum alloys, particularly those of high strength. In this chapter, the fundamental properties of hydrogen in aluminum are reviewed and the role of hydrogen embrittlement in stress corrosion cracking is examined.

II. Properties of Hydrogen in the Perfect Lattice of Aluminum

A. *Solubility*

The solubility of hydrogen in pure crystalline aluminum is relatively small; it is 10^{-6} atomic fraction at the melting point and decreases rapidly with temperature, making it difficult to determine the properties of hydrogen in aluminum accurately. The solubility has been reported by Eichenauer et al. [*1*], Ransley and Neufeld, [*2*] and Opie and Grant [*3*]. Smoothing these results, Talbot [*4*] gave empirical equations for the solubilities in the liquid,

$$\log C_H = (-2761/T) + 2.768, \tag{1}$$

and for the solid

$$\log C_H = (-2580/T) + 1.399, \tag{2}$$

where C_H is in units of cm^3 of hydrogen gas at 0°C and 1 atm per 100 g of aluminum and T is an absolute temperature. Although the data used to determine Eq. (2) were obtained in the high-temperature region (470–660°C), extrapolation of this equation to lower temperatures gives 1.5×10^{-12} atomic fraction as the solubility at 300°K. It should be noted that the presence of traps can increase the equilibrium concentration substantially. It is also possible that a higher concentration of hydrogen can be quenched-in, particularly in light of the low diffusivity of hydrogen in aluminum.

B. Interstitial Site, Partial Molar Volume, and Strain Field

Hydrogen dissolves interstitially and expands the crystal lattice of host metals. Although the interstitial site, partial molar volume, and strain field of hydrogen have been determined in a number of metals, they have not been determined in aluminum.

The location of positive muons has been determined to be the octahedral interstitial site in copper. Since positive muons are regarded as light isotopes of hydrogen, it is generally accepted that hydrogen is on octahedral sites in copper. Unfortunately, a similar approach failed in aluminum because anomalies were observed in the relaxation behavior of positive muons in aluminum [5]. Hydrogen is expected to be accompanied by a significant strain field in aluminum regardless of which interstitial site it is on.

In his review article [*6*], Peisl reported that the partial molar volume of hydrogen, which has been determined in many face-centered cubic metals and alloys, is uniquely 2.9 $Å^3$. Apparently, this volume-constancy rule also applies to body-centered cubic and hexagonal-closed-packed metals. It is also found that the volume remains roughly constant over a wide range of concentrations; relative volume changes from 0.1 to 0.3. No explanation has been given for this unique size of the hydrogen atom in many materials.

C. Diffusivity

There have been a number of investigations on the diffusivity of hydrogen in aluminum since the 1950s [*7–15*]. Ishikawa and McLellan summarized the most recent of these results [*10–14*] along with their own results as of 1986 as shown in Fig. 1 [*15*]. The majority of the diffusivity measurements were made in the high-temperature region and show a significant range of deviation in the diffusion coefficient and the activation enthalpy. In all of these high-temperature experiments, the diffusion coefficients were estimated from thermal evolution of hydrogen from hydrogen-charged specimens except the work by Hashimoto and Kino, where a permeation technique was employed [*14*]. The large spread in the measured diffusivities is attributed to the difficulty in accurately determining small outgassing rates. Csanady et al., who studied the role of the surface oxide layer in the outgassing process of hydrogen, concluded that there was no significant difference in the outgassing behavior of samples with different oxide layers [*11*]. It therefore appears improbable that the surface oxide film plays a major role in causing this spread.

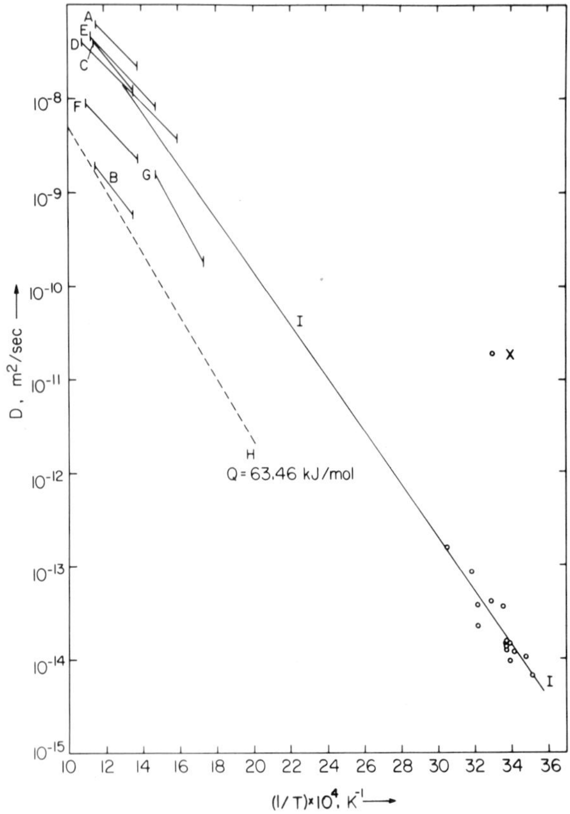

FIG. 1. Diffusion coefficients of hydrogen in aluminum. The letter codes indicate A[12], B[8], C[6], D[10], E[7], F[13] and G[14]. I represents the diffusion coefficients determined around room temperature [15] (shown by open circles). The extrapolation of this line coincides with some of the diffusion coefficients determined near melting points [*6,7,10,12*].

The work of Ishikawa and McLellan [*15*], which was carried out in the temperature range from 285 to 328°K, indicated that a single Arrhenius relation can represent the diffusivities they obtained in this temperature range as well as those determined in the high-temperature range up to the melting point. The best fit was obtained by

$$D(\mathrm{m^2/s}) = 9.2 \times 10^{-5} \exp\,(-55250\ \mathrm{J/mol/R}T). \tag{3}$$

This relation is represented by the line I in Fig. 1. Although a consistency exists between some of the high-temperature diffusivities and the low-temperature diffusivities in terms of the Arrhenius relation, the authors suggest that we allow a significant uncertainty in calculating a diffusivity from their values.

III. Hydrogen-Lattice Defect Interaction

The solubility and the diffusivity of hydrogen can be influenced significantly by trapping of hydrogen by lattice defects. The presence or absence of trapping and the magnitude of the effect on these properties has been debated even for measurements carried out at higher temperatures. The effect can be greater at lower temperatures. Consequently, many hydrogen-induced phenomena must be interpreted with a careful consideration of any hydrogen-trapping effects. Unfortunately, available information is still inadequate for definitive analysis of such effects.

A. Type and Strength of Traps

Lattice defects, such as lattice vacancies, dislocations, and grain boundaries, are potential traps for hydrogen. Some solute atoms and their clusters can also become hydrogen traps. Hydrogen is supersaturation is thought to have a propensity to precipitate as gas bubbles. Therefore, voids, preexisting or formed by hydrogen precipitation, could become hydrogen traps.

The equilibrium concentration of lattice vacancies is approximately three orders of magnitude higher than the solubility of hydrogen at the melting point of aluminum. Thermodynamic calculations also suggest that a significant binding enthalpy should exist between them [*16*]. Therefore, it is reasonable to suspect that lattice vacancies will become significant traps for hydrogen in aluminum. Hashimoto and Kino [*14*] reported the evidence of lattice vacancies trapping hydrogen atoms from their permeation experiment in the temperature range from 573 to 673°K. Ishikawa and McLellan, however, disputed this claim based on their diffusivity measurement around room temperature and their calculation [*15*]. The binding energy between a lattice vacancy and a hydrogen atom was calculated to be 0.5eV (48.1 KJ/mol.) by Besenbacher et al. [*17*].

Foster et al. [*18*] examined the distribution of hydrogen in zone-refined aluminum using the autoradiographic technique. The specimens were charged with tritium at elevated temperatures (typically 500 or 600°C) and then cooled to 0°C. They observed that tritium segregated to dislocations, cell boundaries, and grain boundaries. They also indicated that the presence of impurities, also segregated to these lattice defects, facilitated the precipitation of tritium microvoids along those lattice defects.

The hydrogen trapping at grain boundaries also increases the solubility according to Edwards and Eichenauer [19]. When their 99.999% aluminum specimens were repeatedly used for the thermal extraction experiments, they observed a gradual decline in the solubility of hydrogen. This decline in hydrogen solubility was accompanied by the reduction in grain-boundary area due to grain growth during repeated heating to a temperature near the melting point. By comparison of solubilities between the fine- and coarse-grain states of the specimen, they obtained the grain-boundary solubility and bulk solubility of hydrogen in aluminum as

$$C_{\mathrm{Hg.b.}}\ (\mathrm{mol\ H\ Atm^{-1/2}/m^2 g.b.}) = 1.6 \times 10^{-6} \exp\,(-50640\ (\mathrm{J/mol\ H})/\mathrm{R}T)$$

and

$$C_{\mathrm{Hb.}}\ (\mathrm{mol\ H\ Atm^{-1/2}/mol\ A1}) = 3.35 \times 10^{-3} \exp\,(-64800\ (\mathrm{J/mol\ H})/\mathrm{R}T),$$

respectively. These results indicate that the binding enthalpy between hydrogen and grain boundary is 14KJ/mol (0.15 eV).

Gest and Troiano reported that hydrogen-charging generated two internal friction peaks (at 130 and 170°K) in 7075 aluminum alloy [20]. The 130°K peak was interpreted to be caused by hydrogen-substitutional atom pairing, indicating that significant binding existed between hydrogen and substitutional solute atoms. The 170°K peak was attributed to hydrogen-dislocation pairing. Leger and Piercy [21] further examined the hydrogen-substitutional atom-pairing phenomenon in dilute binary Al-Cu and Al-Mg alloys. They did not observe the hydrogen-induced low-temperature peak observed in 7075 alloy. From the sensitivity limit of their measurements, they concluded that the binding enthalpies between hydrogen and copper and between hydrogen and magnesium were less than 0.05 and 0.06 eV, respectively.

B. Transport of Hydrogen by Moving Dislocations

The diffusivity of hydrogen in aluminum is very small at room temperature. A faster transport mechanism of hydrogen, such as hydrogen transport by moving dislocations, has been speculated to explain phenomena for which a rapid transport of hydrogen is required to rationally explain the observations. To the present authors' knowledge, no direct evidence of this mechanism has been reported in aluminum.

Albrecht et al. [*22*] and Hardwick et al. [*23,24*], examining the effect of cathodic charging in 2124, 7050, and 7075 aluminum alloys, observed that the hydrogen embrittlement was most pronounced when cathodic charging was carried out during slow plastic straining with a strain rate of approximately 10^{-6}/sec. In 7075, which was found to be most susceptible to the hydrogen effect, the hydrogen-induced embrittlement was accompanied by a change in fracture mode from transgranular to intergranular fracture. The authors proposed the enhanced embrittlement effect by concurrent plastic straining during cathodic charging as evidence that dislocations can transport large amounts of hydrogen deep into the interior of the alloys.

Feuerstein and John [*25*] carried out tensile deformation of 99.99% aluminum at room temperature in a UHV chamber while monitoring the outgassing from the specimen. A small amount of hydrogen gas evolved continuously with deformation. At least a portion of this gas was thought to come from the interior of the specimen. The dislocation transport of hydrogen mechanism was suggested as a means for hydrogen to diffuse to the surface.

C. Hydrogen-Induced Mobility of Dislocation

It is still disputed whether hydrogen causes a softening effect or a hardening effect in metals and what mechanism is responsible for this phenomenon. The most extensive works to resolve this question have been carried out in iron [*26*]. The outcome of this dispute will have a profound effect in interpreting various hydrogen effects in several metals including aluminum.

Birnbaum and his coworkers performed a series of in situ deformation experiments in the environmental cell of a high-voltage electron microscope using various metallic specimens [*27–35*]. Apparently, the effect of hydrogen is similar in all the metals examined, including aluminum [*33*] and 7050 and 7075 aluminum alloys [*35*]. They generally observed an increase in dislocation mobility and promotion of highly localized ductile fracture due to the introduction of gaseous hydrogen into the environmental cell. They estimated the hydrogen fugacity in the environmental cell to be between 34.0 + 13.9 MPa and 750 MPa from the blisters formed by gaseous hydrogen between the surface oxide film and aluminum metal [*36*]. The electron beam was suggested to increase the fugacity by dissociating the hydrogen molecules.

IV. Effects of Hydrogen on the Mechanical Properties of Aluminum

Only limited information is available on the effects of hydrogen on mechanical properties of pure aluminum and its dilute solid solutions. This situation in aluminum is in sharp contrast with the extensive information obtained for pure iron and available in analyzing hydrogen effects in iron alloys.

Watson et al. [*37*] examined the effects of cathodic charging on tensile test properties, microhardness, and TEM microstructure, using 99.99% aluminum. The tensile tests using a strain rate of 10^{-4} sec^{-1} revealed that the yield stress increased with charging time, the ultimate tensile stress increased only slightly, and the ductility measured by strain to failure decreased significantly. These effects became observable only after the specimens were precharged for several hours prior to tensile testing. No saturation was evident after up to 80 hours of precharging. In these tests, the cathodic charging was continued until the termination of the test.

The Vickers microhardness tests, using 15 gr. load, carried out on the specimen surface, showed saturation within 10 hours of the cathodic charging. In order to assure that the cathodic charging effect is primarily a hydrogen effect, similar cathodic charging was carried out in an electrolyte that did not contain the recombination poison. This cathodic charging did not produce comparable changes in microhardness measurements, indicating that the observed effects were caused by hydrogen entering into the specimen. The microhardness distribution was also determined as a function of distance from the surface as shown in Fig. 2. The hardness near the surface increased rapidly with charging time. At the same time, the thickness of the surface-hardened layer increased with charging time. The slope of the thickness of this layer squared versus charging time plot yielded 10^{-16} m^2/sec as the room-temperature diffusivity of hydrogen. This value is slightly lower than that measured by Ishikawa and McLellan [*15*] but greater than that measured by Choo and Bernstein [*38*].

The annealing behavior of the cathodic charging-induced hardening was also examined (Fig. 3). The annealing took place in two stages, the second stage coinciding with the annealing stage of cold-rolled aluminum.

Transmission electron microscopy revealed that cathodic charging produced a great increase in the dislocation density near the surface. These dislocations were in two populations: straight parallel dislocations and high-density dislocation bands. The interior regions of the charged foils did not experience the same sort of increase, remaining essentially similar to the annealed condition.

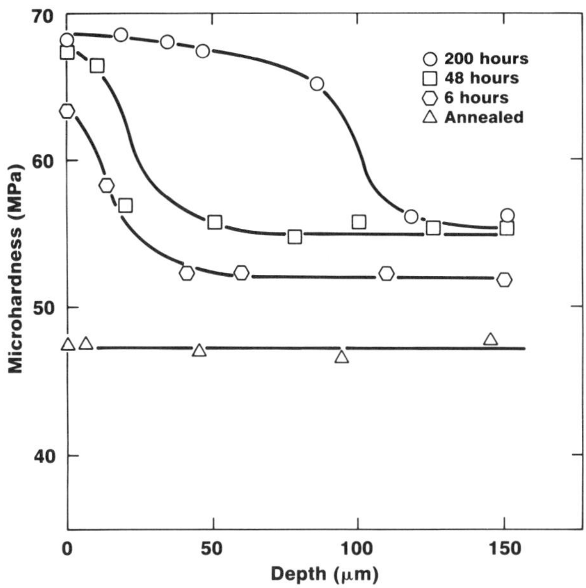

FIG. 2. Microhardness distribution as a function of distance from the surface that had been cathodically charged [37].

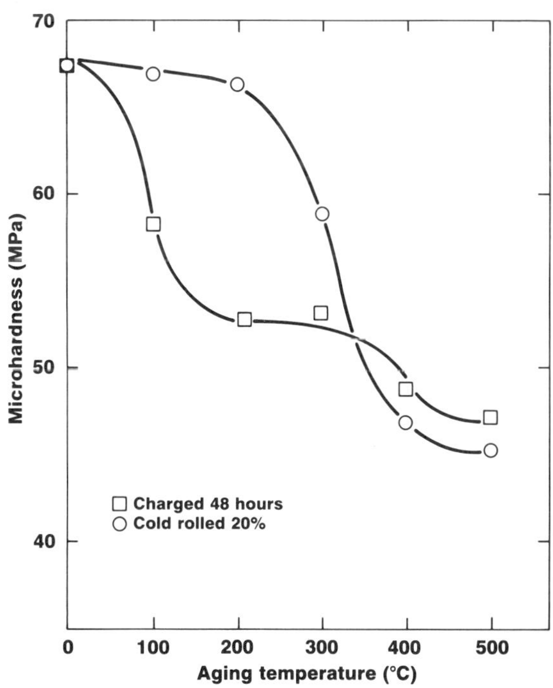

FIG. 3. Isochronal annealing (1 hour) of hardening due to cathodic charging. Hydrogen-induced hardening anneals out in two steps in contrast with work-hardening [37].

V. Stress Corrosion Cracking in Aluminum Alloys

Stress corrosion cracking (SCC) has been investigated extensively since it was first reported in 1919 [*38*]. The joint action of a stress and a corrosive environment produces cracking, often at a stress level below the macroscopic yield stress of the material. The fracture mode under SCC conditions can be different from that observed without the corrosive environment. The subject has been reviewed extensively [*40–47*]. Some reviews are specifically on aluminum alloys [*46,47*]. In spite of extensive investigations, no universal agreement has been reached on operating mechanisms for SCC (see chapter 16). There are two categories of mechanisms dominating the literature: anodic dissolution (AD) and hydrogen embrittlement (HE).

Anodic dissolution is a preferential dissolution mechanism rather than a true hydrogen effect. It is a balance of three electrochemical reactions describing the anodic and cathodic reactions occurring within the growing crack [*48,49*]. There are two anodic reactions: (1) dissolution of bare metal at the crack tip and (2) dissolution of the passive oxide on the crack sides. These anodic currents are balanced by a third cathodic reaction: passivation (oxidation) of the crack sides [*48*]. The stress (applied or residual) aids this process by opening the crack, continuously exposing virgin metal. Although this mechanism is not exclusive to the grain boundaries, the anodic character of the boundaries favors the reactions there. These reactions would be favorable along any high-energy path. Dissolution trenches have been found to lie parallel to the slip planes in specimens of 7075 loaded in mode III [*50*] (Fig. 4). Anodic dissolution dominated popular thinking during the early years of SCC research [*44,46*] and continues to have numerous advocates. It appears, however, that mechanisms in the HE category have been gaining in popoularity [*41,46,47*]. Because our present task is hydrogen in aluminum, we shall primarily discuss the hydrogen embrittlement aspect of SCC. The description of the mechanisms of anodic dissolution can be found elsewhere [*40,43*].

The HE-controlled SSC often invokes a popular mechanism developed by Troiano [*51*] and Oriani and Josephic [*52,53*]. In this mechanism the stresses ahead of an advancing crack tip accumulate an abnormally high hydrogen concentration. This leads to a reduction in the cohesive forces between atoms across the crack plane, enabling the crack to grow through this embrittled region. Once the crack has moved beyond the embrittled region, arrestation occurs. The hydrogen then readjusts to the new stress distribution, diffusing again to the region of high stress ahead of the crack tip, and entire process repeats (See Fig. 5).

Another hydrogen-embrittlement mechanism considers the enhanced susceptibility of the grain boundaries [*54,55*]. Hydrogen is considered to be

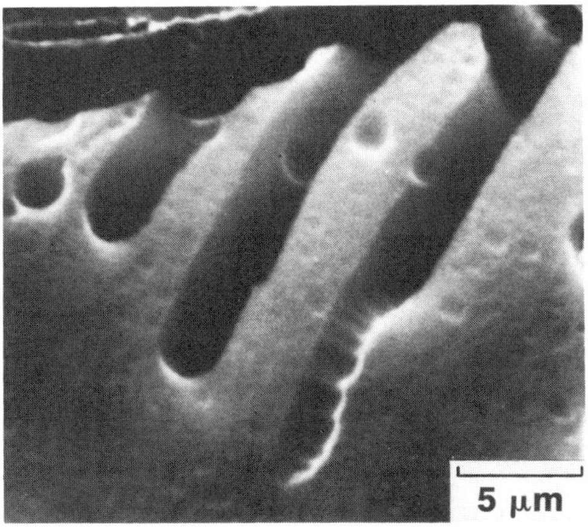

FIG. 4. Apparent dissolution trenches. 7075 in mode-III loading. Sectioned view [50].

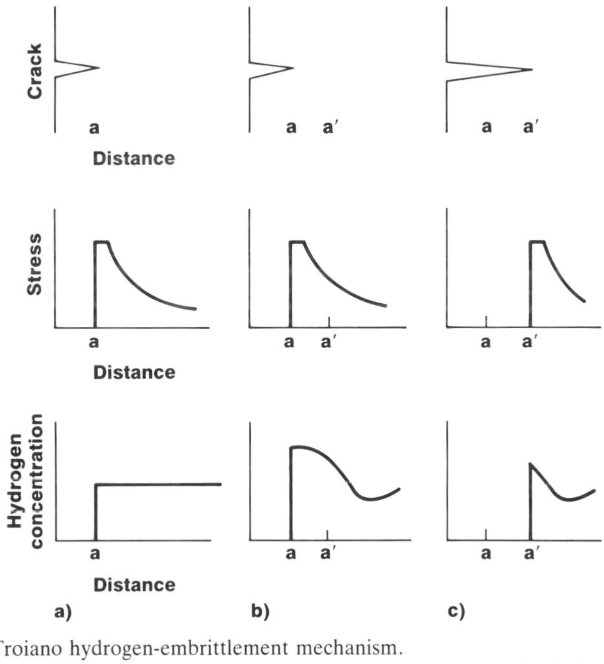

FIG. 5. Troiano hydrogen-embrittlement mechanism.
(a) Initial crack with its stress state and uniform hydrogen concentration below critical level for embrittlement.
(b) Hydrogen diffuses to high-stress region ahead of crack tip, producing an embrittled zone, a-a′.
(c) Crack has advanced through embrittled zone arresting in the ductile material beyond a′. The hydrogen has not had enough time to adjust to the new stress state. When it does, a new embrittled zone will be formed and the crack will continue to grow.

trapped on the grain boundary where it has migrated by diffusion mechanisms: bulk, dislocation-enhanced, and grain-boundary. On the grain boundary, hydrogen decreases the cohesive strength between adjacent grains leading to embrittlement. This can be regarded as an extension of Troiano's model to the grain-boundary region, which is acting as the crack plane [*56*].

Researchers studying primarily liquid metal embrittlement (LME), however, have proposed an alternative mechanism describing HE and SCC phenomena [*57*]. The two phenomena, LME and SCC, share a number of similarities including the effects of microstructure, strength, and temperature and the dependence of crack velocity on stress intensity. Embrittlement is explained in terms of the adsorption of the embrittling species, hydrogen for example, onto the crack surfaces. This then affects dislocation generation from surface sources and hence crack growth. In this context, solute hydrogen is thought to be of less importance than in other hydrogen embrittlement mechanisms. Certainly, absorbed hydrogen may play an important role in some cases of HE, particularly with regards to crack growth. It is well established, however, that in many cases solute hydrogen is essential to explain observed HE phenomena; an adsorption mechanism could not be applicable to these cases.

A. Embrittlement by Precharging of Hydrogen

Hydrogen precharging experiments build the foundation to understand the role of hydrogen in SCC. In these experiments, a series of specimens are precharged with hydrogen by various techniques and are subsequently tested, usually in air. The various phenomena resulting from the precharging treatments are then related to HE mechanisms [*22–24,45,58–61*].

Ductility decreases occur with increasing severity of hydrogen precharging. The ductility loss with hydrogen precharging often recovers with time. An example is shown in Fig. 6, in which the stress–strain relation of an Al-8%Mg age-hardened alloy is shown as a function of cathodic charging time and subsequent aging time [*45*]. Apparently, the presence of hydrogen did not change the shape of the stress–strain curve in this case. The charging reduced the strain to fracture from 20 to 7% as the charging time increased to 1800 seconds. The loss of ductility was recovered by the subsequent aging of the fully charged specimens in dry air. Complete recovery was attained after 10 days. The mode of fracture also changed with hydrogen charging, most frequently from a ductile to intergranular mode. The initial fracture mode prior to hydrogen charging usually returns after recovery.

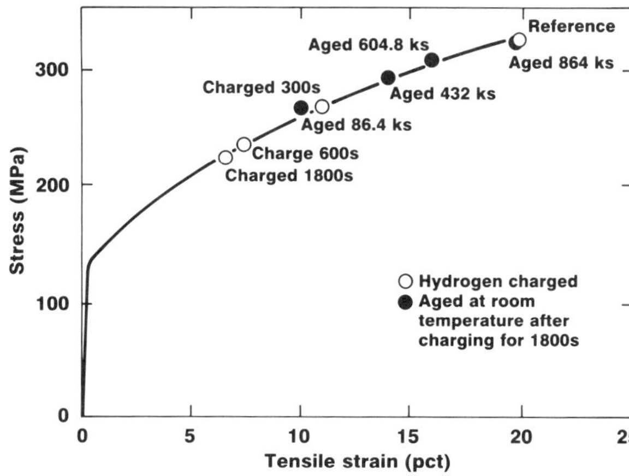

FIG. 6. Effects of cathodic charging and subsequent aging at room temperature on fracture stress and elongation of Al-8%Mg age-hardened alloy. The stress-strain curves all overlap. Open and closed circles indicate the fracture point of specimens that were hydrogen-charged (open circles) and aged at room temperature after hydrogen charging for 1800 seconds (closed circles). The duration of the charging and the aging is indicated along with each point [*45*].

Hydrogen embrittlement shows marked dependence on temperature and strain rate. Figure 7 provides a graphic example of the effects. The highest strain rate in the figure did not produce any loss of ductility. HE increased with decreasing strain rate; the lowest value examined in this study produced the most embrittlement. The temperature dependence of HE is also significant; no HE effect occurred above 320°K. The effect increased rapidly to reach a maximum and then decreased somewhat slowly with decreasing temperature. The temperature for the maximum HE effect increased with strain rate [*45*]. This type of temperature-dependence and strain-rate-dependence can be explained by the transport of hydrogen by moving dislocations. Hydrogen embrittlement also occurs in pure aluminum as discussed in the preceding section.

B. The Role of Hydrogen in Stress Corrosion Cracking

Most of the investigations of the effects of precharging of hydrogen have employed a severe charging condition. Therefore, the observation of hydrogen embrittlement under such conditions does not naturally prove the presence of a hydrogen-embrittlement mechanism under SCC

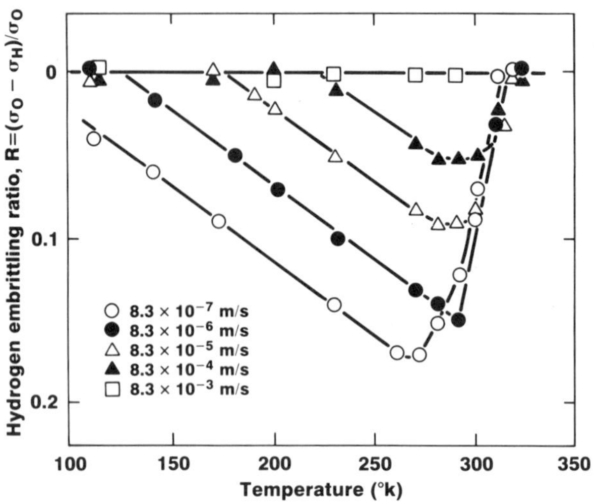

FIG. 7. Effect of temperature and strain rate on hyrogen embrittlement. The largest embrittlement occurs near room temperature. The magnitude of the embrittlement effect increased with decreasing strain rate [*45*].

conditions since technologically important SCC takes place usually under considerably milder charging conditions. A number of careful studies have been carried out providing evidence indicating the presence of a hydrogen-embrittlement mechanism in the SCC of aluminum alloys.

First, the permeation of hydrogen into specimens has been demonstrated by Scamans et al. [*62*] in Al-6%Zn-3%Mg alloys. The specimens were preexposed to water vapor, and the hydrogen evolution during a subsequent tensile test was monitored by a mass spectrometer. Ohnishi and Nakatani [*63*] also examined the permeation of hydrogen in Al-8%Mg alloys. Figure 8 demonstrates the importance of the joint action of stress and environment; the absorption of hydrogen was enhanced if the specimen was immersed with an applied stress (80% of yield stress) in a 3.5% NaCl solution over specimens immersed without any applied stress. The figure also shows the loss of hydrogen by aging at room temperature. Although more extensive studies quantifying the enhanced absorption under different conditions (for example, stress, temperature, and time-dependence) in various specimen materials (varying chemical composition and microstructure) are essential to analyze the SCC in terms of HE mechanisms, it can be concluded that hydrogen is present under SCC conditions and the presence of tensile stress significantly enhanced the absorption of the hydrogen.

Second, loading-mode testing has been proposed to distinguish between

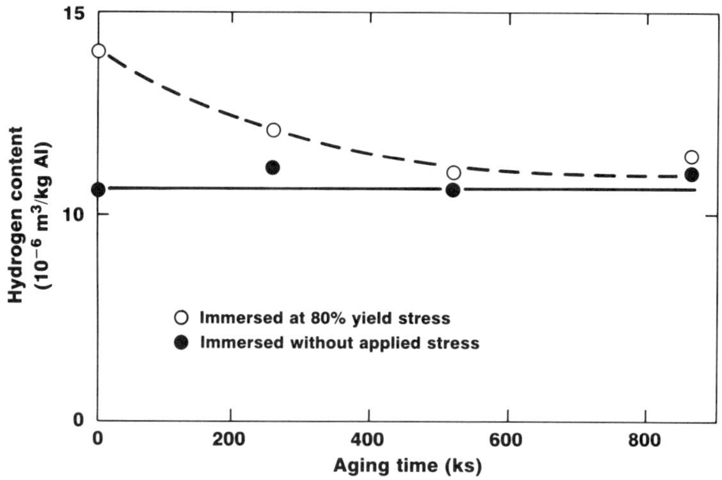

FIG. 8. Effect of tensile stress on hydrogen absorption in Al-8%Mg alloy. Hydrogen absorption was greatly enhanced if a stress was applied during immersion (SCC condition). Progressive aging at room temperature in dry air removed additional absorbed hydrogen [*63*].

anodic dissolution and hydrogen embrittlement mechanisms [*64*]. Hydrogen embrittlement is enhanced by the triaxial stress state that exists at the crack tip of a mode-I crack. Mode III, by contrast, does not produce any triaxial stresses at the crack tip. This line of reasoning provides a means to distinguish between hydrogen embrittlement, which would be affected by a triaxial stress state, and anodic dissolution, which would be relatively unaffected by the stress state. (See Fig. 9). Most of the published results are for tension loading (mode I). The few published results for torsional loading (mode III) seem not to present a clear answer [*50,65,66*]. There seems certainly to be a load dependence on the torsional properties, suggesting a hydrogen effect, but it is generally less than that observed in tensile loading. Mueller et al. [*50*] set out to use loading-mode testing to distinguish between anodic dissolution and hydrogen embrittlement in 7075 alloy. Based on the stress-dependence of the time to failure for mode-I and mode-III loading and the fractography, they concluded that hydrogen embrittlement was responsible for the cracking in mode I while anodic dissolution was responsible for that in mode III. This suggests that a mixed mode is responsible; hydrogen embrittlement is rate-determining but requires the triaxial stress state of mode-I loading, without which anodic dissolution controls cracking. It should be added, though, that loading-mode testing suffers experimentally from the difficulty of achieving pure mode-III loading.

There also are numerous similarities in the mechanical behavior of specimens precharged for HE and those tested under SCC conditions.

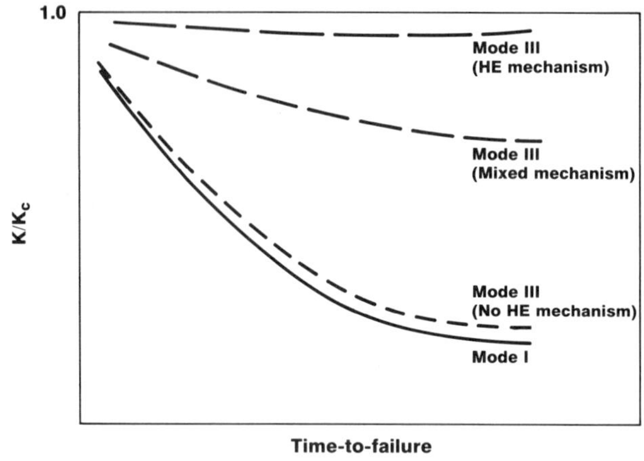

FIG. 9. Mode-III loading should have a strong dependence between K-ratio and time-to-failure if hydrogen embrittlement is not important, a weak dependence if hydrogen embrittlement is important, and an intermediate dependence with a mixed-mode mechanism [*59*].

When SCC or HE predominate, there is a weakening of the grain boundary resulting in intergranular fracture or intergranular secondary cracking [*22,65,67,68*]. Both HE and SCC sensitivities are greatest in underaged microstructures [*22,50,69,70*] and also increase with the strength of the alloy [*45*]. There is also a similarity in the stress-intensity–factor dependence of crack-propagation rate [*44,45,47*] and in the strain-rate dependence [*72,73*]. Additionally, both HE and SCC phenomena are reversible [*20,72*].

Finally, Higashi and his coworkers provide a micrograph (Fig. 10) illustrating the accumulation of hydrogen ahead of a crack tip and its effect [*74*]. They suspended the SCC test of an A1–8% Mg alloy in a NaCl aqueous solution before failure and tensile-fractured the specimen in air. The figure shows four regions of the fracture surface. From left to right, they are the precrack region, the SCC region, the HE region, and the ductile fracture region. On the SEM picture, the second and the third regions appear identical, as both fractured intergranularly. Subsequent EDS analysis could easily distinguish the region fractured in the NaCl solution from that fractured by the subsequent tensile test by the presence of the chlorine element. If a specimen was aged at room temperature for 16 days after removal from the NaCl solution, the third region did not form ostensibly because the hydrogen would have diffused away.

The role of hydrogen in environment-enhanced fatigue of aluminum alloys has also been studied, usually with regard to the effects of water vapor [*75–78*]. These phenomena, though, are complicated by crack clo-

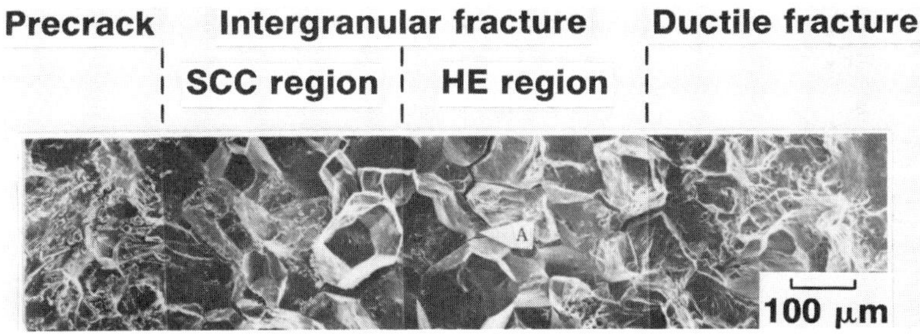

FIG. 10. SEM fractography of Al-8%Mg alloy. There are four regions: the precracked region, the SCC region, the HE region, and the ductile fracture region, from left to right. The second and third regions appear identical on the micrograph. However, the EDS analysis identifies the presence of Cl only on the SCC region [*74*].

sure due to corrosion products [*78*]. The degree of embrittlement is dependent on stress intensity, crack length, exposure-composition strength, and microstructure [*75–78*]. Several mechanisms have been proposed to explain the role of hydrogen in fatigue. Hydrogen could condense in voids lying on active slip planes, resulting in large internal pressures [*75,77*]. Another mechanism describes hydrogen associated with vacancies, slowing the diffusion of the voids, thereby enhancing void clustering [75]. Additionally, there is no reason that the HE mechanisms proposed to explain static-loading phenomena could not be applicable in the case of fatigue.

C. The Structure and Properties of Grain Boundaries

As mentioned previously, the grain boundary seems to be frequently instrumental in HE and SCC processes. The grain boundaries contain precipitates and voids, and they are bounded by relatively soft precipitate-free zones (PFZ). The grain boundary can also act as a trap for hydrogen. All of these can affect the HE or SCC fracture process. Grain-boundary precipitates act as nucleation sites for hydrogen bubbles [*54,55*]. Figure 11 shows ductility recovery of an Al-Zn-Mg alloy at intermediate exposure times observed by Christodoulou and Flower [*54*]. They attribute this recovery to grain-boundary precipitates nucleating hydrogen bubbles, thereby prohibiting hydrogen from wetting the entire boundary. A similar observation, though, has also been explained by a change in the magnesium content and, hence, the reactivity of the boundary [*59*]. Another microstructural feature thought to play an important role in HE and SCC is slip planarity. This mechanism, when accompanied by dislocation transport, can lead to very high local hydrogen concentrations on the grain

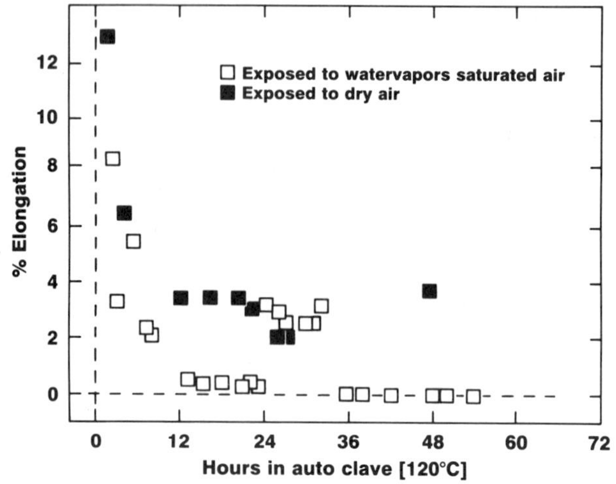

FIG. 11. Ductility recovery of Al-6%Zn-3%Mn at intermediate times (24–36 hours) of exposure to water vapor at 120°C. This is explained in terms of grain-boundary bubbles gettering the hydrogen, preventing wetting of the boundary. Open squares exposed to water-vapor environment. Solid squares exposed to dry environment [*55*].

boundaries [*23*,*24*,*58*,*79*]. Vasudevan et al. [*80*] however, working in the Al-Li-Cu system, found that the SCC behavior is much more dependent on the grain-boundary precipitate spacing than on slip planarity. This is similar to the results of Adler et al. [*69*], who studied the Al-Zn-Mg-Cu system.

D. Hydrogen Embrittlement versus Anodic Dissolution

The aggregate of present SCC results can only be explained by both HE and AD mechanisms; neither mechanism alone can account for all observations. Specifically, how HE and AD mechanisms interact and which will control a given situation is not generally clear. The two mechanisms could operate in parallel or in series. This too may depend on the particular case involved. HE appears to dominate most phenomena in aluminum and aluminum alloys, though AD cannot be generally discounted. The behavior of an aluminum alloy in any specific environment must be investigated to determine the controlling mechanism in that case.

VI. Summary

Aluminum and aluminum alloys have low lattice hydrogen solubilities and diffusivities. This might lead to the erroneous conclusion that these materials are immune to hydrogen degradation. Hydrogen solubility can be enhanced by trapping and hydrogen diffusivity can be increased by processes such as dislocation transport. Unfortunately, these fundamental phenomena have not been exhaustively studied and questions remain. Results as of 1988, though, certainly show that the mechanical properties of aluminum and aluminum alloys are degradated by hydrogen. It has been shown that precharging of hydrogen embrittles aluminum and aluminum alloys. Hydrogen-embrittlement mechanisms apparently control a number of SCC phenomena, although anodic dissolution mechanisms appear to coexist and play an important role in some cases. Here too, though, further study could greatly clarify the present understanding.

References

1. Eichenauer, W., Hattenbach, K., and Pebler, A. (1961). *Zeit. Metall.*, **52**, 682.
2. Ransley, C.E., and Neufeld, H. (1948). *Jour. Inst. Met.*, **74**, 599.
3. Opie, W.R., and Grant, N.J. (1950). *Trans. AIME*, **188**, 1237.
4. Talbot, D.E.J. (1975). *Inter. Met. Rev.*, **20**, 166.
5. Seeger, A. (1978). *Topics in Applied Physics: Hydrogen in Metals I*, G. Alefeld and J. Vokl, eds. Springer-Verlag, Berlin, p. 349.
6. Peisl, H. Ibid., p. 53.
7. Eichenauer, W., and Pebler, A. (1957). *Zeit Metall.*, **48**, 375.
8. Matsuo, S., and Hirata, T. (1967). *Jour. Jap. Inst. Met.*, **31**, 590.
9. Papp, K., and Kovacs-Cseteny, E. (1977). *Scripta Met.*, **11**, 921.
10. Ichimura, M., Imabayashi, M., and Hayakawa, M. (1980). *J. Jap. Inst. Met.*, **44**, 1053.
11. Csanady, A., Papp, K. and Pasztor, E. (1981). *Mat. Sci. and Eng.*, **48**, 35.
12. Papp, K., and Kovacs-Cseteny, E. (1981). *Scripta Met.*, **15**, 161.
13. Outlaw, R.A., Peterson, D.T., and Schmidt, F.A. (1982). *Scripta Met.*, **16**, 277.
14. Hashimoto, E., and Kino, T. (1983). *Jour. Phys. F, Met. Phys.*, **13**, 1157.
15. Ishikawa, T. and McLellan, R.B. (1986). *Acta Met.*, **34**, 1091.
16. McLellan, R.B. (1983). *Scripta Met.* **17**, 1237.
17. Besenbacher, F., Myers, S.M., and Norskov, J.K., cited by T. Ishikawa and R.B. McLellan (1986). *Acta Met.*, **34**, 1093.
18. Foster, L.M., Jack, T.H., and Hill, W.W. (1970). *Met. Trans.*, **1**, 3117.
19. Edwards, R.A.H., and Eichenauer, W. (1980). *Scripta Met.*, **14**, 971.
20. Gest, R.J., and Troiano, A.R. (1974). *Corr.*, **30**, 274.
21. Leger, M., and Piercy, G.R. (1981). *Phil. Mag.*, **43**, 377.
22. Albrecht, J., Bernstein, I.M., and Thompson, A.W. (1982). *Met. Trans.* **13A**, 811.
23. Hardwick, D.A., Taberi, M., Thompson, A.W., and Bernstein, I.M. (1982). *Met. Trans.*, **13A**, 235.
24. Hardwick, D.A., Thompson, A.W., and Bernstein, I.M. (1983). *Met. Trans.*, **14A**, 2517.
25. Feuerstein, S., and John, I.W. (1969). *Jour. Appl. Phys.*, **40**, 3334.
26. Kimura, H., and Matsui, H. (1987). *Script. Met.*, **21**, 319.
27. Eastman, J., Matsumoto, T., Narita, N., Heubaum, F., and Birnbaum, H.K. (1980). In *Proc. of Int. Conf. on Hydrogen in Metals*, I.M. Bernstein and A.W. Thompson, eds. The Metallurgical Society of AIME, Warrendale, Pennsylvania, p. 397.

28. Matsumoto, T., Eastman, J. and Birnbaum, H.K. (1981). *Scripta Met.*, **15**, 1033.
29. Tabata, T., and Birnbaum, H.K. (1983). *Scripta Met.*, **17**, 947.
30. Tabata, T., and Birnbaum, H.K. (1984). *Scripta Met.*, **18**, 231.
31. Robertson, I.M., Tabata, T., Wei, W., Heubam, F., and Birnbaum, H.K. (1984). *Scripta Met.*, **18**, 841.
32. Robertson, I.M., and Birnbaum, H.K. (1986). *Acta Met.*, **34**, 353.
33. Bond, G.M., Robertson, I.M., and Birnbaum, H.K. (1988). *Acta Met.*, **36**, 2193.
34. Shih, D.S., Robertson, I.M., and Birnbaum, H.K. (1988). *Acta Met.*, **36**, 111.
35. Bond, G.M., Robertson, I.M., and Birnbaum, H.K. (1987). *Acta Met.*, **35**, 2289.
36. Bond, G.M., Robertson, I.M., and Birnbaum, H.K. (1986). *Script. Met.*, **20**, 653.
37. Watson, J.W., Shen, Y.Z., and Meshii, M. (1988). *Met. Trans.*, **19A,** 2299.
38. Choo, W.Y., and Bernstein, I.M. (1984). *Met. Trans.*, **15A**, 1953.
39. Rosenhain, W., and Archbutt, S.L. (1919). *Proc. Roy. Soc. Lond.*, **A96**, 55.
40. Speidel, M.O., and Hyatt, M.V. (1972). *Advances in Corrosion Science and Technology*, **vol. 2**. Plenum Press, New York, p. 115.
41. Thompson, A.W., and Bernstein, I.M. (1980). Ibid., **vol. 7**, p. 53.
42. Brown, R.H., Sprowls, D.O., and Shumaker, M.B. (1972). *Stress Corrosion Cracking—A State of the Art*. ASTM STP 518, 87. ASTM, Philadelphia.
43. Latanison, R.M., Gastine, O.H., and Compeau, C.R. (1979). *Environment-Sensitive Fracture of Engineering Materials*, Z.A. Foroulis, ed. The Metallurgical Society of AIME, Warrendale, Pennsylvania, p. 48.
44. Speidel, M.O. (1975). *Met. Trans.*, **6A**, 631.
45. Ohnishi, T. (1987). *Bull. Jap. Inst. Met.*, **26**, 389.
46. Dix, E.H., Jr., (1940). *Trans. AIME*, **137**, 11.
47. Thompson, A.W. (1979). *Environment-Sensitive Fracture of Engineering Materials*, Z.A. Foroulis, ed. The Metallurgical Society of AIME, Warrendale, Pennsylvania, p. 379.
48. Doig, P., and Flewitt, P.E.J. (1981). *Met. Trans.*, **12A**, 923.
49. Doig, P., and Flewitt, P.E.J. (1977). *Proc. Roy. Soc. Lond.*, **A357**, 439.
50. Mueller, M.P., Bernstein, I.M., and Thompson, A.W. (1985). *Corr.*, **41**, 127.
51. Troiano, A.R. (1960). *Trans. ASM*, **52**, 54.
52. Oriani, R.A., and Josephic, P.H. (1972). *Scripta Met.*, **6**, 681.
53. Oriani, R.A., and Josephic, P.H. (1974). *Acta Met.*, **22**, 1065.
54. Scamans, G.M. (1978). *Jour. Mat. Sci.*, **13**, 27.
55. Christodoulou, L., and Flower, H.M. (1980). *Acta Met.*, **28**, 481.
56. Pathania, R.S., and Tromans, D. (1981). *Met. Trans.*, **12A**, 607.
57. Lynch, S.P. (1986). *Aircraft Materials Report 119*. Defense Science and Technology Organisation, Aeronautical Research Laboratories, Melbourne, Victoria, Australia.
58. Albrecht, J., Thompson, A.W., and Bernstein, I.M. (1979). *Met. Trans.*, **10A**, 1759.
59. Tuck, C.D.S. (1985). *Met. Trans.*, **16A**, 1505.
60. Ohnishi, T., Hara, T., Mohri, Y., and Nakatani, Y. (1981). *Jour. Jap. Inst. Light Met.*, **77**, 473.
61. Ohnishi, T., Higashi, K., Inoue, N., and Nakatani, Y. (1981). *Jour. Jap. Inst. Met.*, **45**, 972.
62. Scamans, G.M., Alani, R., and Swann, P.R. (1976). *Corr. Sci.*, **16**, 443.
63. Ohnishi, T. and Nakatani, Y. (1978). *Jour. Jap. Inst. Light Met.*, **28**, 123.
64. Thompson, A.W., and Bernstein, I.M. (1984). *Environment-Sensitive Fracture: Evaluation and Comparison of Test Methods*, ASTM STP 821. ASTM, Philadelphia, p. 114.
65. Sudarshan, T.S., Louthan, M.R., Jr., Place, T.A., and Mabie, M.M. (1985). *Mat. Sci. and Eng.*, **73**, 131.
66. Green, J.A.S., Hayden, H.W., and Montague, W.G. (1976). *Effect of Hydrogen on*

Behavior of Materials. The Metallurgical Society of AIME, Warrendale, Pennsylvania, p. 200.
67. Albrecht, J., McTiernan, B.J., Bernstein, I.M., and Thompson, A.W. (1977). *Scripta Met.*, **11**, 893.
68. Herman, J. (1981). *Mat. Sci.*, **16**, 2381.
69. Adler, P.N., De Iasi, R., and Geschwind, G. (1972). *Met. Trans.*, **3A**, 3191.
70. Maitra, S. (1981). *Corr.*, **37**, 98.
71. Finnegan, J.E., and Hart, W.H. (1976). *Stress Corrosion—New Approaches*, ASTM STP 610, ASTM, Philadelphia, p. 44.
72. Taheri, M., Albrecht, J., Bernstein, I.M., and Thompson, A.W. (1979). *Scripta Met.*, **13**, 871.
73. Hardie, D., Holroyd, N.J.H., and Parkins, R.N. (1979). *Mat. Sci.*, **13**, 603.
74. Higashi, K., Ohnishi, T., Nakatani, Y., and Okabayashi, K. (1981). *Jour. Jap. Inst. Light Met.*, **31**, 386, 644.
75. Broom, T., and Nicholson, A.J. (1960–1961). *Jour. Inst. Met.*, **89**, 183.
76. Wei, R.P. (1970). *Jour. Engg. Fract. Mech.*, **1**, 633.
77. Wei, R.P., Pao, P.S., Hart, R.G., Weir, T.W., and Simmons, G.W. (1980). *Met. Trans.*, **11A**, 151.
78. Vasudevan, A.K., and Suresh, S. (1982). *Met. Trans.*, **13A**, 2271.
79. Rinker, J.G., Marek, M., and Sanders, T.H. Jr. (1984). *Aluminum-Lithium Alloys II*, T.H. Sanders, Jr., E.A. Starke, Jr., eds. The Metallurgical Society of AIME, Warrendale, Pennsylvania, p. 597.
80. Vasudevan, A.K., Liu, J., and Ricker, R.E. (1987). *Proceedings of Environmental Degradation of Engineering Materials III*, M.R. Louthan, Jr., R.D. Simmons, and R. McNitt, eds. Pennsylvania State University Press, p. 321.

19

Oxidation of Aluminum and Its Alloys

D.J. FIELD

Alcan International Limited
Southam Road
Banbury, Oxon England

I. Introduction

Aluminum is an extremely reactive metal and its usefulness as an engineering material is due to the tough, thin, highly protective, self-healing surface oxide film that readily forms when freshly created surfaces are exposed to the environment. At relatively low temperatures (i.e., ≤300°C), the surface oxide can be thought of in terms of a continuous amorphous film that thickens as a function of temperature [*1*].

At high temperatures, duplex oxide films develop that comprise both crystalline and amorphous oxides [*1–3*]. In the situation where reactive alloying elements such as magnesium [*4–6*] and lithium [*7–9*] are present, then selective oxidation occurs leading to the formation of thick, friable, crystalline oxide films.

The oxide films that form on most aluminum alloys during the majority of metal-processing operations are of little consequence; however, there are many instances where the oxide film is of technological importance, e.g., powder metallurgy [*10*], brazing [*11*], scrap recycling [*12*], and all

ISBN 0-12-341831-3

high-temperature metal-processing operations [*12*], where alloys containing magnesium and more recently lithium [*8,9*] are concerned.

Not withstanding the effect of alloying additions, the composition of the oxidizing atmosphere has a significant effect on surface film growth kinetics, composition, surface topography, and crystallography [*6,8,9,13–16*]

This chapter reviews the oxidation of aluminium, including amorphous film growth and the effects of both alloying additions and oxidizing environments.

II. Amorphous Film Growth on Aluminum

At low temperatures, aluminum and its alloys develop thin, protective oxide films which have been described as "amorphous" [*1,4,17*]. Strictly, the term *amorphous* means without structure; however, such films exhibit four strong but diffuse lines corresponding to the cubic spinel γ-Al_2O_3 when viewed in diffraction [*1,17*] and must therefore to some extent be structured.

This film has been the subject of many studies—both experimental [*1,4,18*] and theoretical [*19–21*]. Resistivity studies have shown that films thickened either naturally or thermally can vary in terms of resistance through their thickness. A low-resistance, inner layer exists extending a short distance, ~ 10–15 Å, from the oxide metal interface [*18,22,23*].

Measurements of the rates of film formation with temperature have shown that the film thickens according to logarithmic kinetics reaching a temperature-dependent, ultimate thickness within about 20 hours [*1*]. This film grows by the outward diffusion of cations according to Mott [*19*]:

$$4Al^{3+} + 12e^- + 3O_2 \rightarrow 2Al_2O_3. \qquad (1)$$

From a theoretical viewpoint, it has been proposed that the rate-limiting step in the formation of these films is either electron [*19*] or ion transport [*20*]. The amorphous film structure has been modelled in terms of a glass [*20*] that recrystallizes; however, several weaknesses in this approach have been exposed by Pryor [*21*].

In a later, critical study, Pryor [*17*] considered the possible defect structure of thin, "amorphous" γ-Al_2O_3 and concluded that the film contained an excess of aluminum ions situated in otherwise vacant cation sites, with electrical neutrality being maintained by the provision of three electrons trapped in the vicinity of each excess aluminum ion. Pryor attributed the extremely low dielectric constant (8.5) of "amorphous" γ-Al_2O_3 to the net reduction in cation vacancies. Pryor [*17*] also deduced that the poor, long-range order of these amorphous films on aluminum was due to a small number of aluminum ions captured on anion sites, thereby disrupting the long-range order.

At temperatures above about 425 to 450°C, the "amorphous" γ-Al_2O_3 overlayer on aluminum undergoes a discontinuous change in structure. The rapid migration of oxygen to the oxide–metal interface is readily detectable at these temperatures and evidenced by the nucleation and growth of crystalline γ-Al_2O_3 as a new phase below the amorphous layer [*1–3*]. The initial path of oxygen diffusion is thought to be the preexisting oxygen vacancy structure of "amorphous" γ-Al_2O_3 [*1*].

III. The High-Temperature Oxidation of Aluminum

A. *Dry Oxidizing Environments*

The overall oxidation kinetics for superpurity aluminum in dry oxygen are temperature-dependent; over the temperature range 350–450°C, the kinetics can best be described as parabolic. The reported activation energies for parabolic oxidation show considerable scatter (22.8 [*24*], 41.0 [*25*], and 54.0 [*1*] kcal/mol), which may be due to the differences in sample preparation prior to oxidation. Surface preparation prior to oxidation has a significant effect on the oxidation of aluminum [*13*]. At temperatures above 450°C, the weight-gain curves are roughly sigmoidal in shape, with the initial period of oxidation at a low rate being succeeded by a rapidly rising reaction rate which decreases sharply at longer times as shown in Fig. 1 [*1*].

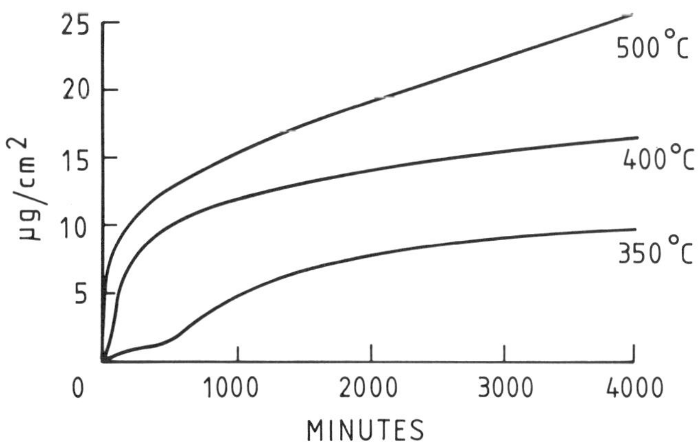

FIG. 1. Oxidation weight-gain curves of aluminum oxidized at various temperatures taken from [*1*].

At such temperatures—i.e. $\geqslant$ 450°C under dry [2–4] and moist [25] oxygen-containing environments—aluminum develops a duplex film of both amorphous and crystalline γ-Al_2O_3. Detailed electron microscopic studies have shown that crystalline γ-Al_2O_3 nucleates as a separate phase at the oxide–metal interface penetrating the metal substrate to a temperature-dependent depth [1,3]. The nuclei, which have a rod-like morphology, expand laterally, until restricted by growth stresses in the metal substrate. New crystals nucleate and grow until coverage of the metal surface is complete. (See Fig. 2.)

Crystalline γ-Al_2O_3 grows expitaxially with respect to the aluminum substrate. The following orientation relationships have been found: $(110)[110]_{Al}/(110)[110]_{OX}$, $(110)[100]_{Al}/(110)[100]_{Ox}$ by Hart and Maurin [3] and Scamans and Butler [5] and $(111)[110]_{Al}(100)[110]_{Ox}$ by Scamans and Bulter [26].

B. *Moist Oxidizing Environments*

The presence of moisture has little effect on the oxidation kinetics of aluminum at high temperatures under oxygen-containing environments. With moist oxygen, the plateau region of the sigmoidal weight-gain curves

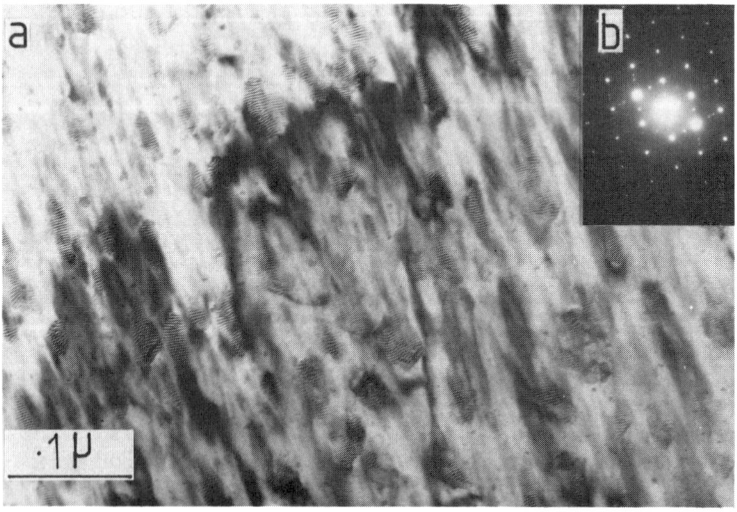

FIG. 2. a. Bright field electron micrograph.
b. Diffraction pattern taken from the same area of aluminum oxidized at 550°C for 15 minutes showing epitaxial crystallites of γ-Al_2O_3.

corresponds with complete coverage of the metal substrate by crystallites of γ-Al_2O_3.

It is known that moisture promotes the formation of subsurface hydrogen blisters in aluminum oxidized at high temperatures [*15,27–29*]. Many oxidation studies have been concerned with the mechanism of blister formation [*13,14,29*], rather than the surface film itself [*28,30*], which may control the transport of hydrogen to the metal substrate [*28*].

In a study of the hydrous surface films on aluminum [*31*], it was proposed that at temperatures above about 360°C, boehmite forms according to the equation:

$$2Al + 4H_2O \rightarrow 2AlOOH + 3H_2O. \quad (2)$$

up to a limiting temperature of about 550 to 590°C, followed by dehydration to γ-Al_2O_3:

$$2AlOOH \rightarrow \gamma\text{-}Al_2O_3 + H_2O. \quad (3)$$

Alwitt [*32*] in a review of the aluminum–water system, proposed that the boehmite dehydration reaction occurred at the lower temperature of 450°C. Poorly recrystallized boehmite is known to dehydrate at temperatures as low as 350°C. Winter and Gruhl [*28*] correlated the enhanced permeation of hydrogen into pure aluminum with the transformation of AlOOH to γ-Al_2O_3 (Eq. 3) and concluded that this mechanism of oxide formation was significant in the formation and permeation of hydrogen to the oxide–metal interface. At high temperatures, aluminum also reacts directly with water to form crystalline γ-Al_2O_3 from both reaction (3) and the environment:

$$2Al + 3H_2O \rightarrow \gamma\text{-}Al_2O_3 + 3H_2. \quad (4)$$

At lower temperatures (i.e., about 100°C), it has been proposed that the transport of hydrogen to the underlying aluminum substrate is controlled by an oxide barrier layer which separates the hydrated oxide film from the metal interface [*33,34*]. Such thin oxide films are known to exhibit significant space-charge separation [*20*]. It has been postulated [*35*] that hydrogen (as protons) from the surface reaction is driven, under the potential and proton-concentration gradient, to the oxide–metal interface. Protons then recombine either at the oxide–metal interface or at cathodic sites (e.g., inclusions and pores) within the metal to form oxide and metal blisters, respectively. The latter process is more applicable to high-temperature reactions where diffusion rates of hydrogen in the metal are rapid. Increased oxidation weight gains due to hydrogen permeation would only be expected if fresh surfaces were exposed by the blistering reaction.

IV. Aluminum-Alloy Oxidation

In terms of oxidation, and for convenience, aluminum alloys can be divided into two groups where the alloying additions either (a) selectively oxidize or (b) do not react. Although alloying additions that fall in the latter group may not react selectively, it does not exclude them from taking part in the general oxidation process.

A. *Nonselective Oxidation*

Elements such as copper [*36*], silicon, and manganese [*37*], when alloyed with aluminum at wt % levels, do not form their own oxides when reacted under oxidizing environments at high temperatures. These alloy systems form duplex surface films of amorphous and crystalline γ-Al_2O_3 in the same manner as pure aluminum.

Copper is known to enter crystalline γ-Al_2O_3 at high temperatures, thereby doping it, reducing the number of cation vacancies, and limiting the transport of Al^{3+} ions to the overlying amorphous film. A.C. resistivity measurements show that doping by copper significantly raises the dielectric constant of the crystalline oxide [*36*].

B. *Selective Oxidation*

Of the alloying additions that are known to react selectively and form their own oxides, magnesium and lithium are the most important. Both Al-Mg and Al-Li alloys are reactive at temperature in excess of about 350°C, lithium alloys being the most reactive [*6*].

1. Al-Mg Alloys

a. *Oxidation Kinetics*

The oxidation kinetics of Al-Mg alloys have been investigated in some detail using vacuum thermogravimetry [*25*]. With this technique it was quickly established that the oxidation kinetics of these alloys are controlled at the oxide–metal interface and not limited by the bulk diffusion of magnesium to the surface.

At temperatures above 350°C, oxidation is rapid. Up to 550°C, the reaction kinetics generally conform to a logarithmic or parabolic rate law,

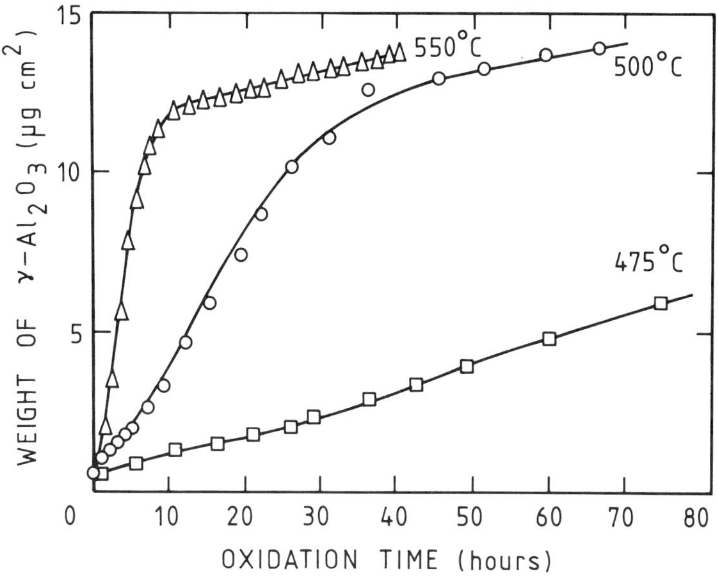

FIG. 3. Weight–time curves of Al-1.55 wt % Mg alloy in oxygen at various temperatures taken from [7].

giving way to approximately linear kinetics. Oxidation weight-gain curves that exhibit such a transition from parabolic type to linear reaction kinetics are generally termed paralinear [*38*]. For Al-Mg alloys, the time interval after which the transition occurs is both temperature- and concentration-dependent. For concentrated alloys (i.e., ~3–4 wt % Mg) and short periods of time, the oxidation kinetics are temperature-dependent, progressing from logarithmic at 350°C to linear at 480°C to paralinear at 520°C and to pseudoparabolic at 550°C [*6*]. With dilute alloys (i.e., Mg contents of the order of 1.5 wt %), both logarithmic and parabolic oxidation kinetics predominate [*7*] (see Fig. 3).

For Al-Mg alloys, the presence of water vapor in the atmosphere has a significant effect on oxidation. Both enhanced [*14*] and in the early stages, reduced rates [*39*] of oxidation due to moisture have been reported. However, this appears to be a function of alloy content, surface preparation, and oxidizing atmosphere dew point.

Under moist environments, the surface blistering of Al-Mg alloys can be severe, and it has been proposed that the enhancement of oxidation by water vapor is due to the generation of new surfaces by the blistering process [*39*]. Information regarding the effect of other gases on oxidation is scant, although it has been reported that "dry" CO_2 atmospheres are beneficial in reducing oxidation [*39*].

b. Oxide Film Structure

For alloys with magnesium concentrations in excess of about 1 wt %, the surface oxide at high temperatures is predominantly MgO [*7*]. This oxide was initially detected using diffraction techniques [*4,14,38*] and has latterly been characterized by electron microscopy [*5–7,26*].

Initially, the MgO-rich surface films were modelled as a scale [*25*]. However, microscopic investigations show that the film is in the form of small magnesia crystallites ranging from 5 to 500 nm. Two modes of MgO formation have been detected using TEM:

1. Direct formation at the oxide–metal interface [*5*].
2. Secondary reduction of the original amorphous overlayer of γ-Al_2O_3 [*5,40*] according to

$$\gamma\text{-}Al_2O_3 + 3Mg \rightarrow 3MgO + 2\ Al \quad \Delta G^\circ_{800K} = -89.5\ \text{kJmol}^{-1}_{O_2}\ . \qquad (5)$$

Oxides formed by mechanisms (1) and (2) have been termed "primary" and "secondary," respectively [*5,8*].

Direct observations of primary MgO formation on Al-5 wt %Mg alloy have been carried out by hot-stage High Voltage Electron Microscopy (HVEM) [*5*]. The nucleation and growth of primary magnesium oxides on Al-Mg alloys is similar to γ-Al_2O_3 formation on pure aluminum in that the primary oxides penetrate the metal substrate; however, MgO also penetrates, thereby disrupting the amorphous overlayer of γ-Al_2O_3 [*8*]. The absence of an induction period and appearance at much lower temperatures of crystalline oxides on Al-Mg alloys (350°C compared to 475°C for pure Al) are due to enhanced oxygen transport through the amorphous overlayer as a result of doping by Mg^{2+} from the alloy substate [*5*].

Primary oxide formation on Al-Mg alloys is pressure-dependent and the spinel oxide $MgAl_2O_4$ predominates at low pressures [*5,26*]. It is known that the oxide-film composition is related to the Mg concentration and an oxide-stability diagram has been constructed, which predicts the oxide composition at the oxide–metal interface as a function of alloy content [*7*]. The oxide-phase regions were defined as follows:

a. $X_{Mg} \geq 10^{-2}$ (i.e., 1.0 at %): MgO is the stable phase at the oxide–metal interface.
b. $10^{-6} \leq X_{Mg} \leq 10^{-2}$ (10^{-4} to 1.0 at %): $MgAl_2O_4$ is stable.
c. $X_{Mg} \leq 10^{-6}$ (10^{-4} at %): γ-Al_2O_3 is stable.

A model for the oxidation of Al-4.2 wt % Mg alloy has been developed which is based on the disruption of the protective, "amorphous" γ-Al_2O_3 overlayer by penetrating magnesium oxides which vary in size with temperature [*8,41*]. Primary MgO can vary in size from about 50 nm at 400°C to 200 nm at 520°C. If oxide penetration into the amorphous film is a function of the oxide size, then there will be situations where primary oxide penetration is sufficient to crack the amorphous overlayer and produce rapid, short-circuit diffusion paths for oxygen to the oxide–metal interface.

Cracking and disruption of the amorphous film is considered to be the event that occurs during the transition from parabolic to linear kinetics in Al-Mg paralinear oxidation. Surfaces that come into contact with the environment after being exposed by primary oxide growth immediately refilm with a layer of Mg^{2+} doped γ-Al_2O_3, beneath which further primary nucleation and growth take place. Due to the development of short circuit diffusion paths, preexisting "primary" oxides become sites for the repeated nucleation and growth of these oxides. The primary oxides form agglomerations of crystals which grow upward from the oxide–metal interface, producing the thick and friable films of magnesia, characteristic of heavily oxidized Al-Mg. At this stage of oxidation, localized regions of the alloy substrate are undercut by oxide growth, and become incorporated into the film. With increasing temperature and correspondingly larger primary oxide size, the oxide–metal interface becomes increasingly more corrugated, giving a larger surface area for oxidation.

c. *Effect of Alloying Additions*

The high-temperature oxidation of Al-Mg alloys is sensitive to certain alloy impurities or additions, particularly Be [*8,42*], Ca [*8,43*], Ag [*8,44*], and Li [*8*]. Although Be at levels as low as 10 ppm is remarkably effective in reducing oxidation, its use is restricted due to the extreme toxicity of the metal and its compounds [*45*]. Both silver [*44*] and calcium additions [*43*] have been proposed as oxidation inhibitors for Al-Mg alloys. In a later study [*8*], the beneficial effect of Ca was confirmed; however, Ag was found to act similarly to Li, enhancing oxidation weight gains over a wide range of temperatures.

There is little information available regarding the oxidation of ternary Al-Mg-X alloys other than Al-2.47 wt % Mg-7.75 wt % Zn [*26*] and Al-3 wt % Li-2 wt % Mg, which will be discussed later (Sections 3a and 3b). It appears that zinc does not participate in the oxidation reaction of ternary Al-Mg-Zn alloys and that oxide films develop with primary and secondary MgO morphology characteristic of the binary Al-Mg alloys.

2. The Oxidation of Aluminum-Lithium Alloys

Early studies of Al-Li alloy oxidation were concerned with the X-ray identification of surface-reaction products following the exposure of alloys to a range of high-temperature environments. Depending upon the oxidation environment employed, a wide range of oxidation products have been reported, such as Li_3N [*46*], $LiAl_5O_8$ [*5*], $LiAlO_2$ [*47*], and Li_2O and Li_2CO_3 [*46*]. In-situ oxidation of Al-1 % Li at low pressures has shown that the crystalline oxide $LiAl_5O_8$ [*5*] forms rapidly at the oxide–metal interface, resulting in complete coverage of the alloy substrate within one minute at 475°C and 10^{-4} τ.

Later systematic studies were carried out investigating the effect of alloy composition [*6*], trace alloying additions and oxidizing environment [*9*].

a. Oxygen-Containing Atmospheres

The reaction kinetics of Al-3 wt % Li and Al-3 wt % Li-2 wt % Mg alloys oxidized under dry and moist 20% O_2/Ar at 480°C are shown in Fig. 4 together with weight-gain curves obtained for Al-4.2 wt % Mg under identical conditions [*6*]. Over the temperature range studied (480–575°C), the weight-gain curves for the lithium alloys exhibited similar features with each environment. Under dry conditions, the weight-gain curves conformed to pseudo parabolic kinetics. However, water vapor promoted breakaway oxidation at all temperatures, typified by an acceleration in the oxidation rate followed by a plateau region of limited weight gain.

Electron microscopic examinations of the oxide films developed on these alloys has shown that oxidation is analogous to Al-Mg (Section IVB1b) and proceeds in the early stages by primary and secondary oxide formation. For Al-3Li, the spinel oxide γ-$LiAlO_2$ was identified as the primary oxide, together with secondary Li_2O formed by the reduction of the γ-Al_2O_3 overlayer by lithium according to

$$6Li + \gamma\text{-}Al_2O_3 \rightarrow 3Li_2O + 2Al \quad \Delta G^{\circ}_{750K} = -79.1 \text{ kJ mol}_{O_2}^{-1}. \qquad (6)$$

For Al-3Li-2Mg, both primary and secondary MgO were detected together with primary γ-$LiAlO_2$ and secondary Li_2O under both wet and dry environments. The free energy of secondary MgO formation at 750K is -89.5 kJ $\text{mol}_{O_2}^{-1}$ (Eq. 5). Where both magnesium and lithium are available for reaction, secondary MgO formation predominates particularly in the presence of water vapor.

In the latter stages of oxidation, primary γ-$LiAlO_2$ nucleation and growth become the dominant mechanism. At this stage of reaction, inter-

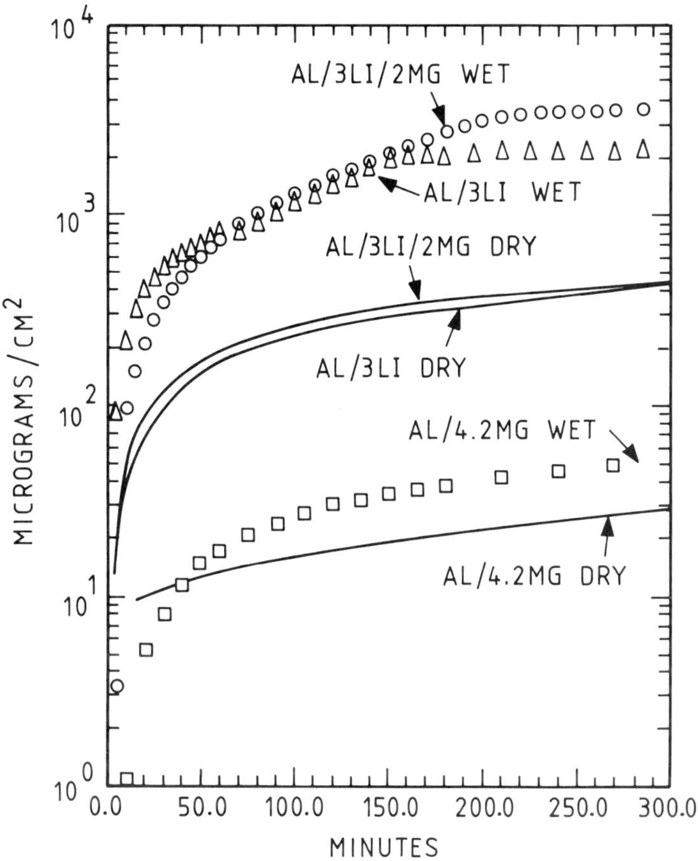

FIG. 4. Oxidation weight-gain curves obtained for Al-3 wt % Li, Al-3 wt % Li-2 wt % Mg and Al-4.2 wt % Mg under "dry" and "moist" 20% O_2/Ar at 480°C, taken from [*4*].

granular oxidation in the wet environment becomes evident. Oxides penetrate down the grain boundaries and, due to the net volume expansion, fresh surfaces are exposed for further reaction, thereby accelerating the oxidation rate as shown in Fig. 4. The plateau region of reduced weight gains corresponds to severe lithium depletion from the thin (0.2-mm) sheet sample. The oxidation rate of Al-3Li is controlled by the nucleation and growth of primary oxides and not limited by the bulk diffusion of lithium to the oxide–metal interface. It is clear that lithium has a dominant effect on oxidation and even magnesium at wt % levels has relatively little effect on the oxidation weight gains experienced by these alloys.

At the time of this writing there was little or no detailed kinetic information regarding the oxidation of other ternary Al-Li alloys (e.g., Al-Li-Cu and Al-Li-Zn) although there was some limited data concerning the quaternary system Al-Zn-Mg-Li [*48*]. It is reasonable to assume that the oxidation characteristics of such alloys will be similar to the equivalent, binary Al-Li alloy.

b. Oxidizing Environments

The reactions of an Al-2.7% Li alloy under both "dry" and "moist" (water vapor saturated at 20°C) 20% O_2/Ar, N_2, and CO_2 over the temperature range 400–575°C have been monitored thermogravimetrically [*9,49*]. The weight-gain curves obtained at 520°C are shown in Fig. 5.

The curves fall into two distinct groups: either dry or wet (except CO_2), in terms of both the magnitude of weight gains and the shapes of the

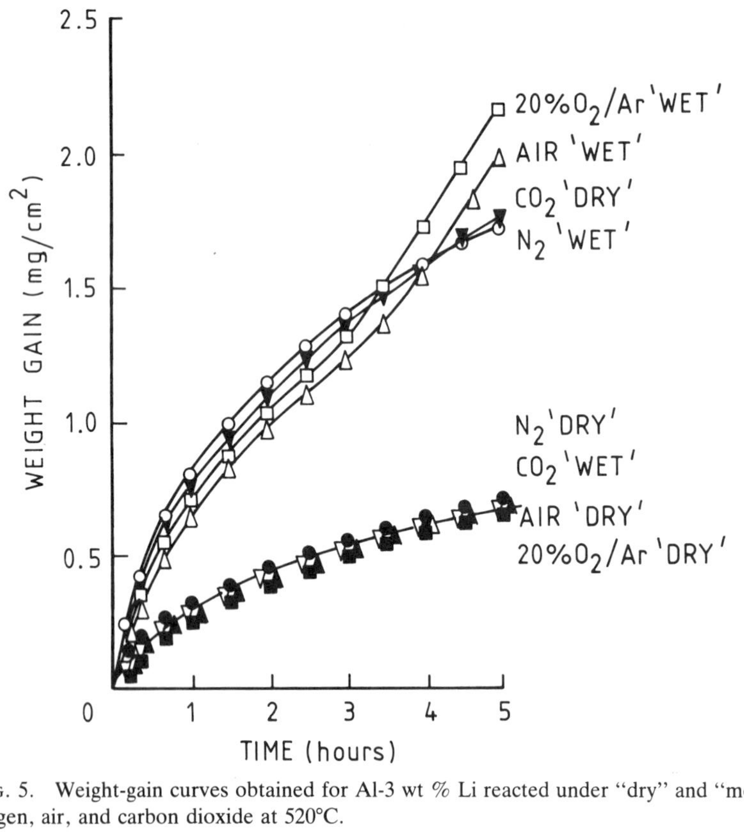

FIG. 5. Weight-gain curves obtained for Al-3 wt % Li reacted under "dry" and "moist" nitrogen, air, and carbon dioxide at 520°C.

curves. Under dry environments (except CO_2) and the wet CO_2 atmosphere, the weight gains conform to pseudoparabolic kinetics and are much lower compared to the wet condition. The presence of water vapor (except CO_2) is dominant and weight gains are high, typified by an acceleration in the reaction rate due to breakaway oxidation during the later stages of reaction. Dried CO_2 gives rise to high oxidation weight gains, with smooth and continuous weight-gain curves conforming to a paralinear kinetics.

Microscopic investigations of the initial stages of the reaction of Al-2.7Li under dry and wet air [*9*] and CO_2 [*50*] have been carried out using an environmental cell and hot stage fitted to a high-voltage electron microscope. Heating was carried out over the temperature range 250–600°C under 30-torr pressure of the selected atmosphere.

The first crystalline oxide phase to develop on the alloy upon heating under dry air was Li_2O. This phase was stable up to about 330°C whereupon it was replaced by Li_2CO_3, which became the predominant crystalline reaction product up to the intermediate temperature of about 450°C. At temperatures in excess of about 450°C, the carbonate was replaced by the tetragonal spinel γ-$LiAlO_2$. An unknown phase was observed at temperatures between about 475 and 525°C.

In the presence of water vapor, the low-temperature reactions were essentially the same except for some other unknown phase, which was detected over a narrow temperature range between 275 and 300°C. The similarity in reaction species between the wet and dry conditions changed in the upper temperature range of 475 to ≥600°C. At 475°C, γ-Al_2O_3 was observed to nucleate, grow, and exist together with the tetragonal spinel. However, at 540°C, the γ-Al_2O_3 disappeared and the cubic spinel $LiAl_5O_8$ formed as a discrete oxide phase in a tetragonal spinel matrix.

Turning to CO_2 environments, the situation changes somewhat. With the dry gas, the first phase to appear is not Li_2O, but a crystalline compound with a structure similar but different to Li_2CO_3, which has been termed a carbonate precursor. This phase exists up to 375°C, when Li_2CO_3 becomes predominant. The carbonate decomposes at about 515°C when the sole reaction product becomes the tetragonal spinel γ-$LiAlO_2$. Water-vapor additions were found to suppress the carbonate precursor and the carbonate itself was strongly detected. From about 320 to 435°C, γ-$LiAlO_2$ was observed while γ-Al_2O_3 was detected from 320 to 557°C. In this temperature range and upward, no carbonate could be detected within the limits of electron diffraction. Following the termination of γ-$LiAlO_2$ formation at 435°C, the cubic spinel formed and became the predominant phase together with γ-Al_2O_3 up to the melting point of the alloy. There is no microstructural or crystallographic information available regarding nitrogen as a reaction environment, although lithium nitride formation in air at high temperatures has been reported by Russian workers [*46*].

References

1. Beck, A.F., Heine, M.A., Caule, E.J., and Pryor, M.J. (1967). *Corr. Sci.*, **7**, 1–22.
2. Doherty, P.E., and Davies, R.S. (1963). *J. App. Phys.*, **34**, no. 3, 619–628.
3. Hart, R.K., and Maurin, J.K. (1970). *Surface Sci.*, **20**, 285ff.
4. de Brouckere, L. (1945). *J. Inst. Metals.*, **71**, 131–147.
5. Scamans, G.M., and Butler, E.P. (1975). *Met. Trans. A*, **64**, 2055–2063.
6. Field, D.J., Butler, E.P., and Scamans, G.M. (1980). *Proc. Ist. Int. Conf. Al-Li Alloys.* TMS-AIME, Stone Mountain, Georgia, pp. 325–346.
7. Grauer, R. and Schmoker, P. (1976). *Werkstoffe und Korrosion*, **27**, 769–774.
8. Field, D.J., Scamans, G.M., and Butler, E.P. (1981). *Proc. 2nd. Int. Conf. Environmental Degradation of Engineering Materials*, Virginia Polytechnic Institute, Blacksburg, September, pp. 393–405.
9. Field, D.J., Scamans, G.M., and Butler, E.P. (1983). *Proc. 2nd. Int. Conf. Al-Li Alloys.* TMS-AIME, Monterey, California, pp. 657–666.
10. Kim, Y.W., Griffith, W.M., and Froes, F.H. (1983). *Proc. ASM Metals Congr. New Developments for Aluminum Applications.* Philadelphia, October, pp. 1–18.
11. Lentz, A.H., Dickerson, P.B., Robinson, P.B., Saunders, H.L., and Sini, D.L. (1979). *Aluminum Brazing Handbook Ed. 3.* The Aluminum Association, Washington.
12. Matienzo, L.J. (1983). *Aluminum*, **59**, no.. 11, E453–E457.
13. Blackburn, P.E., and Gulbransen, E.A. (1960). *J. Electrochem. Soc.*, **107**, no. 12, 944–950.
14. Cochran, C.N., and Sleppy, W.C. (1961). *J. Electrochem. Soc.*, **108**, no. 4, 322–327.
15. Breakspeare, R.J. (1970). *J. Appl. Chem.*, **20**, no. 7, 208–212.
16. Lee, S.T., Shaefer, D.V., and Lockwood, F.E. (1985). *Aluminum*, **61**, no. 7, 504–508.
17. Pryor, M.J. (1971). *Oxidation of Metals*, **3**, no. 6, 523–527.
18. Heine, M.A., and Sperry, P.R. (1965). *J. Electrochem. Soc.*, **112**, no. 3, 359–361.
19. Mott, N.F. (1939). *Trans. Farad. Soc.*, **35**, 1175–1177.
20. Fehlner, F.P., and Mott, N.F. (1970). *Oxidation of Metals*, **2**, no. 1, 59–99.
21. Pryor, M.J. (1971). *Oxidation of Metals*, **3**, no. 3, 271–273.
22. Hunter, M.S., and Fowle, P. (1956). *J. Electrochem Soc.*, **103**, no. 9, 482–485.
23. Heine, M.A., and Sperry, P.R. (1963). *J. Electrochem. Soc.*, **110**, no. 12, 1205ff.
24. Gulbransen, E.A., and Wysong, W.S. (1947). *Phys. Coll. Chem.*, **51**, 1087ff.
25. Smeltzer, W.W. (1958). *J. Electrochem. Soc.*, **105**, no. 2, 67–71.
26. Scamans, G.M., and Butler, E.P. (1975). *Proc. Quatrieme Contres Internatinal Microscopie Electronique Haute Tension*, Toulouse, pp. 341–344.
27. Wilkins, N.J.M., and Wanklyn, J.N. (1959–1960). *J. Inst. Metals.*, **88**, 134–140.
28. Winter, F., and Gruhl, W. (1975). *Aluminum*, **51**, no. 12, 755–757.
29. Winter, F., and Gruhl, W. (1977). *Metall*, **31**, no. 6, 595–599.
30. Greenblatt, J.H. (1962). *J. Electrochem. Soc.*, **109**, no. 12, 139–1142.
31. Altenpohl, D., and Post, W. (1961). *J. Electrochem. Soc.*, **108**, no. 7, 628–631.
32. Alwitt, R.S. (1976). In *Oxide and Oxide Films*, J.W. Diggle, ed., **vol. 4**, Marcel Dekker, New York, pp. 169–253.
33. Vedder, W., and Vermilya, D.A. (1969). *Trans. Farad. Soc.*, **65**, 561–584.
34. Scamans, GM., and Rehal, A.S. (1979). *J. Mat. Sci.*, **14**, 2459–2470.
35. Draley, J.E., and Ruther, W.E. (1957). *J. Electrochem. Soc.*, **104** no. 6, 329–333.
36. Brock, A.J., and Pryor, M.J. (1973). *Corr. Sci.*, **13**, 199–227.
37. Field, D.J., Butler, E.P., and Steward, N.I. (1984). Paper presented at Fall Meeting of TMS-AIME, Detroit. To be published.

38. Kofstad, P. (1966). *High Temperature Oxidation of Metals*. Wiley, New York.

39. Hine, R.A., and Guminski, R.D. (1960–1961). *J. Inst. Metals*, **89**, 417–422.

40. Brock, A.J., and Heine, M.A. (1972). *J. Electrochem Soc.*, **119**, no. 8, 1124–1127.

41. Field, D.J. (1982). *The High Temperature Oxidation of Aluminum Alloys Containing Magnesium and Lithium*. Ph.D. dissertation, University of London.

42. Whitaker, M., and Heath, A.R. (1953–1954). *J. Inst. Metals.*, **82**, 107–116.

43. Muromachi, S., and Minegishi, T. (1960). *Jap. Inst. Light Metals*, **45**, November, 25–28.

44. Brock, A.J., and Pryor, M.J. (1974). Canadian Patent no. 959304, issued December 17.

45. Tolot, F. (1963). *14th. Int. Conf. Occupational Health*, Madrid, **3**, pp. 1098ff.

46. Varakina, L.P., Mironenko, V.N., and Polyanskii, V.M. (1979). *Teknol Legk Splavov.*, **6**, 10–13.

47. Ginsberg, H., and Datta, P.K. (1966). *Aluminum*, **42**, no. 11, 6811-687.

48. Csanady, A., and Kurthy, J. (1981). *J. Mat. Sci. Letters*, **16**, 2919–2922.

49. Field, D.J., Bassett, K.A., and Butler, E.P. (1984). European Patent Application EPO 123 453 Al, filed March 27.

50. Field, D.J., and Butler, E.P. (1983). Unpublished work.

20

Formability and Microstructure

B.A. PARKER

Department of Materials Engineering
Monash University
Clayton, Victoria
Australia

I. Introduction

Formability is a widely used word for describing an alloy's performance in a solid-state shaping operation. Formability is not simply related to fundamental mechanical properties such as strength and ductility and this is readily seen by considering the forming of 3003 H14 canstock. This material has negligible elongation in a tensile test and might be expected, at first sight, to be not very formable and yet it is capable of undergoing the

ISBN 0-12-341831-3

very large shape changes and plastic strains imposed by the drawing and ironing operations used to make a beverage can [*1*].

Formability is, therefore, best defined in terms of the success or otherwise with which an alloy workpiece can be shaped into a component in one or more forming operations. In discussing success, we should also retain some notion of the economics of the material and the process in the broadest sense. Failure of an alloy (i.e., a lack of formability) may occur in a variety of ways:

1. The workpiece may fracture.
2. There may be severe localized deformation (e.g., necking).
3. The material may be too thin in some regions for the end use of the part.
4. The appearance of the product may be unsatisfactory (for example, surface imperfections due to flamboyant Luders bands, galling and wrinkling, or looper lines).
5. The workpiece may not retain the shape imposed by the die (for example, the springback of shallow panels).
6. The formed product may not have the properties desired for the final product.

Some of these failures may be attributed to the design of the part or to the details of the forming process itself, while others clearly arise from the elastic and plastic deformation behavior of the workpiece material. Some aspects of this behavior may be influenced by the microstructure of the alloy concerned. This chapter addresses the relationship between microstructure and the parameters that may affect formability.

Before leaving the broader aspects, however, I wish to emphasize that there may be many ways to solve a forming problem. In other words, process-design modifications, including attention to such aspects as lubrication, may be more important than fine-tuning an alloy composition and microstructure. An alloy may only be described as formable or not in the context of the forming process involved and the forming conditions applying at the time.

The introduction of aluminum alloys into product areas previously exclusively the domain of steel (e.g., the production of car body panels in the late 1970s) led to aluminum alloys acquiring a poor reputation for formability. While there are certainly problems associated with the significantly lower elastic modulus of aluminum, many of the alleged deficiencies may be attributed to the use of processes developed for steel over many decades. There may be a well-founded reluctance to change manufacturing processes, but the notion that aluminum alloys have generally poor formability is untenable.

In a brief chapter of this kind, it is impossible to consider all forming processes and all alloy types. Thus, the majority of the discussion will concern the sheet forming of medium-strength alloys. Many of the topics are relevant to other alloys and other processing operations. Articles on some of the more practical aspects of the press forming of aluminum alloys have been published by Grimes and Wright [*2*] and by Falkenstein [*3*].

II. Failure During Forming

We shall start the discussion by looking more closely at the modes of failure encountered during forming.

A. Fracture

Most medium-strength aluminum alloys break either with a ductile fibrous fracture, where voids are nucleated at second-phase particles, or by a shear mode of failure characterized by a flat, rather featureless, surface crossing many grains and lying approximately parallel to a plane of maximum shear stress. Except in specimens containing very large inclusions, fracture is preceded by plastic-strain localization. This localization may take a number of forms. It may be due to localized work-hardening resulting from plastic instability or to the development of intensive slip bands and shear bands in the microstructure. Alternatively, a high dislocation density may develop around inclusions, dispersoids, or precipitates that do not deform plastically with the matrix. This high local plastic-strain concentration may initiate void formation.

The details of the various modes of fracture are discussed in Chapter 14 by Vasudevan et al. In the context of forming, it can be said that fracture will be delayed to higher strains if the rate of plastic strain concentration can be reduced or, alternatively, if there is a mechanism for the unloading of high local plastic strains by a recovery mechanism.

Final failure modes and macroscopic strain localization are strongly influenced by deformation geometry—for example, theory does not predict necking in a homogeneous workpiece under equibiaxial tension—and the various failure modes may be initiated by local plastic strains or maxima in the shear or normal stresses. The various possibilities for the form of fracture maps in stress and strain space were discussed by Glover, Embury, and Duncan [*4*] and examples were given by LeRoy and Embury [*5*] of the failure modes in a number of aluminum alloys. These geometric effects are among the more important reasons why uniaxial tensile tests often are poor indicators of formability.

B. Forming Limits

As just noted, in medium to low-strength alloys, the end to useful plastic deformation is not associated with fracture but with the localization of plastic strain into a small region of the workpiece. The formation of necks in a tensile test is the most familiar example of such a strain localization and the limits are described over a wide range of strain space in forming limit diagrams [*6*]. Figure 1 includes a forming limit diagram and fracture map drawn in strain space where the axes are the principal strains in the plane of the sheet. The localizations arise when it becomes easier to deform a local region rather than to distribute the strain more uniformly in the workpiece. These instabilities may result from the geometry of the forming process or they may arise from local geometric or material inhomogeneities in the workpiece itself. Local necks in uniaxial tension are a necessary consequence of a work-hardening rate that decreases with increasing strain and there is a potential for necks to occur at a value of strain (ϵ_c) where the stress increment in an infinitesimal geometric im-

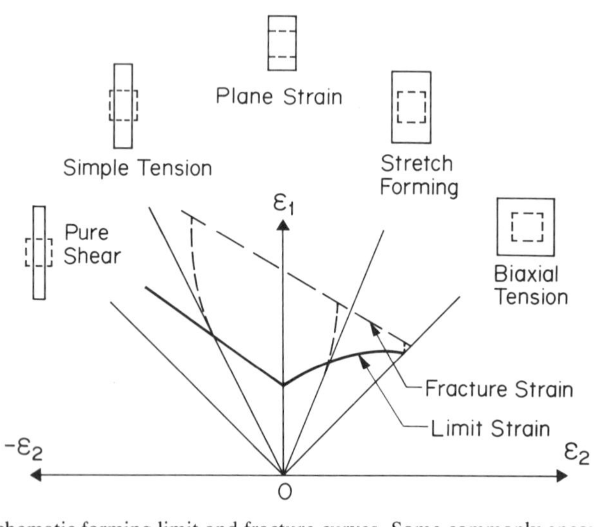

FIG. 1. Schematic forming limit and fracture curves. Some commonly encountered forming geometries are indicated. The dashed lines show the strain trajectories as plane-strain conditions are approached prior to fracture. (Redrawn from LeRoy and Embury [*5*]).

perfection is not balanced by increased work-hardening in the element. For a material deforming according to an idealized work-hardening law, $\sigma = k\epsilon^n$ (σ = stress, ϵ = strain, k is a constant, and n is the work-hardening exponent), this condition is reached when $\epsilon_c = n$. Catastrophic necks form at a somewhat higher strain whose value depends on the shape of the yield surface ($\epsilon_c = n/2$ for isotropic materials) [7].

Not all strain states lead to the formation of necks by simple geometric instabilities and it is necessary to invoke imperfections and heterogeneities to explain the occurrence of catastrophic necks in these geometries. Marciniak and his coworkers [8,9] have provided a valuable description of forming limits in the range of strain states encountered in metal forming. The shapes of forming limit curves arise from the effects of strain geometry on the tendency of the workpiece to develop local necks. A secondary effect, for most materials, comes from the effect of strain geometry on the work-hardening characteristics themselves.

Without going further into the theory of the formation of instabilities, it will be clear that forming behavior will be strongly influenced by the strain and strain-hardening characteristics of the material and by any nonuniformities in structure or geometry that may develop into macroscopic in homogeneities. Low working-hardening rates and low, or negative, strain-rate sensitivities both lead to lower forming limits.

Here I must emphasize that it is the strain and strain-rate-hardening characteristics of the material at strains close to the limit strains that are relevant to the prediction of failure. Many alloys do not conform accurately to a simple constitutive equation (e.g., $\sigma = k\epsilon^n$) and n values calculated by curve fitting over a wide strain range may often give a poor estimate of likely failure strains.

C. Luders Band Formation

Alloys that form macroscopic Luders bands form a special, but important group that includes the 5000 series of Al-Mg alloys and the 2000, 6000, and 7000 series of age-hardenable alloys in the naturally aged (T4) condition. Plastic deformation is generally not uniform in these alloys and takes place in a series of steps in which a small volume of the workpiece deforms rapidly by the propagation of a Luders band. The propagation of the band is accompanied by a load drop in the load extension record. This serrated yielding behavior is discussed in more detail in Chapter 3 by Sanders, Baumann and Stumpf; the concern here is with the appearance of macroscopic bands on the workpiece. The bands are visible because the plastic strain leads to localized thinning, and with increasing plastic strain more bands form and eventually superimpose. The sample thickness does not gradually decrease but, rather, thinning takes place in a series of discontinuous steps. Eventually, the superposition of several of these steps leads to catastrophic failure without the formation of a neck and below the strain at which necking is predicted [*10*].

In addition to alloy composition, the formation of Luders bands is sensitive to the temperature and strain rate of deformation and also to the grain size of the alloy. Further, since the discontinuities in thickness represent a condition of plane strain, their formation is affected by strain geometry in a manner similar to that described for localized necking.

D. Thinning and Plastic Anisotropy

The failure modes discussed so far have been associated with local reductions in thickness. When a metal sheet is stretched, or drawn, there is also a general reduction in thickness. The extent to which a material may be deformed in the plane of the sheet before it reaches some limiting thickness is related to its normal anisotropy. Sheet-metal technology describes this anisotropy in terms of an r-value (the ratio of the width to the thickness strain in a tensile test). A high r-value implies extensive in-plane deformation before a limiting thickness is reached.

Anisotropy is usually also present in the plane of the sheet (planar anisotropy) and can lead to preferential thinning and the formation of ears in, for example, the drawing of cups. The planar anisotropy is described by a Δr-value obtained by measuring the r-values from tensile test samples cut at 0, 45, and 90 degrees to the rolling direction in the sheet.

Plastic anisotropy is closely related to the crystallographic texture of the grains in the sheet. (See Chapter 20 by Fricke and Przystupa.) It is difficult to achieve high r-values in aluminum alloys since appropriate textures (with (111) planes parallel to the rolling plane) cannot be obtained by conventional processing routes. r-values of about 0.7 are typical, considerably below those of deep drawing quality steels, but not much lower than those of High-Strength Low Alloy (HSLA) steels. Limiting draw ratios (limiting ratio of the blank diameter to the cup diameter for a successful draw) have been correlated with r-values [11], but there is some suggestion [*12*] that the relationship does not hold for r-values less than unity. (See also Hosford [*13*].)

Planar anisotropy may be eliminated if a random texture can be formed. Again, this is difficult to achieve in conventional processing and the tendency toward earing is reduced by having a balance between two texture components—the cube texture that leads to the formation of ears at 0 and 90 degrees to the rolling direction and the retained rolling texture that gives ears at the 45-degree position [*14*]. The origin and control of textures is discussed in more detail by Fricke and Przystupa.

E. Plastic Strain Gradients

Microscopic strain gradients occur at particles and grain boundaries through an accumulation of geometrically necessary dislocations. Their effect on formability is discussed in Section III.D. Macroscopic strain gradients are developed in most formed parts simply by the geometry of the forming process. Examples include drawn cups where the base of the cup is essentially underformed and the walls are extensively deformed and consequently a region exists between the two where the strain gradient changes rapidly. A second example is that of simple bending where a strain gradient is developed through the thickness of the sheet. The maximum value of the local strain will depend upon the hardening characteristics of the material and any tendency to develop strain inhomogeneities and high local strains will be spread, somewhat, by high levels of strain and strain rate hardening.

In a situation involving large macroscopic strain gradients, there are, however, other important considerations and the analysis by Hosford [*13*] of the important parameters in the drawing and ironing of cans is worthy of note. Hosford regards the important parameters as the r- and Δr-values of the material and attaches little significance to work-hardening provided the material has an n-value ≥ 0.2. Indeed, he emphasizes that in redrawing and ironing operations, low work-hardening rates are desirable since the top of the cup wall will be strengthened relative to the base if the material work hardens significantly in the initial cupping operation. During redrawing, the bottom of the new wall is formed from material that was not hardened in the initial operation; this material has to carry the increased forces exerted by the redrawing of the work-hardened region near the top of the wall. In this example, high levels of strain-hardening lead to large strain gradients and the possible failure of the unhardened material.

F. Product Appearance and Performance

A number of parts may be successfully formed but nonetheless be unacceptable owing to the presence of markings that mar the appearance of the end product. Examples of such markings include wrinkles, flamboyant Luders bands, and orange-peel effects. The formation of wrinkles is controlled by the geometry of the forming process and also by the plastic anisotropy of the material [*15*]. Luders bands occur in a number of alloys and are discussed elsewhere in this chapter. Flamboyant, or type-A bands, can be controlled by grain size or by temper-rolling alloys before forming [*16*]. Orange-peeling is generally caused by an overlarge grain size. An interesting effect, known as looping, occurs in some alloys that have a nonuniform grain structure, giving rise to a structural texture parallel to the rolling direction [*17*].

G. *Elastic Modulus and Springback*

As a final topic in this section, I shall briefly consider the effect of the elastic modulus a metal forming. The elastic modulus of aluminum is low (approximately one-third that of steel), resulting in relatively large elastic deflections for a given load. When the load is removed after a forming operation, there may be significant springback creating problems in the retention of the imposed shape. The majority of these problems may be overcome by careful attention to component and tooling design, but difficulties certainly remain, particularly when forming shallow panels.

Higher-modulus aluminum-lithium alloys and metal matrix composites have been developed, largely for aerospace applications, but it is unlikely that these materials will find their way into general forming aplications.

III. Microstructural Variables in Formability

A. *Work-Hardening of Aluminum Alloys*

The fundamental aspects of this topic are discussed in Chapter 3 by Sanders, Baumann, and Stumpf. My purpose here is to describe the development of the deformed microstructure in various alloys and to relate this to the rate and homogeneity of strain-hardening.

The stress–strain curves for a number of commercial alloys are shown in Fig. 2. The data was collected by hydraulic bulge testing in order to minimize geometric effects on necking and serrated yielding and also to provide data over a wide strain range. The alloys were all tested in the annealed condition. Typical compositions for these commercial alloys will be found in Chapter 3. However, the trend in Fig. 1 reflects an increasing alloying addition from commercially pure aluminum (1145) at the bottom of the figure to alloy 5182 at the top of the figure. The main effect is that of increasing magnesium level in solution.

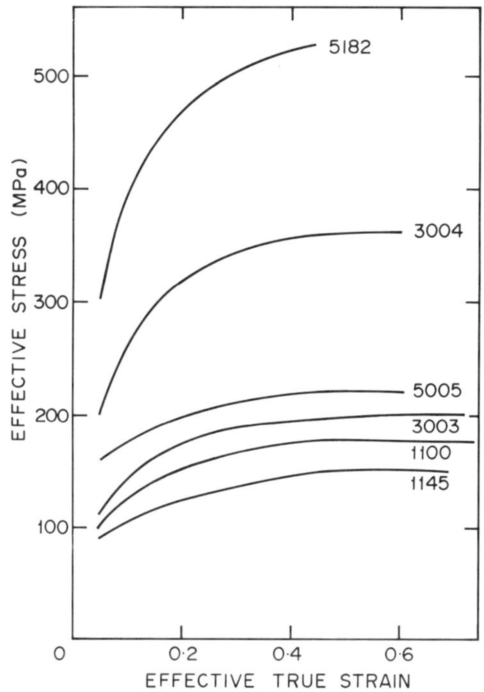

FIG. 2. Stress–strain curves for commercial aluminum alloys obtained by hydraulic bulge testing. The data are not compensated for the effects of normal anisotropy [34]. However, the *r*-values for the alloys are similar.

The initial flow stress of alloy 1145 is quite low and work-hardening is sustained to quite high strains. At low strains, the microstructure (Fig. 3) shows the development of dislocation tangles particularly in the vicinity of grain boundaries and triple points With increasing strain (Fig. 4), the tangles develop into well-defined cell walls. The cells are essentially equiaxed near the grain boundaries but may be elongated in the grain interiors where the number of operating slip systems may be rather less than the five required to maintain compatability at the grain boundaries.

The dislocation cells become smaller with increasing strain, but appear to reach a limiting size of a little less than one micron at a true strain of around 0.15. Additional plastic strain leads to a gradual increase in the misorientation across the cell walls and there is a gradual sharpening of the walls indicative of the relaxation of the dislocation structure into a more stable arrangement.

There is increasing evidence for the formation of additional microstructural features at high strains. Features within grains are described

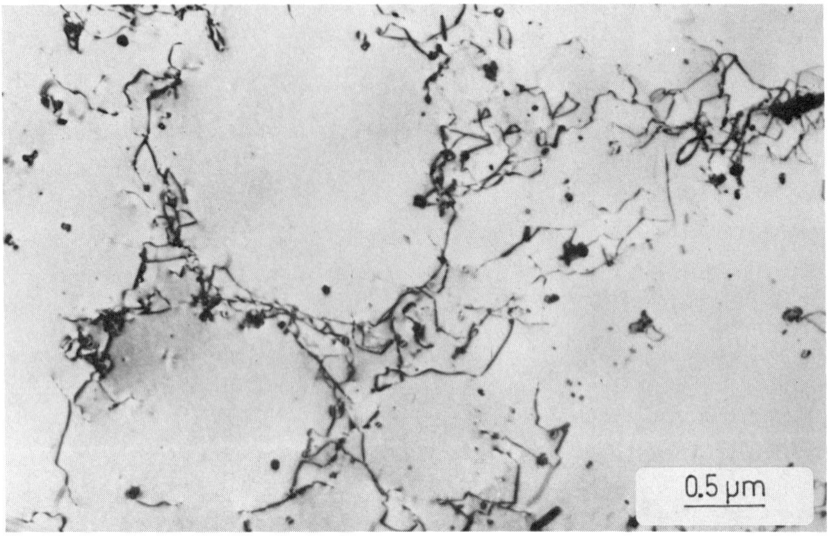

FIG. 3. Alloy 1145 deformed to a strain of 0.02.

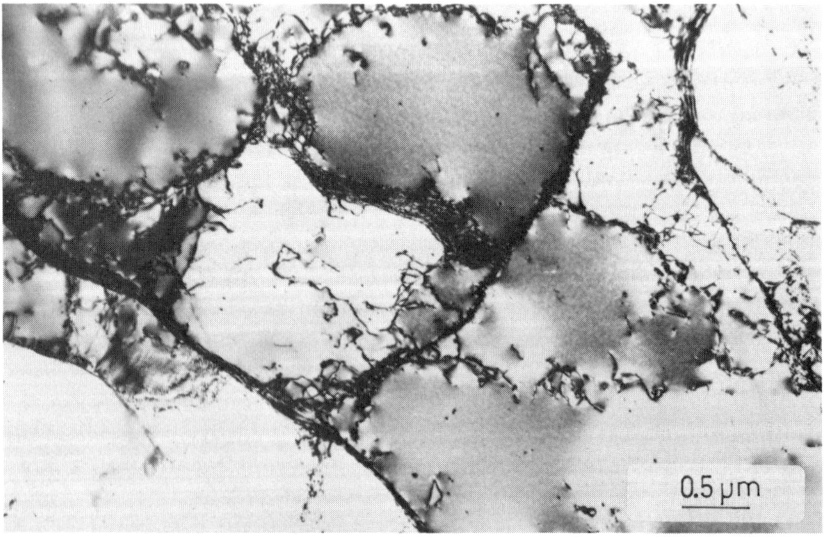

FIG. 4. Alloy 1145 deformed to a strain of 0.2.

as microbands and the more macroscopic features, which may develop from microbands, are called shear bands [*18*]. The features have seen extensively studied in copper and its alloys [*19*] and some observations have also been made in aluminum alloys [*20*]. See also Fig. 23 of Chapter 3 by Sanders, Baumann and Stumpf. These components of the deformed microstructure can extend over several grains and can result in large differences in the local crystallographic orientation with consequent effects on local mechanical properties. As structural inhomogeneities, they may be sites for the initiation of failure in a sample and so the conditions under which they form may be important.

Dillamore et al. [*21*] have suggested that the features are a form of a plastic instability and so occur when the work-hardening characteristics of adjacent elements are not able to compensate for local variations in stress. Indeed, the large shear bands may represent a form of work-softening. I previously noted the propensity of some aluminum alloys to fail in a catastrophic shear mode and I shall discuss the contribution of various microstructures later. There is insufficient evidence to suggest that shear instabilities precede all failures, but clearly the features of heavily deformed microstructures in complex alloys require further attention.

B. Dynamic Recovery

The formation of the dislocation cell structure and its refinement with increasing strain is a consequence of dynamic recovery by which the overall elastic strain field of the dislocation structure is reduced. The rearrangement of the dislocation structure involves, predominantly, the cross-slip of screw dislocations. Al-Haidary et al. [*22*] have suggested that there are two identifiable ways in which cross-slip is involved in dynamic recovery. First, there is the relaxation of the dislocation network into a low energy configuration. Second is the annihilation of dislocations of opposite sign. The first process appears to occur over a very wide temperature range whereas the second is restricted to rather higher deformation temperatures. The annihilation mechanism leads to rather narrower cell walls and a more stable dislocation array reminiscent of static recovery. This paper considers the effect of strain rate, solute, and temperature on the two processes and reaches some useful conclusions on the development of the deformed microstructure in aluminum alloys.

Cross-slip is a process that may be aided by thermal activation. Its rate depends on the local stress and the temperature. Consequently, it leads to strain-rate and temperature effects in the deformation behavior that are considered shortly. Cross-slip is, however, not the only factor that gives rise to temperature and strain-rate effects during cold-work and there is increasing interest in the contribution of dislocation intersections, the formation of jogs, and the generation of point defects to the thermally activated component of plastic flow (although these factors are now regarded as having little effect on strain-hardening) [*23*]. The effect of dynamic recovery in lowering the work-hardening rate is clearly detrimental to formability. However, as I shall note later, dynamic recovery can lead to the unloading of local strain concentrations with beneficial effects on the failure of higher-strength alloys. Dynamic recovery also leads to a positive strain-rate sensitivity, the advantages of which are also discussed shortly.

C. *Solute Strengthening*

Relatively few elements have a high equilibrium solubility in aluminum at room temperature. The only solute elements important in the common forming alloys are magnesium and, to a lesser extent, copper. These elements are mobile at ambient temperature and tend to diffuse toward dislocations so that there is a mutual reduction in strain energy. The segregation of solutes has a marked inhibiting effect on dynamic recovery and a well-defined cellular dislocation structure is not observed in alloys with a significant solute content (Fig. 5).

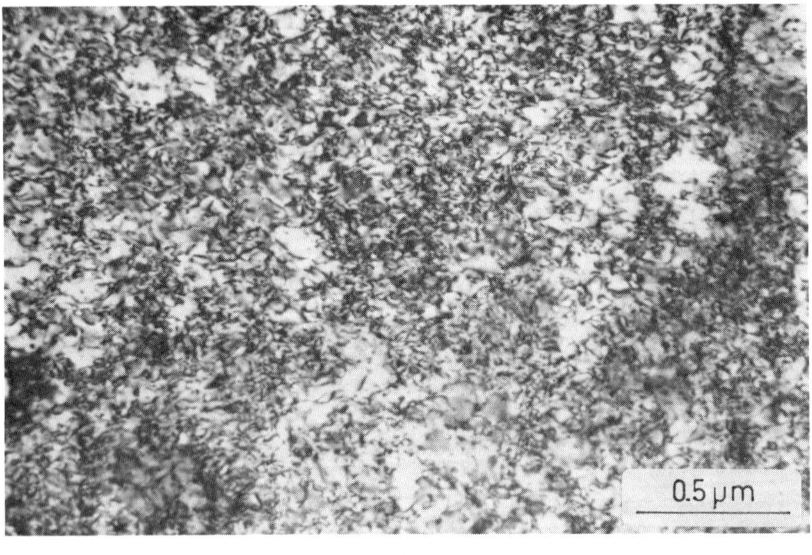

FIG. 5. Alloy 5182 deformed to a strain of 0.2.

The absence of dynamic recovery leads to a significantly higher rate of work-hardening and, with the added advantages of solid solution strengthening itself, has led to the development of the 5000 series of Al-Mg alloys, which are widely used for forming. The increased work-hardening does not come without problems. The segregation of solute atoms to dislocations leads to a reduced mobile-dislocation density and a discontinuous flow behavior. The discontinuities are associated with localized strain and thinning in the form of Luders bands, which can lead to surface irregularities and premature failure due to the local strain concentrations caused by the interaction of Luders bands [*10*].

The serrated nature of plastic flow in these alloys arises from an apparent negative rate sensitivity. In other words, a rise in strain rate in an element of the workpiece is accompanied by a lower flow stress, and this may in turn lead to the rapid development of geometric inhomogeneities. The strain-rate effects will be discussed further. Temperature also has important effects in these alloys. At low temperatures, solutes are immobile and serve to strengthen without the formation of Luders bands, whereas at higher temperatures, the solutes move easily with the dislocations and have little retarding effect on the cross-slip process. It would appear that, at least as far as magnesium is concerned, that room temperature is about the most unsuitable temperature for the forming of solute-strengthened alloys and some attractive results have been obtained for alloys formed at mildly elevated temperatures [*24*].

The development of rapid solidification processes and rapid heat-treatment techniques allows the development of alloys supersaturated with other solutes such as manganese, iron, and zirconium. These elements do not diffuse rapidly at room temperature and may provide solute-strengthening and sustained work-hardening without the disadvantages of discontinuous yielding.

D. Particle-Hardening

The particles present in medium-strength alloys may be divided into two classes depending on whether or not the particles are plastically deformed along with the matrix during deformation. The separation into classes depends not only on the inherent mechanical properties of the particles, but also on their size and spacing [*25*].

If particles are cut by the moving dislocations, then there is a tendency to concentrate deformation into a number of persistent slip bands in the microstructure since the passage of dislocations through the precipitates

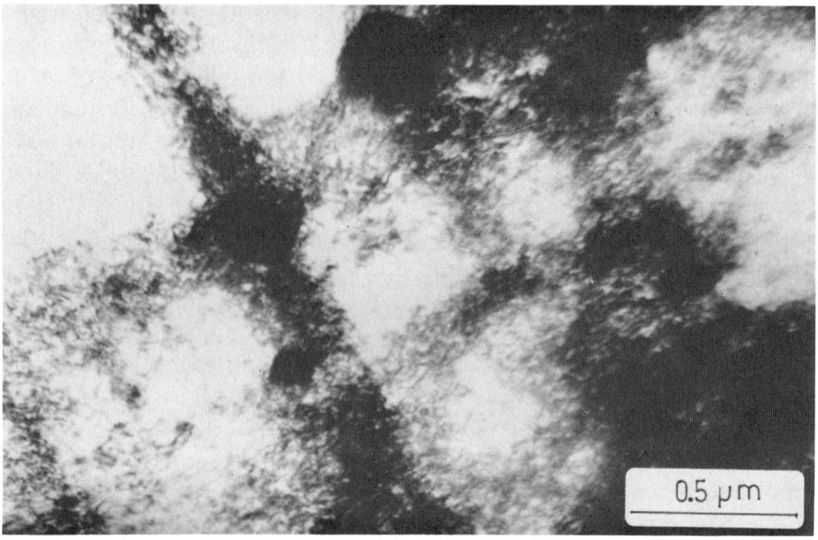

FIG. 6. Alloy 2036-T4 deformed to a strain of 0.2. Note the zones of higher dislocation density developed around the dipersoid particles.

will reduce their effectiveness to resist further deformation. The work-hardening rate of these structures is quite low and the concentration of slip deformation may provide a mechanism for the initiation of a premature shear failure or assist in the nucleation of grain-boundary cracks.

Structures containing nondeformable particles on the other hand have high work-hardening rates since dislocation debris is left around the particles by dislocations moving in the slip plane or cross-slipping around the particles (Fig. 6). The presence of the additional dislocations increases the effective size of the obstacles and adds an extra component to the hardening. The advantages gained through the increased work-hardening are offset, to some extent, by the local plastic-strain concentration near the particles, which can lead to the initiation of fracture by decohesion or cracking of the particles. Some mitigation of this effect may, in turn, result from localized dynamic recovery.

A critical dispersion of particles can result in a change in the deformed structure. This occurs when the structure consists of a nondeformable particle population dispersed with a particle spacing around one micron. A cell structure forms quickly at low strains and is stabilized by the particles (Fig. 7). Increasing strain causes only an increase in misorientation across the cell walls together with a sharpening of the cell-wall structure. Such structures are capable of large plastic strains in certain

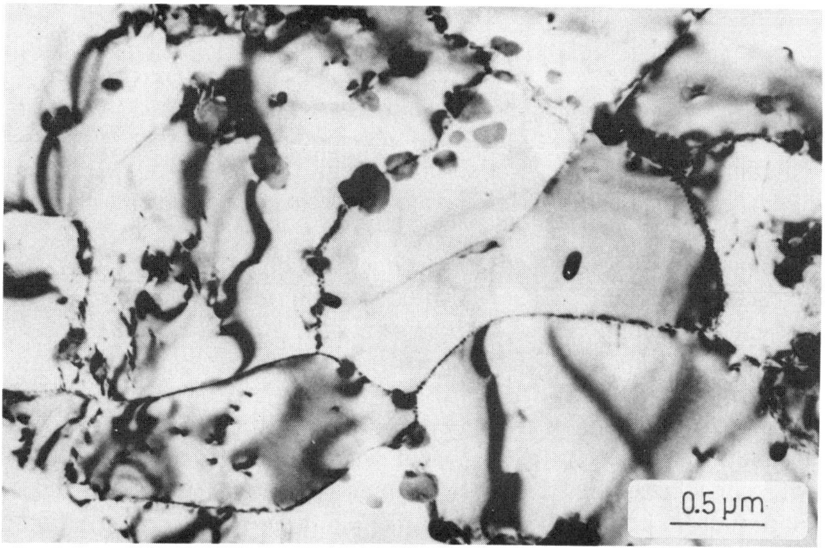

FIG. 7. Fine-grained 1.8% Mn alloy deformed to a strain of 0.3 showing a dislocation cell structure, with sharp boundaries, stabilized by fine manganese-containing particles.

deformation geometries, but they do appear to suffer from having a low work-hardening rate [*26*].

Alloys can, of course, be produced with more than one population of particles. It may be desirable for a structure to have a fine precipitate for strength (particle size around 0.01 μm) and to have, in addition, a population of larger particles (around one micron) to break up the slip pattern and prevent the formation of persistent slip bands. Such duplex particle structures have been shown to have useful combinations of strength and ductility in titanium alloys [*27*] as well as in aluminum alloys [*28*]. The ideas have been extended to medium-strength formable alloys with some success [*29*].

Summarizing the effects of particle structure, we can say that, from the point of view of formability, fine particles are not desirable because of their low work-hardening rates and tendency to form intensive slip bands. Intermediate sized particles in the range from one to five microns provide alloys with some strength and have beneficial effects on formability, whereas still larger particles (greater than five microns) may lead to fracture initiated by failure of the particles or their interface with the matrix.

E. The Effects of Strain Rate and Temperature

Some of the effects of strain rate and temperature were considered in discussing other aspects of the deformed microstructure. A rise in temperature accelerates the rate of thermally activated processes such as cross-slip and climb and decreases the strain-hardening rate. In alloys containing solutes, serrated yielding is eliminated and recovered dislocation structures are seen.

When thermally activated processes are involved, the effects of strain rate and temperature cannot be treated separately. Thermally activated processes have finite relaxation times that bear little relation to the rate of concurrent deformation. In simple structures a higher imposed strain rate will lead to a higher flow stress as the dislocation structure becomes harder due to a reduced rate of dynamic recovery. This behavior is described as positive strain-rate sensitivity. A consequence of positive rate sensitivity is that an element of the workpiece straining faster than an adjacent element will harden more and the strain-rate difference will decrease, reducing a tendency for strain localization [*30*].

It will now be clear that high levels of strain-rate-hardening and high strain-rate sensitivities will help to distribute plastic deformation more uniformly in a deforming workpiece. It will also be recognized, however, that high strain-rate sensitivites are associated with processes that lead to lower rates of strain-hardening. We might ask how we might make the best use of the two phenomena in the design of forming alloy. Strain-rate sensitivity is a function of many variables including stress, strain, microstructure, and temperature; although its value is never high, it is altered by quite small changes in these parameters. Work-hardening, on the other hand, tends to decrease with increasing temperature. In alloy design, therefore, an attempt should be made to make the best use of strain-rate-hardening at strains where strain-hardening is becoming less effective (i.e., at higher strains).

To make the point in another way, consider an alloy that deforms according to the constitutive equation

$$\sigma = \mathrm{k}\epsilon^{n}\left(\frac{\dot{\epsilon}}{\dot{\epsilon}_0}\right)^{\mathrm{m}}$$

where, for the sake of our present argument, the strain-rate exponent, m may be strain-dependent and evaluated close to the limit strain. $\dot{\epsilon}_0$ is a reference strain rate introduced to make the term dimensionless. Applying the usual condition for diffuse necking $\left(\frac{1}{\sigma}\frac{\partial\sigma}{\partial\epsilon} = 1\right)$, we find that the limit strain is given by [*31*]

$$\epsilon_c = \left[\frac{n}{1 - \dfrac{dm}{d\epsilon} \ln \left(\dfrac{\dot{\epsilon}}{\epsilon_0} \right)} \right]$$

Although the magnitude of the rate-dependent term is not easily evaluated, it is clear that, for a given value of n, the limit strain increases with the slope ($dm/d\epsilon$). Formable alloys might therefore be designed with as high a work-hardening as possible and a strain-rate sensitivity that is increasing with strain at strains close to the limit strain. We shall look further at how this might be achieved a little later.

F. Grain Size and Texture

A fine grain size is important in many forming alloys, not necessarily because of an improvement in properties, but rather because large-grained alloys can have an unacceptable finish after forming, (e.g., orange-peel effects). Fine grain sizes also assist the homogenization of slip, limiting the length of slip bands formed in any one grain. Excessive reduction in grain size may however lead to problems. As the grain size approaches one micron, dislocation substructures do not develop in the grains and the work-hardening rate may be quite low. These structures may also fail by unstable shear, in an unpredictable manner, by the cooperative action of a number of grains [*26*].

The crystallographic texture of sheet-forming alloys has long been a topic of interest; Section II.D noted its relationship to plastic anisotropy described by the r and Δr-values. An important characteristic of most forming alloys is that they should have high r-values and low Δr-values although there are special situations (e.g., the drawing of square cups) where some anisotropy may be desirable. Considerable effort over many decades has led to the development of deep drawing steels with r-values up to 2 and these have excellent deep-drawing properties. r-values can be calculated from crystallographic textures and so desirable textures can be predicted [*32*].

Aluminum alloys develop a number of textures during deformation and annealing. These are discussed in Chapter 20 by Fricke and Przystupa, but essentially the more widely used alloys have combinations of three common textures:

1. The cube texture {100} ⟨001⟩ that forms on annealing more heavily deformed alloys at higher temperatures.

2. The deformation texture {123} ⟨412⟩ (with other components forming a tube in orientation space) formed by cold-rolling.
3. An annealing texture, similar to the deformation texture, formed in many alloys by either continuous recrystallization or by discontinuous recrystallization at relatively low temperature. This texture is frequently designated the R-texture.

The r-value of alloys containing different amounts of these texture components are all around 0.7, less than 1.0, (the value for a randomly oriented grain structure) Random structures are not obtainable by conventional processing and indeed little can be done to increase r-values above 0.7. For the pure cube texture, Δr is quite high and ears are formed on drawn cups in the sheet-rolling direction and at 90° to it. High Δr values are also found for the R-texture, but in this case the ears are formed at angles of 45° to the principle directions. Minimum Δr-values are obtained by having a balance of the cube and the retained rolling texture so that eight small ears are formed and the effects of planar anisotropy minimized [*14*].

In commercial-purity aluminum and the more dilute aluminum alloys, careful attention is paid to composition and thermomechanical treatment, particularly for alloys to be deep drawn, in order to ensure minimum earing. Since some of these alloys are drawn in the cold-rolled condition, a strong cube texture is needed in the hot-rolled slab.

IV. Ideal Microstructures for Formable Alloys

From the discussion so far, it is possible to suggest an ideal microstructure for a formable aluminum alloy. Such an alloy must not fracture prematurely in any strain geometry, it must not have a tendency toward persistent slip band or shear band formation, it should not thin unduly during forming, and it should have the capacity to spread locally occurring high plastic strains; further, it should be capable of unloading highly localized plastic strains such as those occurring around inclusions. The properties required are:

1. A high work-hardening rate, sustained to as high a strain as possible.
2. A high, positive strain-rate sensitivity, particularly at high strains.
3. No particles larger than about three to five μm.
4. No deformable precipitation-hardening particles or, alternatively, a duplex particle structure with fine particles for strength and large particles to inhibit the formation of intensive slip bands.

5. A crystallographic texture with (111) planes parallel to the rolling plane and negligible planar anisotropy.

How can we get such an ideal material? Requirement (3) can be achieved, at some cost, by careful control of solidification and subsequent processing. Requirement (5) does not seem to be achievable at realistic cost with processing technology available as of early 1988. A high work-hardening rate at high strains is best achieved with a structure in which dynamic recovery is inhibited by a mobile solute. Localized plastic flow in such structures has to be inhibited by the presence of a dispersion of moderately sized particles. These particles also help to increase the work-hardening rate, but they also cause another form of local plastic deformation due to the collection of dislocation debris around them. The ideal alloy must therefore have sufficient solute to sustain work-hardening up to the average strain but not so much that dynamic recovery is inhibited at the higher local strains around particles. An example of dynamic recovery occurring in the strain field near a large particle is shown in Fig. 8. The notion is that dynamic recovery will unload the high local plastic strain and prevent the initiation of fracture at the particles. A valuable additional advantage of dynamic recovery at high strains is that it causes positive rate sensitivity and helps prevent the development of thickness defects.

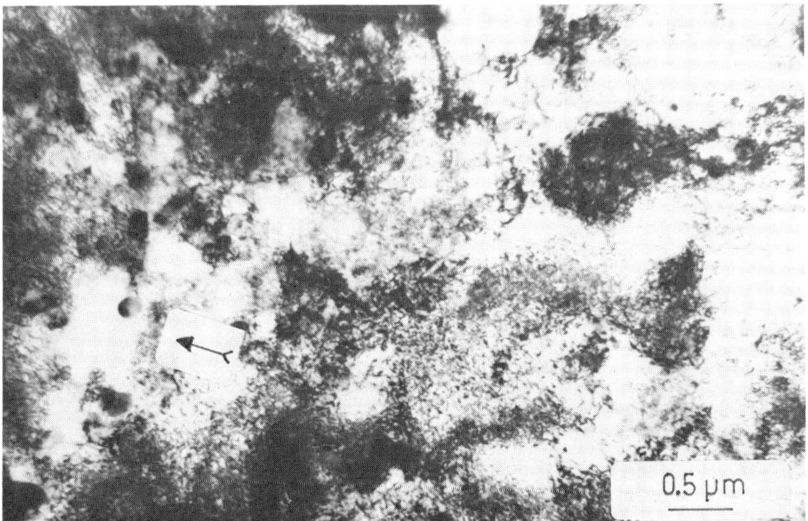

FIG. 8. Alloy 2036-T4 deformed in biaxial tension to an effective strain of 0.55. The formation of a dislocation-free "cell" at a dispersoid particle may be seen.

V. Current Trends in Forming Alloy Development

Table I summarizes the range (as of 1988) of medium-strength forming alloys and tempers developed largely, but not exclusively, for automotive-body-panel applications. Figure 9 shows the forming and fracture limits for typical samples of these alloys where the information is available. The more recently developed 6009 and 6010 alloys perform better than the earlier 2036 alloy largely because the strain-rate sensitivity is less negative at higher strains. The strain-rate sensitivities of the alloys are shown in Fig. 9 as a function of strain and it may be seen that for a given strength, the forming limits correlate quite well with the higher rate sensitivity. Figure 10 and 11 take the point further and compare the strain-rate sensitivities and forming limits of an Al-Mg alloy formed at elevated temperature and an experimental 2036 base alloy containing manganese to provide a dispersion of moderately sized particles. These two alloys show significantly higher forming limits and demonstrate, on the one hand, the advantages of warm forming and, on the other hand, the value of the kind of microstructure described in Section IV. Figure 8 shows evidence for dynamic recovery at dispersoid particles observed close to the fracture surface in the 2036 alloy. The development of medium-strength forming alloys designed from a microstructural viewpoint has been described by Beaver and Parker [*33*].

TABLE I

MECHANICAL PROPERTY DATA FOR MEDIUM-STRENGTH, FORMABLE ALUMINUM ALLOYS

	MPa					
Alloy Temper	σ_y	σ_u	% El.	*r*	*n*	*m*
2036-T4	190	340	24	.7	.23	−.005
6061-T4	145	240	22	.6	.22	—
6009-T4	130	235	25	.7	.23	~0
6010-T4	180	290	25	.7	.22	−.001
5182-0	130	280	25	.8	—	−.01
5182-0 (200°C)	150	220	67	.8	—	+.04
3003-0	42	110	30	.6	—	+.01
3004-0	70	170	20	.4	—	−.01
E-T4	190	340	19	.5	—	+.005
A-K Steel	180	300	40	1.6	.23	+.015

Note: Data obtained on commercial samples (except for experimental alloy E) may not be *average* data.

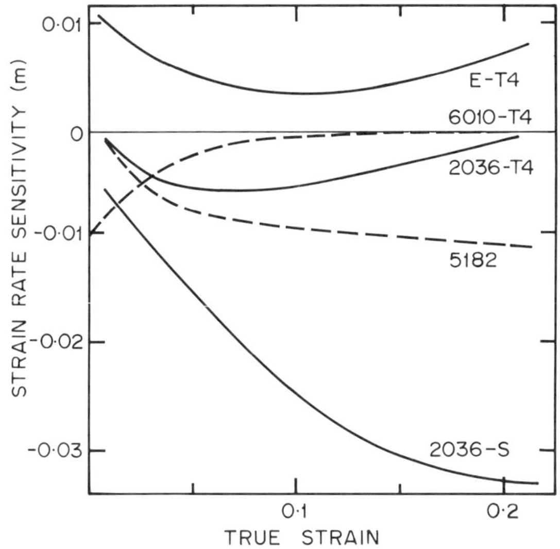

FIG. 9. The variation of the strain-rate sensitivity (m) with true strain for commercial alloys and for experimental alloy E in indicated tempers.

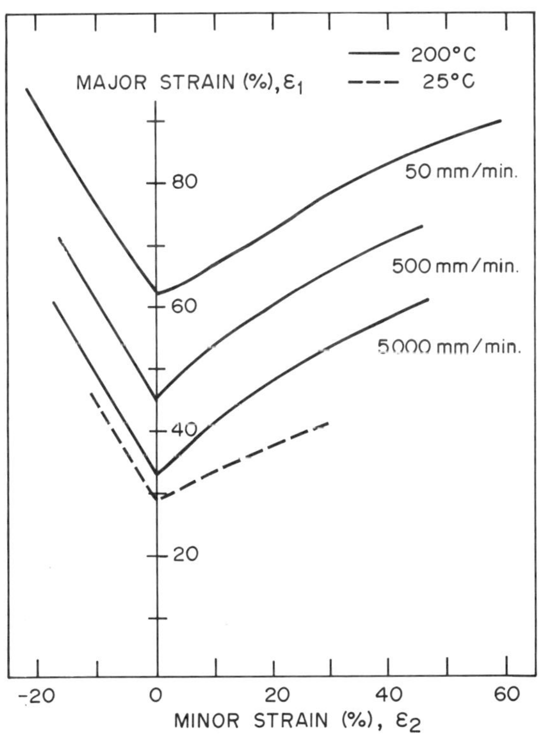

FIG. 10. The effect of temperature and deformation rate on the forming-limit curve for alloy 5182-0. (The forming limits were not rate sensitive at 25°C.) [*24*]

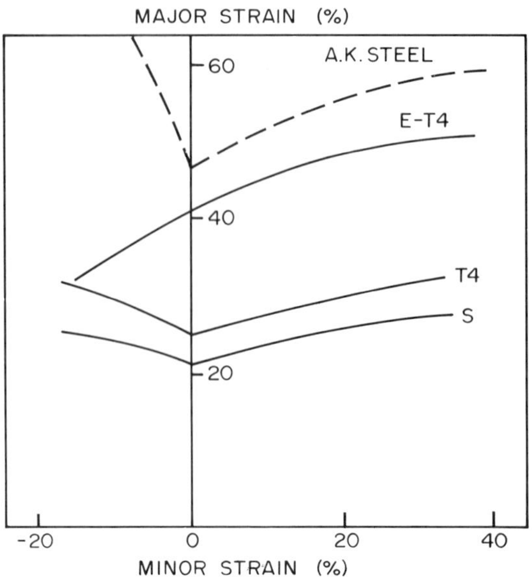

FIG. 11. Forming-limit diagram for alloy 2036 in the solution-treated (s) condition and the naturally-aged (T4) condition. The curve marked E-T4 is for an alloy similar to 2036 but with an additional particle distribution achieved by adding manganese. The curve for a typical deep-drawing steel is included as a reference.

VI. Conclusion

I have described a number of observations providing insight into the relationship between microstructure and formability in aluminum alloys. As the chapter is not a comprehensive review, the interested reader is referred to the following references. New processing technologies will bring new microstructures with different properties and different forming behaviors. An understanding of the basic parameters describing formability will aid their introduction into the sheet-metal industry.

Acknowledgments

Work on the development of formable aluminum alloys has been supported by Comalco Ltd., The Australian Research Grants Scheme, and the Harold Armstrong Memorial Fund at Monash University. The contributions by P.W. Beaver and R.G. O'Donnell are gratefully acknowledged.

References

1. Sowerby, R., and Sareen, B.K. (1978). in *Formability Topics—Metallic Materials*, A.K. Schmieder and J.R. Newby, eds. ASTM STP 647. American Society for Testing and Materials. Philadelphia, Pa. 49–64.
2. Grimes, R., and Wright, J.C. (1967). *Sheet Metal Ind.* **44**, 391–416.
3. Falkenstein, H.P. (1982,3). *Aluminium* **58**, E227–231, E246–257; **59**, E23–24, E47–48, E104–105, E125–126, E179–181, E244–247, E310.
4. Glover, G., Duncan, J.L., and Embury, J.D. (1977). *Metals Tech.* **4**, 153.
5. LeRoy G.H., and Embury, J.D. (1978). In Formability: Analysis, Modelling and Experimentation, S.S. Hecker, A.K. Ghosh, and H.L. Gegel, eds. The Metallurgical Society Warrandale, Pa. 183–207.
6. Keeler, S.P. (1965). *Sheet Metal Ind*, **42**, 683–691.
7. Backhofen, W.A. (1972). *Deformation Processing*. Addison-Wesley, Reading, Mass.
8. Marciniak, Z., and Kuczynski, K. (1967). *Int. J. of Mech. Sci.* **9**, 609–620.
9. Marciniak, Z. (1974). *Aspects of Material Formability*. McMaster University, Hamilton, Ontario, Canada.
10. Beaver, P.W. (1982–1983). *J. Mech. Work. Tech.* **7**, 215–231.
11. Riggs, B.A. (1973). *Sheet Metal Ind* **50**, 620–625.
12. Dillamore, I.L. (1978). *Proc. 10th Int. Congr. of I.D.D.R.G.* Portcullis Press, Redhill, pp. 347–354.
13. Hosford, W.F. (1978). In *Formability: Analysis, Modelling and Experimentation*, S.S. Hecker, A.K. Ghosh, and H.L. Gegel, eds. The Metallurgical Society, Warrandale, Pa. 78–95.
14. Blade, J.C., Bryant, A.J., and Thomas, A.T. (1976). *Metals Tech.* **3**, 380–386.
15. Szacinski, A.M., and Thomson, P.F. (1984). *J. Mech. Work. Tech.* **10**, 87–102.
16. Brindley, B.J., and Worthington, P.J. (1970). *Met. Reviews* **4**, No. 145, 101–111.
17. Althoff (1976) *Aluminium* **52**, 632–635.
18. Hatherly, M. (1982). *Proc. 6th Int. Conf. on Str. of Metals and Alloys,* Gifkins, R.C. ed. Pergamon, Oxford. 1181–1195.
19. Malin, A.S., and Hatherly, M. (1979). *Metal Sci.* **13**, 463–472.
20. Brown, K. (1972). *J. Inst. Metals* **100**, 341–345.
21. Dillamore, I.L., Roberts, J.G., and Bush, A.C. (1979). *Metal Sci.* **13**, 73–77.
22. Al-Haidary, J.T., Petch N.J., and De Los Rios, E.R. (1983). *Phil Mag A* **47**, 869–890.
23. Kuhlmann-Wilsdorf, D. (1985). *Met. Trans.* **16A**, 2091–2108.
24. Ayres, R.A., and Wenner, M.L. (1979). *Met. Trans.* **10A**, 41–46.
25. Martin, J.W. (1980). *Micromechanisms in Particle Hardened Alloys.* Cambridge University Press, Cambridge, England.
26. Westengen, H. (1982). *Proc. 6th Intl. Conf. on the Str. of Metals and Alloys*, Gifkins, R.C. ed. Pergamon, Oxford. 461–466.
27. Lutjering, G., and Weissman, S. (1970). *Acta Metall.* **18**, 785–795.
28. Prince, K.C., and Martin, J.W. (1979). *Acta Metal.* **27**, 1401–1406.
29. Beaver, P.W., and Parker, B.A. (1982). *Proc. 12th. Biennial Congr. I.D.D.R.G., Associazione Italiano di Metallurgra*, Milan, Italy pp. 85–96.
30. Ghosh, A.K. (1977). *J. Eng. Mats. and Tech. (Trans ASME, H)*, **100**, 264–274.
31. Parker, B.A. (1979). *Proc. 5th Int. Conf. on the Str. of Metals & Alloys*, Hassen, P., Gerold, V., and Kostorz, G. eds. Pergamon, Oxford.

32. Da Costa Viana, C.S., Davies, G.J., and Kallend, J.S. (1978). In *Textures of Materials*, G. Gottstein and K. Lücke, eds., vol. II, Springer-Verlag, Berlin. 447–453.
33. Beaver, P.W., and Parker, B.A. (1986). *Mater. Sci. and Eng.* **82**, 217–224.
34. Wagoner, R.H. (1981). *Met. Trans. A.*, **12A**, 235–241.

Additional References

General

Backhofen, W.A. (1972) *Deformation Processing*. Addison-Wesley, Reading, Mass.

Hecker, S.S., Ghosh, A.K., and Gegel, H.L., eds. (1978). *Formability: Analysis, Modelling and Experimentation* A.I.M.E., New York.

Schmieder, A.K., and Newby, J.R., eds. (1978). *Formability Topics: Metallic Materials*. A.S.T.M. STP 647.

Sachdev, A.K., and Embury, J.D., eds. (1987). *Formability and Metallurgical Structure* The Metallurgical Society, Warrandale Pa.

Aluminium Alloys

Embury, J.D. (1978). "Some Aspects of the Formability of Aluminium Alloys" in *Aluminium Transformation Technology and Applications*, Pampillo, C.A. Biloni, H. and Embury, J.D. eds. American Society for Metals, Metals Park, Ohio.

Embury, J.D., Iricibar, R., Morris, L.R., and Duncan, J.L. (1981) "Formability of Aluminium Sheet Alloys" in *Aluminium Transformation Technology and Applications. 1981*, Pampillo, C.A. Biloni, H. Mondolfo, L.F. and Sacchi, F. eds. American Society for Metals, Metals Park, Ohio.

21

Texture

WILLIAM G. FRICKE, JR.

Alloy Technology Division
Alcoa Laboratories
Alcoa Center, Pennsylvania

MAREK A. PRZYSTUPA

Consultant
Redondo Beach, California

I. Introduction

Because single crystals have different properties along the various crystallographic directions, polycrystalline materials are anisotropic if the crystals are not randomly distributed. Experience teaches us that the distribution of grain orientations seldom approaches a random distribution because any deformation process tends to align grains, and new grains produced by recrystallization also tend to align them.

Real materials are, therefore, inherently anisotropic. Mechanical properties vary somewhat with angle to the rolling direction; drawn products, for example, deform more easily in certain directions. The differences are not always significant, but the phenomenon is important enough for the metallurgist to consider. In this regard, it should be remembered that not all anisotropy is due to preferred orientation. Manufacturing processes can also align intermetallics and voids, elongate grains, and cause preferential precipitation, all of which can produce differences in properties in different directions.

ISBN 0-12-341831-3

II. Orientation of Crystals

The basic evidence for preferred crystallographic orientation is in the pole figure measured by X-ray diffraction techniques. Poles are defined as imaginary lines drawn perpendicular to particular selected crystallographic planes. The pole figure is a drawing of how the planes in a particular material are facing.

Figure 1 shows two pole figures for a typical hot-rolled aluminum sheet. The view is looking down on the sheet with the rolling direction running vertically. Cold-rolled sheet and the interior of extrusions with a large width-to-thickness ratio—in fact, any material subject to plane strain conditions—would have a similar texture. These pole figures are of the same material; they present the same information in different ways. When the pole figure is made with diffraction off of (111) planes (the left side of Fig. 1), the data characteristically cluster in a lip-shaped area running along the transverse direction (TD). There are also clusters of poles slightly off the rolling direction (RD). When the pole figure is made with 200 diffraction (the right half of Fig. 1), two stirrup-shaped areas are characteristically seen.

For many purposes, a pole figure is enough. The type of texture is suggested by the shapes of the contour lines and its strength is measured by the maximum intensity. When one wants to know what particular orienta-

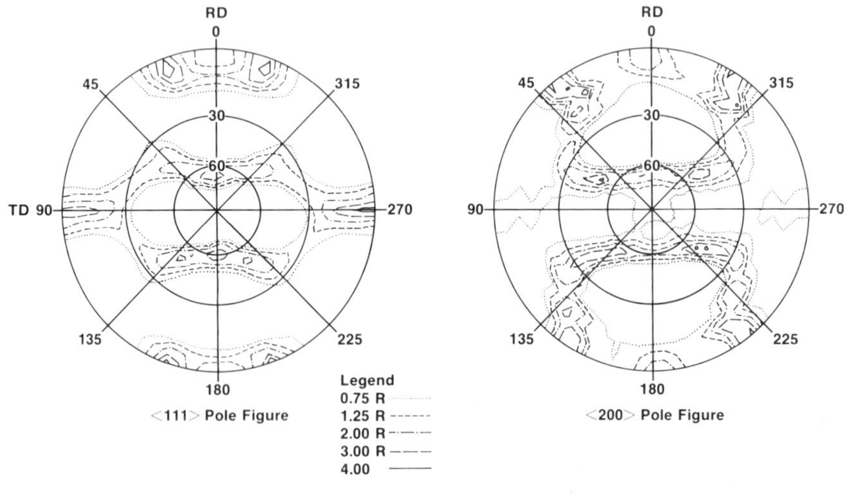

Hot Rolled Aluminum Alloy

FIG. 1 Typical pole figures of hot rolled aluminum.

tions are present, the task becomes more difficult. Each grain in the sample contains four ⟨111⟩ poles and three ⟨200⟩ poles, and the pole figure does not tell which set of poles belong to a particular grain. When more than one orientation is present, as is usually the case, mental interpretation is very difficult.

Fortunately, there exists a technique for removing the ambiguity. Crystallite orientation distribution function (ODF) analysis allows the separation of a texture into its individual components [*1*,*2*]. This is possible because, although the two pole figures of Fig. 1 represent the same data, the overlap of poles from different orientations occurs in different places. The process is analogous to taking two dimensional photographs of a subject from two different angles and reconstructing a third dimension from them.

The ODF result is an equation describing the relative amount of material that has each of the possible orientations, where orientation now means what crystallographic plane is parallel to the surface of the sheet and what crystallographic direction in that plane is pointed in the rolling direction. Three variables are needed to describe the angles between the crystallographic axes and the specimen axes. Figure 2, which is a graph of the ODF equation for the material of Fig. 1, represents three-dimensional Euler angle space. For convenience, slices have been made through this solid at selected values of one of the angles, phi, in order to show data as a function of the other two angles: psi and theta. The clusters of data representing the rolling texture have been numbered 1 through 6 for identification, but it should be remembered that these are part of a continuous distribution.

The question now arises, how do you interpret these results? There exist charts that can be laid on the ODF sections to give those locations in Euler space where particular plane-direction combinations of orientation should appear [*3*]. These charts necessarily represent orientations with simple Miller indices. Equations also exist for translating Euler angles into Miller indices. The trouble is that nature itself does not particularly prefer simple orientations, so one ends up saying the data are "close to" certain ideal orientations and various investigators have selected different indices.

One way to represent the ODF results is to use the definitions of the Euler angles to plot the relations between the crystallographic and specimen axes. Figure 3 shows how the normal direction of the sheet and the rolling direction lie in relation to the crystallographic reference frame for each of the six orientations of the ODF result.

The normal direction (ND), which represents what crystallographic planes tend to be parallel to the surface of the sheet, forms a series along a line from ($\bar{1}01$) to somewhere between ($\bar{1}12$) and ($\bar{1}13$). Meanwhile, the rolling direction favored varies from [$1\bar{2}1$] to near [$1\bar{1}1$].

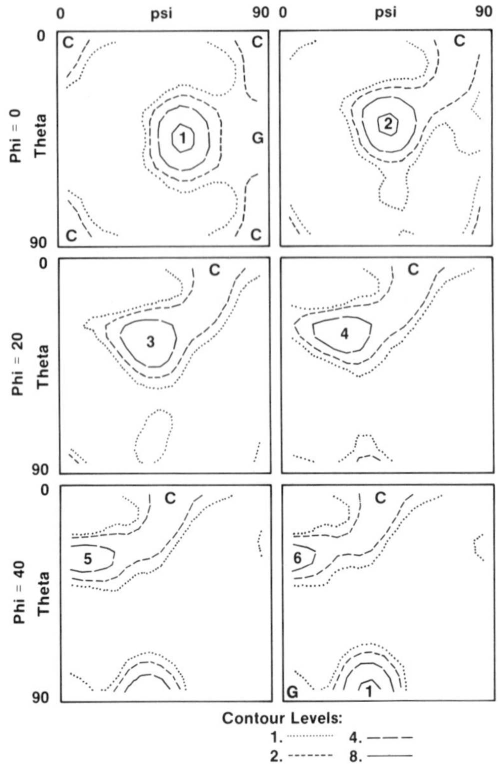

Hot Rolled Aluminum Alloy

FIG. 2 Graphic representation of the orientation distribution function for the same material. Components of the rolling texture are numbered 1 through 6. Remnants of the recrystallization texture are labeled C for cube (100) [001] and G for Goss (110) [001].

Now that the relations between the crystallographic axes and specimen axes have been established for a number of frequently occurring grains, it is instructive to go back and plot where the poles for these grains would appear on the pole figures. This is done in Fig. 4 for the four ⟨111⟩ poles of each orientation. In other words, these are the places where the four triangles of Fig. 3 end up when the ND of each orientation is rotated to the center of the drawing and the RD is placed on top.

In the actual sample, there are grains that have equivalent orientations to the ones plotted, grains in which the Miller indices are interchanged or have a different sign. The locations in Fig. 4 therefore have equivalent positions on the opposite side of the vertical and horizontal lines.

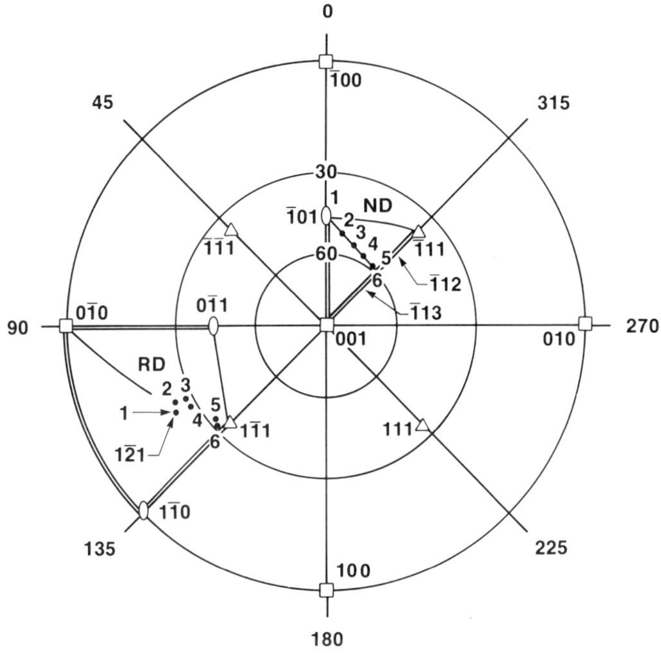

Hot Rolled Aluminum Alloy, Inverse Pole Figure

FIG. 3 Location of the orientations 1 through 6 in crystallographic space. Planes parallel to the sheet surface are indicated by plotting how the sheet normal (ND), and the rolling direction (RD) fall in relation to the crystal lattice.

When the points have been replotted into one quadrant, they form clusters predicting where data would fall in a pole figure for the orientation series 1 through 6. This is done for both ⟨111⟩ and ⟨200⟩ poles in Fig. 5. In other words, when these locations, derived from the ODF interpretation of the specimen, are compared to the original pole figures (Fig. 1), it can be seen which orientations contribute most to particular pole figures. In fact, since the textures in specimens tend to be similar, it is often not necessary to do the actual ODF to compare strengths of components of texture. For example, the pole figures show that this particular sample has more of components 1, 2, and 3, and less of 5 and 6.

There are some data in the pole figures, however, particularly the ⟨200⟩ one, that are not accounted for by orientations 1 through 6. These are present in the ODF plot of Fig. 2 in locations C and G and represent a small amount of cube and Goss orientations not completely destroyed by the rolling. These minor components are evident in the ODF results, but

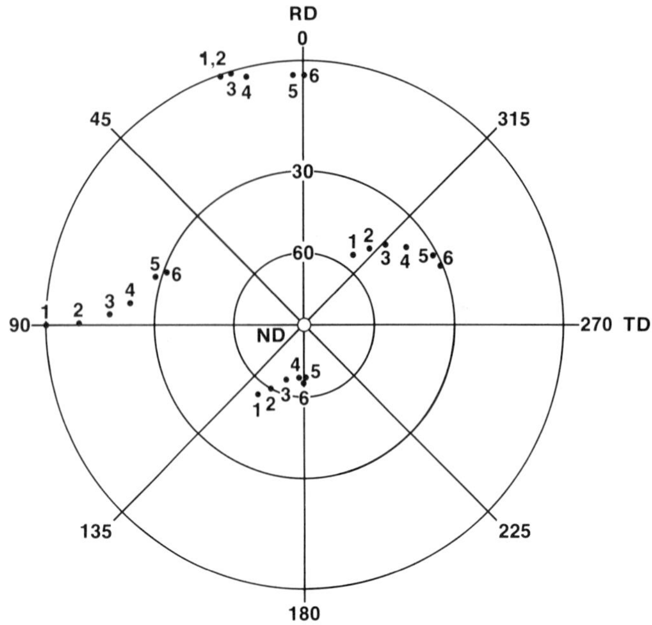

Hot Rolled Aluminum Alloy
<111> Poles for Grains of Selected Orientations

FIG. 4 Location of the 111 poles for the six orientations of Fig. 3 after the normal direction and rolling direction have been rotated to the positions of the pole figure.

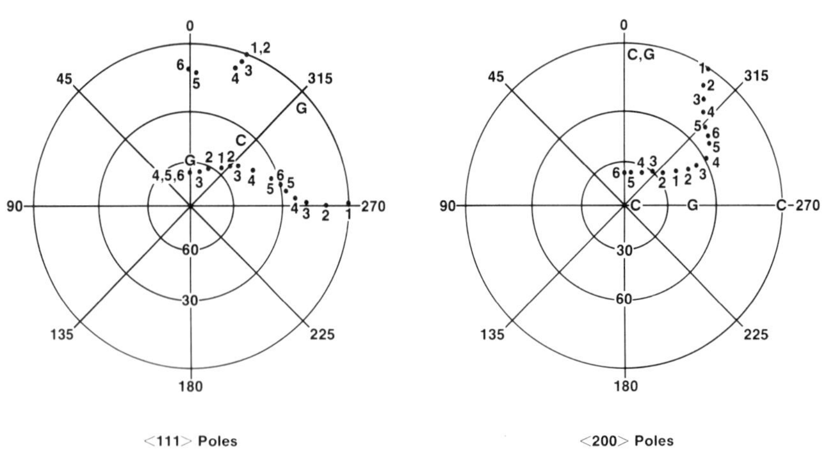

FIG. 5 Locations of poles for the six orientations after accounting for equivalent grains. Compare these locations with the original pole figures of Fig. 1.

tend to be obscured in the pole figures themselves. They should appear in the C and G locations of Fig. 5. In the actual ⟨111⟩ pole figure (Fig. 1), the cube component appears only as slight "wings" tending to draw the data out along the 45° diagonals. It is more evident in the ⟨200⟩ pole figure. The Goss component is weak enough to be lost in the pole figures.

Certain of the texture components have earned names. These names imply not only the orientation of the unit cell in relation to the rolling plane, but how that cell is rotated with respect to the rolling direction. The cube and Goss components occur in recrystallization textures. They can be visualized in terms of how the cubic unit cell lies in relation to the sheet or how a tetrahedron with (111) surfaces lies in that unit cell (Fig. 6). The tetrahedron is important because its surfaces are slip planes and its edges are slip directions in aluminum. Further plastic deformation, whether during fabrication or mechanical testing, would occur on these planes in these directions.

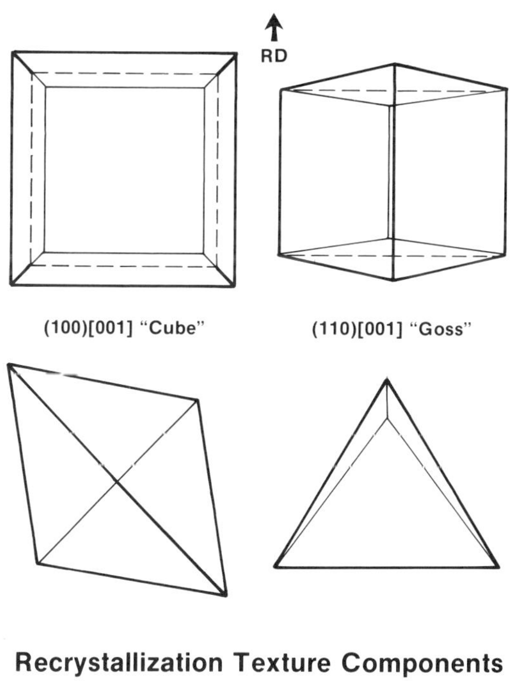

FIG. 6 Perspective drawings of the cubic unit cell and the Thompson tetrahedron for two recrystallization components. The view is looking down on the sheet surface with the rolling direction vertical. The dashed line is the trace of a plane through the center of the cube and parallel to the surface. Edges of the cube above the surface are accentuated.

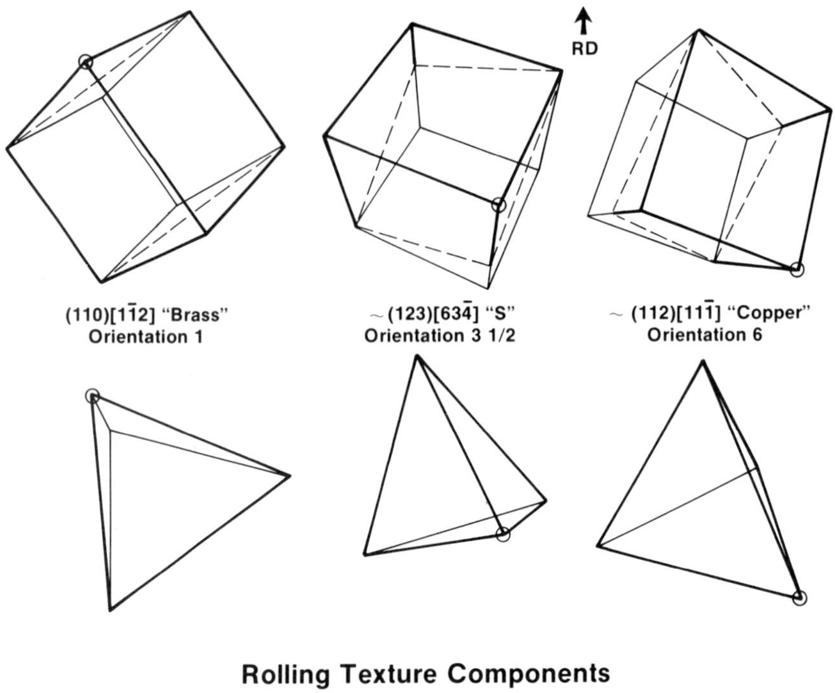

FIG. 7 Perspective drawing of the cubic unit cell and Thompson tetrahedron for three rolling texture components. The components can be imagined as a continuous series of orientations where the unit cell rotates between the positions at the left and right. One corner of the cell is marked with a circle.

As the ODF plot and pole figures suggest, the rolling texture contains a continuous series of orientations and it is arbitrary whether one considers this a large number of true orientations or relatively few orientations with imperfection in achieving the ideals. In thinking, one often picks the end points where the series begins repeating itself, along with an intermediate point (Fig. 7). The intermediate point, sometimes called S, has been described in a variety of Miller indices.

III. Average Values of Physical Properties

A full description of texture in a polycrystalline material can be given either in terms of fractions of the ideal texture components or by an ODF equation. Both descriptions are equivalent, and both can serve as a starting point in an evaluation of the average properties. For example, for texture described in terms of the volume fractions of the idealized components, each of the components can be represented as a single crystal grain of an appropriate orientation, and average values of the property of interest can be calculated as a weighted average based on the amount of that component. For texture described by an ODF equation, the procedure is more complex. Since ODF functions are continuous, the directional properties also have to be represented by continuous functions. Any reasonable approximation should serve the purpose, but since in an ODF equation spherical harmonics are used, it is also convenient to use harmonics in approximating single crystal properties. As a result, the averaging operation simplifies to the summation of the products of appropriate coefficients [*1*,*4*]. Both methods are then simple to implement and although they assume pure superposition (e.g., no interactions between grains and no grain shape effects), they have been quite accurate in predicting both elastic and plastic behavior in many materials [*1*,*4*,*5*]. The remainder of this chapter will discuss in more detail some of the methods related to the prediction of plastic anisotropy from crystallographic texture.

A. *Plastic Anisotropy*

Several different models of plastic behavior of FCC polycrystals have been proposed over the years. The Sachs model is historically the oldest [*6*]. It assumes that the real material can be represented by an aggregate of parallelepiped grains of the same height but with different crystallographic orientations, and that the cross-section of each grain is proportional to the volume fraction of the orientation it represents. It further assumes that there is only one active slip system in each grain and that this system is the one with the highest Schmid factor. If such an aggregate is subjected to a tensile deformation, the equal strain condition in the tensile direction is preserved, and the tensile stress component in each grain is related to the critical resolved shear stress through the Schmid factor [*7*,*8*]. The yield stress in the polycrystal can then also be easily evaluated through an appropriate averaging procedure. In spite of the fact that this model is physically appealing, it violates compatibility conditions, as it tacitly assumes that the lateral deformations in all grains are different. It is nevertheless simple to use and has proven useful as a lower bound approximation for yield strengths [*8*,*9*], yield loci [*10*], and, to some extent, also in the prediction of earing [*11*].

The Taylor [*12*] and Bishop-Hill [*13*,*14*] model, often referred to as a full-constraint model, represents an improvement over the Sachs approach. It assumes that all grains deform the same (e.g. compatibility condition is fulfilled, but requires five active slip systems in each grain). In order to find which slip systems operate, Taylor proposed a criterion of the minimum sum of the total shears, while Bishop and Hill utilized an equivalent criterion of the maximum external work. In a case then when all external strains are known, the procedure requires calculation of the

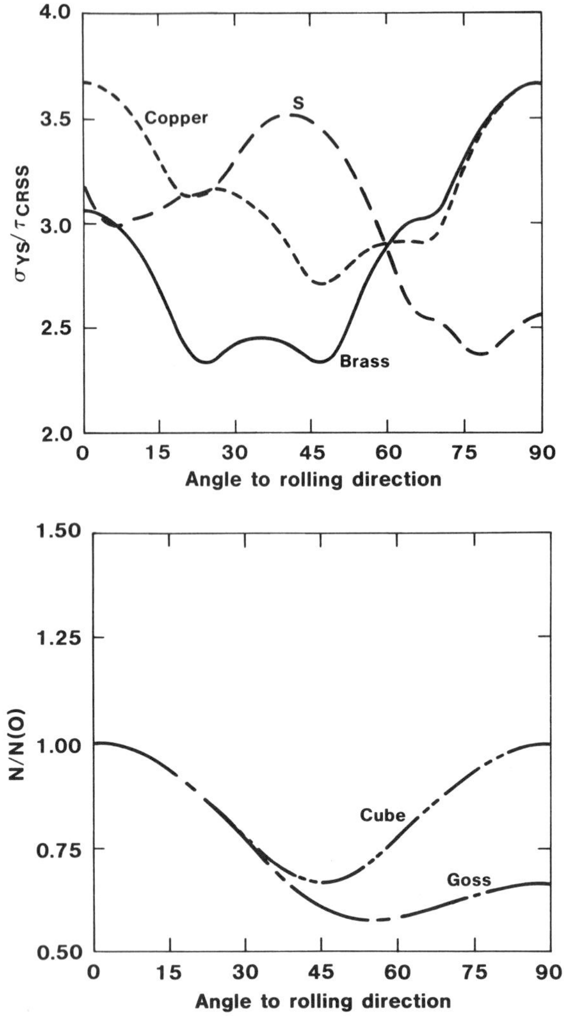

FIG. 8 Variation of the yield strength between rolling and transverse direction for major rolling (a) and recrystallization (b) textures calculated using the Taylor/Bishop-Hill model.

Taylor factors for each of the representative crystal orientations and then calculation of the average Taylor factor [*7,8*]. For the case when not all of the strain components are known (as, for instance, in tensile deformation), it is necessary to use the minimization procedure, resulting in determination of not only average Taylor factors, but also r values [*15*]. The Taylor/Bishop-Hill model is then both more realistic and more versatile than the Sachs approach. It is also quite simple to use as both the values of the Taylor factors [*7,8,16*], and the coefficients of spherical harmonics representation of the Taylor factors [*1,4,17–19*] are readily available.

As an example of the application of the Taylor analysis, a change of the yield strength with the angle to the rolling direction in the plane of the sheet for major aluminum texture components and axisymmetric deformation is shown in Fig. 8. In all instances, the yield strength has been normalized by the critical resolved shear stress of a single crystal; e.g., it is given in terms of Taylor factors. Note that for all rolling texture components, the curves have minima; while for recrystallization components, maxima are observed. In order to further illustrate the effects of rolling and recrystallization on sheet anisotropy, the curves for an arbitrary rolled material with 25% brass, 25% S, 25% copper, and 25% random texture and for another recrystallized material with 33% Goss, 33% cube, and 34% random-texture components are plotted in Fig. 9. The curves suggest

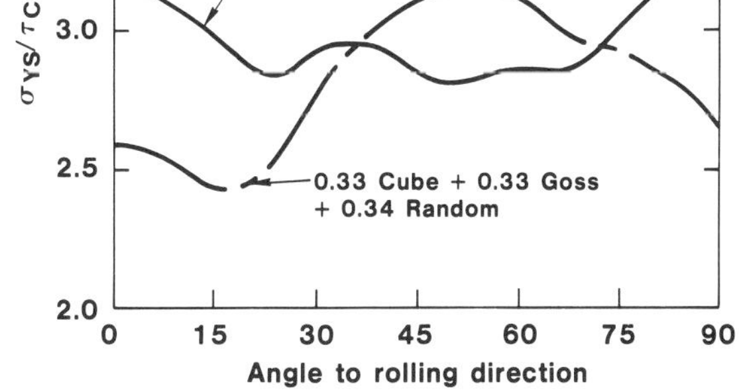

FIG. 9 The change of the yield strength between rolling and transverse direction for hypothetical rolled material with 25% brass, 25% S, 25% copper, and 25% random textures and for recrystallized material with 33% Goss, 33% cube, and 34% random textures calculated using the Taylor/Bishop-Hill model.

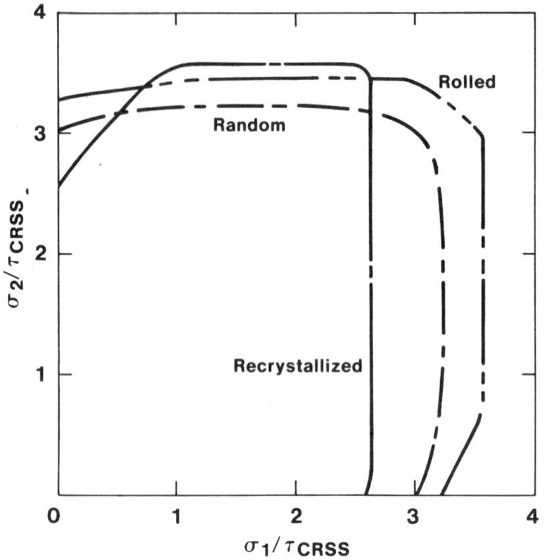

FIG. 10 Yield loci for random texture and for the same hypothetical rolling and recrystallization textures as shown on Fig. 9, calculated according to the Taylor/Bishop-Hill model.

that the combined effect of the rolling is a yield strength minimum in the 30 to 60 degree range, while recrystallization results in a maximum in approximately the same range. This effect is consistent with the experimental observations and is the main reason why it is possible to achieve plastic isotropy through appropriate thermomechanical treatment [*20,21*]. Taylor analysis has also been instrumental in predicting yield loci for known textures [*10,13,14*], and an example of such predictions for random and for arbitrary rolling and recrystallization textures specified above is shown in Fig. 10.

In spite of its power, the Taylor model is not without some drawbacks. The most restrictive one is a nonuniqueness of solution when several combinations of the five slip systems can accomplish the required deformation. Prediction of the texture development is in such a case nonunique, and some kind of averaging and/or arbitrary choice is necessary. Fortunately this ambiguity can be eliminated by imposing a condition of the compatibility of the strain rates between grains, as proposed by Assaro et al. [*22*]. Another problem with the Taylor model is that it requires five slip systems to operate, although in order to support some types of deformations, fewer than five systems are necessary. A relaxed constraints model [*23,24*] proposed in the early 1980s takes that aspect of slip geometry into account. In general, the full (e.g., Taylor) and relaxed constraints models are complementary. The first one works better for equiaxed grains and small strains, while the second one seems to be more appropriate for highly elongated grains and when strains are large [*25*].

As already indicated, the Taylor/Bishop-Hill and Sachs models are also quite effective in predicting both earing and r values. There have been several studies addressing the relationships between texture and earing in aluminum alloys and it seems that both models give very reasonable results [*11,26–29*]. It has been also shown that it is possible to correlate the amount of earing with the values of the fourth-order coefficients in the ODF equations. Unfortunately, the success in applying this correlation to aluminum alloys has been very limited [*1,4,26,30*]. In general, the predictions of the anisotropy ratios (r values) from the plastic behavior models have been quantitatively correct, although predicted vallues are usually larger than the experimental ones [*31,32*]. The methods for calculating r values directly from the ODF equation coefficients have also been proposed [*1,4,33*].

B. Precipitation Effects

A very important question related to the texture in aluminum alloys is how precipitation affects texture-induced anisotropies. It is reasonable to expect that the spherical and/or incoherent precipitates should have no effect, while plate and rod-like precipitates, confined to specific crystallographic planes, should result in some changes in the plastic anisotropy. For aluminum alloys, the most common directional precipitates are those on either {100} or {111} habit planes [*34*]. Hosford and Zeisloft [*35*], Agrawal and Hosford [*36*], and subsequently Bate et al. [*37,38*] studies those effects and showed that {100} precipitates will act to eliminate texture-induced plastic anisotropy while {111}-type precipitates will tend to reinforce it. In order to illustrate the magnitude of these effects, a variation of the yield stress component resulting from {100} plate-like precipitates, calculated using the Hosford and Zeisloft method [*35*], has been plotted as a function of the angle to the rolling direction in Fig. 11. The calculations have been performed for the same textures as those shown in Fig. 8. Since in all instances the corresponding curves have opposite trends, the {100} habit plane precipitates will have a tendency to suppress the texture effect. Experimental observations for Al-Cu alloys conform these predictions [*35,37*]. They show that the magnitude of the change depends on the aspect ratio and volume fraction of the precipitates and that in the peak age conditions {100} habit plane precipitates can almost totally eliminate texture effect [*37*].

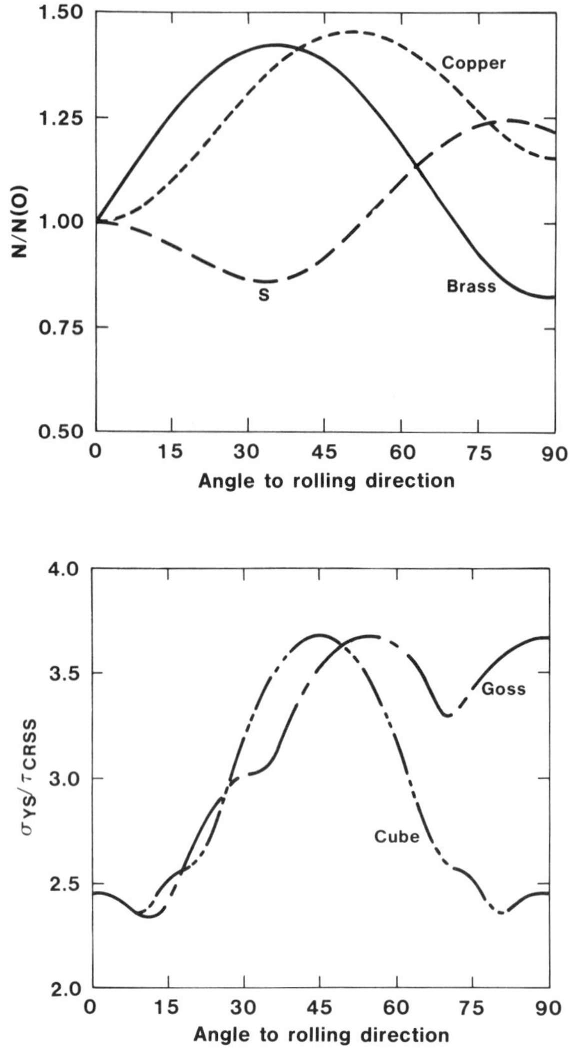

FIG. 11 Variation of the yield strength component index for {100} plate-like precipitates in (a) major rolling and (b) recrystallization textures calculated by the Hosford and Zeisloft method [35]. All values are normalized by the values of the yield strength component index in the rolling direction.

IV. Summary

There are currently several excellent ways of characterizing texture in the aluminum alloys. They are very accurate and allow for the determination of the effects of the compositional and processing variables on the development of the individual texture components. Texture in aluminum alloys can be described either by an ODF equation or by the amount of cube, Goss, brass, S and copper texture components. At the same time, a number of reliable models allowing for the prediction of the mechanical as well as physical properties from the texture data are currently available. The models predict that for the typical aluminum rolling texture components, a minimum in the yield strength, between rolling and transverse directions, should be observed, while for recrystallization components, a maximum in the same region is expected. It is also possible to predict precipitation effect, and the directional precipitates on {100} habit planes are expected to weaken texture-induced plastic anisotropy, while {111} habit plane precipitates tend to reinforce it. The mechanical behavior models are quite versatile and ultimately should be instrumental in predicting both the consequences of the different processing routes on texture evolution, and the interactions between texture and directional precipitations during processing.

References

1. Bunge, H.J. (1982). In *Quantitative Texture Analysis*, H.J. Bunge and C. Esling, eds. Deutsche Gesellschaft fur Metallkunde, Adenauerallee 21, 6370 Oberursel 1, pp. 383–406.
2. Roe, R.J. (1966). *J. Applied Phys.*, **37**, 2069–2072.
3. Davies, G.J., Goodwill, D.J., and Kallend, J.S. (1971). *J. Appl. Cryst.*, **4**, 67–70.
4. Bunge, H.J. (1982). *Texture Analysis in Material Science: Mathematical Methods*. Butterworth, London.
5. Davies, G.J., Goodwill, D.J., and Kallend, J.S. (1972). *Met. Trans.*, **3**, 1627–1631.
6. Sachs, G. (1928). *Z. Verein Deut. Ing.*, **72**, 734.
7. Reid, C.N. (1973). *Deformation Geometry for Materials Scientists*. Pergamon Press, New York.
8. Kocks, U.F. (1970). *Met. Trans.*, **1**, 1121–1143.
9. Kallend, J.S., and Davies, G.J. (1971). *J. Inst. Metals*, **99**, 257–313.
10. Viana, C.S. da C., Kallend, J.S., and Davies, G.J. (1979). *Int. J. Mech. Sci.*, **21**, 355–371.
11. Tucker, G.E.G. (1961). *Acta Met.*, **9**, 275–285.
12. Taylor, G.I. (1983). *J. Inst. Metals*, **62**, 307–324.
13. Bishop, J.F.W., and Hill, R. (1951). *Phil. Mag.*, **42**, 414–427.
14. Bishop, J.F.W., and Hill, R. (1951). *Phil. Mag.*, **42**, 1298–1307.

15. Hosford, W.F., Jr., and Backofen, W.A. (1964). In *Fundamentals of Deformation Processing*, (W.A. Backofen, eds.) Syracuse, University Press, pp. 259–292.
16. Chin, G.Y., and Mammel, W.L. (1967). *Met. Trans.*, **239**, 1400–1405.
17. Bunge, H.J., Grzesik, D., Ahrndt, G., and Schulze, M. (1981). *Arch. Eisenhuttenwes.*, **52**, 407–411.
18. Bunge, H.J. (1985). In *Texture in Non-Ferrous Metals and Alloys*, H.D. Marchant and J.G. Morris, eds. Metallurgical Society of AIME, Warrendale, Pennsylvania, pp. 145–171.
19. Royer, F., Mohr, J.M., and Travard, C. (1975). *C. R. Acad. Sc. Paris*, **281**, 555–558.
20. Hatch, J.E. (1984). *Aluminum*. American Society for Metals, Metals Park, Ohio.
21. Wilson, D.V. (1969). *Met. Rev.*, **14**, 175–188.
22. Assaro, R.J., and Needleman, A. (1985). *Acta Met.*, **33**, 923–962.
23. Honneff, H., and Macking, H. (1978). In *Texture of Materials*, G. Gottstein and K. Lucke, eds., **vol. I**. Springer-Verlag, Berlin, pp. 265–275.
24. Kocks, U.F., and Canova, G.R. (1981. In *Deformation of Polycrystals: Mechanisms and Microstructures*, N. Hansen, A. Horsewell, T. Leffers, and H. Lilholt, eds. Riso National Laboratory, Roskilde, Denmark, pp. 35–44.
25. Canova, G.R., Kocks, U.F., and Jonas, J.J. (1984). *Acta Met.*, **32**, 211–226.
26. Welch, P.I., Kallend, J.S., and Davies, G.J. (1978). In *Texture of Materials*, G. Gottstein and K. Lucke, eds., **vol. II**. Springer-Verlag, Berlin, pp. 355–369.
27. Pochettino, A., Pernot, M., and Penelle, R. (1981). In *Deformation of Polycrystals: Mechanisms and Microstructures*, N. Hansen, A. Horsewell, T. Leffers, and H. Lilholt, eds. Riso National Laboratory, Roskilde, Denmark, pp. 205–209.
28. Sowerby, R. (1985). In *Texture in Non-Ferrous Metals and Alloys*, H.D. Marchant and J.G. Morris, eds. Metallurgical Society of AIME, Warrendale, Pennsylvania, pp. 99–115.
29. Rodrigues, P.M.B., and Bate, P.S. (1985). In *Texture in Non-Ferrous Metals and Alloys*, H.D. Marchant and J.G. Morris, eds. Metallurgical Society of AIME, Warrendale, Pennsylvania, pp. 173–187.
30. Hirsch, J., Musick, R., and Lucke, K. (1978). In *Texture of Materials*, G. Gottstein and K. Lucke, eds., **vol. II**. Springer-Verlag, Berlin, pp. 437–446.
31. Welch, P.I. (1982). *Z. Metallkde*, **73**, 381–386.
32. Lequeu, P., Montheillet, F., and Jonas, J.J. (1985). In *Texture in Non-Ferrous Metals and Alloys*, H.D. Marchant and J.G. Morris, eds. Metallurgical Society of AIME, Warrendale, Pennsylvania, pp. 189–212.
33. Royer, F., Mohr, J.M., Travard, C., and Parniere, P. (1977. *Mem. Sc. Rev. Metallurgie*, **4**, 449–455.
34. Lorimor, G.W. (1977). In *Precipitation Process in Solids*, (K.C. Russell and H.I. Aaronson, eds.) Metallurgical Society of AIME, Warrendale, Pennsylvania, pp. 87–119.
35. Hosford, W.E., and Zeisloft, R.H. (1972). *Met. Trans.*, **3**, 113–121.
36. Agrawal, S.P., and Hosford, W.F. (1976). *Met. Trans.*, **7A**, 1857–1871.
37. Bate, P. Roberts, W.T., and Wilson, D.V. (1981). *Acta Met.*, **29**, 1797–1814.
38. Bate, P., Roberts, W.T., and Wilson, D.V. (1982). *Acta Met.*, **30**, 725–737.

22

Strengthening Mechanisms in Aluminum Alloys

J.D. EMBURY

McMaster University
Hamilton, Ontario, Canada

D.J. LLOYD

Alcan International Limited
Kingston Research & Development Centre
Kingston, Ontario, Canada

and

T.R. RAMACHANDRAN

Indian Institute of Technology
Kanpur, India

I. Introduction

In attempting to review the strengthening mechanisms in aluminum alloys, it is important to recognize that an important technological transition is in progress regarding structure-property relationships. Traditionally, alloy technology has relied on the control of the grain size and morphology by the action of dispersoids, and strengthening was achieved

ISBN 0-12-341831-3

by taking advantage of reduced solid solubility at lower temperatures. However, a variety of new processes (such as rapid solidification, mechanical alloying, and vapor deposition [1–3] have been developed which extend the kinetic range of processing and substantially decrease the intrinsic scale of the microstructure. As a result, the degree of microstructural flexibility and resulting properties have been greatly increased.

Further, there is an increasing recognition that the final use of the material will dictate the design and processing method for the material. This has an important impact for strengthening mechanisms from several points of view. Large-scale fabrication processes, such as rolling and extrusion, can result in degradation of the microstructure and the final properties are controlled by the thermal and mechanical stability of the initial microstructure. In composite materials, the metal matrix is a vehicle for carrying the load-bearing reinforcement. The level of mechanical properties arises from controlling the scale and dispersion of the reinforcement, together with the strength of the matrix-reinforcement interface, rather than the interaction of dislocations with obstacles in the matrix. Also, the selection of materials for applications involving forming and fabrication is dictated not by strength level, but by a more complex combination of properties involving the ability to distribute strain and to avoid local damage processes. Thus, it must be recognized that the understanding of strengthening mechanisms is only one part of the mosaic of knowledge needed to optimize the processing-structure–properties relationship.

II. Structural Aspects of Aluminum Alloys

Prior to discussing the basic methods of strengthening, it is of value to summarize the scale of microstructural constituents and the basic mechanisms for producing second-phase particles in commercial alloys.

During solidification, coarse intermetallic compounds are formed. These are retained, to some extent, through homogenization and fabrication to the final product. They can have a significant influence on recrystallization and fracture. These compounds, or inclusions, typically contain Al, Fe, Mn, and Si. Westengen [4] has detected as many as nine binary (Al-Fe) and ternary (Al-Fe-Si) phases in commercial alloys. In the Cu-, Mg-, Zn-, and Mn-containing alloys, additional phases are formed. Some typical examples, together with their scale, are given in Table I. Coarse intermetallics affect the recrystallization of the alloys. It is now well established that the region surrounding particles develops large misorientations during cold-working, and recrystallization nucleates in these regions [7,8]. As a

TABLE I

THE NATURE AND SCALE OF INTERMETALLIC COMPOUNDS FORMED DURING SOLIDIFICATION

Alloy	Nature of the phase	Volume fraction (%)	Average particle diameter	Center to center spring	Reference
2014-T6	$Q(Al_4CuMgSi_4)Al_{12}(Fe, Mn)_3Si$	2.9	5.8 μm	8.4 μm	5
2024-T851	$Al_2Cu(Mn, Fe)_3CuAl_2Mg_2Si$	4.4	5.6 μm	7.4 μm	5
2124-T851	$Al_{20}Cu_2(Mn, Fe)_3$	1.0	4.5 μm	9.2 μm	5
7075-T7351	$(Fe, Mn, Cu)Al_6Mg_2SiCuAl_2$	2.6	5.7 μm	11.6 μm	5
7079-T651	$(Fe, Mn, Cu)Al_6Mg_2Si$	0.9	4.2 μm	9.4 μm	5
3003	$Al_{12}(Fe, Mn)_3Si$		~1 μm		6

result, coarse particles have a major influence on the grain size of the alloy [*9,10*]. The particles may also provide sites for heterogeneous nucleation of precipitates and thereby lead to the formation of a precipitate-free zone around them. Such zones can develop high strain concentrations resulting in fracture. These large particles may fracture shortly after the onset of plastic flow. Experiments indicate that as many as 25–50% of the constituent particles are cracked after 7% plastic strain in uniaxial tension [*11*].

Commercial alloys also contain a finer distribution of dispersoid particles. Dispersoids are incoherent, nonshearable particles containing Mn, Fe, Cr, and Zr. (In metastable alloys produced by rapid solidification, they may contain other elements such as Co and V.) Dispersoids are produced during any high-temperature soaking and hot-working, and they are usually in the 10–100-nm size range. They are primarily used in retarding recrystallization and grain growth. They also promote slip homogenization and can be valuable additions to alloys tending to deform by planar slip, such as peak aged alloys and Al-Li alloys [*12,13*]. However, dispersoids can also introduce quench-sensitivity by aiding heterogeneous nucleation of phases during the quenching from solution-treatment.

Precipitate particles with a size below 10 nm form during the age-hardening treatments and maximize the strength of the alloys. Precipitation behavior in aluminum alloys has been widely studied. The results are reported in several reviews [*14,17*]. The precipitation sequences in some of the alloys are shown in Table II. The important parameters in terms of alloy strength are the density, size, and homogeneity of the precipitate-particle distribution and the nature of the precipitate–matrix interface. An important point to note is the development of precipitate-free zones (PFZ) at grain boundaries during aging. Localization of strain in these regions can lead to fracture. The origin of the PFZ has been discussed in detail in the

TABLE II

THE SOLID-STATE PRECIPITATION SEQUENCES FOR VARIOUS AL ALLOYS

Al-Cu	Supersaturated solid solution → GP zones → θ'' → θ' → θ (equilibrium) $CuAl_2$
Al-Cu-Mg	Supersaturated solid solution → GP zones → S^1 Al_2CuMg → S (equilibrium) Al_2CuMg
Al-Mg-Si	Supersaturated solid solution → GP zones → β^1 Mg_2Si → β Mg_2Si
Al-Zn-Mg	Supersaturated solid solution → GP zones → η^1 hexagonal $MgZn_2$ → η hexagonal $MgZn_2$
	In alloys with high Mg-Zn ratios, T′ may form instead of η; T′ on further aging is replaced by T, Mg_{32} (Al, Zn)$_{49}$.
Al-Cu-Mg	Supersaturated solid solution → GP zones → S″ → S′ → S (Al_2CuMg)
Al-Li	δ' + supersaturated solid solution → δ' Al_3Li → δAlLi
Al-Li-Cu	δ' + supersaturated solid solution → GP zones → θ'' → θ' → θ → T_1 Al_2CuLi
Al-Li-Mg	δ' + supersaturated solid solution → δ' Al_3Li → Al_2LiMg
Al-Li-Cu-Mg	δ' + supersaturated solid solution → GP zones → S″ → S′ + δ' → S + δ

Note: δ' usually cannot be suppressed during the quench from solution-treatment temperature. Further precipitation of δ' occurs during the GP zone and normal precipitation sequence of the Al-Cu and Al-Cu-Mg alloys. It should also be noted that other metastable T phases can form in the Li-containing alloys.

literature [*18,19*]; multistage heat-treatments [*20*] and trace element additions [*21*] can sometimes be used to minimize the width of the PFZ.

Two other parameters that will influence the alloy strength are the grain size and the scale of the dislocation substructure. Both of these parameters can be strongly affected by the dispersoid-particle distribution.

The size, density, and shape of all particles can be varied by altering the rate of cooling from the melt. In this context, both rapid solidification processing (RSP) and vapor deposition may assume considerable importance [*1,3*]. These processes offer the advantages of refined grain size and increased solid solubility of potential dispersoid-forming elements, which can be subsequently precipitated as fine particles. The concentration of coarse insoluble particles can be minimized and their morphology controlled. As these factors favorably affect the strength of the alloy. Another interesting process development is mechanical alloying [*2*], by which dispersoids of carbides and oxides can be introduced into the matrix. Using this technique, it is possible to obtain strengths in conventional solid-solution 5XXX aluminum alloys comparable to those in 7XXX-series precipitation-hardened alloys. A typical example is the IN-9052 (Al-4% Mg-0.8% O-1.1% C), which has an attractive combination of tensile strength, fatigue strength, toughness, and corrosion-resistance.

III. Strengthening Mechanisms

For clarity, specific strengthening mechanisms are considered together with examples of their commercial utilization and future technological potential.

A. *Grain-Size Effects*

Aluminum and its alloys exhibit strengthening due to grain-size refinement according to the well-known Hall-Petch [*22*] equation:

$$\sigma_F = \sigma_1 + kd^{-1/2} \tag{1}$$

where σ_F is the flow stress, σ_1 is a frictional stress, d is the average grain diameter, and k is a constant that characterizes the difficulty of transmitting slip across the grain boundary for a given alloy system. Early work indicated that grain-size strengthening was not very strong in aluminum because the value of k was some five times smaller than in ferrous alloys [*23,24*]. However, although k is small, the effects of grain sizes can become important at very fine grain sizes. By utilizing particle-pinning in high volume fraction binary and quasi-binary eutectic alloys, Morris [*25*] has shown that grain sizes approaching one micron can be achieved. An example is the eutectic Al-6% Ni system. Lloyd [*26*] has shown that the Hall-Petch relationship remains operative at these fine grain sizes to give considerable strengthening. Even finer-scale structures, 100–500 nm in diameter, can be achieved by techniques such as sputtering and vapor deposition. Using this technique, Al-Al_2O_3 laminates have been made by Alpas [*58*]. The TEM micrograph in Fig. 1 shows the structure of the laminate in cross section. The oxide layers are thin, linear features delineating the aluminum layers. The influence of the interlamellar spacing in the strength is shown in Fig. 2 and it obeys the Hall-Petch relationship. The slope of the curve, $k = 0.092$ Mpa$\sqrt{m}$, is intermediate between $k = 0.065$ for Al and $k = 0.12$ for find-grained Al-6% Ni [*26*].

It is of interest to note that in other systems, based on Fe [*27*], Mo [*28*], and Cu-Nb [*29*], cold-drawing can be used to refine the scale of two-phase

FIG. 1 Microstructure of Al/Al_2O_3 laminate produced by vapor deposition.

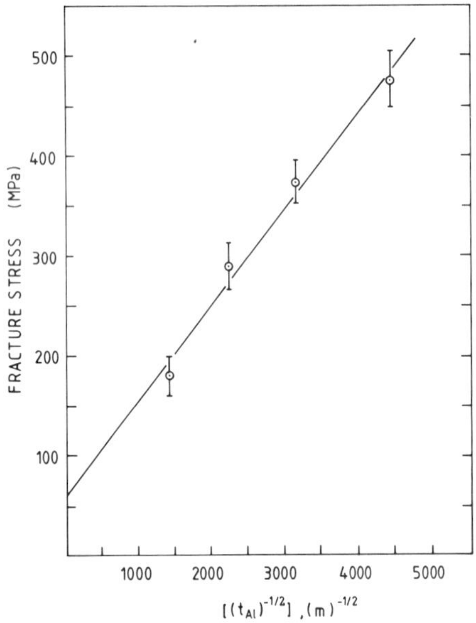

FIG. 2 Dependence of the fracture stress on the lamellar spacing.

systems and to give strengths within a factor of five of the theoretical strength of the matrix. Since the refinement is produced by drawing, the scale of the microstructure is dependent on the initial interlamellar spacing (λ_0) and the drawing strain (n), and the Hall-Petch relationship is expected to be of the form [27,29]:

$$\sigma = \sigma_i + \frac{\mathrm{k}}{\sqrt{\lambda_0}} \exp.\left(\frac{n}{4}\right). \tag{2}$$

As of the late 1980s, this has not been explored in aluminum-based systems. However, the mechanism is attractive because by combining a fine initial starting structure (obtained from rapid solidification or sputtering) with extensive wire or strip drawing, a variety of high-strength products might be produced.

B. Solid-Solution Strengthening

Solid-solution strengthening in aluminum alloys can be considered using the approach developed by Fleischer for copper alloys [30]. Strengthening from solute atoms arises either from differences in atomic size between the solute and solvent atoms or from differences in elastic modulus.

The modulus effects can be analyzed by considering the variation in flow stress with concentration, $d\sigma_F/dc$, with a modulus factor

$$\frac{1}{G}\left(\frac{dG}{dc}\right)\left[1 + \frac{1}{2G}\left(\frac{dG}{dc}\right)\right] \tag{3}$$

where G is the shear modulus and c is the solute concentration.

The size effects can be considered in terms of Ω_A and Ω_S^*, the atomic volumes of aluminum and the solute atom, and expressed as

$$\frac{\Omega_s^* - \Omega_A}{\Omega_A} \times 100\% . \tag{4}$$

The work of Noble et al. [31] is illustrated in Fig. 3 and indicates that the size effect is the dominant source of solute-strengthening in aluminum. The large solute effects in systems such as Al-Cu make the analysis of strengthening due to other factors (such as precipitation-hardening) difficult to analyze unambiguously.

Solutes also introduce other effects such as dynamic strain aging with its resulting serrated flow in Al-Mg alloys. Detailed descriptions of solute-dislocation reactions of this type are beyond the scope of the present chapter.

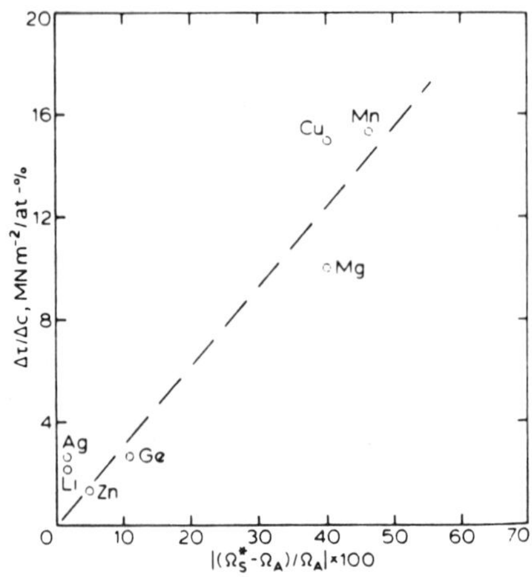

FIG. 3 Strengthening due to atomic size effects (Noble et al. [31]).

One important role of solutes concerns not the initial yield stress, but the modification of the work-hardening behavior at large strains [32]. Thus, solutes such as magnesium are able to increase the flow stress achieved after large plastic strains far above that of pure aluminum, as shown in Fig. 4. This behavior can be analyzed following the seminal work of Kocks and Mecking, by considering the variation in work-hardening rate $\Theta = d\sigma/d\epsilon$ with the current flow stress, σ. An example taken from the work of Rollet et al. [33] is shown in Fig. 5. This illustrates two very important points of high strain deformation in alloys: the presence of magnesium solute clearly inhibits dynamic recovery and gives rise to a much slower decrease in the hardening rate than in pure Al., while the Al-Cu system containing hard particles actually shows a slight increase in the hardening rate at high stresses. This increase has been attributed to geometric hardening due to constrained flow between closely spaced particles, analogous to the laminate situation discussed previously.

C. *Strengthening by Second-Phase Particles*

The treatment of strengthening by second-phase particles is usually considered in terms of the interaction of a dislocation with an array of point obstacles [34–36]. This provides a useful description of the different ways second-phase particles can influence dislocation motion. However, it does not include some important aspects of strengthening that may arise due to the compatibility of the second-phase particle and the matrix during the

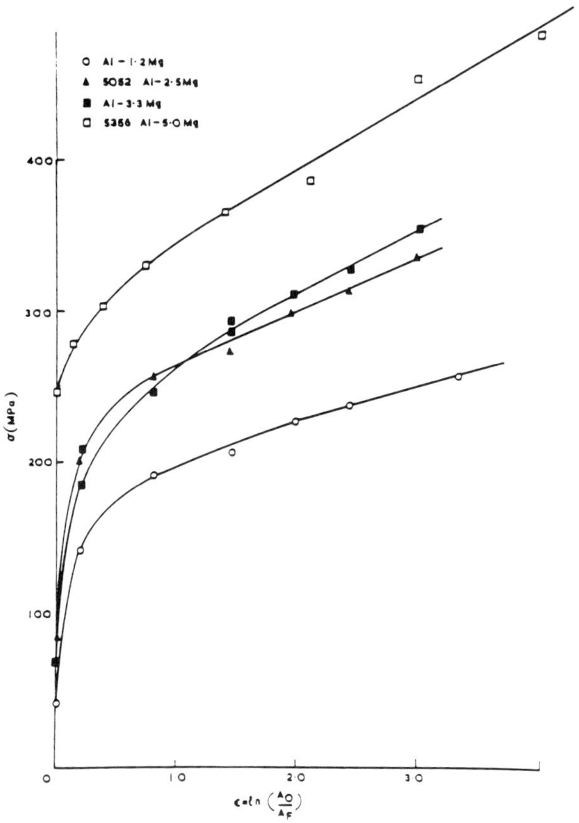

FIG. 4 Influence of Mg content on the stress–strain behavior.

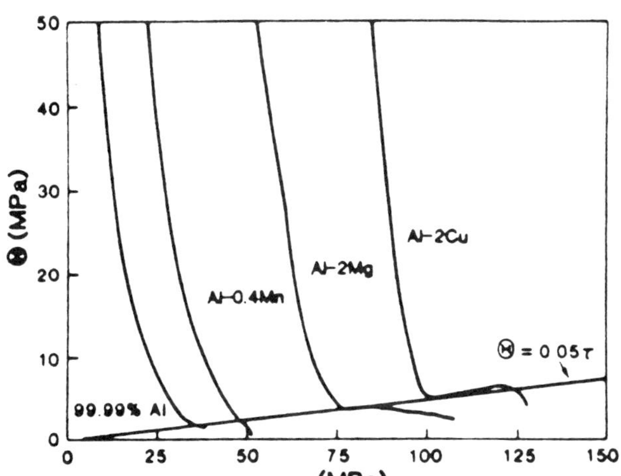

FIG. 5 Work-hardening rate versus resolved shear stress (Rollett et al. [33]).

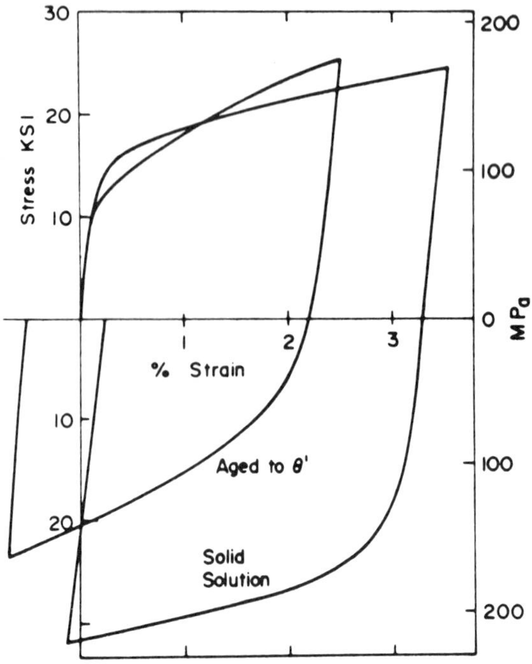

FIG. 6 Bauschinger effect in Al-Cu alloy in the solution-treated and solution-treated and aged conditions.

plastic flow. If the particle supports elastic stresses produced by the plastic flow of the matrix, during reversed flow these stresses can give rise to a Bauschinger effect (i.e., a marked asymmetry in mechanical response when the direction of stressing is reversed). Bauschinger effects have been reviewed [*37*,*38*] and an example is shown in Fig. 6 for Al-Cu in the solution-treated and solution-treated and aged conditions. The precipitation-hardened alloy has a much larger Bauschinger effect, and the interaction between particles and dislocations will be strain-path–dependent. Therefore, treatments based on monotonic flow and interaction with point obstacles may neglect such contributions to the strength.

Further, it must be realized that in commercial alloys, many strengthening mechanisms are operating simultaneously. Weak obstacles, such as solute atoms and dislocations, can be considered as linearly additive components of the flow stress, as can grain–boundary and sub–grain-boundary strengthening. However, for multicomponent precipitation systems, more complex addition laws may be more appropriate Koppenaal and Kuhlmann-Wilsdorf [*59*] have suggested a sum of squares or Pythagorean addition rule:

$$\tau_c^2 = \tau_1^2 + \tau_2^2 \tag{5}$$

where τ_1 and τ_2 are strengthening contributions from obstacles 1 and 2, respectively.

IV. Models of Initial Yield Stress

The initial yield stress of the material can be modelled, in an idealized way, by considering the particles as point obstacles and calculating the point forces exerted on the dislocation, as shown diagrammatically in Fig. 7. When the dislocation is bent through a critical angle, Θ_c, the obstacle strength, F_c, will be reached and

$$F_c = 2T\cos\left(\frac{\Theta_c}{2}\right) \tag{6}$$

where T is the line tension of the dislocation, which can be assumed to be $\simeq Gb^2/2$.

The local value of the radius of curvature of the dislocation loop, R, is related to the applied stress, τ, by

$$\tau b = \frac{T}{R}. \tag{7}$$

Thus, the contribution to the flow stress is of the form:

$$\Delta\tau = \frac{F}{bL}. \tag{8}$$

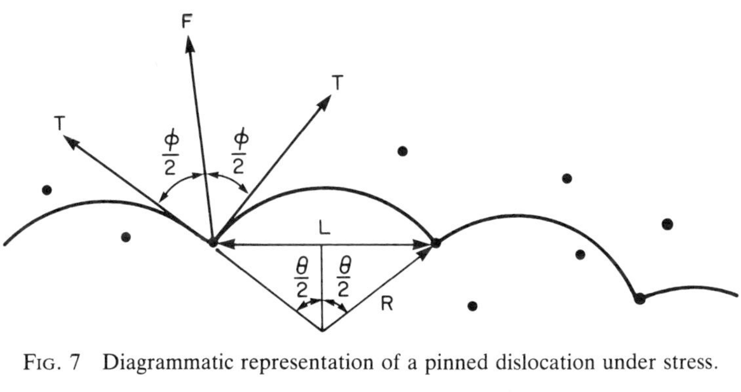

FIG. 7 Diagrammatic representation of a pinned dislocation under stress.

This involves both the intrinsic strength of the obstacle and its effective spacing, L.

The effective spacing depends on the way in which the dislocation contacts the array of obstacles. For a square array of obstacles, the spacing can be written as $L_s = N_s^{-1/2}$, where N_s is the number of obstacles per unit area. If the dislocation bends (i.e., if the obstacle strength is increased), L becomes equal to $L_s/(\cos \Theta/2)^{1/2}$. Thus, the increase in flow stress is then given as

$$\Delta\tau = \frac{2T\left(\cos\dfrac{\Theta}{2}\right)^{3/2}}{bL_s}. \tag{9}$$

This expression is valid for weak obstacles, where $\Theta > 130°$. But for strong obstacles, the dislocation is forced to bow extensively between the obstacles, and the strength is better described by an expression of the form

$$\Delta\tau = \frac{0.8}{bL_s}\left(2T\cos\frac{\Theta}{2}\right). \tag{10}$$

Eq. 10 is valid for obstacles with $\Theta < 100°$.

In considering the application of strengthening models to aluminum alloys, several problems arise. The interaction of dislocations and particles involves consideration of a detailed treatment of dislocation-line tension as the dislocation contacts the obstacle and the change in L_s due to particle size. In addition, aluminum precipitates can occur in a range of shapes, depending on the alloy system and the state of aging.

The specific nature of the various types of dislocation–particle interactions were reviewed by Brown and Ham [*34*]. However, in order to characterize the interaction, it is necessary to quantify a variety of parameters (such as the elastic misfit of the particle, the surface energy, and the ordering energy) or establish the detailed structure of dislocations within the particles. In aluminum alloys, this is made difficult by the complex sequence of precipitation reactions that occur in the solid state. Thus, it is extremely difficult to evaluate the required parameters, due to both the small scale and the nonequilibrium nature of the phases.

With these limitations in mind, the strengthening in aluminum alloys is considered to arise from one of the following mechanisms:

1. Creation of increased surface energy due to particle shear (chemical or surface strengthening).
2. Stacking-fault–strengthening due to the difference in stacking-fault energy between the particle and the matrix.

3. Modulus-strengthening due to differences in the elastic moduli between the particles and the matrix.
4. Coherency-strengthening due to elastic misfit of the particle in the matrix.
5. Order-strengthening, where the particle is forced to create an antiphase boundary due to the passage of a glide dislocation.
6. Orowan-strengthening due to dislocation looping around the particle in the case of impenetrable obstacles.

Each of these mechanisms has been considered at length in other reviews [*35,36*] and will not be discussed in detail here. Table III gives the basic formulations involved for the different mechanisms. It will be noted that several of the mechanisms have volume fraction (f) and size-dependence (r) of the form $\tau\alpha(fr)^{1/2}$, which means that a straightforward volume fraction and size-dependence will not distinguish between the mechanisms.

Chemical-strengthening has been suggested as the hardening mechanism for the Al-Zn system [*39*], although this can only be the case for short aging times because the strength is inversely dependent on particle radius and requires an increasing particle volume fraction to provide strengthening.

Stacking-fault strengthening has been used to explain the strength in dilute Al-Ag [*40*] and for GP-zone strengthening in Al-Zn-Mg alloys [*41*].

Strengthening in Al-Zn-Mg alloys has also been attributed to modulus strengthening [*42*]. There is also considerable modulus-strengthening possible in Al-Li [*31*].

Order-strengthening can account for the strengthening in binary Al-Li alloys, where the strengthening phase is ordered Al_3Li [*31*]. As noted previously, Al-Li can exhibit significant modulus-strengthening. Figure 8 from Noble et al. [*31*] shows that order-hardening overestimates the strength and modulus-hardening underestimates it. However, the modulus of Al_3Li is not accurately known and the antiphase boundary is only known to be in the 0.15–0.2 J m^{-2} range. So even in this simple system, there is still uncertainty. In commercial Al-Li alloys, elements such as Cu, Mg, and Zr are also present, giving rise to additional precipitates, which complicate the situation even further.

Overaged alloys and alloys strengthened by fine oxide particles obey Orowan-hardening. A comparison between experimental and calculated Orowan-strengthening in Al-Al_2O_3, using the results of Hansen [*43*], is shown in Fig. 9. This form of strengthening is somewhat easier to analyze because it is independent of the properties of the particle and only dependent on interparticle spacing. Dispersion-strengthened alloys produced by rapid solidification and mechanical alloying are expected to have considerable Orowan-strengthening.

TABLE III

PARTICLE-STRENGTHENING MECHANISMS

Strengthening Mechanism	Formula	Symbols	Reference
Chemical-surface	$\tau_{cs} = \left(\frac{6\gamma_s^3 \ bf}{\pi T}\right)^{1/2} \frac{1}{r}$	γ_s = surface energy	44
Stacking fault	$\tau_{SF} = \Delta\sigma^{3/2} \left(\frac{3\pi^2 fr}{32T\, b^2}\right)^{1/2}$	$\Delta\sigma$ = difference in stacking-fault energy between matrix and particle	40
Modulus	$\tau_G = \frac{\Delta G}{4\pi^2} \left(\frac{3\Delta G}{G_M b}\right)^{1/2} \left[0.8 - 0.143 \ln\left(\frac{1}{r}\right)\right]^{3/2} r^{1/2} f^{1/2}$	ΔG = difference in modulus between matrix and particle	45
	$\tau_G = 0.9\,(fr)^{1/2} \frac{T}{b} \left(\frac{\Delta G}{G_M}\right)^{3/2} \left[2b \ln\left(\frac{2r}{f^{1/2} b}\right)\right]^{-3/2}$		42
Coherency	$\tau_{cs} = k(\epsilon G)^{3/2} \left(\frac{rfb}{T}\right)^{1/2}$	ϵ = misfit parameter $\epsilon = \frac{1}{3}\left(\frac{1+v}{1-v}\right)\frac{\Delta a}{a} \approx 2/3\ \delta$ where δ = fractional difference in lattice parameter k is a numerical factor, $k = 3$ for edge, and $k = 1$ for screw dislocations.	46
Order	$\tau_0 = \frac{\sigma_{APB}}{2b} \left[\left(\frac{4\ \sigma_{APB} r f}{\pi T}\right)^{1/2} - f\right]$ for $r < \frac{T}{\sigma_{APB}}$ and $\tau_0 = \frac{\sigma_{APB}}{2b} \left[\left(\frac{4f}{\pi}\right)^{1/2} - f\right]$ for $r > \frac{T}{\sigma_{APB}}$	σAPB = antiphase-boundary energy	34
Orowan	$\tau = \frac{0.8\ Gb}{2\pi(1-\nu)^{1/2} L} \ln\left(\frac{2r}{r_0}\right)$	L = clear interparticle spacing r_0 = inner cut-off radius $\approx b$	47

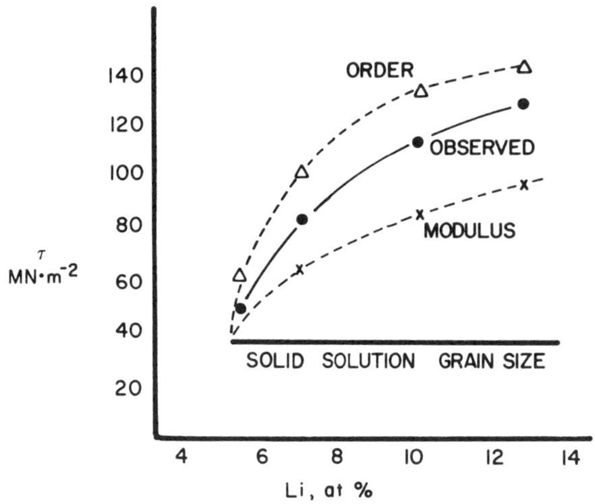

FIG. 8 Strengthening in Al-Li (Noble et al. [31]).

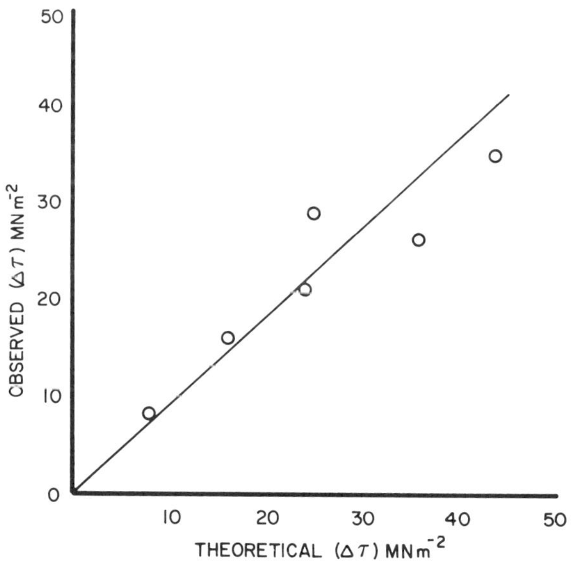

FIG. 9 Orowan-strengthening in $Al\text{-}Al_2O_3$ (Hansen [43]).

V. Composite Strengthening

In the 1980s, processing routes have been developed that make metal-matrix composites (MMCs) viable commercial materials. Most of the initial development of MMCs concentrated on continuous-fiber reinforcement, using C, B, and SiC fibers. While these reinforcements, in principle, can achieve the optimum strength levels, they suffer from the facts that continuous fibers are expensive and that the resulting composite has no formability. The effort in MMCs has concentrated on whisker and particulate reinforcements, which are much cheaper reinforcements, and the resulting composite can be extruded and rolled. Both powder-metallurgy and molten-metal routes have been developed for making MMCs, and these are becoming available in semicommercial quantities. The properties of particular interest are high modulus and strength, high wear-resistance, and low coefficients of thermal expansion, although there is a significant loss in ductility. Aluminum is a particularly attractive matrix because of its low density, and the main reinforcement has been SiC, but other reinforcements (such as Al_2O_3, B_4C, and Si_3N_4) are also being considered [*48*].

The possible strengthening mechanisms of composites are:

1. Orowan-strengthening
2. Shear-lag–type strengthening
3. Eshelby-type strengthening

Orowan-strengthening will obey the equation in Table III. Particulate and whisker reinforcement usually involve particles in the 1–15-μm size range, with volume fractions of 0 to 30%. While the particles are not uniformly distributed, making it difficult to know a realistic value for L in the Orowan equation, a realistic estimate for Orowan-strengthening is about 10 MPa, which is far less than that observed, particularly in whisker-reinforced material.

In the shear lag analysis [*49*], shear forces at the interface between the reinforcement and the matrix develop tensile stresses in the reinforcement. For the particles to be fully loaded, they have to have an aspect ratio greater than S_c, given by

$$S_c = \frac{\sigma_p}{\tau_i} \tag{11}$$

where σ_p is the particle strength and τ_i is the interfacial shear strength. The interfacial shear strength is not known, but should approach the shear strength of the matrix (i.e., $\sigma_m/2$) where σ_m is the matrix tensile strength. For SiC, $\sigma_p \simeq 2000$ MPa and $\tau_i \simeq 150$ MPa to give $S_c \simeq 13$.

For $S > S_c$, the shear lag analysis gives a composite strength, σ_c, of

$$\sigma_c = \tfrac{1}{2}V_p\sigma_p + V_m\sigma_m \tag{12}$$

where V_p and V_m are the particle and matrix volume fractions and σ_p is the particle strength.

In both whisker- and particulate-reinforced composites, $S < S_c$. For whiskers, Arsenault [50] obtained $S = 4$. If $S < S_c$, the shear lag analysis gives the following expression for composite strength: for particulates,

$$\sigma_c = \left(\frac{V_p\sigma_m S}{4}\right) + V_m\sigma_m, \tag{13}$$

and, for whiskers,

$$\sigma_c = \left(\frac{V_w\sigma_m S}{2}\right) + V_m\sigma_m. \tag{14}$$

These equations assume no stress transfer at the particulate or whisker ends. Nardone and Prewo [51] have suggested that end loading can provide additional strengthening and the composite strength then becomes, for particulate,

$$\sigma_c = \left(\frac{V_p\sigma_m(S + 4)}{4}\right) + V_m\sigma_m \tag{15}$$

and, for whiskers,

$$\sigma_c = \left(\frac{V_m\sigma_m(S + 2)}{2}\right) + V_m\sigma_m. \tag{16}$$

Therefore, end loading results in an effective increase in the reinforcement aspect ratio. It should be noted that tensile loading across the long face of the reinforcement is not included in this modified analysis. Since the reinforcement is often not well aligned, tensile loading across the long face may also be operating.

Figure 10 shows a plot of σ_c/σ_m versus V_f, for different aspect ratios of particulate, predicted by Eq. 13 and 15. Also shown on this plot are experimental results for a range of particulate composites produced by Dural Aluminum Composite Corporation. Clearly, a modified form of the shear lag analysis is required to account for the composite strengths achieved, because in all these composites the aspect ratio is only 1–2. It must also be noted that a homogeneous distribution of the particulate in the composite is assumed. This is not the case, as seen from the microstructure of extruded 6061-15v/o SiC shown in Fig. 11.

Eshelby strengthening [52] arises from the fact that an elastic inclusion

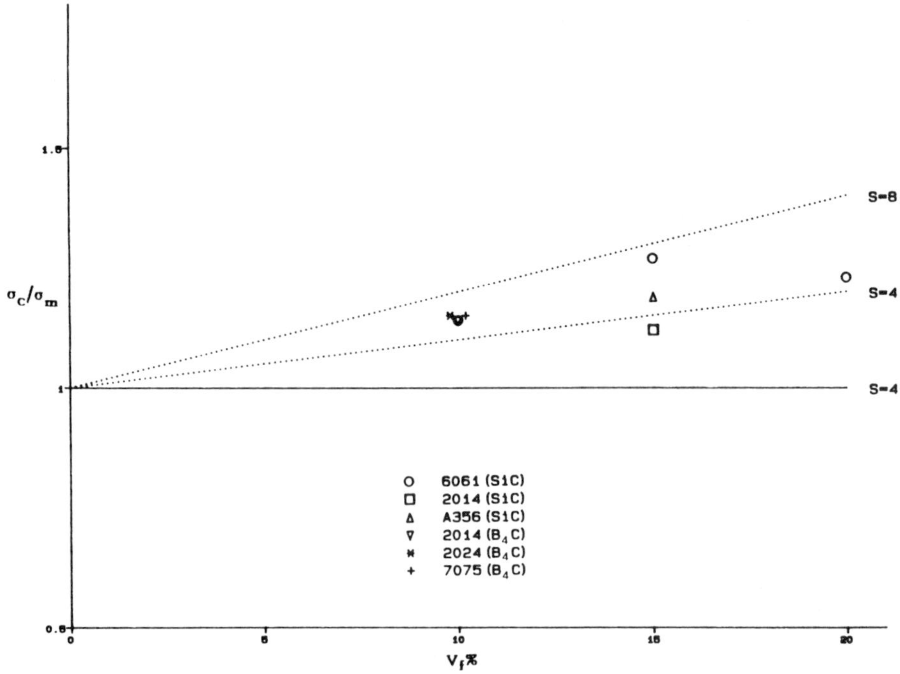

FIG. 10 Predictions of strength according to shear lag analysis (continuous line) and modified shear lag (dotted lines).

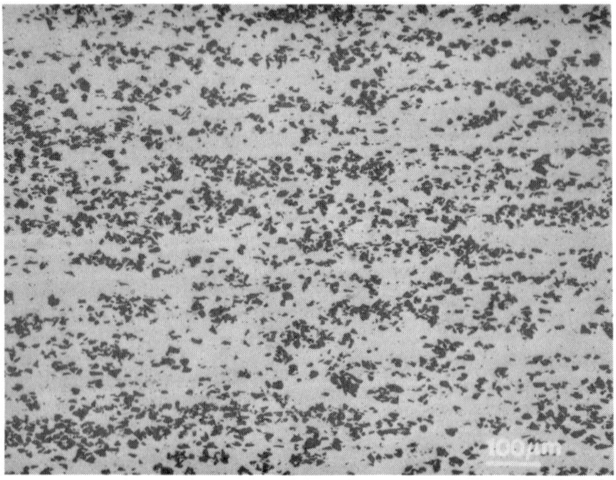

FIG. 11 Microstructure of extruded 6061-15v/o SiC.

embedded in a matrix with different elastic properties induces a stress field in and around the inclusions. This stress field gives rise to the Bauschinger effect referred to earlier [*38*]. In the original treatment, only a single inclusion was considered, but Mori and Tawaka [*53*] have accounted for the interaction between inclusions and the model has been applied to metal-matrix composites [*54*].

The yield stress σ_{yc} of the composite is given by [*55*]

$$\sigma_{yc} = c_1\sigma_{ym} + c_2\epsilon_p \tag{17}$$

where σ_{ym} is the yield stress of the matrix, ϵ_p is the plastic strain, c_1 is a nondimensional yield stress raiser obtained from calculating the eigenstrain, and c_2 is the work-hardening rate.

The model can also be used to predict the Young's modulus of the composite, E_c, which is given by [*55*]

$$\frac{E_c}{E_m} = \frac{1}{1 + V_f(E_m/\sigma_0)\epsilon^*_{33}} \tag{18}$$

where E_m is the modulus of the matrix, V_f the volume fraction of reinforcement, and ϵ^*_{33} a strain computed from the model.

For aluminum reinforced [*54*] found that the Eshelby approach gave much better agreement than the conventional shear lag analysis for predicting elastic modulus, as shown in Fig. 12.

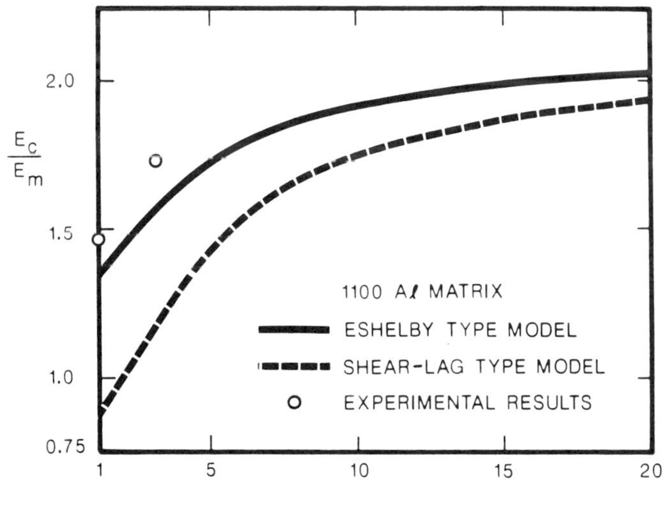

FIG. 12 Predictions of the Eshelby analysis for the elastic modulus against aspect ratio for SiC-whisker–reinforced aluminum (Arsenault and Taya [*54*]).

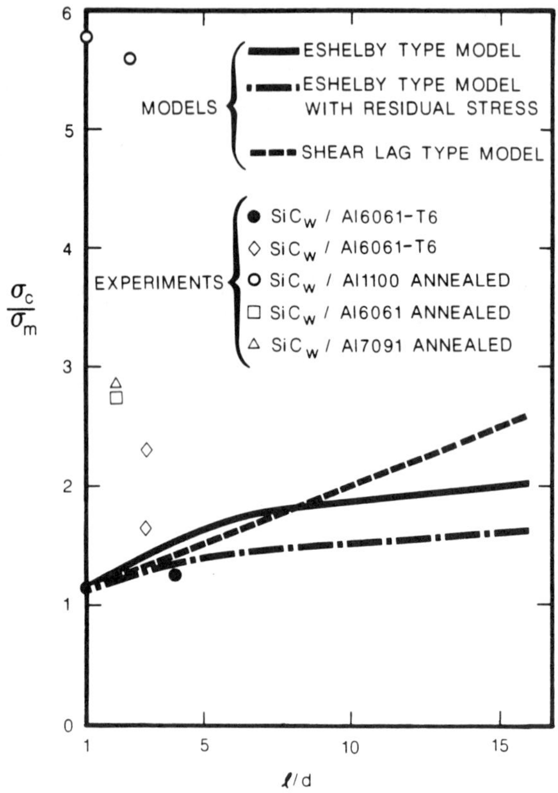

FIG. 13 Strength predictions from the Eshelby analysis for SiC-whisker–reinforced aluminum (Taya and Arsenault [*54*]).

For the yield stress, however, both the Eshelby and shear lag analyses tend to greatly underestimate the strength, as shown in Fig. 13. This is still the case when residual stresses, due to the differential thermal expansion between the matrix and the reinforcement, are included in the model. One possible explanation for this underestimate is that the yield stress of the matrix is increased in the composite due to an enhanced dislocation density. The difference in coefficients of expansion between the reinforcement and the matrix develop stresses upon cooling (these being the stresses incorporated into the model in the residual stress modification noted previously) which may be relaxed, to some extent, by dislocation punching. This results in an increased dislocation density locally in the region of the reinforcement.

When the strengths of different particulate-reinforced MMCs are being considered, account has to be taken of differences in aspect ratios between the particulates and the degree of particle alignment produced by the processing. In general, whisker-reinforced materials are stronger than particulate, and this is expected from the shear lag equations. In the Eshelby model, this difference would reflect differences in aspect ratio and degree of alignment.

The difficulty in assessing all the strengthening mechanisms is the recurring problem of correctly accounting for the inhomogeneous particulate distribution and the range of particle shapes. The results are still too limited to decide whether the shear lag or Eshelby approach is preferable, but the latter is attractive because it is capable of predicting strength anisotropy, residual stresses [*56*] and thermal conductivity [*57*].

VI. Summary

The strength of aluminum alloys involves a wide range of factors, many of which are difficult to model and, in regards to precipitation-strengthening, are often poorly understood. In the commercial context, the strengthening mechanisms used can be generalized as follows:

1. In the 1000-series alloys, which are essentially pure aluminum, there is weak solution-and dislocation-hardening, depending on alloy and heat-treatment.
2. In the 3000-series Al-Mn and 5000-series Al-Mg alloys, there is strong solution-and dislocation-hardening, since these are often used in a worked temper. Additions of Mg and Si are sometimes made to the Al-Mn alloys to introduce weak precipitation-hardening.
3. The heat-treatable 2000 (Al-Cu), 6000 (Al-Mg-Si), and 7000 (Al-Zn-Mg (Cu)) alloys are strongly strengthened by solution- and precipitation-hardening. These alloys are sometimes used in the unrecrystallized or thermomechanically worked conditions and then have considerable dislocation and substructure strengthening.

In the future, new processes under development will provide material strengthened by fine grain size, dispersoid hardening, and reinforcement.

References

1. Cahn, R.W. (1982). *Ann. Rev. Mater. Sci.*, **12**, 51.
2. Gilman, P.S., and Benjamin, J.S., (1983). *Ann. Rev. Mater. Sci.*, **13**, 279.

3. Bickerdike, R.L., Clark, D., Easterbrook, J.N., Hughes, G., Mair, W.N., Partridge, P.G., and Ranson, R.C. (1984–1985). *Inter. Jour. of Rap. Sol.*, **1**, 305.
4. Westengen, H. (1982). *Z. Metallkde*, **73**, 360.
5. Low, J.R., Jr., Van Stone, R.H., and Merchant, R.H. (1972). NASA Tech. Rep. no. 2, Res. Grant NGR 38-087–003. Carnegie-Mellon University, Pittsburgh. Quoted by G.T. Hahn and A.R. Rosenfield (1975). *Met. Trans*, **6A**, 653.
6. Mathew, E.V., Ramachandran, T.R., Gupta, K.P., and Das, S. (1984). *J. Mater. Sci. Lett.*, **10**, 605.
7. Humphreys, F.J. (1977). *Acta Metall.*, **25**, 1323.
8. Lloyd, D.J. (1986). In *Microstructural Control in Aluminum Alloys: Deformation, Recovery and Recrystallisation*, E. Henry Chia and J.J. McQueen, eds. Met. Soc. AIME, p. 45.
9. Nes, E. Ibid, p. 95.
10. Wert, J.A. Ibid, p. 67.
11. Cox, T.B., and Low, J.R., Jr. (1972). NASA Tech. Rep. no. 3, Res, Grant NGR 39-087-003. Carnegie Mellon University, Pittsburgh.
12. Dowling, J.M., and Martin, J.W. (1973). *I.C.S.M.A.-3*, Cambridge, England, **2** 170.
13. Starke, E.A., Jr., Sanders T.H., Jr., and Palmer, I.G. (1981). *J. Metals*, **33**, 24.
14. Kelly, A., and Nicholson, R.B. (1963). *Progr. in Mater. Sci.*, **10**, 151.
15. Brown, L.M., and Ham, R.K. (1971). *Strengthening Methods in Crystals*, A. Kelly and R.B. Nicholson, eds. Elsevier, New York, p. 9.
16. Polmear, I.J. (1981). *Light Alloys—Metallurgy of the Light Metals*. Edward Arnold, London, p. 26.
17. Hatch, J.E. ed. (1984). *Aluminum: Properties and Physical Metallurgy*. Amer. Soc. for Met., p. 134.
18. Lorimer, G.W., and Nicholson, R.B. (1969). *The Mechanism of Phase Transformations in Crystalline Solids*. Inst. Met. Micrograph no. 33. p. 36.
19. Jacob, M.H., and Pashley, D.W. Ibid, p. 43.
20. Hehemann, R.F. (1971). *Met. Trans.*, **2A**, 39.
21. Polmear, I.J. Op. cit., p. 21.
22. Hall, E.O. (1951). *Proc. Phys. Soc., London*, **B64**, 747.
23. Carreker, R.P., and Hibbard, W.R. (1957). *Met. Trans. A.I.M.E.*, **209**, 1157.
24. Phillips, V.A. (1952). *J.I.M.*, **81**, 623.
25. Morris, L.R. (1976). *Proc. I.C.S.M.A.-4 Nancy, France*, **2**, 649.
26. Lloyd, D.J. (1980). *Met. Sci.*, **14**, 193.
27. Embury, J.D., and Fisher, R.M. (1966). *Acta Met.*, **14**, 47.
28. Wilcox, B.A., and Gilbert, A. (1967). *Acta Met.*, **15**, 601.
29. Spitzig, W.G., Pelton, A.R., and Laabs, F.C. (1987). *Acta Met.*, **10**, 2427.
30. Fleischer, R.L. (1963). *Acta Met.*, **11**, 203.
31. Noble, F.W., Harris, S.J., and Dinsdale, K. (1982). *Met. Sci.*, **16**, 425.
32. Lloyd, D.J., and Kenny, D. (1982). *Met. Trans.*, **13A**, 1445.
33. Rollett, A.D., Kocks, U.F., and Doherty, R.D. (1986). In *Formability and Metallurgical Structure*, A.K. Sachdev and J.D Embury, Met. Soc., p. 211.
34. Brown, L.M., and Ham, R.K. (1971). In *Strengthening Mechanisms in Crystals*, A. Kelly and R.B. Nicholson, Halsted Press, John Wiley & Sons, New York.
35. Lloyd, D.J. (1985). *I.C.S.M.A.-7, Montreal, Canada*. Pergamon Press, p. 1745.
36. Ardell A.J. (1985). *Met. Trans.*, (1986). **16A**, 2131.
37. Bate, P.S., and Wilson, D.V. (1986). *Acta Met.*, **34**, 1097.
38. Embury, J.D. (1987). *Materials Forum*, (1970). **3**, 27.
39. Harkness, S.D., and Hren, J.J. (1970). *Met. Trans.* **1A**, 43.
40. Gerold, V., and Hartmann, K. *Trans. Jap. Inst. Met.*, **9** (supplement), 509.

41. Kovacs, I., Lendvai, J., Ungar, T., and Groma, G. (1977). *Acta Met.*, **25**, 673.
42. Melander, A., and Persson, P.A. (1978). *Acta Met.*, **26**, 267.
43. Hansen, N. (1970). *Acta Met.*, **18**, 137.
44. Kelly, A., and Fine, M.E. (1957). *Acta Met.*, **5**, 365.
45. Knowles, G., and Kelly, P.M. (1971). *Effect of Second-Phase Particles on the Mechanical Properties of Steel.* Iron and Steel Inst., London, England, p. 9.
46. Gerold, V., and Haberkorn, H. (1966). *Phys. Stat. Sol*, **16**, 675.
47. Ashby, M.F. (1966). *Acta Met.*, **14**, 679.
48. Nair, S.V. Tien, J.K., and Bates, R.C. (1985). *Inter. Met. Rev.*, **30**, 275.
49. Piggot, M.R. (1980). *Load-Bearing Fibre Composites.* Pergamon Press.
50. Arsenault, R.J., (1984). *Mat. Sci. Eng.*, **64**, 171.
51. Nardone, V.C., and Prewo, K.M. (1986). *Scripta Met.*, **20**, 43.
52. Eshelby, J.D. (1957). *Proc. Roy. Soc. Lond.*, **A241**, 376.
53. Mori, T., and Tawaka, K. (1973). *Acta Met.*, **21**, 571.
54. Taya, M., and Arsenault, R.J. (1987). *Scripta Met.*, **21**, 349.
55. Arsenault, R.J., and Taya, M. (1987). *Acta Met.*, **35**, 651.
56. Takao, Y., and Taya, M. (1985). *J. Appl. Mech.*, **107**, 806.
57. Hatta, H., and Taka, M. (1985). *J. Appl. Phys.*, **58**, 2478.
58. Alpas, A. Unpublished work. McMaster University, Hamilton, Ontario, Canada.
59. Koppenaal, T.J., and Kuhlmann-Wilsdorf, D. (1964). *J. Appl. Phys.*, **4**, 59.

Part VI

Competitive Materials to Aluminum

23

Fiber-Reinforced Plastic Composites

F.R. JONES

School of Materials
Division of Ceramics, Glasses and Polymers
University of Sheffield
Sheffield, England

I. Introduction

A composite is defined as a mixture of physically distinct components. This definition clearly refers to a number of artifacts, but for engineering materials, it usually applies to fiber- or particulate-reinforced materials. An early example is the straw-reinforced clay bricks of the Old Testament. Modern examples are car tires from nylon and steel-fiber–reinforced rubber as well as fiberglass boat hulls and car bodies. The most important class is the so-called fiber-reinforced plastics (FRP), but recent advances in fiber

ISBN 0-12-341831-3

technology and materials processing have made fiber-reinforced metals (FRM), ceramics (FRC), and cements commercially practical. Much of the competition to aluminum comes from the more established fiber-reinforced plastics—in particular, those reinforced with glass fibers (GFRP or GRP), carbon fibers (CFRP), or aramid fibers (AFRP). On economic grounds, industrial composites are mostly based on GRP, while aerospace applications are generally based on the significantly more expensive CFRP. The aramid fibers are replacing glass fibers in a number of applications where their superior specific modulus enables lighter parts to be produced. By careful choice of fiber type, orientation, and matrix, the properties of the material can be engineered to cater for the service stresses encountered in a particular application. This chapter is not intended to be a complete discourse, but will concentrate on the materials science of fiber-reinforced plastics composites. FRM, FRC, and related materials are not discussed in depth, but much of the argument is equally relevant. Laminate theory has been developed to predict the performance of a complex angle ply composite from the mechanical and thermomechanical properties of the components. Computer software is now universally available to support these calculations and an appropriate discussion is beyond the scope of this chapter. (See for example, Tsai [*1*].)

II. Fiber Reinforcements

The theoretical strength of a solid has been shown by Orowan [*2*] to be directly related to the separation of the atoms and the surface energy of the material

$$\sigma_t = (E\gamma_a/a)^{1/2} \tag{1}$$

where E is the Young's modulus, γ_a = the equilibrium surface-free energy, and a = interplane spacing in the crystal. These properties are all interrelated so that the highest-strength materials are obtained when a dense crystalline arrangement occurs with strong interatomic forces. The covalently bound crystals and metallic crystals (the transition metals in particular) yield the highest theoretical strengths because of the presence of cleavage planes in ionically bonded crystals. For the ideal structural materials, a high strength-to-weight ratio is required. The property is counter to the requirements for isotropic high-strength–high-modulus materials but high modulus can be achieved in the less dense material by orientating the molecular bonds in the direction of the applied stress. This has been the practice in fiber technology over the decades. It follows that high specific strength (σ/ρ) and modulus (E/ρ) will be found in highly orientated anisotropic materials. If one compares the data for polyethylene in Table I, the effect of drawing the randomly coiled isotropic polymer allows the theoretical modulus to be approached. Similar technology has also been applied to polyacrylonitrile (PAN) precursors for the production of high-stength carbon fibers [*13*].

Another aspect of strength is the inevitable presence of Griffith flaws, which makes the practical strength significantly less than theoretical. Thus, relatively isotropic materials such as glass can be strengthened by reducing the number and size of flaws in the surface. Fiber technology has a role to play in the production of high-strength glass. Bartenev and Izmailova [*4*] have argued that the strength of glass fibers arises from the presence of an oriented quenched layer on the surface of the fiber. They concluded this from the observation that contrary to plate glass, etching reduced the strength of "defect"-free glass. There is no disagreement with the statistical distributions of strengths illustrated in Fig. 1, but with the explanation of the characteristic values for the fibers in each group. The object of this brief discussion is to illustrate the statistical nature of the failure processes, which applies equally to carbon fibers as to the other reinforcing fibers. Table I compares the mechanical performance of typical commercial reinforcing fibers for fabrication into composite materials. The strength of these fibers is largely defect-controlled so that the reported values are gauge-length–dependent and in many cases are averages obtained from the failure of either unimpregnated or resin-impregnated bundles.

The other advantages of fiber technology are (1) the use of resin coatings to protect the fiber surface during processing and fabrication and (2) the use of extensible matrices which allow the use of brittle ceramics in applications that would otherwise be impractical.

The advantages of different fiber types are best assessed by comparing their specific moduli and strength as illustrated in Fig. 2. It is clear that carbon fibers, Kevlar 49, and Kevlar 149 are strong contenders for use in applications where light weight and high strength are required, such as aerospace and sports projects. The relative low cost of E-glass fibers makes them the preferred reinforcing agent in industrial application. These fibers are therefore also discussed in further detail.

A. *Carbon Fibers*

High-strength carbon or graphite fibers can be manufactured by the so-called PAN route developed at the Royal Aerospace Establishment, Farnborough, U.K., by W. Watt and coworkers or from mesophase pitch.

TABLE I

TYPICAL MECHANICAL PROPERTIES OF FIBERS AND MATRICES

	Relative density	Young's modulus (GPa)	Tensile strength (GPa)	Elongation at break (%)	Fiber diameter (μm)
Fibers for composites					
E-glass	2.6	72	3.0	4.0	10–16
S-glass	2.5	87	3.5	4.0	12
S-2 glass	2.49	86	4.0	5.4	10
Mesophase carbon	2.02	380–820	2.0–2.4	0.5–0.2	~10
(PAN) HM carbon	1.9	370	1.7	0.5	8
(PAN) HT carbon	1.8	250	2.8	1.2	8
(PAN) HE carbon	1.8	230	4.5	2.0	8
SiC continuous (Nicalon)	2.5	200	3.0	1.5	10–15
SiC Whisker	3.2	480	7.0	—	1–50
S*i*TiCO (Tyranno)	2.35	200	3.0	1.5	8–10
Boron	2.6	410	3.4	0.8	100
α-alumina (FP)	3.9	380	1.7	0.4	20
β-alumina (Saffil)	3.3	300	2.0	0.5	3
$Fe_{80}B_{20}$ (metallic glass)	—	100–150	3.6 (yield)	—	—
Steel (high C)	7.8	210	2.8	—	250
Kevlar 149 (PPT)	1.47	180	3.45	1.9	12
Kevlar 49	1.45	128	2.65	2.4	12
Kevlar 29	1.44	60	2.65	4.0	12
Nomex (PmPT)	1.4	17.3	0.7	22.0	12
PBT	1.5	250	2.4	1.5	20
Polyamide 66	1.44	5.0	0.9	13.5	10
Polyethylene: (theory)	—	200	—	—	—
(solution spun)	1.0	100–150	—	—	—
(drawn)	~1.0	6–80	—	—	—
Conventional materials					
Steel	7.8	210	0.34–2.1	—	—
Aluminum alloys	2.7	70	0.14–0.62	—	—
Glass	2.6	60	—	—	—
Resins					
Phenolic	1.4	7	—	~0.5	—
Epoxy	1.2–1.4	2–3.5	0.05–0.09	1.5–6	—
Polyester	—	2–3.9	0.04–0.085	1–25	—
Polymers					
Nylon 66	1.14	2.0	0.07	60	—
HDPE*	0.96	1.3	0.03	~500	—
LDPE*	0.92	0.25	0.01	~600	—

* Polyethylene
HDPE High density polyethene or polyethylene
LDPE Low density polyethene or polyethylene

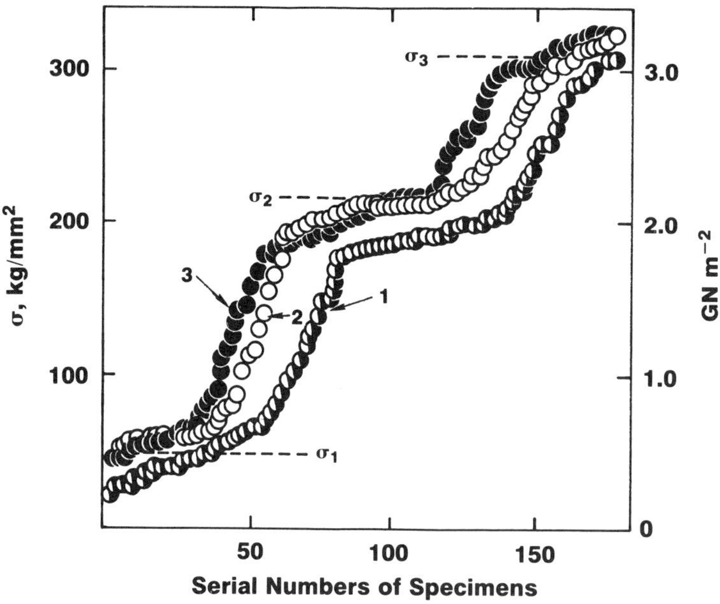

FIG. 1 Distribution of strength values for three series of industrial glass-fiber specimens 10 μm in diameter, in order of serial number. (1) Average strength 1.48 GN m^{-2}, specimen length 40 mm. (2) 1.76 GN m^{-2} and 25 mm. (3) 1.89 GN m^{-2} and 25 mm. (after Bartenev and Izmailova [*4*]) Reproduced from McCrum [5] with permission of the controller, Her Majesty's Stationery Office, London.

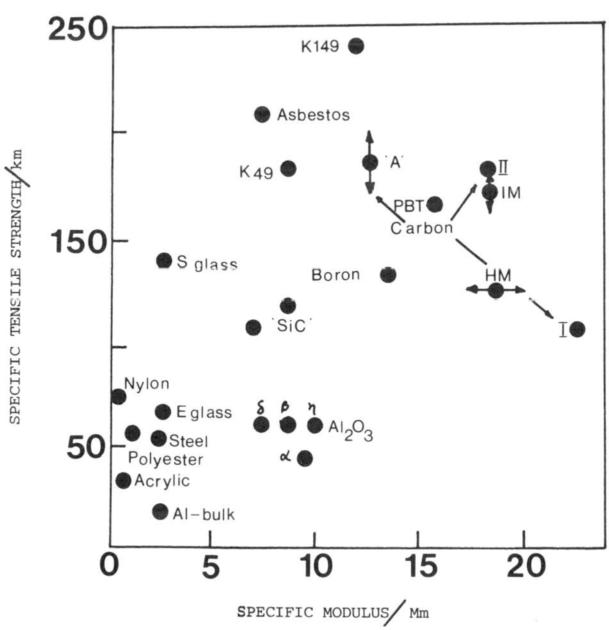

FIG. 2 Specific properties of carbon fibers compared with other filamentary materials. The value for aluminum is based on a typical aerospace alloy such as 2024.

The former is by far the most important commercial route although pitch-based fibers could be potentially more economic when the technology for the production of consistent fibers has been developed.

1. Production of Carbon Fibers from Polyacrylonitrile

A description of the production of carbon fibers is given schematically in Fig. 3. The process can be discussed in terms of the following stages:

The *precursor fiber* is obtained by wet-spinning a carefully filtered polyacrylonitrile copolymer dope (solution). Hot-stretching of the fiber to orientate the macromolecules has a profound effect on the properties of the final carbon fiber.

Oxidation of the precursor fiber is carried out under restraint in oxygen atmosphere at 200 to 220°C so that the molecular orientation induced during stretching can be maintained. Polyacrylonitrile is the favored precursor because of the self-induced thermal polymerization of the pendant nitrile groups, to produce a ladder polymer. (This is schematically illustrated in Fig. 4.) Because of the limited number of adjacent isotactic nitrile groups, the cyclized sequences involve approximately 5–15 groups. In addition, in areas where the molecules are not so well orientated, cyclization reactions between syndyotactic nitrile groups will cause the ladder polymer to take on a curved rather than planar ribbonlike form. The cyclized rings of the thermally modified PAN are also puckered. It is now known that these reactions precede the oxidation reaction and that the role

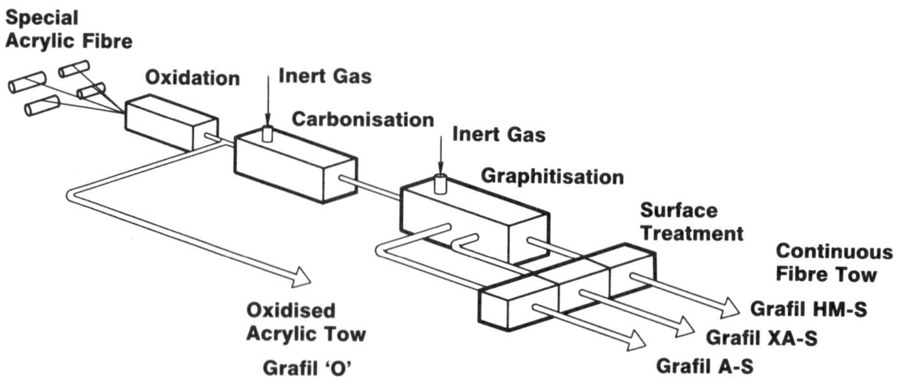

Fig. 3 Carbon-fiber–manufacturing process polycrylonitrile. [after Hysol-Graphil Co. Ltd]

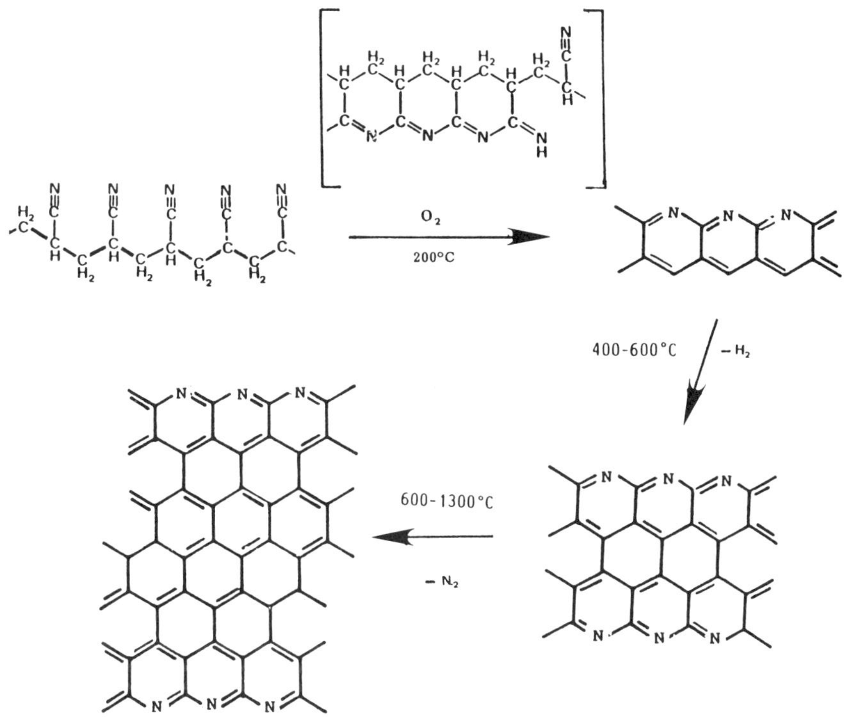

FIG. 4 Schematic representation of the chemistry of carbon-fiber preparation from PAN.

of oxygen is to aromatize the cyclized polymer into a planar aromatic structure that facilitates the condensation between graphitic nuclei in the *carbonization* and *graphitization* stages, which are carried out in an inert atmosphere over the temperature range of 400 to 2500°C. The tensile strength of the fiber reaches a maximum at 1500°C, the so-called type-II or high-strength (HS) fiber. Graphitization in a temperature range of 2500°C produces the type-I or high-modulus (HM) fiber. The distribution of strength values are approximately the same for both fibers, only differing in the mean value. In recent years, the production technology has been improved and a wide range of fibers are now commercially available (for example, the so-called type-A fiber, which has been subjected to 1000°C). The properties of the commercial Grafil fibers are given in Table II. The high-strain fibers are favored for the majority of applications. For special use, a fiber with a modulus approaching 800 GPa (i.e., 80% theoretical) can be produced by hot-stretching techniques.

TABLE II

TYPICAL PERFORMANCE OF COMMERCIAL CARBON FIBERS

Fiber type	Typical strength		Typical modulus		Typical strain
	GPa	ksi	GPa	msi	(%)
XA (high-strength)	3.0–3.3	435–480	220–240	32–35	1.3–1.4
XA (high-performance)	3.3–3.6	480–520	220–240	32–35	1.4–1.5
XA (high-strain)	3.7+	535+	220–240	32–35	1.5–1.7
HM (high-modulus)	2.3–2.6	335–375	330–350	48–51	0.7
IM (intermediate-modulus)	2.9–3.2	420–465	280–300	41–43.5	1.0

Source: From Hysol-Graphil Ltd.

2. STRUCTURE AND SURFACE TREATMENT OF CARBON FIBERS (FROM PAN PRECURSORS)

The relatively small size of the graphite nuclei and the curved ribbonlike molecular structure of the oxidized acrylic fiber result in a turbostratic graphite structure to the fiber. A schematic representation of the transverse and longitudinal structures is given in Fig. 5. The type-I fibers have also been shown to have the skin-core structure indicated in Fig. 5a,

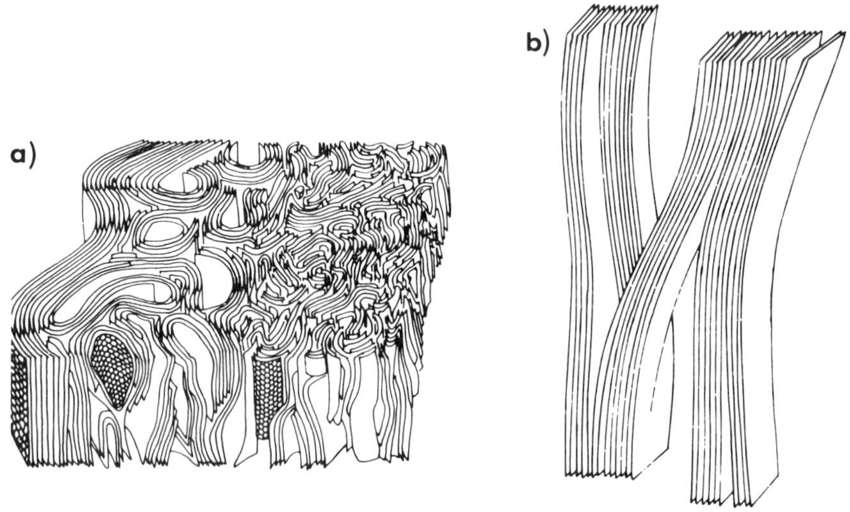

FIG. 5 Schematic carbon-fiber structure: (a) three-dimensional model. (b) longitudinal section showing misoriented crystallite. (after Bennett and Johnson [*6*])

whereas for the type II and type A, these differences are not observable in transmission electron microscopy. The misalignment of the graphite crystallite ribbons as shown in Fig. 5b is not considered to be responsible for the strength-reducing flaws in graphite fibers. However, the presence of impurities which can catalyze the graphitization locally result in a high degree of misorientation and crystallite continuity where a prospective flaw of critical size can form [*7*,*8*].

An essential stage in the production of carbon fibers is the postcarbonization surface oxidation. This process can be carried out chemically, thermally, or electrolytically, with the latter being the favored route. The treatment is essential to the optimization of the adhesion of the fiber to the resin matrix, which can drastically affect the performance of the fiber composite. In the early production of carbon fibers, the surface treatment was too severe and caused a serious reduction in the toughness of the composite. In recent times, a better understanding of the relationship between the type and degree of surface treatment and composite performance has been achieved. Figure 6 demonstrates the need for a carefully controlled surface oxidation to optimize the performance of the composite. The interlamina shear strengths (normalized to the normal industrial treatment and denoted 100%) at different degrees of treatment are compared to the notched tensile strengths (also normalized to the value for 100% treatment) for varying surface treatments. The impact strength of a multidirectional composite reaches a maximum at commercial treatment levels.

The nature of the adhesive bond to carbon fibers is still the subject of discussion, but it has been shown that the surface of the oxidized type-I fiber has a micropore volume of $3–14 \times 10^{-5}$ cm^3 g^{-1} (from BET and ESCA studies), which can be interpreted as micropores of 0.7-nm diameter. Surface analysis has demonstrated that the main functional groups present on the surface of the treated fiber are of the —COOH, $>C=O$, —OH family. The relative proportions of each would appear to be determined by the surface-treatment technique. However, ~10% of the surface atoms on a commercial fiber are found to be oxygen, and a third of these are in the form of carboxylic (COOH) acid groups. Removal of these groups by vacuum desorption does not appear to reduce the interlamina shear strength of the composite. Furthermore, the final surface was found to be highly active, which led Denison and coworkers [*10–12*] to postulate that the surface consisted of micropores at interruptions in the graphite basal planes and that the functional groups were located at these points. Desorption of these groups was considered to produce activated sites which could contribute equally to the adhesion of the resin. A schematic view of the top surface layers is given in Fig. 7.

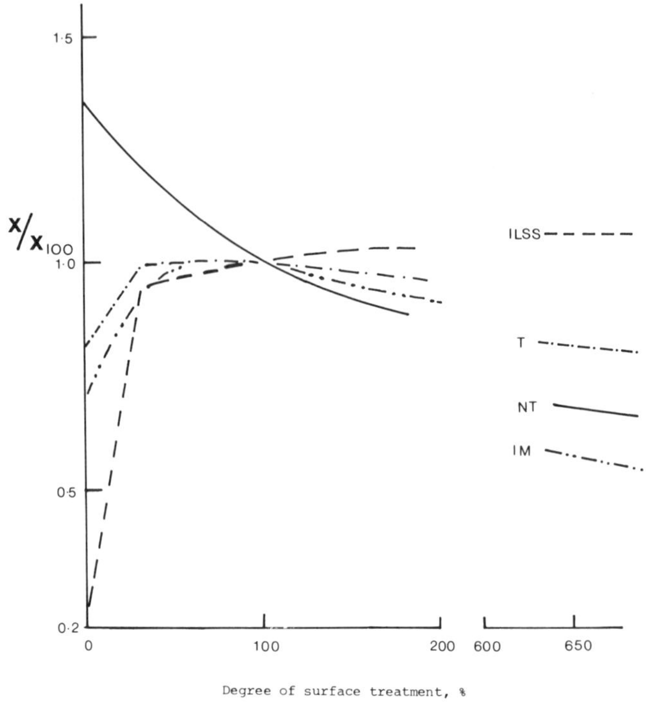

FIG. 6 Effect of surface treatment (normalized to commercial level at 100%) on the mechanical properties of carbon-filter–epoxy composites. X = ILSS, interlaminar shear strength; T, tensile strength; NT, notched tensile strength; IM, Impact strength. X_{100} = property at 100% (or commercial) degree of surface treatment. Redrawn from data given by Dunsford et al. [9] with permission of the controller. Her Majesty's Stationery Office, London.

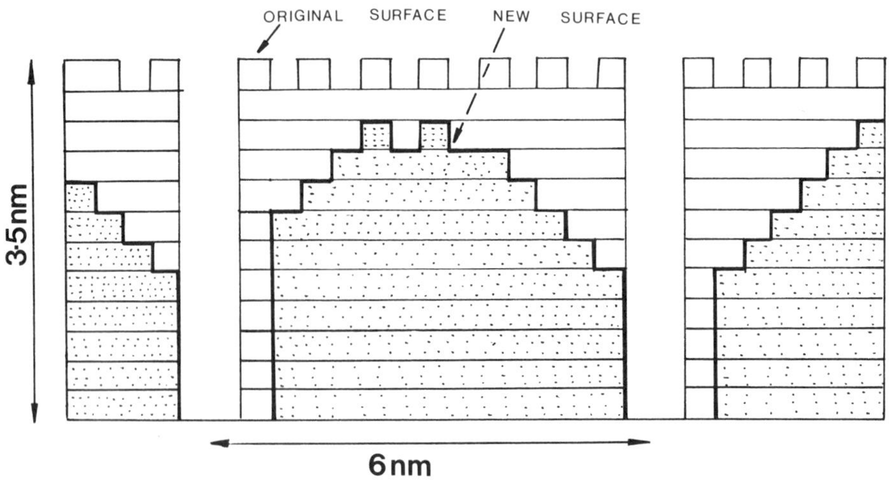

FIG. 7 Schematic model of the surface of type-I carbon filter. (after Denison et al. [*12*])

The structure and analysis of the surface of the type-II fibers is more complicated and consistent with electron microscope studies, which have shown the absence of a surface skin. The treated surface is saturated with carboxylic groups and labelling studies suggest that they are spaced at <0.3 nm. This can only be understood in terms of a microporous surface where a significant proportion of the graphite basal planes are exposed in order to accommodate the large number of functional groups present [*10*].

Because of their brittle nature, the fibers are consequently dressed with an epoxy resin size (i.e., coating) to minimize handling damage and facilitate fabrication and processing.

3. Carbon Fibers from Pitch Precursors

High-strength carbon fiber can also be manufactured from petroleum pitch. This involves the polymerization of the condensed aromatic molecules to form a mesophase or liquid crystal solution at temperatures above 350°C for extrusion melt spinning into fibers for thermosetting in air and carbonization at temperatures around 2000°C. The main difference between the two types of fiber lies in differing microstructures. The pitch fibers have a radiating crystal structure as illustrated in the fracture surfaces given in Fig. 8. The difference arises from the fact that the crystal nuclei are formed prior to spinning where the molecules are collectively

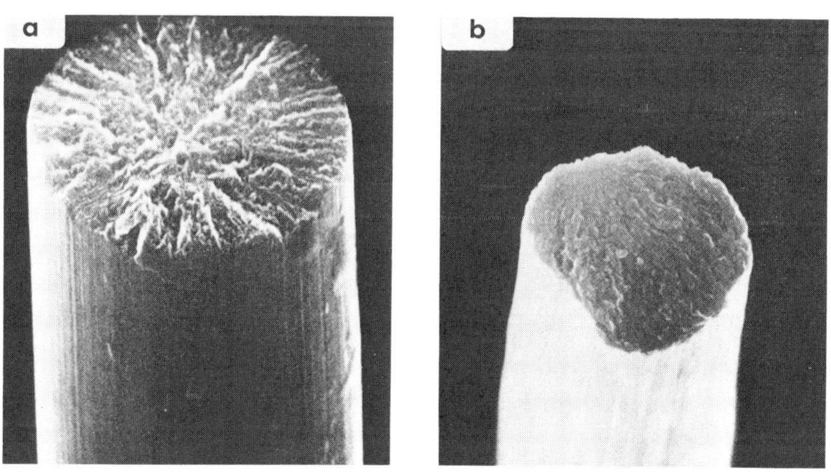

Fig. 8 Fracture surfaces of high-modulus carbon fibers. (a) mesophase pitch fiber precursor. (b) PAN fiber precursor. (after Bacon [*14*])

orientated. This effect results in the fibers having a lower transverse strength than PAN-based fibers. Furthermore, variations in the precursor pitch can lead to inconsistencies in the final fiber. These disadvantages have limited the impact of this technology, but advances since the late 1970s have demonstrated that high performance can be achieved ecnomically [*11*].

B. Aramid Fibers

The two commercially important fibers in this class are Du Pont products Kevlar 49 (poly 1,4-phenylene terephthalamide) and Nomex (poly 1,3-phenylene isophthalamide). A higher-performance fiber, Kevlar 149, was announced in 1987 [*15*]. Other manufacturers have competitive products near to commercialization. The fibers are produced by an analogous technology to pitch-based carbon fibers in that liquid crystal or nematic spinning dopes are employed.

In solution, these polymers with highly aromatic molecular skeletal backbones take on rigid rod conformations, which at concentrations of ≈ 5 to 10% in polar solvents orientate into liquid crystals The phase change from isotropic to anisotropic behavior is accompanied by a reduction in solution viscosity. These observations enable spinnable solutions from high–molecular-weight polymer to be obtained. This situation is illustrated in Fig. 9, where the phase diagram for poly(p-phenylene terephthalamide) in concentrated sulfuric acid shows that in 100% H_2SO_4, a higher solids content than 10% could be achieved. It was further observed that at concentrations greater than 12%, the melting point of the liquid crystals increased linearly with temperature (Fig. 10). Thus at 90°C, a solids content of 20% could be achieved. By careful control of the spinning conditions ($\approx$20% in 99.8% H_2SO_4 at 90°C, the temperature at which the anisotropy begins to decrease and polymer degradation is minimized), a good-quality fiber can be wet-spun into an aqueous coagulant bath and washed with a weak alkali to remove acidic impurities. In fact, residual sodium sulfate is a reported impurity and considered to be responsible for the observed variations in fiber-moisture absorption (3 ± 0.5%). A spin-stretch factor of between 1 and 14 is used in the production of high-modulus fibers.

1. Heat Treatment

The final morphology is achieved by a subsequent but brief heat treatment for a few seconds in an inert atmosphere at temperatures between

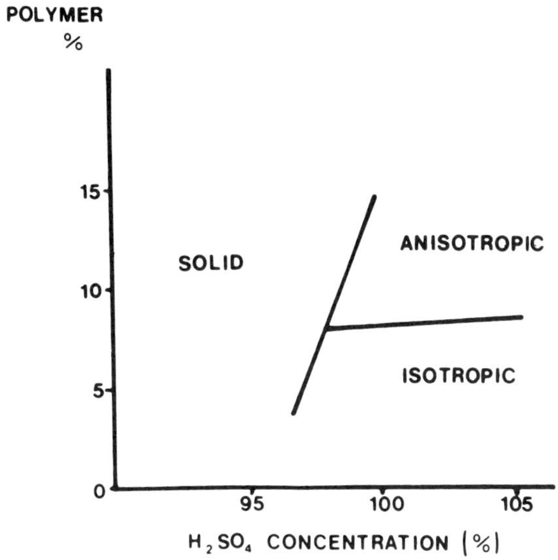

FIG. 9 PPD-T/H_2SO_4 phase diagram (PPD-T η=2.7). (after Magat [*16*])

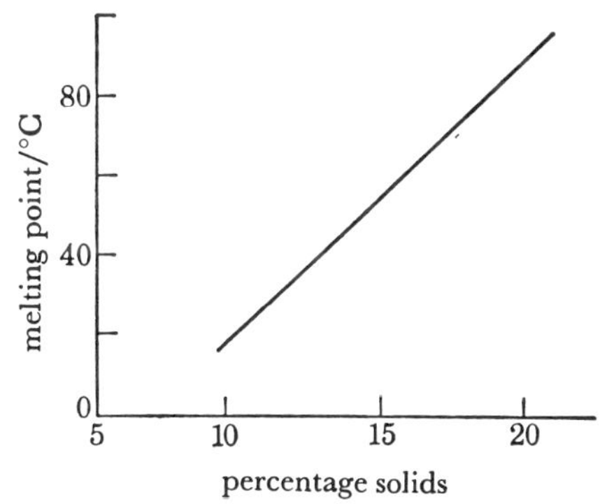

FIG. 10 Effects of solids concentration on the melting point of PPD-T/H_2SO_4 solutions. (after Magat [*16*]).

150 and 550°C when recrystallization occurs. For the high-performance Kevlar 49, heat treatments at temperatures of the order of 450°C lead to a perfection in the lateral order of the polymer. The improvement in modulus is shown in Fig. 11. The treatment times are critical since an undesirable drop in fiber strength can result from excessive polymer degradation. For the higher-modulus Kevlar, a hot-stretching technique is presumably employed to produce a more highly oriented crystal structure.

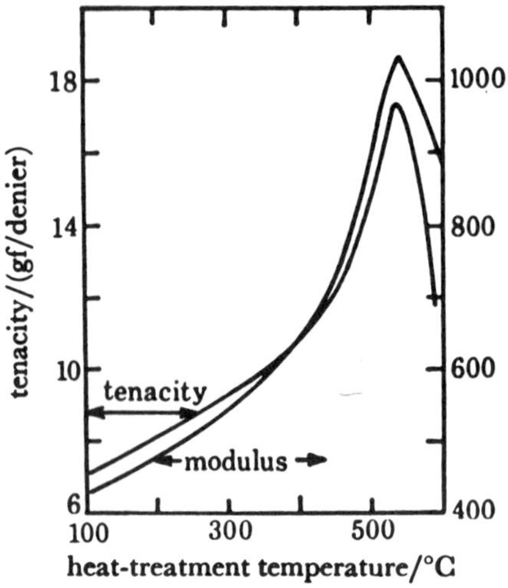

FIG. 11 Effect of heat treatment on properties of poly(1,4-benzamide) fibers. (after Magat [*16*])

2. STRUCTURES AND PROPERTIES OF ARAMID FIBERS

The spinning conditions result in a radial crystallite structure similar to that shown in Fig. 8a for pitch C-fibers. Because of the conformational arrangement of the polymer chains, a pleated structure is formed as illustrated in Fig. 12.

Kevlar 29 is the lower-modulus 1,4-Aramid with less lateral order and is

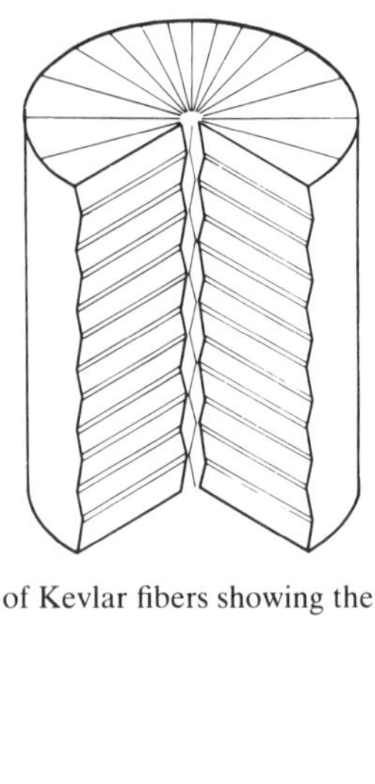

FIG. 12 Schematic diagram of Kevlar fibers showing the radially arranged pleated sheets. (after Dobb et al. [*17*])

primarily used for tire cord and ropes. Kevlar 49 is the high-modulus fiber obtained by high-temperature recrystallization. The irregular chain conformation of the 1,3-Aramid is not suitable for high stiffness, so that Nomex fibers have a significantly lower modulus.

The highly aromatic molecular structure gives rise to high thermal stability for an organic polymer. However, ultraviolet (UV) light can sensitize the oxidative degradation process. Although the degradation products are self-screening, it is essential that in applications where UV environments are encountered that protective films (which may be GRP for composites) are employed.

The polar nature of the molecule means that the equilibrium moisture content is significant and steps are required to dry the fibers and to prevent moisture ingress during use.

The crystalline morphology just described is responsible for the poor fiber transverse strength and the tendency to fibrillate. Consequently, the fibers have poor compressive strengths. The composites also have relatively poor transverse properties in compression or in shear because of the limited adhesion chemistry between fiber and matrix and between fiber crystallite planes. These properties can be utilized in applications where impact resistance is critical, such as ballistic protection.

Attempts have been made to improve the fiber-resin adhesion by using plasma-type surface treatments and increase fiber integrity by γ-irradication cross-linking, but commercial treated fibers are not available. The fiber surface can be contaminated with textile-processing aids, which may also contribute to the poor transverse properties of Kevlar composites.

C. Glass Fibers

A number of glass compositions are commonly used for the production of glass fibers. The most important is E-glass, whereas for specialist applications others can be used: for example, S-glass where a higher modulus is required, ECR glass for acid resistance, and AR glass for alkali resistance (i.e., cement matrices). C-glass is usually produced only in the form of veil for chemically resistant GRP linings.

The fibers are prepared by drawing from the bushing below a molten glass reservoir as illustrated schematically in Fig. 13. Since the strength of glass fibers is determined by the distribution and size of surface defects [indeed, the virgin strength can fall from 3.5 to GPa; (see Fig. 1)], the most important and commercially sensitive stage in the process is the surface treatment or size which is applied by rubber roller prior to assembly into rovings. The rovings consist of strands, each containing 2000

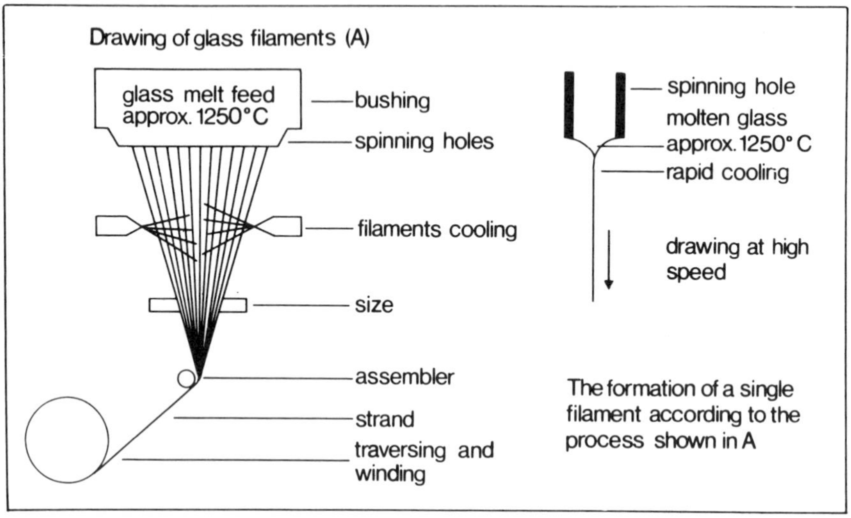

FIG. 13 Schematic diagram of glass-fiber spinning. (after Klunder [*18*])
(a) Drawing of glass filaments.
(b) Formation of a single filament according to the process shown in (a).

filaments. The rovings can be either used directly, rewound onto a bobbin (a small twist being put in to improve handling) for further textile processing, or chopped into short fibers for the production of chopped strand mat. (See Fig. 14.)

The surface "finish" is a critical stage in the process, ensuring compatibility with the resin or matrix and high-speed conversion into composite materials with minimum catenation or fluffing of the strands, fiber breaks, and damage. The "finish" consists of an adhesion promoter called a coupling agent, a protective polymeric coating or size, binder, and other additives which ensure good handleability.

D. Fiber-Reinforcement Form

The "finish" on glass fibers and the form of the reinforcement is dictated by the composite fabrication route required for the particular application. This is an important advantage of fiber-composite materials since the reinforcement can be tailored to the complexity of the molding operation. The production engineering can also be tuned to short-run or high production volumes of components. The fibers take the form of discontinuous and continuous reinforcements as illustrated in Fig. 14. The major forms of

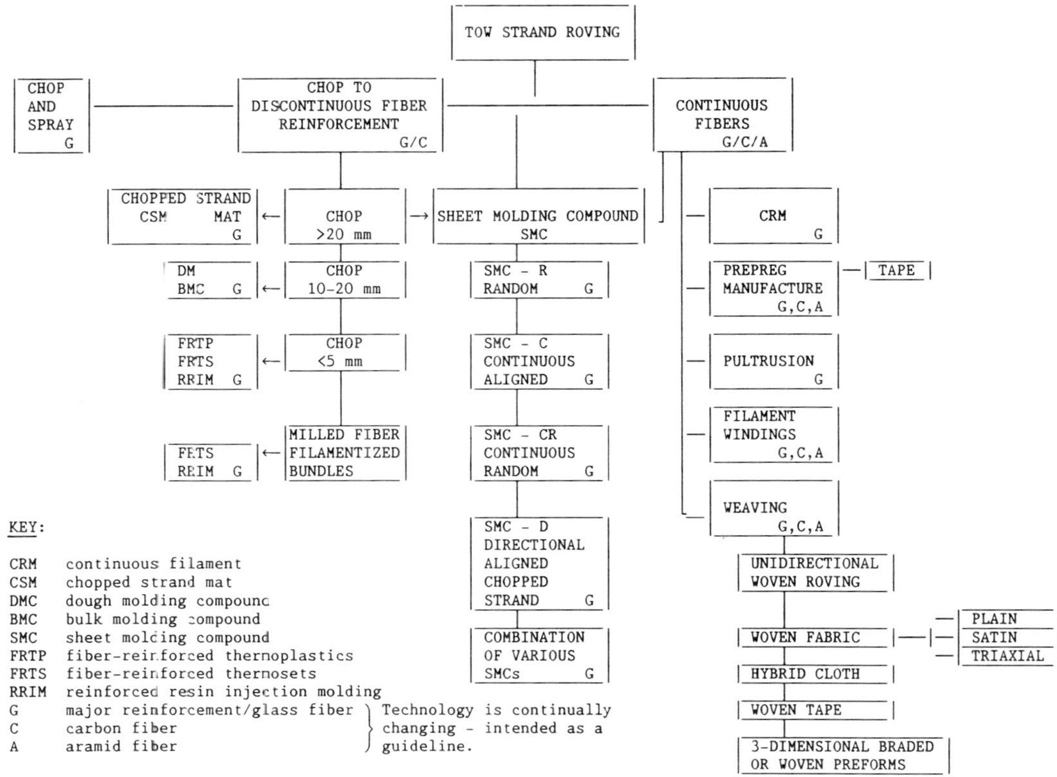

FIG. 14 Types and forms of glass-fiber products. R = Random; D = Directional; C = Continuous; CR = Continuous Random.

reinforcements used for carbon, glass, and Aramid fiber composites are indicated but because of changing technology and new applications these are given as a guideline only. Furthermore, the use of textile technology in composite materials has expanded significantly since the early 1980s with the use of hybrid woven cloths (e.g., carbon–glass) and three-dimensional braided or knitted reinforcements.

III. Fabrication Processes

There are many fabrication routes for fiber composites and these are reviewed in some detail in Table III. This review is by no means extensive since subtle combinations of these routes are often used to either produce a complex shape or to automate an otherwise labor-intensive process. Another example of recent innovation is the use of polyimide foam cores as mandrels for tape laying or filament winding. The resin-curing process for the thermosetting materials or thermal cycle for the crystallizable thermoplastic matrices is critical so that the fabrication route needs careful selection as does the molding conditions. As indicated previously, the performance of the composite largely emanates from the volume fraction and orientation of the fibers. The various techniques therefore produce composites with a wide variation in properties as illustrated in Fig. 15.

A. Prepreg Moldings

Much of the competition for aluminum alloys from composites comes from the use of high-performance carbon-fiber and glass-fiber laminates fabricated by prepreg molding using autoclave or pressclave techniques. A "prepreg" consists of collimated fibers or woven cloth preimpregnated with an epoxy resin (or other thermoset or high-performance thermoplastic) modified with a thermoplastic or elastomeric toughening agent and containing a latent hardener. (Prepreg from high-volume thermoplastics such as polypropylene and nylon are just becoming commercially available.) For the epoxy system, a latent hardener such as the insoluble dicyandiamide (DICY) is micronized in the resin solution prior to wet-impregnation into the collimated fibers and woven cloth and drying. The curing temperature and the kinetics of cross-linking can also be adjusted by incorporating a promoter or accelerator. The resin viscosity is adjusted to give the prepreg the appropriate tack by adding thermoplastic polymers or "advancing" the resin with partial reaction with the hardener (not with

TABLE III

FABRICATION PROCESS FOR FIBER COMPOSITES

Processes	Reinforcement	Vol %	Method	Common resin	Typical application
1. Hand lay-up	CSM or woven mat (G)	13–30	Resin impregnation of dry fiber with brush and roller	Polyester	GRP boat hull
2. Spray-up	Chopped rovings (G)	15–30	Direct application of chopped rovings and resin	Polyester	Similar to item 1
3. Autoclave	Continuous (G, C, A)	60	Forming of stacked prepreg in vacuum bag	Epoxy Bismalemides PMR PEEK FRTP	Aircraft wing or rotor blade
4. Filament winding	Continuous rovings, occasionally particulate fillers included (e.g., sand) (G, C)	50	Wet resin carried by fibres onto mandrel at predetermined angles. Variants: prepreg tape, woven tape, dry preform manufacture	Polyester Vinyl ester Epoxy Phenolic	Pipes and containers
5. Pultrusion	Continuous, occasional CSM and fillers (G)	60–80	Wet resin-impregnated roving or cloth formed through die	Polyester Epoxy Vinyl ester	I-beam or rods
6. Centrifugal casting	Chopped rovings, occasional particulate fillers (e.g., sand (G)	40–60	Chopped rovings and resin sprayed into rotating mold	Polyester	Pipes

TABLE III (*Continued*)

Processes	Reinforcement	Vol %	Method	Common resin	Typical application
7. Compression molding	Chopped or continuous and fillers (G, A, C)	10–60	SMC, DMC, wet resin plus CSM preform, prepreg, phenolic molding powder. Pressed to shape in a closed mold between matched dies.	Polyester Epoxy Phenolic	Medium-sized complex moldings (e.g., lorry body panels)
8. Hot-stamping thermoplastics	Continuous and discontinuous (G)	15–50	Forming of FRTP sheeting with fully impregnated fibers	Polypropene Nylon Advanced thermoplastics PEEK PES	Medium-sized thermoplastic moldings (e.g., car fascia)
9. Resin injection or transfer molding	Continuous or discontinuous plus polymer foam inserts (G, C, A)	25–30	Wet resin injected into woven or CSM preform within a closed mold	Polyester Vinyl ester	GRP car body panels
10. Reinforced reaction injection molding (RRIM)	Short-milled glass fiber or fiber preform, (G) similar to item 9	5 20	Fiber–polyol dispersion and isocyanate mixed at rapid mixing head and injected into mold. Chemistry controlled to give rapid reaction.	Polyurethane (Nylon 6)	Large complex parts (e.g., auto parts for impact resistance but with low stiffness)
11. Injection molding	Short fibers	5–20	Conventional injection molding of precompounded granules DMC, BMC, FRTP, FRTS	Polyester resin Several thermoplastics or phenolics and related molding powders	Numerous small-to-medium domestic moldings

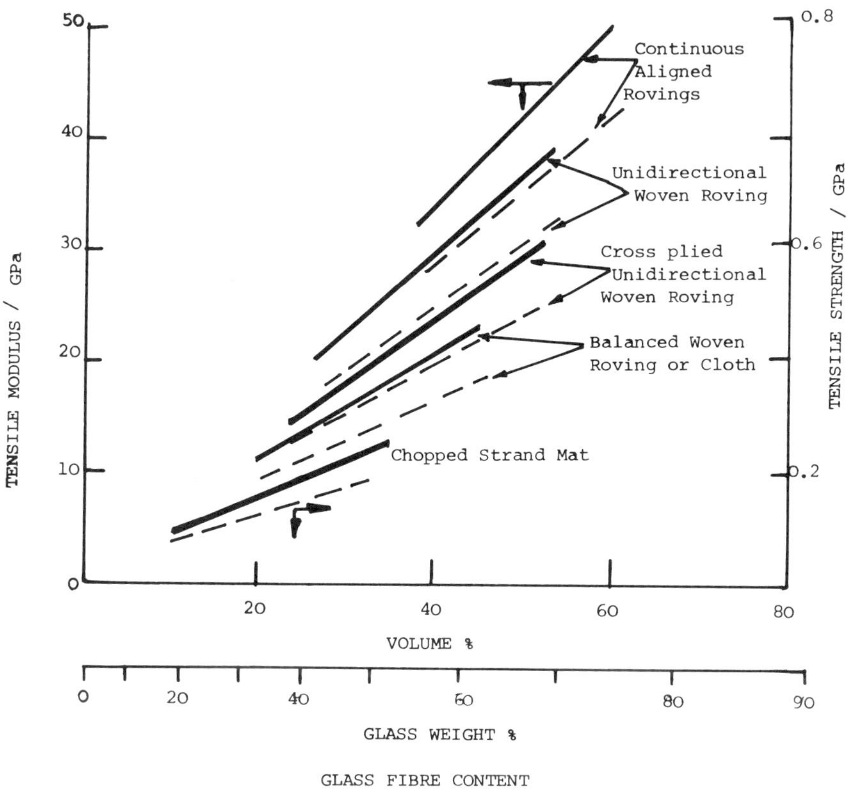

FIG. 15 Stiffness and tensile strength of polyester–glass laminates, showing effect of different forms of reinforcement over practical range of glass content.

DICY). The prepreg is stored between plastic release films in a sealed container in a freezer. After thawing and temperature equilibration to prevent moisture condensation and ingress, the unidirectional sheet is cut to appropriate dimensions and stacked at predetermined angles to produce a balance laminate (e.g., $0^{\circ}_{2}/\pm45^{\circ}/90^{\circ}_{2}/\pm45^{\circ}/0^{\circ}_{2}$; the subscripts define the number of prepreg lamina in each ply; the choice of stacking sequence is determined by the required laminate properties). The prepreg stack is laid up between release cloths and "blotting paper or similar" bleeding pack to soak up the excess resin in the prepreg. (See Fig. 16.) The bleeding of excess resin facilitates the removal of entrapped air and residual volatile impurities during consolidation of the prepreg into a void-free laminate. The curing schedule is also carefully controlled since the exothermic nature of the chemical reactions can lead to a temperature build-up in thick

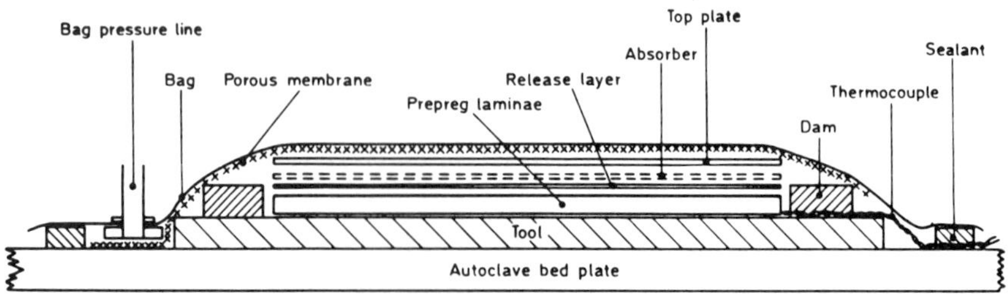

FIG. 16 Prepreg molding details. (after Purslow and Childs [*19*])

sections. Some of these processing difficulties can be overcome using thermoplastic matrices such as polyethersulphone (PES) and polyether etherketone (PEEK), but these have their own processing problems. Prepreg molding techniques have a complicated technology, but offer the mechanical engineer the opportunity to orientate fibers in the direction of the stresses of complex structures.

IV. Plastic Matrices for Fiber Composites

Plastic matrices for fiber composites generally can be divided into thermoplastics, chemosets, and thermosetting polymers. The latter two broadly fall into the same category of thermosetting resins, but for the purposes of this discussion we differentiate between those requiring additional hardeners, curing agents, or catalysts (the chemosets) and those that are quenched at a partially reacted state and continue to cure on heating. In both cases, curing involves the generation of a three-dimensional cross-linked molecular structure. Although it is generally assumed that the cross-linking process is homogeneous, there are notable examples where an abundance of indirect evidence points to a heterogeneous amorphous molecular structure. In the unsaturated polyester resins, this can arise from the statistical nature of free-radical addition polymerization, whereas for the epoxy resins, it is more likely to result from the heterogeneous nature of some of the curing agents.

Figure 17 gives the various types of matrices, but for this brief discourse, the discussion is limited to the main types: the epoxy resins and unsaturated polyester resins. The polyester resins are low polymers (molar mass $\approx$2000 g mol^{-1}) dissolved in a reactive diluent (styrene monomer), while the epoxy resins are relatively small molecules. These two systems satisfy

Thermoplastics
Saturated polyester eg.
poly(butyleneterephthalate) PBT

Polyamide (eg. Nylon 6)

Polyurethane

Polypropene

Polycarbonate

Polyphenylene ether PPO
-polystyrene blend

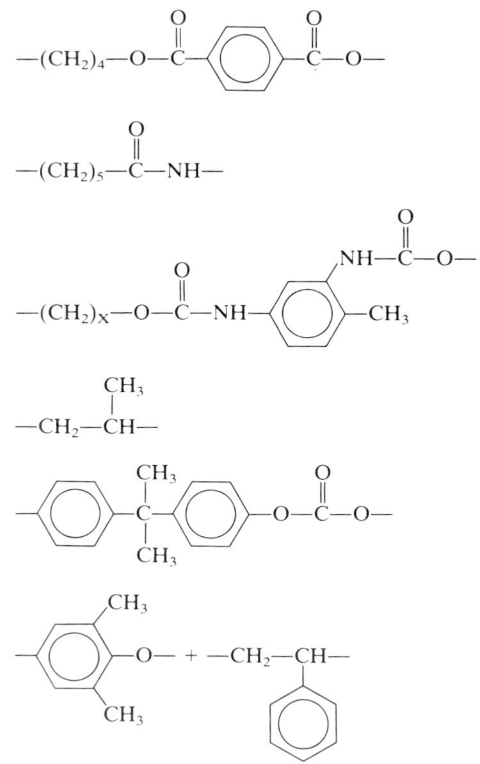

FIG. 17 Matrices for plastics composites.

Polyethersulphone PES

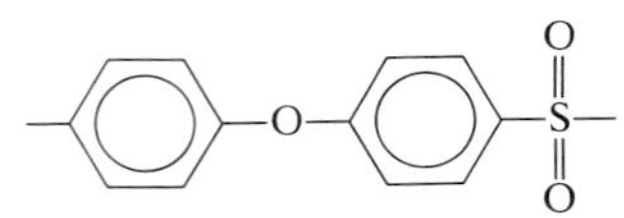

Polyetheretherketone PEEK

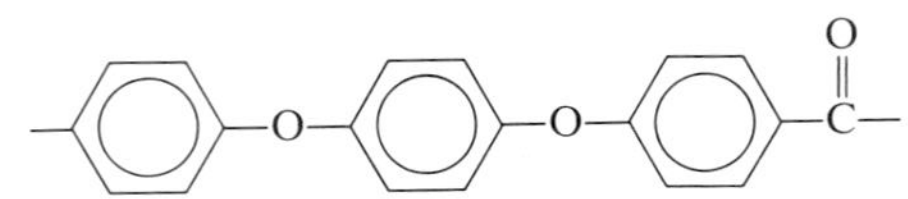

Thermosets

Phenolics (phenolformaldehyde resole)

Polyimides—conventional condensation
—PMR 15
—Bisaleimide

Polyurethane

Chemosets

Unsaturated polyester resin
(typical example is a styrenated
isophthalic polyester resin)

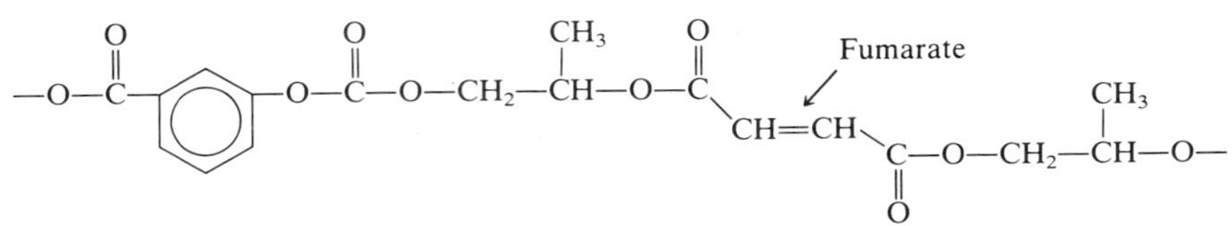

Vinyl ester resin
(also styrenated)

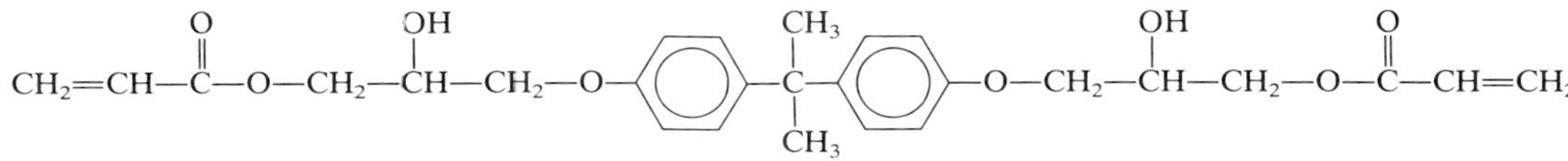

Epoxy resins

Difunctional:
eg. Diglycidyl ether of
Bisphenol A ($n = 0$)
(Epikote 828)

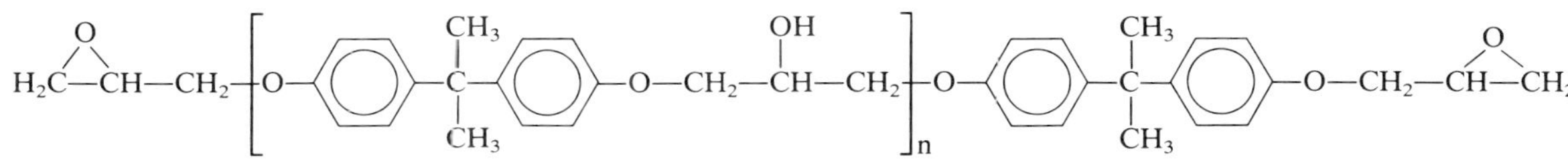

Trifunctional:
eg. MY 0510

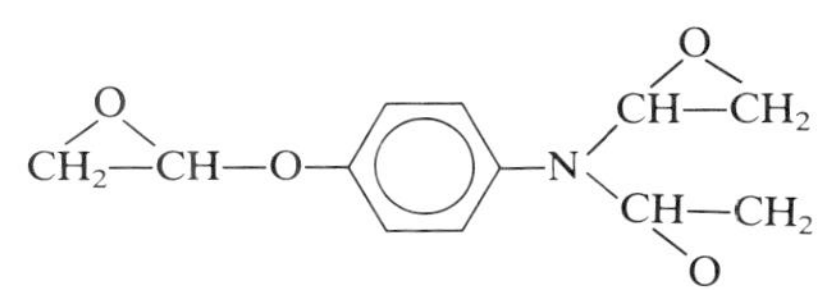

Tetrafunctional:
eg. MY 720

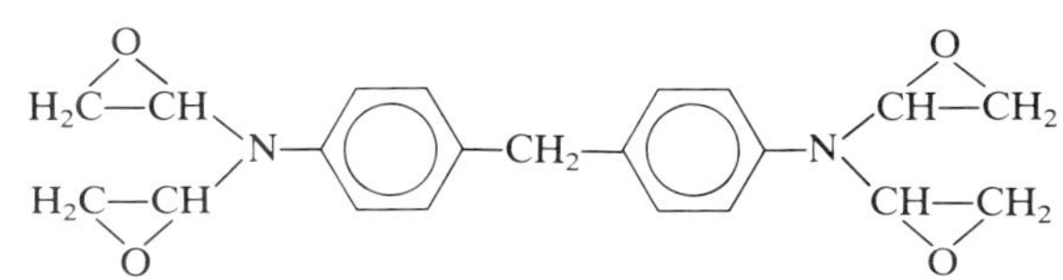

FIG. 17. (*Continued*)

the requirements of an appropriate matrix:

1. Low viscosity for fiber impregnation
2. High reactivity on curing
3. Chemical control of cure without volatile formation
4. Good mechanical properties of the cured resin

Criterion (3) is not readily achieved by the thermosets and more complicated processing is required. The PMR-15 and bismaleimides contain thermally polymerizable unsaturated groups and give significantly higher temperature performance (up to 400°C) than the advanced epoxy matrices (up to 200°C) with the potential of void-free laminates. These advanced systems are beyond the scope of the text and extend the potential for fiber composites but also pose different production-engineering problems which are being researched.

A. *Epoxy Resins*

The difunctional diglycidyl ether of Bisphenol A (e.g., Epikote 828) and related oligomers of degrees of polymerization up to four, form the basis of the general-purpose resins for both adhesives and composites. The glycidyl amines have a higher functionality and potentially higher cross-link densities and form the basis of the resins for aerospace composites.

1. Curing of Epoxy Resins

The epoxide or epoxy $-\overset{O}{\overbrace{\mathrm{CH-CH}}}$ functional group is highly reactive and forms the locus for the curing reactions through homopolymerization and copolymerization with a hardener. The properties of cured epoxy resins are therefore strongly determined by the curing system chosen and can range from the hardeners that require stoichiometric equivalent quantities with the reactive epoxy functional groups, say 80 parts/hundred of resin (phr) by weight, through to catalytic curing agents used at the level of a few phr.

Hardeners are molecules that have active "hydrogen" atoms which can add to the epoxy group. There are two main types: the acid anhydrides and the amines and polyamides.

The reaction with amines and polyamides can be described by the following chemical equations: Of note is the formation of pendant hydroxyl groups during the curing process, which is particularly valuable for promoting hydrogen bonds with substrates. These resins find potential uses as adhesives for this reason. With the polyamides, the reactive groups are well spaced and the cured resin has a higher fracture-toughness.

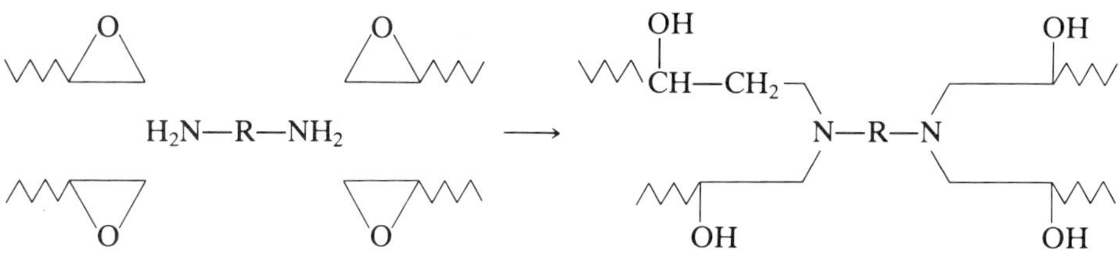

(2)

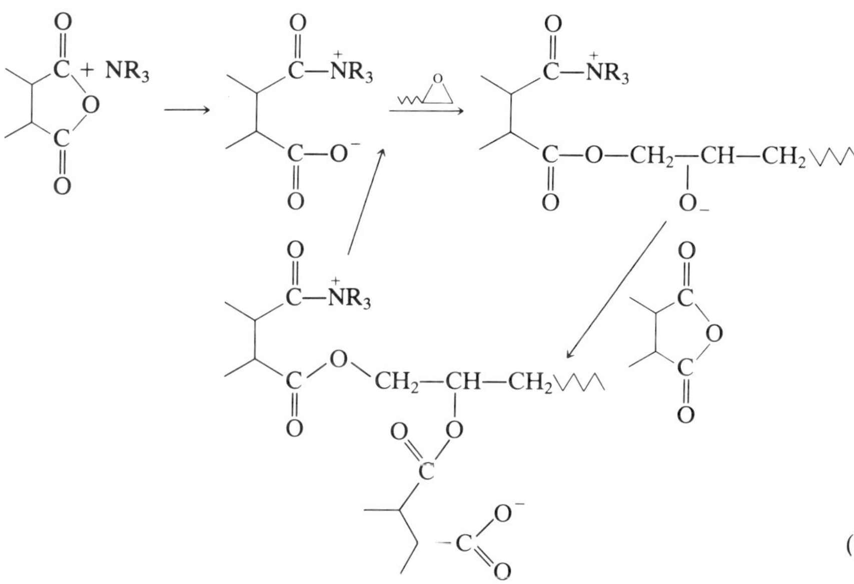

(3)

With the anhydride curing agents, several reactions are possible, giving a cured resin structure consisting of ester linkages with some pendant hydroxyl groups at irregular intervals along the polyester skeleton. The number of hydroxyl groups and the final polarity will depend on the relative rates of the individual reaction and on the presence of adventitious moisture. For composite fabrications, the rates of reaction are too slow and the hardening process is catalyzed by 0.5 to 2 phr of a tertiary amine which initiates an alternating copolymerization between the anhydride and epoxy groups (Eq. 3). The resulting resin is much less polar since hydroxyl groups will only form in side reactions and the molecular structure is essentially that of a polyester.

The catalytic curing agents such as tertiary amines and boron trifluoride complexes (at 2 to 6 phr level) initiate an anionic or cationic addition homopolymerization of the epoxy groups, respectively, to produce essentially a polyether molecular structure (Eq. 4). Dicyandiamide is also used in catalytic quantities but the chemistry is rather complex and still uncertain and produces a molecular structure akin to that expected of a polyfunctional amine.

$$BF_3:NH_2Et \longrightarrow H^+ + [BF_3NR_2]^-$$

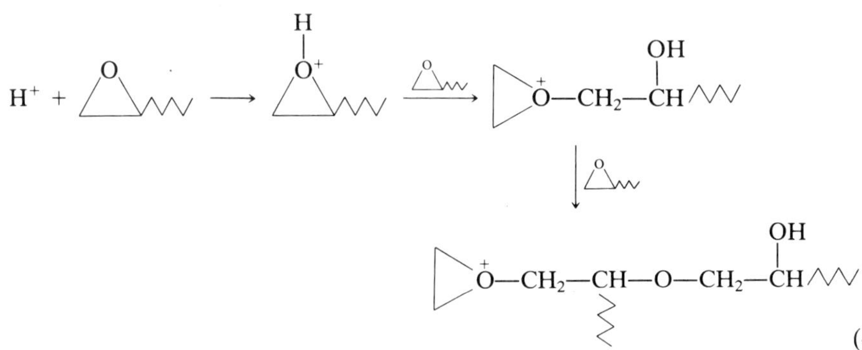

(4)

Both $BF_3 \cdot NH_2Et$ and DICY are latent curing agents, becoming active at 140°C and being ideal for prepreg manufacture. Commercial systems use mixtures of curing agents to optimize the temperature and rates of curing for appropriate applications. For example, the former is often used in conjunction with the amine DDS (see Table IV) and DICY with the accelerators "Monuron" and "Diuron" to give, for example, low-temperature processing.

It should also be remembered that the complete reaction of the epoxy groups is not normally achieved except after long times at high temperature so that higher-heat-distortion temperatures and chemical resistance can only be achieved by appropriate postcuring. The mechanical properties of epoxy resins are therefore determined by the hardener or curing systems and time-temperature profile. The typical range of resin properties that can be achieved from one base epoxy resin are given in Table IV. These can be generalized by appreciating that the heat-distortion temperature and Young's modulus increase with cross-link density and the concentration of aromatic groups within the network.

TABLE IV

EFFECT OF CURE MECHANISM ON THE PROPERTIES OF AN EPOXY RESIN (DIGLYCIDYL ETHER OF BISPHENOL 'A' [DGEBA])

Type	Agent	Level (phr)	HDT max (°C)	σ_{mu} (MPa)	E_m (GPa)	ϵ_{mu} (%)
Anhydride hardener	NMA	80	218	90	3.3	2.7
	THPA	60	120	81	2.7	4.5
	PA	50	110	82	3.0	4.1
Amine Hardener	DDS	33	193	78	3.1	6.0
	DTA	8	95	70	3.4	5.3
	TETA	13	96	62	3.6	2.6
Catalyst	BF_3 ·MEA	4	176	58	—	1.8
	DICY	4	125	—	—	—

DDS = diamino diphenyl sulphone.
DICY = dicyandiamide.
DTA = diethylene triamine.
HDT = heat-distortion temperature (maximum achievable value given).
NMA = nadic methylenetetrahydrophthalic anhydride.
PA = phthalic anhydride.
TETA = triethylene tetramine.
THPA = tetrahydrophthalic anhydride.

B. Unsaturated Polyester Resins

These are low-viscosity solutions of a linear unsaturated polyester resin in a reactive diluent which is invariably styrene. The base resin is prepared by polycondensation of maleic anhydride or fumaric acid with appropriate glycols and aromatic acids. The reactivity of the resin is determined by the number of fumarate groups in the polymer skeleton (Fig. 17). These trans-unsaturations are more likely to have been formed by cis-trans isomerization of maleic anhydride during the polycondensation process. The degree of isomerization can be varied by careful choice of the glycols and the reaction conditions. In contrast to the epoxy resins, a wide range of properties is achieved by varying the aromaticity and the fraction of reactive fumarate groups in the base resin. The curing process is a free radical addition copolymerization of the reactive diluent (usually styrene) with the polyester resin, with an average of two styrene units per polyester molecule. The reaction is initated by free radicals formed by thermal decomposition of an organic peroxide (e.g., benzoyl peroxide or methyl ethyl ketone peroxide) (called the catalyst) or redox reaction with an aromatic amine or more usually a cobalt soap (called the accelerator).

TABLE V

TYPICAL PROPERTIES OF CURED POLYESTER RESINS

	Cast resin properties					Laminate properties			
Polyester-resin type	Flexural strength (MPa)	Tensile strength (MPa)	Tensile modulus (GPa)	Elongation (%)	HDT (°C)	Glass (%)	Flexural strength (MPa)	Tensile strength (MPa)	Tensile modulus (GPa)
Orthophthalic	100	65–75	3.2	2.0–4.0	55–110	30	150	90	7
Isophthalic	140	70–85	3.5	3.5	75–130	30	230	120	8
Neopentyl glycol	130	70	3.4	2.4	110	30	170	90	7
Isophthalic/neopentyl glycol	130	60	3.4	2.5	90–115	30	160	90	7
HET acid	80	40–50	3.2	1.3–4.0	55–80	30	150	85	7
Isophthalic/HET acid	85	55	3.2	2.9	70	30	150	90	7
Bisphenol A	130	60–75	3.2	2.5–4.0	120–136	30	170	90	7
Chlorinated paraffin	110	50–60	3.4	1.2–4.8	55–80	30	140	90	7
Isophthalic/chlorinated paraffin	90	60	2.0	4.8	50	30	140	100	7

Source: From Weatherhead [20].

Resins can be preaccelerated during manufacture by incorporating a predetermined level of cobalt. But normally the "accelerator" comes in the form of a styrene solution and the catalyst as a solution or dispersion in an ester plasticizer. Thus, a wide range of resin reactivity (i.e., curing time and temperature) can be achieved by choice of chemical structure and concentration of catalyst and accelerator. These range from hot-curing formulations for compression molding of SMC, for example, or for cold-curing hand lay-up materials. In addition, the induction period before the onset of polymerization and therefore working times in hand lay-up systems or for factory impregnation of fibers prior to on-site installation of, for example, *in situ* pipe linings (the Insituform process) can be adjusted by incorporating known levels of free radical inhibitors such as hydroquinone.

The properties of the cured resin can be obtained by postcuring above the maximum-heat-distortion temperature which is typically $\approx$100°C. The overall performance of the resin can be varied over a wide range by careful formulation of the base polyester resin as shown in Table V.

The vinyl esters are similar resins, differing by the presence of terminal vinyl groups rather than fumarate links within the molecular backbone. These resins find application where corrosion resistance is specified.

V. Mechanical Properties of Unidirectional Fiber Composites

Fiber composites are, in general, highly anisotropic such that the fibers must be arranged at angles to one another to counteract multidirectional service stresses. In general, the characteristic failure behavior is one of damage accumulation over a wide range of stress and strain, culminating in failure or collapse. In addition to the anisotropic nature of the fiber-reinforced materials, both differences in the physical properties of the components and the statistical variability of fiber strength contribute to performance. Thus, under stress, debonding and microcracking occur which under most conditions is comparatively innocuous but can act as nuclei for more severe failure processes such as environmental attack. For example, as with some metals, environmental stress corrosion of E-glass GRP can occur as a result of interfacial failure [*21*]. But effects such as these are beyond the scope of this brief introduction, which considers the performance of unidirectional composites with brief reference to angle ply laminates. The science of discontinuous fiber composites is complicated by the need to transfer stress between the fiber and matrix, the stress concentrations at fiber ends, and the effects of flow-induced orientation. For a formal discussion, the reader is referred to the textbook by Folkes [*22*] and the article by Bader [*23*].

A. *Unidirectional Laminae*

1. Longitudinal Stiffness

When a tensile load is applied parallel to the perfectly bonded continuous fibers of a unidirectional lamina, the strain in the matrix ϵ_m will be equal to the strain in the fiber ϵ_f. Since $E_f > E_m$, the stress carried by the fibers σ_f is greater than that in the matrix σ_m. This is the basis of fiber reinforcement and

$$\begin{aligned} \epsilon_f &= \epsilon_c = \epsilon_m \\ \sigma_c &= \sigma_f V_f + \sigma_m V_m \\ \sigma_f &= E_f \epsilon_c \\ \sigma_m &= E_m \epsilon_c . \end{aligned} \tag{5}$$

By equating the volume fraction of fibers V_f with the fractional fiber area and the volume fraction of matrix V_m with the fractional matrix area, it can be shown that

$$E_l = E_f V_f + E_m V_m \tag{6}$$

and, since $V_m = (1 - V_f)$,

$$E_l = E_f V_f + E_m (1 - V_f). \tag{7}$$

This is known as the law of mixtures and assumes that both the fibers and matrix are perfectly elastic. At high V_f, any deviations due to viscoelastic behavior of the matrix are relatively insignificant. Similarly, the additional stresses due to differences in the Poisson's ratios of the fibers and matrix only lead to an error of $<2\%$. Experimental verification of Eq. (7) has been established.

2. Longitudinal Tensile Strength

Under a rising tensile stress parallel to the fibers, either the fibers or the matrix will fail depending upon the relative failure strains of the fibers (ϵ_{fu}) and matrix (ϵ_{mu}). Two cases need to be considered.

a. Case 1: $\epsilon_{fu} < \epsilon_{mu}$

With uniformly strong fibers at high volume fractions, when the fiber fails, all the load is thrown onto the weaker matrix. The effective cross-sectional area is reduced and the composite will immediately fail.

The strength of the composite is given by

$$\sigma_{lu} = \sigma_{fu} V_f + \sigma'_m (1 - V_f) \tag{8}$$

where σ'_m is the stress in the matrix when the fiber breaks as described in Fig. 14.

At low volume fractions (typically <~0.006–0.03 for GRP and CFRP), the matrix can carry the load leading to multiple fracture of the fibers. The strength is given by

$$\sigma_{lu} = \sigma_{mu} (1 - V_f). \tag{9}$$

These opposing strength mechanisms are described graphically in Fig. 18 and demonstrates clearly that a critical fiber volume fraction exists before reinforcement occurs. Figure 18 also demonstrates that the matrix contributes insignificantly to the failure strength of a unidirectional composite parallel to the fibers so that the strength is approximately given by Eq. (10):

$$\sigma_{lu} = \sigma_{fu} V_f. \tag{10}$$

In practice, however, the fibers have a variable failure strain (as shown in Fig. 1); a cumulative weakening process needs to be considered. Thus, fiber breaks will occur randomly at their weakest points. The first fiber break may lead to immediate failure if the flaw that is formed has a

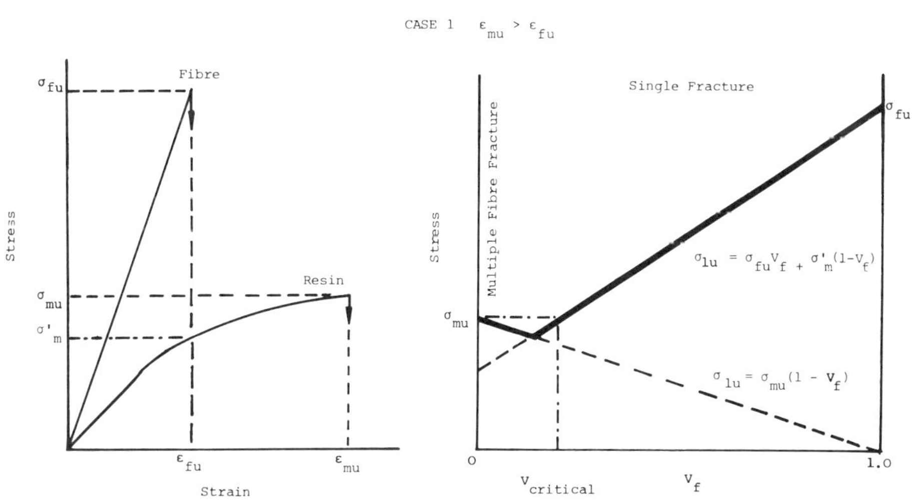

FIG. 18 (a) Stress–strain curves of fiber and resin for $\epsilon_{mu} > \epsilon_{fu}$.
(b) Variation of fracture strength of unidirectional laminae σ_{lu} with V_f for $\epsilon_{mu} > \epsilon_{fu}$ (case 1).

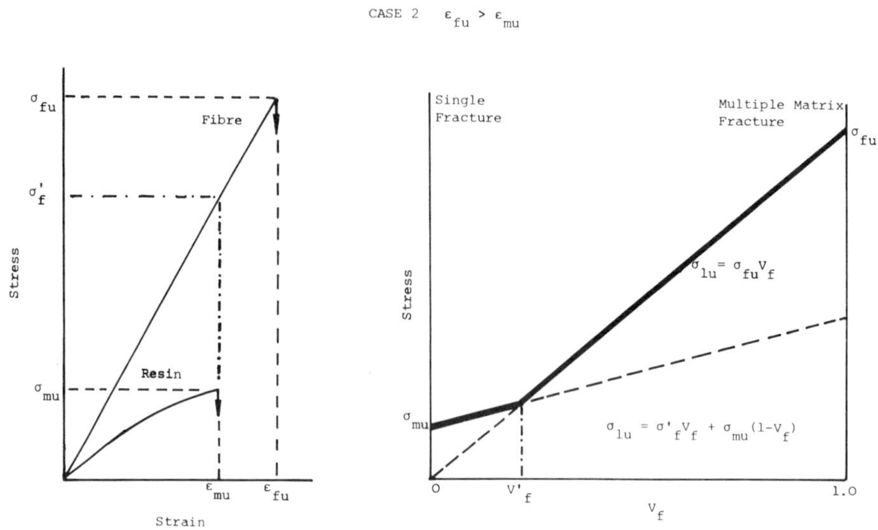

FIG. 19 (a) Stress–strain curves of fiber and resin for $\epsilon_{fu}>\epsilon_{mu}$. (b) Variation of fracture strength of unidirectional laminae σ_{lu} with V_f or $\epsilon_{fu}>\epsilon_{mu}$ (case 2).

sufficient intensity to cause the crack to propagate. This situation only arises when the fibers are very strongly bound to the matrix which itself has a low fracture toughness. In practice, however, the stresses at the fiber ends are modified and fiber-resin debonding occurs. The process of cumulative fiber damage is expected to occur until the ineffective lengths of the multiple fractured fibers is reached. However, there is an increased probability of fracture of adjacent fibers due to the redistribution of stresses that occurs at a fiber failure. This leads to a sequential failure and the production of a Griffith flaw which can result in composite failure. The number of fibers in an impregnated tow over which the stress is distributed and the critical number of adjacent fiber breaks are still the subject of research, but it does mean that dispersion of a high-failure-strain fiber (such as glass) into a bundle of low-failure-strain fibers (such as carbon) can lead to an increased composite strength. This effect has been called the hybrid effect.

The tensile fracture surface of a unidirectional composite has a brushlike appearance with varying degrees of fiber fracture and pullout (Fig. 20), making meaningful strength predictions difficult. For further discussion of this problem, the reader is referred to the texts by Hull [*24*] and Kelly [*25*].

FIG. 20a Scanning electron micrograph of fracture surface of carbon fiber–epoxy resin lamina tested in longitudinal tension. Fracture surface is relatively smooth and consists of a network of blocky outcrops of fibers and resin at different levels. (from J.D. Grundy, unpublished data). Reproduced from Hull [24] with permission.

FIG. 20b Scanning electron micrograph of fracture surface of glass fiber–polyester resin lamina tested in longitudinal tension, illustrating brushlike appearance associated with extensive fiber pullout. Reproduced from Hull [*24*] with permission.

b. Case 2: $\epsilon_{fu} > \epsilon_{mu}$

Under these circumstances, the matrix fractures first, throwing the additional load onto the fibers. At low V_f, as illustrated in Fig. 19, the fibers are unable to support the additional load and the composite breaks, giving

$$\sigma_{lu} = \sigma_f' V_f + \sigma_{mu} (1 - V_f). \tag{11}$$

σ_f' is the stress in the fibers when the matrix fractures. (See Fig. 19.)

At higher V_f when $V_f > V_f'$ $(= \sigma_{mu}/\sigma_{fu} - \sigma_f' + \sigma_{mu})$, the fibers are able to support the extra load when the matrix fails. Additional loads cause multiple cracking of the matrix. This phenomenon has been described in detail by Aveston and Kelly [*26*], but is also relevant to the fracture of laminates, which is discussed in Section VI. The strength of the laminate is clearly determined by the average strength of the fibers and

$$\sigma_{lu} = \sigma_{fu} V_f. \tag{12}$$

A practical example of the damage-accumulation process in these types of composites is the increased toughness of cement and concrete that has been reinforced with fibers.

3. Transverse Stiffness

A simple model for predicting the transverse Young's modulus of a unidirectional composite assumes a continuity of stress across the fiber–resin interface such that

$$\sigma_c = \sigma_f = \sigma_m \tag{13}$$

then

$$E_t = (E_f E_m)/[E_f (1 - V_f) + E_m V_f]. \tag{14}$$

This equation is also given in the following form:

$$1/E_t = V_f/E_f + V_m/E_m. \tag{15}$$

The fibers are noticeably much less effective in raising the modulus of the composite. As shown in Fig. 21, this equation is a poor estimate of the actual transverse modulus and a number of attempts have been made to estimate the nonuniform distribution of strain. The most rigorous approach is that of Hashin and Rosen [*27*], for which upper and lower bounds are given in Fig. 21. The Halpin and Tsai [*28*] equations are generally applicable and allow for variations in packing geometry and

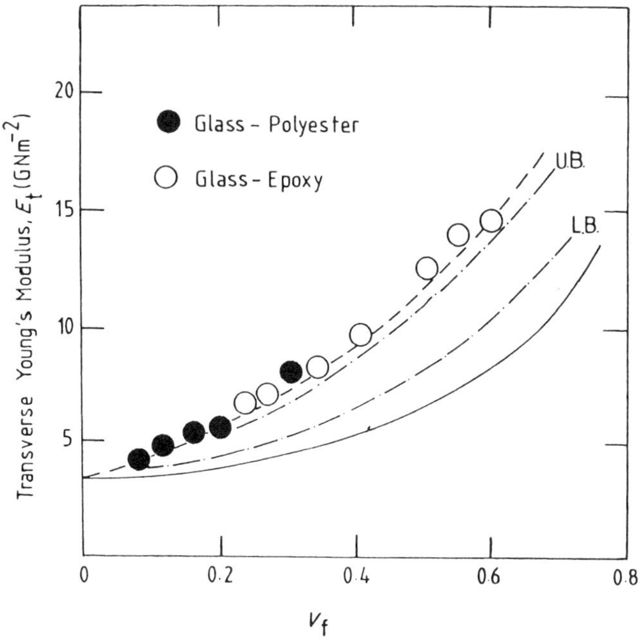

FIG. 21 Comparison of experimental and theoretical values of transverse Young's modulus of the unidirectional composites as a function of fiber volume fraction, (——) Rule of Mixtures (Reuss bound), (-··-) Hashin-Rosen upper (UB) and lower (LB) bounds, (----) Halpin-Tsai estimate. (after Bailey and Parvizi [*31*]).

regularity with the factor ξ

$$E_t = E_m(1 + \xi\eta V_f)/(1 - \eta V_f) \tag{16}$$

where $\eta = [(E_f/E_m) - 1]/[(E_f/E_m) + \xi]$.

4. Transverse Strength

The transverse strength of a unidirectional composite will be determined on a simple model by the strength of (1) the fibers transversely, (2) the interfacial bond, or (3) the matrix, whichever is weakest. This model is unacceptable since it fails to take into account stress concentrations in the matrix surrounding the completely embedded fibers.

The composite will therefore fail at a stress σ_{tu}, which is significantly less than σ_{mu} such that

$$\sigma_{tu} \simeq 0.5\ \sigma_{mu} \tag{17}$$

The matrix-failure strain in the composite ϵ_{mu} is much reduced by the strain magnification that arises from the presence of the much stiffer fiber. In practice this means that for a stress continuity, the strain in the matrix resin will be proportionally higher than the composite strain by the ratio of the moduli E_m/E_f. Kies [29] used a regular fiber array to predict the effective volume fraction of fibers and defined a strain-magnification factor (SMF) such that

$$SMF = \frac{\text{strain in the resin between the fiber}}{\text{strain in the transverse ply}} \tag{18}$$

and that

$$SMF = \frac{[2R + D]}{[2RE_m/E_f + D]} \tag{19}$$

where R = fiber radius and D is the interfiber spacing. For a square array,

$$D = R\,[\pi/V_f)^{1/2} - 2]. \tag{20}$$

Substituting Eq. (20) into (19) and putting $SMF = \epsilon_{mu}/\epsilon_{tu}$, then

$$\epsilon_{tu} = \epsilon_{mu}/\{1 + 2[(\pi/V_f)^{1/2} - 2]^{-1}\}. \tag{21}$$

In real laminates, the fibers are not arranged in a square array and Eq. (22) was considered by Garrett and Bailey [30] to give a better representation:

$$V_f(\text{true}) \simeq \pi R^2/(D + 2R)^2. \tag{22}$$

As shown in Fig. 22, the Kies equations do not accurately predict the transverse strength. Since debonding of the fibers prior to failure had been observed, Bailey and Parvizi [31] considered the stress concentrations in the interface for well-bonded and nonbonded fibers and derived the following equations:

(a) Well-bonded fibers:

$$\epsilon_{tu} = \epsilon_{mu}E_m/KE_t \tag{23}$$

(b) Nonbonded fibers:

$$\epsilon_{tu} = \epsilon_{mu}V_mE_m/K'E_t. \tag{24}$$

K and K' are maximum stress concentrations at the appropriate fiber matrix contact points for cases (a) and (b), respectively. These equations are plotted in Fig. 22 with K and K' taken as 1.0 and give an improved fit to the transverse strength but represent an oversimplification of the actual performance. It should of course be borne in mind (as with any strength predictions) that true values reflect small variations in the quality of the

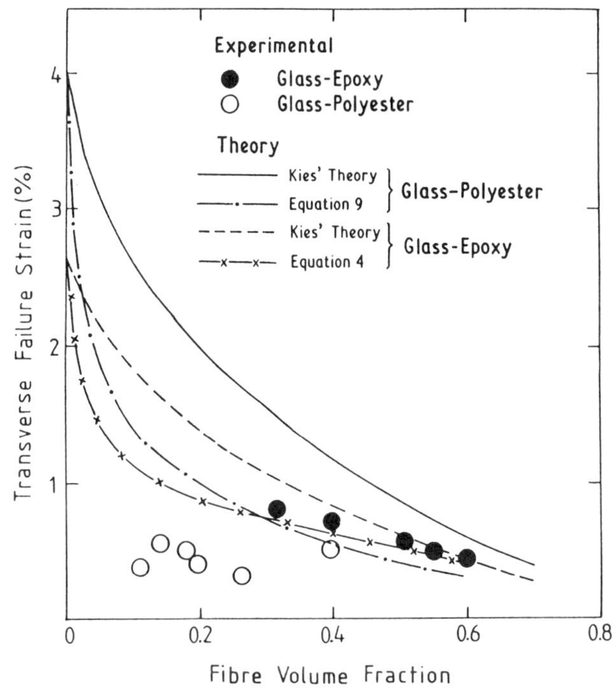

FIG. 22 Comparison of the theoretical predictions and experimental results of the transverse failure strain of the unidirectional composites with fiber volume fraction. (after Bailey and Parvizi [*31*])

product. However, the composite failure strains and stresses are low in the transverse direction, and lamination (or weaving) techniques need to be used in the design of fiber composites.

Table VI gives typical properties of 90° and 0° composites from carbon- and glass-fibers in epoxy resin. At other angles, to a first approximation σ_{cu} falls rapidly from 0° properties over the first 5° when fiber fracture dominates, through 5 to 20° when intralaminar shear occurs; at 20 to 45°, both intralaminar shear and transverse failure occur; and at 45 to 90°, the strength falls of more slowly when transverse failure is dominant.

VI. Failure Processes in Laminates

In practice, fiber composites are constructed with fibers at angles to one another. For advanced composites, this invariably involves the stacking of unidirectional plies (e.g., prepreg) at appropriate angles. These constructions need to be balanced to prevent warpage. The nomenclature describes

TABLE VI

MECHANICAL PROPERTIES OF UNIDIRECTIONAL LAMINATES

Laminate	0° CFRP	0° GFRP	0° AFRP
E/GPa	127 ± 3	42 ± 1	83
σ_{cu}/GPa	1.7 ± 0.1	0.92	1.31
ϵ_{cu}/%	1.16 ± 0.04	2.2	—
ν	0.29 ± 0.004	0.27	0.34
V_f	0.62 ± 0.02	0.55 ± 0.05	0.62 ± 0.03
Laminate	**90° CFRP**	**90° GFRP**	**90° AFRP**
E/GPa	8.3 ± 0.3	14 ± 0.5	5.6
σ_{cu}/GPa	0.039 ± 0.003	0.056	0.039
ϵ_{cu}/%	0.48 ± 0.05	0.5	—
ν	0.019 ± 0.002	0.09	—
V_f	0.02 ± 0.02	0.55 ± 0.05	0.062 ± 0.03

AFRP = Kevlar 49/epoxy composite.

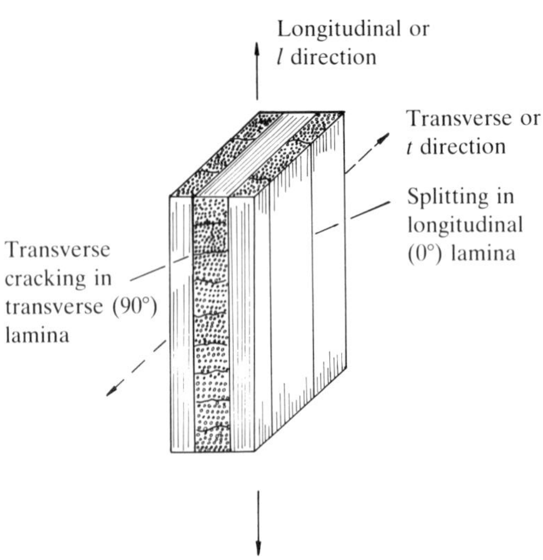

FIG. 23 Representation of model 0°/90°/0° cross-ply laminate with regularly spaced cracks in transverse lamina and splits in longitudinal laminae.

the number of plies of each type, for example, a simple eight-ply cross-play laminate $0_2°/90_4°/0_2°$ or more complex angle ply laminate, such as $0_2°/90_2°/\pm45°/90_2°/0_2°$. From the foregoing, it will be appreciated that the failure processes are highly complex. In particular, the free edge stresses in angle ply laminates can significantly modify the sequence of failure in individual laminae. A detailed discussion of these mechanisms is beyond the scope of this chapter. For an introduction to this topic, the reader is referred to the textbook by Hull [*24*]. Figures 23 and 24 show the transverse cracking and longitudinal splitting of the model 0°/90°/0° cross-ply laminate and illustrate the accumulation of damage that occurs in these systems. Because of the constraint afforded by the 0° plies, for optimum performance the 90° plies should be finely dispersed and those with a large

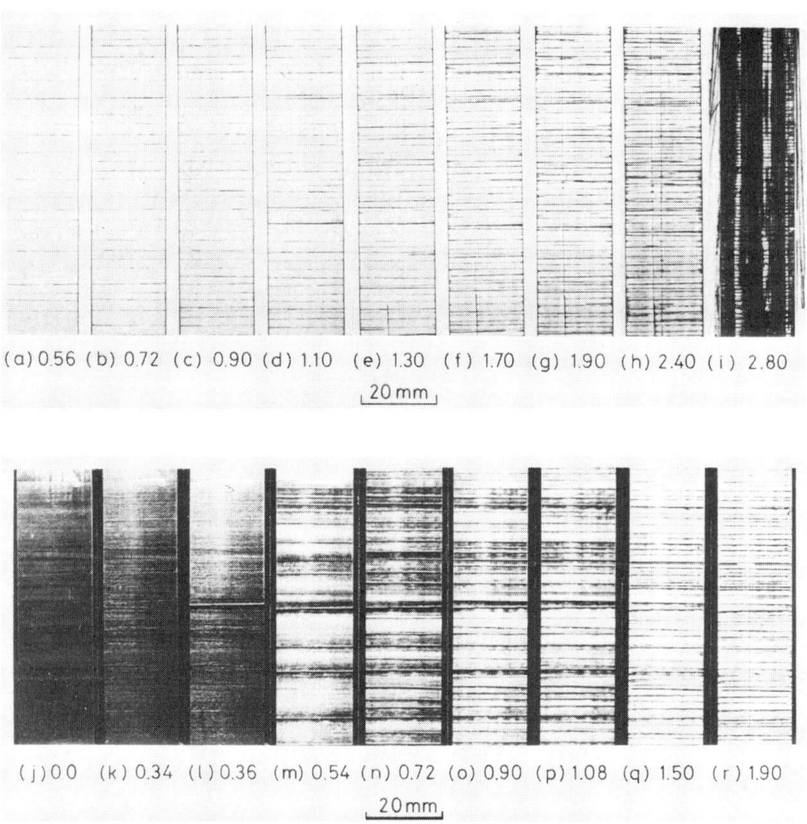

FIG. 24 Photographs of 0°/90°/0° specimens at the indicated strain levels (%) under (a to i) bright-field and (j to r) dark-field illumination. Note multiple transverse cracks in 90° ply stress "whitening" and longitudinal splitting in the 0° plies. (after Manders et al. [*32*])

Poisson's ratio (i.e., ±45° plies) should be located in the laminate mid-plane. The stacking sequence also needs to be selected to minimize the free edge stresses.

The important result of these studies and of fatigue studies is that microcracking is likely to occur in service and that design practice needs to take into account the presence of benign cracks in composite structures.

VII. Concluding Remarks

Peel et al. [33] have compared the design-specific strengths of carbon-fiber composites and aluminum alloys (Fig. 25) where it is seen that the major competitive thrust comes from CFRP applications where tensile stresses are expected. The weight savings of about 15% can readily be achieved using appropriate design considerations. Although the performance in compression is apparently not so beneficial, improvements in the retained compressive strength after impact can be achieved with composites fabricated from selected resins, with controlled fiber-resin interface and laminate configuration. It should also be pointed out that replacement of metal components in aircraft has been achieved using a conservative

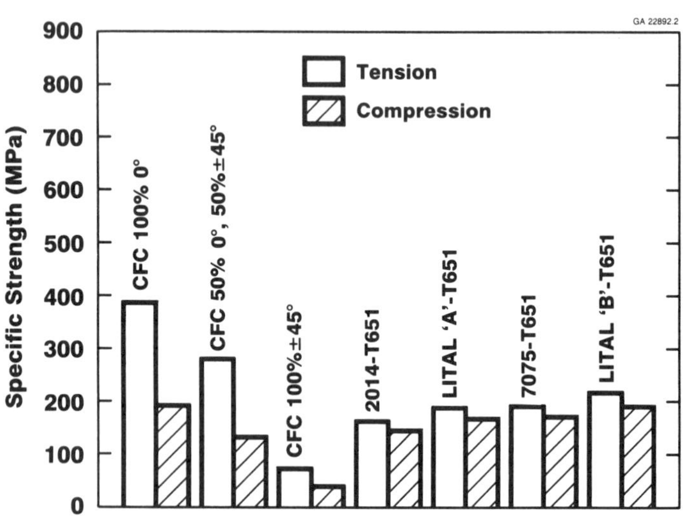

FIG. 25 Specific values of design-allowable strengths in tension and compression for selected carbon-fiber–composite lay-ups and aluminum alloys. Reprinted with permission from *Aluminum-Lithium Alloys II* [33].

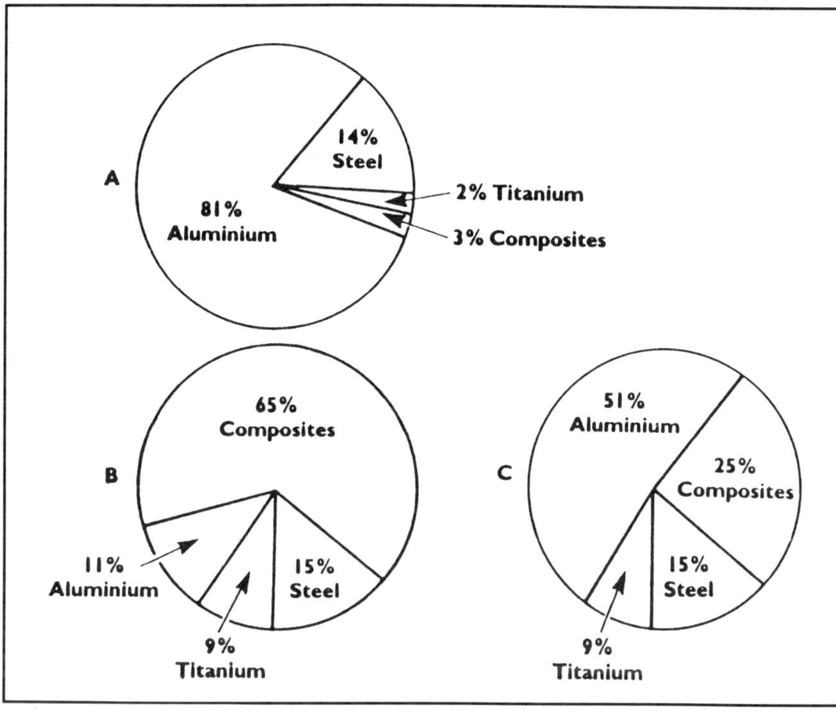

FIG. 26 Boeing prediction of future materials use in large commercial, advanced-technology subsonic aircraft, showing materials weight distribution: (a) 767; (b) composite scenario (1985–1995 subsonic aeroplane); (c) aluminum scenario (1985–1995 subsonic aeroplane). Reprinted with permission from *Aluminum Lithium Alloys II* [*34*].

design philosophy. The use of fiber composities has been fully justified by the in-service durability obtained and further advances can be expected. In addition, GRFP has proved an attractive competitor for industrial applications where ease of fabrication and low energy costs are important. Figure 26 gives the composites and aluminum scenarios for materials in large commercial aircraft, where the growing importance of fiber composities is clearly demonstrated [*34*].

Acknowledgments

The author wishes to thank J.E. Bailey for introducing him to the science of fiber composites. It is also a pleasure to acknowledge the contributions of the Composites Research Groups at the Universities of Surrey and Sheffield. The publishers and authors cited are also thanked for permission to reproduce their figures and tables.

References

1. Tsai, S. W. (1985). *Composite Desig*ı. (1985) USAF Mat. Lab., Think Composites, Dayton, Ohio.
2. Orowan, E. (1949). *Rep. Prog. Phys.*, **12**, 185.
3. Watt, W. (1985). In *Handbook of Composites, Vol. 1: Strong Fibers*, (W. Watt and B.V. Perov, eds.) Elsevier Science, Amsterdam, The Netherlands, pp. 327–387.
4. Bartenev, G.M., and Izmailova, L.K. (1984). *Soviet Physics Sol. St.*, **6**, 920.
5. McCrum, N.G. (1971). *Review of the Science of Fiber Reinforced Plastics*. Her Majesty's Stationery Office, London.
6. Bennet, S.C., and Johnson, D.J. (1978). In *Proc. 5th London Carbon and Graphite Conf.*, **vol. 1**. Society of the Chemical Industry, London, p. 377.
7. Bennet, S.C., Johnson, D.J., and Johnson, W. (1983). *J. Matter. Sic.*, **18**, 3337.
8. Reynolds, W.N., and Sharp, J.V. (1974). *Carbon*, **12**, 103.
9. Dunsford, D.V., Harvey, J., Hutchings, J., and Judge, C.H. (1981). "The Effect of Surface Treatment of Type 2 Carbon Fibres on CFRP Properties." RAE Technical Report 81096. Her Majesty's Stationery Office, London.
10. Denison, P., Jones, F.R., and Watts, J.F. (1985). *J. Mater. Sci.*, **30**, 4647–4656.
11. Denison, P., Jones, F.R., and Watts, J.F. (1986). *Surface and Interface Analysis*, **9**, 431–435.
12. Denison P., Jones, F.R., and Watts, J.F. (1988). *Surface and Interface Analysis*. 12,455.
13. Bacon, R. (1980). *Phil. Trans. R. Soc. London*, **A294**, 437–442.
14. Rand, B. (1985). In *Handbook of Composites*, *Vol. 1: Strong Fibers*, W. Watt and B.V. Perov, eds. Elsevier Science, Amsterdam, The Netherlands, pp. 495–575.
15. Riewald, P.G., Dhingra, A.K., and Chern, T.S. (1987). In *ICCMVI and ECCM 2*, **vol. 5**, F.L. Matthews, N.C.R. Buskell, J.M. Hodgkinson, and J.J. Morton, eds. Elsevier Applied Science, London, pp. 5.362–5.370.
16. Magat, E.E. (1980). *Phil. Trans. R. Soc. London*, **A294**, 463–472.
17. Dobb, M.G., Johnson, D.J., and Saville, B.P. (1979). *Phil Trans. R. Soc. London*, **A294**, 483–485.
18. Klunder, J. (1970). *Silenka Service Manual on Glass Fibers*. Silenka B.V., Amsterdam, The Netherlands.
19. Purslow, D., and Childs, R. (1986). *Composites*, **17**, 127–136; RAE Technical Report 85039 (1985), Her Majesty's Stationery Office, London.
20. Weatherhead, R.G. (1980). *Fiber Reinforced Polymer Technology*. Applied Science, London.
21. Jones, F.R., and Rock, J.W. (1983). *Composites*, **14**, 262–269.
22. Folkes, M.J. (1982). *Short Fibre Reinforced Thermoplastics*. Wiley, Chichester, England.
23. Bader, M.G. (1983). In *Handbook of Composites*, *Vol 4: Fabrication of Composites*, A. Kelløy and S.T. Mileiko, eds. Elsevier Science, Amsterdam, The Netherlands, p. 177.
24. Hull, D. (1981). *An Introduction to Composite Materials*. Cambridge University Press, England.
25. Kelly, A. (1973). *Strong Solids*. Clarendon Press, Oxford, England.
26. Aveston, J. and Kelly, A. (1973). *J. Mater. Sci.*, **8**, 352.
27. Hashin, Z., and Rosen, B.W. (1964). *J. Appl. Mechanics Trans. ASME*, **31**, 223.
28. Halpin, J.G., and Tsai, S.W. (1967). "Environmental Factors in Composite Materials Design." U.S. Air Force Materials Laboratory Technical Report AFML-TR67-423.
29. Kies, J.A. (1962). "Maximum Strains in the Resin of Fiberglass Composites." U.S. Naval Research Report NRL-5752.

30. Garrett, K.W., and Bailey, J.E. (1977). *J. Mater. Sci.*, **12**, 157, 2189.
31. Bailey, J.E., and Parvizi, A. (1981). *J. Mater. Sci.*, **16**, 649.
32. Manders, P.W., Chou, T.-W., Jones, F.R., and Rock, J.W. (183). *J. Mater. Sci.*, **18**, 2876–2889.
33. Peel, C.T., Evans, B., Baker, C., Bennett, D.A., Gregson, P.J., and Flaver, H.M. (1983) In *Aluminum-Lithium Alloys II*, (T.H. Sanders and E.A. Starke, eds.) Metallurgical Society of AIME, Warrendale, Pennsylvania, p. 328.
34. Quist, W.E., Norayanan, G.H., and Wingert, A.L. (1983). In *Aluminum-Lithium Alloys II*, (T.H. Sanders and E.A. Starke, eds. Metallurgical Society of AIME, Warrendale, Pennsylvania. p. 313.

Suggested Reading

Composites

Chawla, K.K. (1987). *Composite Materials—Science and Engineering*. Springer-Verlag, New York.

Hull, D. (1981). *An Introduction to Composite Materials*. Cambridge University Press, Cambridge England.

Jones, R.M. (1973). *Mechanics of Composite Materials*. McGraw-Hill, Washington. D.C.

Kelly, A. (1973). *Strong Solids*. Clarendon Press, Oxford, England.

Kelly, A., and Rabotnov, Yu. N. (1985). *Handbook of Composites*, **Vols. 1–4**. Elsevier Science, Amsterdam, The Netherlands.

Tsai, S.W. (1985). *Composites Design 1985*. Think Composites, Dayton, Ohio.

Tsai, S.W., and Hahn, H.T. (1980). *Introduction to Composite Materials*. Technomic, Westport, Connecticut.

Fibers

Donnet, J.B., and Bansal, R.C. (1984). *Handbook of Fiber Science and Technology*, **Vol. 3**. *Carbon Fibers* Marcel Dekker, New York.

Plueddeman, E.P. (1982). *Silane Coupling Agents*. Plenum, New York.

Plastic Materials

Brydson, J.A. (1981). *Plastics Materials*. Butterworth, London.

Reinforced Plastics Technology

Weatherhead, R.G. (1980). *Fiber Reinforced Polymer Technology*. Applied Science, London.

24

Titanium Alloys

J.C. WILLIAMS

General Electric Company
Cincinatti, Ohio

and

G.W. KUHLMAN

Alcoa Forging Division
Cleveland, Ohio

I. Introduction

Titanium alloys have been used for critical load bearing applications in fixed and rotary wing aircraft (both military and commercial) and in aircraft turbine engines for the past 20 years. These alloys have experienced growing acceptance and their usage has increased with each new generation of aircraft and engine until recently when, in combination, the rising cost of Ti alloys and the alleged benefits of advanced polymer matrix composites have caused the usage to level off. Yet, because of the potential for weight savings, the improved durability, the relative ease of fabrication, the superior temperature capability, and repairability in the field, Ti alloys remain an attractive choice for critical applications.

In this chapter, we shall discuss the properties and attributes of Ti alloys and draw some comparisons between them and competing advanced structural materials, most notably Al and polymeric matrix composites

ISBN 0-12-341831-3

and the new Al alloys, many of which are made by powder-metallurgy methods. We also shall attempt to separate promise from reality since this always is a troublesome issue when the pros and cons of materials of varying degrees of maturity are discussed.

II. Properties of Ti Alloys

The properties of Ti alloys have been the subject of numerous reviews [*1–9*]. References 1 through 9 are a representative but not exhaustive list of such reviews and summaries. The purpose of this chapter is not to provide a lengthy review of the properties and mechanical behavior of Ti alloys. However, it is important to demonstrate that Ti alloys as a class of materials can have a wide range of properties; this is shown in Table I. While it is instructive to illustrate this range, it also is important to emphasize that this range spans the properties of Ti alloys as a class. Thus the entire range of properties shown in the table cannot be realized in a single alloy. If the properties of Ti alloys based on the ordered intermetallic compounds Ti_3Al and TiAl are included in the table, the range becomes even greater. This is shown in Table II for purposes of comparison. The reason for including these tables is to demonstrate the versatility of Ti alloys.

In addition to the attractive properties of Ti alloys, they also have excellent general, pitting, and exfoliation corrosion resistance. Ti alloys also exhibit low galvanic currents when placed in contact with graphite-reinforced composites and Al alloys so that cathodic attack in multimaterial structures is generally not a problem. This is not the case for

TABLE I

PROPERTIES OF TITANIUM ALLOYS (WITHOUT TI ALUMINIDES)

Property	Units	Range of values
Young's Modulus	million psi [MPa]	11–20 [76–138]
Density	pounds per cubic inch [gm/cc]	0.15–0.19 [4.2–5.3]
Yield strength	thousands of psi [MPa]	30–200 [207–1380]
Ultimate tensile strength	thousands of psi [MPa]	130–210 [897–1449]
Tensile ductility	percentage	0–40
Fracture toughness [K_{Ic}]	thousands of psi $\sqrt{in}$ [MPa$\sqrt{m}$]	20–110 [22–121]
Fatigue strength [@ 10^6 cycles]	thousands of psi [MPa]	30–100 [207–690]
Fatigue crack growth threshold	thousands of psi $\sqrt{in}$ [MPa$\sqrt{m}$]	5–20 [6–22]
Stress corrosion cracking threshold	thousands of psi $\sqrt{in}$ [MPa$\sqrt{m}$]	20–100 [22–110]

TABLE II

PROPERTIES OF TITANIUM ALLOYS (WITH TI ALUMINIDES)

Property	Units	Range of values
Young's Modulus	million psi [MPa]	11–25.5 [76–176]
Density	pounds per cubic inch [gm/cc]	0.14–0.19 [3.9–5.3]
Yield strength	thousands of psi [MPa]	30–200 [207–1380]
Ultimate tensile strength	thousands of psi [MPa]	130–210 [897–1449]
Tensile ductility	percentage	0–40
Fracture toughness [K_{Ic}]	thousands of psi $\sqrt{in}$ [MPa$\sqrt{m}$]	15–110 [17–121]
Fatigue strength [@ 10^6 cycles]	thousands of spi [MPa]	30–100 [207–690]
Fatigue crack growth threshold	thousands of spi $\sqrt{in}$ [MPa$\sqrt{m}$]	5–20 [6–22]
Stress corrosion cracking threshold	thousands of psi $\sqrt{in}$ [MPa$\sqrt{m}$]	20–100 [22–110]

Al-alloys–graphite composites, which works to the advantage of Ti alloys. With a few exceptions, they also exhibit relative immunity to stress corrosion cracking (SCC) and at high strength levels are considerably better than high-strength steels or high-strength Al alloys with regard to SCC [*9*].

The properties of a particular Ti alloy are also very dependent on the processing history, both thermal and mechanical. Some typical properties for three selected alloys are shown in Table III to illustrate this point. From this table it can be seen that not only can the strength of an alloy be altered by heat treatment, but the toughness and ductility also can be significantly altered by thermomechanical processing. In fact, the $\alpha + \beta$ Ti alloys such as Ti-6Al-4V were probably the first commercially available materials in which the guaranteed properties were achieved through thermomechanical processing. This is true for both flat-rolled products and forgings. As of the late 1980s, high-performance applications of Ti alloys require microstructures and properties that only can be attained through complex thermomechanical processing; yet, these have become practical to the extent that they are routinely reproduced in a wide range of commercial product forms. In competition with high-strength Al alloys, Ti alloys typically offer the designer higher specific and absolute strengths, excellent fatigue-and fracture-related properties, and superior elevated-temperature capability. It is these combinations of properties for Ti alloys that have been and will continue to be exploited in the manufacture of critical structures.

The properties of Ti alloys shown in Tables II and III suggest that these alloys compete very favorably with Al and graphite composites in terms of

TABLE III

Selected Properties of Three Titanium Alloys

	Yield strength (ksi) [MPa]	Ultimate strength (ksi) [MPa]	Tensile elongation (%)	Fracture toughness (ksi$\sqrt{in}$) [MPa$\sqrt{m}$]
Ti-6Al-4V [references 5, 7, 8][a]				
Heat treatment/condition[a]				
β-processed	128 [883]	144 [993]	11	73 [80]
β-quenched	142 [979]	156 [1076]	10	86 [95]
Annealed	141 [972]	155 [1069]	14	69 [76]
Solution treat. and aged	137 [945]	151 [1041]	13	73 [80]
Ti-6Al-2Sn-4Zr-6Mo [reference 3]				
Processing and heat-treatment condition				
β-process + STA {1625°F/AC + 1100°F/4h}	152 [1048]	174 [1200]	7	52 [57]
High α + β process + STA {1625°F/AC + 1100°F/4h}	162 [1117]	176 [1213]	13	31 [34]
α + β process + STA {1650°F/AC + 1000°F/4h}	167 [1151]	180 [1241]	14	24 [26]
α + β process + anneal {1300°F/AC}	154 [1062]	164 [1131]	13	24 [26]
Ti-10V-2Fe-3Al [reference 6]				
Processing/heat-treatment condition				
High strength				
β process-STA (1400°F/WQ + 930°F/8h)	183 [1262]	194 [1337]	8	42 [46]
β process-DA (930°F/8h)	179 [1234]	188 [1296]	9	48 [53]
α + β process-STA (1400°F/WQ + 930°F/8h)	177 [1220]	184 [1269]	11	51 [56]
Intermediate strength				
β process-STA (1400°F/WQ + 1000°F/4h)	166 [1158]	174 [1200]	11	58 [64]
β process-DA (1000°F/4h)	157 [1083]	168 [1158]	14	56 [62]
α + β process-STA (1400°F/WQ + 1000°F/4h)	164 [1130]	171 [1179]	14	58 [64]
Low strength				
β process-STA (1400°F/WQ + 1100°F/4h)	137 [945]	144 [993]	17	93 [102]
β process-DA (1100°F/4h)	131 [903]	159 [1096]	14	88 [97]
α + β process-STA (1400°F/WQ + 1100°F)	146 [1007]	158 [1089]	17	71 [78]

[a] Yield strength and fracture toughness are quite sensitive to oxygen content.

stiffness and strength. In addition, the notch toughness of Ti alloys is decidedly superior to that of composites. It can be seen from the tables that the toughness and tensile ductility of Ti alloys exhibit inverse trends. That is, microsctructures that exhibit the highest toughness often have corresponding low tensile ductility. While explanation of this point is not the purpose of this chapter, it is appropriate to say that fracture strain during a tensile test is dominated by the strain to nucleate microvoids, whereas the toughness derives a major contribution from crack-path roughness. Roughness is typically greatest in microstructures that have the lowest strains for void nucleation. A more important point is that the toughness and tensile ductility of Ti alloys can be altered at a constant strength level. At specific strength levels (strength/density) in the range (2.5–3.3) $\times$ 10^4 meters, Ti alloys compete very well with composites. This becomes even more apparent when the relative cost of precision forgings and castings of Ti alloys are compared to comparable composite shapes. Composites, on the other hand, become attractive for large components where the size exceeds the plan area that can be conveniently produced in single forgings or castings and the shapes are complex relative to those that can be readily fabricated from plate (e.g., by superplastic forming and diffusion bonding). However, while such large composite structural members can be fabricated, the cost of such components is high and the ability to nondestructively inspect these is limited at present. Thus, the immediate application of composites for primary structure does not appear to be without its own set of problems.

III. Economics of Ti-Alloy Applications

Discussion of the economics of Ti-alloy applications is more complicated than comparing and discussing the properties of Ti alloys, Al alloys, and composites because the competitiveness is more dependent on the product form required for a particular application. For example, superplastic forming (SPF) and simultaneous superplastic forming and diffusion bonding (SPFDB) of Ti alloys have emerged as important fabrication technologies for complex Ti-alloy sheet structures with high section moduli, but the cost of Ti-alloy sheet is considerably higher than plate or forging billet. Thus, the economic aspects of materials selection for a particular application becomes component- and even detailed-design–specific. This is more the case for Ti alloys than for Al alloys because the increase in the cost of sheet with decreasing gage is more rapid for Ti alloys. The reason for this is related to the reactivity of Ti and the resulting production techniques used to make sheet. Since an analogous situation does not exist for Al-alloy sheet, the cost of sheet for these two materials diverges with decreasing thickness.

Recently, it has been shown that special processing of high-strength–ingot metallurgy Al alloys can render them superplastic. These alloys now can also be formed into complex geometries and therefore become more cost-competitive. Thus, for ambient-temperature applications, it is unlikely that Ti alloys can compete with these new-generation high-strength, superplastic Al alloys. However, at present the Al alloys that can be superplastically formed have very limited elevated temperature capability, and SPFDB Ti-alloy structures are still attractive for applications above ≈121°C e.g., for such structures as aft body skins of aircraft where the engine exhaust plume would overheat an Al-alloy structure. Compared to composites, Ti alloys are generally more economical for fabrication of complex shapes, especially nonaxisymmetric ones which are typical of airframe components.

The economics of Ti alloys have been driven by their relatively high material and fabrication cost in comparison with Al alloys, with resulting emphasis by both producers and users on development of net/near-net shape-making technologies that provide improved cost-effectiveness, as measured by "buy-to-fly" ratio or other value analysis techniques. In addition to SPF and SPFDB technologies reviewed previously for sheet, there have been significant developments in Ti-alloy precision forging and precision investment casting technologies to produce complex three-dimensional structures demanded by the aerospace industry, with little or no subsequent machining. In Ti-alloy precision forgings, Hot Die Forging is the critical process technology, wherein superalloy dies are heated at a close to the metal temperature and the strain-rate–sensitivity of Ti alloys is captured to provide the necessary deformation control to achieve very complex, thin web/rib shapes. Today, an equivalent degree of precision forging shape design and tolerance capability is available in Ti alloys as has been typical for precision Al-alloy forgings. In fact, in the six-year period after 1980, Ti-alloy precision forging capabilities emerged that required nearly 20 years to develop for high-strength Al-alloys, reflecting the emphasis placed on Ti-alloy–part cost reduction. Lost-wax, vacuum precision investment casting is not a new technology; yet, for Ti alloys its broad-based application has been recent and paced by the commercialization of hot isostatic pressing (HIP) for upgrading the quality of very large and very complex precision Ti castings used primarily in jet engines and recently as primary airframe structure in competition with forgings and other fabrication techniques. HIP has been shown to be particularly effective in upgrading Ti-alloy castings due to Ti alloys' to scavenge their own oxides, although HIP is also being used to upgrade Al-alloy castings as well.

Recently, emerging RST powder-metallurgy (PM) Al alloys based upon Al-Fe-X systems with improved elevated-temperature capability in the 177–316°C range are providing both mechanical-property and cost challenges to Ti alloys in some of their traditional applications. PM fabrication techniques are also candidates for reduction in Ti-alloy final-part costs and/or development of improved property characteristics. Ti-alloy PM processes are to two general types: blended elemental (BE) alloys consolidated using cold-hot isostatic pressing and sintering or pre-alloyed systems made by several RST techniques, consolidated by HIP with or without subsequent hot-working. The BE technique is used commercially to produce conventional pure Ti, $\alpha + \beta$ Ti alloys, and β Ti alloys in shapes that compete effectively with castings for noncritical components where corrosion resistance is the primary property sought. RST pre-alloyed Ti-alloy PM, however (while providing the potential for unique alloy formulations, including rare-earth dispersoids and titanium aluminides for improved elevated temperature capabilities and/or property enhancement of traditional alloys), has not found broad commercial acceptance due to the very high cost associated with powder production. On the horizon are Ti-alloyed powder-making techniques that may effect significant reductions in Ti PM cost and make them more competitive with Al-alloy PM materials.

Thus, in considering the economics of Ti-alloy applications in competition with Al alloys or composites, it is evident that manufacturing technologies (e.g., SPF-SPFDB, precision forging, precision casting, and PM) play a critical role in the decision-making process in competitive materials selection and exploitation of the range of mechanical properties Ti alloys offer the user. A further complication to the economics of Ti-alloy applications has been the behavior of the Ti market and the feast-or-famine supply–demand relationships that have typified the Ti industry since its inception in the mid-1940s. With aerospace as the largest single consumer of Ti alloys, the Ti market has been paced by aerospace demand, marked by periods of high demand and thus high cost, followed by periods of low demand and cost due to decline in aerospace business and or entry of new producers. Currently, free-world Ti market capability is less than 50% utilized and users are enjoying a period of relatively low-cost Ti-alloy product forms. Conversely, in 1978–1980, the last major aerospace surge, the cost of various Ti-alloy product forms was two to four times that prevalent today and, in this period, the extremely high cost of Ti alloys forced users to seek substitutions of lower-cost Al or Fe alloys. Growing commercial application of Ti alloys has served to broaden the market base and provide a more stable Ti industry with a more stable pricing structure over the long term.

IV. Ti-Alloy Applications

Traditionally, Ti alloys have been inexorably linked with critical aerospace applications that demand the superior characteristics these alloys possess and can tolerate the cost associated with their manufacture. A classic example is the turbine engine, for without Ti alloys the jet engine as we know it would not exist. Alpha and $\alpha + \beta$ Ti alloys are extensively used in the fan and compressor sections of such engines due to their light weight, excellent specific properties, fatigue-and fracture-related properties, and elevated-temperature characteristics. Ti alloys are used in turbine engines at temperatures up to 538°C and, with the emergence of alloys based upon Ti_3Al or TiAl, usage at 704–816°C, in competition with some Ni/Co-based superalloys, is probable. With Ti alloys, the weight of this propulsive system became practical, and state-of-the-art turbine engines making extensive use of advanced Ti alloys have thrust-to-weight ratios of 4:1. Application of Ti aluminides is one of the pacing technologies to further increases in thrust-to-weight ratio and engine performance. Jet engines have been and will continue in the foreseeable future to be the largest and most demanding consumer of Ti alloys in several product forms, primarily forgings and castings.

Major use of Ti alloys in highly critical fixed and rotary wing airframe structures, the second largest consumer, did not begin until the early 1970s with the construction of wide-body commercial and transport aircraft such as the Douglas DC-10, Lockheed L1011, and Boeing 747; highly sophisticated military aircraft such as the USAF F-15 and B1A/B and USN F-14; and advanced helicopters such as the USA UH-60. Military airframes continue to be a larger consumer of Ti alloys than commercial airframes, primarily due to cost. In competition with Al alloys or composites, Ti-alloy usage in airframes is driven by their elevated-temperature capability, specific strength and/or modulus, and fatigue-and fracture-related properties. For example, the workhorse $\alpha + \beta$ Ti alloy Ti-6Al-4V is extensively used in such structures. With compositional control and thermomechanical processing, specifically a beta-solution treatment and anneal, this alloy provides superior fracture toughness, e.g., 105 $MPa\sqrt{m}$ and superior fatigue crack propagation resistance. With the advent of sophisticated fracture-mechanics design techniques now prevalent in most advanced aircraft (particularly military programs), we must anticipate even more extensive airframe use of Ti alloys, thermomechanically processed to enhance or tailor fatigue-or fracture-related properties such as fatigue crack propagation. Airframe applications of Ti alloys include sheet, plate, forging, casting, and extrusion product forms.

The final aerospace application of Ti alloys is in spacecraft and missile systems, wherein all available product forms, including Ti BE PM and Ti composite parts, are exploited. Ti alloys have proven particularly effective in such highly weight-critical and cryogenic applications and can be found in components ranging from cruise missile wings to ordnance missile empennage to pressure bottles to critical space shuttle primary structure and tankage.

Commercial application of Ti alloys has received extensive industry focus in order to improve market stability and reduce the market's dependency on the aerospace sector. Most, if not all, of the major commercial applications of Ti alloys exploit the excellent corrosion resistance of Ti materials in a variety of environments. For example, welded commercially pure Ti tube has been developed to compete effectively with stainless steels and/or Cu/Ni-based alloys in powder-generation heat exchangers and evaporators in desalinization plants. In these applications, while Ti components require higher initial costs, Ti alloys have proven through many years of field experience to extend the lives and drastically reduce maintenance costs of such systems. In addition to applications in chemical and food-processing equipment primarily for castings, recently Mo-bearing Ti alloys such as Ti-6Al-2Sn-4Zr-6Mo and Beta C (Ti-3Al-8V-6Cr-4Mo-4Zr) have been demonstrated to be effective materials for use in down-hole, sour gas oil field applications, especially those requiring large-diameter, seamless extruded pipe. Other commercial applications for Ti alloys range from Al-alloy anodizing fixtures to specialized sporting goods to prosthetic devices. In this latter application, precision casting, BE powder metallurgy, and/or forging fabrication techniques for Ti-6Al-4V alloy are being utilized in combination with special coating technology to effectively compete with existing Co-Cr alloys and provide devices with significantly improved durability and reliability.

V. Conclusions

Modern designers, in the conduct of their tasks, have available a number of competing materials (and processes that produce them) including high-strength Al alloys, Al-graphite composites, and Ti alloys, many of which offer similar or overlapping capabilities. In selection of the best material and/or process for any given application, of necessity the designer must consider a number of factors, not all of which are complementary. Frequently trade-offs in properties, cost, fabricability, etc. are required. Ti alloys have been shown to offer a range of properties including excellent specific and absolute strength/modulus, elevated temperature performance, fatigue- or fracture-related properties, and corrosion resistance that make them formidable competitors with high-strength Al alloys or composites.

The mechanical properties of Ti alloys are highly dependent upon processing history; yet, this dependency has been exploited in the development and proven commercial application of combined thermal and mechanical processing techniques for these alloys to tailor their characteristics to final-part–specification requirements. Due to the relatively high cost of Ti alloys, sophisticated fabrication technologies such as superplastic forming, superplastic forming and diffusion bonding, precision forging, precision investment casting, and powder-metallurgy techniques are available on a commercial scale to provide the most cost-effective final part possible consistent with design requirements. Such technologies have proven in many cases to make Ti alloys quite cost-competitive with Al-alloy–fabricating techniques and/or composites.

Since the Ti industry's inception in the mid-1940s, Ti-alloy applications have been predominantly in critical aerospace applications such as turbine engines, fixed and rotary wing airframes, and missiles and spacecraft that exploit the excellent mechanical property combinations (particularly elevated-temperature and fatigue- and fracture-related properties) Ti alloys display. In recent years, the Ti industry has gained stability through broader-based commercial Ti-alloy applications in powder generation, desalinization, oil field, chemical process, prosthetic devices, etc., which have served to reduce the cost-volatility of Ti materials.

While the growth of Ti alloys' usage appears to have leveled due to user-cost reduction pressures and the advent of improved Al alloys, particularly superplastic and RST PM Al alloys, and/or composites with mechanical properties comparable to existing Ti alloys, it appears that Ti alloys will continue to play a major role in advanced gas turbines through development of titanium aluminides and enhanced fabricating technology for existing alloys; in advanced airframe structure designed by stringent fracture mechanics techniques through advanced alloys or enhanced existing alloys and fabrication techniques, including precision forgings and castings and lower-cost pre-alloyed Ti PM; and in the commercial sector, where Ti alloys still display superior corrosion resistance and total life-cycle cost-effectiveness.

References

1. Paton, N.E., Williams, J.C., Chestnutt, J.C., and Thompson, A.W. (1976). "The Effects of Microstructure on the Fatigue and Fracture of Commercial Titanium Alloys," AGARD Conference Proceedings, **185**, p. 4–1.

2. Duerig, T. W., Terlinde, G. T., and Williams, J. C. (1980). "Phase Transformations and Tensile Properties of Ti-10V-2Fe-3Al," *Metall. Trans.*, **11A**, 1987.

3. Chestnutt, J.C., Rhodes, C.G., and Williams, J.C. (1976). "The Relationship between Mechanical Properties, Microstructure and Fracture Topography in $\alpha + \beta$ Titanium Alloys," ASTM STP 600. ASTM, Philadelphia, p. 99.

4. Margolin, H., Chesnutt, J.C., Luetjering, G., and Williams, J.C. (1981). "Fracture Fatigue and Wear: Critical Review," *Titanium '80, Science and Technology*, H. kimura and O. Izumi, eds., **vol.1**. TMS-AIME, Warrendale, Pennsylvania, p. 169.

5. Starke, E.A., and Williams, J.C. (1984). "The Role of Thermomechanical Processing in Tailoring the Properties of Aluminum and Titanium Alloys." In *Deformation, Processing and Structure*, George F. Krauss, ed. ASM, Metals Park, Ohio, p. 279.

6. Kuhlman, G.W., Pishko, R., Kahrs, J.R., and Nelson, J.W. (1984). "Isothermal Forging of Beta and Near-Beta Titanium Alloys." In *Beta Titanium Alloys in the 1980's*, R. Boyer and H. Rosenberg, eds., p. 255.

7. Sparks, R.B., and Long, J.R. (February 1974). "Improved Manufacturing Methods for Producing High Integrity More Reliable Forgings" Technical Report AFML-TR-73-301. Air Force Materials Laboratory, Wright-Patterson, Ohio.

8. Chesnutt, J.C., Thompson, A.W., and Williams, J.C. (February 1978). Technical Report AFML-TR-78-68. Air Force Materials Laboratory, Wright-Patterson, Ohio.

9. Blackburn, M.J., Smyrl W.H., and Feeney, J.A. (1972). In "Stress Corrosion Cracking in High Strength Steels and in Titanium and Aluminum Alloys," B.F. Brown, ed. Naval Research Laboratory, Washington, D.C.

25

Magnesium Alloys

ROBERT BUSK

International Magnesium Consultants, Inc.
Hilton Head, South Carolina

I. Physical Metallurgy

In contrast to aluminum, magnesium has a close-packed hexagonal crystal structure, with parameters of $a = 3.202$, $c = 5.199$ angstroms, and $c/a = 1.624$ (which is very close to the ratio of 1.633 obtained by piling spheres in the same arrangement). This structure is basic to much of the physical metallurgy of magnesium and magnesium alloys.

At room temperature, slip occurs mainly on $(0001)\langle 11\bar{2}0\rangle$, with a small amount somtimes seen on pyramidal planes such as $(10\bar{1}1)\langle 11\bar{2}0\rangle$. As the temperature is raised, pyramidal slip becomes easier and more prevalent. However, note that the slip directions, whether associated with basal or the pyramidal planes, are coplanar with (0001), a general observation for all observed slip in magnesium and magnesium alloys. Therefore, it is impossible for a polycrystalline piece of magnesium to deform without cracking unless deformation mechanisms other than slip are available. These mechanisms are twinning, banding, and grain-boundary deformation.

Twinning occurs principally on the $(10\bar{1}2)\langle 10\bar{1}1\rangle$ system, resulting in a basal plane rotation around the $(10\bar{1}2)$ plane of 86°. Thus, the reoriented

ISBN 0-12-341831-3

basal plane within the new twin allows slip at an angle to the original plane. The geometry is such that, after twinning, there is an extension in the *c* direction of the parent lattice and a shortening in the *a* direction. Twinning, therefore, occurs most easily when a compressive stress is applied parallel to the basal plane. When the parent grain is under restraint (as is the case in a polycrystalline material), accommodation is accomplished by kinking on either the $(10\bar{1}0)$ or $(11\bar{2}0)$ planes.

After compression deformation, such as during rolling, large numbers of bands inclined to the normal of the compression axis are formed. These arise from a complex twinning and slip-deformation mechanism. Twinning first occurs on the $(10\bar{1}1)$ plane, then retwinning takes place on the $(10\bar{1}2)$ plane. Slip on the reoriented basal planes also occurs. This mechanism is important for the rolling of magnesium alloys.

Movement at the grain boundaries is an important mechanism for the deformation of magnesium. Shear at the grain boundary occurs anelastically, followed by grain-boundary migration to stabilize the shear. The. mechanism becomes more important as the grain-boundary area increases and as the temperature increases.

One important result of the deformation mechanisms peculiar to magnesium is that grain size is particularly important to the strength and ductility of magnesium alloys. As the grain size decreases, the grain-boundary area increases and grain-boundary shear becomes more probable, resulting in an increase in ductility. As the grain size decreases, the path length for the dislocation movements required for either slip or twinning decreases, resulting in an increase in strength. Thus, the finer the grain size, the higher are both strength and ductility. It is considerably more important for magnesium than for aluminum that all forms of alloys have as small a grain size as possible.

When magnesium alloys are deformed (as by rolling, extrusion, or forging), the basal planes of all the crystals in a polycrystalline aggregate tend to become parallel to the plane of deformation. Since twinning occurs when a compressive stress is applied parallel to the basal plane, extension of a deformed piece of magnesium in the direction of working must rely on deformation principally by slip and grain-boundary shear, with some twinning due to grain-boundary constraints. On the other hand, compression in the direction of working can be accomplished by slip, grain-boundary shear, and extensive twinning. The result is that wrought magnesium products, in contrast to wrought aluminum products, can have a lower compressive than tensile-yield strength. Since a fine grain size inhibits twinning and encourages grain-boundary sliding, the compressive yield strength approaches the tensile yield strength as the grain size decreases and becomes equal to it at a diameter of about 0.005 mm.

Magnesium forms significantly extensive solid solutions with a limited number of elements, principally those in groups Ib, IIb, and IIIb, plus the metals in groups IIIa, IVa, and Va, and the high–atomic-number members in the lanthanide series. Several elements (namely Si, Mn, Zr, low–atomic-number members of the lanthanide series, and Th) are important in spite of limited solid solubility. Because of its electropositive nature, magnesium forms high-melting intermetallic compounds with many elements, an extreme example being Mg_3Sb_2, which has a melting point of 1228°C. Because of its high vapor pressure (a boiling point of 1100°C), magnesium does not form alloys at normal pressures with the very-high-melting metals such as Mo, Ta, and W.

The strength and ductility of magnesium alloys are controlled by using the same mechanisms as those used for aluminum alloys: solid-solution hardening, precipitation hardening, cold-work, and grain-size refinement. As just noted, grain-size refinement is much more important to the strength and ductility of magnesium alloys than of aluminum alloys. The principle elements that form solid solutions that will precipitation-harden are aluminum, silver, yttrium, and zinc. The most common of these is zinc, especially in conjunction with the addition of zirconium for grain refinement.

II. Corrosion

As with aluminum, the corrosion properties of magnesium are controlled by the nature of the films that form on the metal and by the presence or absence of other metals that can lead to galvanic attack.

In the absence of condensed water, a thin, tightly adhering film of MgO forms. Protection is excellent, with a corrosion rate of about 10^{-5} mm/yr. In the presence of condensed water, MgO reacts with the water to form $Mg(OH)_2$, which has an equilibrium pH of 10.4 and is, therefore, stable in the presence of bases. However, it is not stable in the presence of acids and will break down when these are present. Strong acids, such as hydrochloric, destroy the film and rapidly dissolve magnesium. The carbon dioxide in the atmosphere dissolves in any condensed water to form carbonic acid, which reacts with the magnesium hydroxide to form hydroxy-carbonates. The same is true of other acid-producing compounds, such as sulphur dioxide in industrial atmospheres, the latter forming sulphates. If aluminum is present in the alloy, the aluminum-containing hydrotalcite ($MgCO_3 \cdot 5Mg(OH)_2 \cdot 2Al(OH)_3 \cdot 4H_2O$) forms in the presence of carbonic acid. Hydrotalcite is more stable than the aluminum-free hydroxy-carbonates.

Although the films that form on magnesium are quite stable, they are slightly soluble in water and, therefore, will not provide complete protection over long periods of time if exposed to water. They will also break down in the presence of certain activating ions such as chloride, bromide, sulphate, and chlorate. Salt water, which contains chlorides, is, therefore, a harsh environment for magnesium alloys.

Two anions, namely chromate and fluoride, produce very protective films on magnesium. A small quantity of sodium fluoride in water produces a film of magnesium fluoride on magnesium which is far more protective under normal conditions than magnesium hydroxide. Magnesium has even been used for constructing handling equipment for concentrated HF. Chromates are the basis for many protective chemical treatments for magnesium.

Thus, magnesium, unlike aluminum, is stable in alkaline and chloride-free environments; it is less stable in acidic and chloride-containing conditions. Since the world is in general acidic, and chlorides are ubiquitous, magnesium is less corrosion-resistant than aluminum.

Magnesium is the most electropositive of the commonly used structural metals and is, therefore, the anion in any pair of metals. If the pair are electrically connected and there is an electrolyte present, the magnesium will dissolve while protecting the metal to which it is connected. It is important in any structure containing both magnesium and other metals to protect the composite from galvanic corrosion either by interrupting the electrical circuit or by removing any electrolyte. The procedures used for joining aluminum to other metals are also those used for magnesium. It is especially important to guard against galvanic corrosion between magnesium and aluminum. The magnesium hydroxide formed as the magnesium corrodes results in an alkaline environment that can seriously attack the aluminum member. Both the anode and the cathode can be destroyed. High-purity aluminum and aluminum alloys of the 5000 series polarize sufficiently that no galvanic current is produced in contact with magnesium. 5000-series aluminum alloys can be used as washers in magnesium-containing assemblies.

The susceptibility of magnesium to galvanic corrosion means that the presence of second phases in the microstructure will result in poor corrosion resistance of the alloy in an electrolyte such as salt water if any second phase present is cathodic to magnesium and if the hydrogen overvoltage is such that hydrogen can be discharged at the second phase. It is especially important to control the amount of Fe, Ni, and Cu in magnesium alloys. Both the amount of the impurity and the distribution of the impurity in the alloy are important. For best corrosion-resistance, the amount of impurity

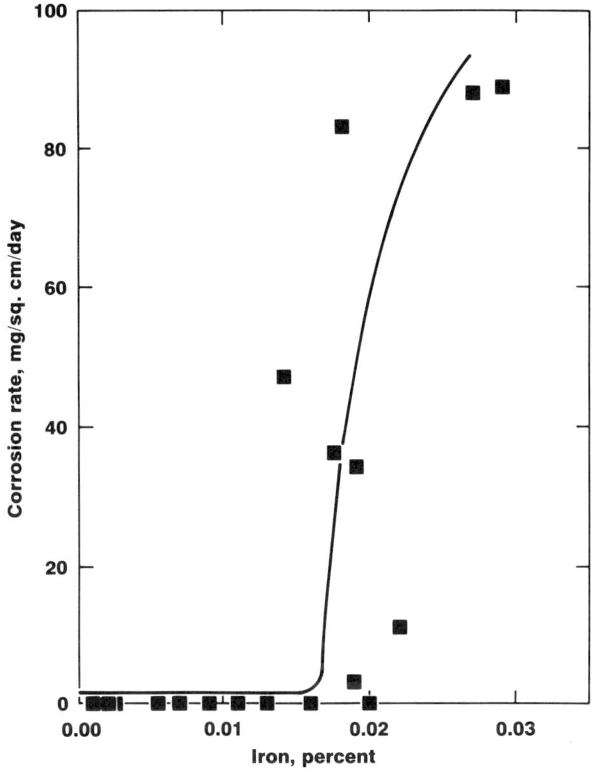

FIG. 1 Effect of iron on corrosion of pure magnesium (alternate immersion in 3% NaCl). Hanawalt, J.D., Nelson, C.E., and Peloubet, J.A. (1942). "Corrosion Studies of Magnesium and Its Alloys," *Transactions of the American Institute of Mining and Metallurgical Engineers*, **147**, 273–299.

should be low, and the cooling rate from the molten state should be rapid. Figure 1 is an illustration of the effect of composition on the corrosion of pure magnesium in salt water, using iron as the impurity. Below about 0.017% Fe, the corrosion rate is low. Above this amount, the rate climbs rapidly to very high values. Other impurities (for example, nickel and copper) show this same "tolerance-limit" behavior.

The corrosion properties of magnesium alloys are well illustrated in Table I. In this table, the rural environment is near Midland, Michigan, where the average SO_2 content of the air was 7 ppb. The industrial site was within the Dow Chemical plant in Midland, Michigan, where the average SO_2 concentration was 20–25 ppb. The marine site was 24 meters from the ocean at Kure Beach, North Carolina. The salt-water test was alternate

TABLE I

Typical Corrosion Rates of Four Magnesium Alloy Types (Loss of Thickness in Microns/Year)

Alloy	Form	Atmospheric exposure			Salt water
		Rural	Industrial	Marine	
AZ31B-F	Extrusion	14	25	20	24
AZ91A-T6	Gravity-cast	4	15	22	24
AZ91D-F	Die-cast	3	14	6	16
ZK60A-T5	Extrusion	14	36	25	59
For comparison: 2024-T3	Al sheet	0.1	2	2	
Steel (0.27% Cu)		15	25.4	150	

immersion in 3% NaCl. Within the magnesium alloys, the two that contain aluminum (AZ91A and AZ91D) have lower rates in all four environments than the alloy that does not contain aluminum (ZK51A). This is the effect of the formation of hydrotalcite on the surface of alloys containing aluminum. Although the impurity contents of both alloys are the same, the corrosion rate of the rapidly cooled alloy (die-cast AZ91D) is significantly lower than the slowly cooled alloy (sand-cast AZ91A) in both the marine and the salt-water environments. All three magnesium alloys have higher rates than 2024, but lower rates than the steel. The corrosion rate of the die-cast alloy AZ91D is not greatly different from 2024 in the marine environment.

III. Alloys

Commercial magnesium alloys can be considered in two major groupings: those that contain zirconium as an alloying element and those that do not. Zirconium is added in small quantities (0.5–1.0 wt. %) because it is a potent grain refiner of cast magnesium. Other elements are then added with zirconium to produce a variety of different alloys. Zinc and silver are added for room-temperature strength; rare-earth elements, thorium, and yttrium are added for elevated-temperature strength. Since aluminum and zirconium cannot coexist in molten magnesium, almost all of those alloys that do not contain zirconium contain aluminum. Zinc is added to aluminum-containing alloys for further strength, and manganese is added in small quantities for corrosion resistance. Silicon is added to die-cast alloys to improve elevated-temperature strength through dispersion-hardening

TABLE II

NOMINAL COMPOSITIONS OF COMMERCIAL MAGNESIUM ALLOYS, Wt %

Name	Composition	Remarks
AM60A	6 Al; 0.15 Mn	Die-cast
AM100A	10 Al; 0.10 Mn	Gravity-cast
AS41A	4 Al; 1 Si	Die-cast
AZ31B	3 Al; 0.2 Mn; 1 Zn	Wrought; Fe, Cu, Ni at low levels
AZ31C	3 Al; 0.2 Mn; 1 Zn	Wrought
AZ61A	6 Al; 0.15 Mn; 1 Zn	Wrought; Fe, Cu, Ni at low levels
AZ80A	8 Al; 0.12 Mn; 0.5 Zn	Wrought; Fe, Ni at low levels
AZ81A	7.5 Al; 0.13 Mn; 0.7 Zn	Gravity-cast
AZ81E	8 Al; 0.2 Mn; 0.6 Zn	Gravity-cast; Fe, Cu, Ni at low levels
AZ91A	9 Al; 0.13 Mn; 0.7 Zn	Gravity-cast; Cu at low levels
AZ91B	9 Al; 0.13 Mn; 0.7 Zn	Die-cast
AZ91C	9 Al; 0.13 Mn; 0.7 Zn	Gravity-cast
AZ91D	9 Al; 0.13 Mn; 0.7 Zn	Die-cast; Fe, Cu, Ni at low levels
AZ91E	9 Al; 0.13 Mn; 0.7 Zn	Gravity-cast; Fe, Cu, Ni at low levels
AZ92A	9 Al; 0.13 Mn; 2 Zn	Gravity-cast
EQ21A	2 Didymium; 1.5 Ag; 0.7 Zr	Gravity-cast
EZ33A	3 Mischmetal; 3 Zn; 0.7 Zr	Gravity-cast
HK31A	3 Th; 0.7 Zr	Wrought; gravity-cast
HM21A	2 Th; 1 Mn	Wrought
HZ32A	3 Th; 2 Zn; 0.7 Zr	Gravity-cast
K1A	0.7 Zr	Wrought; gravity-cast
M1A	1 Mn	Wrought
QH21A	2 Ag; 1 Didymium; 1 Th; 0.7 Zr	Gravity-cast
WE54A	4 Nd + heavy RE; 5 Y; 0.5 Zr	Gravity-cast
ZE41A	4 Zn; 1 Mischmetal; 0.7 Zr	Gravity-cast
ZE63A	6 Zn; 3 Mischmetal; 0.7 Zr	Gravity-cast
ZK30A	3 Zn; 0.7 Zr	Gravity-cast; wrought
ZK40A	4 Zn; 0.7 Zr	Wrought
ZK51A	5 Zn; 0.7 Zr	Gravity-cast
ZK60A	6 Zn; 0.4 Zr	Wrought
ZK61A	6 Zn; 0.8 Zr	Gravity-cast
ZM21A	2 Zn; 1 Mn	Wrought

by the finely divided Mg_2Si produced by the rapid solidification characteristic of pressure die casting.

Table II gives the composition of the principle commercial magnesium alloys in common use together with the form in which each is usually used. The naming system is the one given in ASTM Recommended Practice B275, where the first two letters specify alloying elements, the first two numbers show the wt % of those elements, and the final letter is a serial designation to separate otherwise identical names.

IV. Physical Properties

Some of the more important physical properties for magnesium are given in Table III. The low density, only about two-thirds that of aluminum, is the principle incentive for the structural use of magnesium. The highest density for any alloy is still only 1.87 g/cm^3; in many applications, a lighter structure can be built of magnesium than of any other material.

The specific heat is also a reason for using magnesium in many applications. The specific heat of a unit mass of magnesium is higher than that of other structural metals:

Metal	Specific heat (J/kg °K)
Mg	1025
Al	900
Ti	532
Ni	471
Fe	448
Cu	386
Zn	382
Mo	276
Ag	235
Ta	139

TABLE III

PHYSICAL PROPERTIES OF MAGNESIUM

Property	Value	Temperature, °C
Boiling point	1100°C	
Density	1.738 g/cm^3	20
	1.692 g/cm^3	400
	1.58 g/cm^3	650 (liquid)
Electrical resistivity	4.10 ohm meter × 10^8	20
Heat of fusion	368 kJ/kg	
Melting point	650°C	
Specific heat	1025 J/kg K	20
	1150 J/kg K	300
	1327 J/kg K	600
Thermal conductivity	154 W/m K	20
	148 W/m K	100
	130 W/m K	400
Thermal expansion	25.2 unit strain × 10^{-6}/°C	20
	26.9 unit strain × 10^{-6}/°C	100
	32.5 unit strain × 10^{-6}/°C	400
Vapor pressure	3.6 × 10^2 Pa	650 (liquid)

When magnesium is substituted for an equal mass of another metal, it acts as a better heat sink and stays cooler for a longer period of time. As an example, missile parts subjected to rapid but short-duration heating will remain cooler when made of magnesium rather than aluminum.

By contrast, the specific heat of a unit volume of magnesium is lower than that of other metals:

Metal	Specific heat (J/cc °K)
Mg	178
Ta	231
Ti	240
Al	243
Ag	247
Zn	272
Mo	282
Cu	346
Fe	352
Ni	419

If an equal volume of magnesium is substituted for another metal, the magnesium will heat up or cool down more rapidly. If it is acting as a radiator, it will more readily dissipate heat from the part. The preceding comparisons are made with the properties for pure metals; however, specific heat is not much affected by alloying, so that the conclusions hold for alloys as well as for the pure metals. This property is utilized in the die casting of magnesium alloys. Because of good mechanical properties, magnesium die castings can be made of the same wall thickness as those of aluminum. The lower-volume specific heat of magnesium means that the part cools more quickly in the die and it is possible to cast more parts per hour of magnesium than of aluminum.

The crystal structure of magnesium affects some physical properties by introducing anisotropy. This is true, for example, for electrical resistivity, thermal conductivity, and thermal expansion, where this value is different parallel to the *c* direction in the lattice than in the *a* direction. The effects are small and seldom of practical significance.

V. Mechanical Properties

Table IV gives some representative properties of a few of the more common magnesium alloys. The modulus of elasticity of magnesium and magnesium alloys is 45 GPa at room temperature. The specific strengths

TABLE IV

Representative Mechanical Properties of Magnesium Alloys (Stress in MPa)

						Creep strength			Rotating beam fatigue			
		Tensile strength (25°C)				Stress for 0.2% TE in 100 hrs			Stress for failure at cycles			
Alloy	Temper	% E	TYS	CYS	TS	95°C	200°C	250°C	10^5	10^6	10^7	5×10^7
Gravity castings												
AZ91C	F	2	95	95	165	—	—	—	110	105	95	90
	T4	14	85	85	275				135	115	105	100
	T6	5	130	275	—	—	—	—	125	105	90	85
AZ92A	F	2	95	95	165				115	105	95	92
	T4	9	95	95	275				125	110	105	100
	T5	2	110	110	180				125	110	105	100
	T6	2	145	145	275	55	10	—	130	115	100	97
EQ21A	T6	4	195	195	261	—	95[a]	36	—	100	98	96
QE22A	T6	4	205	205	275	—	85[a]	30	150	125	115	105
WE54A	T6	4	200	200	275	—	—	56	—	102	99	97
ZE41A	T5	5	140	140	205				124	97	96	94
ZK51A	T5	8	165	165	275	65	30	—	105	75	70	75

Die castings												
AM60A	F	8	130	130	220							
AS41A	F	6	140	140	210		40[a,c]		110	90	90	—
AZ91A or D	F	3	160	160	230		25[a,c]					
Extrusions												
AZ31B	F	15	200	95	260	50	—	—	180	160	145	140
AZ80A	T5	7	275	240	380	—			190	175	160	155
HM31A	T5	10	270	160	305	—	75	70	125	105	90	90
ZK30A	F	18	239	213	309		28[a]	—	152	131	124	124
ZK60A	T5	12	295	215	360	20	5	—	210	170	150	145
ZM21A	F	11	162	—	255							
Sheet												
AZ31B	0	21	150	110	255	55	—	—	—	—	—	—
	H24	15	220	180	290	30	—	—	155[b]	120[b]	103[b]	103[b]
HM21A	T8	11	160	130	250	—	80	50	—	—	—	—
ZK30A	—	8	185	154	270	—						
ZM21A	0	13	131	—	232	—						

Notes: % TE = percent total elongation.
[a] Creep extension only.
[b] Cantilever bending fatigue with $R = -1$.
[c] 150°C.
%E = Percent elongation
TYS = Tensile yield strength
CYS = Compressive yield strength
TS = Tensile strength

(strength/density) are such that structures of magnesium castings can frequently be lighter than the same structures in aluminum. Structures made of magnesium wrought products cannot be made lighter than aluminum if the critical design criterion is yield or tensile strength. In bending, however, the stiffness is proportional to the modulus and to the cube of the thickness; the bending strength is proportional to the yield strength and to the square of the thickness; and the weight is proportional to the density and the first power of the thickness. When the critical design criterion is bending, there can be weight savings realized with magnesium. The point is illustrated by the following comparison, using relative numbers, between extruded magnesium and aluminum alloys.

For plates of equal stiffness at room temperature:

Alloy	Bending strength	Weight	Stiffness
AZ31B-F	50	76	100
ZK60A-T5	98	78	100
2024-T4	100	100	100
7075-T6	119	102	100
Al-Li	116	89	100

For plates of equal weight at room temperature:

Alloy	Bending strength	Weight	Stiffness
AZ31B-F	88	100	233
ZK60A-T5	162	100	215
2024-T4	100	100	100
7075-T6	115	100	95
Al-Li	147	100	142

For plates of equal strength at room temperature:

Alloy	Bending strength	Weight	Stiffness
AZ31B-F	100	106	280
ZK60A-T5	100	78	103
2024-T3	100	100	100
7075-T73	100	93	76
Al-Li	100	83	80

For plates of equal stiffness at 315°C:

Alloy	Bending strength	Weight	Stiffness
HM31A-T5	436	72	100
2024-T4	100	100	100
7075-T6	146	113	100
C110M (Ti)	988	138	100
17-7PH (steel)	1258	185	100

For plates of equal weight at 315°C:

Alloy	Bending strength	Weight	Stiffness
HM31A-T5	843		
2024-T4	100		
7075-T6	113		
C110M (Ti)	517		
17-7PH (steel)	368		

This same effect is seen when comparing the dent-resistance of alloys. Figure 2 compares the size of the dent in sheet of equal weight when a hemispherical tup is dropped on a clamped sheet from various heights. The larger the ordinate number, the greater the size of the dent produced. Since many times a permanent dent means a service failure (for example, in an automobile fender or a suitcase), dent-resistance is an important service property. The greater dent-resistance of magnesium in comparison to other metals is related not only to the better resistance to bending as noted, but also to a greater tendency to anelastic deformation. There is greater dent-recovery after the blow in magnesium than in the other metals.

The anelasticity of magnesium is manifested in another important property, namely damping. Damping is the anelastic absorption of vibrational energy with production of heat. The higher the damping capacity, the more rapidly are vibrations absorbed. A high damping capacity tends to make a part run more quietly, and, since vibrations are absorbed efficiently, vibrational stress is transmitted inefficiently. This property is utilized by using magnesium alloys for vibration fixtures and for platforms for mounting equipment sensitive to vibrations. Since vibrations are transmitted poorly, a higher fatigue strength in an assembly than would be predicted from laboratory fatigue data is often realized.

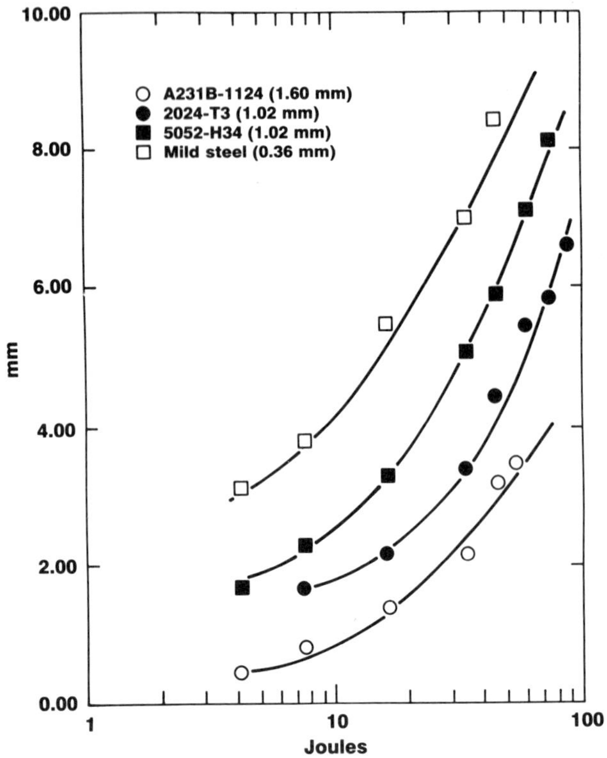

FIG. 2 Equal-weight comparison for dent-resistance.

TABLE V

DAMPING CAPACITY

		Specific damping capacity, %, at MPa				
Alloy	Form	5	15	20	25	35
AZ31B-F	Extrusion	1.0	1.6	2.2	3.1	4.0
AZ91A or D	Die-cast	0.8	1.6	2.5	—	—
AZ92A-T6	Gravity-cast	0.4	0.7	1.6	3.1	4.8
EZ33A-T5	Gravity-cast	—	4.9	12.6	18.2	22.4
K1A-F	Cast or wrought	40.0	49.0	56.0	62.0	66.0
A380 Al	Die-cast	1.0	—	—	—	—
355-T6 Al	Gravity-cast	—	0.5	0.7	1.0	—
356-T6 Al	Gravity-cast	0.3	0.5	0.6	0.8	1.2
Cast iron	Gravity-cast	—	5.0	12.0	14.0	16.0

Specific damping capacity is defined as the ratio of the vibrational energy absorbed per cycle of vibration to the total energy of vibration at the maximum amplitude of the cycle, expressed as a percentage. Table V lists the specific damping capacities of some magnesium alloys plus cast iron and a few aluminum alloys for comparison.

VI. New Developments

Three of the newer developments in metallurgy may have a significant impact on the future course of magnesium alloys.

Ion implantation is a method of injecting individual ions of an element or compound into a lattice. Because the injection is individually accomplished, "solid solutions" not otherwise obtainable can be formed. Thus, for example, extensive amounts of iron, chromium, or tungsten can be put into solid solution in magnesium only in this way. Since iron, for example, is very deleterious for the corrosion resistance of magnesium when it is present as a second phase, it has not been possible to determine whether or not it has any beneficial effects on the films that form on magnesium. Since iron in solid solution will not have any galvanic corrosion effect, the possibility of corrosion improvement can be studied. This approach is applicable to a large number of elements. P.L. Hagans at the Dow Chemical Company has already shown that iron in particular has some beneficial effects on corrosion when added to magnesium in this way. Although the effect produced is only in a very thin film on the surface, this may be enough to be of practical interest.

Metal-matrix composites are receiving attention for advanced structural applications. Magnesium is probably the preferred material as the metal matrix for several reasons. It has the lowest density. Within broad limits, the properties are determined only by the fiber used, not the matrix. For the best specific strength, the lowest possible density for the matrix is required. Magnesium wets, but does not react with, most of the fibers being studied. Thus, strong bonds can be formed easily between magnesium and the fiber. An exception is graphite, which must be surface-treated to be wet properly by magnesium. Molten magnesium reacts readily with both oxygen and nitrogen to produce solid reaction products. Therefore, when magnesium is infiltrated into a bundle of fibers in dry air, complete infiltration is obtained without the need for any external pressure. It is very easy to prepare complex castings with chosen emplacement of fibers using an infiltration technique. Use of fibers to strengthen selected areas of complex magnesium castings should become common using techniques already common for producing tubeless passageways.

Rapid solidification of metal alloys is being used to produce extreme nonequilibrium structures and ultrafine grain size. Because many elements form complete liquid solutions but limited solid solutions with magnesium, the potential for improved alloys based on supersaturated solutions is large. As previously noted, the properties of magnesium alloys are particularly sensitive to grain size, and improvement should be better for magnesium than for other materials. As early as the mid-1950s, the Dow Chemical Company was producing an extruded version of ZK60 made from atomized powder rather than from ingot. With the atomizing technique used, the cooling rate was of the order of $10^{4\circ}$K/s. More recently, Santosh Das and coworkers at Allied-Signal Corporation [*1*] have been studying the effects of rapid solidification on a number of magnesium alloys, using a technique that has a cooling rate of the order of $10^{6\circ}$K/s. Some comparisons illustrate the property improvements that have been obtained:

Alloy	% E	TYS	CYS	TS	Remarks
ZK60A	12	295	215	360	Extruded from ingot
ZK60B	15	310	310	380	Extruded from powder ($10^{4\circ}$K/S)
Mg-5Al-5Zn-7Y	5	455	455	510	Extruded from strip ($10^{6\circ}$K/S)
Mg-9Y-0.5Zr	8	370	370	415	Extruded from ingot

Note that the finer grain size produced by cooling at about $10^{4\circ}$K/s not only improves the strength but equalizes the compressive and tensile yield strengths of ZK60. The ingot-cast alloy containing Y is included in order to judge the degree of improvement that may be due to the rapid cooling rate used on the Mg-Al-Zn-Y alloy [2]. One of the effects of Y in solid solution in magnesium is the inhibition of the normal twinning mode. For these alloys, the compressive and tensile yield strengths are equal regardless of the grain size.

References

1. Das, S.K. Chang, C.F. and Raybould, D. (1986). "High Performance Magnesium Alloys by Rapid Solidification Processing." Proceedings of the International Magnesium Association.

2. Cooling, S.L. (1968). "Development of High Strength Magnesium-Base Sheet Alloy." Contract No. DAA-25-67-C-0160 (Frankford Arsenal), Philadelphia, Pa.

Bibliography

1. Busk, Robert S. (1986). *Magnesium Product Design*. Marcel Dekker, New York. A compilation of the engineering and fabrication properties of magnesium and magnesium alloys and of applications for the metal.
2. Emley, E.F. (1966). *Principles of Magnesium Technology*. Pergamon Press, London. An excellent treatise on the practical handling of magnesium alloys by producers of magnesium products.
3. Godard, Hugh P., Jepson, W.B., Bothewell, M.R., and Kane, Robert L. (1967). *The Corrosion of Light Metals*. John Wiley & Sons, New York. The section on magnesium by Bothwell is an excellent summary of the corrosion behavior of magnesium and its alloys.
4. Raynor, G.V. (1959). *The Physical Metallurgy of Magnesium and its Alloys*. Pergamon Press, London. An excellent review of the physical metallurgy of magnesium, especially as related to alloy theory.
5. Roberts, C. Sheldon (1960). *Magnesium and Its Alloys*. John Wiley & Sons, New York. A thorough review of the physical metallurgy of magnesium and its alloys, especially as related to deformation.

Appendix 1

*Nominal Chemical Compositions of Wrought Alloys**

Percent of Alloying Elements—Aluminum and Normal Impurities Constitute Remainder								
Alloy	Silicon	Copper	Manganese	Magnesium	Chromium	Nickel	Zinc	Titanium
1050	..	..	99.50 percent minimum aluminum			..	..	..
1060	..	..	99.60 percent minimum aluminum			..	..	..
1100	..	0.12	99.00 percent minimum aluminum			..	..	..
1145	..	..	99.45 percent minimum aluminum			..	..	..
1175	..	..	99.75 percent minimum aluminum			..	..	..
1200	..	..	99.00 percent mininum aluminum			..	..	..
1230	..	..	99.30 percent minimum aluminum			..	..	..
1235	..	..	99.35 percent minimum aluminum			..	..	..
1345	..	..	99.45 percent minimum aluminum			..	..	..
1350[f]	..	..	99.50 percent minimum aluminum			..	..	..
2011[a]	..	5.5	..	..	..	..	..	..
2014	0.8	4.4	0.8	0.50	..	..	..	..
2017	0.50	4.0	0.7	0.6	..	..	..	..
2018	..	4.0	..	0.7	..	2.0	..	..
2024	..	4.4	0.6	1.5	..	..	..	..
2025	0.8	4.4	0.8	..	..	..	..	..
2036	..	2.6	0.25	0.45	..	..	..	..
2117	..	2.6	..	0.35	..	..	..	..
2124	..	4.4	0.6	1.5	..	..	..	..
2218	..	4.0	..	1.5	..	2.0	..	..
2219[b]	..	6.3	0.30	..	..	..	..	0.06
2319[b]	..	6.3	0.30	..	..	..	..	0.15
2618[c]	0.18	2.3	..	1.6	..	1.0	..	0.07
3003	..	0.12	1.2	..	..	..	..	..
3004	..	..	1.2	1.0	..	..	..	..
3005	..	..	1.2	0.40	..	..	..	..
3105	..	..	0.6	0.50	..	..	..	..
4032	12.2	0.9	..	1.0	..	0.9	..	..
4043	5.2	..	..	..	..	..	..	..
4045	10.0	..	..	..	..	..	..	..
4047	12.0	..	..	..	..	..	..	..
4145	10.0	4.0	..	..	..	..	..	..
4343	7.5	..	..	..	..	..	..	..
4643	4.1	..	..	0.20	..	..	..	..
5005	..	..	..	0.8	..	..	..	..
5050	..	..	..	1.4	..	..	..	..
5052	..	..	..	2.5	0.25	..	..	..
5056	..	..	0.12	5.0	0.12	..	..	..
5083	..	..	0.7	4.4	0.15	..	..	..
5086	..	..	0.45	4.0	0.15	..	..	..
5154	..	..	..	3.5	0.25	..	..	..
5183	..	..	0.8	4.8	0.15	..	..	..
5252	..	..	..	2.5	..	..	..	..
5254	..	..		3.5	0.25	..	..	..
5356	..	..	0.12	5.0	0.12	..	..	0.13
5454	..	..	0.8	2.7	0.12	..	..	..

* Reprinted with permission from the Aluminum Association, Washington, D.C.

Percent of Alloying Elements—Aluminum and Normal Impurities Constitute Remainder								
Alloy	Silicon	Copper	Manganese	Magnesium	Chromium	Nickel	Zinc	Titanium
5456	..	..	0.8	5.1	0.12	..	..	..
5457	..	..	0.30	1.0	..	..	..	..
5554	..	..	0.8	2.7	0.12	..	..	0.12
5556	..	..	0.8	5.1	0.12	..	..	0.12
5652	..	..	..	2.5	0.25	..	..	..
5654	..	..	..	3.5	0.25	..	..	0.10
5657	..	..	..	0.8	..	..	..	..
6003	0.7	..	..	1.2	..	..	..	..
6005	0.8	..	..	0.50	..	..	..	..
6053	0.7	..	..	1.2	0.25	..	..	..
6061	0.6	0.28	..	1.0	0.20	..	..	..
6063	0.40	..	..	0.7	..	..	..	..
6066	1.4	1.0	0.8	1.1	..	..	..	..
6070	1.4	0.28	0.7	0.8	..	..	..	..
6101	0.50	..	..	0.6	..	..	..	..
6105	0.8	..	..	0.62	..	..	..	..
6151	0.9	..	..	0.6	0.25	..	..	..
6162	0.6	..	..	0.9	..	..	..	..
6201	0.7	..	..	0.8	..	..	..	..
6253	0.7	..	..	1.2	0.25	..	2.0	..
6262[e]	0.6	0.28	..	1.0	0.09	..	..	..
6351	1.0	..	0.6	0.6	..	..	..	..
6463	0.40	..	..	0.7	..	..	..	..
6951	0.35	0.28	..	0.6	..	..	..	..
7005[d]	..	..	0.45	1.4	0.13	..	4.5	0.04
7008	..	..	..	1.0	0.18	..	5.0	..
7049	..	1.6	..	2.4	0.16	..	7.7	..
7050[g]	..	2.3	..	2.2	..	..	6.2	..
7072	..	..	..	..	..	..	1.0	..
7075	..	1.6	..	2.5	0.23	..	5.6	..
7108[h]	..	..	..	1.0	..	..	5.0	..
7178	..	2.0	..	2.8	0.23	..	6.8	..
8017[i]	..	0.15	..	0.03	..	..	..	..
8030[j]	..	0.22	..	..	..	..	..	..
8170[i]	0.09	..	..	..	..	..	..	..
8177[k]	..	..	..	0.08	..	..	..	..

Note: Listed herein are designations and chemical composition limits for some wrought unalloyed aluminum and for wrought aluminum alloys registered with the Aluminum Association. This list does not include all alloys registered with the Aluminum Association. A complete list of registered designations is contained in the "Registration Record of International Alloy Designations and Chemical Composition Limits for Wrought Aluminum and Wrought Aluminum Alloys." These lists are maintained by the Technical Committee on Product Standards of the Aluminum Association.

[a] Lead and Bismuth, 0.40 percent each.
[b] Vanadium 0.10; Zirconium 0.18.
[c] Iron 1.1.
[d] Zirconium 0.14.
[e] Lead and Bismuth, 0.6 percent each.
[f] Formerly designated EC.
[g] Zirconium 0.12.
[h] Zirconium 0.18.
[i] Iron 0.7.
[j] Boron 0.02.
[k] Iron 0.35.

Index

B

F

G

H

I

Contents of Previous Volumes